RESERVOIR GEOPHYSICS

EDITED BY

ROBERT E. SHERIFF

Co-editors

Alistair R. Brown
Thomas L. Dobecki
Bob A. Hardage
David H. Johnston
James D. Robertson

Marc H.F. DeBuyl
O. Ed Gilbert, Jr.
James H. Justice
Wulf F. Massell
Joseph E. Warren

Project Editor Alistair R. Brown

SEG gratefully acknowledges the support of this publication by the SEG Foundation through the generous donations of Amoco Production Company.

Society of Exploration Geophysicists
P.O. Box 702740, Tulsa, OK 74170-2740

Library of Congress Cataloging-in-Publication Data

Reservoir geophysics / edited by Robert E. Sheriff; project editor
    Alistair R. Brown.
        p.   cm.—(Investigations in geophysics; no. 7.)
    Includes bibliographical references and index.
    ISBN 1-56080-057-7: $79.00
    1. Oil reservoir engineering.   2. Gas reservoirs.   3. Prospecting—
Geophysical methods.   I. Sheriff, Robert E. II. Brown, Alistair
R. III. Series.
TN871.R465   1992
622′.338—dc20   92-17973
                                                       CIP
ISBN 0-931830-46-X (IG Series)
ISBN 1-56080-057-7 (Volume)

Published 1992
Second Printing 1994
Printed in the United States of America

CONTENTS

PREFACE

Historically, geophysicists fed the results of their work to exploration geologists, who then fed their conclusions to development geologists, who in turn fed the results of their studies (maps) to petroleum engineers, who used the maps, in conjunction with measurements in wells and on samples taken from wells, to develop and produce oil and gas fields. Only rarely was the direction of information flow reversed. The knowledge and capabilities of exploration people generally did not contribute to development-production decision making except for the parts of their information that had been included in the hardcopy maps and reports. The upstream data were not reviewed in the light of the information found out downstream. This obviously inefficient method of communication has persisted throughout most of petroleum exploration and development history.

The 1983–84 Society of Exploration Geophysicists (SEG) Executive Committee authorized formation of a permanent committee, the SEG Development and Production Committee, and charged the committee with improving communication between geophysicists, development geologists, and reservoir engineers.

The SEG Development and Production Committee, now in its seventh year, is composed of volunteer members. Two meetings of the full committee have been held each year and four chairmen have served: 1984–86, Robert E. Sheriff; 1986–88, David H. Johnston; 1988–90, Wulf F. Massell; and 1990–date, Danny Melton.

Also about 1983–84 the American Association of Petroleum Geologists (AAPG) set up a committee on Development Geology and the chairmen of the respective SEG and AAPG committees became members of each other's committees. Geophysicists and geologists then had a number of communication vehicles, but not so between geophysicists and engineers; thus the SEG Development and Production Committee concentrated its efforts on communication between geophysicists and engineers.

At its first meeting the Committee determined that its first objective should be educating geophysicists and engineers in each other's disciplines, that is, telling geophysicists what engineers need to know and telling engineers what geophysicists can do for them. Several earlier attempts had been made to improve communication between geophysicists and engineers but there had been little follow-through. Getting geophysicists and engineers to talk with each other turned out to be more difficult than originally anticipated. Some engineers thought that geophysicists were simply trying to worm their way into the production/ development budgets, which were much larger than the exploration budgets, and there was some truth in such charges. There was no broad-scale recognition that the different disciplines had much to contribute to each other. Accordingly, the committee has emphasized communicating case histories that show how the disciplines have contributed significantly. This book is a further step toward this end.

A number of communication vehicles have been developed:

Sessions of geophysical papers are given at the Society of Petroleum Engineers (SPE) Annual Meetings, and sessions of petroleum engineering papers at SEG Annual Meetings;

The practice of appointing liaison representatives to the SPE and the Society of Professional Well Log Analysts (SPWLA) was begun;

A permanent Intersociety Coordinating Committee including Presidents and President-elects of the SEG, AAPG, SWPLA and SPE Societies has been formed;

Joint workshops and forums where practitioners of the disciplines participated include:

The Midwest SEG meeting in 1987,
The SPWLA symposium in London in 1987,
SEG/Los Alamos Workshop on Interwell Surveying in 1988,
Joint Meeting with China Petroleum Society in 1988,
The AAPG Annual Meeting in San Antonio in 1989,
New Mexico Technology meeting in Socorro in 1989,
SPE Forum in 1989,
Cosponsorship of Maracaibo meeting on Reservoir Characterization with Sociedad Venezuelana de Ingenieros Geofisicos in 1990,

International Symposium on Borehole Geophysics for Petroleum, Hydrology, Mining, and Engineering Application at the University of Arizona in 1990,

The AAPG Annual Meeting in San Francisco in 1990, and

The first Archie Conference in Houston in 1990.

Distinguished lecturers for the SEG, SPE, and AAPG have made reservoir geophysics the subject of their lectures;

Technical Development and Production Geophysics luncheons are included at the SEG Annual Meetings.

This present book, *Reservoir Geophysics*, is the latest effort. Reservoir Geophysics is very much a committee effort; it has been underway since 1986. Most decisions regarding the book have been made at the numerous committee meetings. The Committee thanks Margaret Sheriff, who hosted the many breakfast meetings. I personally express my appreciation to the members of the Reservoir Geophysics Book Committee, which include Alistair Brown, Marc Debuyl, Tom Dobecki, Ed Gilbert, Bob Hardage, Jim Justice, Dave Johnston, Wulf Massell, Jamie Robertson, and Joe Warren.

R. E. Sheriff,
Committee Chairman and Editor,
Houston, December 1990

Chapter 1: Reservoir Management
Introduction

*David H. Johnston**

Geophysics and Reservoir Development

This book describes the recent application of geophysical technology to oil and gas field development and exploitation. Historically, most applications of geophysical technology have been limited to exploration. We normally think that the objectives of the geophysicist, geologist, and engineer are quite different. However, the complete description of a reservoir from an engineering perspective requires measurements over many length scales. These scales range from core analyses of the effects of pore microstructure on fluid flow, to the mapping of large faults which define the limits of a reservoir. Geological, geophysical, and engineering measurement technologies span much of the production scale, from the centimeter vertical resolution in some well logs, to the 10–100 m lateral and vertical resolution of three-dimensional (3-D) seismic surveys, and finally, to the scale of the entire reservoir in well interference tests.

Many companies now recognize that an integrated approach to reservoir development planning is most cost effective. Teams of geophysicists, geologists, and engineers plan the acquisition of data from the best sources and analyze and integrate all of the information into a consistent description of the reservoir. This team approach requires that each member must understand the technology involved in obtaining each source of data and the data accuracy so that the best possible information is used to estimate reservoir properties.

This book, therefore, serves two purposes: (1) For the development geologist or reservoir engineer, the technical background (in non-mathematical terms) to evaluate geophysical data as a source of information for reservoir description is provided. (2) For the practicing geophysicist, as well as the geologist and engineer, case studies which illustrate the proven and potential value of geophysical data to solutions of reservoir production and development problems are presented. The book emphasizes application, seeking whatever geophysical technologies are appropriate to solve a specific problem. The case studies reprinted in

*Exxon Production Research Co.

this volume are by no means an exhaustive survey of reservoir geophysics. They should, however, outline the wide range of applications where geophysical data are being used in field development and set the stage for the introduction of new geophysical technologies in the future.

Integrated Reservoir Development

Why is there a move toward a team approach to reservoir description and development? There are two reasons. First, the decision to develop a new field offshore, or a deep play onshore, or a field in a remote location requires accurate appraisals of oil and gas in place, potential production rates, and ultimate recovery. If developed, economic pressures further require that these high-cost fields be brought on stream quicker and that recovery be increased. In addition, these new fields tend to be smaller and more complex than those found in the past. Second, in a climate of uncertain oil and gas supplies and unstable prices, many companies are focusing on increasing reserves by more precise definition and detailed characterization of mature fields. To illustrate the importance of this trend, analyses by the Bureau of Economic Geology (Fisher, 1988) indicate that for the U.S., excluding Alaska, up to 80 billion barrels of oil and 180 TCF of gas can be added to current onshore reserves by infill drilling, extension drilling, and well recompletion, but only if this work is based on detailed reservoir characterization. In addition, enhanced recovery is estimated to be able to increase U.S. oil reserves by another 28 billion barrels with the application of advanced technology (at $30 per barrel, National Petroleum Council, 1984).

The trend is further illustrated by examination of oil company expenditures in the U.S. for production and development compared with exploration, shown in Figure 1. The data clearly suggest an increase in the importance of production and development activities. The ratio of development to exploration expenditures (Figure 2) has risen steadily since 1979, even during the industry downturn beginning in 1986. Because of the increasing importance in developing mature fields,

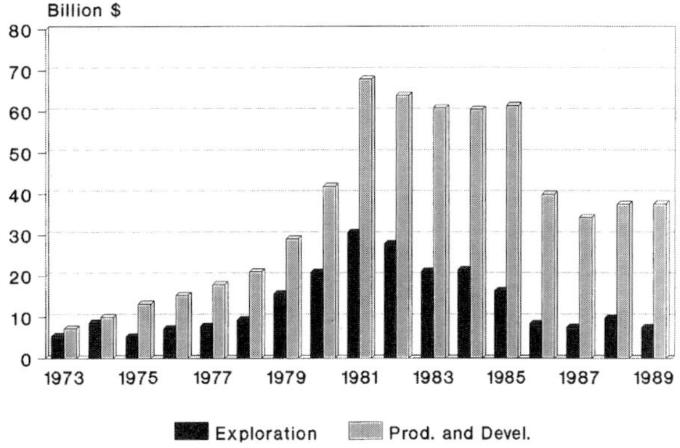

Fig. 1. U.S. estimated expenditures for exploration, production, and development from 1973 through 1989 (American Petroleum Institute statistics).

detailed production strategies must be devised that incorporate all available geological, geophysical, and engineering data.

The Role of Geophysics in Reservoir Management

J. D. Robertson (1989) defines reservoir management as maximizing the economic value of a reservoir by optimizing recovery of hydrocarbons while minimizing capital investments and operating expenses. Thus, reservoir management is the economic process of raising the worth of a property to its highest possible level. Economic value generally increases when more reserves are proved or when the reservoir's producing rate increases. Capital investments (drilling, seismic shooting, lease bonuses, etc.) and operating expenses (lease rentals, staff costs, taxes, etc.) must be incurred to find and develop reserves and these expenditures offset value.

Given this definition of field management, development strategies must meet these five basic objectives:

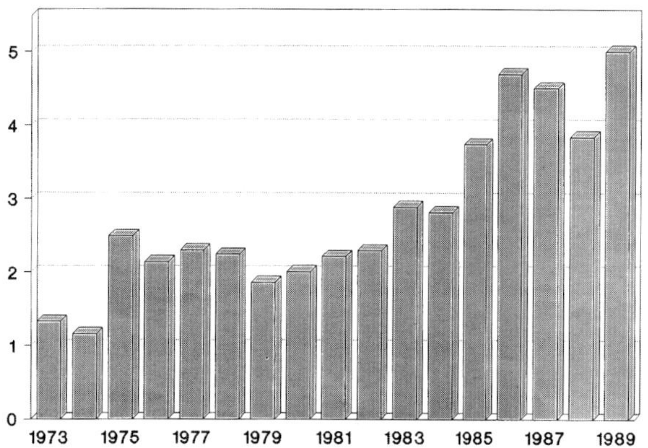

Fig. 2. U.S. estimated ratio of production and development to exploration expenditures from 1973 through 1989 (American Petroleum Institute statistics).

(1) reduce the cost of field development, which often translates to minimizing the number of wells,
(2) optimize total reserves,
(3) optimize production recovery,
(4) reduce operating costs of the developed field, and
(5) enhance recovery if economically justified.

Expenditures that drain present worth of a field must be traded off against the chance of increasing present worth by adding reserves and/or increasing production. This process is a continuous balancing act.

Robertson suggests that geophysics impacts reservoir management in two ways. First, geophysical analyses can lead to the identification of reserves that may not be produced by the existing development plan. Second, geophysics can save costs by minimizing dry holes and poor producers. Thus, geophysical data can contribute to reservoir economics by adding reserves or by lowering costs or by either. Either result can be sufficient justification for the expense of a surface or borehole seismic survey.

In addition to these economic incentives, Jim Jordan, head of Petroleum Engineering Research at Shell, has identified two significant technical challenges that reservoir management must face by the end of the century (Jordan, 1986). The first is the early and accurate characterization of the reservoir in terms of volumetrics, fluid properties, lithology, and continuity. The second challenge is to improve reservoir surveillance techniques so that fields under production may be accurately monitored and efficiently managed.

Reliance upon conventional engineering data, such as core analyses, well logs, and production history for reservoir characterization or surveillance, cannot provide the complete information required to meet these challenges. The case studies presented illustrate that a key to improved characterization and surveillance is the use of high resolution geophysical measurements integrated with conventional data within a geological model of the reservoir. Geophysical methods can, therefore, provide quantitative information to enhance or constrain reservoir simulation models.

In order to gain a perspective as to the specific applications of geophysics in production and development, consider that field management may be separated into four phases: (1) predevelopment, (2) initial development, (3) operating, and (4) enhanced recovery. In each phase, specific tasks are required in reservoir management to meet development or production goals. Also in each phase, geophysical data can play a key role in accomplishing these tasks.

Predevelopment Phase

After assessment of the discovery well, the primary goal of the predevelopment phase is to maximize reserves by delineation of the reservoir. Tasks include characterizing the trap and determining its structural nature, limits, fluid flow boundaries, and volume. Reservoir limits, thickness distribution patterns, and the lateral and vertical extent of hydrocarbon accumulation must be determined. In some cases, it may also be necessary to determine the drive mechanism and, if appropriate, delineate the extent of the aquifer. Studies early in the predevelopment phase can help identify potential problems of pressure interference among reservoirs producing from a common aquifer. The commercial value and profitability of the field is then assessed, based in large part on the reservoir volumetrics derived from delineation.

Initial development planning is focused on the placement of offset wells. With little available well control, seismic data often provide the primary information on reservoir or aquifer extent. High resolution seismic data have greatly improved the ability to define structures and continuity of pay and nonpay zones and, in some cases, to indicate the presence of gas or oil.

In particular, 3-D seismic data have proven effective in reducing the risk of dry holes and in reducing the costs of delineation wells by the prediction of drilling hazards such as faults and overpressure. Images of the subsurface from 3-D seismic data provide such a detailed structural picture of the reservoir that many oil companies now acquire such data early in the predevelopment phase of a field. 3-D seismic data is such a powerful tool that its use pervades many of the case studies presented in this book. The paper, "Reservoir Management Using 3-D Seismic Data" by J. D. Robertson, presents an excellent summary of the range of field development applications that have been found for 3-D seismic data.

In some fields, borehole seismic techniques such as offset-source vertical seismic profiles (VSP) are gaining increasing credibility as tools for delineating reservoir extent or locating faults and other geological discontinuities in the vicinity of a well. By placing seismic sources, sensors, or both in the well, within or near the reservoir, the subsurface can be imaged with greater detail than can be obtained from surface seismic data. These borehole seismic data can be acquired, processed, and interpreted within a time-frame that can significantly impact a development drilling program.

Initial Development Phase

The primary goal in the initial development phase is to plan production well locations to maximize recovery. Reservoir and aquifer description, rather than delineation, becomes the primary task of the engineer. Lateral facies and porosity variations must be estimated to determine if barriers to flow are present, and to determine if all units of the reservoir are continuous. Fluid contacts must be mapped in detail, formation pressures and controls on fluid saturations determined, and reservoir fractures analyzed. A complete geologic model of the reservoir must be built, analyzed, and validated. The drilling strategy, guided by reservoir simulations based on this model, is then formulated.

Geophysical data are beginning to play an active role in reservoir description and characterization. Seismic stratigraphy provides information on depositional patterns and lateral facies variations. In addition, other seismic analysis tools provide methods for mapping fluid contacts, estimating lithology, and characterizing reservoir fractures.

A primary objective in reservoir description during initial field development is to estimate rock properties, such as lithology and porosity, for input into reservoir flow simulation models. Seismic data can be used to improve the spatial description of these properties by taking advantage of its dense areal coverage compared to well data. Methodologies for integrating seismic data to constrain and optimize geologic models derived from geologic and well-log data are being developed by a number of companies. Results from these methods can be used to estimate rock properties away from well control, while retaining much of the vertical resolution of the well data. These methods of integrating seismic and well data increasingly are being employed in the field development setting for reservoir description.

In addition, geophysical data can be used to evaluate bed continuity and reservoir heterogeneities either indirectly with geostatistical methods or directly with cross-borehole measurements.

Operating Phase

The goals in the operating phase of field development are to reduce operating costs by efficient reservoir management and to ensure maximum hydrocarbon recovery. The key to a successful operating phase is reservoir surveillance. Differences between actual and predicted performance are used to update the geologic model of the reservoir and to revise the depletion strategy. Infill drilling based on detailed reservoir characterization may be used to further increase reserves. Secondary recovery processes may be used to enhance recovery. The detection of permeability barriers, the estimation of reservoir continuity, and the determination of sweep efficiency are all critical tasks in the operating phase.

Geophysical data, closely integrated with detailed core, well, and reservoir performance data, are beginning to impact reservoir surveillance. In addition, the detection of reservoir heterogeneities that exist on a distance scale between well spacing can help locate bypassed oil and enable the engineer to more effectively plan an infill drilling program. With the drilling of development wells, this phase of field development may be where we will see the greatest growth of applications for cross-well geophysics.

Enhanced Recovery Phase

If the return on investment warrants, enhanced oil recovery (EOR) processes may be used to increase recovery efficiency. Reservoir surveillance during EOR processes is recognized as a critical tool for the evaluation of EOR efficiency. Real-time mapping of steam, fire, or CO_2 fronts can provide the opportunity to control or modify the EOR process, thus reducing operating costs. In addition, reservoir heterogeneities and anisotropy have been shown to be the most common cause for EOR failure. Therefore, detailed reservoir characterization is essential before embarking upon such a project.

EOR processes induce changes in the physical properties of reservoir rocks that are detectable by geophysical methods. Several case studies have shown that 3-D surface seismic and cross-well seismic data can be used to monitor EOR progress. In addition, these studies illustrate the importance of geophysical reservoir characterization prior to the design of the recovery process.

While the four phases of reservoir management described may appear distinct, this is rarely the case in the real world. As Robertson (1989) notes, the process of reservoir management is iterative. Data acquired early in the life cycle of a reservoir are constantly being evaluated to form the basis for development and production decisions (such as locating production and injection wells, siting and designing platforms, setting flow rates, managing pressure maintenance, performing workovers, planning waterflood and EOR strategies, etc.). When implemented, these development and production activities in turn generate new information that revises or refines the reservoir model.

Geophysics is one of many tools used in the reservoir evaluation loop. Taking 3-D seismic data as an example, an initial interpretation of the survey impacts the original development plan. As development wells are drilled, the added information from logs, cores, drill stem tests, pressure tests, etc., is used to revise and refine the original interpretation. Elements of the 3-D data that may have been initially ambiguous can begin to make sense. Thus, Robertson suggests that both the seismic interpretation and the reservoir model become more detailed and sophisticated.

As a result, while geophysical data can contribute to reservoir economics during any phase of the develop-

ment process, the usefulness of the data lasts for the life of the reservoir.

Chapter Summaries

Chapter 1. Reservoir Management

Integrated reservoir studies and the team approach to reservoir development and management are such important concepts that the remainder of Chapter 1 consists of papers that expand upon this theme. J. G. Richardson and R. M. Sneider begin by examining the advantages and disadvantages of a synergistic team approach to reservoir development. They also discuss the key technical questions and problems to be solved during reservoir development. H. H. Haldorsen and T. V. Golf-Racht clearly define a reservoir management philosophy and the main elements of reservoir description. For geophysicists and geologists, this paper provides the engineering perspective of field development. The authors discuss the data available for reservoir description and the strategies used for the integrated application of the data. Finally, J. D. Robertson summarizes how 3-D seismic data fit among the many tools that a reservoir management team employs to plan and monitor the development and production of a field.

Chapter 2. Petrophysical and Geophysical Background

While the geophysicist, geologist, and engineer tend to think in different terms, the ultimate objective of all three disciplines is to obtain a detailed understanding of the reservoir's internal and external structure. Indeed, the data sets unique to each profession originate from the same geology. Chapter 2 discusses the basic petrophysical properties used by engineers and geologists to describe the reservoir as well as how those properties are related to the measurements made in geophysics.

For the geophysicist, Chapter 2 explains what the reservoir engineer and development geologist need to know in order to construct a model of the field. For the engineer and geologist, Chapter 2 summarizes the basic principles of seismic wave propagation and addresses important issues such as the resolution and limitations of the seismic method.

Chapter 3. Geophysical Technology

A variety of technologies can be applied to reservoir geophysics. Since this book emphasizes application, not methodology, Chapter 3 reviews the seismic tools that are applied in the case studies that follow. Because of the importance of 3-D seismic data for reservoir management, the bulk of Chapter 3 is devoted to a discussion of data acquisition, processing, and interpretation. In addition, methods for the use of seismic data acquired in the borehole are also discussed.

Chapter 4. Reservoir Delineation

Prior to the application of high resolution seismic techniques to reservoir delineation, the location of offset and infill wells was guided primarily by conceptual geologic models of the reservoir. Often these models were poorly constrained—even in ''well understood'' fields. The analyses of seismic data can provide strong constraints on the geologic model and even result in new prospects within an old field.

Chapter 4, on characterizing the trap, presents a number of case studies in which both 3-D surface seismic and borehole seismic data are used to aid in: the development decision, locate offset wells, planning development wells, locating drilling hazards such as faults, and assessing the risk of subsidence as a result of production.

Chapter 5. Reservoir Description

Until recently, the use of seismic data has been limited to providing a structural model of the reservoir. However, processing and interpretation techniques have evolved so that such data can help define the reservoir features internal to the trap. Tools are being developed to improve the spatial description of reservoir rock properties by taking advantage of the dense areal coverage of seismic data as compared to well data.

These techniques relate seismic attributes, which may be as simple as the amplitude of a reflected seismic wave, or as complicated as velocities and density obtained from a full inversion of the seismic data, to rock properties derived from well log analyses. Chapter 5 describes and illustrates seismic methods that provide control on reservoir parameters such as net and gross thickness, net porosity and porosity thickness, lithology, and Poisson's ratio.

Chapter 6: Reservoir Surveillance

Research in petrophysics suggests that seismic data can be used to image the changes in the physical properties of a reservoir rock as the rock undergoes enhanced recovery processes. Chapter 6 reviews this physical basis of geophysical methods for reservoir surveillance and presents case studies in which seismic data are used to monitor EOR processes.

These applications of geophysics to reservoir surveillance make use of proven technology like 3-D seismics but are, at the same time, stimulating the development of new technologies such as cross-well tomography. Time-lapse geophysical images of the reservoir may be able to monitor changes induced by production operations and aid in planning future field development.

Chapter 7. Alternative and Emerging Technologies

Seismic applications dominate reservoir geophysics. However, there are circumstances where nonseismic applications can be used effectively. These methods, described in Chapter 7, can monitor changes in physical properties such as in situ strain, temperature, fluid salinity, bulk density, magnetic susceptibility, porosity, fluid pressure, and fluid distribution. In many situations, seismic methods are not sensitive to these changes or, to isolate the influence of one property from the others, may not be possible.

Chapter 7 also discusses emerging and future technologies that will impact reservoir geophysics. These include multicomponent 3-D seismic surveys which can provide information on reservoir fracture orientation and density, downhole seismic sources, interactive interpretive processing, and parallel processing.

Chapter 8. Economics

The final chapter, Chapter 8, discusses the economics of reservoir development. In particular, the papers address issues such as the economics of decision making, the value of geophysical information compared with other types of information, and the impact of oil prices on the value of that information.

References

Fisher, W. L., 1988, An assessment of the natural gas resource base of the United States: Bureau of Economic Geology Report to the U. S. Department of Energy.

Jordan, J., 1986, The role of the petroleum engineer in the year 2000: Presented at the 1986 Ann. Soc. Petr. Eng. Mtg., New Orleans.

National Petroleum Council, 1984, Enhanced oil recovery: Report to the U.S. Department of Energy.

Robertson, J.D., 1989, Reservoir management using 3-D seismic data: Geophysics, The Leading Edge of Exploration, 8, No. 2, 25.

Synergism in Reservoir Management[1]

Joseph G. Richardson and Robert M. Sneider**

Introduction

Modern reservoir management requires teamwork and close coordination among geologists, geophysicists, and engineers at all stages of a reservoir's life. Technological and business changes in the petroleum industry demand that this technical and management team find and develop fields in more efficient and cost-effective ways. Unprecedented challenges face industry in every facet of the petroleum exploration and development business. Much of the search for oil and gas is in high-cost, frontier, and hostile environments such as deeper water, the arctic, and remote desert terrain. Economic pressures require that fields be brought on stream quicker and that recovery be increased. At the same time, new fields are often smaller and more complex. Many fields are marginal and economically sensitive because of their geographic location.

In addition, we see rapid-to-explosive advances in new exploration and development technology, with increased emphasis on engineering. Primary and enhanced recovery processes require more accurate and complete reservoir description to reduce engineering risks within tolerable bounds. All of these factors are coupled with the fact that the world demand for petroleum continues to increase, and the industry continually fails to replace reserves consumed. Given the recent worldwide decline in exploration and drilling, the world reserves of petroleum will continue their downward trend with the prospects for a supply crisis in the 1990s.

Reservoir Description—The Key to Reservoir Management

A comprehensive and detailed picture of the reservoir rocks and fluids and the aquifer is essential to optimize hydrocarbon recovery and maximize income. The need for reservoir description starts when a discovery is made and is appraised to obtain the best estimates of hydrocarbons in place, recoverable reserves, and rates of production. In fact, usually as a field or reservoir goes through the typical "life cycle" of appraisal, planning, development, and reservoir management (surveillance), an ever more complete description is both necessary and possible.

Some key questions and problems to be solved during field or reservoir exploitation follow.

What does the reservoir look like? What is its external geometry and what is the continuity of the internal pore space and fluids?

Will the reservoir have an effective natural water drive? If so, what is the aquifer geometry, continuity, and strength?

Where should wells and platforms be located?

How should the wells be completed and perforated?

Will recoveries be better by water or gas displacement?

Will water or gas injection be needed and when?

Will enhanced recovery processes be needed and when?

In order to answer these questions, detailed data about the three-dimensional (3-D) distribution and continuity of the pore-fluid system of the reservoir and aquifer system is required. Experience indicates that many times and in many fields these key problems in reservoir analyses are studied by engineers without the complete understanding of available geological and geophysical knowledge of the reservoir and aquifer. A detailed reservoir description is not asked for until a critical technical and economic project failure is in the making or has happened.

Geological and geophysical data are essential elements of most aspects of reservoir description. A good example is the determination of reservoir continuity. A critical first step in the reservoir description process is the recognition of any correlative reservoir subzones or layers and the intervening dense, imperme-

[1]This paper was edited by David H. Johnston, Exxon Production Research Co., Box 2189, Houston, TX 77252.
*Richardson, Sangree, and Sneider, 11767 Katy Freeway, # 330, Houston, TX 77079.

able, or low permeability strata. Knowledge of the depositional and diagenetic processes controlling reservoir and nonreservoir rock is essential to determine one's ability and degree of confidence in correlating these units.

For displacement recovery processes, the determination of pay continuity and the estimation of the percent pay that will be floodable based on the well spacing have a major impact on recovery estimates and on the decision to infill.

Seismic facies analyses and well-documented outcrop studies can add significantly in establishing interwell correlations with confidence. Flow test data dovetailed with knowledge of the reservoir and nonreservoir framework based on geology and geophysics provides the best description of reservoir continuity and discontinuity.

Maps, structural and stratigraphic cross sections, and fence and block diagrams are used to convey the 3-D geometry, distribution, and continuity of the reservoir, nonreservoir, and aquifer. Isopach maps without accompanying detailed correlation sections have been the pitfall of many projects. Net pay isopach maps, drawn to provide the basis for determining hydrocarbons in-place, have tricked many petroleum engineers into believing a reservoir is more continuous, more homogeneous, and less stratified than it actually is.

Timing of Data Acquisition

A coordinated data acquisition program, developed by the reservoir management team, can greatly improve the probability of correct assessment in the appraisal, planning, development, and management stages of a reservoir's life. Although the quantity of data which can be obtained is limited by the number of wells that have been drilled up to any time, the type and quality of information developed on each well should be planned carefully. As a field evolves from the initial appraisal to the planning, development, and management phases, the data acquisition program should be continued. Surveillance of reservoir performance coupled with analysis of a complete data base can improve probabilities of correct responses to production problems.

Figure 1 illustrates the communication among team members that occurs during the appraisal phase following discovery of a major accumulation of oil and gas or of both. Geophysicists now have high-resolution seismic data from the use of higher frequencies and improved computer processing. These data can better define structures and the continuity of pay and nonpay zones. In some cases, seismic data can indicate the presence of gas or oil. Geologists study the seismic data and assess the depositional environment of various intervals from examinations of cuttings,

Appraisal Phase

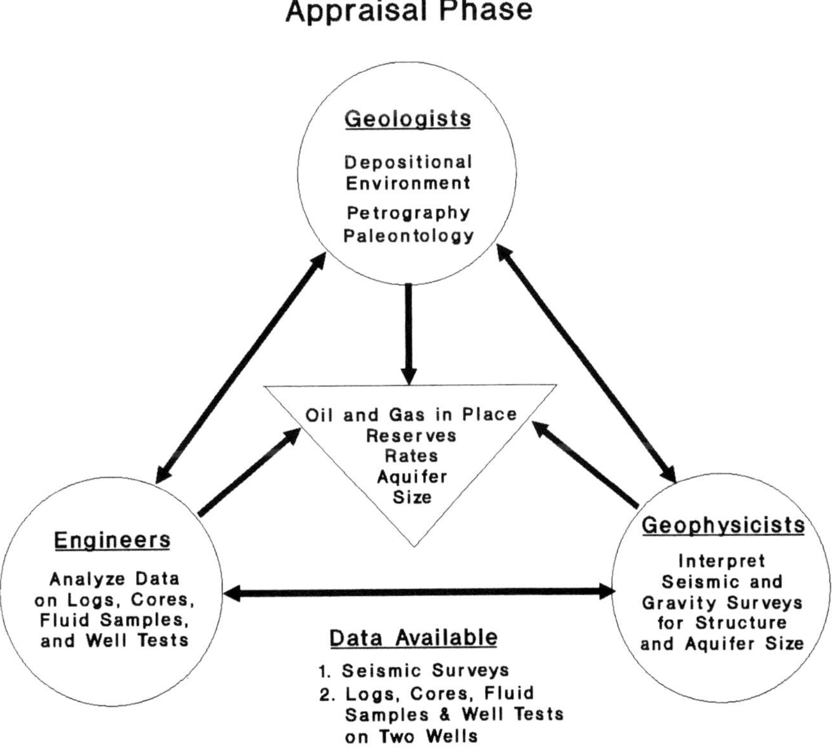

Fig. 1. Communication among team members during appraisal phase.

cores, and logs. Conceptual models based on studies of recent sediments and outcrops permit the extension of well data in preliminary correlations of the various zones. Engineers analyze data from cores, logs, and well tests which, together with results from seismic and geologic interpretations, permit preliminary estimates of oil and gas in place, aquifer size, reserves, and producing rates.

As an example, core analysis results on a foot-by-foot basis are used to develop a correlation of permeability versus porosity. When such correlations can be developed on the first well or two, porosities read from logs on subsequent wells can be used to estimate permeabilities for each foot in the absence of core data. (Calibration of logging parameters by plotting porosities from logs versus porosities from cores is a necessary part of this procedure.) The averaging procedure used on foot-by-foot values from core data and core/log correlations should be consistent with well tests and geologic data on the depositional environment to yield valid reservoir descriptions.

Figure 1 illustrates another aspect of the appraisal phase. The engineer must work with the geologist and geophysicist to estimate the strength of potential water drive. Simple calculations using rock and water compressibilities can show that the aquifer pore volume needs to be much larger than the reservoir pore volume for the aquifer to be of potential significance in furnishing a water drive. Thus, early collaboration can

establish priorities for detailed mapping of aquifers. Team studies can also identify potential problems of pressure interference among reservoirs producing from a common aquifer.

If the appraisal phase results encourage development of a prospect, additional teamwork is necessary to plan for optimum development of a field (see Figure 2). High costs in offshore waters require that the number of platforms and wells be optimized. Detailed geophysical work includes interpretation of fine-grid seismic data for improved mapping of the reservoir and aquifer by geologists. Additional appraisal wells drilled during the planning phase and accumulation of basic reservoir data on rock and fluid properties are necessary for developing sound development plans. Improved reservoir description is made possible by analysis of data from up to 10 wells drilled for appraisal and evaluation of the prospect. The best possible geologic description is vital to assess potential reservoir performance under alternate modes of operation. Engineers use these maps in computer studies of reservoir and well performance to choose the optimum fluid injection program and optimum number and location of wells.

As operations move into the development phase, plans for completing each well to drain the entire oil zone with a minimum number of workovers and with a minimum production of gas and water require close interaction among engineers and geologists. As seen in

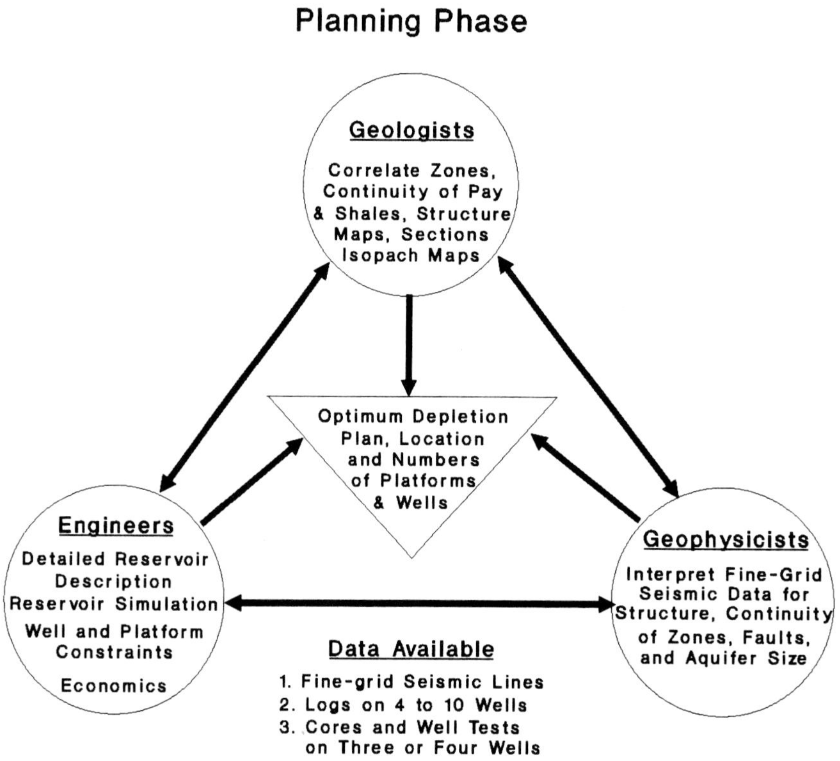

Planning Phase

Fig. 2. Communication among team members during planning phase.

Figure 3, each discipline has a role in developing the best policies for well completions and workovers. Critical decisions involving policies on completion intervals can be evaluated using reservoir simulators. Recoveries and times to flood the reservoir can be compared by operating the reservoir under alternate modes to aid in choosing the best plan. For example, a plan for opening several intervals in both injection and production wells and flooding to high water cuts will often prove optimum. Selective injection and production from each potentially separate interval could require a prohibitive number of wells—or time—if the zones were flooded in sequence.

Close surveillance of reservoir performance by teams of engineers and geoscientists is important in the management of a reservoir operation (see Figure 4). Study of the pressure data and performance of various wells often reveals the need to alter injection plans. Sometimes all intervals have not taken water or produced as planned and sometimes studies reveal a different zonation or fault pattern than those used in the original plan. In the surveillance phase of operation of large reservoirs, 3-D reservoir simulators can be most useful in quantifying the benefits of additional wells and workovers in recovery of bypassed oil.

Synergism and Organization

An organizational approach to help solve some of the reservoir management problems and challenges that lie ahead in the next decade is the use of a variety of generalists and specialists in a task-force approach—a synergistic team. The dictionary defines synergy as "the cooperative action of discreet agencies so that the total effect is greater than the sum of the effects taken independently." In the context of the petroleum business, synergy means geologists, geophysicists, and engineers work together more effectively and efficiently as a team or task force than working as a "group" of individuals. The synergistic approach among members from different disciplines yields better and quicker results than can come from individual members working alone.

For synergism to be a significant force in an organization and to be truly effective in exploration and production, all participants must be open to new ideas and both the individuals and management must want and be able to work and share. There are barriers to synergism and the synergistic team. Some participants have personalities that are not suitable, others lack good communication skills (an essential ingredient), and some fear that by working in a team their contributions will not be fully recognized. Some managers lack the understanding of the concept or do not wish to share the glory, defeats, or problems with other managers. Some organizational structures inhibit synergistic decision-making.

The advantages of a synergistic team approach to reservoir management are

Development Phase

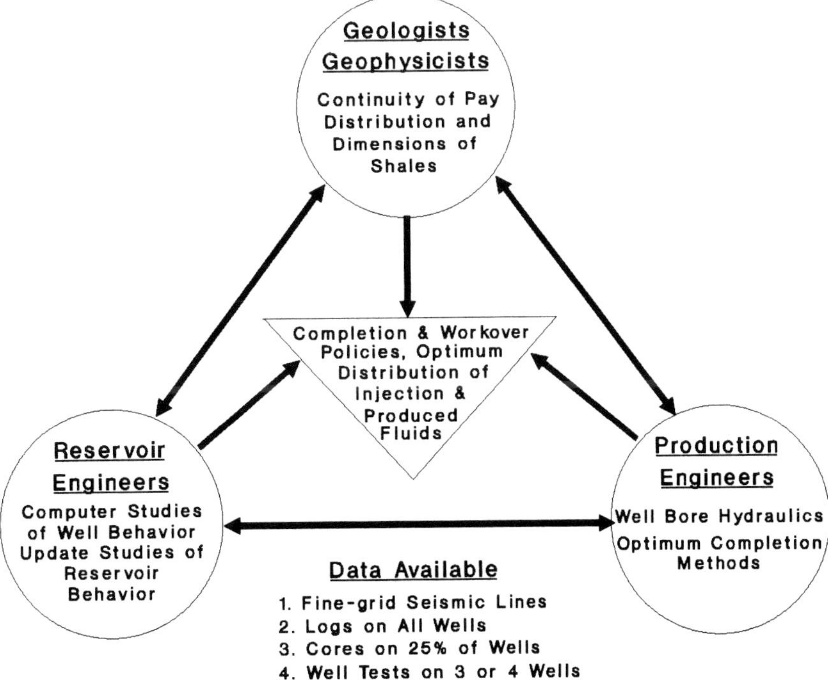

Fig. 3. Communication among team members during development phase.

Members work toward a common goal with specific objectives.

The team can focus early on key problems.

Team members have diverse, specialized training which allows (1) awareness and use of proved and new technology, (2) development of a comprehensive data base, and (3) selection of the best data and methods to solve the problems.

Time and cost required to complete the project is reduced.

However, there are potential disadvantages of working as or on a team:

Training and experience is necessary to become an effective team. Members need skills in communication, an understanding of other disciplines' technology, and a strong desire to work with others.

An individual member's contributions may not be recognized by management.

The chances for raises and the promotion of a member may be reduced if the team reports to a manager from a different discipline or function. This problem can be mitigated by self-evaluation of the team members.

Management must be convinced that the advantages of better and cost-effective solutions outweigh the disadvantages caused by personnel problems.

Emergencies often necessitate the initiation of a team approach. However, the full benefits of synergism are seldom realized since the team is dissolved when the crisis is over.

The proper organizational structure and management style to make the synergistic team most effective is not yet known. Several major oil companies and both small and large independent companies are experimenting. One large oil company set up a small "independent division" of about 35 professionals and support staff; they operated as a separate company and competed in the same geographical area as the company's traditional 175 employee operating division. As reported by Sneider (1990), the small company was organized by plays or projects and had a very flat organization. Individual team leaders reported directly to the president who had a great deal of technical and monetary authority. In contrast, the larger group had four levels of management and review when the experiment started.

During the five years of the experiment, the staff experience, budgets, technical data bases, and the economic and risk criteria were essentially the same for the two groups. The results of the competition are amazing. The small synergistic team approach found

Reservoir Management Phase

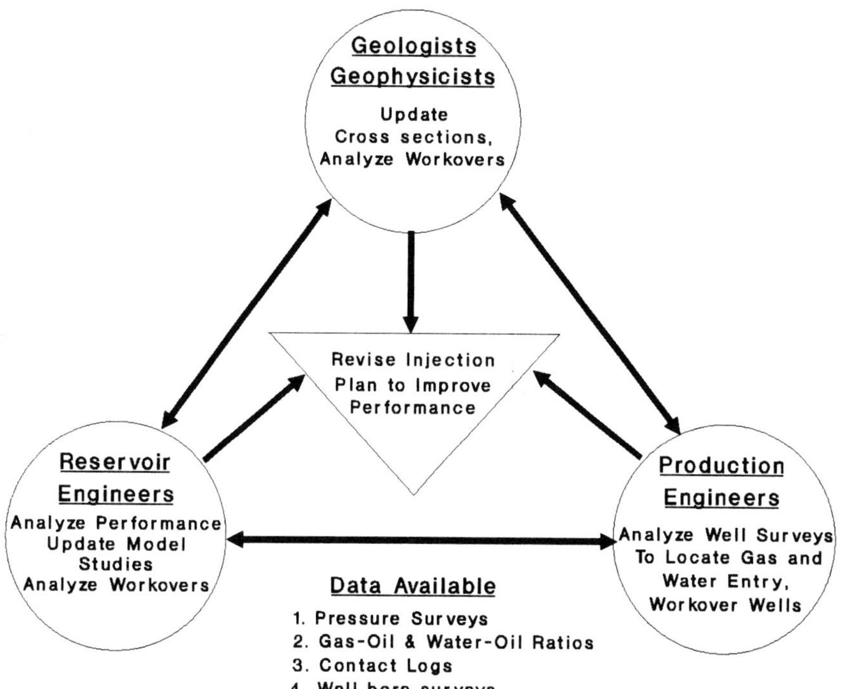

Fig. 4. Communication among team members during reservoir management phase.

about 2.8 times the reserves at about half the finding costs compared to the traditional approach. The development costs were significantly lower for the smaller group. Based on these results, the parent company reorganized the larger group, reducing the number of management levels and increasing the authority of lower levels.

While the use of synergistic teams is appropriate for both large and small companies, the approach may actually have a greater chance for success in smaller organizations. In large companies synergism is, and can be, employed although often diluted and hindered by bureaucracy and size. However, whatever problems there may be with the synergistic team concept, in the next decade economics, the compelling need for future petroleum reserves, and the explosion in technology demand that the petroleum industry develop interdisciplinary, cooperative teams.

Conclusions

A good reservoir description designed to answer key reservoir performance questions is a fundamental tool for appraisal, development, planning, and reservoir management. The incremental cost to obtain an adequate well or seismic data for a reservoir description is very small compared with its value in improved recovery. The time to complete a reservoir description is before significant expenditures are planned and spent. It is not in the best interest of the company (or client) to wait to start a reservoir description program until a crisis or disaster is upon the project.

A detailed reservoir description program should be started during the planning stage as a way to follow and modify depletion programs. Optimum reservoir management requires teamwork and close coordination among geologists, geophysicists, engineers, and managers through all stages of the life of a reservoir.

Reference

Sneider, R. M., 1990, The economic value of a synergistic organization: Presented at the 1990 Archie Conference, Houston.

Reservoir Management Into the Next Century[1]

Helge H. Haldorsen and Theodor van Golf-Racht**

Introduction

Reservoir management is the process of synthesizing available knowledge about a reservoir into an economically optimum field development and production program.

The successful end of an exploration venture is really only a beginning; decades of production will dictate whether a technical-economic success, or a disaster, is in the making. Safe development schemes which yield the forecasted production rates as a function of time are a company's responsibility. The company's task is to manage the reservoir in such a manner that rate forecasts are met and a profitable operation is possible even with the lowest envisioned oil price scenario.

The decision to develop a hydrocarbon field (or to implement a field project) is purely economic. The project's net present value (NPV) must be positive before platforms, facilities, and pipelines can be ordered. The NPV is given by an integral over time (t):

$$NPV = \int \left[\frac{\begin{array}{c} OR(t) \times OP(t) - CAPEX(t) \\ - OPEX(t) - TAX(t) - MISC(t) \end{array}}{(1 + \text{discount rate})^t} \right] dt, \tag{1}$$

where $OR(t)$ is oil rate, $OP(t)$ is oil price, $CAPEX(t)$ is capital expenditures, $OPEX(t)$ is operating expenses, $TAX(t)$ is taxes, and $MISC(t)$ is other expenses; all are functions of time (see Figure 1).

The main tasks of the underground sciences are

(1) to contribute toward optimizing development and production strategy,
(2) to accurately predict future field production rates with a high degree of precision,
(3) to maximize the recovery factor, net present value, and production rates,
(4) to reduce capital expenditures and operating expenses, and

(5) to quantify uncertainty in predictions of future production rates.

Timing is clearly important since equation (1) is a function of time.

Reservoir Management Philosophy

Reservoir management philosophy is a system for guiding the execution of judgement, cooperation, and responsibility with regard to hydrocarbon reservoir evaluation and the technical-economic optimization of recovery schemes. The proposed philosophy is shown schematically in Figure 2. This list is to use as a check at all future field development discussions.

The first questions asked about a discovery are

(1) How much oil and gas may be available?
(2) What are the characteristics of the rock and its fluids?
(3) What description can be given of the reservoir system.

Often the weakest link in a development plan is the reservoir system description.

The Reservoir System

A professional working with a hydrocarbon reservoir builds within his mind and on paper and computer a conceptual model of the reservoir, a subjective representation of a complex and largely unknown reality (Figure 3). The model will contain some elements that approximate reality along with others that are distorted or inaccurate. System understanding grows as static and particularly dynamic information accumulate. With time the representativeness of our model of the real system grows and the precision of estimated future rates and recoverable reserves increases (Figure 4). Unfortunately, major decisions are often made when we know least about the field. Sometimes major decisions should be delayed until more is known, i.e., additional appraisal, long-term testing, and/or phased development should be considered.

More aggressive use of exploration wells should be considered:

[1] Abridged by R. E. Sheriff from paper NMT 890023, New Mexico Tech Centennial Symposium, Socorro, NM, October 16–18, 1989.
*Norsk Hydro Produjksjon A.S., Kjorbokollen, Box 490, 1301 Sandville, Norway.

Production testing of longer duration (creation of transients, better rate-versus-time information, depletion effects, areal continuity);

Injectivity testing (if critical for the development strategy);

Horizontal well creation and testing to determine its benefit over a vertical well;

Testing of temporarily abandoned wells;

Pressure measurements in plugged and abandoned wells with a permanent pressure recorder left behind to monitor pressure throughout the field life.

The main elements in reservoir system description are (1) data acquisition, (2) reservoir description for static computations, (3) reservoir description for dynamic computations (rates versus time), and (4) assessment of risk and levels of uncertainty.

While a reservoir in the subsurface is intrinsically deterministic, uncertainty is introduced because of

(1) Incomplete information about the reservoir's dimensions, internal architecture, and rock property variability on all scales;
(2) Complex spatial disposition of reservoir blocks;
(3) Rock property and structure variability with spatial position and direction;
(4) Unknown relationships between property values and the volume of rock being averaged over; and
(5) Relatively few dynamic values known (time-dependent effects, how rock architecture affects a recovery process).

Reservoir description is a combination of all observations and educated deductions, interpolations, and extrapolations according to an inferred geologic model.

Reservoir description must be associated with a "volume scale" (Figure 5). The "microscopic scale" involves the study of grain and pore-size distributions, pore-wall roughness, packing arrangements, clay lining of pore throats, porosity types, mineralogy, cementation effects, and other features discernible in scanning electron microscope (SEM) and thin sections.

The "macroscopic scale" is the scale associated with a typical core, the scale at which we usually determine reservoir rock properties: porosity, permeability, dispersivity, compressibility, relative permeabilities, and capillary pressure-saturation relation-

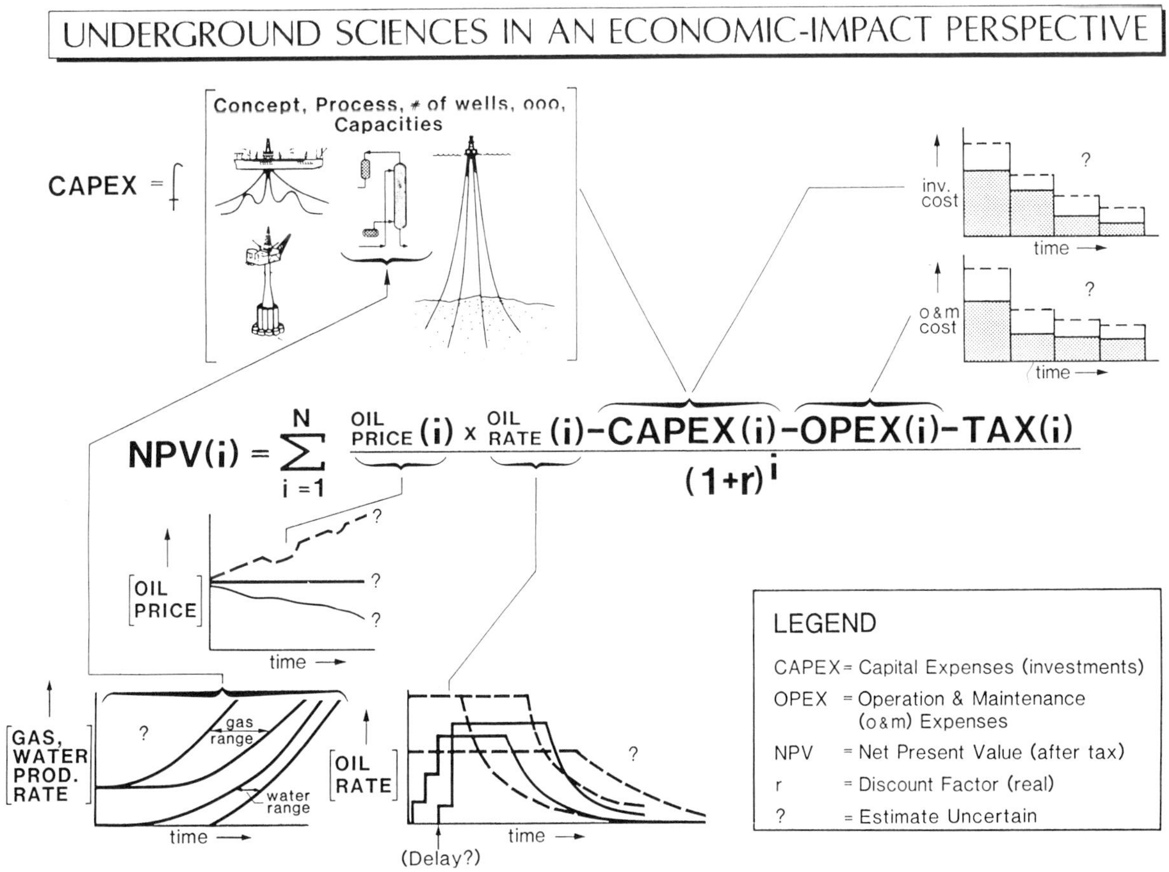

$$NPV(i) = \sum_{i=1}^{N} \frac{\text{OIL PRICE}(i) \times \text{OIL RATE}(i) - CAPEX(i) - OPEX(i) - TAX(i)}{(1+r)^i}$$

Fig. 1. The net present value equation.

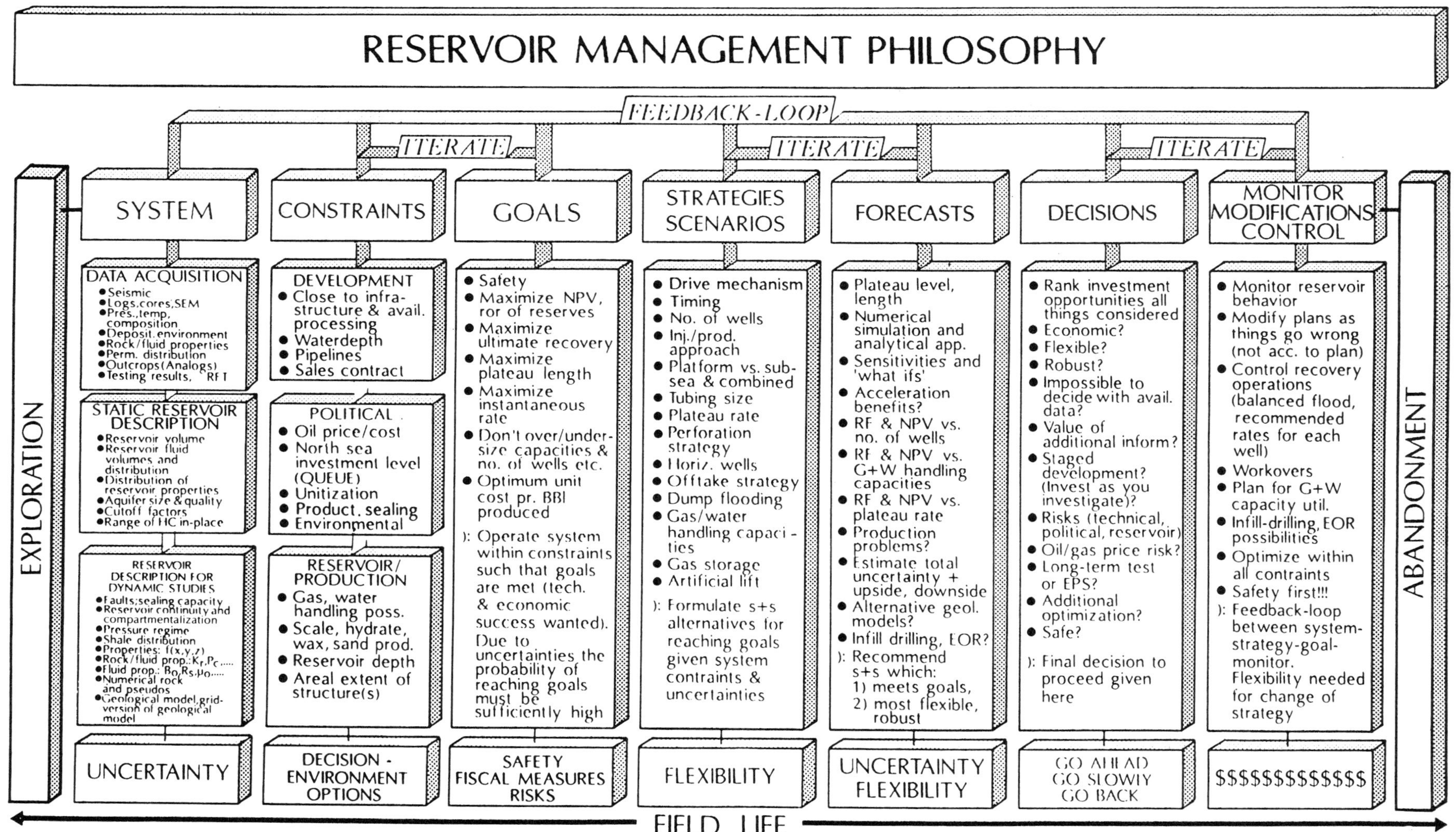

Fig. 2. Reservoir management check-list. Scanning electron microscopy (SEM), Reservoir formation testing (RFT), Hydrocarbons (HC), Net present value (NPV), rate of return (ror) per (pr), Injection/production (Inj/prod), strategies and scenarios (s + s), recovery factor (RF), gas plus water (G + W), enhanced oil recovery (EOR), early production system (EPS), according (acc).

ships. Heterogeneity is evidenced in every well cored; properties vary significantly with position and direction. Macroscopic behavior is generally very difficult to predict from microscopic data.

The "megascopic scale" is that of hydrocarbon volume calculations and numerical simulation studies.

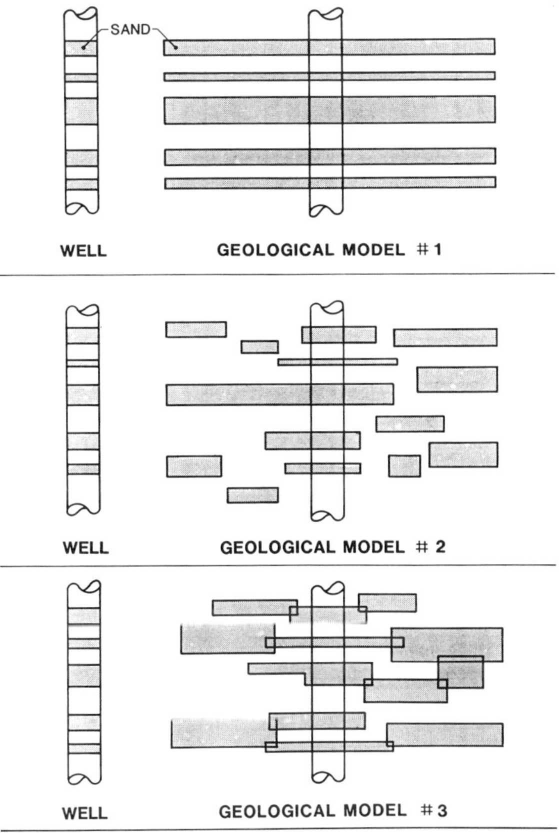

Fig. 3. Three interpretations of how well data relate to the surroundings.

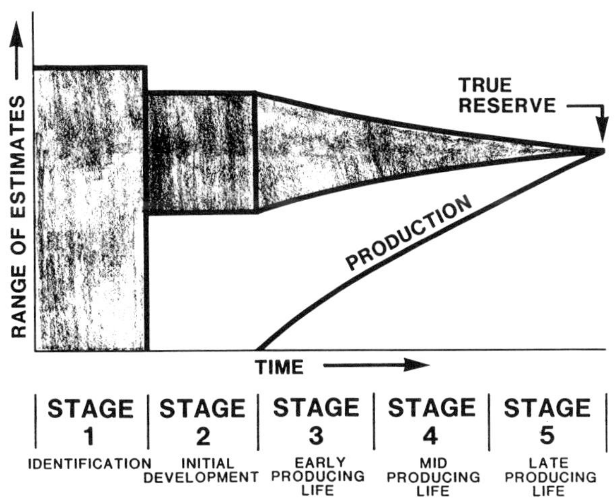

Fig. 4. Expected range of estimates as function of time of estimate.

This scale involves the blocks in the computational grid that represent a discretized version of the continuous geologic domain. At this scale we describe the internal architecture (or "plumbing system") of the reservoir to delineate hydraulic units and the major obstacles to fluid flow. We are interested in the dimensions, shapes, orientations, and spatial disposition of reservoir building blocks and the severity of compartmentalization due, for example, to faults and isolated sands. The difficulties in representing megascopic values by macroscopic averages is comparable to the difficulties in going from the microscopic to the macroscopic. Some reservoir engineering measurements (pressure tests, tracer tests, logs) average over domains of megascopic size.

The "gigascopic scale" concerns regional geologic events such as subsidence rates, basin size, and regional tectonic events. These concern stress fields which may have influenced the stratigraphy and configuration of the reservoir (faults, fractures, stress conditions).

The acquisition of seismic data finds fields, tells their areal and vertical extent, gives the reservoir surface topography, supplies parameters for calculating hydrocarbon pore volume, and tells where faults are.

There is no substitute for a high-quality, three-dimensional (3-D) seismic interpretation. Maps derived from two-dimensional (2-D) seismic data are poor substitutes since fields appear different when constructed from 2-D and 3-D data. Seismic resolution has improved over the years and by use of attribute techniques, vertical seismic profiling (VSP) data, well-

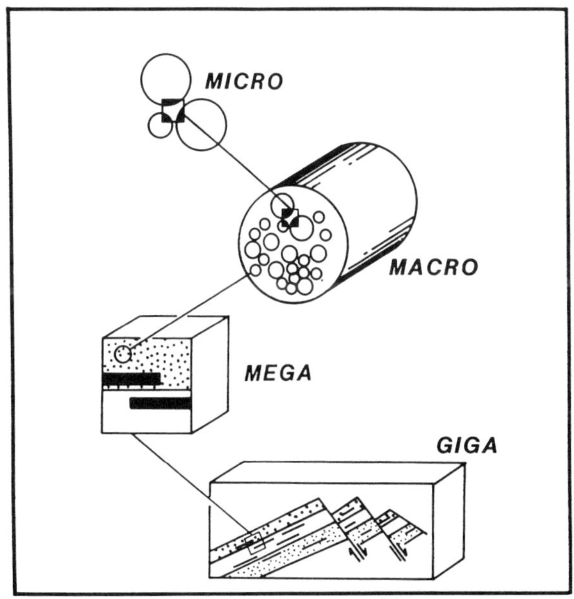

Fig. 5. Scales associated with porous media averages.

to-well seismic measurements, time-lapse seismic monitoring, etc. Seismic interpretations aid in:

(1) reservoir architecture (facies) mapping (e.g., of a channel sand),
(2) mapping reservoir properties such as porosity,
(3) locating fluid contacts, and
(4) monitoring fluid fronts.

Interwell seismic may soon yield preproduction information about lateral continuity.

Initial reservoir pressure, among the most important system parameters, indicates the potential energy of the reservoir. The main problems to decide are (1) which primary recovery process or engineered drive is most appropriate, (2) if and when pressure maintenance should be initiated, and (3) if and when artificial lift is needed. Even in overpressured reservoirs undergoing waterflooding, maintaining overpressure provides steady total produced liquid rates at high water cuts toward the end of field life. Formation pressure studies in exploration, appraisal, and development wells are invaluable for delineating pressure continuity among neighboring formations, finding fluid contacts, indicating areal pressure continuity, and delineating the sealing capacity of faults. Distributed pressure measurements provide both static and dynamic reservoir data. To monitor the field-wide reservoir pressure over time is important. This information directly affects the offtake strategy, aids in history-matching of models, and informs about the degree of compartmentalization and aquifer support.

Measurements on cores yield many of the parameter values needed for reservoir simulation. Obtaining cores is an arduous task that involves drastic treatment of the cores which may change the properties to be determined. In drilling a well a core is subjected to higher pressures than its natural habitat, flushed by drilling mud, and put under different stresses in the core barrel. On its way to the surface pressure and temperature are reduced by enormous amounts, gas comes out of solution, retrograde condensation occurs, everything expands, and the core may fracture. More drastic treatment occurs in the laboratory where smaller plugs are drilled from the core and the mud is flushed out to arrive at a ''fresh'' or ''restored state'' core. To determine realistic rock-fluid interaction values, the core needs to be saturated so that various phases are correctly proportioned in the correct pore spaces, and put under pressure, stress, and temperature conditions to match in-situ conditions, with live hydrocarbons and gas-saturated brine. Even with the best procedures, representative and valid data involve major uncertainties.

Obtaining a sample of unsaturated hydrocarbons is usually straightforward. Possible pitfalls involve oils close to saturation (or bubble point) pressure and gas/condensate systems. Representativeness is the main concern; no component should escape and the right mol and weight proportion of components must be maintained.

The main rock properties to be measured are Klinkenberg-corrected absolute permeability (vertical and horizontal), porosity, compressibility, and the effects of overburden stress on these. Usually the first two are measured on plugs every 25 cm, but bias is often introduced because samples are selected from available intervals of 50 cm. The list of properties every 25 cm is the basis for selecting layers in the simulation model (hydraulic units).

The fluid data to be measured depend on the type of field. For undersaturated oil systems, measurements needed are of viscosity, density, formation volume factor, compressibility, solution gas/oil ratio, all as a function of pressure (at constant reservoir temperature), and bubble-point pressure. For gas and gas-condensate reservoirs, needed measurements are composition, viscosity, dew point, liquid drop-out curve, gas-formation factor, and compressibility. Black-oil reservoir models generally require different data than full composition models. Those fluid parameters which could make or break field development should receive the greatest attention.

Data that reflect the interaction between rock and fluids include capillary pressure curves, relative permeabilities, residual saturations, hysteresis effects, adsorption isotherms, and capillary desaturation curves. These data are essential for subsequent analytical studies and numerical simulation and they are often the most difficult, costly, and time-consuming to acquire.

How is the most representative data set selected? Steady-state and unsteady-state experiments on the same system seldom give the same relative permeability. This and similar relevant questions seldom have clear-cut answers and appreciable uncertainty is always present. On the other hand, the performance of an oil or gas field is rarely ever a disaster solely due to wrong laboratory data. Most often surprises are linked with erroneous assessments (and modeling) of the large-scale geologic architecture or the reservoir's plumbing system, sand accessibility and interwell floodability problems, high-permeability (thief) zones, inadequate or excessive aquifer support, fractures, sealing faults, shales, and so on. Nonrepresentative laboratory data can, however, lead to over- or under-dimensioning of water-handling capacity, miscalculation of injectivity, and in some situations may lead to a wrong choice of recovery mechanism.

Insight into values to use in a simulation and problems that might develop often can be obtained by comparing what happened with what was predicted to happen in older fields that are analogs. Post-mortem reviews should focus on understanding serious discrepancies.

Estimates of the amount of hydrocarbons in place often exhibit a nonsymmetrical distribution around the "base value", the 50 percentile value. Usually high and low values appropriate to the 20- and 80- percentile points are also calculated. Monte Carlo or conditional property-field simulation techniques are utilized to determine the overall distribution, and an uncertainty of ±25 percent is not uncommon.

The very important net-to-gross (N/G) cutoff factor, i.e., when a unit should be eliminated from the definition of "pay", is a complex function of recovery process, size, shape, and position of a member in question, the contrast in properties, time, etc. Arbitrary cutoff definitions (e.g., 0.6 mD permeability or 5 percent porosity) may be in great error.

The system description also involves quantification of uncertainty about in-place volumes, compartmentalization due to faults, shale continuity, aquifer size and quality, etc. (Figure 6).

Constraints and Goals

Many constraints, which can be very restrictive, are imposed on the development of a hydrocarbon field: development criteria, political considerations and regulations, limitations about the field itself. Given the system, system uncertainties, and the constraints, the problem is how to formulate realistic technical-economic goals and how to reach them.

The ultimate goals may be to maximize the ultimate recovery (national interest), or to maximize the net present value (company interest), or to keep a plateau rate as long as possible (sales department interest), or some other measure of economic success. Optimal exploitation also requires the prevention of noneconomic developments and maintaining a flexible response to arising opportunities. In a climate of reduced and fluctuating oil prices and smaller, more complex fields, emphasis has shifted from pure optimization to risk reduction.

At every point in a decision process questions to be asked include:

Is it useful?
Is it cost effective?
Can it be achieved by alternate means? and

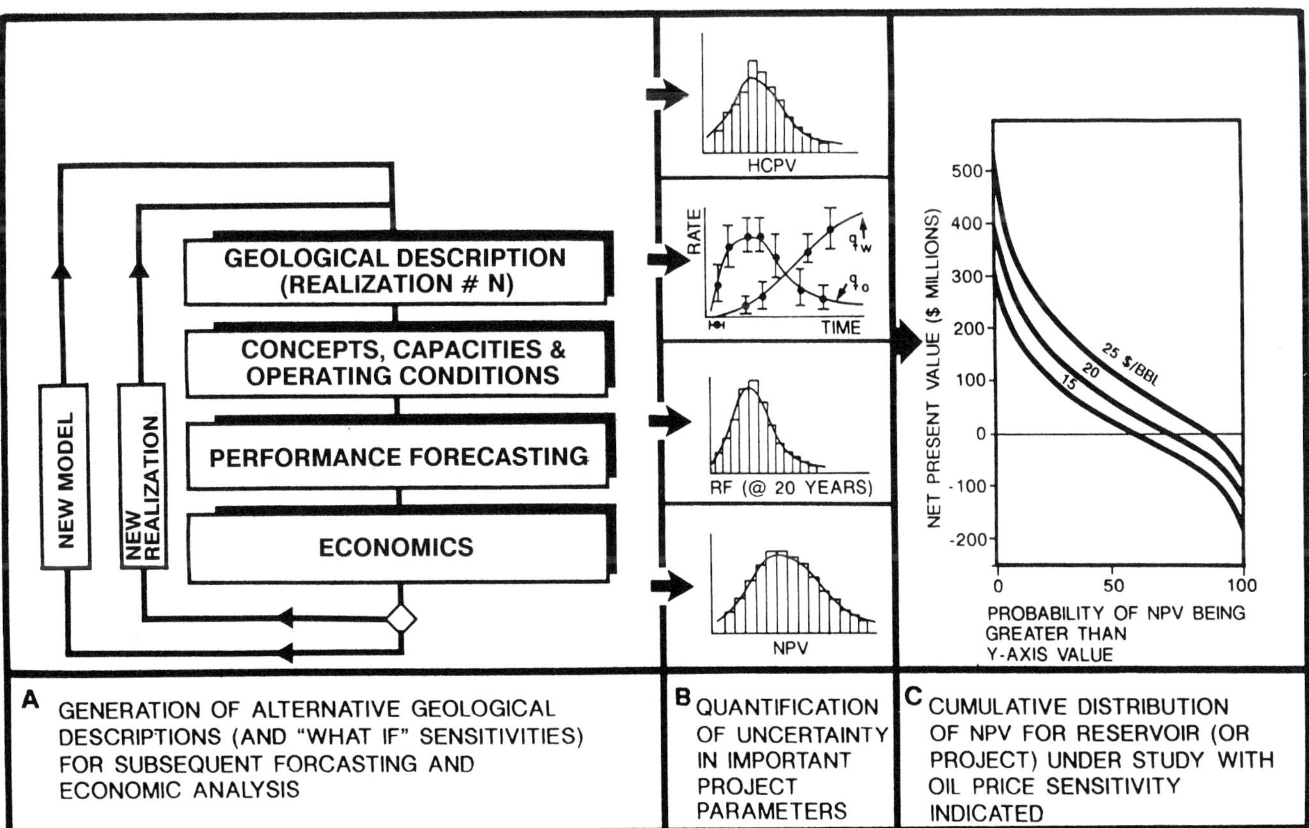

Fig. 6. Exposing the uncertainties in project parameters. Hydrocarbon pore volume (HCPV), recovery factor (RF), net present value (NPV).

What are the constraints and long-term consequences?

The better the integrating of disciplines, the greater the chance of formulating realistic goals and meeting them.

Reservoir Simulation

Today numerical simulation studies form the basis of almost all decisions on commerciality, maximum effective rate, overall offtake plans, and initiation of secondary and enhanced oil recovery projects, as well as the major business decisions such as the location and number of platforms, depletion strategy, the number and location of injection/production wells, and gas and water handling capacities.

Reservoir simulation is the process for inferring reservoir behavior from the behavior of a mathematical grid model. Simulation models have evolved from single, homogeneous "sugar-cubes" to sophisticated, 3-D, multiphase, numerical simulation models with thousands of grid blocks. Some have become "black boxes" that give little physical understanding or geological insight. Modeling combined with synergy among disciplines reduces the probability of failure, allows the study of various strategies and "what ifs", allows investigating sensitivities, and permits the design of robust and flexible development plans.

Clearly a model is no better than the data on which it is based and data invariably involve uncertainties. Most studies include up- and down-side cases as well as a base case, so that parameter uncertainty can be translated into rate-forecast uncertainty. Flexibility guards against the adverse effects of uncertainty; if a field is not behaving as forecasted, a back-up scenario may be implemented. However, flexibility is costly.

While an extensive menu of recovery options (Figure 7) is theoretically available, guidelines for their use often disregard timing, cost, equipment, and operational considerations. The selection of recovery mechanism should only be decided after screening candidate strategies with respect to system characteristics, constraints, goals, and technical/economic feasibility studies.

Prior to numerical simulation an attempt should be made to predict the reservoir behavior by analytical approaches, such as material-balance, fractional flow theory, decline curve analysis, etc., to gain insight into the problem.

Grid blocks in simulators (Figure 8) are considered homogeneous regardless of their size. Hence geological architecture and property variability at a scale smaller than the grid block must be handled by using pseudo-functions and effective properties. Larger-scale features such as areally extensive shales and sealing or partially sealing faults can be embedded as red traffic lights. The use of millions of grid blocks could preserve local behavior around wells, maintain sharp fluid fronts, and correctly represent viscous, capillary, dispersive, and gravity effects. However, a compromise must be made between the desire for detail and computational cost/efficiency standpoints. Even if data are of excellent quality and fully representative, predictions may still be unrepresentative because of numerical dispersion, grid orientation, or other numerical effects (Waldren, 1988), or because

POSSIBLE GRIDS

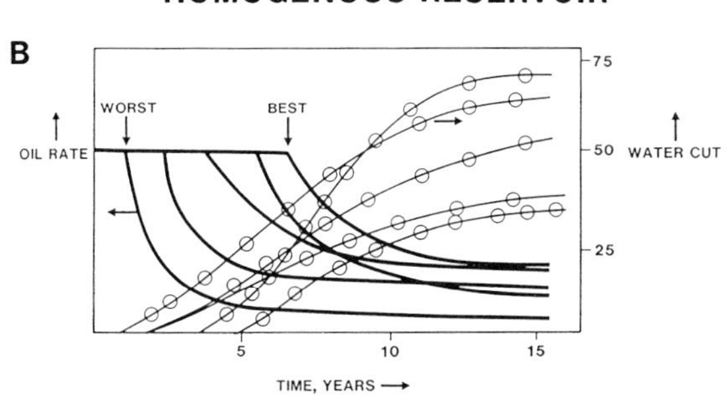

HOMOGENOUS RESERVOIR

Fig. 8. Gridding for reservoir simulation. (A) (1)–(5), grids of different sizes; (6) combination of grid sizes; (B) effect of grid on production.

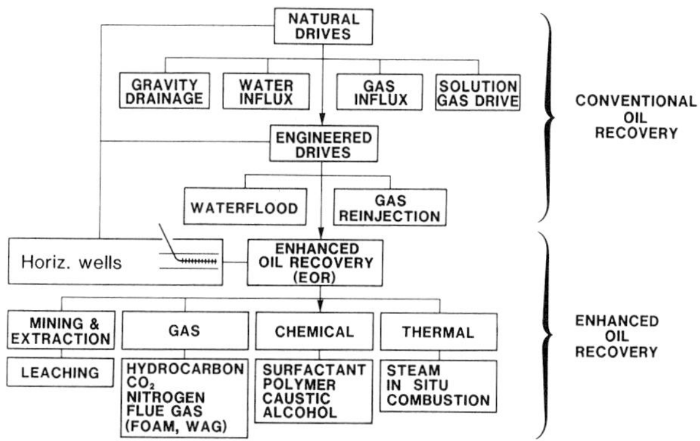

Fig. 7. Recovery mechanisms.

shale continuity, fault positions, aquifer size, or other geologic factors affect breakthrough times and recoveries.

Practical problems include (1) the identification of hydraulic units, (2) the assignment of average physical properties to hydraulic units at wells, (3) the interpolation and extrapolation of properties away from wells, and (4) the incorporation of geologic architecture such as sand and shale continuity, aquifers, faults, etc. Saleri et al. (1988) note that standards for building models do not exist, and a standard methodology is clearly needed.

To build up a full-field model, many companies study the problem at increasing scales (Figure 9):

(1) 1-D modeling and matching of laboratory flooding experiments on cores; capillary pressure and oil-water relative permeability may be reinforced by matching pressure changes versus time and saturation distributions during a flood.

(2) Single-well simulation models to match drill-stem tests; modification of values from laboratory measurements may be necessary. Long-term testing should see increasing application; the value of the information obtained should be compared with the costs of total field failure rather than the cost of the testing.

(3) Cross-sectional models through selected parts of the field; these investigate the vertical flow of phases and vertical sweep efficiency (Table 1). Perforation strategies, reservoir descriptions (e.g., shale continuity), layering schemes, rates, relative permeabilities, etc. are studied. A large number of layers must be modeled (Figure 10a); many of these will be combined into coarser layers (Figures 10b,c) in sector and full-field models using pseudo-relative permeabilities and pseudo-capillary pressures. Different recovery mechanisms should be investigated. For example, for a dipping oil reservoir with initial pressure well above the bubble point, production by depletion followed by secondary gas cap formation and gravity drainage might be compared to waterflooding.

(4) Sector models of relatively confined areas of the field.

(5) Full-field models and history-matching; these are the models that really count. Grid block dimensions are often hundreds of meters with only a few layers; the scale ratio between these blocks and laboratory core measurements is often 10^{12}.

NUMERICAL MODEL STUDIES: FORECASTING METHODOLOGY

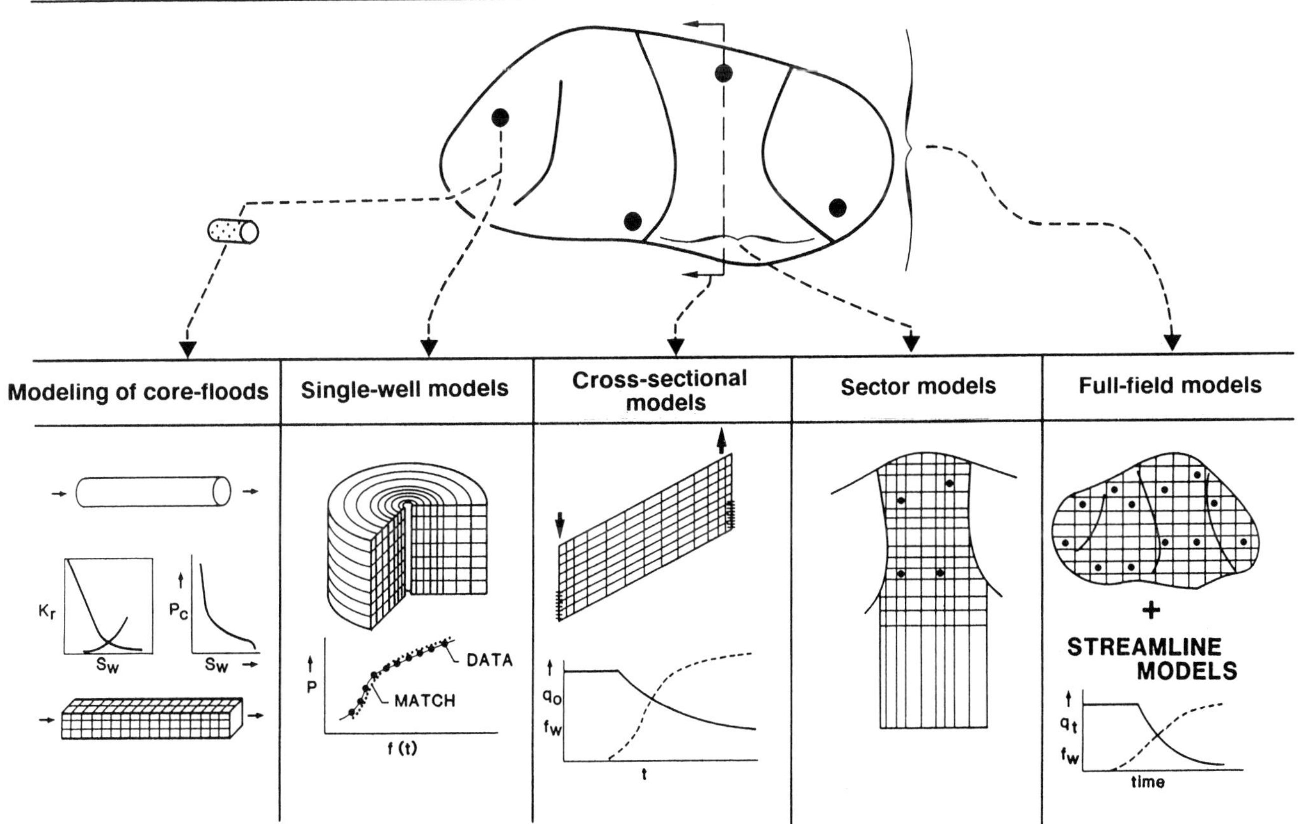

Fig. 9. Numerical modeling for reservoir simulation. Relative permeability (K_r), water saturation (S_w), capilary pressure (P_c), well pressure (P), production history f(t), oil production (q_o), formation water production (f_w).

Grid effects can result in serious distortions. A base case simulation and sensitivities to input data changes are run using a "one-factor-at-a-time approach" (which may be misleading because of interactions). The recovery factor associated with each sensitivity run is associated with an estimate of the probability that the parameter would take on that value. Results are then combined in a Monte Carlo approach. The final result is a range of values along with associated probabilities for taking on values within the range.

Strategies, Scenarios, and Forecasting

Many scenarios, sensitivities, and "what ifs" should be investigated with reservoir simulators (Figure 6):

Sensitivity of recovery factor and project economic indicators to the number of wells used;

The influence of aquifer size, fault compartmentalization, shale continuity, vertical permeability, etc. on plateau duration and ultimate recovery;

The influence of plateau rate on ultimate recovery and net present value;

The influence of relative permeabilities and residual saturations on production rates through time;

The effects of injection-production well locations; a streamline model often can cut computer costs (Hewett and Behrens, 1989);

The influence of offtake strategy on plateau duration.

Reservoir simulation models do not themselves optimize recovery operations; a skilled engineer must interpret the results of simulation runs and institute changes in the model (well positions, perforation intervals, injection rates, production rates, etc.) that will result in increased performance.

Because many unverifiable assumptions are inherent in forecasts, cheap flexibility is necessary when actual and predicted outcomes differ. For example,

Over dimensioning of water and gas handling capacities has been a common (but expensive) means of providing flexibility regarding errors in relative permeability values;

If aquifer support is stronger than expected, water injection wells may be converted to producers;

If a reservoir is more discontinuous than assumed, extra platform slots should be planned. Pulse, interference-testing, or reservoir formation testing (RFT) should be initiated early to reduce uncertainty; and

If water injectivities are less than expected, fracturing of injection wells may be required.

Table 1. Factors affecting sweep efficiencies and residual oil saturation.

	Sweep efficiencies			Residual oil Saturation
Quantity	E_D	E_A	E_V	
Throughput	strong	strong	strong	slight
Mobility ratio	strong	strong	strong	slight
Gravity	moderate	slight	strong	slight
Capillary forces	slight	slight	moderate	strong
Heterogeneity	slight	strong	strong	moderate
Relative permeabilities	strong	slight	moderate	strong
Well pattern	slight	strong	moderate	slight
Well spacing	slight	moderate	moderate	slight
Dip/structural well position	slight	strong	strong	slight
Faults (positions, lengths, sealing capacities, throws, architecture)	slight	strong	moderate	slight
Offtake strategy	slight	moderate	moderate	slight
Perforation policy	slight	moderate	strong	slight
Shut-in criteria	slight	moderate	strong	slight
Fracturing during waterflooding	slight	strong	moderate	slight
Anisotropy in permeability	slight	strong	moderate	slight
Zonal allocation of prod./inj. wells	slight	strong	strong	slight
Future application of profile control agents	slight	slight	strong	slight
Future application of mobility control agents	strong	strong	strong	slight
Future application of S_{OR}-reducing agents	strong	moderate	moderate	strong
Workover strategy	slight	slight	strong	slight
Gas/water handling capacities	slight	moderate	moderate	slight

There is an intimate relationship between the choice of recovery mechanism (number of wells/slots, need for injection or processing, etc.) and the choice of development concept (steel or concrete structures, subsea solutions, floaters, ships, facilities, etc.). Massive engineering effort may be undertaken too early, before the optimum concept is known. A too-early sub-optimal solution may, or may not, make more money than a late-optimum solution. Companies generally want to recover costs and earn money as soon as possible.

Offtake strategy involves determining which wells/areas should be produced on a given day, and at what rates, through the life of the field. Gas-oil ratio (GOR) and/or water-oil ratio (WOR) increasing with time and limited gas-water handling capacity is an important constraint in many fields. The end-of-plateau is often when the capacity can no longer handle all the gas and/or water produced. Offtake strategy determines how production from different wells is controlled to meet management goals. The optimum allocation usually means maximizing the net present worth.

As an example of offtake strategy decisions (Kirwan, personal communication, 1983), assume a field with gas/water injection (gravity drainage) where cusping/coning is a problem. Initially all wells have low GOR/WOR and excess capacity exists in the gas/water handling system. The major objective of offtake strategy during the plateau period is to use the excess capacity to maximize ultimate recovery while maintaining the plateau rate. Options are to

Maximize offtake from areas with low cumulative production;

Maximize production from current high GOR/WOR are as to maximize production from these areas prior to the decline point;

Limit production from areas with high cumulative offtake;

Limit production from potential cusping/coning wells to less than the critical coning rate;

Develop oil and gas production targets by well.

Offtake targets should be reviewed regularly to account for changes in reservoir performance. Other constraints that might influence the decision include the efficiency of recovery might be strongly rate dependent, GOR may be more a function of rate than of cumulative offtake, and a high withdrawal rate may cause uneven depletion. We may expect cybernetic offtake strategies ("brains on simulators") to be used in the future rather than arbitrary strategies.

Half of all wells are expected to be horizontal wells by the mid 1990s.

Two 1500 ft horizontal wells on the crest of an oilfield with gas cap pressure maintenance will replace clusters of 3 to 4 gas injectors.

The number of wells or slots needed in gas fields will be reduced by a factor of four. Reduced draw-down implies less liquid drop-out and fewer production problems.

In low permeability gas reservoirs the long well-bore will compensate for the low permeability and therefore less hydraulic fracturing will be required.

Both peripheral and pattern-style water injection wells will be used. Wells that traverse faults will eliminate many fault compartmentalization effects. More channel-sand reservoirs may be hit by horizontal wells.

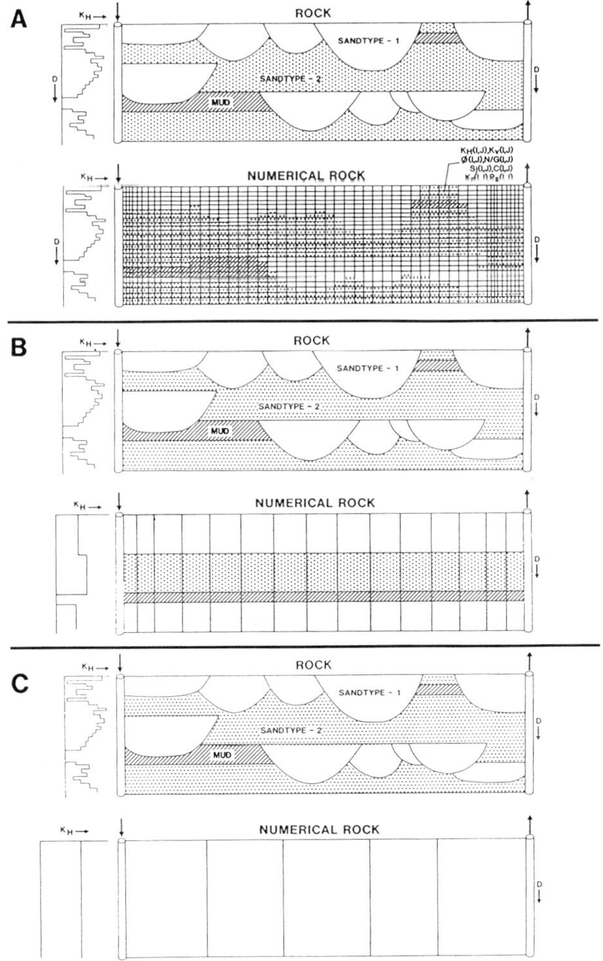

Fig. 10. Assigning property values to grid cells for reservoir simulation: Horizonal permeability (K_H).

Many applications will accelerate oil production, reduce the number of wells, and reduce coning and cusping. Sand production will be less because of reduced draw-down.

For thin oil rims the lowering of draw-down will reduce coning and cusping.

Marginal fields close to existing processing can be developed.

For naturally fractured reservoirs both injection and production will be enhanced since a well will intersect many more fractures. The lower draw-down will also enhance the production and injection since fractures tend to close as the fluid pressure decreases.

After a field has been found, horizontal appraisal wells will give better sampling of reservoir properties and more complete evaluation.

No strategy-scenario-forecast cycle is complete before horizontal wells have been considered.

Decisions and Decision Making

Usually capital investments are approved only after a comprehensive review by senior executives at which all classical economic indicators must be presented, risks exposed, timing issues discussed, and the investment must be compared to other investment opportunities. In particular, a project must be technically viable, economically feasible, flexible and robust, and safe.

The "Go slow" and "Go back" flags (Figure 2) frequently are not used enough. Most reservoir data are static and the dynamic (time-dependent) data urgently are needed. Long-term testing or staged development or an early production system should be considered as the rule rather than as the exception. Often additional information should be gathered before an investment decision is made. For example, long-term testing of one or more wells, enhanced recovery pilot projects, drilling and testing of a horizontal well, etc. Additional information should be gathered as long as the information contributes to improving the net present value and reduces the investment risk.

Reservoir-related decisions always depend on estimates based on very limited and uncertain information. Therefore, the decision must be made if further information gathering is justified. The estimates on which decisions are made are almost always biased; people usually overestimate project performance (Brush et al., 1982; Campbell, 1986; Castle, 1986). Quantifying the probability of alternative events will indicate if expectations are too optimistic or pessimistic. The value of information is the probability weighted outcome if information is gathered, minus the outcome if information is not gathered. Lohrenz (1988) described a method for evaluating the value of information; his decision-tree (Figure 11) method is explained by Warren in Chapter 8 of this book. This method is also useful in indicating which factors are critical. One of the difficulties in applying the method is how to establish the probabilities of alternative events. Application is limited to cases where the outcome of alternative events can be quantified as net present value or similar economic measures.

As an example of estimating the value of information, assume that it is not known if there will be aquifer support for a newly discovered field, but if there is no aquifer support, development will be marginal. Should a long-term test be made of the discovery well? If the long-term test indicates support, additional wells will be drilled; if no support, the well will be produced as long as possible to minimize the loss. Both high and low production cases, assumed to be equally probable, are to be evaluated, but the probability of aquifer support (f_a) is not known. The analysis (Figure 12) yields Figure 13. The criteria for inaugurating the long-term test is that the value of the information ($C_{\max}$) is positive, and in this instance this is true if the probability of aquifer support exceeds 13 percent. Often only a small probability of aquifer support will justify long-term testing.

Monitor, Modify, and Control

In offshore fields a new choice of recovery strategy is practically impossible once production has begun. Flexibility is usually limited to selective production, workover or recompletion, infill drilling, horizontal wells, redrilling (sidetracking), artificial lift, capacity utilization, offtake strategy, etc.

RFT measurements should be run in new development wells to check for differential depletion between

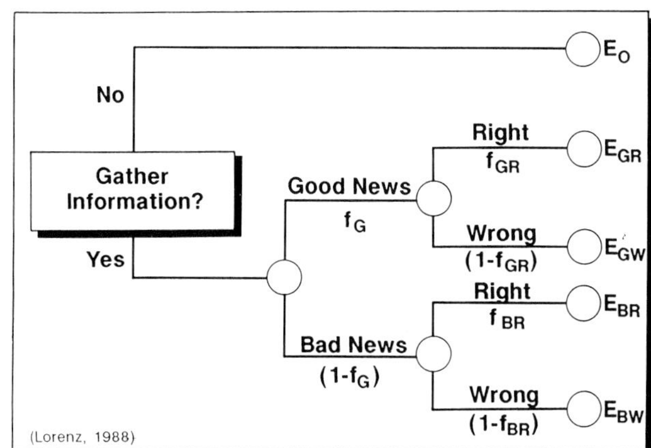

Fig. 11. Value of information decision tree. Probabilities (f_G, f_{GR}, f_{BR}); end result (E).

layers. RFT measurements which indicate vertical permeability need to be reconciled with petrophysical and geological evidence. An integrated reporting system is needed with proper custodianship for data integrity. A complete record of reservoir and individual well behaviors is warranted. The geophysical and geological mapping system which contains the static reservoir description should be used in conjunction with the production reporting system through which reservoir performance is monitored. History matching should yield the dynamic data to resolve uncertainties about the geologic model. Decisions should be based on all data available and a consistent interpretation which explains all phenomena. Inaccuracy should not be used as an excuse to explain facts away.

LONG TERM TEST DECISION TREE

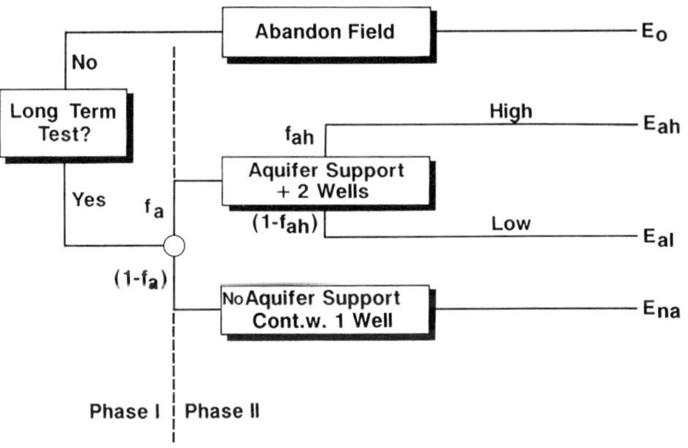

Fig. 12. Decision tree for aquifer support example. Probability of aquifer support (f_a). End result (E).

LONG TERM TEST:

VALUE OF INFORMATION

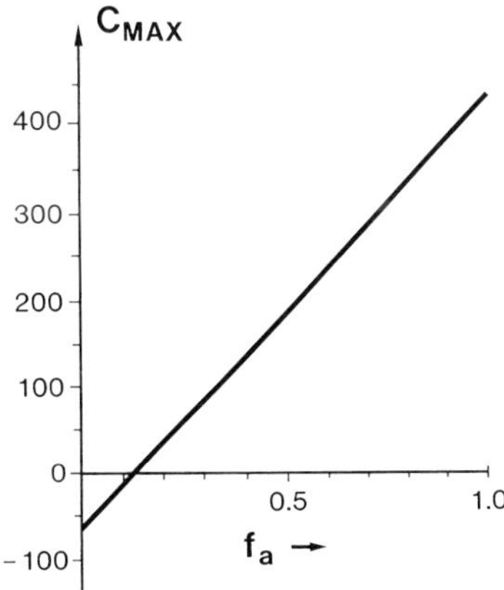

Fig. 13. Net value of information (C_{max}) versus probability of aquifer support (f_a).

Summary and Recommendations

A philosophy for managing reservoirs from exploration to abandonment has been presented (Figure 2). Reservoir description (and the adequacy of its representation in forecasting tools) is the foundation upon which field development plans rest; the amount of truth in forecasted production rates, the degree to which selected facilities will fit, and the amount of reality in economic projections are all intimately related to the reservoir description. Integration of disciplines is required to ensure that all data are considered in all contexts and that global rather than local optimality (within each discipline) is achieved.

Some selected summary points follow:

(1) Due to the large up-front exposure of capital in field developments, the risk regarding future cash flow must be quantified a priori. If things are too uncertain, management should raise a ''go slow'' sign and consider long-term testing to collect dynamic data. Data are, however, only useful if they affect decisions. The petroleum industry needs standards for forecasting (simulation) and in particular for uncertainty estimation. Uncontrolled and arbitrary practices in reservoir description and simulation are prevalent today. 3-D seismic data acquired early can help make valid up-front decisions as well as be useful throughout the field life (see following paper by Robertson).

(2) A large potential exists for horizontal wells for injection, production, and even appraisal. Next generation field developments may employ mostly horizontal production/injection wells with only a third the number of wells. Larger (9⅝-inch) tubing will be necessary to inject and/or produce larger volumes per horizontal well.

(3) Post-mortem reviews can give valuable empirical information about the ability to predict and they may directly impact the development of an analog field with similar recovery mechanism.

(4) Concept optimization should go on concurrently with reservoir recovery mechanism optimization. If development concepts are frozen too early (because detailed engineering has been done too early), the overall optimization is really fictious because the number of wells, capacities, etc., are fixed and significant changes are not really possible. Allow ample time for screening concepts, and postpone detailed engineering studies until a clearly superior concept emerges. Consider horizontal wells. If risks are too high, go slowly and consider phased development or long-term testing or both. Underground scientists and concept selectors must work closely together.

(5) Reservoir engineers must recommend offtake strategy and production rates. If the goal is to keep a production plateau as long as possible, the selection of well production rates should change through time. The

best solution may not be to control rates in such a manner that gas or water breaks through at all wells at the same time; this might result in costly gas and water handling capacity remaining under-utilized for years. Longer plateau periods and higher oil recovery may be achieved by producing high GOR/WOR wells to the maximum possible extent while handling capacity is available, and saving low GOR/WOR wells for later, an approach that is contrary to the normal instinct to shut-in/choke-back high GOR and WOR wells.

(6) Many fields need dynamic data. If (a) analysis indicates that a field can be developed with a positive net present value, (b) the stand-alone economics of a long-term test is positive, (c) the computed value of dynamic information is large, and (d) the worry is the reservoir risk, then long-term testing seems appropriate to prevent a large potential loss. Fields may fall into this category where oil rates are not high enough to yield attractive stand-alone economics. In offshore reservoirs with particularly high geologic or reservoir risk or both, second generation tests with injection-production facilities should be utilized up-front to investigate the most critical parameters or assumptions prior to full-field development.

(7) Leaving continuous measurement sensors in discovery and appraisal wells when they are plugged and abandoned would be helpful. With subsequent field development these observation wells will yield valuable interference and water-front arrival data on a continuous basis.

(8) Petroleum engineering research should be focused on real field evaluation, production problems, and opportunities eventually leading to applications in the field.

References

Brush, R. M. and Marsden, S.S., 1982, Bias in engineering estimation: J. Petr. Tech., February.

Campbell, J. M., 1986, Nontechnical distortions in the analysis and management of petroleum investments: J. Petr. Tech., December.

Castle, G. R., 1986, North Sea score card: Soc. Petr. Eng. Paper 18358.

Hewett, T. A. and Behrens, R. A., 1989, Scaling laws in reservoir simulation and their use in a hybrid finite difference/streamtube approach to simulating the effects of permeability heterogeneity: 2nd NIPER/DOE Conference.

Lohrenz, J. 1988, Net values of our information: J. Petr. Tech., April.

Saleri, N. G. and Toronyi, R.M., 1983, Engineering control in reservoir simulation, Part 1: Soc. Petr. Eng. Paper 18305.

Waldren, D., 1988, Discretisation techniques for multiphase flow, in Edwards, S., (Ed.), Mathematics in Oil Production : Oxford Univ. Press Inc.

Reservoir Management Using 3-D Seismic Data[1]

*James D. Robertson**

Introduction

The geologic detail needed to properly develop most hydrocarbon reservoirs substantially exceeds the detail required to find them. This obvious, but compelling, precept has fueled the steadily increasing application of three-dimensional (3-D) seismic analyses to reservoir management. A measure of the increase is that 3-D surveys now account for half the seismic activity in the offshore Gulf of Mexico and North Sea, and the percentage has risen yearly since commercial 3-D surveys were first shot in these areas in 1975 (Figure 1). Use of the technology in other offshore areas and on land likewise is growing rapidly.

A 3-D seismic survey is one of the many tools that a reservoir management team employs to plan and monitor the development and production of a field. The seismic information is integrated with well logs, pressure tests, cores, and other engineering/geoscience data from the discovery and delineation wells to formulate an initial field development plan. As more wells are drilled, logged, and tested, and production histories are recorded, the interpretation of the 3-D data volume is revised and refined to take advantage of the new information. Aspects of the interpretation that were initially ambiguous become clear as an understanding of the field builds, and inferences from the seismic data become more detailed and reliable. The 3-D data volume evolves into a continuously utilized and updated management tool that impacts reservoir planning and evaluation for years after the seismic survey was originally acquired.

Types of 3-D Seismic Analyses

The interpretations that a geophysicist might perform with 3-D seismic data can be grouped conveniently into those that examine the geometric framework of the hydrocarbon accumulation, those that analyze rock properties, and those that try to monitor fluid flow and pressure in the reservoir (Figure 2).

These analyses affect and, one hopes, significantly improve decisions that must be made about volume of reserves, well/platform locations, and recovery strategy.

Geometric Framework

I use "geometric framework" as a collective term for such spatial elements as the attitudes of the beds that form the trap, the fault and fracture patterns that guide or block fluid flow, the shapes of the depositional bodies that make up a field's stratigraphy, and the orientations of any unconformity surfaces that might cut through the reservoir. A 3-D seismic data volume samples the geometric framework on a regular 3-D grid (generally 10 to 30 m laterally and 5 to 15 m vertically).

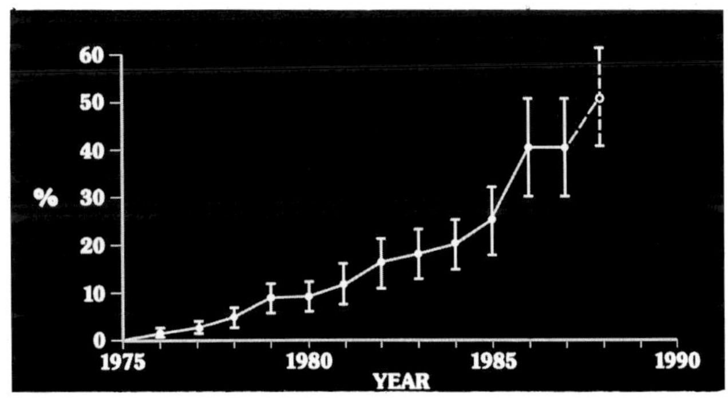

Fig. 1. 3-D seismic as a percentage of total seismic in the Gulf of Mexico and North Sea.

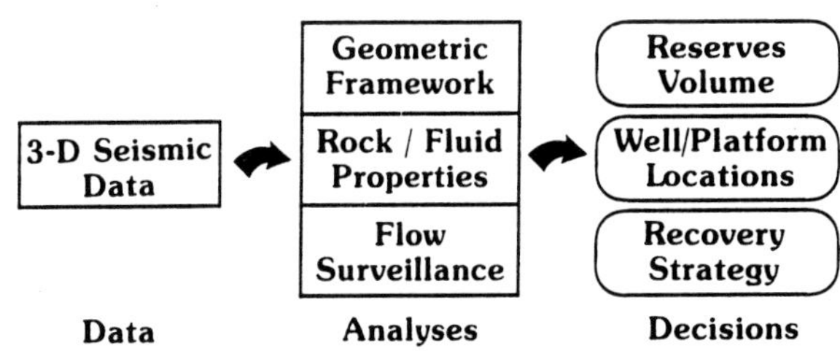

Fig. 2. Applications of 3-D seismic data to reservoir management.

[1] Presented as the Fall 1988 Distinguished Lecture, Society of Exploration Geophysicists. An earlier version published in The Leading Edge, **8**, No. 2, 25–31.

*ARCO International Oil and Gas Company, 2300 W. Plano Parkway, Plano, TX 75075, USA.

By mapping traveltimes to selected events, displaying seismic amplitude variations across selected horizons, isochroning between events, noting event terminations, slicing through the volume at arbitrary angles, compositing horizontal and vertical sections, optimizing the use of color in displays, and using the wide variety of other interpretive techniques available on a computer workstation, a geophysicist can synthesize a coherent and quite detailed 3-D picture of a field's geometry.

An example of mapping geometric framework is shown in Figures 3 and 4. Figure 3 is an early structure map of the Prudhoe Bay Field on the North Slope of Alaska (Morgridge and Smith, 1972). It shows the basic elements of the trap at Prudhoe Bay: the dip to the south and southwest, the boundary fault to the north, and the erosional truncation of the Sadlerochit reservoir to the east. Figure 4 is a time slice from one of the 3-D seismic volumes that now exist over the field (location shown in Figure 3). The reds and blues are seismic peaks and troughs, and the time slice cuts the volume at about the level of the unconformity truncating the Sadlerochit. One can clearly see the Sadlerochit subcrop imaged as a red-blue-red event traversing north-northwest to south-southeast across the right side of the image. The northern boundary fault appears as a sharp, arcuate lineament crossing the top of the time slice. More subtle elements of the geometric framework, such as additional east-west and northwest-southeast faulting that disrupts the continuity of the red and blue seismic events, are also evident.

A time slice, which is an example out of the general category of regularly gridded two-dimensional (2-D) digital images, can be analyzed with processing techniques developed for imagery such as satellite pictures. An example is shown in Figure 5. An artificial sun is assumed to exist above and to the south of a stack of time slices, and the texture of the time-slice stack is highlighted as a light and shadow pattern by illumination from this sun. One can clearly see the

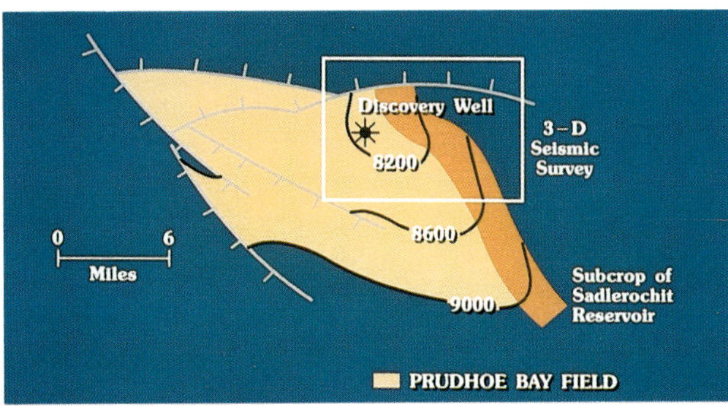

Fig. 3. Prudhoe Bay Field, Alaska—1971 Sadlerochit structure map (Morgridge and Smith, 1972).

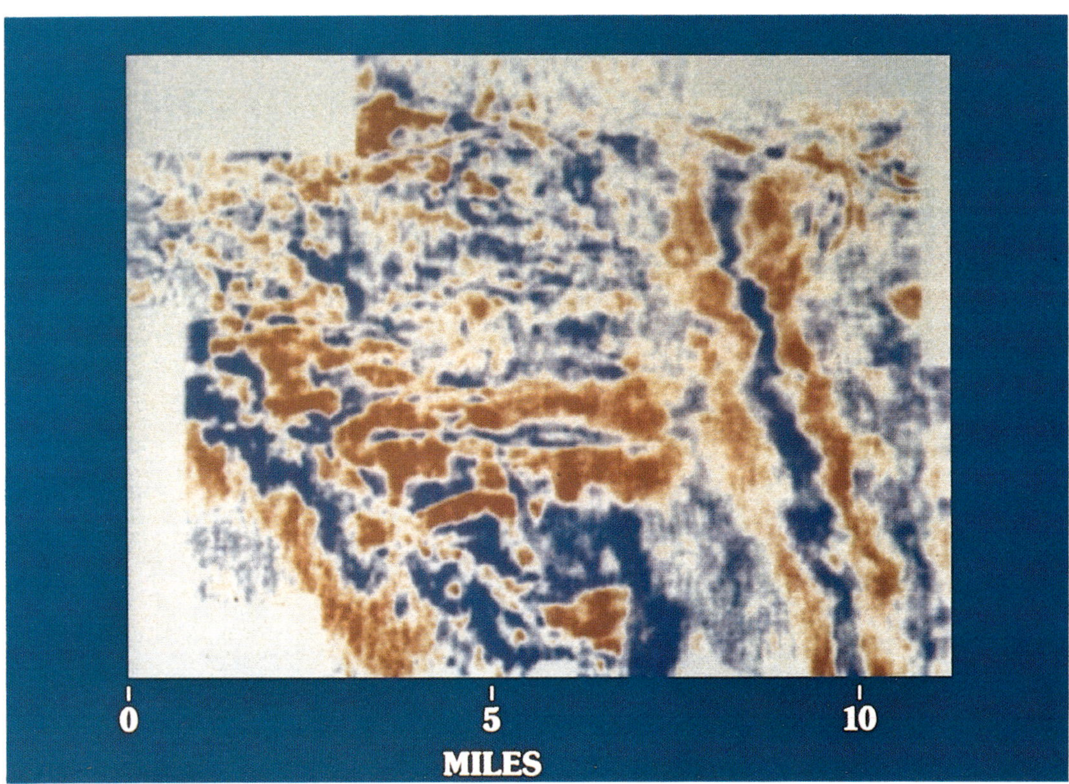

Fig. 4. Prudhoe Bay Field, Alaska—time slice (courtesy of D. A. Fisher).

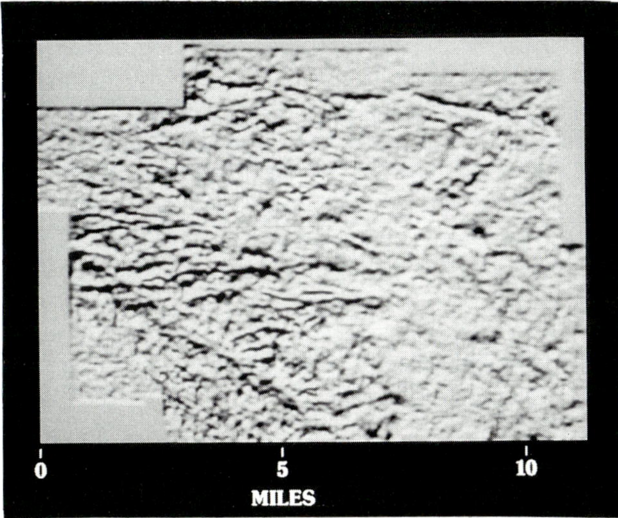

Fig. 5. Prudhoe Bay Field, Alaska—sun-shaded stack of time slices (courtesy of D. A. Fisher).

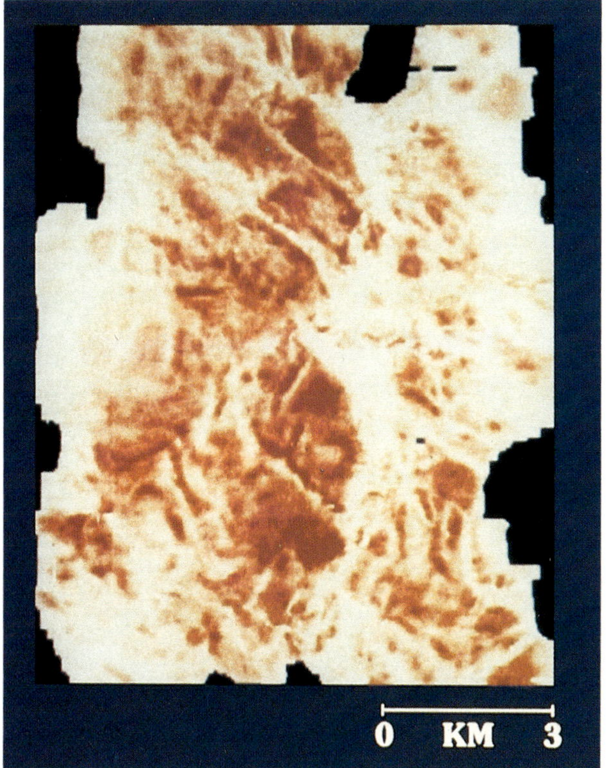

Fig. 6. Gas Field, North Sea—horizon slice (courtesy of D. A. Fisher).

northern boundary fault as a sharp, white-to-black transition, and the fine detail of the remainder of the faulting in the field also is visible. A sun placed to the south best illuminates the east-west faults. An interpreter obviously can move the sun to a different compass point to optimize the highlighting of faults at other orientations.

Figures 6 and 7 are a second example of using 3-D seismic data to map geometric framework. This field in the southern North Sea produces gas from traps in which Rotliegendes sandstones are sealed by overlying Zechstein evaporites. The traps are rotated normal-fault blocks, and the 3-D seismic volume assists in the delineation of these structures. Figure 6 is a display called a horizon slice. An interpreter has picked the base of the Zechstein throughout the seismic volume, color-coded the strength of the seismic reflection from this horizon, and displayed the seismic amplitude as a 2-D image. The areas colored white have a weak seismic reflection, and progressively darker reds indicate strengthening of the seismic amplitude. The display shows the normal faults as white lineaments, and an interpreter can use the image to produce a detailed map of the individual fault blocks. A large fault runs north-south approximately above the zero tick on the scale bar. The fault blocks to the west are rotated and dip off to the west, and these are the blocks that form the field.

Figure 7 is a display called dip magnitude. Seismic traveltimes have been measured from the surface to the base of the Zechstein, and the traveltime differences among adjacent grid points have been converted to the time dip of the horizon. The magnitude of this time dip is color-coded, red and yellow for steep dips (normal fault planes) and green and blue for flatter dips

(the tops of the fault blocks). The representation clearly shows where the faults are located and aids in precisely mapping the size and continuity of individual traps.

Figures 8, 9, and 10 are a third example of using 3-D seismic data to map geometric framework. This example, also from the North Sea, is the Cormorant Field,[2] which produces from sandstones in the Middle Jurassic Brent Group that are tilted, unconformably eroded, and sealed by overlying Upper Jurassic shales. A geologic cross-section (Figure 8) and colocated seismic line (Figure 9) from Stiles and McKee (1986) depict the attitudes of the beds and the trapping mechanism. The major objective of the development plan was to locate producing wells as far updip as possible (to delay water production and minimize attic oil) without actually crossing the truncation edge (and thereby missing the productive sandstones at the top of Brent stratigraphic section). The Brent truncation can be clearly picked on time slices through the 3-D seismic volume over the field (the broad black and gray event running north-northeast to south-southwest somewhat left of center in Figure 10), thus providing an accurate map of the truncation edge for positioning development wells and planning recovery strategy.

[2]See also pages 150 to 163.

The Prudhoe Bay and North Sea examples illustrate that 3-D seismic data map the gross, controlling, geometric elements of a field. This type of analysis is the traditional use for the 3-D seismic technique and has contributed very significantly to the geologic characterization of many hydrocarbon traps.

An example of imaging stratigraphic shapes with 3-D seismic data is shown in Figures 11 and 12. The data volume comes from the Matagorda area of the offshore Gulf of Mexico, and Riese and Winkelman (1989) discuss the details of the example. A single seismic section (Figure 11) in the 3-D volume contains a variety of nearly flat events. Some appear and then disappear and others vary laterally in amplitude; the stratigraphic significance of any particular reflector is not obvious on the two-dimensional (2-D) display. Note the short black event at about 0.7 s located directly above the zero tick on the scale bar. A time slice through the 3-D volume (Figure 12) reveals that the event is a transverse cut through a meandering stream channel, and the stratigraphic situation becomes clear when the lateral spatial sampling of the 3-D volume is used. (The green bands crossing the

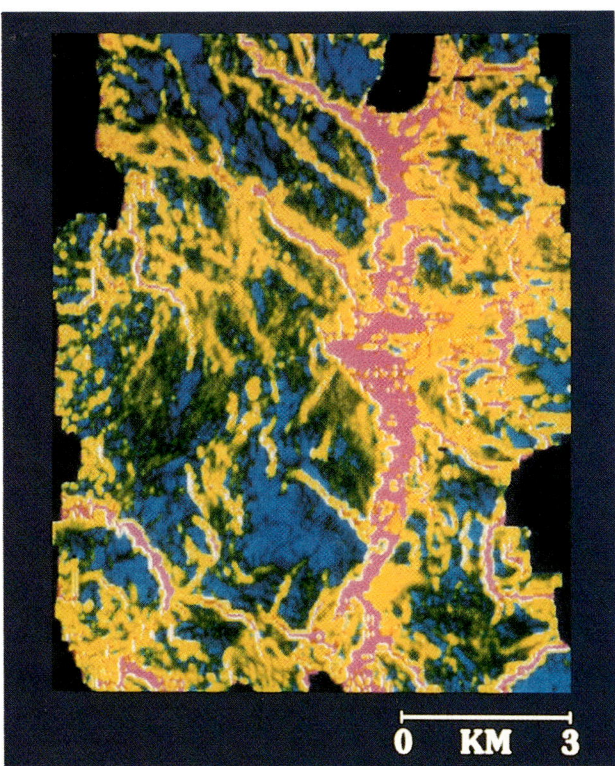

Fig. 7. Gas Field, North Sea—dip magnitude (courtesy of D. A. Fisher).

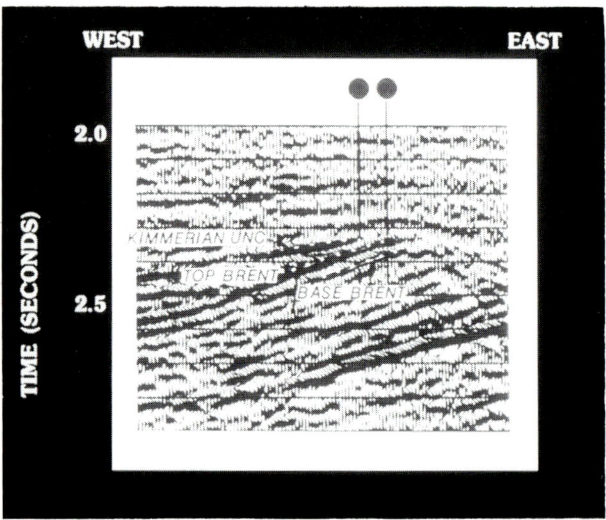

Fig. 9. Cormorant Field, North Sea—seismic section (Stiles and McKee, 1986).

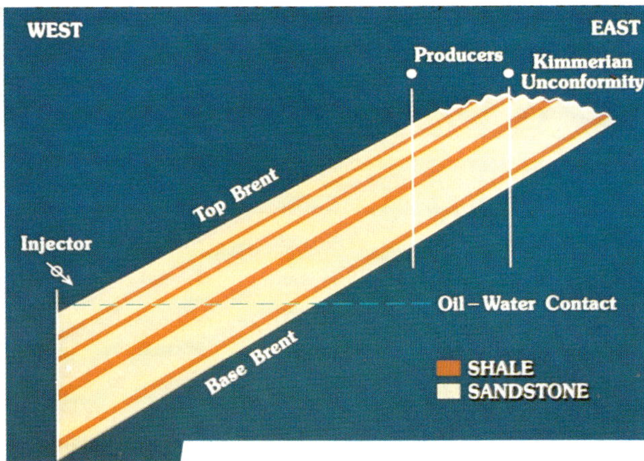

Fig. 8. Cormorant Field, North Sea—structural cross-section (Stiles and McKee, 1986).

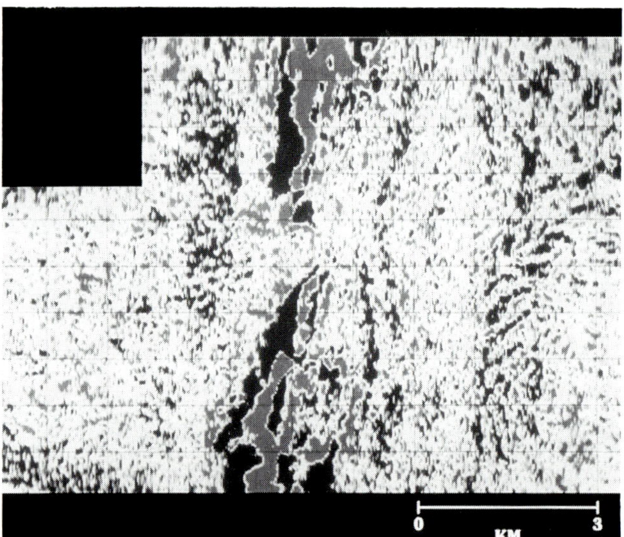

Fig. 10. Cormorant Field, North Sea—time slice (courtesy of J. H. Stiles, Jr.).

northern part of the slice are fault cuts coming up through the data). The example illustrates that minor character changes in 3-D seismic data tend to correlate with real geologic changes. The seismic subtleties generally are not noise or acquisition/processing errors, and the challenge is to deduce correctly their geologic significance.

Rock Properties

The second general grouping of 3-D seismic analyses (Figure 2) involves the qualitative and quantitative definition of rock properties. Amplitudes, phase changes, interval traveltimes between events, frequency variations, and other characteristics of the seismic data are correlated with porosity, fluid type, lithology, net pay thickness, and other reservoir properties. The correlations usually require borehole control (well logs, cuttings, cores, etc.) both to suggest initial hypotheses and to refine, revise, and test proposed relationships. An interpreter develops a hypothesis by comparing a seismic parameter in the 3-D volume at the location of a well to the well's information, often through the intermediary of a synthetic seismogram or 2-D or 3-D seismic model. The hypothesis is then used to predict rock properties between wells, and subsequent drilling validates (or invalidates)

the concept. Gas saturation in sandstone reservoirs is probably the rock property that has been most successfully mapped by 3-D seismic surveys. The presence of free gas typically lowers sharply the seismic velocity of relatively unconsolidated sandstones and creates a strong acoustic impedance contrast with surrounding rock. The contrast produces a seismic amplitude anomaly. Since the early 1970s, this "bright spot" effect has been widely exploited to detect gas saturation with standard 2-D seismic sections. When the effect occurs in 3-D volumes, gas-saturated sandstones can be accurately mapped laterally across fields at multiple producing horizons.

Figure 13 is an example from the Pleistocene Trend of the offshore Gulf of Mexico. The display is a horizon slice through one of the producing intervals in the field, and the red areas are strong seismic amplitude anomalies (the straight black lines are lease boundaries overlaid on the display). Drilling has confirmed that the anomalies correlate with gas pay, and the seismic data clearly can be used by the reservoir management team to locate wells and estimate the areal dimensions of productive pools.

A less common type of analysis is illustrated in Figure 14. This example is located on the North Slope of Alaska, and the formation of interest is the Lis-

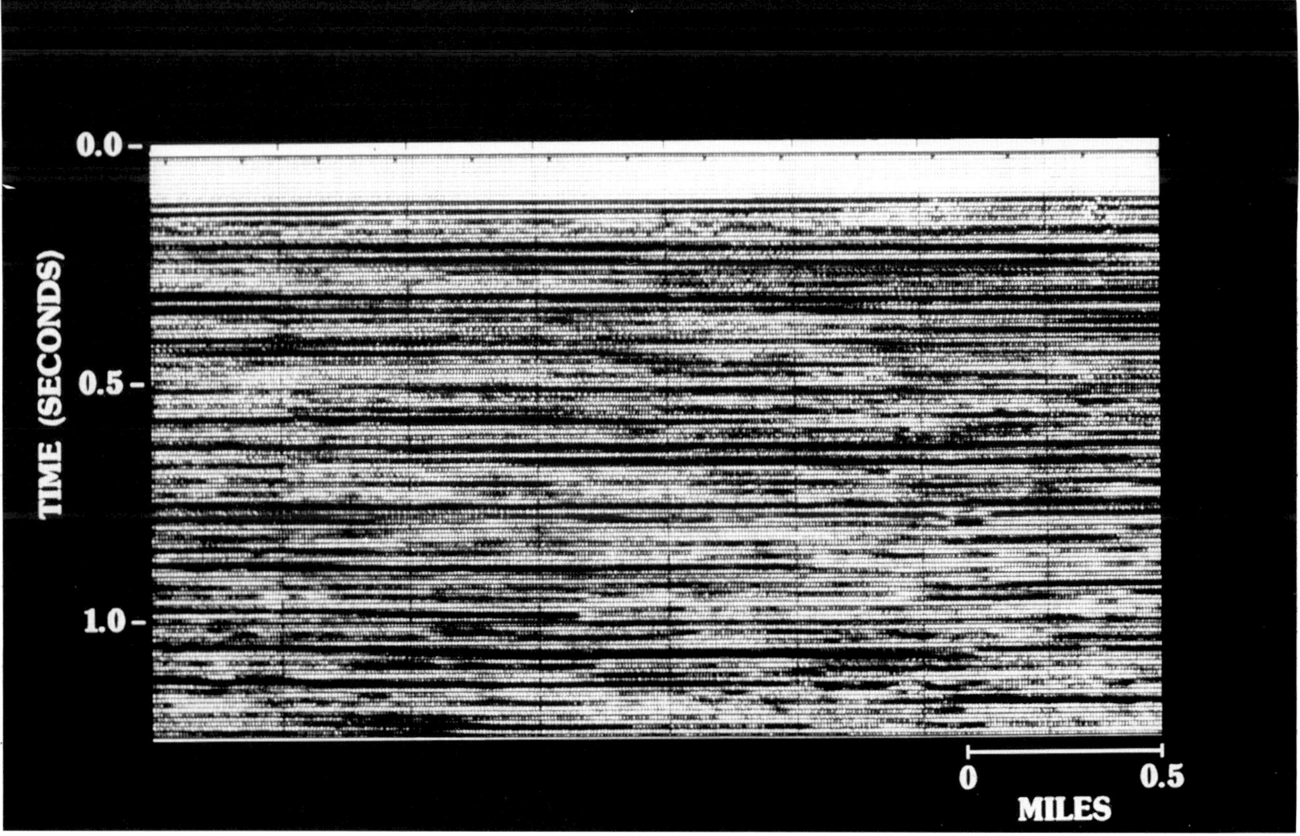

Fig. 11. Matagorda Block 668, Offshore Texas—seismic section (Riese and Winkelman, 1989).

burne, a carbonate that produces from below the Sadlerochit at Prudhoe Bay. The amplitudes of the Lisburne reflection were determined at points of well control and compared to Lisburne porosity-thicknesses measured in the same wells. The comparison produced transform functions that were used to convert seismic amplitudes directly into porosity-thickness values at seismic grid points between wells. Figure 14 is the Lisburne reflection scaled to porosity-thickness. A number of wells drilled after this analysis have tested the quality of the porosity-thickness pre-dictions, and the tests have matched the predictions to within a few units.

The Gulf of Mexico and Lisburne examples demonstrate that 3-D seismic data guide interwell interpolations of reservoir properties. Given a 3-D survey, one does not have to settle for crude, linear interpolations of reservoir parameters between wells. The reservoir manager can use the seismic volume to pinpoint and understand nonlinear lateral changes, an approach that nearly always results in lower costs, fewer surprises during development, and better production.

Flow Surveillance

The third general grouping of 3-D seismic analyses (Figure 2) consists of those designed to monitor the actual flow of the fluids in a reservoir. Such flow surveillance is possible if one (1) acquires a baseline 3-D data volume at a point in calendar time, (2) allows fluid flow to occur through production and/or injection with attendant pressure/temperature changes, (3) acquires a second 3-D data volume a few weeks or months after the baseline, (4) observes differences between the seismic character of the two volumes at the reservoir horizon, and (5) demonstrates that the differences are the result of fluid flow and pressure/temperature changes. Of course, one must be careful not to vary seismic acquisition and processing parameters between surveys and thereby introduce differences that can be mistaken for fluid flow effects. One expects that the seismic character of horizons above

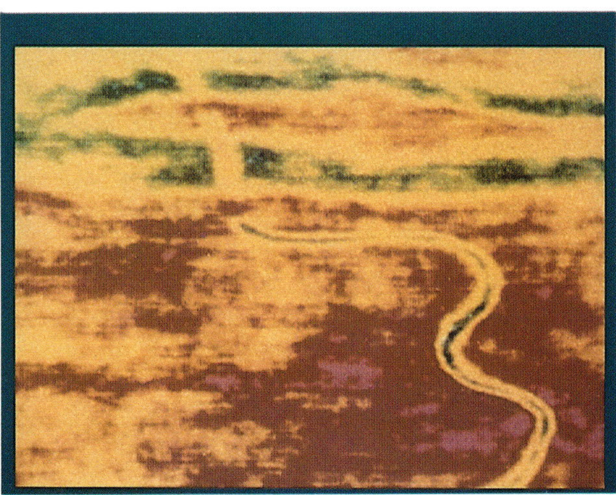

Fig. 12. Matagorda Block 668, Offshore Texas—time slice (Riese and Winkelman, 1989).

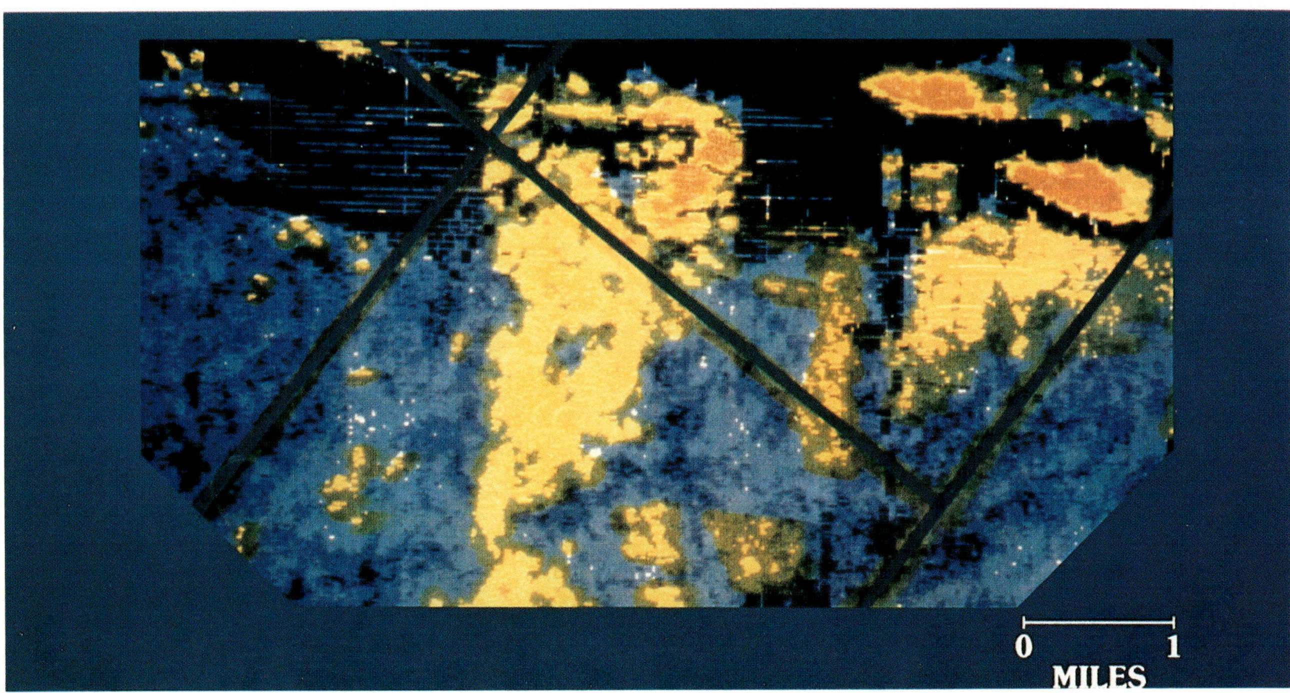

Fig. 13. Ship Shoal Block 332 Field, Offshore Louisiana—horizon slice (courtesy of G. G. Chong and G. J. Mitch).

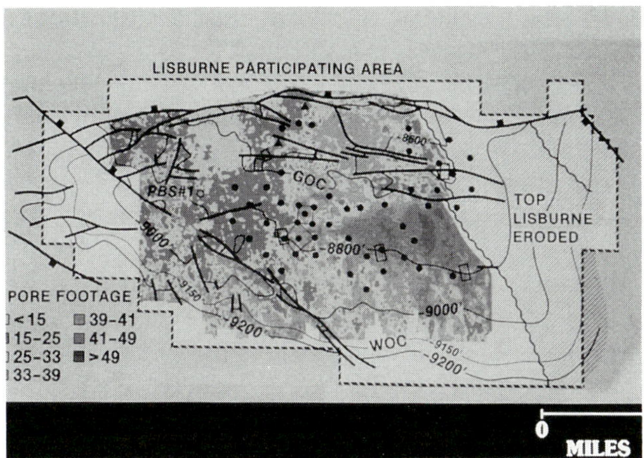

Fig. 14. Prudhoe Bay Field, Alaska—Lisburne porosity-thickness from 3-D seismic data (courtesy of S. F. Stanulonis and N. Kumar).

the reservoir would be virtually identical (geology generally changes much more slowly than fluid flow!). Hence, an interpreted flow-induced difference can be validated indirectly by verifying that the difference occurs at the reservoir event, but not elsewhere in the 3-D volume. Of course, one can acquire a third, a fourth, etc. survey and continue the surveillance by computing additional 3-D difference volumes.

Flow surveillance with multiple 3-D seismic surveys is at a very early stage of research and development, but its potential impact on reservoir management is enormous. Most current practice of the technique has been directed toward monitoring enhanced oil recovery (EOR) processes. An example[3] is shown in Figure 15 from Greaves and Fulp (1987). An experimental, oxygen-driven, thermal EOR pilot was performed on a depleted oil field, the Holt Sand Unit, in north Texas. The top section in Figure 15 is a line through a 3-D data volume acquired before the start of the pilot. The Holt Sand is the event identified by the white triangles, and its seismic amplitude is low. In this color-coded display of envelope amplitude, the bright red and yellow event is a limestone lying a hundred meters below the Holt and not associated with the reservoir. The middle section in Figure 15 lies in the same spatial position in the 3-D data volume as the top section, but was acquired a few months after the start of oxygen injection/thermal combustion. Likewise, the bottom section is the same line reshot about a year after startup. One can observe that the oxygen injection/thermal combustion process has produced a dramatic increase in the strength of the Holt Sand reflection, and that the formation is increasingly affected as calendar time passes. The combination of oxygen injection and creation of combustion gases increases

gas saturation in the reservoir (in effect, the experiment is creating an artificial bright spot), the thermal process is altering the rock, and the multiple seismic data sets are monitoring the changes.

One also can observe changes in the amplitude of that portion of the limestone reflection that lies directly below the Holt bright spot. The interpretation is that the seismic energy passing through the burned Holt is attenuated by partial gas saturation in the pores, so a weakened seismic wave is transmitted down to the limestone. Mapping the dimming of the limestone event thus is an alternative to mapping the Holt bright spot. The two indicators are making a somewhat different measurement: the gas saturation/thermal alteration specifically at the top of the Holt in the case of the bright spot versus gas saturation/thermal alteration throughout the Holt interval, not just at the top, in the case of the dimming. Second-order differences in the indicators permit one to infer how the process is affecting the top of the reservoir as opposed to the middle and lower portions. The seismic monitoring in this example does map an apparent vertical override of the gas some months after the start of the pilot; this and other details of the monitoring are described fully in Greaves and Fulp (1987).

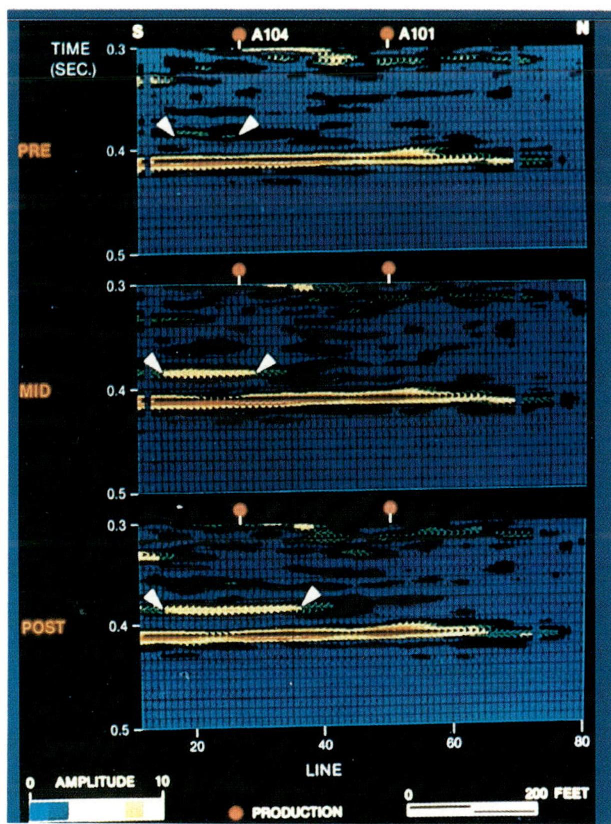

Fig. 15. Holt Sand Unit EOR pilot—seismic sections (after Greaves and Fulp, 1987).

---

[3] See pages 309 to 320.

Figure 16 displays horizon slices at the top of the Holt. An acoustically homogeneous layer overlies the formation, so the preburn horizon slice on the left is actually a map of reservoir porosity, the rock property that most strongly varies amplitude before injection of the oxygen. The dark reds to the northeast and southwest of the injection well indicate that the reservoir is more porous in these directions than elsewhere. One might predict a priori that the fireflood would preferentially propagate in these directions. That is exactly what happened as monitored by the amplitude changes in the midburn and postburn slices and by temperature/fluid measurements made in production and observation wells.

The lesson here is that 3-D seismic snapshots map changes in fluid/pressure/flow regimes. Although not yet as commercialized as mapping geometric frameworks or estimating rock properties between wells, this application of 3-D seismic surveying may eventually become just as important. It has the potential to measure reservoir performance directly, to provide timely feedback that can change development/production plans, and to be more spatially specific than pressure tests. In the future, the technique might be used to monitor gas cap movements, control production and injection rates for optimum recovery, map pressure/temperature distributions, and even decipher stress patterns, particularly in tectonically active areas.

Roots

The first 3-D seismic survey ever shot appears to have been an experimental survey acquired by Exxon Production Research Company in 1967 at Friendswood Field near Houston, Texas. Walton (1972) describes the experiment and presents some of the data. This first 3-D volume was not migrated and also not corrected for normal moveout. The interpretation consisted of tracing emerging wavefronts and looking for wavefront discontinuities that might indicate faulting (Figure 17).

During 1967–72, various petroleum companies carried out other experimental surveys, including some with transducers in water tanks performed by William S. French and coworkers at Gulf Oil Corporation about 1970. It was not until 1973, however, that the first commercial 3-D seismic program was conducted a land survey in Lea County, New Mexico, shot by Geophysical Service Inc. for a six-company consortium (Amoco, ARCO, Chevron, Mobil, Phillips, Texaco). A time slice from this survey (Figure 18) clearly images two anticlines separated by a saddle and flanked by north-south faults. The first commercial marine survey was shot in 1975 by Geophysical Service Inc. for Sun Oil Company in the High Island area of the Gulf of Mexico.

Tables 1 (for land) and 2 (for marine) are comparisons of typical seismic acquisition parameters used in early versus recent surveys. The actual areal extents of the surveys have not changed dramatically, not surprisingly since the sizes of the fields targeted for mapping are not changing much. Of course, there are some modern surveys that do cover very large areas; the tabulated comparisons are for typical field-size surveys, not these regional ones. The comparisons between recorded samples show that recent surveys

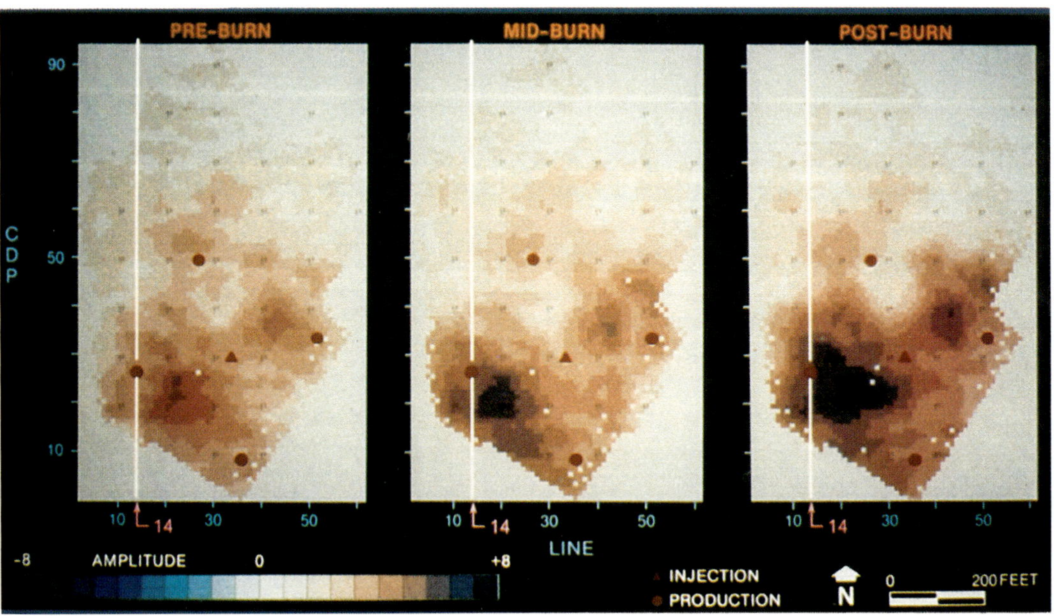

Fig. 16. Holt Sand Unit EOR pilot—horizon slices (courtesy of R. J. Greaves and T. J. Fulp).

are acquiring in excess of an order of magnitude more data than the early ones. The additional data capacity is being used primarily to shrink grid sizes, thereby improving lateral spatial resolution, rather than to increase fold.

Use of 3-D seismic technology grew steadily from the mid-1970s onward, particularly in the marine environment, as acquisition and processing improved. The best available evidence is that more than a hundred 3-D surveys were shot worldwide by 1980. Ongoing innovations in acquisition and processing (streamer tracking, real-time binning, 3-D stacking, 3-D migration, 3-D velocity analysis, etc.), the advent of supercomputers, and the explosive growth in computer graphics workstations for interpretation continue to fuel the use of 3-D seismology. It is virtually certain that more than a thousand 3-D surveys have now been acquired worldwide by the petroleum industry.

Future

A very safe prediction is that the application and sophistication of 3-D seismic technology will grow steadily for the foreseeable future. The following are some specific areas in which progress is occurring.

Acquisition, Processing, and Interpretation Methods

At sea, operators are shooting surveys with various combinations of multiple streamers, multiple source arrays, and multiple boats; are experimenting with towing streamers in circles around targets like salt domes to improve the imaging of radial faults; and are shooting into receivers fixed on the ocean bottom.

On land, where one is free from the constraints of towing a streamer, operators are deploying many innovative acquisition geometries that make full use of the multichannel capabilities of modern seismic sys-

tems. The geometries account for terrain and cultural obstacles while optimizing subsurface coverage and mixes of offsets and azimuths, all at the lowest possible cost. Some experimental work is underway to acquire 3-D three-component surveys, thus adding shear and converted wave data volumes to the standard compressional wave volume.

Advances in supercomputing will continue to speed up processing and permit the inclusion of more sophisticated algorithms in processing schemes. Infusion of analytical techniques from remote sensing and other image processing disciplines is beginning to affect 3-D seismic interpretation. Automated information extraction (for example, algorithms that pick events after a few control points are specified) is becoming a routine part of interpretation, and many facets of 3-D seismic analysis are amenable to being impacted eventually by artificial intelligence technology.

Routine Use in Exploration

A recent innovation in 3-D seismic surveying has been acquisition along widely spaced lines followed by filling in of the data volume by numerical interpolation before performing 3-D migration. This 3-D scheme (known variously as reconnaissance, exploration, or wide-line 3-D) depends on a good interpolation algorithm to be successful, and even then some steep-dip information is lost. However, the technique has the potential to lower acquisition costs to a point where it is feasible to shoot 3-D for exploration, and these types of surveys are now penetrating the seismic market.

3-D Seismic With Downhole Sources and Receivers

The standard 3-D seismic data volume is acquired with sources and receivers at the Earth's surface. It is

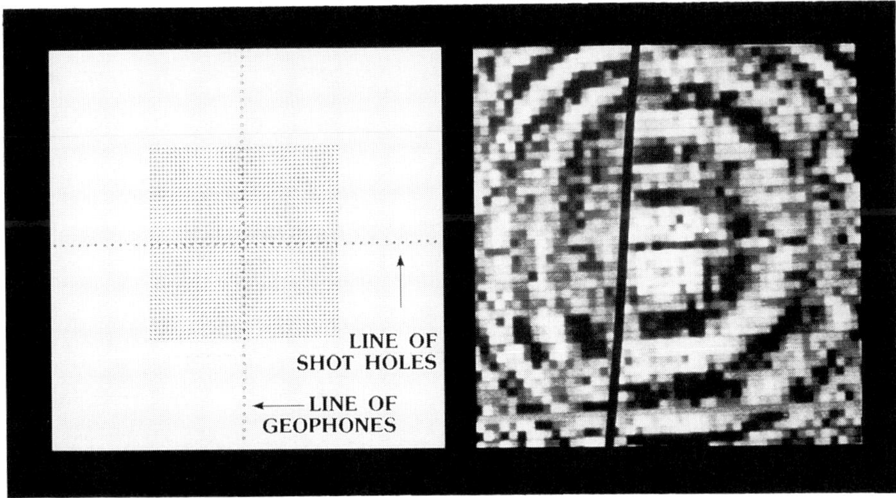

Fig. 17. 1967 Exxon experimental 3-D survey—Friendswood Field, Texas
(Walton, 1972).

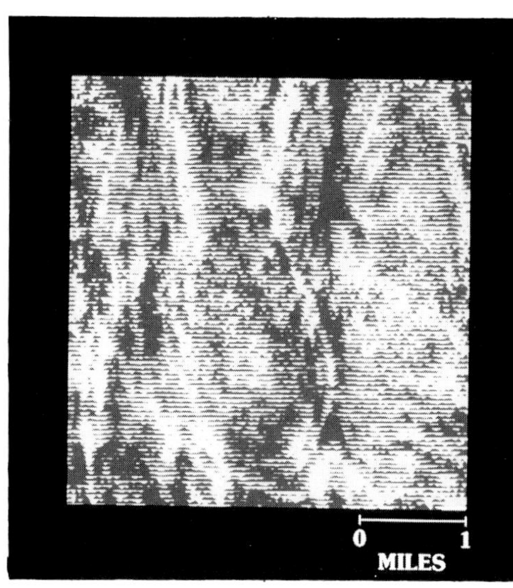

Fig. 18. 1973 GSI commercial land 3-D survey, Lea County, New Mexico—time slice (courtesy of W. R. Cotton and R. J. Graebner).

logistically possible to put sources and/or receivers in boreholes and to record part or all of the 3-D data volume with this downhole hardware. This approach is an active area of research. Depending on the acquisition configuration, one records various kinds and amounts of reflected and transmitted seismic energy, which can then be sorted to provide information on geometric framework, rock properties, and flow surveillance, just like surface surveys. Advantages of downhole placement are that higher seismic frequencies generally can be recorded, thereby improving resolution, and that surface-associated seismic noise and statics problems are lessened or avoided. The main disadvantages are that source and receiver plants are constrained by the physical locations of available boreholes; borehole seismology can be affected by tube waves and the like, so downhole placement is not noise-free; a borehole source cannot be so strong as to damage the well; and the logistics and economics of operating in boreholes are complex, though not necessarily always worse than operating on the surface. One can imagine a time when borehole seismic sources

Table 1: Land acquisition parameters for 1973 New Mexico 3-D survey versus 1988 survey.

Parameter	1973	1988	Ratio
Areal coverage	26 mi^2	35 mi^2	1.4
Acquisition time	28 days	40 days	1.4
Live traces	96	480	5
Number of vibrator points	3700	10 000	2.7
Recorded traces	3.5×10^5	4.8×10^6	14
Trace length	5 s	5 s	1
Sampling interval	4 ms	2 ms	2
Recorded samples	4.4×10^8	1.2×10^{10}	27
Number of CMPs	1.5×10^4	1.4×10^5	10
CMP grid size	220 × 220 ft	82.5 × 82.5 ft	0.15
Fold	12 × 2	15 × 2	1.25
Sweep frequency	8-55 Hz	10-100 Hz	2

Table 2: Marine acquisition parameters for 1975 Moray Firth 3-D survey versus 1988 Gulf of Mexico survey.

Parameter	1975	1988	Ratio
Areal coverage	43 mi^2	36 mi^2	0.8
Acquistion time	30 days	35 days	1.2
Line spacing	200 m	50 m	0.25
Cable length	2400 m	3000 m	1.25
Number of groups	48	120	2.5
Group interval	50 m	25 m	0.5
Source-point interval	25 m	25 m	1
Trace length	5 s	7 s	1.4
Sampling interval	2 ms	2 ms	1
Recorded samples	2.7×10^9	3.3×10^{10}	12
Line miles	350	1225	3.5
CMP grid size	50 × 100 m	12.5 × 25 m	0.06
Fold	48	60	1.25
Source strength	1450 in^3	4000 in^3	2.75

and receivers might be standard components of the hardware run into wells and accepted as routine and valuable devices for reservoir characterization and flow surveillance.

Summary

The petroleum industry's twenty-year experience with 3-D seismic surveying is an example of a technological and economic success. Today, the investment in a 3-D survey typically results in fewer development dry holes, improved placement of drilling locations to maximize recovery, recognition of new drilling opportunities, and more accurate estimates of hydrocarbon volume and recovery rate. These outcomes improve the economics of development and production plans and make the surveys cost-effective. More skillful reservoir management will be a theme of the 1990s, and 3-D seismic technology will be part of the advancement.

References

Greaves, R.J., and Fulp, T.J., 1987, Three-dimensional seismic monitoring of an enhanced oil recovery process: Geophysics, **52**, 1175–1187.

Morgridge, D.L., and Smith, W.B., Jr., 1972, Geology and discovery of Prudhoe Bay Field, Eastern Arctic Slope, Alaska: Am. Assn. Petr. Geol., Memoir 16, 489–501.

Riese, W.C., and Winkelman, B.E., 1989, Atlas of seismic stratigraphy: Am. Assn. Petr. Geol., Vol. **3**, 59–71.

Stiles, J.H., and McKee, J.W., 1986, Cormorant: development of a complex field: Soc. Petr. Eng., Paper 15504.

Walton, G.G., 1972, Three-dimensional seismic method: Geophysics, **37**, 417–430.

Chapter 2: Petrophysical and Geophysical Background
Basic Petrophysics and Geophysics

*R. E. Sheriff**

Engineer-Geophysicist Cross-Communication

Reservoir geophysics involves a blending of disciplines that historically have had little to do with each other. Selecting the location for a discovery well usually involves a geophysicist working closely with an exploration geologist, but, once hydrocarbons are found the geophysicist usually has not been further involved. When testing to determine if a discovery is commercial and planning a cost effective way to develop and produce the hydrocarbons, the responsibility shifts from exploration to development organizations, involving people with different backgrounds and outlooks. The development geologist and petroleum engineer often know only the structural picture developed from the exploration geophysical data without appreciating the ambiguities or alternative interpretations of the picture. Much that could be done to refine the structural and stratigraphic picture is often not done. Exploration groups try to keep costs down until a hydrocarbon accumulation is established and thus they do not do the extra work that can refine the picture of the reservoir. Many geophysicists who have not dealt with petroleum engineers do not appreciate what engineers need to know, and many petroleum engineers do not realize how geophysical technology could help them achieve their objectives.

A more synergetic approach is now developing where petroleum engineers, geologists, and geophysicists work together as a team, sharing their knowledge and each applying their specialized knowledge to help solve the other's problems. This chapter briefly reviews some principles of petroleum engineering that need to be understood by a geophysicist and principles of geophysics that need to be understood by an engineer.

The Nature Of Reservoirs

A petroleum engineer is mainly interested in whether hydrocarbons are present and whether (and

*Geoscience Dept., Univ. Houston, Houston, TX 77204–5503.

how) they can be produced at a profit. More specifically, he is interested in

the pore spaces of a rock (porosity),

how the pore spaces are interconnected (permeability),

the nature of the fluids filling the pore spaces (fluid saturation),

the energy or pressure that may cause the fluids to flow (drives),

the vertical and areal distribution of reservoirs and pore-connected spaces, and

barriers to fluid flow (sealing and nonsealing faults, stratigraphic barriers, etc.).

These facts have to be determined from available information, which probably consists of

surface seismic, gravity, magnetic, and other geophysical data,

borehole logs of various types,

cores taken in boreholes,

analyses of fluids recovered in drillstem tests,

production and pressure data,

VSP and perhaps other specialized geophysical measurements,

occasionally tracer data, and

drilling rate logs, mud logs, and other well data.

Well logs, geologic background, and well-to-well log correlations supplemented by seismic character studies give an overall picture of the stratigraphy and stratigraphic changes across the reservoir, and production and pressure data (and occasionally tracer data) give information about the connectivity of reservoir members between wells. Surface geophysical data, while lacking the vertical resolution of borehole logs and cores, provides the only data source that gives detailed information about areal distributions.

Porosity and Permeability

The important aspects of reservoir rocks are the void spaces (porosity) and the interconnection of the porespaces (permeability). Porosity is the volume fraction of space not occupied by the rock matrix. Not only average porosity is important but also porosity distribution, both vertically and horizontally. Reservoir porosity is determined from measurements on cores and well logs using relationships that are somewhat empirical. Porosity is one major factor in determining seismic velocity and reflectivity so that velocity measurements and (sometimes) seismic amplitude measurements can yield porosity information.

The fluid flow rate q (volume of fluid per unit time, in mL/s) in the x-direction for the one-dimensional (1-D) flow of a single fluid is expressed by the empirical Darcy's Law:

$$q = \left(\frac{kA}{\mu}\right)\frac{\Delta P}{\Delta x} \qquad (1)$$

where k is permeability in darcys, A is the cross-sectional area in cm^2, μ is the viscosity in centipoises, ΔP is the pressure differential across the thickness Δx in atmospheres per centimeter. Commercial reservoirs generally have 5–30 percent porosity and 0.005 to 5 darcy permeability. Flow equations become more complicated when dealing with several fluids simultaneously, when the fluids are compressible, when flow rates are high, and when flow is not linear (for example, near a producing well where the fluids converge toward the well).

Permeability is measured on cores but is not measured directly by borehole logs. Permeability often correlates with porosity, or measurements that correlate with porosity, such as resistivity, and these measurements are used to predict permeability. Permeability may differ areally across a reservoir and from zone to zone within a reservoir, and permeability often varies directionally. Permeability is qualitatively indicated by comparing measurements from logs that differ in their effective penetration, or by the buildup of mudcake at permeable zones.

Rock Structure

Most rocks are composed of grains, with voids between the grains. The porosity remaining from that present at deposition (primary porosity) sometimes is added to by cavities that develop subsequent to deposition (secondary porosity) because of solution, chemical replacement, or other processes. Original pore space or the connections between pore spaces may be destroyed by cementation or other diagenetic processes. Fractures or joints sometimes provide connections between pore spaces.

The trapping of fluids requires seals to prevent fluid escape. A seal or cap rock is one in which the capillary displacement pressure is sufficiently high to prevent entry of gas or oil into the pores of the caprock. Fluid flow through a seal is usually so small that even during long geologic time only a small portion of the trapped fluids will have passed through. Seal rocks are characterized by very small connections (throats) between pores. Plastic cap rocks are better than brittle ones, which fracture more easily. Typical cap rocks are clays, shales, salt, and anhydrite, and sometimes other rocks such as dense limestone act as cap rocks. Cap rocks have not only very low permeabilities (of the order of $10^{-6}-10^{-8}$ darcies) but also very high capillary pressures. Capillary pressure, the pressure difference across the curved interface between two immiscible fluid phases jointly occupying the interstices of a rock, is due to the tension of the interfacial surface.

Most information about rocks is obtained from samples (such as full-diameter cores, sidewall plugs, or samples brought up with the returning drilling mud) or logs of physical measurements made in boreholes. Physical property measurements determined from samples and logs have intrinsic limitations—the rock may have been altered in the process of drilling or by exposure to borehole fluids, or the samples may not be representative of the entire formation. Properties such as porosity and permeability may change within short distances, and they also may be directional, that is, anisotropic. Measured values often vary considerably with different samples and consequently calculations based on these measurements may have large uncertainties. Because the anisotropy generally is not known, calculations usually assume that properties are the same in all directions (isotropy). Initial estimates of reserves based on a discovery well are apt to be in error, perhaps by a factor of two. Estimates of reserves usually change considerably as production progresses and as the values used in initial calculations are refined.

A geologic picture develops from the core, log, and seismic data. Cores and thin sections cut from cores are examined by X-ray diffraction, photomicrographs, scanning electron micrographs, etc. The environment of deposition is assessed based on the core, log, well test, and seismic data supplemented by paleontologic depth indicators and isotope ratio measurements that give the temperature at the time of deposition.

Shales and their distribution throughout a reservoir rock play an especially important role in how fluids flow through a rock. Shale distribution is determined from examination of cores and logs and well-to-well correlations. Determination of the depositional environment is very important in interpreting shale distribution.

Since most rocks are somewhat inhomogeneous and anisotropic, fluids flow more easily in one direction than in another. A reservoir bed usually includes shale or other types of laminations that separate the bed into subbeds (zones) which behave somewhat independently. Flow may occur within one part of a bed while bypassing other parts of the bed's porespaces. Clearly both horizontal and vertical variations of porosity and permeability vitally affect fluid flow.

Reservoir Fluids

The fluids in a reservoir tend to separate into layers because of density differences; gas forms the uppermost layer, oil with dissolved gas the middle layer, and brine the lowest. The boundaries between the layers may not be sharp but rather transition zones in which the fluid proportions change gradually. Water in amounts of 10 to 30 percent generally coexists with the gas and oil in the upper zones. Most reservoir rocks have mixed wetability and are only wetted by water in the smallest pore spaces and at grain-to-grain contacts. The rounded surfaces of grains are often coated with asphaltenes adsorbed on their surfaces. Not all of the pore space is available to mobile fluids. The specific gravity of oils is usually 0.80 to 0.95. Hydrocarbon gas is mainly methane with some ethane, propane, and other hydrocarbons; the proportions of the higher hydrocarbons in the gaseous state depend on the pressure (especially) and temperature.

Gas usually is quite soluble in oil, especially under the pressures and temperatures in a reservoir. Much of the gas comes out of solution as the pressure is lowered to atmospheric pressure in the course of production. The liquid volume shrinks, up to 50 percent, as the gas is evolved, and the liquid becomes more viscous. The volume of a volatile oil at reservoir temperature and pressure may be more than twice that at stock tank conditions. The behavior of a mixture of hydrocarbons as pressures and temperatures change can be quite complicated. Consider the phase diagram shown in Figure 1. At pressures and temperatures above the critical point the distinction between gas and liquid no longer exists. If the starting reservoir conditions should be at point A and if the pressure is lowered over time without significantly affecting the temperature, then gas will begin to come out of solution when conditions reach B (the bubble point). When conditions reach C the volume will be 25 percent gaseous. If the starting reservoir conditions should be at the point D, then liquid will begin to condense when conditions reach E (retrograde condensation), and when conditions reach F the maximum amount will have liquified and further lowering of pressure will cause the oil to vaporize. If the starting conditions are G, no condensation will occur until the temperature is lowered.

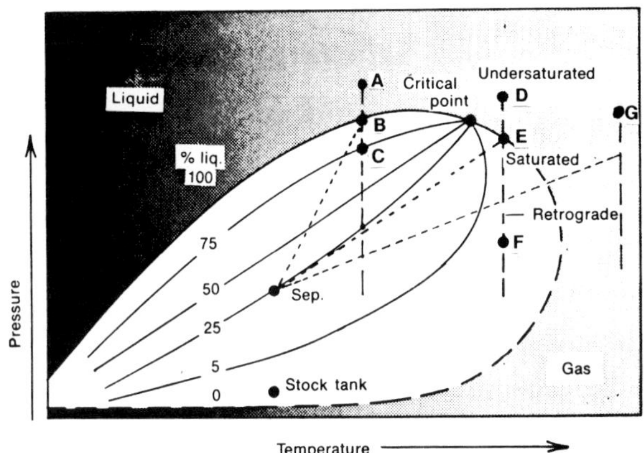

Fig. 1. Phase diagram for a mixture of hydrocarbons. The long dashed lines indicate isothermal changes (as might happen in a reservoir) as pressure lowers, for three possible initial conditions A, D, and G. The short-dashed lines indicate reductions of both pressure and temperature, as might happen to fluids flowing to the surface and into a separator, where proportions are appreciably different than in the reservoir. Fluid volumes will also change appreciably as pressures-temperatures are changed. (After Dickey, 1979: 192)

The effective permeabilities of fluids depend on their relative saturations and the nature of other fluids present. Where gas and liquid phases coexist, the gas inhibits the flow of the liquids.

Oil field brines usually contain six major ions, Na^+, Ca^{++}, Mg^{++}, Cl^-, SO_4^{--}, HCO_3^-. Sometimes ions of K, Sr, Ba, Fe, I, Br, or other elements are present in significant amounts.

Estimating Reserves

Initially a field's recoverable reserves are estimated based on the reservoir volume, porosity, saturation values, and recovery factor estimates. The specific tools for determining the hydrocarbons originally in place are structure and isopach maps, porosity and hydrocarbon saturation maps, and cross sections based on log correlations and seismic data. Seismic hydrocarbon indicators (see pages 47–48) may play an important role in reservoir calculations. Reserve figures are usually modified after some production history provides produced-volume and pressure-decline values. Seismic measurements play a major role in defining reservoir volume and how reservoirs are broken up into separate pools.

Energy to Drive Hydrocarbons to Production Wells

Hydrocarbons are pushed from reservoir pore spaces by water or gas. The gas for gas drive may come from the expansion of the gas in a gas cap as the pressure declines, or from the gas dissolved in the oil, which comes out of solution as the pressure declines. Water drive requires an understanding of the aquifer

Table 1. Efficiencies of different drives

Drive	Percent oil recovered	Percent gas recovered
Solution gas (depletion)	5–15	> 90
Gas cap expansion	20–40	70–90
Water invasion	35–70	50–70

surrounding an accumulation since that is where the water to displace the hydrocarbons must come from. Effective water drives generally require connected aquifiers 100 times the volume of the reservoir. The order of magnitude of drive efficiencies is indicated in Table 1.

The upper values are generally achieved when reservoirs are homogeneous, isotropic, and permeable and when oil has low viscosity. Water drive is not as effective as gas drive for gas production because water tends to trap more gas in unrecoverable places.

The amount of recoverable hydrocarbons depends on a number of factors, including the interstitial porosity; pore sizes and shapes and interconnectivities; saturations of oil, gas, and water; interfacial tensions and wetability; and fractures—their spacing, width, and orientations. These factors affect not only primary recovery but also secondary and enhanced recovery.

Primary, Secondary, Enhanced Oil Recovery

The natural energy present in a reservoir drives the fluids to the producing boreholes and results in primary recovery. After primary energy and production decline, the natural energy is often supplemented by artificial means such as gas or water injection, which is called secondary recovery. Gas or water injection may be initiated soon after production begins.

More exotic means of stimulating production involve injection of miscible fluids, surfactants or other chemicals, injection of steam, and setting fire to the hydrocarbons in a reservoir. These methods are called enhanced oil recovery (EOR). Residual oil may be left in a reservoir as discontinuous globules trapped by interfacial tension, which tends to vanish when the driving liquid is miscible. Miscible fluids, which are expensive, are often injected as a slug that is then pushed by injected water or gas. EOR floods tend to flow through the more permeable beds or channels and thus may bypass portions of a reservoir. Seismic methods may help monitor the changes in a reservoir that result from production and oil recovery efforts where gas is involved (e.g., as a result of a fire-flood) or where the seismic velocity is changed significantly (for example, decreased as a result of the temperature increase resulting from steam injection). Obviously we need to know how fluids actually flow through a reservoir in order to optimize (and, often, to make economical) EOR projects.

Pressure Measurements

Inhomogeneities and anisotropy may cause the channeling of water or gas phases, by-passing pockets of oil (and leaving them as unproducible). Sometimes by-passed regions may be invaded if sufficient fluid is injected or by-passed oil can be recovered by infill drilling or reperforating wells. Evidence as to reservoir continuity is obtained mainly from pressure histories, observing the pressure changes in one well when conditions are changed in another well. When production in a well is stopped, the early build-up of pressure mainly depends on the immediate vicinity of the well and the influence of more remote parts of the reservoir affects the pressure curve later. Proximity to faults or other barriers to fluid flow affect pressure changes. Complications arise when two or more reservoirs are connected to the borehole. Leakage into other formations because of channeling behind the casing also affects shut-in pressure tests.

Changes During Producing History

Multiphase situations result in complex sequences of reservoir conditions as production progresses. May (1984) gives the following production scenario for a very simple situation, a uniform porous sand containing only oil with dissolved gas.

"As the pressure is reduced, small amounts of gas evolved from solution accumulate in the pores and increase the resistance to flow of oil without the gas itself becoming mobile, but, as the concentration increases, the gas bubbles eventually join up to form a continuous phase, and from then on both fluids move through the formation. . . . the permeability of the porous formation to gas increases, and the permeability to oil decreases. Hence, with continued production, the produced gas/oil ratio tends to rise, this tendency being accentuated both by shrinkage of the liquid-phase volume and by the increase in its viscosity with loss of dissolved gas."

Dickey (1979) describes several case histories that did not perform as expected. Sometimes additional wells were drilled to investigate reservoir conditions in more detail to find out the reasons. Dickey states

"this sort of geological data ought to be obtained early in the life of the pool, and the progress of any enhanced recovery scheme should be closely monitored. . . . One can't help wondering, however, whether [more] careful environmental reconstruction could have been made on the original wells, and at least the possibilities of inhomogeneities in the reservoir could have been predicted. On a more positive note, it is now becoming more generally recognized that the geology of a petroleum reservoir should be worked out in detail during the primary development, and taken into account in planning secondary and enhanced recovery.''

The distribution of porosity, permeability, and other properties over the area of a reservoir are incorporated in reservoir simulation models. The models are updated as additional information becomes available. Scenarios of production programs are run on a model to determine how to optimize the economic return. Obviously the most realistic maps and data are required to achieve this end result.

The Seismic Method

Simplistically, a seismic source injects mechanical energy into the earth, the energy travels as seismic waves down to where rock properties change, where a portion of the energy is reflected back toward the surface where it is detected by sensors (geophones). The big advantage of the seismic method over other methods is fine horizontal resolution; seismic measurements are usually made every 25–50 m and often closer.

The parameters measurable from seismic data are (1) the arrival times of seismic waves (called events), (2) amplitudes of the events, (3) the character of events (which tells something of the fine structure of reflecting interfaces), and (4) the patterns of events (which sometimes tells about local structure and depositional environment). A geophysicist routinely calculates several things from examination of seismic records:

1. From the elapsed traveltimes (and velocity information)—the depths to interfaces that separate rock types;

2. From differences in the traveltimes with surface location—the dip;

3. From differences in the traveltime with source-to-geophone distances—the velocity of seismic wave travel;

4. From measurements of reflection amplitudes—the contrast in rock properties at interfaces;

5. From discontinuities in reflection patterns—the locations of faults and stratigraphic changes.

Under certain circumstances, interepretation can go further and determine:

6. From velocity values—conclusions about the lithology, abnormal pressure, fluid content, or temperature;

7. From lateral changes in amplitude measurements—the locations of hydrocarbon accumulations and changes in stratigraphy, porosity, or thickness;

8. From patterns in the seismic data—the depositional environment in which the rocks were laid down.

9. From changes with measurement direction—velocity anisotropy and inferences as to fracture orientation or permeability anisotropy.

10. From changes between measurements made at different times—the locations of the changes.

Every determination involves uncertainties and ambiguities. Values of porosity or porosity-thickness sometimes can be calculated from seismic measurements with reasonably good accuracy, while at other times only crude estimates are possible. While we cannot calculate directly from seismic data some reservoir parameters such as permeability, permeability usually correlates with porosity, which we often can estimate. The differences between the effects of oil and water in pore spaces may be too small to measure but gas-oil, gas-water, and oil-water contacts can be seen sometimes. Because seismic measurements usually are made at the surface, some distance above a reservoir, we can see detail for shallow reservoirs that cannot be resolved for deep ones.

How Seismic Waves Propagate

The seismic waves which concern us mostly are *P*-waves (compressional or acoustic waves) that involve alternate compressions or expansions of inter-molecular distances (Figure 2). The sense of motion travels through the earth until eventually it reaches geophones, where the mechanical wave energy is converted into the electrical signals that are recorded. Surfaces in the earth that undergo the same sense of motion at any given time are called wavefronts; lines perpendicular to the wavefronts are called raypaths (in isotropic media), and we (rather loosely) think of the seismic energy as traveling along raypaths. Wavefronts and raypaths bend where the seismic velocity changes, in the same manner as sound and light wavefronts and raypaths bend when their velocity changes, that is, according to Snell's Law.

Where the velocity and/or density change, such as at the interfaces separating different kinds of rocks, wave energy partitions among reflected and transmitted waves. The ratio of the amplitude of a reflected wave to that of the incident wave is called the reflectivity. The reflectivity is smaller than 0.1 for most interfaces in the earth and only for particularly strong reflectors does it reach values as high as 0.3. The fraction of energy reflected is given by the square of the reflectivity, and usually less than 1 percent of the energy is reflected at any interface, most often considerably less.

At perpendicular incidence, that is, when a raypath strikes a reflecting interface at right angles, the reflectivity is given by the simple expression,

$$R = \frac{\rho_2 V_2 - \rho_1 V_1}{\rho_2 V_2 + \rho_1 V_1} \, , \qquad (2)$$

where ρ_1 and V_1 are the density and velocity in the incident medium and ρ_2 and V_2 are those beyond the interface. The situation is slightly more complicated at gradational interfaces (where a change of waveshape also occurs) or at rough scatterers (which are rarely a problem). Relation (2) becomes much more complex and dependent on angle and changes in Poisson's ratio

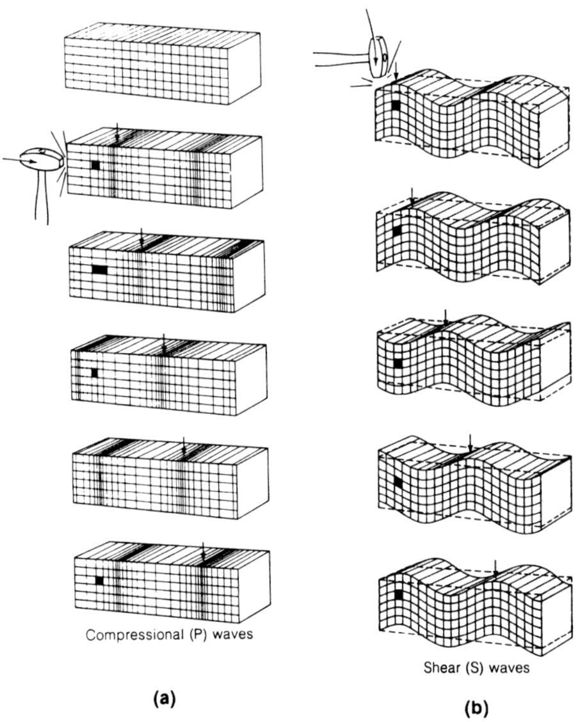

Fig. 2. The two types of seismic waves that can travel through the body of an isotropic material. (a) *P*-waves involve particle motion in the direction in which the wave is traveling, indicated by the alternate compressions and rarefactions. (b) *S*-waves involve particle motion perpendicular to the direction in which the wave is traveling. The type of wave depends on the initial stress changes, here indicated by the hammers. The initially undeformed block is shown at the top left. The arrows indicate successive locations of the same phase [a compression in (a), an upward deflection in (b)] at successive equal time intervals. (After Philipps, 1968.)

when incident raypaths are not perpendicular to the interface, as is discussed later.

The Convolutional Model and the Nature of Reflections

The earth acts linearly with respect to seismic waves in almost all practical situations and as a consequence the principle of superposition applies. Each reflected wave causes its own effect at each geophone (or hydrophone) independent of what other waves may be affecting the geophone at that time. The geophone response is simply the sum (linear superposition) of the effects of all the waves striking the geophone. The mathematical expression of this is that a seismic trace is the convolution of an embedded wavelet with the reflectivity of the earth, with noise added. This concept, the convolutional model of a seismic record, is at the heart of almost everything we do with seismic data.

A synthetic seismogram shows the seismic data that would be obtained for a line run across a geologic model. A synthetic seismogram is based on the convolutional model and is useful for identifying the reflection events from particular interfaces, for studying how geologic changes should appear, and to assess the likelihood that some specific feature can be detected. Its applicability depends on how closely the geologic model (usually a simple layered model) matches the actual geology. The reflectivity values in its manufacture are usually based on well-log data. Synthetics usually (but not necessarily) assume vertical ray travel and that only primary reflections are of concern, and they often ignore density variations. They can be made to allow for multiples (waves that are reflected more than once), nonvertical travel, nonnormal incidence, raypath bending (by ray-tracing through a model), and wave-theory effects such as diffractions and mode conversion.

The embedded wavelet in field seismic data is usually a reasonably long wavetrain. However, in processing we can (within limits) replace the field wavelet with a shorter wavelet that is easier to interpret; this will be discussed later. The replacement wavelet is almost always a zero-phase symmetrical wavelet because such a wavelet minimizes timing problems and maximizes resolution. While there are problems with wavelet replacement, modern processing generally accomplishes this reasonably well.

How Rock Properties Affect Velocity

Elasticity theory gives us expressions for the velocity of seismic *P*- and *S*-waves (V_S and V_P) in terms of rock elastic constants for simple cases, and the elastic media can be determined from velocity and density ρ measurements. With isotropic media,

$$V_P = \sqrt{[(\lambda + 2\mu)/\rho]}, \text{ and} \qquad (3a)$$

$$V_S = \sqrt{(\mu/\rho)}, \qquad (3b)$$

$$\frac{V_S}{V_P} = \left[\frac{0.5 - \sigma}{1 - \sigma}\right]^{1/2} \qquad (3c)$$

where λ and μ are the Lamé constants and σ is Poisson's ratio. V_P, V_S, and V_S/V_P also can be expressed in terms of other elastic constants. S-waves (also called shear and transverse waves) involve particle motion perpendicular to the direction of wave travel (Figure 2). For some purposes S-waves are the main subject of study.

Seismic velocity and density depend strongly on porosity and also on other rock properties such as lithology, fluid content, depth, pressure, age, compaction, cementation, and temperature. Of course, these are not all independent; for example, porosity, compaction, pressure, temperature, and age usually depend on depth. Velocity generally increases nonlinearly with depth, but local lithology alters this general rule.

Seismic velocity depends strongly on porosity and often the increase of velocity with a decrease of porosity is the principal factor controlling velocity. Theory leads to equations such as those of Gassmann (1951) and Biot (1956) to relate seismic velocity to porosity and the rock and fluid properties. We often use the simpler empirical time-average equation to calculate porosity from velocity measurements:

$$\frac{1}{V} = \frac{\phi}{V_f} + \frac{1 - \phi}{V_m}, \qquad (4a)$$

where ϕ is the porosity, V_f is the velocity of the interstitial fluid, and V_m is the velocity of the rock matrix material. This equation is also written in terms of specific transit times for the rock, fluid, and matrix materials, $\Delta t, \Delta t_f, \Delta t_m$:

$$\Delta t = \phi \Delta t_f + (1 - \phi)\Delta t_m \qquad (4b)$$

Since reflectivity depends mainly on velocity (the changes in density being generally smaller), we sometimes interpret amplitude variations as indicating variations in porosity, especially with limestones and dolomites. Of course, such interpretations involve ambiguity because of our inability to separate the effects of porosity, thickness, fluid content, and other variations, especially for thin reservoirs, but with the addition of independent information about a reservoir we can sometimes do this.

If gas is the interstitial fluid, the velocity is often markedly lower than when water fills the pore spaces. This fact forms the basis of most of the hydrocarbon indicators that are discussed subsequently. Oil in the pore spaces also lowers the velocity compared to water and occasionally oil produces enough velocity change to detect. The effect of gas on velocity is large and nonlinear; a very small amount of gas may have a larger effect than complete gas saturation.

Velocity depends strongly on the differential pressure, that is, the difference between the lithostatic pressure (the pressure of the overlying column of rock) and the interstitial fluid pressure (normally the hydrostatic pressure, the pressure of a column of interstitial fluid extending to the surface). This fact is used to predict abnormal pressure situations from seismic velocity measurements.

Resolution

Reflecting interfaces are generally spaced much more closely than the wavelength of the embedded wavelet so that successive reflections overlap. In fact, most reflections are the interference composites of a number of component reflections and separating the effects of changes in the individual components is not easy. Resolution—the ability to separate effects—ultimately limits what can be done with seismic data, especially in reservoir studies.

The natural unit of measure for seismic waves is the seismic wavelength (usually indicated by the symbol λ), which can be related to velocity (V) and dominant frequency (f) (or to the dominant period (T), which can be approximated on seismic sections):

$$\lambda = V/f = VT. \qquad (4)$$

For shallow hydrocarbon reservoirs the velocity is usually small and the dominant frequency rather large, say 6000 ft/s (1800 m/s) and 60 Hz, so the wavelength is of the order of 100 ft (30 m). For deep reservoirs the velocity is usually much larger and the dominant frequency much smaller, say 15 000 ft/s (4500 m/s) and 15 Hz, so the wavelength is of the order of 1000 ft (300 m). Most hydrocarbon reservoirs are thinner than a seismic wavelength and often thinner than a quarter wavelength, which is an important value in seismic interpretation.

While there is subjectivity involved in specifying a resolvable limit, a value of about one quarter-wavelength is generally used. For shallow reflecting interfaces where the velocity is small and wavelet frequency content relatively high, a quarter-wavelength is of the order of 25 ft (or 8 m). Resolution decreases rapidly with depth because velocity generally increases with depth and frequency content decreases with depth. Thus the resolvable limit for deep reflectors is much larger than for shallow ones and the ability of seismic methods to resolve reservoir details decreases rapidly with depth. Incidentally, the resolvable limit applies to the definition of structural features—such as the detection of faults—as well as to stratigraphic aspects.

Some aspects of reservoirs can be distinguished even where they are thinner than the resolvable limit. For example, a bed of contrasting acoustic impedance may yield enough reflection to be seen even where its thickness is only 1/30th wavelength. The values for this detectable limit are about 5 ft (1.5 m) and 50 ft (15 m) for the shallow and deep reflectors in the wavelength calculations given.

Obviously specifications of resolvable and detectable limits are not rigid. The quality of the data, how much noise is present and of what type, whether the reflection of interest is much stronger than interfering events, and the experience and knowledge of the interpreter are clearly factors in determining resolution.

One especially instructive synthetic seismogram is of a wedge pinchout, such as shown in Figure 3. The interference of the reflections from the top and base of the wedge produces an amplitude buildup when the wedge thickness is one-quarter of the dominant wavelength; this thickness is the tuning thickness. Where a reservoir is thicker than λ/4, the wedge thickness is determined by timing the trough and peak (or vice-versa) associated with the reflections from the top and base of the reservoir. However, at and below the tuning thickness the trough-to-peak interval cannot be

used for determining thicknesses The amplitude measurements then will depend on the thickness.

Where a reservoir involves only two lithologies, for example, interbedded sand and shale embedded in shale, and where the total thickness is about a quarter-wavelength, the reflection amplitude is proportional to the net thickness of the reservoir—the net thickness of sand in this example. Furthermore, the amplitude is not sensitive to how the lithology is distributed, that is, on whether the sand is in one relatively thick layer or in several thin layers. Where we have a calibration point such as a well, we sometimes use this to map the net pay thickness.

The areal portion of a reflecting interface responsible for contributing significantly to a reflection is a large area often spoken of as the "Fresnel zone". The Fresnel zone is generally much larger than the trace spacing, so that on unmigrated seismic sections the effects of subsurface changes are spread out over many seismic traces, tending to obscure the changes. However, migration collapses the Fresnel-zone so that subsurface changes can be seen sharply at their correct locations. Migrated seismic data are especially good at showing edges such as the edges of reservoirs, stratigraphic bodies, fault locations, etc. Constraints on migration are discussed later.

The ideal embedded wavelet is a short, sharp, known wavelet that bears a simple, direct relationship to the interfaces which reflections involve. The wavelet, which should have a distinct timing point to indicate the traveltime to the reflector, should also create minimum confusion in interfering with nearby events. For maximum information content, we require a broad bandpass (at least 1.5 octaves) with good high-frequency content, and faithful amplitude values.

The wavelet generated by most seismic sources (Vibroseis being the most notable, but not the only, exception) is nearly minimum phase, that is, the energy builds up as rapidly as possible after the source activation. Most field data are nearly minimum phase. A minimum-phase wavelet (Figure 4) is not the easiest

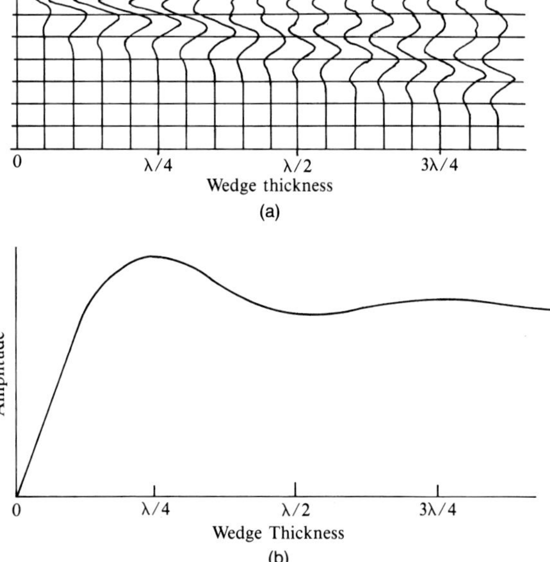

Fig. 3. Reflections from the top and bottom of a wedge that is pinching out. The medium above and below the wedge is the same so that there is no reflection when the wedge has zero thickness. (a) Seismic section across the wedge, whose thickness is specified in fractions of the dominant wavelength. (b) Maximum amplitude as wedge thickness varies. (From Sheriff, 1991, Figure T-13.)

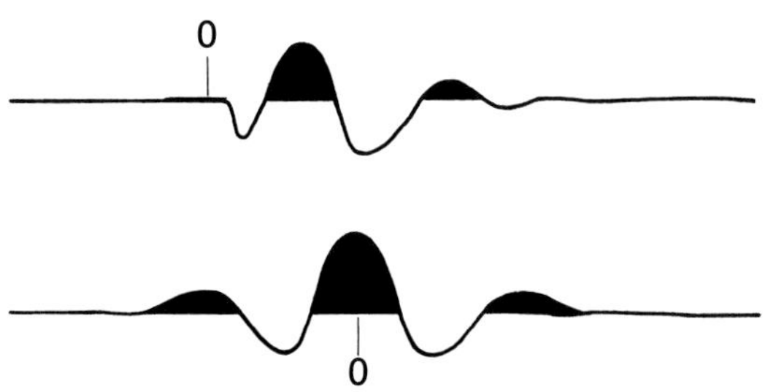

Fig. 4. A minimum-phase wavelet (above) and zero-phase wavelet (below), both for positive reflectivity and displayed with standard polarity. (From Sheriff, 1991, Figure P-5.)

to interpret. However, we can change the shape of the embedded wavelet in processing.

A waveshape can be synthesized by the superposition of cosine waves of different frequencies and amplitudes; this is the "Fourier theorem". The amplitudes and phases of the component cosine waves determine the sum waveshape and where the wave will be located. For a wave to have its maximum value at time zero, each component must have zero phase since each cosine curve has maximum amplitude at zero. When the phase of all components is zero, a zero-phase wavelet results; such a wavelet is symmetrical (Figure 4). Where the bandpass is limited, there will also be buildups at locations other than zero, which we call "side lobes". Phase shifting of any component subtracts from the peak value and produces some waveshape asymmetry. A zero-phase wavelet has minimal sidelobes and comes as close as possible to being an ideal wavelet. Most seismic sections today approximate zero-phase embedded wavelets.

If the amplitude of each frequency component is the same within a passband and zero for all frequencies outside (a boxcar frequency spectrum), the number and magnitude of side lobes depend on the band of frequencies passed and on the sharpness of the cutoff at the edges of the bandpass. A narrow bandpass results in sizeable sidelobes, which can interfere with and distort other events. Bandpass width is measured in octaves; an octave is a frequency ratio of two; thus from 15 to 30 Hz is one octave, from 30 to 60 is another, from 60 to 120 still another. About 1.5 octaves is the minimum acceptable bandwidth. In order to have the peak (or trough) of a wavelet to be picked to be sharp, the bandpass also needs to include high frequencies. Thus we require a broad bandwidth including high frequencies. Neither high frequencies without broad bandwidth nor broad bandwidth without high frequencies will achieve a good wavelet shape.

Data Acquisition

The nature of the seismic source is not very important as long as the source provides enough energy that reflections are detectable above the background noise. Many source types are available and the selection as to which source to use in any particular situation is usually based on cost, location, accessibility, legal, and other considerations. Seismic sources for land operations are about evenly divided between explosives and vibrators, but other sources are also occasionally used. Air guns are used for most marine work, but water guns and other sources are available for certain situations.

The Vibroseis method (Figure 5) is now the most used land source. The vibroseis injects energy into the ground at low intensity but over a long period of time (7 to 35 s) and subsequent correlation processing effectively shortens the embedded wavelet into a compact one. The correlated vibroseis trace is mixed phase, having both the characteristics of a zero-phase autocorrelation and those of the nearly minimum-phase earth through which the energy has passed. The Vibroseis usually creates less damage than explosive sources because the intensity is low, but the results after correlation are apt to be comparable.

The earth restricts the bandwidth and high frequency content. High frequencies are attenuated more in passage through the earth. In an attempt to compensate for the increased loss of high frequencies, we

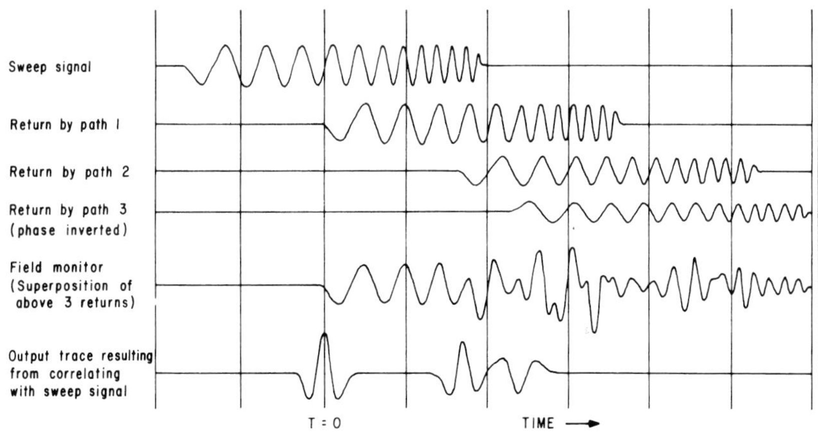

Fig. 5. The Vibroseis method. A sweep of continuously varying frequency (top trace) is reflected by three reflecting interfaces (next 3 traces), yielding their sum as the geophone output (5th trace). After correlation with the sweep the zero time and individual reflections can be seen. (Courtesy Conoco.)

sometimes inject more high frequencies into the ground with our source by using small sources or a nonlinear vibroseis sweep, but there are practical and economic limits to these procedures.

Land detectors are usually moving-coil geophones that respond to vertical motion. Usually a number of detectors (6 to 24) are hard-wired together to form a "geophone group" at each detecting station, and typically 96 or more (up to 1024, and sometimes even more) detector groups are used simultaneously. The use of groups discriminates against random energy and waves that are traveling with large horizontal components. The most important noise contributor on land is ground roll, a horizontally traveling surface wave that is apt to be especially strong when surface sources (such as the vibroseis) are used. The use of groups also discriminates against steeply dipping reflection events, so that field recording is usually a compromise between attenuating the ground roll sufficiently while not seriously damaging dipping reflections.

Marine detectors are usually pressure-sensitive hydrophones located in a streamer, typically about 2 miles (3–5 km) long, that is towed behind a seismic ship. Usually a number of hydrophones are connected together to form each hydrophone group and 96 to 500 hydrophone groups are distributed along the streamer. Stringent positioning controls are imposed so that the streamer's location is known correctly despite horizontal drift due to currents.

Sometimes the analog signals from the detector groups are transmitted directly to the recorder where they are digitized, and sometimes the signals are sampled and digitized near the detectors. Virtually all recording today is digital.

Almost all seismic data are acquired by the common-midpoint (CMP) method (Figure 6). Each source is recorded at a number of geophone or hydrophone locations and each geophone or hydrophone location is used to record the response from a number of source locations. The midpoint for any source-detector combination is the surface location midway between the source and detector locations. For horizontal reflectors with horizontally uniform velocities, the data for which the midpoints are common involve the same reflecting point. The CMP method provides data redundancy that permits considerable attenuation of several types of noise, the measurement of stacking velocity, and statistical corrections for near-surface variations.

Seismic Data Processing

The primary objective of seismic data processing is to emphasize primary seismic reflections (signal) in comparison with other energy (noise) received by the geophones or hydrophones. The principal method of accomplishing this task is combining data elements,

i.e., stacking. The redundant data recorded in the field are sorted according to common midpoints (CMP), corrected for geometrical differences in travel paths (normal-moveout correction, NMO), and combined in a CMP stack; this increases the magnitude of primary reflections more than it increases random noise and systematic noise (such as multiples). Dip-moveout (DMO) processing corrects for the fact that the component traces of a CMP gather for dipping reflections do not involve the same reflecting point.

An ideal seismic section should be as noise free as possible. We especially do not want noises to change horizontally, because we want to be able to attribute horizontal changes to changes within the earth. Toward this end we try to treat all data in the same way.

Reflections on seismic sections may be obscured because the long downgoing wavetrain causes reflections to overlap each other. The embedded wavelet involves the specific waveshape imposed by the source, as modified by the geophones and recording instruments and passage through the earth. Wavelet processing is a type of deconvolution that involves (1) attempting to ascertain the shape of the embedded wavelet, (2) finding an operator to change it to a more interpretable wavelet, and (3) applying this operator to

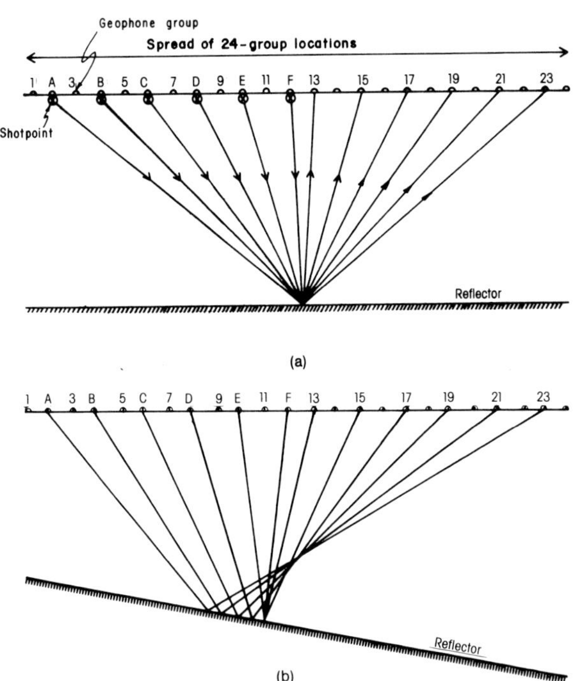

Fig. 6. Common-midpoint method. (a) In six-fold surveying with 24 geophone groups, the same reflecting point is involved for source A into geophone 23, B into 21, C into 19, D into 17, E into 15, and F into 13. (b) The reflecting point is smeared when the reflector dips. (From Sheriff, 1991, Figure C-8.)

the data. Wavelet processing is also used to make sure all the embedded wavelets in a gather are the same, so that high frequency elements are not lost in stacking. Wavelet processing is also used to improve the tie between data recorded in different ways or to tie synthetic seismograms to field data.

The faithful preservation of amplitudes in acquisition and processing is essential. Amplitudes vary for a multitude of reasons. Some unwanted amplitude variations can be avoided by careful acquisition and processing practices, some can be corrected for based on theory, and some do not significantly affect relative amplitudes. While we hardly ever know true amplitudes, relative amplitudes can be estimated closely enough that useful results can be obtained.

Reflecting points are generally not located vertically underneath midpoints and data are migrated to place reflection events at the locations appropriate to the reflecting points. Migration is not only important in displaying seismic structure but also in making faults and subtle features sharper and more evident. A wide variety of more-or-less equivalent methods can be used for migration. Conventional 2-D migration involves relocating seismic events based on observed in-line dip components; components of dip perpendicular to the lines are not measured and thus are not allowed for. One result is that migrated lines may not tie at line intersections. This defect, and most other migration defects, are remedied by 3-D migration, where 3-D data are available. Migration also assumes that all data represent primary reflections or diffractions. Migration smears out noise and the quality of a final migrated section strongly depends on how much noise was present in the unmigrated data.

Accurate migration requires knowledge of the velocity distribution, but the migration process generally is tolerant of small errors in velocity. Lateral changes in velocity usually are small over any single reservoir. If DMO processing has been applied, steep and even conflicting dips do not create significant problems. Components of dip perpendicular to the seismic lines do create some problems if only 2-D data are available. Migration often produces major improvement in lateral resolution.

The vertical scale on seismic sections, including most migrated sections, is usually traveltime and so is nonlinear in depth, and time-to-depth conversion (as well as migration) requires knowledge of the local velocity. This ordinarily does not impose serious problems in reservoir applications because well control is available to give independent velocity information. However, sizable changes in velocity can occur, especially across major faults, and interpretations made before the well control became available can present distorted pictures of a reservoir.

Processing also involves measurements and calculations of various types (attributes such as interval velocity, envelope amplitude, instantaneous phase or instantaneous frequency) that can communicate information to an interpreter. Another function of processing is to display the data to facilitate communication. The primary function of attribute displays is to make more evident subtle features which otherwise might be overlooked. Parameters such as scale, scale ratios, linearity in traveltime or depth, display amplitude, polarity, wavelet shape, display mode, color, etc. are very important in making features visible, for an interpreter cannot interpret features he does not notice. The importance of display is often under appreciated.

Seismic Interpretation

Seismic interpretation involves deriving a simple, plausible geologic model that is compatible with the observed seismic data. The model is never unique and an interpretation involves a sequence of somewhat arbitrary choices. An interpretation is also never complete but is simply discontinued after decisions are made. As more data become available, for example as the result of drilling wells, an interpretation should be reviewed and modified. For example, if pressure tests indicate that certain wells are not in communication, reexamination of the seismic data may turn up an explanation not previously proposed. Occasionally a completely new interpretation should be made inasmuch as an accumulation of modifications is apt to cause inconsistencies.

Most 3-D seismic work is now interpreted at interactive work stations and work stations are also being used increasingly for 2-D interpretation. Interpretations move quickly once data are in a form accessible to the work station; this provides a big advantage for reinterpretations, as only the new data have to be entered when the old data are already accessible. Work stations make it feasible to do a lot of checking that otherwise never gets done because of time limitations, and thus they result in more consistent interpretation.

Hydrocarbon Indicators

The presence of hydrocarbons in a rock's porespace lowers the seismic velocity and density and increases the absorption. Under favorable circumstances local accumulations create seismic effects that can be detected; these are called hydrocarbon indicators (see Table 2). The lowering of velocity and density changes reflection amplitude and waveshape in various ways depending on the contrast with the surrounding rock. Where a water-saturated reservoir rock has a lower acoustic impedance than the surrounding rock, hydrocarbons increase the contrast and a high-amplitude

Table 2. Hydrocarbon indicators. All of the indicators can have causes other than hydrocarbon accumulation. (From Sheriff, 1991, Figure H-7.)

Structural crest or against a fault?	**trapping location**
Local increase in amplitude?	**bright spot**
Local decrease in amplitude?	**dim spot**
Discordant flat reflector?	**flat spot**
Local waveshape change?	**polarity reversal** or local **phasing**
Reservoir limits consistent?	(if reservoir base and top separately visible)
Polarities consistent?	(if zero-phase data)
Low-frequencies underneath?	**low-frequency shadow**
Time sag underneath?	**velocity sag**
Lower amplitudes underneath (and sometimes above)	**amplitude shadow**
Increase in amplitude with offset?	**AVO anomaly**
P-wave but no *S*-wave anomaly?	***S*-wave support**
Data deterioration above (and perhaps minor bright spots)	**gas chimney**

bright spot results; this is the most common situation, especially with Tertiary clastics. Sometimes, however, the water-saturated reservoir rock has an appreciably higher acoustic impedance than the surrounding rock, in which case the hydrocarbons decrease the contrast and a low-amplitude dim spot results. Sometimes the water-saturated reservoir rock has an acoustic impedance slightly higher than that of the surrounding rock and then hydrocarbons change the contrast from positive to negative so that a reversal of polarity results. The reflection from the reservoir rock is often only one component in a composite reflection and changing the one component may change the waveshape. Where a reservoir has a sharp fluid contact, a reflection from the resulting contrast may produce a flat spot. Where a bright spot is present, the amplitude of reflections beneath and above the reservoir may be weakened as a result of amplitude balancing in processing. Where reservoirs are thick, the increased time in transiting the low-velocity reservoir may cause a sag in underlying reflections. A low-frequency shadow is sometimes seen immediately underneath a reservoir. Gas leaking from a reservoir may lower the velocity in the overlying section, producing velocity distortion and masking effects. The variation of amplitude with offset (see next section) is another hydrocarbon indicator.

Where hydrocarbon indicators can be seen, they are useful in mapping the reservoir and sometimes they can be used quantitatively to outline the reservoir and define the volume of hydrocarbons. Sometimes reflection amplitudes correlate nicely with quantities such as porosity-feet and hydrocarbon producibility. Hydrocarbon indicators have been associated with reservoirs of all ages and types, including both gas and oil reservoirs, but they are most apt to be useful with gas reservoirs in Tertiary clastic sections. False indications are possible; a small gas saturation may produce larger effects than a sizeable gas accumulation. The magnitude of hydrocarbon indicator effects may be so small that they are not obvious. Thus the confidence

that can be placed on hydrocarbon indicators (or their lack) varies with the situation, but generally a case is strengthened where several indicators point to the same conclusions.

S-waves, AVO, and Anisotropy

Almost all seismic work is based on study of *P*-waves, waves in which particle motion is parallel to the direction of wave travel. *S*-waves involve particle motion perpendicular to the direction of wave travel. There are two such directions, in consequence of which there can be two independent modes of *S*-waves; *S*-waves that involve just one of these are polarized. *P*- and *S*-waves depend on different elastic moduli and hence measure of both gives more information about the media through which they travel. Measurements with both *P*- and *S*-waves permit determining elastic moduli.

Where a seismic ray is incident on an interface at an angle other than perpendicular, both *P*- and *S*-waves are generated. Then the reflectivity is no longer given simply by equation (2) but depends also on the ratios of *S*- to *P*-wave velocities (V_S/V_P), or, what is equivalent, to Poisson's ratios (σ). V_S/V_p and σ are related by equation (3c). Shuey (1985) approximates reflectivity *R* by an expression of the form,

$$R \approx (\Delta V_P/2V_P + \Delta\rho/2\rho)\{1 + [A_o + \Delta\sigma/R_o(1 - \sigma)^2]$$
$$\times \sin^2\theta + B(\tan^2\theta - \sin^2\theta)\} , \quad (5)$$

where θ is the average angle involved, $\Delta V_P/V_P$ and $\Delta\rho/2\rho$ are the fractional changes in *P*-wave velocity and density, $\Delta\sigma$ is the change in Poisson's ratio σ, and A_o, R_o, and *B* involve the terms in the reflectivity at normal incidence. The term in brackets becomes more important as the angles increase and the term involving *B* becomes important only at very large angles and is usually ignored. This approximation is reasonably good for most reservoir rocks under the usual measurement conditions. V_S/V_P and σ do not change very much except where gas fills the pore spaces; consequently the term in brackets in equation (5) is usually unimportant except where gas is involved. Amplitude-variation-with-offset (AVO) methods measure how amplitude changes with offset and are used as an indicator of gas.

Seismic wave velocity depends on direction in media that are anisotropic. An S-wave traveling through such a medium may split into two waves that travel at different velocities (birefringence). The two polarizations of S-waves are usually thought of with respect to the measurement orientations but anisotropy imposes preferred directions (called principal directions). Fractures produce anisotropy and where the effects are sufficiently large seismic waves may be used to determine fracture orientation and intensity.

Other Velocity Effects

The major factor involved with hydrocarbon indicators is the effect of hydrocarbons on velocity, but velocity changes sometimes indicate other factors relevant to reservoir development and evaluation. The decrease in velocity associated with an increase in porosity is sometimes used to map porosity, and the velocity decrease associated with overpressured formations is sometimes used to map them and predict the degree of overpressuring.

The velocity of the intervals between reflections can be calculated from the variation of arrival time with source-to-geophone distance. Velocity can also be calculated from reflection amplitude where amplitudes have been preserved faithfully. This calculation, a type of inversion, essentially involves solving the reflectivity equation (equation 2) for the acoustic impedance $Z(t)$ as a function of the variation of amplitude with time, $A(t)$:

$$Z(t) = Z_o e^{k \int A(t) dt} \qquad (6a)$$

Usually a velocity-density relationship is assumed and the equation is solved for the interval velocity $V(t)$:

$$V(t) = V_o e^{k' \int A(t) dt} \qquad (6b)$$

The solution requires independent velocity data to evaluate the terms Z_o, k, V_o, and k' and to constrain the solution. The result is called a seismic log. Seismic log displays sometimes nicely outline hydrocarbon accumulations and the numerical values sometimes correlate with net porosity-thickness values or well productivity.

Acknowledgment

I thank Joseph G. Richardson who helped prepare this chapter.

References and Further Reading

Biot, M. A., 1956, Theory of propagation of elastic waves in a fluid-saturated porous solid: J. Acoust. Soc. Am., **28**, 168–191.

Brown, A.R., 1988, Interpretation of three-dimensional data (2nd Ed.), Am. Assn. Petrol. Geol. Memoir **42**.

Dickey, P. A.,1979, Petroleum development geology: Petroleum Publ. Co.

Dobrin, M. B. and Savit, C. H., 1988, Introduction to geophysical prospecting (4th Ed.): McGraw-Hill Book Co., New York.

Gassmann, F., 1951, Elastic waves through a packing of spheres: Geophysics, **16**, 673–685.

Hardage, B. A., 1985, Vertical seismic profiling (2nd Ed.): Geophysical Press, London.

May, C. J., 1984, Reservoir engineering, *in* Hobson,G.D., Ed., Modern petroleum technology, 5th ed.: John Wiley & Sons, Inc., 165–198.

McQuillin, R., McEvoy, R., and Steele, R., 1984, An introduction to seismic interpretation, 2nd Ed.: Graham & Trotman.

Phillips, O. M., 1968, Heart of the earth: Freeman & Co.

Press, F., and Siever, R., 1978, Earth, 2nd Ed.: W. H. Freeman and Co.

Robinson, E. S. and Coruh, C., 1988, Basic exploration geophysics: John Wiley & Sons.

Sheriff, R. E., 1989, Geophysical methods: Prentice-Hall.

Sheriff, R. E., 1991, Encyclopedic dictionary of exploration geophysics, 3rd Ed.: Soc. Expl. Geophys.

Sheriff, R. E. and Geldart, L. P., 1982–83, Exploration seismology (2 vols): Cambridge University Press.

Shuey, R. T., 1985, A simplification of the Zoeppritz equations: Geophysics **50**, 609–614.

Telford, W. M., Geldart, L. P., and Sheriff, R. E., 1990, Applied geophysics, 2nd Ed.: Cambridge University Press.

Waters, K. H., 1987, Reflection seismology, 3rd Ed.: John Wiley & Sons.

White, J. E. and Sengbush, R. L., 1987, Production seismology: Geophysical Press.

Chapter 3: Geophysical Technology
Technologies of Reservoir Geophysics

*Alistair R. Brown**

Introduction

Reservoir geophysics, also referred to as development and production geophysics, addresses the delineation, description, and surveillance of hydrocarbon reservoirs. The subject is new, broad, and as yet poorly defined. Many technologies are encompassed and their common thread is the extraction of more subsurface detail than traditional geophysical techniques. The vast majority of the applicable technologies are acoustic and most fall within the broad definition of seismology. Today's most important and most widely used technique is the three-dimensional (3-D) seismic survey. In fact, collection, processing, and interpretation of 3-D seismic data probably constitute 70–80 percent of all reservoir geophysics, and borehole acoustic techniques constitute most of the remainder.

Figure 1 reviews the spectrum of acoustic techniques and compares them in terms of resolution, signal frequency, and range. Adequate resolution of course is critical to all reservoir studies and optimum collection and processing of 3-D seismic data achieve the best possible resolution for data collected on the Earth's surface. Further improvement can be achieved only at the expense of reduced range by placing the source and/or receiver below ground closer to the reservoir under study. This placement leads to the techniques of vertical seismic profiling (VSP) with the receiver downhole; crosswell tomography with the source and receiver both downhole but using different holes; and sonic logging, sonic imaging, and other types of well logging with the source and receiver down the same hole. Real projects involve the integration of many geophysical technologies in order to achieve synergism between them in addition to the synergism that can be achieved between geophysics, geology, and engineering. This chapter outlines the applicable geophysical technologies starting with 3-D seismic.

*Consulting Reservoir Geophysicist

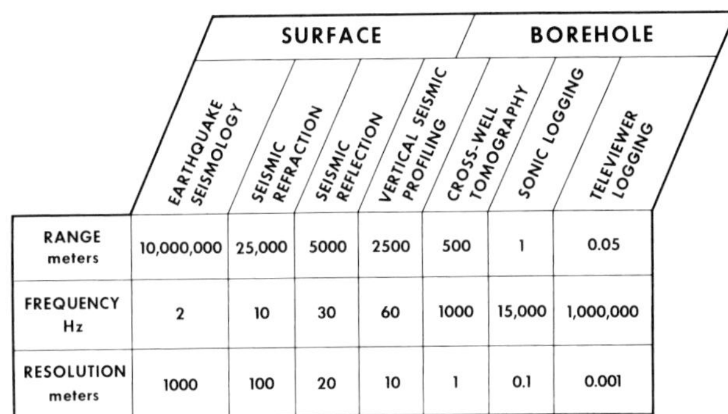

	SURFACE				BOREHOLE		
	EARTHQUAKE SEISMOLOGY	SEISMIC REFRACTION	SEISMIC REFLECTION	VERTICAL SEISMIC PROFILING	CROSS-WELL TOMOGRAPHY	SONIC LOGGING	TELEVIEWER LOGGING
RANGE meters	10,000,000	25,000	5000	2500	500	1	0.05
FREQUENCY Hz	2	10	30	60	1000	15,000	1,000,000
RESOLUTION meters	1000	100	20	10	1	0.1	0.001

Fig. 1. Comparison of subsurface acoustic techniques.

Collection and Processing of 3-D Seismic Data

Seismic waves are reflected, refracted, and diffracted by subsurface structure and stratigraphy, according to the principles discussed in Chapter 2, to yield recorded seismic data. Converting this recorded data into an interpretable form is the task of seismic processing. Common-midpoint (CMP) stacking attenuates multiple reflections and improves signal-to-noise (S/N) ratio. For reflectors with significant dip the assumption that CMP traces are reflected from the same point on the reflector is inadequate. Dip-move-out correction (DMO) is required in addition to normal-moveout correction (NMO). DMO correction and stacking is effectively prestack partial migration and serves two functions: (1) DMO moves finite-offset events updip and removes reflection-point smear, and (2) DMO removes the dip-dependence of stacking velocities. Wavelet processing attempts to understand and remove the signature of the energy source, the response of the detectors and recording instruments, and the filtering effects of the subsurface. Then the embedded wavelet, which is the shape of the relection from an individual interface, is changed to aid interpretability. Many other processes attack specific noises in order to enhance S/N ratio and hence the visibility of geologic detail. Seismic migration, normally near the end of the processing sequence, images the subsurface by undoing the defocussing effects of the earth. Specifically, migration performs three major functions: (1) reflections recorded out-of-place because of dip are repositioned; (2) diffraction patterns from discontinuities are collapsed; and (3) energy spread over Fresnel zones is focussed. Structural configuration thus is clarified significantly and interpreted structural anomalies will be located correctly. Stratigraphic resolution is also greatly improved as dispersed amplitudes are imaged. Because the earth is 3-D, the aberrations in recorded seismic data are 3-D and thus can be correctly compensated only by performing migration in three dimensions. 3-D migration is commonly partitioned into two orthogonal components which is economically efficient and technically valid where the velocity field is fairly uncomplicated. Understanding the 3-D velocity field is critical for accurate migration. In areas of very complex structure the difficulties of velocity determination may limit significantly the benefits achievable by 3-D migration.

A successful 3-D survey must be diligently designed with the objectives in mind and with some knowledge of the geology. The subsurface sampling intervals must be small enough to avoid aliasing during processing and interpretation. The Nyquist theorem states that every signal must be sampled at least twice per wavelength. The highest frequency of interest, F_{max}, provides the shortest wavelength and thus determines the maximum permissible subsurface spacing. The limiting value of the subsurface spacing D is determined from the formula

$$D \leq 1000/(2\, F_{max}\, DIP_{max}),$$

where DIP_{max} is measured on the unmigrated seismic section in milliseconds per unit distance. To design with some margin of safety is prudent because generally it is difficult to determine F_{max} and DIP_{max} exactly. Hence we often write the above formula as

$$D \leq 1000/(3\, F_{max}\, DIP_{max}),$$

implying that the design goal is three samples per shortest wavelength. The surface area over which data are collected in order to image the desired subsurface area must also be computed. Considering the movement of structural events around the edge of the prospect during migration, data must be collected in an additional fringe area if the dip is upward toward the center of the prospect. The width of this fringe area, or migration distance M, is given by

$$M = TV^2DIP/4,$$

where T is two-way record time and V is velocity. In order to properly focus seismic amplitudes spread over Fresnel zones for stratigraphic objectives, an additional Fresnel zone radius R around the edge of the prospect must be considered. The width of the additional fringe area to satisfy this requirement is given by

$$R = (V/2)\sqrt{(T/F)} = \sqrt{(Z\lambda/2)},$$

where Z is depth and λ is wavelength. In practice, the above design parameters are calculated and compromises and adjustments are made based on the hardware available for the survey, the nature of the objective, how close the principal area of interest is to the edge of the prospect, etc. Subsurface sampling intervals computed according to the above design considerations range from 10 to 100 m and may not be the same in dip and strike directions. Vertical sampling is determined by the digital sample interval of 2 or 4 ms. Thus the dimensions of the grid cell of seismic data points are typically around $5 \times 15 \times 50$ m. Hence the subsurface detail present in 3-D seismic data normally is very great. Interpreters should bear this in mind during their interpretation and mapping, and ensure that they make use of all the data. The sampling requirements for interpretation are the same as for processing.

3-D seismic data are collected over the designed area and with the designed subsurface spacings by means of a variety of techniques. At sea a boat pulling a hydrophone streamer and an array of air guns steams back and forth across the prospect normally collecting one line of data at a time. If the required spacings in dip and strike directions are different, the boat normally steams along the dip direction. In this way the lesser spacing in the dip direction is provided by the hydrophone separation in the streamer and the wider spacing in the strike direction is provided by the separation of parallel boat tracks. Navigation accuracy is of the utmost importance as maximum probable error must be significantly less than the required spacings. Positioning of the boat can be accomplished best by a trilateration method from fixed base stations. Positioning of the hydrophones relative to the boat generally is measured from a series of compasses in the streamer. The processing of these two sources of navigation data determine which recorded traces fall into which "bins". A bin is a small subsurface area which defines one set of recorded traces to be stacked together in processing. Collecting closely spaced 3-D seismic data in an area of producing platforms (Figure 2) generates special problems in obstacle avoidance. Often separate source and recording boats are used so that the area adjacent to each platform can be undershot. 3-D seismic data on land generally are collected with several parallel geophone lines (commonly four or six) laid out simultaneously. Several shots, or other sources of energy (commonly six or eight), are executed across one end of these geophone lines (Figure 3). After moving along the number of geophone groups appropriate to the intended fold of multiple coverage, the series of six or eight shots is repeated. In this way data are collected over a "swath" or strip of the

Fig. 2. Seismic data collection vessel in an area of producing petroleum platforms. (Courtesy Halliburton Geophysical Services Inc.)

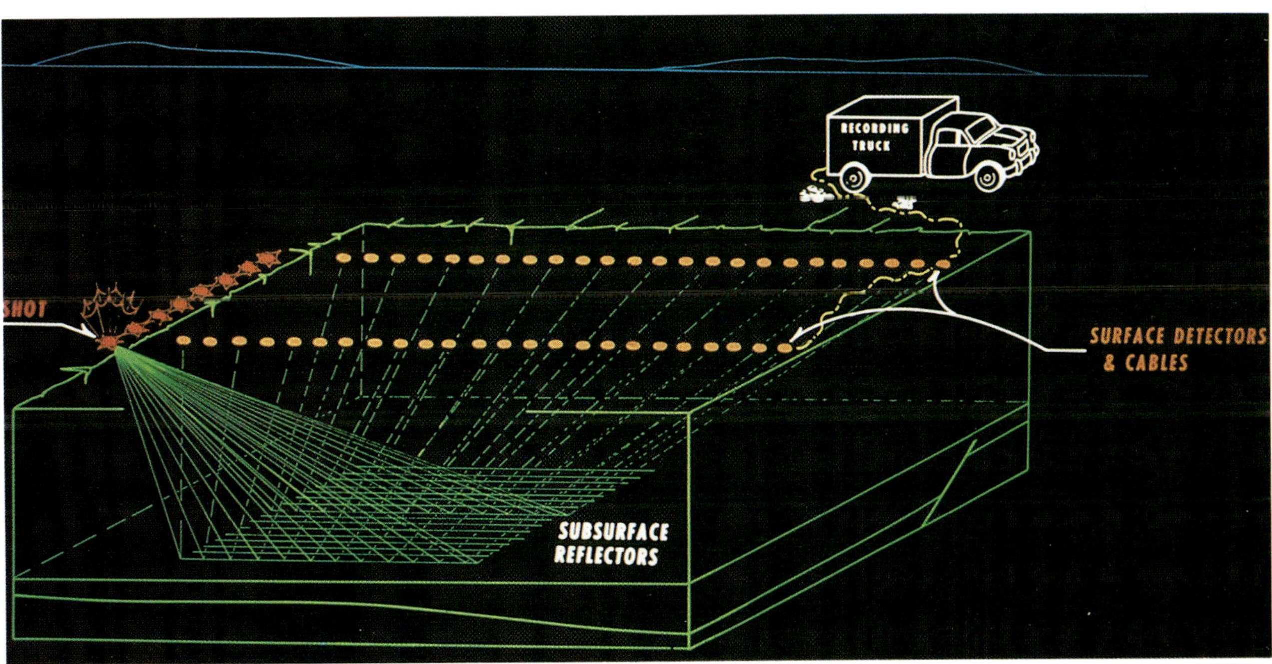

Fig. 3. Collecting land 3-D seismic data by the swath technique (Courtesy Halliburton Geophysical Services Inc.)

subsurface. Successive adjacent parallel swaths then cover the prospect. Multiple fold in both line and crossline directions is necessary for static correction determination, particularly at swath boundaries. An alternative acquisition technique is to lay out a ''patch'' of geophones and then to shoot around and through the patch. In areas of limited access land 3-D data are sometimes collected with sources and receivers placed only around squares or loops. This method, however, gives low-fold and poorly determined corrections, so generally provides poorer data. Another low-fold method involves a single line of shots and a single perpendicular line of geophones. These latter techniques are avoided wherever swath or patch methods are possible and economic.

In the last few years there has been an increasing demand for less expensive 3-D surveys. Boats pulling two (or three) streamers now collect two lines of data simultaneously; sometimes the air guns are divided into two separate arrays for further efficiency. Two boats, each towing two streamers and two air gun arrays collect 12 lines of data with one pass. Boats also collect 3-D data in circles, either successively overlapping circles of the same radius or in a spiral; this pattern gives great operational efficiency as recording is continuous with no line turnarounds. Great areas of marine survey have been covered with a widely spaced form of linear 3-D data collection known as "exploration 3-D". Here lines are spaced four or five times more widely than the sampling theorem demands and the omitted lines are interpolated during processing. This arrangement depends critically on "intelligent interpolation", an interpolation process based on trace attribute recognition and propagation which mimics the interpreter's ability to jump correlate with seismic character. This approach prevents aliasing during 3-D migration and provides a data volume with all its associated interpretive advantages (see following discussion). However, this widely spaced 3-D data has significantly lower resolution than full 3-D data, resolution which is related to the spacing of the lines as collected rather than that after interpolation. The data are good and useful but different from full detailed 3-D coverage. There is also a trend toward the derivation of 3-D volumes from irregular grids of 2-D coverage. This trend permits 3-D migration of 2-D data by filling in the under-sampled grid loops by Intelligent Interpolation. This approach yields data of even further reduced resolution, which again is related to the original data density incorporated into the processing. An interpreter newly confronted by a "volume of 3-D data" should appreciate how the data were collected and processed so that he can adapt his expectations of subsurface detail accordingly. Generally these types of widely spaced 3-D data are inappropriate for reservoir studies.

Velocity, which pervades the seismic method, is the speed with which seismic energy travels through rocks and thus controls arrival time. Velocity is also in the reflection coefficient formula and thus controls reflection amplitude. The curvature in time of a seismic reflection from a single shot as a function of source-to-receiver offset is NMO and is the basis of absolute seismic velocity measurements (Figure 4). This velocity is used directly in NMO correction prior to CMP stacking, but the interrelated effects of the near-surface and ray path complexity need to be understood simultaneously. 3-D dip correction of stacking velocity is accomplished by ray tracing and yields the velocity values needed for DMO correction and seismic migration. The accuracy of 3-D migration processes in imaging the subsurface and providing the interpretive benefits previously discussed depends critically on the accuracy of the velocity field used. In structurally straightforward areas velocity fields commonly are determined with adequate accuracy but in highly complex areas, like overthrust belts, understanding of the velocities is a major difficulty. Dix's formula converts

stacking velocity to interval velocity which can be interpreted in terms of lithology, stratigraphy, pressure, porosity or fluid content. However, this conversion usually is possible only in a gross sense as moveout-derived interval velocity is unreliable when calculated for small intervals. The limit is typically an interval of about 150 ms (approximately 250 m). If we consider this limit in terms of the frequency content of interval velocity measurement, it converts to an upper limit of 3 Hz. The product of velocity and density, making acoustic impedance, occurs in the reflection coefficient formula. Inversion of the seismic trace thus converts seismic amplitudes to estimates of acoustic impedance, and hence velocity, over the seismic bandwidth. This bandwidth is typically about 8–60 Hz so there remains a gap of about 3–8 Hz in the frequency content of velocity determination which can be filled only by interpretive intervention and well data. Average velocity, the final item on Figure 4, is the velocity needed for converting seismic time to depth and is thus critical for determining true dip attitudes and structural closure. Very small lateral velocity changes can cause significant changes in dip and closure, so sophis-

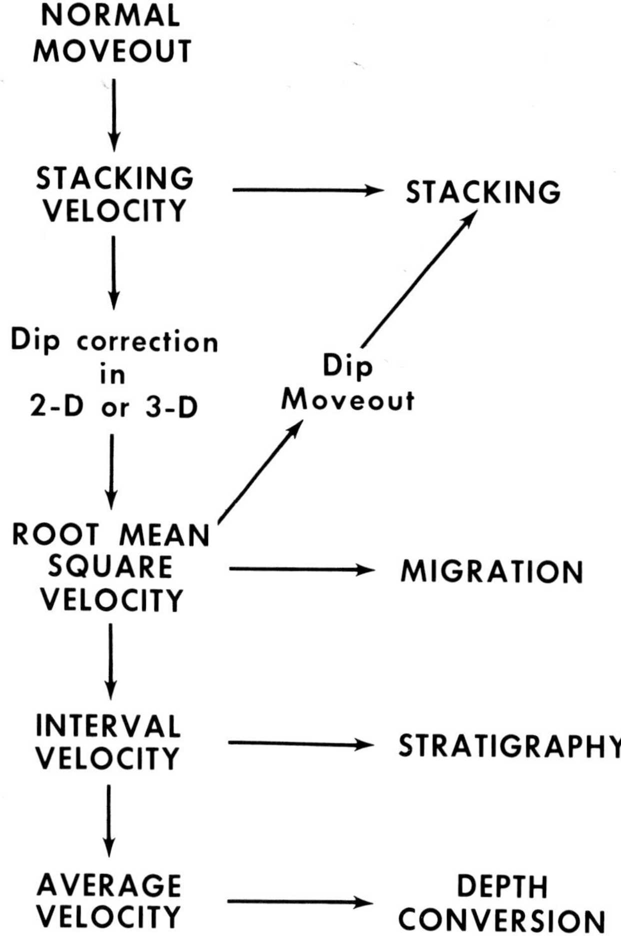

Fig. 4. Seismic velocity measurements and their use.

ticated normalization, quality control, smoothing, and well tying of velocities on a layer by layer basis are generally necessary. Velocity anisotropy introduces errors which require further correction for some layers.

Interpretation of 3-D Seismic Data

A 3-D seismic survey provides a volume of data for interpretation. Seismic projects have always sought a 3-D interpretation, so 3-D data has interpretive benefits which must be exploited fully. Because of the subsurface spacing design discussed previously, each data point in a 3-D volume is less than half a wavelength from the adjacent data point in every direction. Consequently, there are no grid loops around which the interpreter must tie his horizons and thus no areas over which he must interpolate his interpretation. Therefore, a more accurate and consistent interpretation should result, but the methods of 3-D interpretation differ from those of 2-D interpretation. The 3-D array of regularly and closely spaced data points permits the volume to be sliced preferentially along three orthogonal planes (Figure 5) yielding "lines" (vertical sections), "crosslines" (vertical sections), and "time slices" (horizontal sections, also known as Seiscrop™ sections). Complete interpretation, which must comprehend all the data, requires the use of all of these types of sections and perhaps certain other special purpose slices also. Vertical sections along diagonal lines connecting wells or in the plane of a deviated well ("arbitrary lines") assist in well tying. Slices along one structurally interpreted horizon, thus showing the spatial distribution of acoustic properties over one bedding plane, are "horizon slices" and have value for studying stratigraphy and reservoir properties. "Fault slices" are parallel to a major fault and are used for studying secondary faulting, fault-related structure, and sealing at faults. Many kinds of composite displays permit the 3-D seismic interpreter to appreciate the three-dimensionality of the subsurface and interpret more meaningful information from the data. Hence the 3-D interpreter must learn to use horizontal sections and various special-purpose slices in order to get the most out of the data at each stage of interpretation.

The time slice or horizontal section shows all the seismic events for a particular seismic time or depth. Figure 6 shows a horizontal section through a Gulf of Mexico gas reservoir; the high amplitudes are reflections from the reservoir interfaces. The attitude of any feature on a time slice indicates the feature's own local strike. The picking of a seismic reflection provides the

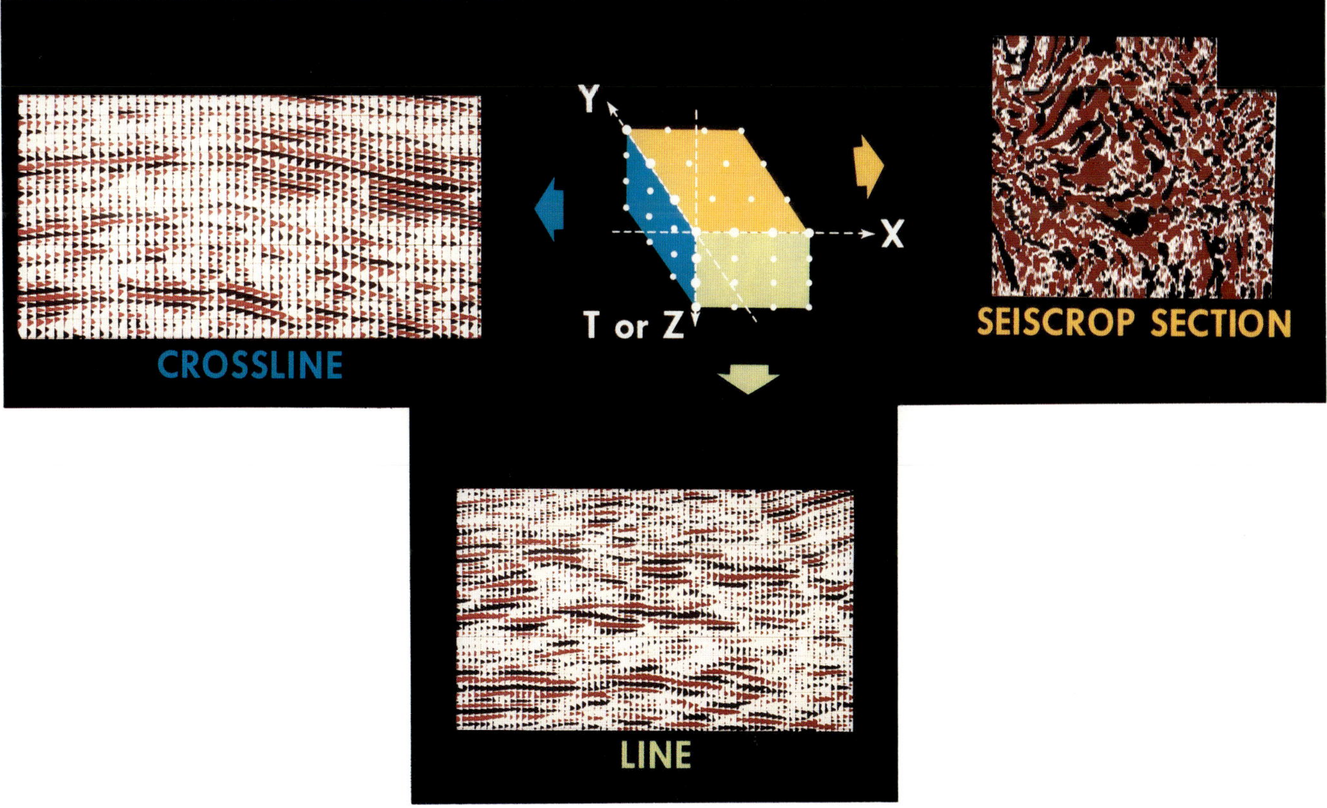

Fig. 5. Slicing a 3-D data volume in three orthogonal directions to yield lines (vertical sections), crosslines (vertical sections), and time slices (horizontal or Seiscrop sections).

strike of that reflecting surface and thus a contour at the time at which the time slice was generated. Direct contouring from time slices thus yields a structural contour map directly without the traditional intermediate stages of timing and posting. Picking of a fault on a sequence of time slices yields a fault plane map directly. It is important that a 3-D interpreter makes use of all the data in the volume so that maximum subsurface detail is extracted. The proper use of a combination of vertical, horizontal, and other sections, coupled with the use of an interactive system, facilitates this; the picking of every vertical section would be excessively time-consuming and also less effective. Structural interpretation of 3-D seismic data generally reveals enormously greater subsurface detail than was believed present on the basis of the previous 2-D data interpretation. Typically the number of interpretable faults increases significantly, maybe ten-fold. Also the shape of the structural contours reveals fine detail in horizon configuration not present on previous

maps. These changes should be considered normal and structural interpreters of 3-D data should not expect contours to be as smooth as their maps showed previously. Various special approaches can be used to help reveal especially subtle faults. In areas of low dip a subtle fault may show more discontinuity in strike than dip, and thus be more visible on a horizontal section than on a vertical one. Automatically tracked structural contour maps may be manipulated to reveal subtle lineations interpretable as faults. These manipulations include surface differentiation, dip magnitude and azimuth plotting, and map smoothing and subtraction to yield a high-spatial-frequency residual. Fault slices parallel to the interpreted position of a growth fault have revealed details of splinter faults close to the growth fault and caused by differential movement on the faults.

The most striking features revealed by time slices are stratigraphic because the map-style view often provides distinctive shapes readily relatable to geo-

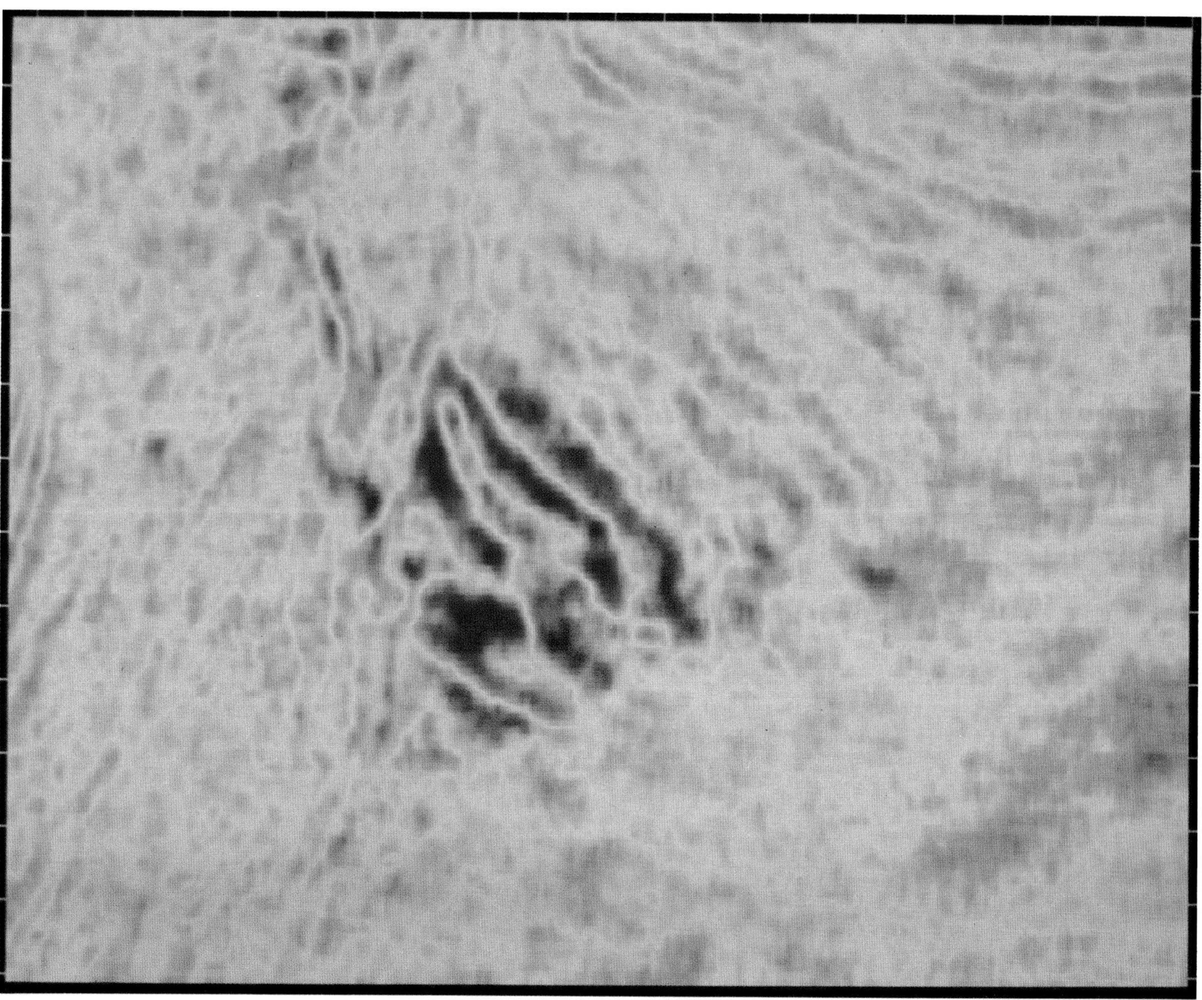

Fig. 6. Time slice or horizontal section from Gulf of Mexico 3-D data showing high amplitude reflections from a gas reservoir. (Courtesy CNG Producing Company.)

logic experience of depositional systems. In flat-lying beds, horizontal sections slice along bedding planes and thus provide views of ancient deposition which look just like the views of modern deposition that one gets out of an airplane. In the presence of structure, however, the same map-style view of the bedding plane is obtained by slicing through the data volume along a structurally interpreted horizon, thus reconstituting a depositional surface. Figure 7 is a horizon slice showing a loop of a meandering river channel, the inside of which has been filled with point bar deposits. The value of a horizon slice for stratigraphic interpretation critically depends on the correctness of the structural interpretation on which it was based. Sometimes the continuity of an observed stratigraphic feature can indicate the correctness or otherwise of the structural interpretation. Horizon slices are made simply and routinely today by interactive interpretation systems. Guided semiautomatic horizon tracking or fully automatic spatial horizon tracking starting from input control is used to follow the selected horizon along its maximum amplitude; the times generated by this tracking provide the time structure map, and the amplitudes provide the horizon slice. Horizon slices made in this way have been used to recognize and study channels of various kinds, offshore bars, point

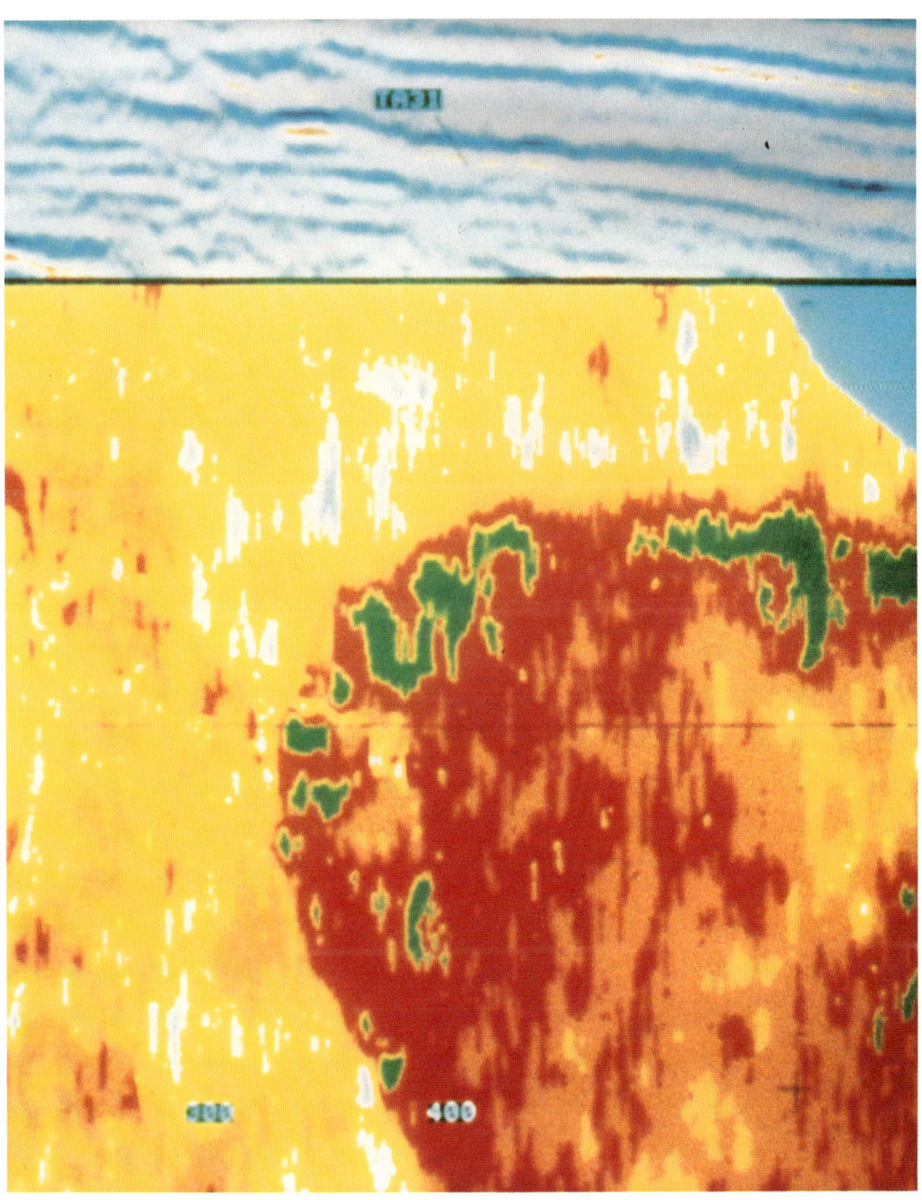

Fig. 7. Horizon slice showing spatial distribution of seismic amplitude over one reflection. The high amplitudes, shown in red and green, indicate one loop of a meandering river channel. (Courtesy Halliburton Geophysical Services Inc.)

bars, crevasse splays, depositional edges, pinchouts, and other stratigraphic features. Clearly for this approach, shape is the critical diagnostic. In carbonate areas reef buildups can be delineated and erosional features such as sink holes observed. Horizon slices through a reservoir zone have been used to delineate areas of maximum porosity. Horizon slices along unconformities reveal the patterns of subcrop below, and have been used to delineate subcropping porous zones and zones of fluid saturation. Bright spots from hydrocarbon accumulations have been mapped in detail and the exact position of downdip reservoir limits have been determined by amplitude extrapolation. The utility of horizon slices has been demonstrated on land and marine data for depths to 4000 m. The limiting factors are the data's resolution, S/N ratio, and amplitude and phase preservation.

Amplitude, Phase, and Color

Seismic amplitude is full of subsurface information about the reservoirs we wish to study but great care must be exercised in data collection and processing for the amplitudes to be of sufficient fidelity. Automatic gain adjustment or short-gate scaling must not be used and proper, accurate corrections must be applied for ground coupling, instrument sensitivity, near-surface irregularities, moveout, variations in stack fold, etc. Hydrocarbon fluids reduce the acoustic impedance of reservoir rocks. Often, particularly in young unconsolidated rocks, this reduction produces a "bright spot" or anomalously high amplitude reflection. If, in addition to the high amplitude, the characteristics of the reflection satisfy certain criteria, the bright spot can be positively identified as a hydrocarbon reservoir. Figure 8 shows very high amplitude reflections from the top and base of pay in the upper reservoir and less strong reflections from the lower reservoir. The base of pay reflection for the upper reservoir in Figure 8 is a fluid contact reflection or "flat spot". Reservoir reflections may show a "polarity reversal" between the hydrocarbon zone and the water zone. Under

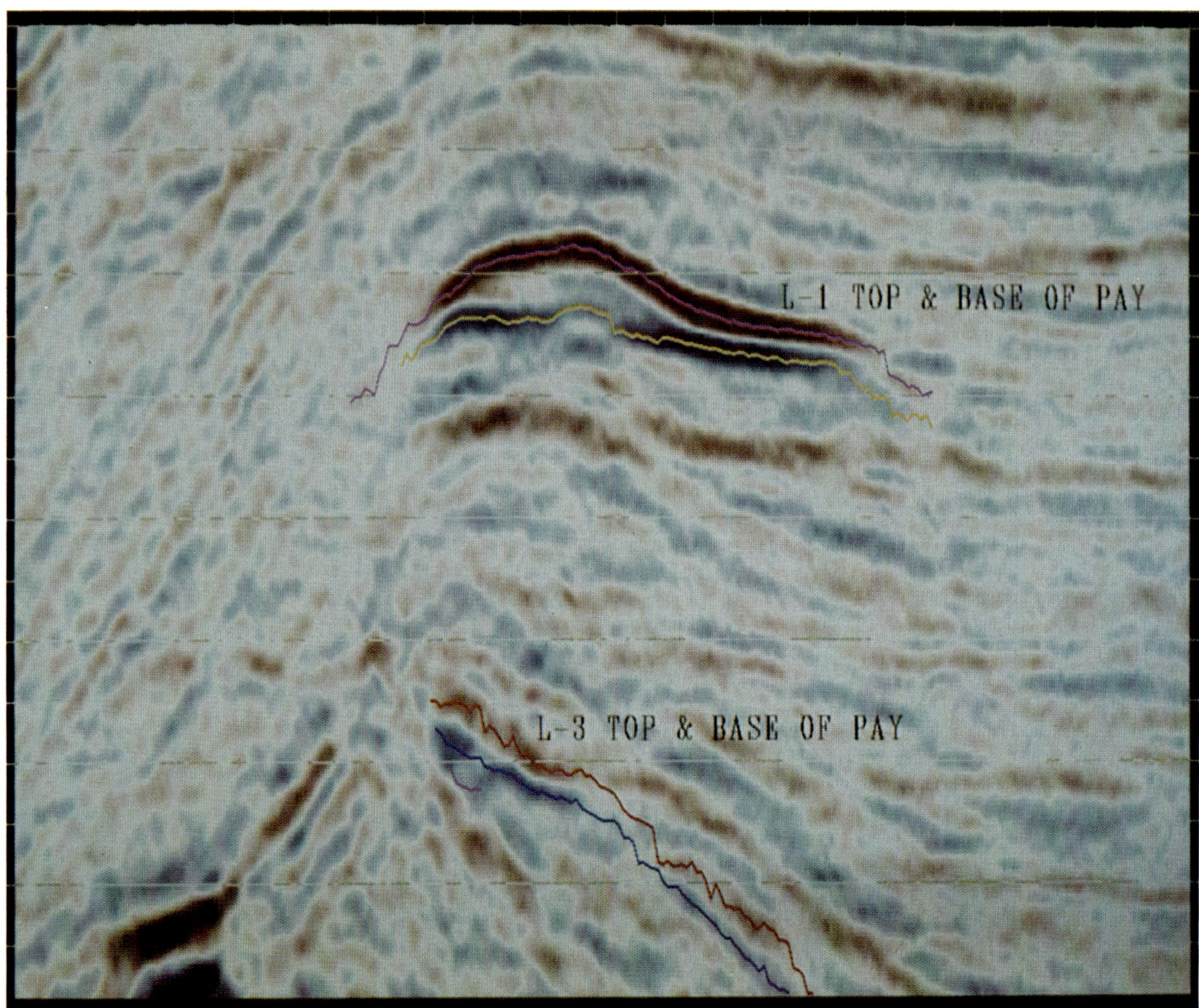

Fig. 8. Auto-tracked high amplitude zero-phase seismic reflections from two Gulf of Mexico gas reservoirs seen on a vertical section.(Courtesy CNG Producing Company.)

some circumstances, generally for older rocks, reservoir reflections may show a reduced amplitude or "dim spot". Once reservoir reflections of sufficient S/N ratio have been identified, an attempt can be made to interpret them in terms of reservoir properties. Such properties affecting amplitude include lithology, nature of fluid, pressure, hydrocarbon saturation, porosity, net-to-gross ratio, and the tuning effects of reservoir thickness. Without well control these effects are unseparable, but often those properties affecting the reservoir as a whole, such as the first three or four just discussed, can be taken as known from the first well or wells drilled. Lateral variations in reservoir amplitude can often be ascribed reliably to variations in porosity and/or net-to-gross ratio. Tuning effects between reservoir reflections confuse these effects but can be removed; these subjects are discussed separately below. For the common type of clastic reservoir where the acoustic impedance of the reservoir sand is lower than that of the embedding shales, higher amplitude indicates better reservoir quality regardless of whether that better quality is higher porosity, higher net-to-gross ratio, or even higher hydrocarbon saturation (tuning effects excepted). In general, therefore, high amplitude is usually good. Lateral variations in amplitude are best observed on seismic sections displayed in color. These variations are best related to reservoir interfaces if the data are zero phase.

Studies of amplitude and time thickness for thin beds are complicated by tuning phenomena. For a reservoir whose thickness is close to the tuning thickness or less, the reflections from the top and bottom will interfere with each other both constructively and destructively. The tuning thickness is the same as the limit of separability of two closely spaced reflections, namely one-quarter of the dominant seismic wavelength. The best way to visualize tuning phenomena in amplitude and time is with a thinning wedge (Figure 9). For a given acoustic contrast and a reservoir thickness well above tuning, the amplitude is normal. As the tuning thickness is approached, the amplitude increases, reaching a maximum at the tuning thickness. The magnitude of the amplitude increase depends on the size of the wavelet's side lobes. For thicknesses less than tuning, the amplitudes of the reservoir reflections decrease until they are lost in the background noise. This point is the "limit of detectability" ("visibility"). Because the tuning thickness is also the closest separation possible for the wavelets from the two reservoir interfaces, the wavelets retain that time separation as the reservoir thins. For zero-phase wavelets the resultant time distortion is equally divided between the top and bottom reservoir reflections as shown in Figure 9, the seismic top being displaced earlier in time and the seismic base later. (These effects are modified where the reflections from top and

base are not of equal amplitude and opposite polarity.) The seismic expression of Tertiary clastic reservoirs is often a strong reflection from the top and an equally strong one of opposite polarity from the fluid contact. As the downdip reservoir limit is approached these two reflections form a thinning wedge and all the tuning phenomena discussed here occur and can in many cases be observed. The downdip reservoir limit is not at the limit of visibility but beyond the limit. If the tuning thickness and the limit of visibility can both be recognized, if they are well separated, and if the amplitude of the reservoir top and the fluid contact are judged to be of the same amplitude, then the amplitude gradient between tuning and the limit of visibility can be extrapolated outward to zero. In this way the actual downdip reservoir limit can be inferred. For lower S/N reservoir reflections and irregular internal reservoir structure, tuning phenomena will be more complicated and modeling may be necessary to understand the observed effects.

The embedded seismic wavelet is the shape of the reflection from a single isolated reflecting interface, and a normal seismic trace is the composition of thousands of superimposed wavelets reflected from thousands of interfaces. The effective length in time of the wavelet is determined by its frequency content and its amplitude by the acoustic contrast of the interface causing the reflection, but the wavelet's shape is determined by the phase. Most seismic energy sources are minimum phase, meaning that the amplitude of the recorded signal reaches a maximum after a minimum delay from the first receipt of energy (Figure 10). The response of the geophones, the recording system, and the near-surface are also minimum phase, but they compound the length and complexity of the wavelet

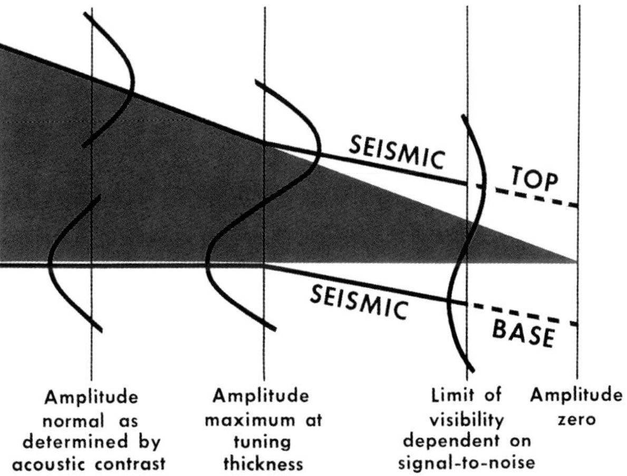

Fig. 9. Tuning phenomena in amplitude and time which occur between reflections from the top and base of a reservoir as the reservoir thins. (The reflectivity at the top and base of the reservoir are assumed to be equal in magnitude and opposite in polarity.)

embedded in recorded seismic data. The objective of wavelet processing is to unravel these natural processes in a controlled manner and shape the resultant wavelet for maximum interpretability. Most interpreters today who wish to relate individual seismic amplitudes to individual subsurface interfaces in pursuit of a detailed interpretation, prefer zero-phase data (Figure 10). There are many reasons why zero-phase is an aid to stratigraphic and reservoir interpretation. The zerophase wavelet is symmetrical with most of its energy concentrated into the central lobe; hence one particular wavelet from one interface of interest is easier to recognize in the presence of the other energy and noise on the seismic trace. Compared to a minimum-phase wavelet, zero-phase gives one principal trace excursion rather than two, so there is less ambiguity in understanding the polarity and identity of the reflection under study. For a zero-phase wavelet, the traveltime of the reflection from an interface is that of the center of the wavelet coincident with the amplitude maximum. Thus the time and peak amplitude for the reflection from one subsurface interface are at the same position on the seismic trace. This position is very important when extracting time and amplitude over reservoir surfaces in the pursuit of reservoir description.

Color increases the visual dynamic range of seismic data. We cannot interpret subsurface detail that we cannot see and color display at a large scale certainly permits us to see more. Typically a gradational color scheme is used in which two distinctly different colors or ranges of colors are applied to positive and negative amplitudes with the intensity of each color proportional to amplitude values. Figure 8 is an example using a popular blue and red scheme. A black-and-white, variable area and wiggle trace conventional display has three serious shortcomings for observing details of reservoir amplitudes: peaks are black and troughs are white, and thus look very different. The interpreters eye is biased toward the peaks and the troughs are relatively neglected. It is very difficult to compare the relative amplitude of peak and adjacent

trough because they look so different. Peaks, whose amplitude on the section exceeds the trace spacing, are clipped or optically saturated, rendering the highest amplitudes invisible. The correct horizontal position for each trace is normally at a zero-crossing; the excursion of the trace carries the highest amplitudes away from this correct position. Color display corrects all these shortcomings by coding color intensity directly to trace amplitude. By seeing adjacent peaks and troughs with equal clarity in gradational blue and red (for example), reflections from the top and bottom of a reservoir can often be recognized because their amplitudes vary in unison. For high S/N ratio reflections, observation of the relative amplitude of peak and adjacent trough can often indicate the phase of the data. Interpretive recognition of zero-phase is an important check that an interpreter should apply to his data before embarking on a detailed reservoir study.

Interactive Interpretation

Interactive interpretation systems, or workstations, have developed rapidly since their introduction in 1980. More interpreters convert to interactive systems each year particularly for the interpretation of 3-D seismic data and detailed reservoir studies. The data are held on magnetic or optical disk and are viewed and interpreted on the screen of a color monitor. The benefits of working interactively are many. Much time is saved by eliminating the bookkeeping and paper-handling associated with conventional methods. The interpreter can compose exactly the data slice or combination of slices needed for the study of the problem at hand. He has color as the standard display and flexibility of color to suit individual requirements and data problems. Once learned the system permits moving through the data quickly, without distraction, and the interpreter can maintain a productive idea flow. With the incorporation of well and other nonseismic data into interactive system data bases, more information can be reviewed easily by the interpreter, so that the final interpretation should be more consistent with all the available evidence and hence be a better interpretation. Because of the speed and efficiency of the many different data operations, interpreters are able to extract more information from 3-D seismic data than is normally possible by traditional techniques. Interpretations, therefore, are not necessarily accomplished more quickly, but they are certainly more complete and accurate. The central function of an interactive system is horizon tracking and several modes are generally available. Manual tracking, also known as point or stream tracking, depends only on the judgement of the interpreter as he moves the cursor over the data. Automatic tracking by the computer of a chosen reflection seen on the interactive screen is efficient in most cases, but autotracking is

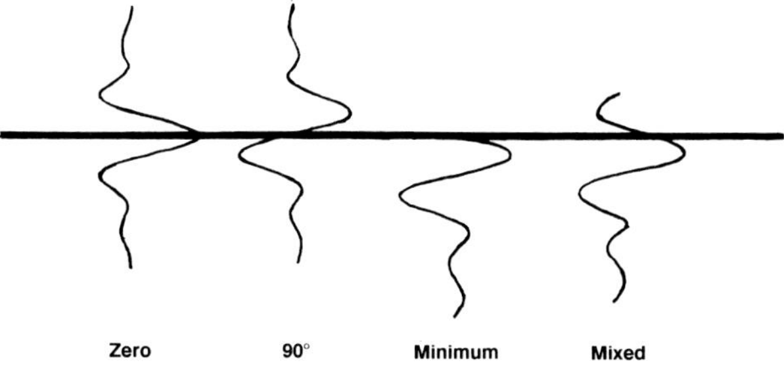

Fig. 10. Wavelets of four different phases showing the relative relationships between the amplitude maximum and the time of the reflecting interface.

also highly desirable for detailed amplitude studies because the computer can track the maximum amplitude in the seismic waveform. The time and amplitude stored in the digital data base are then accurately crestal amplitude values. These times and amplitudes relate directly to the same reservoir interface if the data are zero-phase, the reservoir reflection has good S/N ratio, and the same event has been tracked everywhere. Thus there is an important synergism between amplitude, phase, and interactive horizon tracking for detailed reservoir studies. An important development of horizon tracking is fully automatic spatial tracking where the computer attempts to track through the whole 3-D volume starting from input control.

Automatic horizon tracking of zero-phase 3-D seismic data correctly following an identified reservoir reflection yields a horizon slice over a reservoir interface. A horizon slice is a powerful product for studying reservoir properties. Color display is again valuable for observing trends and patterns in the spatially varying amplitude. Figure 11 shows a horizon slice over the reflection from the top of a Gulf of Mexico gas reservoir. The dark reds are the highest amplitudes and thus the zones of better reservoir quality. The low amplitude lineations running east-northeast to east are faults. The high amplitude lineation running approximately west-northwest is caused in part by tuning

between the reflections from the top of the reservoir and the fluid contact, as the reservoir thins to the north. Typically horizon slices are generated over both the top and base reservoir reflections. Similarity of the amplitude patterns confirms that they are caused principally by lateral variations within the reservoir. There may be some lateral variations in the embedding material above and below the reservoir but the effect of these can be suppressed simply by adding together (in absolute value) the two horizon slices. This is a kind of interpretively constrained inversion, focusing on the amplitude variations caused by the reservoir rather than by the embedding material. The amplitude patterns on the individual or composite horizon slices can be interpreted in terms of the extent of reservoir fluids, of variations in reservoir quality, or of tuning. A high amplitude lineation is interpretable as a tuning phenomenon, indicating where the reservoir has a thickness equal to the tuning thickness, if the lineation follows structural strike or is parallel to the downdip reservoir limit (e.g., Figure 11). With knowledge of the reservoir time thickness and an understanding of the tuning mechanism in the data, the tuning effects can be edited out to produce horizon slices or composite horizon slices in detuned amplitude. For a low acoustic impedance reservoir, a high in detuned amplitude means better reservoir quality. In order to be more quantitative, calibration to well data is essential. De-

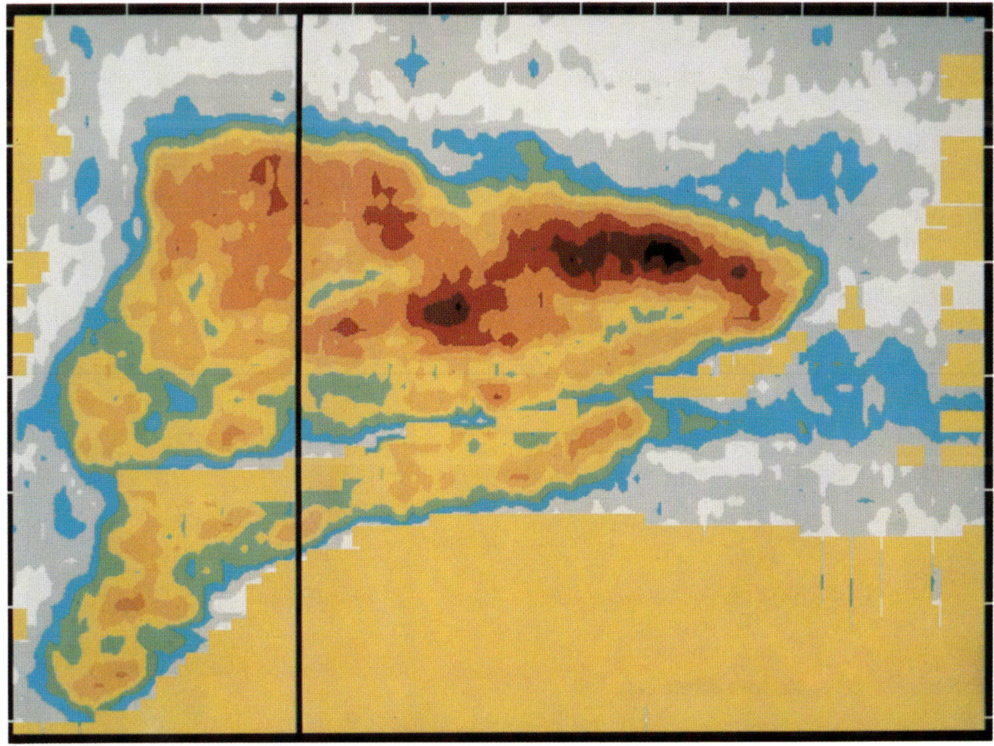

Fig. 11. Horizon slice along the reflection from the top of a Gulf of Mexico gas reservoir. (Courtesy CNG Producing Company)

tuned amplitude can then be expressed in terms of net-to-gross ratio, porosity, or the product of the two. Calibration to well data in a field of many wells can be performed either with a single calibration factor, which implies a high confidence in the seismic relative to the well data, or local calibration factors at individual wells with spatial interpolation between, which implies a high confidence in the well data relative to the seismic. In either case better input is provided to reservoir simulation studies than well data alone.

Seismic Inversion and Shear Properties

The well known formula for reflection coefficient RC, in terms of the velocity V and density ρ, is

$$RC = (\rho_2 V_2 - \rho_1 V_1)/(\rho_2 V_2 + \rho_1 V_1),$$

and the reflection coefficient is related to seismic amplitude through knowledge of the wavelet. The simplest form of seismic modeling is the generation of synthetic seismograms: velocity and density functions from well data are combined using the above formula and the resulting reflection coefficient function is convolved with the wavelet believed to match the one in the data. 2-D modeling involves the generation of a synthetic seismic section based on postulated subsurface geology. By comparison of the synthetic section with an actual section, the geologic postulate is tested. "Seismic inversion" attempts to drive the reflection coefficient formula backward in order to infer acoustic impedance, or velocity and density separately, from seismic amplitude. To do this with reasonable accuracy, the wavelet in the data must be known and assumptions must be made about the presence or absence of multiple reflections and other nonprimary energy. Furthermore seismic inversion can only operate over the seismic bandwidth, approximately 8–60 Hz. The upper frequency limit imposes the seismic resolution restriction on the acoustic impedance outputs. The lower limit, because it does not reach zero, indicates that inversion cannot generate absolute values of acoustic impedance but only relative ones. Absolute values of interval velocity are often determined from NMO and combined with inversion velocity for a final output. However, NMO velocity is only accurate up to frequencies of about 3 Hz, as mentioned previously. More sophisticated forms of inversion have proliferated in an attempt to address these issues. Many inversion approaches are now actually based on forward modeling. 2-D synthetic seismic sections are generated from an initial subsurface model and compared to the real data; then the model is changed and the synthetic section is updated and compared to the real data again. If after several iterations no further improvement is achieved, the updated model is the inversion output. Some approaches incorporate a priori information in order to improve the completeness and uniqueness of the output. The fundamental interpretive benefit of any form of seismic inversion is that "interface" information is converted to "interval" information. Hence the final presentation used by the interpreter is closer to geology (which is the study of rocks rather than of rock boundaries), is more readily tied to well logs, and is more directly relatable to reservoir properties. We should, however, be careful to distinguish between the interpretive benefits of inversion and those of display, in particular the use of color. Figure 12 shows a reservoir horizon slice in acoustic impedance. The brown colors represent low acoustic impedance values and indicate the producing zone of the reservoir. Here

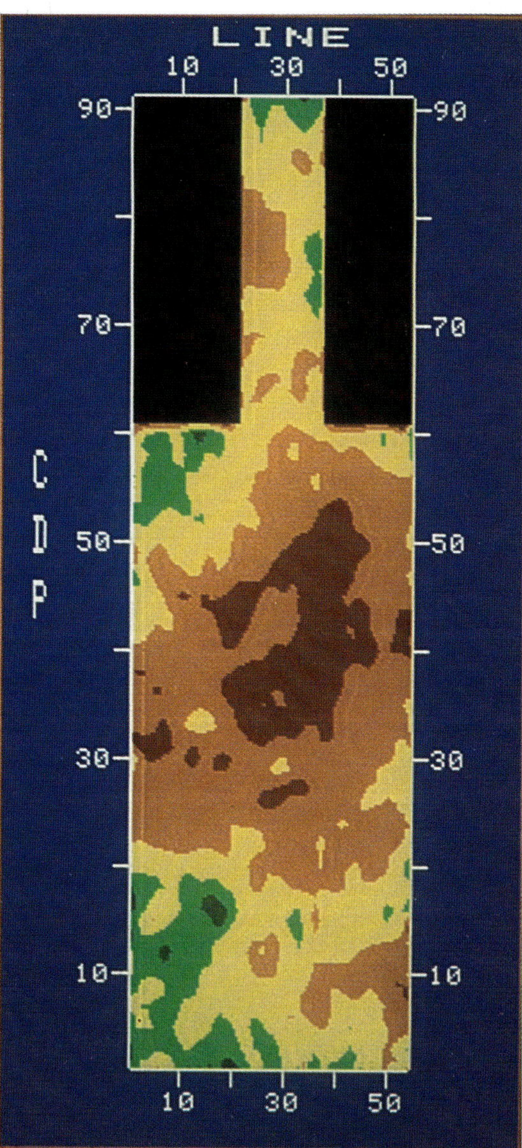

Fig. 12. Horizon slice in acoustic impedance over a producing reservoir in Peru. Brown indicates low acoustic impedance. (Courtesy Occidental Exploration and Production Company.)

the visibility of the reservoir fluids has been increased by the amplitude integration in the inversion process.

If seismic energy is reflected at an angle θ, the reflection coefficient formula becomes approximately

$$RC = (\rho_2 V_2 - \rho_1 V_1)/(\rho_2 V_2 + \rho_1 V_1)]\cos^2\theta$$
$$+ 2.25(\sigma_2 - \sigma_1)\sin^2\theta,$$

where σ is Poisson's ratio for each medium. For $\theta = 0$, this formula reduces to the normal incidence formula in the previous paragraph. The formula here is an approximation of the very complicated Zoeppritz equations made by Shuey to aid comprehension of the factors affecting variations of reflection amplitude with incident angle. For no Poisson's ratio contrast across a reflecting interface, the second term is zero and the variation with angle is simply the cosine factor, which causes a decrease of amplitude with increasing angle. If there is a significant Poisson's ratio contrast, as normally occurs at the boundary of gas sands, then the second term becomes important and amplitude generally increases with increasing angle. This increase in amplitude with increasing angle of incidence, or recording offset, can be used as a diagnostic in the identification of gas reservoirs. Studies of amplitude variations with offset (AVO) necessarily involve the use of prestack, single-fold data and are, therefore, subject to both random and systematic noise. Preconditioning and correction of the individual traces are thus essential, often followed by partial stacking to increase S/N ratio. Partial stacks displayed in different colors can be combined by color mixing to yield color sections whose variations indicate the varying domi-

nance of different offsets ranges. Figure 13 displays the gradient of amplitude variations in color, so red indicates increases with offset and blue indicates decreases. In general there is a lot of lithologic detail in amplitude-versus-offset data, but it is difficult to interpret; prestack seismic modeling is often required.

The study of shear wave (S-wave) seismic data in conjunction with normal longitudinal wave data provides additional information on the elastic properties of rocks similar to that provided by amplitude-versus-offset studies. Velocity ratio between P-waves and S-waves is also a function of Poisson's ratio, according to the following formula:

$$V_p/V_s = \Delta t_s/\Delta t_p = \sqrt{[2(1 - \sigma)/(1 - 2\sigma)]}.$$

Much S-wave data on land has been collected by a separate field operation using a S-wave vibrator. Modern trends prefer recording with three-component geophones, which in the same data collection operation collect P-waves and two polarized S-waves from the same seismic source. S-waves are not transmitted through water, so no equivalent operation is possible in the marine environment. However, where the water bottom is hard and for large angles of incidence, there is significant mode conversion of P- to S-waves at the water/rock interface. P-waves and S-waves can be separated in processing by sophisticated types of velocity filtering. S-waves are not affected by gas, so the amplitude phenomena of bright spots, flat spots, dim spots, and polarity reversals do not occur on S-wave sections. Their absence can be used as another identifying characteristic of hydrocarbon reservoirs.

Fig. 13. Ampliude-variations-with-offset section where red indicates increases with offset and blue indicates decreases. Increase of amplitude with offset often can be interpreted as an indication of gas. (Courtesy Western Atlas International.)

Velocity ratio is the most useful measurement for reservoir studies. Reflections must be accurately correlated between the *P*-wave and *S*-wave sections, a pair of which are shown in Figure 14, so that transit times can be measured for the same interval. This would desirably require independent identifications from *P*- and *S*-wave synthetic seismograms or *P*- and *S*-wave VSP. The transit time ratio is then the reciprocal of the velocity ratio, as indicated by the above formula. Velocity ratio is higher for increased porosity, for increased carbonate/sand ratio, and for decreased sand/shale ratio. It is also affected by pore shape. If porosity is known from wells and is cross-plotted against velocity ratio, the resultant distribution can be interpreted in terms of pore aspect-ratio spectra, that is, the relative abundance of spherical pores and thin cracks.

S-waves involve particle vibrations perpendicular to the direction of propagation of the seismic energy. Clearly there are two such directions which are themselves mutually perpendicular. In an isotropic medium these two *S*-waves travel at the same velocity, which is much less than the velocity of *P*-waves. However, in an anisotropic medium where elastic properties depend on direction, the two velocities are different. A polarized *S*-wave, that is an *S*-wave vibrating in one preferred direction, traveling through an anisotropic medium will be split into two mutually perpendicular *S*-wave vibrations, parallel to and perpendicular to the anisotropic axis. The most common form of anisotropy of petroleum interest is oriented fractures. An *S*-wave vibrating parallel to the fracture direction sees unfractured rock and so travels with the normal *S*-wave velocity. On the other hand, an *S*-wave vibrating perpendicular to the fracture direction sees the full effect of the fractures and its travel is slowed down accordingly. The extent to which this *S*-wave is slowed down is a measure of the fracture density. If the fracture direction is known, an oriented *S*-wave source can be used to generate *S*-waves polarized parallel to and perpendicular to this direction. By comparing the measured interval velocities of these two *S*-waves in the zone of interest, a velocity reduction and thus a fracture density can be determined. Vertical seismic profiles provide reliable velocity measurements for this kind of study.

Borehole Studies

Zero-offset vertical seismic profiling (VSP) involves an energy source near the top of a borehole and a geophone in the borehole recording at a succession of depths. The resultant profile is a display of time traces

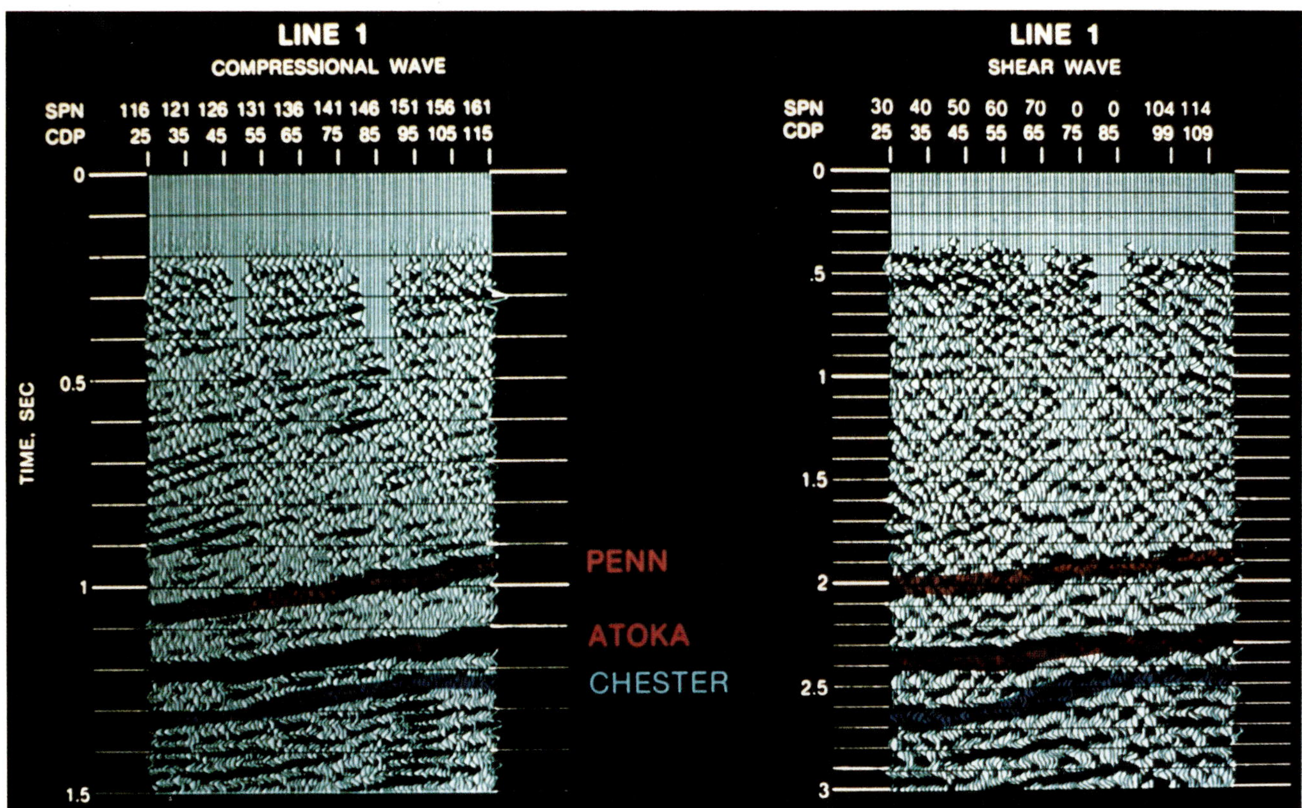

Fig. 14. Comparative *P*-wave and *S*-wave sections. (Courtesy Arco Oil and Gas Company.)

for a sequence of well geophone depths (Figure 15). The well geophone receives downgoing energy directly from the source and upgoing energy reflected from below. For a well geophone depth of zero, all the energy is from below so this trace approximates the surface seismic data. When the well geophone is located at a particular reflecting interface, the travel-times for the downgoing direct energy and the upgoing reflected energy from that interface are the same. In this way positive identification of seismic reflections with rock boundaries or features in the well logs is possible. Thus the basic task of a zero-offset vertical seismic profile is for calibration of seismic to borehole data, that is, as an upgraded synthetic seismogram. Because the depth of the well geophone is accurately known and the corresponding reflection time can be determined in the manner just described, time-depth relationships are established and thus reliable subsurface seismic velocities are obtained. Because upgoing and downgoing energy have opposing dips on a VSP

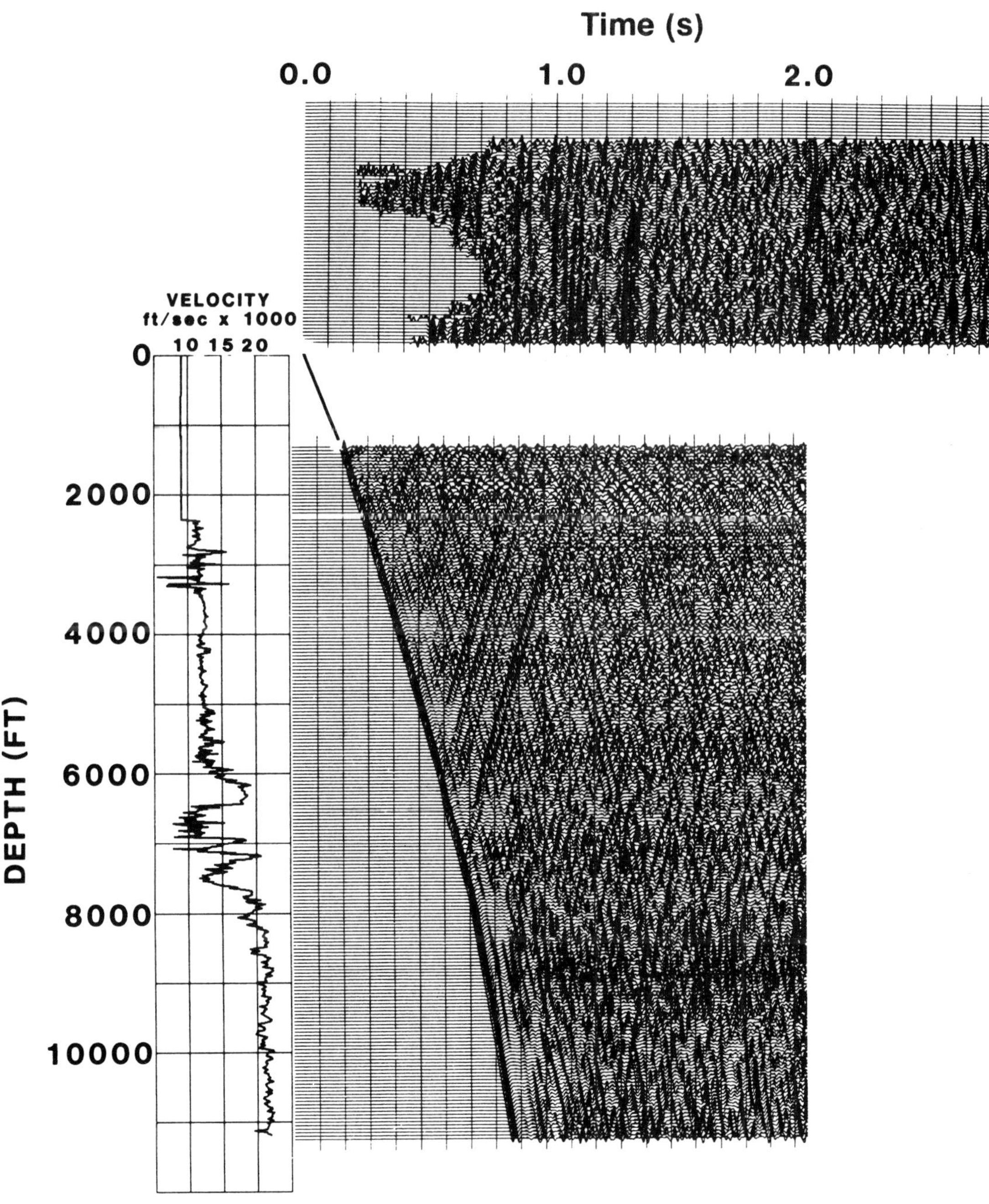

Fig. 15. Zero-offset vertical seismic profile tied to surface seismic data and well log. (Courtesy Phillips Petroleum Company.)

(Figure 15), they can be separated in processing. The upgoing energy is reflected only and, in the early part of each VSP trace, no significant multiple reflections occur. This permits the generation of a truly multiple-free trace. If a VSP is recorded with a multicomponent geophone, an *S*-wave VSP can also be constructed. This recording provides reflection identification, time-depth relationships, and velocity for *S*-wave studies.

Because VSP involves a source on the surface and a receiver in the borehole, or vice versa, the travel is one-way only and hence the data are potentially higher resolution than normal surface seismic data. Various types of offset VSP and walkaway VSP record lateral coverage in the vicinity of the borehole, taking advantage of one-way-only travel. When a well is thought to have just missed the target, an offset VSP can provide local high resolution coverage to indicate which way to move. For a deviated well, vertical incidence VSP coverage can be recorded with the source vertically above the down-hole receiver. For a vertical well, lateral coverage can be obtained either by moving the receiver up the borehole or by moving the receiver along the surface. The former is usually preferred because the multiple receiver positions in the borehole permit the separation of upgoing and downgoing energy during processing. Figure 16 shows an offset VSP; the strange shape of the coverage results from the midpoint ray path geometry for a source on the surface and a receiver at different depths in a vertical hole. Conversion from seismic traces for different geophone depths to traces for different offset positions, as in Figure 16, involves an important processing step called the VSP to CMP transform. This process is difficult in the presence of complex structure or lateral velocity variation. Walkaway profiles generally involve movement of both the source and receiver. For example, the source is moved along the surface in a direction perpendicular to the direction of deviation of the borehole, and this move is repeated for several receiver positions in the hole in order to separate upgoing and downgoing energy. Successes of borehole profiling include the delineation of small faults and stratigraphic changes, and the mapping of thin reservoir sands.

The basic notions of offset VSPs and walkaway VSPs are being extended in a variety of novel ways. By placing an array tool with multiple sensors at fixed positions in the borehole and by using multiple source positions at various azimuths on the surface, a 3-D VSP can be generated and still have the separation of upgoing from downgoing waves. This procedure has been done in the marine environment. However, it is clearly faster and more economical to capture data at a large number of surface positions with receivers than it is with seismic sources, whether or not the objective is 3-D data. This desire has led to the "reverse VSP"

in which an energy source in the borehole is recorded by several receivers on the ground. A key requirement of this technology is a nondestructive borehole energy source and one of the more promising ones is the resonant swept-frequency oscillator. Alternatively, the vibrations of the active drill bit have been used; this procedure involves continuous recording for several minutes and cross-correlation in processing, an approach similar in principle to Vibroseis processing. This permits a reverse VSP to be recorded while the well is being drilled which has obvious economic advantages. In fact, measurement while drilling (MWD) is an important technological trend that permits downhole logging and other measurements to be made without loss of rig time. The reverse VSP technology is expected to facilitate the routine implementation of 3-D VSP in all environments in the near future.

Cross-well tomography is a further extension of borehole profiling where source and receiver are in different boreholes. Tomography is a general term meaning reconstruction of a section through a body

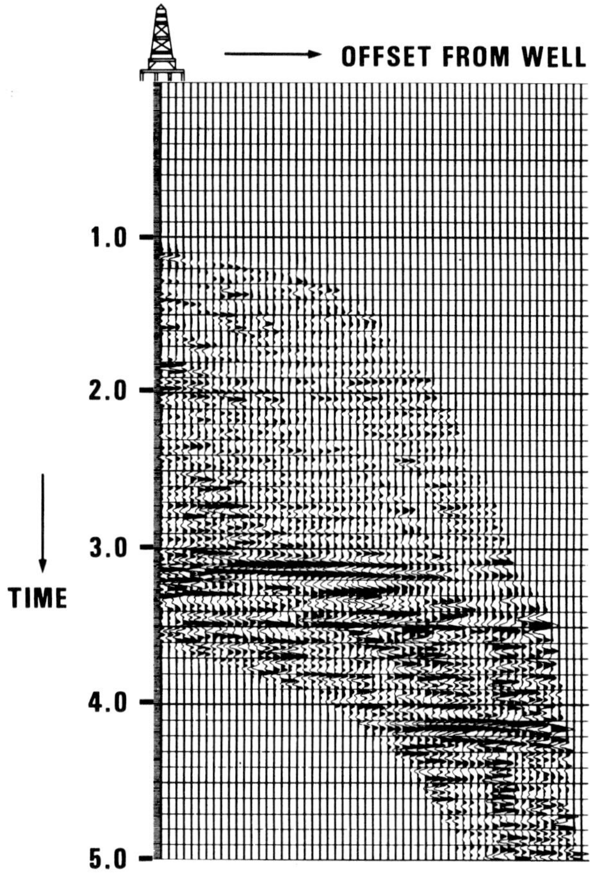

Fig. 16. Offset vertical seismic profile showing subsurface seismic coverage as a function of offset distance from the well. (Dillon, P.B., and Thomson, R.C., 1984, Geophys. Prosp. **32**, 790–811)

from a series of measurements around the body. Tomography is an established technique in medicine and material testing. In seismic the recorded observations can in general be borehole-to-borehole, borehole-to-surface, or surface-to-surface and can use either transmitted or reflected energy. Processing of traveltimes to determine velocity can be for interpretive mapping or for improved seismic data processing. One major application for reservoir geophysics is the transmission of energy from one borehole to another directly through the reservoir zone. For closely spaced wells, improvements in resolution are possible. Processing of multiple recordings for different downhole source and receiver positions in a pair of wells provides a tomogram in acoustic velocity from the arrival times and a tomogram in acoustic transparency from the recorded amplitudes. Interpretation involves the extension of features penetrated by the wells by means of the tomographic patterns in velocity and transparency. Another major application is the study of high resolution reflections recorded from one borehole to another within, and adjacent to, the reservoir zone. For any of these approaches the great practical difficulty is the operation of a nondestructive downhole energy source with adequate range. Great developmental effort is currently being directed toward this problem; vibrators, air guns, and piezoelectric transducers have all been used.

Sonic logging involves a source and receiver in the same borehole and has been in use for many years. Modern improvements are many. In order that the sonic measurement of transit time, and thus velocity, is for the unaltered rock formation, the separation of source and receiver is much longer than it used to be. The resultant increased penetration reaches behind mud cake, alteration, and hole irregularities. Instrumental difficulties, including cycle skipping and triggering on noise bursts, are circumvented by sonic waveform logging, which is the recording of the whole trace of energy arriving at the receiver rather than just the first arrival. The sonic waveform log is an extension of the regular sonic log just as the vertical seismic profile is of the well velocity survey. A recorded waveform log section (Figure 17) is quality controlled and its first arrival tracked to give a reliable transit time measure. In addition to this measure of the P-wave arrival, an S-wave arrival can be observed and tracked (Figure 17) to generate a shear transit-time log. This additional valuable information permits the calculation of velocity ratio V_p/V_s and Poisson's ratio, and from these ratios some lithologic identification. Also an S-wave synthetic seismogram can be generated for event identification on S-wave sections. A sonic waveform log in a deviated well can record reflections from nearby acoustic interfaces; these events will generally be recognized as those not parallel to the borehole. These sonic images are also recorded with special tools.

The borehole televiewer does not seek penetration of the rock formations but simply takes pictures of the borehole wall. The tool is held centrally in the wellbore and an acoustic transducer operating at megahertz frequencies spins on a vertical axis recording from a compass each time it passes through north. The traveltime of the recorded signal measures the borehole radius and the amplitude measures the wall's reflectivity. Figure 18 shows televiewer sections displaying both of these parameters in color against geographic direction and depth in the borehole. Anomalously long traveltimes or anomalously low amplitudes indicate fractures and one of the most important functions of the televiewer is to identify fractures and measure their orientation. Fractures are clearly visible in Figure 18. We can deduce that they are striking east-to-west and dipping steeply to the north. The formation microscanner, another borehole tool incorporating dipmeter and microresistivity measurements, also is used to see fractures.

Time-Lapse Measurements and Other Techniques

In principle any of the discussed techniques could be repeated on the same reservoir at different times during a reservoir's production history in order to monitor the manner in which the reservoir changes. Repetitive 3-D surveys are being used for reservoir surveillance and, because they give measurements as a function of elapsed time, they are sometimes referred to as "4-D surveys". Data collection and processing must be identical from one survey to the next; so, among other precautions, geophones are commonly fixed at the bottom of boreholes throughout the surveillance operation. Resultant sections, both vertical and horizontal, are compared between the different time-lapsed surveys and often subtracted one from another to give "difference sections". Enhanced-oil-recovery (EOR) operations that have been geophysically monitored include fire flood, steam flood, water flood, and carbon dioxide flood. In a fire flood, residual oil in place is burned with the injection of oxygen, and the expansion of an artificial seismic bright spot caused by the combustion gases is observed on reservoir horizon slices. The paper by Greaves and Fulp included in Chapter 6 gives a clear example of asymmetric expansion of a bright spot indicating that the reservoir was distinctly anisotropic. Studies of attenuation on the reservoir clearly show that uniform vertical fire flood propagation had only occurred during the first part of the burn. Thereafter, the combustion gases had streamed to the top of the reservoir. In a steam flood, steam is injected to mobilize heavy oil which is too viscous to be producible by normal means. Laboratory studies show that an increase in

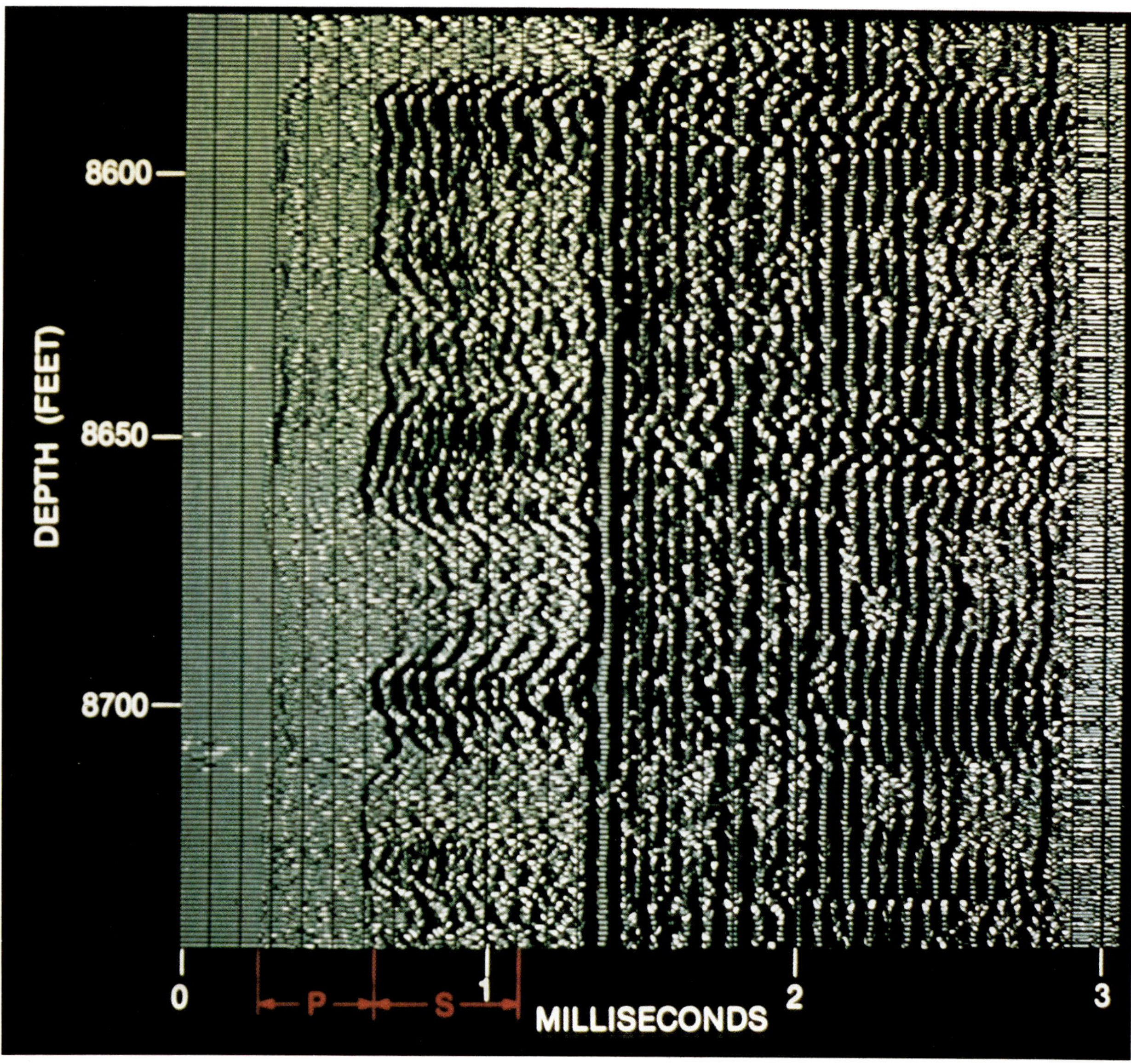

Fig. 17. Sonic waveform log showing *P*-wave, *S*-wave arrivals. (Courtesy Arco Oil and Gas Company.)

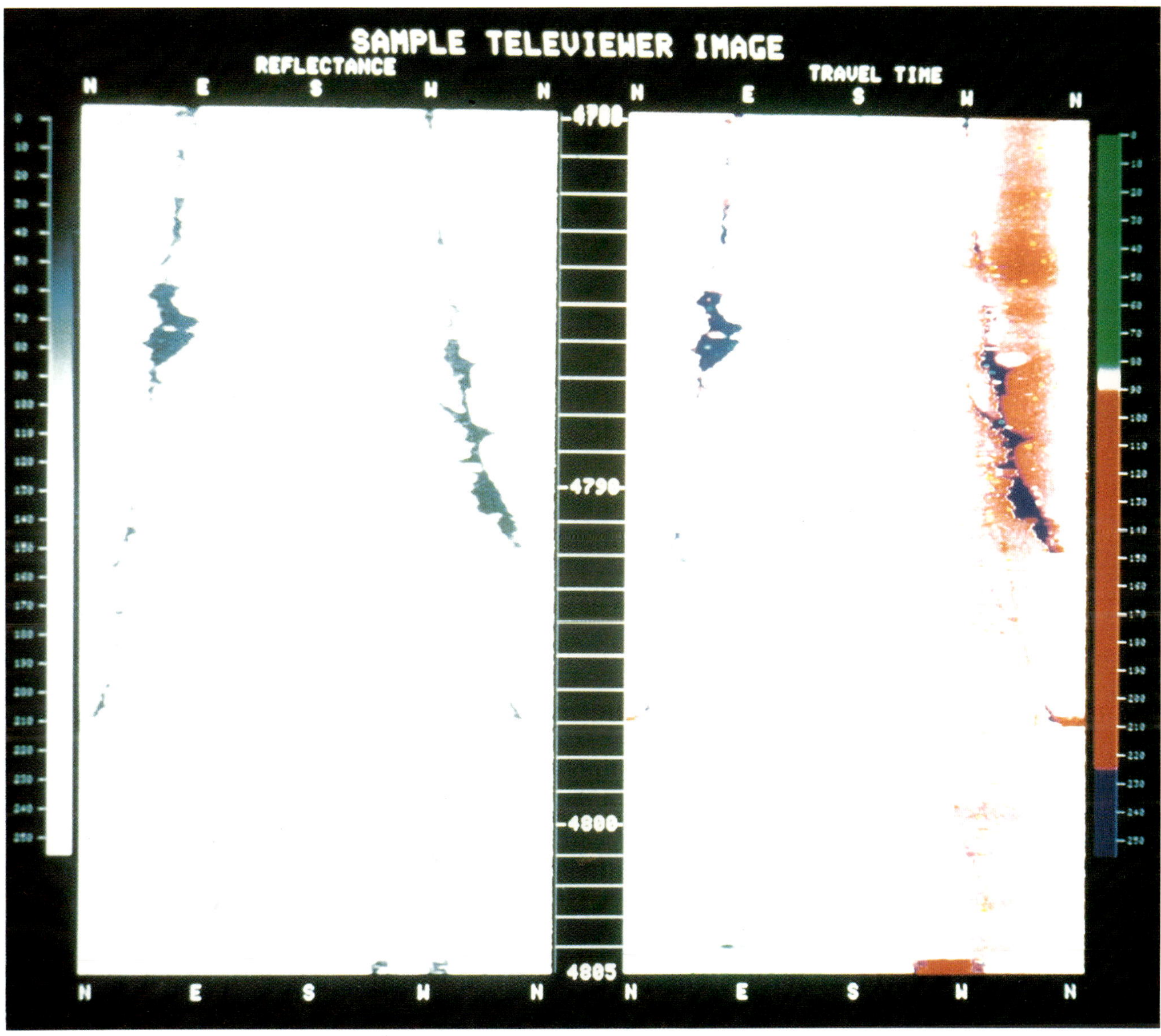

Fig. 18. Borehole televiewer images in traveltime and reflectance. The gray and blue lineations are fractures with east-west strike and north dip. (Courtesy Arco Oil and Gas Company.)

temperature of heavy oil significantly decreases the acoustic velocity. The effect can be observed on vertical seismic sections as an increase in amplitude and also as a push-down in underlying, previously flat, reflections. Time slice difference sections, such as Figure 19, show colored areas of difference interpreted as zones of heat. Expanding zones of heat resulting from steam injection have also been monitored using cross-well velocity tomograms, as have carbon dioxide flood fronts.

There are a variety of other techniques relevant to reservoir geophysics. Salt proximity surveys are a form of VSP survey in a well drilled in, or close to, a salt dome. The energy source is placed over the salt and is recorded by a geophone at several positions in the well. For each geophone position, an aplanatic curve, which is the locus of all possible refraction points between salt and sediments that satisfies the measured arrival time, is constructed. The envelope of the aplanatic curves is then the interpreted position of the salt face. Passive microseismic listening is used in the study of fractures induced for reservoir stimulation. Hydrofracturing causes stresses in the rock near the well. These stresses gradually relieve themselves with the release of microseismic murmurs. By regarding the induced fracture as a waveguide, processing of the recorded microseismics can determine the orientation of the fracture and hence the direction from which the reservoir will be drained. The borehole gravimeter is used to measure gravity at the top and

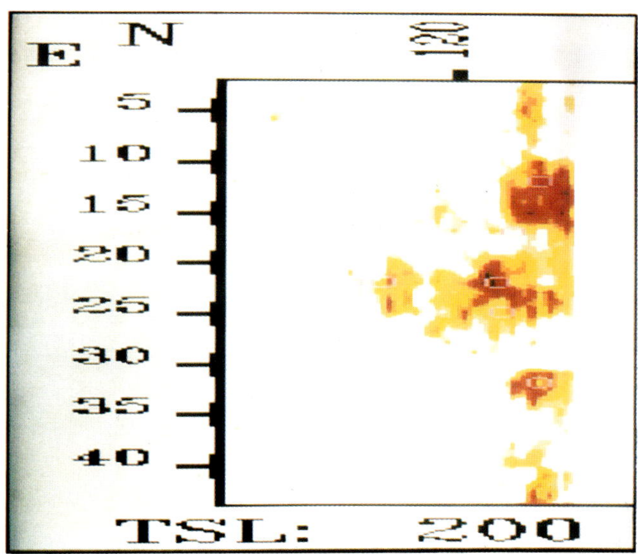

Fig. 19. Time slice difference section indicating the changes due to heat between two 3-D data sets recorded at different stages of a steam-flood project. (Courtesy Western Atlas International and Amoco Canada Petroleum Company Limited.)

bottom of zones of interest in a borehole. From these measurements the density, and hence porosity, of the intervening formation can be computed; this tool has much greater penetration into the formation than conventional density logging techniques.

Chapter 4: Reservoir Delineation
Characterizing the Trap

Alistair R. Brown and O. Ed Gilbert[‡]*

Accurate determination of the physical boundaries and internal segmentation of a reservoir has significant impact on development and production operations, beginning with the original volumetric calculations and continuing through the final stages of reservoir depletion. The enormous amount of detailed data provided by modern geophysical methods and their low cost relative to the overall cost of most drilling programs make them spectacularly effective as reservoir delineation tools.

This chapter provides nine case histories of the application of geophysical methods to reservoir delineation. The three-dimensional (3-D) seismic survey is the premier tool of the reservoir geophysicist, and is prominently featured in this chapter; six of the case histories deal with applications of 3-D seismic.

In the life cycle of most fields discovered since the early twentieth century, the discovery phase has been driven primarily by geophysical data, usually two-dimensional (2-D) seismic surveys. In the development and production phases, however, seismic data have been utilized far less often as a major tool, despite the considerable technical and economic advantages. Mature fields often have not been considered good candidates for seismic surveys, because the well data base is usually (and as we shall see, erroneously) considered adequate for delineating the traps, siting infill wells, and planning and guiding enhanced recovery operations. In addition, the logistic difficulties for seismic operations posed by existing surface facilities often have been perceived as a problem.

Early delineation drilling, infill drilling, exploitation of small internal subaccumulations, and reservoir flow modeling usually have been driven by a small number of one-dimensional (1-D) samples (the wells) and a geologic field model. The geologic field model is itself inferred from the limited well and seismic data, comparison to experience in other fields, and conceptual geologic models of structural styles and sediment body geometries derived from studies of outcrops and mod-

ern environments. Most such conceptual geologic models have serious limitations. Too often the field model is unavoidably influenced by the interpreter's training and experience, and perpetuation of older ideas which the interpreters may have inherited from their predecessors. Most of our conceptual geologic models also are incorrect or inadequate; the Mississippi is not the paradigm for all deltas, and many of our ideas about faulting in the vicinity of salt domes are naive, to name only two examples. The great advantage of 3-D seismic is that it allows the geophysicist-geologist team to develop the most comprehensive and objective geologic field model possible.

The reservoir flow model is of ultimate importance to the complete and economic depletion of a reservoir. Yet the flow model is obviously only as good as the geologic field model upon which it is based. An inadequate reservoir flow model can seriously impact the development program, resulting in the drilling of non-economic wells into isolated fault blocks, premature water or gas influx, premature pressure depletion, and a host of other potential problems. The importance of the geologic field model to the reservoir model and the development plan can be illustrated by a simple example.

One of Exxon's discoveries was first recognized in 2-D seismic data and interpreted as a simple stratigraphic trap (primarily because the interpreter's past experience had been with stratigraphic traps). After reprocessing the data, the interpreter was able to recognize a fault bounding one flank of the accumulation. After drilling a discovery well, it was decided to acquire and interpret a 3-D seismic survey, largely because of high well costs. The 3-D survey revealed 125 faults, 20 of which caused significant segmentation and pressure isolation of blocks within the reservoir. Economic analysis indicated that several of the smaller fault-bounded reservoir blocks could not be economically developed, resulting in major savings in drilling costs. The cost of the survey was about one-fourth that of a single well, a good investment by any measure.

*Consulting Reservoir Geophysicist.
[‡]Production Department, Exxon USA.

This is a typical example of the economic leverage of 3-D seismic. It also is common that, in most new structural-trap fields, the number of significant faults is underestimated by at least an order of magnitude, and that these additional faults are largely revealed by 3-D seismic.

The potential also is recognizable for older fields. ARCO recently reported that a 3-D survey over the aging High Island 24L field in the Gulf of Mexico found additional reserves of 50 million ft^3 of gas and 300 000 bbl of oil; the 3-D survey cost less than one-half as much as a shallow well in the field, and extended the life of the field by eight years.

Several of the case histories in this chapter deal with the use of 3-D seismic to map complex fault patterns, the best-known and most common application. Others describe such diverse uses as reservoir juxtaposition across faults and distribution of sealing gouge within the fault zones, two of the newer and more exciting applications of the data.

Development and production operations are, above all, exercises in economics. Therefore the case histories presented here in general tell economic stories. The first describes how step-out wells were used in an attempt to delineate a reservoir, and how a subsequent 3-D program revealed an areally limited and complex reservoir that could never have been understood using well data alone. The second case history deals with a mature and "fully understood" field (over 60 wells) for which an infill and outpost drilling program was planned. A 3-D survey intended for research purposes was shot because the subsurface was so "well known". Results of the survey revealed that the faulting was actually poorly understood, and the geologic model was wrong in several critical details. More important, it revealed a major unknown trap in the mature field, and turned a money losing drilling program into a money maker.

Although 3-D seismic is the most versatile item in the reservoir geophysicist's toolbox, it is by no means the answer to every problem. In one of our case histories the vertical seismic profile (VSP) method was used to delineate quickly and inexpensively a small stratigraphic trap. In this case the thin reservoir, the need for rapid interpretation results, and difficulty in obtaining surface access made the VSP the tool of choice. The VSP also can be used efficiently to accurately delineate the geometries of faults, and our last case history deals with the application of this method to guiding a complex drilling program in the vicinity of potentially troublesome faults.

The utility of geophysical methods in development and production operations is now clearly demonstrated. For maximum advantage, geophysical studies, either 3-D seismic or borehole surveys, should begin as early as possible in the life of the field, and be continually updated throughout the history of the field. The value of geophysical methods to field development can be seen in the practice of some companies that acquire a 3-D survey as a follow up to any discovery well. In recent years, some companies have also initiated the practice of acquiring 3-D surveys prior to drilling exploration wells in environments, such as the offshore, characterized by high drilling costs and complex traps. In general, these are companies that recognize the value of 3-D seismic through extensive experience

The geophysical interpreter should have a good working knowledge of geophysical data processing techniques, and should be involved in any data acquisition program from the beginning. The interpreter should not rely upon processing which he does not understand, because the data collection and processing exert a strong control over the final data quality and information content. Nor can the interpreter work in isolation. The ideal is a synergistic team of geologist, geophysicist, and engineer to ensure comprehensive geologic and engineering models of the reservoir.

Management of this team also is important. Our experience clearly demonstrates the value of geophysics to reservoir development and management. But the benefits are not easily or magically unleashed simply by acquisition of the data; painstaking and lengthy processing and interpretation is generally required. Acquisition and analysis of VSP data may take a few months, whereas acquisition and processing of a 3-D survey often requires six months or more, followed by a lengthy and detailed full interpretation of the data. Several years ago a number of interpreters were polled to determine the duration of an "average" 3-D interpretational project. The answers ranged from three weeks in the case of mapping one simple horizon, to over sixteen months for a particularly complex field. The average interpretation time, for mapping and studying several stratigraphic horizons and a moderately complex fault patterns, was about six months. New display and interpretation technology makes the interpreter's task easier and faster, but there is still a tradeoff of speed versus accuracy and completeness. In the end, the message of the following case histories is that geophysics can provide valuable contributions to understanding and developing the reservoir.

Segmentation and Distribution of a Problem Reservoir, Arauca Field, Llanos Basin, Colombia

O. Ed Gilbert, Jr. and Esso Colombian Exploration Staff*

Introduction

Sometimes we learn more from adversity than from outright success. The Arauca Field was under development at the time when 3-D seismic was first coming into general use as a development and production tool, but at a time when wells were still regarded as the primary delineation tool. Our experience with the Arauca Field provides a case study of the economic impact of geophysics as a production and development tool, and provides a capsule summary of the influence of major advances in 3-D seismic application during the 1980s.

Several very important lessons were learned from Arauca. The first and most important was that seismic is an extremely powerful delineation tool, and spectacularly cost-effective, particularly where well costs are high. The cost versus risk ratio is now sufficiently favorable that in many of Exxon's operating areas a 3-D survey is shot as the immediate follow-on to a discovery well. A second lesson was that it is often profitable to reprocess and reinterpret older surveys, to take advantage of advances in technology. In our reevaluation of the Arauca Field, lowered costs of seismic processing made it economically attractive to reprocess the data, and advances in 3-D migration and statics correction software resulted in an improved data quality. But most important, the advance from first-generation paper section and light table display technology to a computer-driven interactive display and interpretation system allowed us to perform far more detailed analyses of the data than was possible in the original interpretation. The interpretation techniques used are not described in detail, because they are not particularly sophisticated and are relatively widely used today.

Arauca—A Brief History

The Arauca prospect was initially identified as an elongated domal culmination on a southwest-plunging structural high in the Llanos Basin of eastern Colom-

*Exxon USA, Alaska Interest Organization, P.O. Box 2180, Houston, Texas 77252-2180. Formerly Exxon Production Research Company.

bia (Figure 1). The primary target reservoir was the Mirador Sand of Eocene age, although many structures in this basin also produce from Paleocene and Cretaceous reservoirs. However, the quality of the 2-D seismic data was such that structure at the Paleocene-Eocene level could not be directly mapped. The lowest mappable horizon, the mid-Miocene "near-top-of-Carbonera" seismic marker, indicated a simple domal structure (Figure 1).

The Arauca 1 well was spudded near the crest of the structure as defined by the 2-D seismic data. In July 1980, the Arauca 1 was completed in the Mirador, and a 19 m section of oil-filled Mirador reservoir was encountered. The Arauca 1 also encountered secondary reservoirs in the Paleocene and Cretaceous. Previous experience in the basin suggested that the Mirador would be a blanket nonmarine sand. The Arauca 2 was spudded in May, 1981 just off the crest of the structure (Figure 1), and completed in August, 1981. There was production from the Paleocene, but the Mirador reservoir, essential to the economic success of the well, was missing. The Paleocene top also was unexpectedly high by 170 ft (51 m). The structure was obviously more complex than anticipated, and at a per well cost of about US $20 million, it was apparent that stepout wells would be a very expensive method for reservoir delineation in this environment.

In September 1981, INTERCOL spudded the Arauca 3 well, targeted for a bottomhole position (Figure 1) which appeared on the 2-D seismic data to be structurally lower than the Arauca 2 well. When completed in September 1982, the bottomhole elevation was even higher than that of the 2 well. The well had some production from the Paleocene reservoir, but the economically essential Mirador reservoir was, once again, missing.

Concurrently with drilling the Arauca 3 well, INTERCOL contracted with GSI to acquire and process a 3-D seismic survey over the Arauca structure. The data were acquired in early 1982. For INTERCOL to acquire a 3-D survey over an undeveloped field was, at that time, an unusual move. The prevailing "common wisdom" was that 3-D seismic was such an expensive tool that it could only be justified for delineation and development of proven productive fields. However, in the Llanos Basin, the cost of the entire survey, including processing, was 30 percent that of a single well. Clearly, if the survey was of any use at all as a delineation tool, it would be cost-effective.

Because of drilling commitments the Arauca 4 well was spudded in April, 1982. In January 1983, the Arauca 4 was completed. The Mirador was missing.

The 3-D seismic survey was a qualified success. The 3-D migration successfully imaged deeper stratigraphic horizons, including the top of the Paleocene and the top of the Cretaceous, both of which could be neither recognized nor mapped from the existing 2-D data. Unfortunately, the usefulness of the 3-D data set

to the INTERCOL staff was limited by the available interpretation technology. At that time the most widely used interpretation tool was the rear-projection light table; vertical seismic sections were plotted as variable-area wiggle-trace paper displays, and tied to time-slice displays projected from a film onto the frosted glass of a light table. Because line-to-slice ties were difficult, this system could not take full advantage of the increased resolution available from the seismic data, nor could very small displacement faults be effectively recognized and mapped. There was no provision for quantitative mapping of amplitude or other attributes, and of course the wiggle-trace display provided little amplitude information useful in correlating across faults. In short, most of the more pow-

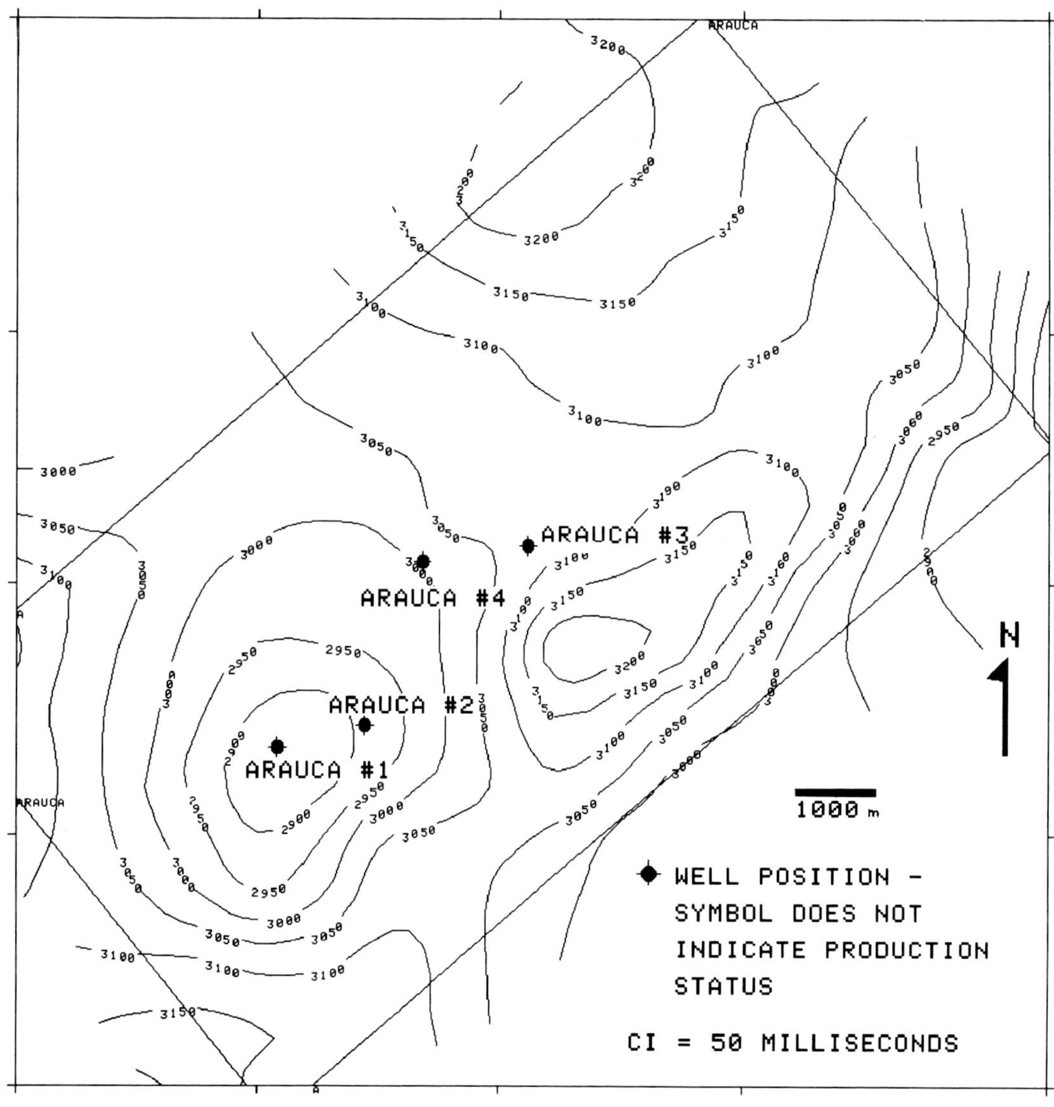

Fig. 1. Time-structure map of the Arauca structure on the mid-Miocene "near-top-of-Carbonera" seismic mapping horizon. In the original 2-D seismic data, this was the deepest mappable horizon. Note that this map is constructed from the 3-D data, but the main feature, the unfaulted domal culmination centered at Arauca 1, is essentially identical to the exploration interpretation.

erful attributes of modern interactive interpretation systems were not available.

The original interpretation did reveal that structure at the reservoir level was far more complex than first imagined, with a semisystematic fault pattern believed to be indicative of wrench faulting. In addition, the structural crest at the reservoir level was discovered to be about 400 m to the east of the crest at the mid-Miocene Carbonera marker level. Unfortunately the Mirador was far too thin to resolve seismically, so the critical question, that of the reservoir extent, remained unanswered. Several geologic models, based on onlap onto a growing dome, were developed, but all interpretations of Mirador distribution continued to be based more on conceptual models than on actual data.

Given the uncertainties in mapping reservoir distribution, INTERCOL management elected not to drill additional wells. Although we did not realize it at the time, the data necessary to solve the problem of Mirador distribution were already in hand, but the display and interpretation technology just wasn't advanced enough to allow us to fully utilize the data.

In retrospect, how might the results of the 3-D program have affected the drilling program had the interpreted survey been available before the Arauca 3 and Arauca 4 wells were drilled? Certainly the structural complexity would have alerted us to potential problems. Even granted the velocity and related depth-conversion problems in this area, it is unlikely that the Arauca 3 well would have been drilled up-plunge from and structurally higher than the disappointing Arauca 2 well. And it is virtually certain that the Arauca 4 well would not have been drilled in a position on the crest of the structure, given that the earlier wells indicated that the Eocene would likely be absent and the Paleocene a poor producer on this structure. Twenty-twenty hindsight is a wonderful thing, but it would probably not be too far off the mark to speculate that the 3-D seismic, even limited by the available display and interpretation technology, might have resulted in significant cost savings in the drilling program.

Arauca Revisited

In early 1986 INTERCOL was faced with a decision whether or not to relenquish certain leases in the Arauca area, and contracted with Exxon Production Research to reprocess and reinterpret the 3-D data. Improved processing technology did make significant improvements in the data quality, with the greatest improvements resulting from removal of line-to-line statics problems and from the use of true 3-D migration as opposed to two-pass x/y migration. During the reprocessing of the data the interpreters worked with the processors on a daily basis to help insure that the final product would meet the interpreters' needs.

However, the greatest improvements in our 1986 re-examination of the Arauca data resulted from major advances in interactive interpretation technology achieved in the four years following the original interpretation.

Structural Interpretation

Several of the features of Exxon's in-house interactive seismic interpretation system—the ability to scan rapidly through large volumes of data, to accurately tie together lines and time-slices, and to recognize very small offsets—allowed us to map a more realistic and systematic fault pattern. Mapping interval thickness changes across faults and mapping how high into the section various faults reached revealed two distinct sets of normal faults. Several northwest-trending faults on the crest of the main structure were active in Oligocene (and possibly in Eocene) time (see time slice in Figure 2). East- and northeast-trending en echelon normal faults along the south and southeast flanks of the structure are mid-Miocene to Pliocene in age, and represent the shallow manifestation of minor right-lateral extensional strike-slip faulting in the basement (Figure 3).

In addition to the detailed analysis of fault patterns and ages, we were able to use other capabilities of the interactive interpretation system to help unravel the complex structural and stratigraphic history of the Arauca structure. Interactive datumming in particular allowed us to determine areal changes in unit thickness, and more important, to quickly visualize how the structure looked at specific times in its history. The dense areal coverage of the 3-D data allowed us to recognize subtle stratigraphic onlap surfaces and unconformities which could not be recognized from paleontological or stratigraphic criteria, but which were keys to the broad-scale tilting or arching of the structure. As we shall see, the structural history of the Arauca structure greatly influenced reservoir distribution.

Stratigraphic Interpretation

The paper displays used in the original interpretation did not allow quantitative interpretation of amplitude information contained in the 3-D data, and in any case the thin Mirador is seismically invisible. However, seismic modeling indicated that variations in amplitude on the underlying top-of-Paleocene reflection might provide an indirect means of mapping the Mirador distribution. The ability to map reflection amplitudes in an areal sense is, of course, unique to 3-D seismic data.

In the Arauca 1 well a thickened Eocene to lowermost Oligocene section overlies the Paleocene. Destructive interference between wavelets from the top of Paleocene surface and a marker unit within the lower Oligocene result in lower amplitudes on the composite reflection (Figure 4). Where the Eocene and lowermost Oligocene are absent, the intra-Oligocene marker lies close to the top of the Paleocene, and constructive interference results in higher amplitudes on the top of Paleocene reflection.

Figure 5 is an areal display of amplitudes on the top of Paleocene reflection, with well penetration positions plotted. Red areas indicate the extent of the thickened lowermost Oligocene section (and by infer-

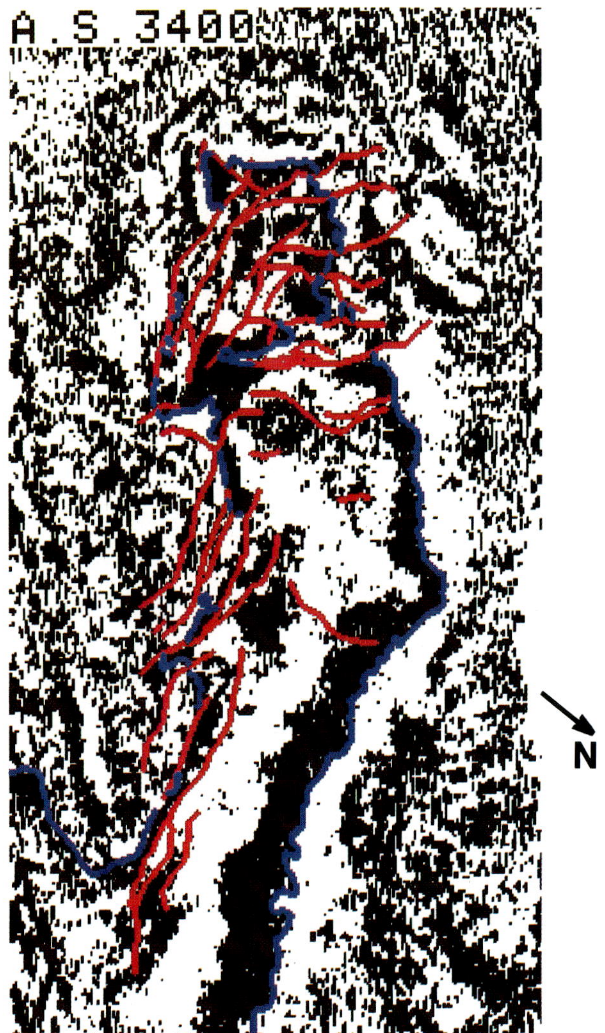

Fig. 2. Time slice or horizontal seismic section through the Arauca structure at the Eocene/Paleocene reservoir level (blue). The geometry of the structure, the precise location of the culmination, and the intensity of faulting (red) are in sharp contrast to the structure at the mid-Miocene level. Note particularly the numerous faults along the southeast flank and at the southwestern terminus of the structure. Higher data density and better migration frequently, but by no means invariably, result in improved imaging of complex structure. Section is oriented as shown by the box in Figure 1.

ence the probable maximum extent of the Eocene) and blue the areas where the corresponding section is thin and the Eocene most likely missing.

Geological History and Reservoir Distribution

Based on our analysis of the data from Arauca, and unpublished data on regional tectonics, we were able to construct a geologic model to explain the complex structure and unexpected reservoir distribution. Although the Arauca structure has undergone almost continuous deformation from Late Cretaceous through Pleistocene time, two "episodes" are particularly important.

In Early Eocene through Early Oligocene time the small crustal block which would eventually become the Arauca structure, bounded on the southeast side by a basement-involved fault, was gently tilted toward the northwest. At the same time it was elongated along a northeast-southwest direction by subsidence of the developing Andean foredeep to the southwest. This elongation resulted in development of northwest-trending faults localized near the crest of the present structure (Figure 3).

Figure 6 depicts the areal extent of thickened Eocene (?)—Early Oligocene section in the immediate vicinity of the Arauca 1 and 2 wells, superimposed on the fault pattern map shown in Figure 3. The likely extent of the thickened section is clearly limited to the area immediately surrounding the older Eocene (?)—Oligocene grabens. We considered several geologic models to explain the apparent reservoir distribution. The most promising were (1) the Mirador was deposited as a nonmarine blanket sand, and later eroded away except where preserved in the downthrown blocks associated with the northwest-trending faults prior to deposition of the Oligocene section, or (2) the northwest-trending faults were active as early as Eocene time, the Mirador was deposited as updip nonmarine valley fill sands in the old grabens, and sediments eventually overlapped the graben boundaries during the Oligocene. In a practical sense there is no real advantage of one model over the other, although we prefer the second, which implies a slightly greater volume of reservoir overlapping the graben boundaries. In either case, the areal extent of the potential reservoir is far less than implied by any prior geologic model.

In later Oligocene and Miocene time continued subsidence into the growing Andean foredeep resulted in continued tilting of the Arauca structure to the southwest. In mid-Miocene time the inferred basement fault along the southeastern boundary of the structure was reactivated as a strike-slip fault with minor right-lateral displacement, as determined from fault geome-

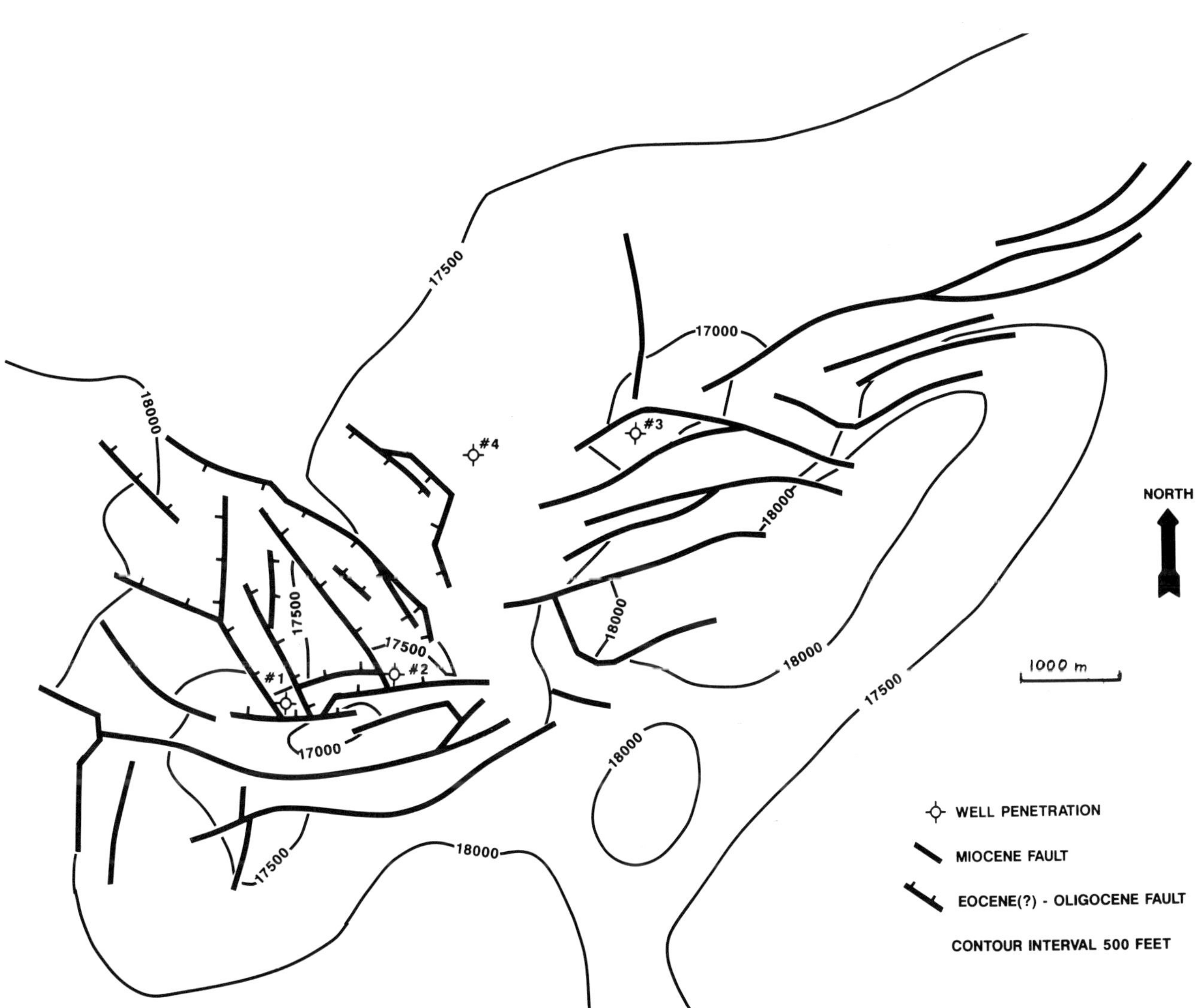

Fig. 3. Depth structure map of Arauca at the reservoir level, as interpreted from the 3-D seismic data. Note that there are two distinct sets of faults, an older Eocene(?) to Oligocene age system which trends northwest and dies out upward into unfaulted lower Oligocene section. The second set is younger and penetrates much higher, into the Miocene section, and associated folding is seen as high as the Pliocene section. This set is interpreted to be the shallow manifestation of minor strike-slip faulting at the basement level.

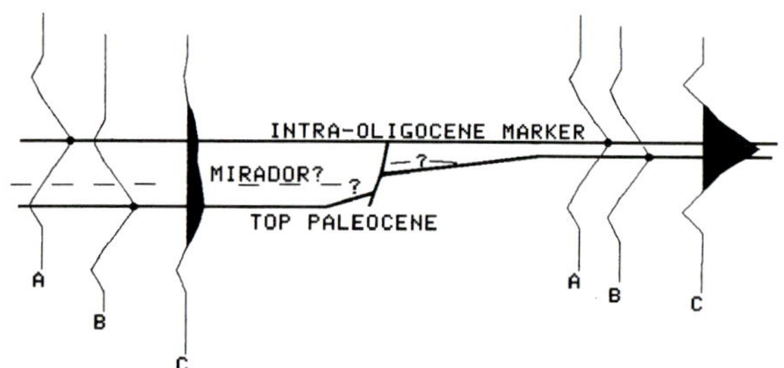

Fig. 4. Amplitude variations caused by the constructive and destructive interference of reflection energy from the Top Paleocene (near reservoir) and a seismic reflector internal to the overlying Oligocene section. Where the Eocene reservoir is present and the lower Oligocene is thicker (left side of figure), reflected energy from the Paleocene top (B) and the intra-Oligocene marker (A) experiences destructive interference, resulting in lower amplitude and "splitting" of the composite reflection (C). Where the Eocene is absent and the lowermost Oligocene is thinner (right side of figure), reflected energy from the two markers undergoes constructive interference, resulting in a much higher amplitude for the single-peaked composite reflection (C). Stratigraphic implications of the amplitude variation cannot be directly inferred from the seismic data, but must be determined through one-dimensional seismic modeling of wells.

tries at the Miocene level. It was this Miocene deformation, and compressional deformation continuing at least into the latest Pliocene, which formed the present gross structural geometry at Arauca, and it is clear why reservoir distribution models of Eocene onlap onto a growing dome were unsuccessful.

The ultimate test of this new interpretation would of course be new drilling. Unfortunately, the reserve figures implied by the new interpretation make drilling here a less attractive proposition than ever, and there is no possibility of further development at all under the prevailing depressed oil prices.

Summary

Our experience at Arauca clearly demonstrates that the true nature of the structure and the distribution of the reservoir would never have been understood using only delineation wells and 2-D seismic data. In addition, it clearly demonstrates that the additional data twice forced fundamental revisions of the geologic models.

At the time of acquisition, the cost of the 3-D survey was about 30 percent that of a single well; with modern

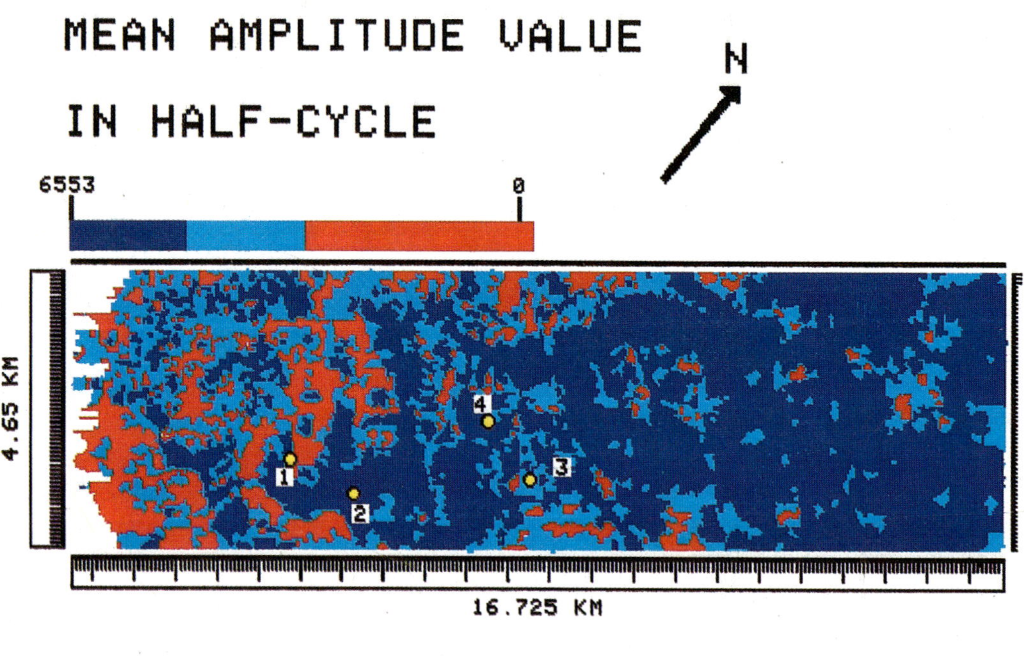

Fig. 5. Seismic amplitude map on the "top Paleocene" reflection. Note that the Arauca 1 well position lies well within the lower-amplitude red/pale blue zone, while the Arauca 2, 3, and 4 wells lie in the zone of higher blue amplitudes. Note that the Eocene reservoir is seismically invisible, but its maximum likely areal extent can be inferred from preservation of the thickened section. The area covered is approximately the northwestern two-thirds of the box shown in Figure 1.

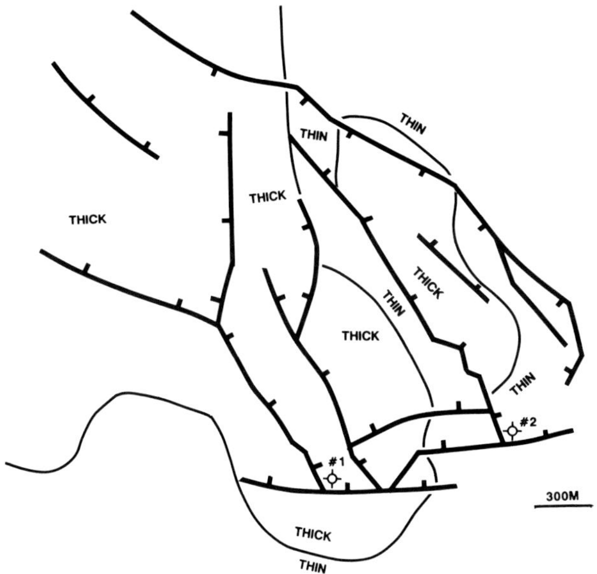

Fig. 6. Distribution of thickened Eocene-lowermost Oligocene section (from seismic amplitudes) superimposed on fault pattern at crest of structure. The Eocene Mirador reservoir is likely to be absent altogether in the areas marked "THIN", as indicated by its absence in the Arauca 2 well. Several alternative geologic models may account for the absence of the Mirador, as described in the text, but the distribution of thickened areas of the Eocene-lowermost Oligocene section constrains all of the models.

acquisition techniques and faster processing capabilities, the cost of a similar survey would likely be less than 20 percent that of a single well. Early utilization of 3-D seismic at Arauca would by no means have solved all of the problems encountered, but it is not unfounded speculation to assume that early utilization to revise the drilling plan would have resulted in substantial cost savings. In addition, it is clear that utilization of modern display and interpretation technology plays a major role in the effective utilization of 3-D seismic data.

Acknowledgments

I would like to thank the management of ESSO (formerly INTERCOL) and Exxon Production Research Company for permission to publish the results of this study. I would particularly like to acknowledge my project co-workers, Carlos Rodriguez (formerly of INTERCOL), and Elijah White Jr., Karen Romine, J. A. Klingshirn, M. G. Purifoy Jr., and Pinar Yilmaz (Exxon Production Research Company). T. R. Bultman and T. P. Harding of Exxon Production Research Company provided advice on the structural aspects of the reinterpretation.

New Prospects in an Old Field
The Northeast Viboras Field, Brooks County, Texas

Dana Butters, John Dickinson,[‡] Stuart W. Fagin,[§] Gordon H. Weisser,***
and Ed Gilbert Jr.[‡‡]

Introduction

A recurring argument against three-dimensional (3-D) seismic surveys for field development and production is that with sufficient well data the seismic survey provides no additional understanding of the field. Our collective experience with 3-D seismic indicates that such surveys invariably result in recognition of unexpected structural complications, and often require reinterpretation of the major structural framework as well.

Exxon's experience at the Northeast Viboras Field provides one such example. Three 3-D seismic surveys over this field were collected in 1983 to evaluate different data acquisition techniques (Dickinson, 1990). Northeast Viboras was selected as a test site because there is extensive subsurface control (40 wells), existing two-dimensional (2-D) seismic, and a well-documented structural and stratigraphic model of the geology. Northeast Viboras was to serve as a "control" case, because the subsurface was so well-known. What actually happened was that the 3-D interpretation forced a reevaluation of the structural model, and a revision of the drilling program. In economic terms, the revision of the drilling program resulted in a change from a potentially money-losing well program to a profitable program. The results of the Northeast Viboras program have been used within Exxon as an illustration of the economic advantages of 3-D seismic data in mature fields.

*Exxon USA, Offshore/Alaska Exploration, P.O. Box 4279, Houston, Texas 77210. Formerly Exxon USA, Production Department, South Texas Division.
[‡]Exxon Production Research Company.
[§]Exxon USA, Offshore/Alaska Exploration, P.O. Box 4279, Houston Texas 77210. Formerly Exxon Production Research Company
**8411 Brae's Meadow, Houston, Texas 77071. Formerly Exxon Production Research Company.
[‡‡]Exxon USA, Alaska Interest Organization, P.O. Box 2180, Houston, Texas 77252-2180. Formerly Exxon Production Research Company.

Geologic Setting

The Northeast Viboras Field is one of several fields along the Vicksburg—South May trend of south Texas (Figure 1). The Northeast Viboras Field proper is located on the downthrown side of the listric Viper Fault, one component of a north-trending down-to-basin imbricate growth fault system of Eocene-Miocene age. The primary reservoirs are in fluvial-deltaic sands of the lower and middle Frio Formation. The primary trap is a north-south trending anticlinical rollover on the low side of the main Viper Fault (Figure 2). The gross geometry of the trap is related to movement on the underlying listric Viper Fault.

The large anticlinical closure is broken by numerous smaller faults of the system, which generally strike northeastward (Figure 2). The original interpretation of these smaller faults was as a series of throughgoing synthetic and antithetic faults which merged with the Viper Fault both along strike and at depth (Figure 2). In this model, smaller hydrocarbon accumulations within the main trap are established by closure against sealing faults.

Early Interpretation

Prior to 1985 the structural interpretation of the field was based on well data and 2-D seismic lines across the structure. Based on a structural model developed from prior experience in this trend, the faults revealed by the various 2-D seismic lines were connected to form a splaying pattern of synthetic and antithetic faults, as shown in Figure 2. In this structural model, all of the synthetic faults were considered as splays from the main Viper Fault, and connected to it in the subsurface. At this stage of interpretation, potential traps consisted of small fault-controlled closures against the smaller faults, wholly within the downthrown block of the Viper Fault.

Revised Interpretation

In 1985 the interpretation of the 3-D seismic data forced a reevaluation of the local structural setting, with major consequences for the drilling program.

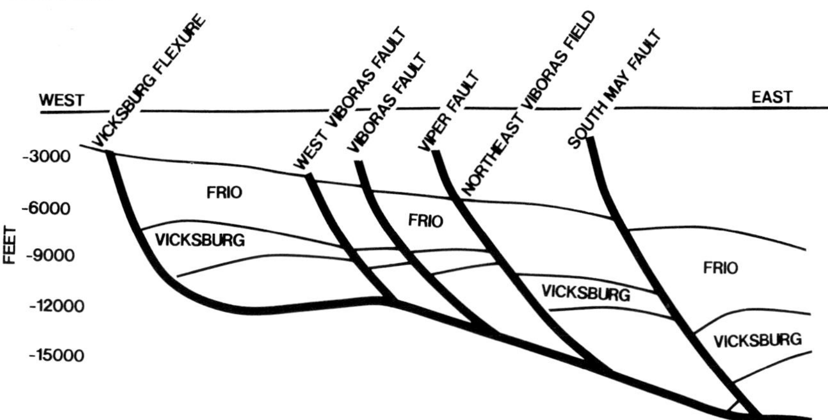

Fig. 1. Schematic northwest-southeast cross-section through the growth fault trends of south Texas. The primary trap at Northeast Viboras is formed by the Frio-level rollover between the Viper and South May faults. Not to horizontal scale.

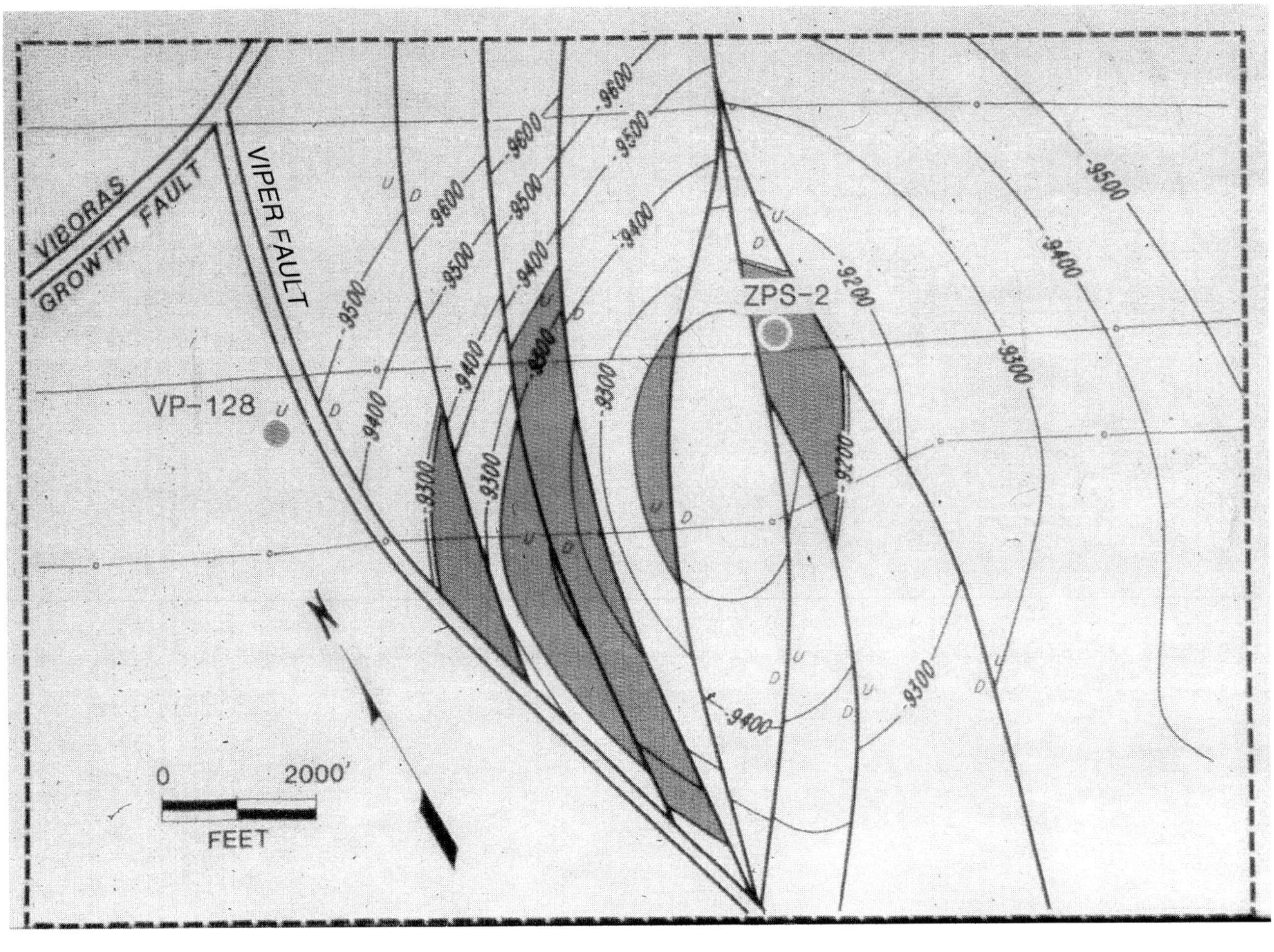

Fig. 2. Original structural interpretation, in depth, of the Northeast Viboras area, based on 40 wells (not shown) and three 2-D seismic lines (shown as light lines). Note the connection of the fault cuts revealed by the seismic lines, the throughgoing secondary faults, and the smooth map trend of the main Viper Fault.

Interpretation of the 3-D data volume indicated that many of the smaller faults within the hanging wall of the main fault did not directly connect to the main fault (Figure 3). This phenomenon of small isolated faults is commonly seen in 3-D seismic data, but is usually not considered in model-based structural interpretations.

The good reflection continuity and low dips meant that the time slices were particularly useful in interpretation of the fault pattern at Northeast Viboras.

Figure 4 shows a typical time slice through the main reservoir level. The prominent red reflection marks the top of the main reservoir sand (actually a zone of several sands). The offsets caused by a major set of synthetic faults are obvious just to the upper left center of the figure. These faults had been recognized from the older 2-D seismic data, but note that they have a much more pronounced easterly trend than previously interpreted. In the prior interpretation,

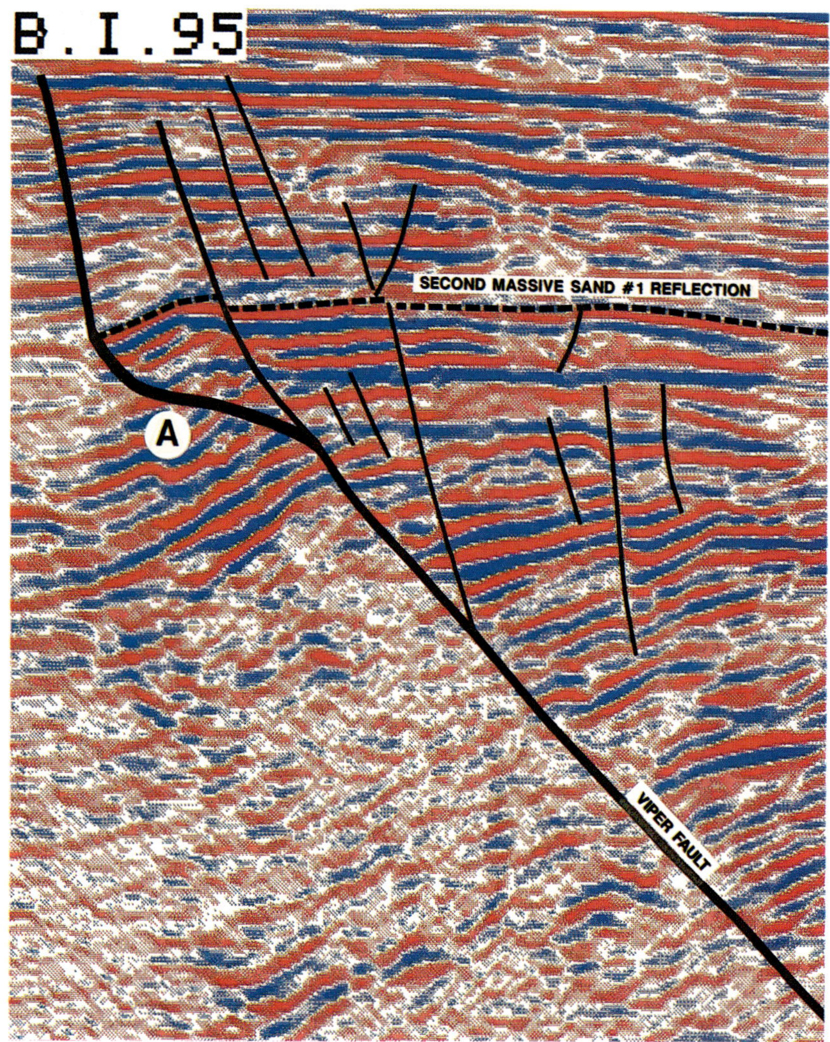

Fig. 3. A representative northwest-trending inline from the 3-D survey. In this and the following figures, the top of the productive interval is indicated by the dashed line. The productive zone also includes several sands below the second massive sand which are not resolvable as separate reflections in the seismic data. Closures are in part fault controlled; the lateral limits of proven or potential production are shown in Figure 6. The increased resolution of the 3-D data reveals many more mapable faults, most of which are restricted to specific stratigraphic intervals and are not throughgoing (do not connect to the Viper Fault) in either cross-section or plan view. The low-angle curved segment of the Viper Fault at ''A'' is apparent dip; along this segment the Viper Fault swings sharply to the northwest (Figure 4), and is therefore dipping into the plane of this cross-section.

lacking any control in the area between the widely separated 2-D lines, the wrong fault cuts had been connected.

A second, completely independent set of synthetic faults is somewhat less obvious; this set appears as branches off the main Viper Fault trace in the bottom center of Figure 4. In prior interpretations, these small faults, although not actually detected by a chance 2-D seismic line, were linked to the larger synthetics to the north. However, the areal coverage provided by the time slices clearly demonstrates that there are two separate and unconnected fault sets, with different trends. The recognition of the more southerly faults as a separate set is important because it is these faults which were the dominant structural control on several of the smaller traps detected by the new seismic data.

The most important feature of the new interpretation was the revision in the trend of the Viper Fault. Based on the existing 2-D seismic data, the Viper Fault had been interpreted as having a gently curving northwesterly trend. However, the 3-D seismic data revealed that the Viper Fault actually turns sharply to the west (Figure 4), indicating that a much larger area than originally anticipated lay within the prospective downthrown block of the fault system.

Immediately north of the offset in the Viper Fault the time slices indicate a triangular structural closure, bounded on the south by the westerly trending segment of the Viper Fault, on the southeast by a synthetic "splay", and on the northwest by the dip of the reservoir section (Figures 4 and 5). This trap was of particular interest, since it lay in an area not previously considered prospective.

Economic Impact

Prior to interpretation of the 3-D data, Exxon's planned drilling program consisted of seven wells, but it became apparent that the results of the seismic interpretation warranted a reexamination of the drilling program.

Under the revised drilling plan, the program was reduced to three wells, one of which was a test of the

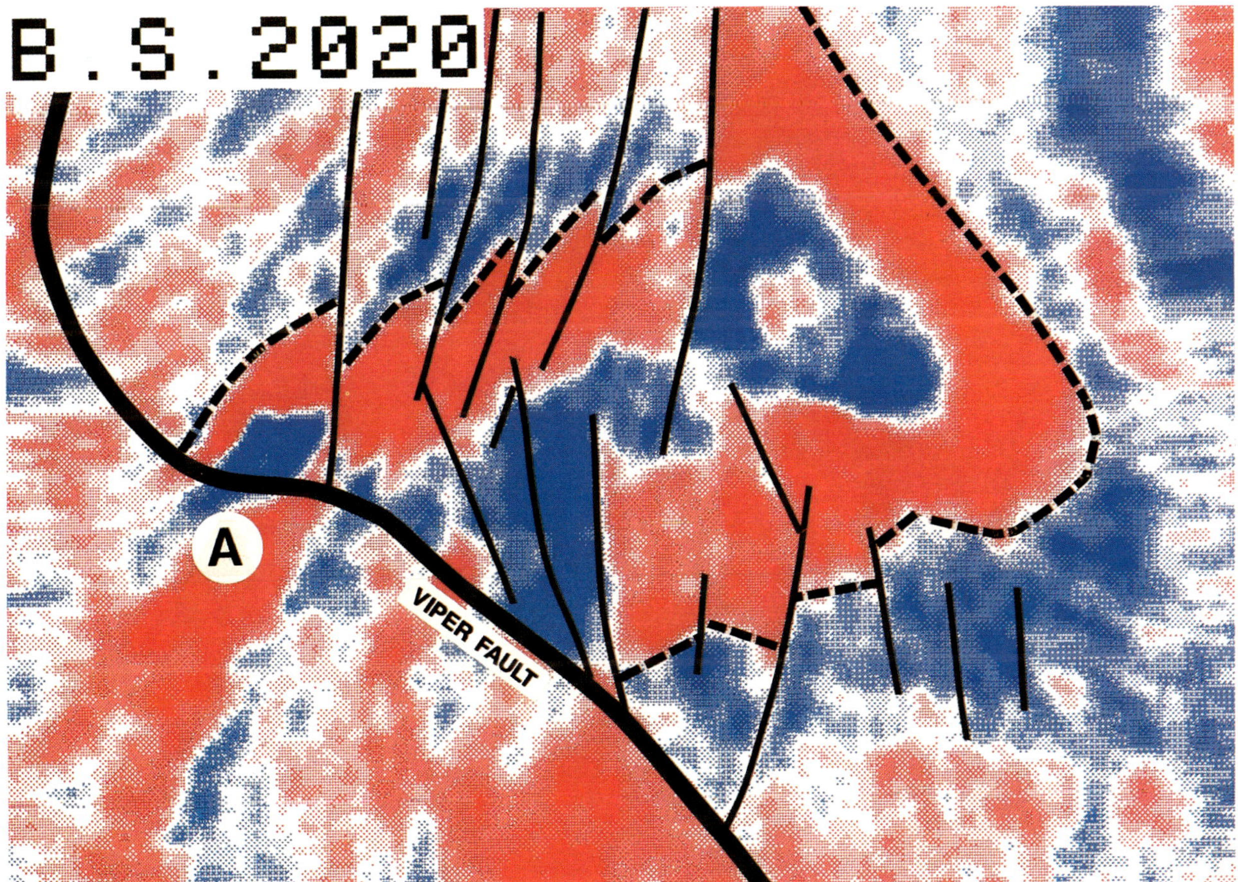

Fig. 4. A representative time-slice through the top of the reservoir interval at 2020 ms. The variable trends and set relationships of the smaller faults are apparent from the disruption of the reservoir top reflection; compare the pattern to Figure 2. The abrupt turn in the trace of the main Viper Fault at "A" is not readily apparent from this section, but must be mapped by tying together inlines, crosslines (CDP lines), and time-slices.

newly discovered closure in the bend of the Viper Fault. Analysis of the original seven-well drilling plan in light of the revised interpretation indicated that the program would have resulted in four dry holes, two non-profitable wells, and one marginal producer. The net result would have obviously been a money-losing program.

The first of the three wells in the revised program, the Viboras Pasture 128, tested the newly discovered closure formed by the bend in the trace of the Viper Fault (Figure 6). This well encountered significant originally-pressured gas reservoirs. If the seismic survey (which was originally shot as an experimental survey) is included in operational expenses, the reserves booked from this single well still resulted in an anticipated investor's rate of return sufficient to cover the total estimated costs of the revised three-well program.

A second well, the Viboras Pasture 130 was also successful, but the third well was not an economic success, as it encountered an unmapped fault. The reason the fault was not detected and mapped is still problematical, but is likely the result of the interpretation technique used. (The fault trends are oblique to

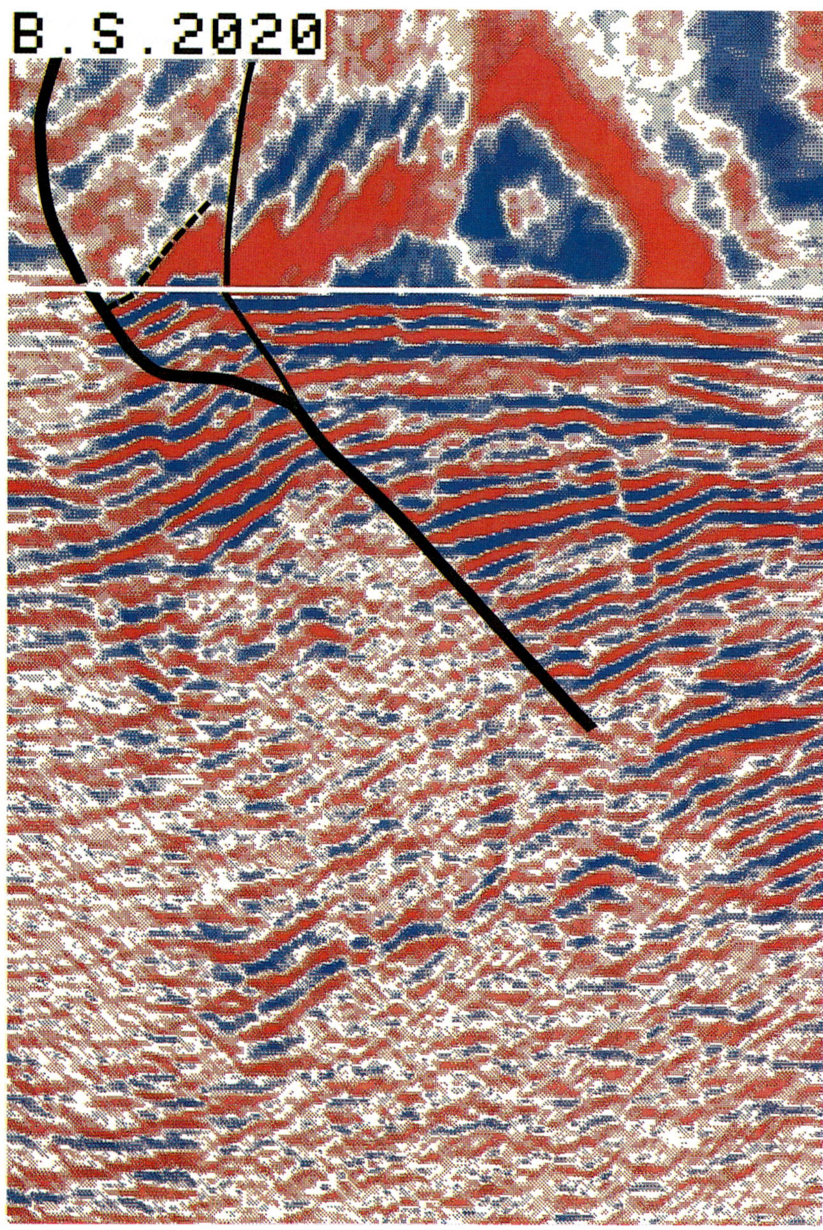

Fig. 5. Tie between inline 95 and the 2020 ms time-slice, showing the previously unrecognized trap in the bend of the Viper Fault.

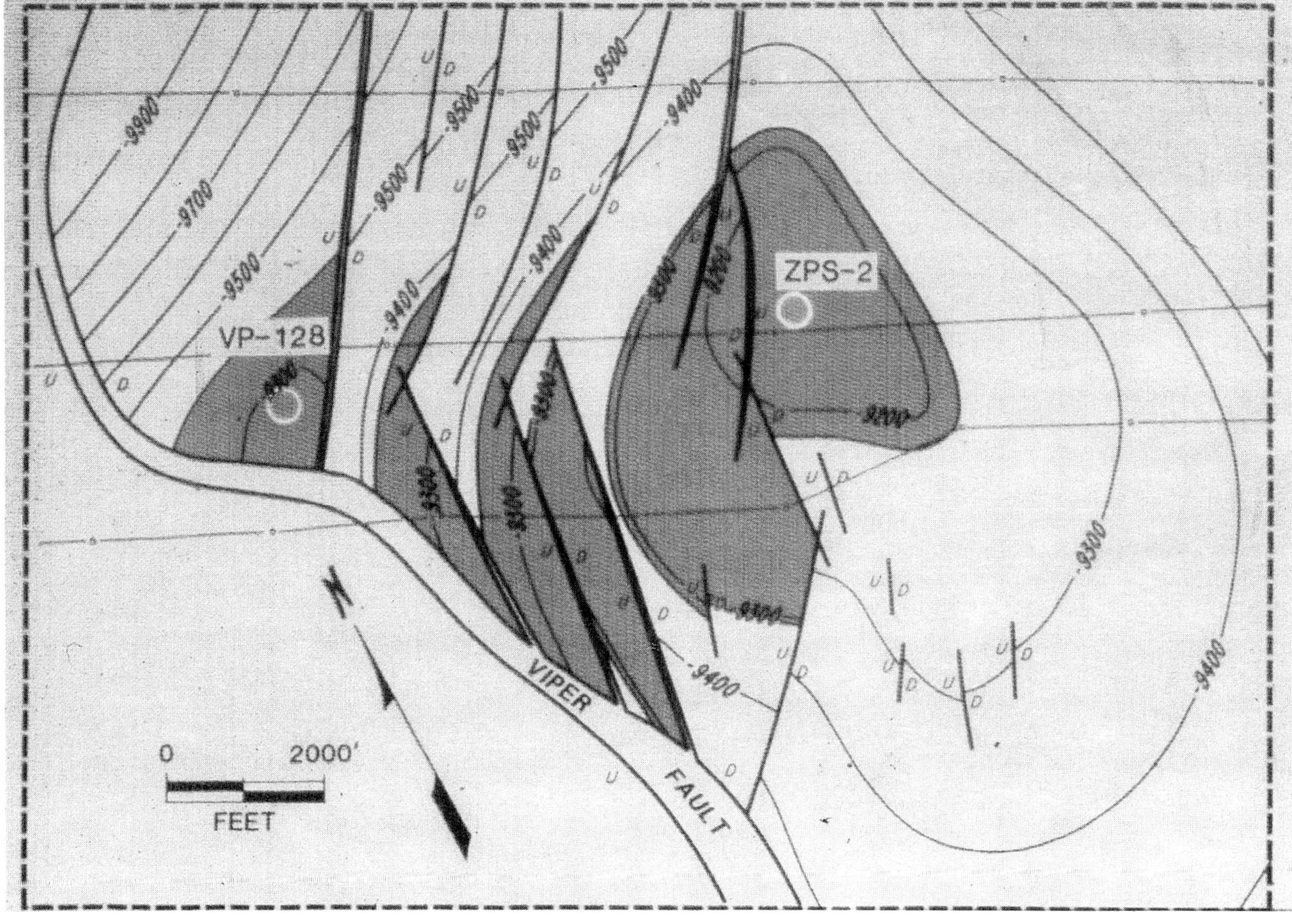

Fig. 6. The new interpretation, in depth, as revised from the 3-D survey, and the location of the Viboras Pasture 128 well (VP128). Note also that the fault gaps are much more precisely mapped, and that the structure is less simplistic than the interpretation shown in Figure 2.

both northeast and southwest oriented lines of the survey and are therefore hard to recognize on either set; however, they are recognizable on specially generated northeast trending lines). Despite the problem with the third well of the program, the program is an overall success, and this success is directly attributable to the better structural interpretation made possible by the 3-D seismic survey.

The 3-D seismic data have also proven useful for other purposes, for which the economic benefits in dollars are not directly quantifiable. These uses include determination of which fault blocks are in pressure communication for production monitoring, determining which reservoir horizons should be perforated in specific fault blocks for maximum drainage efficiency, and economic justification of well workovers.

Summary

In this case, availability of the 3-D seismic data resulted in a revision of the drilling program which converted a potential money losing program into a profitable program. The success of this example has encouraged Exxon's production offices worldwide to make much more extensive use of 3-D seismic, and has been used within Exxon USA and its affiliates as an example of the potential technical and economic benefits of such surveys in field development.

Acknowledgments

We acknowledge the South Texas Production Division of Exxon USA for allowing us to publish the results of the Northeast Viboras program. F. K. Levin, T. H. Crawford, J. W. Stelzig, M. A. Harper, and J. M. Scheel of Exxon Production Research were instrumental in planning and executing the seismic program. Ross A. Ensley of Exxon Production Research also participated in the original interpretation of the data.

Reference

Dickinson, J. A., 1990, Comparison of 3-D Seismic Data Acquisition Techniques on Land: 60th Ann. Internat. Mtg., Soc. Expl. Geophys., Expanded Abstracts, 913–916.

Reservoir Development Using Offset VSP Techniques in the Denver-Julesburg Basin[1]

*P. W. Cramer**

Introduction

Efficient and economical development of reservoirs is a primary goal of production earth scientists and engineers. In small fields typical of many of today's discoveries, achieving this goal is becoming increasingly difficult because of declining prices and the small reserves involved. Channel-sand reservoirs are among the most difficult to develop efficiently, yet they frequently have excellent production characteristics. The difficulties arise from the depositional mechanisms involved in channels that create sand bodies whose thickness and quality can vary rapidly over short distances, perhaps as short as a few hundred feet. These rapid variations make it difficult to use conventional seismic data successfully for mapping, and the cost of three-dimensional (3-D) surveys may not be justifiable, given the small size of the field. In such cases, VSPs can often be used as a cost-effective, accurate means to delineate reservoirs and to optimize the number and location of development wells. A case history is presented showing the successful development of a channel-sand field by use of VSP techniques. Background information on VSP methodology is provided, as well as a thorough presentation of the planning of the VSP survey, its acquisition and processing, and the interpretation of the data. Results of subsequent drilling are shown to agree closely with those predicted by the VSP. Finally, the acquisition of additional VSP data is shown to allow further mapping of the reservoir and refinement of locations for future drilling.

Initial Field Development

Wattenberg field is located in the Denver-Julesburg basin in northeastern Colorado. A portion of the field stretches across acreage operated by Berge Exploration and Sierra Energy in Weld County, as shown in Figure 1. A map of the acreage and a typical lithologic column for the area are shown in Figure 2.

[1]©1988 Society of Petroleum Engineers, Journal of Petroleum Technology, February 1988, 197–205.
*Shell Offshore, Inc. Formerly Western E&P Inc.

Initial production on this acreage was gas from the Sand J series at depths of between 7900 and 7950 ft (2408 and 2423 m). The third well (Well 34-3), however, also encountered a thick (26 ft [8 m]) Sand D section at 7820 ft (2384 m), which was oil-bearing. As Plybon and Oldham (1985) described, Sand D is a member of the Dakota group and is believed to have been deposited in a channel system as part of a westward-prograding coastal-plain system. The productive channels can range in size up to 1000 ft (300 m) across and as much as 40 ft (12 m) thick. Outside these channels, regional thickness is usually under 10 ft (3 m). After Well 34-3 was completed, Wells 34-4, 34-5, and 34-6 were drilled (see Figure 2 for locations), but none encountered the Sand D channel found in Well 34-3. Figure 3 shows a cross-section through the field at Sand D level.

Following these unsuccessful wells, the operators decided that some form of seismic control was necessary to map the Sand D channel before trying another well. Key concerns were turnaround time for acquisition and processing, cost, and, of course, reliability of the data obtained.

Turnaround time was important because of the desire to increase production from the Sand D reservoir as quickly as possible. Because conventional 2-D seismics and a 3-D survey were ruled out as a result of time and permitting factors, a VSP was selected as an alternative method to satisfy the rapid turnaround criterion and to provide high-resolution, detailed data for precise delineation of the reservoir.

Background on VSP Methods

The traditional VSP is acquired with a surface energy source placed close to the well location and a downhole tool containing a geophone package that is clamped to the borehole wall to record the seismic signals from the energy source. This procedure is repeated with the tool moved to a different depth in the well each time. Given small dips and a straight well, the information gathered comes from the immediate vicinity of the wellbore itself, and directly below the current total depth (TD) of the well. However, if the

"

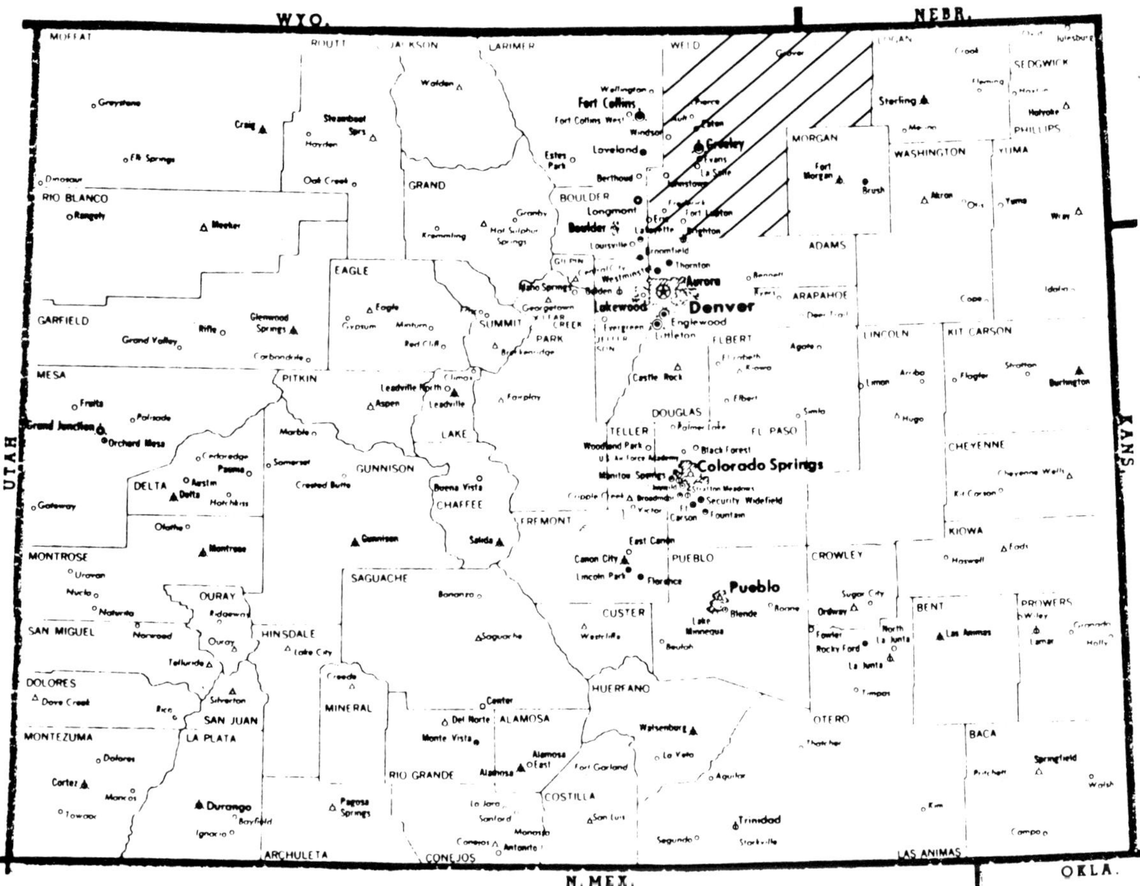

Fig. 1. State of Colorado.

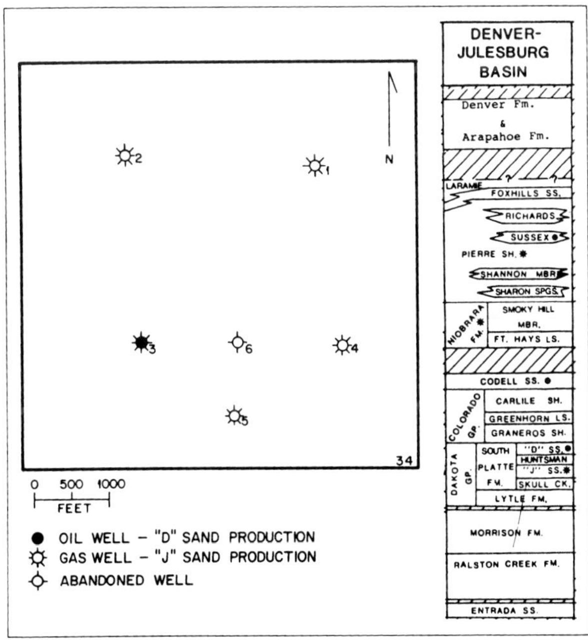

Fig. 2. Plot of acreage operated by Berge and Sierra with stratigraphic column (courtesy Berge Exploration).

source is moved some distance away from the well, as in Figure 4, information is collected along a lateral profile away from the well in the direction of the source location. By moving the source location or by simultaneously using more than one source at different locations, we can generate a series of lateral profiles to provide information away from the well in several different directions. In this fashion, then, multioffset VSPs can provide critical information about the quality and lateral extent of reservoir rock around a well.

Offset VSPs have several inherent advantages over traditional surface seismic lines and 3-D surveys. Chief among these is improved resolution, both laterally and vertically, because of the unique geometry of the VSP, which results in a much shorter energy travel path and therefore superior data quality. Further, VSPs can typically be acquired and processed in a matter of hours or a few days, instead of weeks or months for a 2-D surface seismic grid or a 3-D survey. Hence, when time is critical, the offset VSP is a logical choice.

Offset VSPs have two primary disadvantages. First, in complex structural situations, the proper acquisition

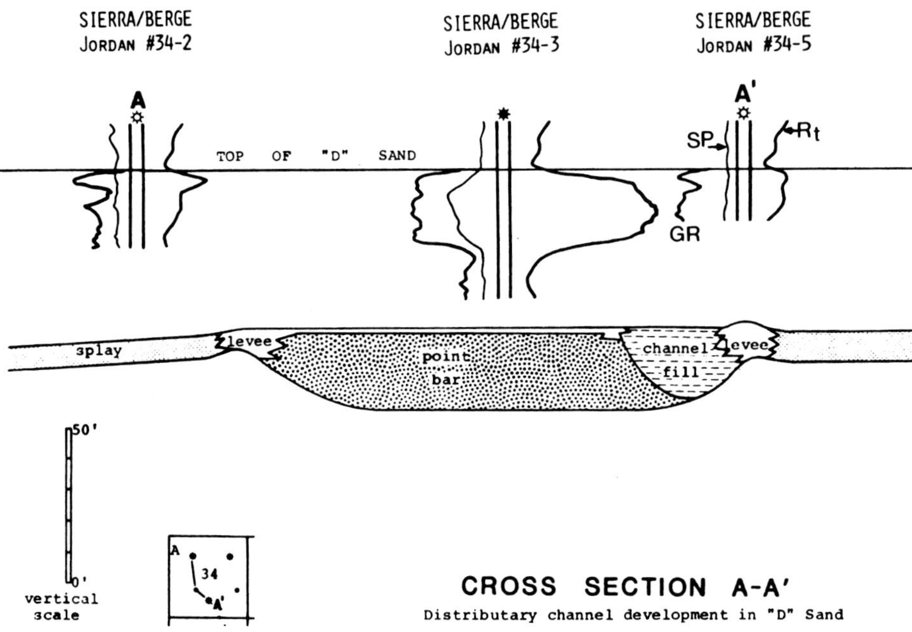

Fig. 3. Cross-section through Berge and Sierra acreage (courtesy Berge Exploration).

and processing of VSP data can be quite complicated and may not provide the best information to solve problems related to the complex structure. Therefore, 3-D surveys may be better suited to these cases. Second, the extent of the lateral profiles generated by offset VSPs is limited. In some cases, the information provided by offset VSPs may not extend over sufficient distance to be of optimum use.

In an offset VSP, the distance over which information is collected is governed by the source-to-geophone offset distance and the depth of the geophone. Practical limits, however, are placed on source offset distances by such factors as depth of objective, dip, vertical velocity variation, and well deviation. Excessively long source offsets usually result in poor data quality because large amounts of coherent noise are generated through refraction and other undesired energy-propagation effects.

Survey Modeling

Before the VSP data were acquired, detailed modeling was performed to determine whether seismic character changes could be correlated to thickness changes in Sand D and thus be used to delineate the reservoir. Modeling was also used to test and to verify possible offset-VSP survey configurations. The data for this model, as shown in Figure 5, consisted of layer thicknesses and interval velocities taken from well logs in the immediate vicinity. Sand D was allowed to vary in thickness from 7 to 30 ft (2 to 9 m). The modeling was performed with the QUIK®[2] software package. The output of the modeling package was then entered into our VSP processing software and processed to simulate the data that would be generated from an actual VSP survey. Figure 6 shows the results of this process. Clearly, a distinct character change can be correlated with the known changes in Sand D thickness across the model. Modeling also demonstrated that reasonable survey parameters (source

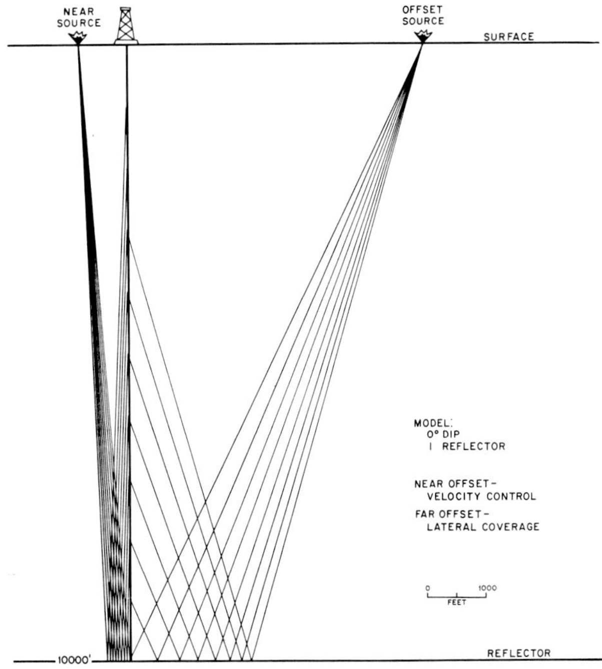

Fig. 4. Ray-path geometry of zero-offset and far-offset VSP's.

[2]Trademark of Sierra Geophysics.

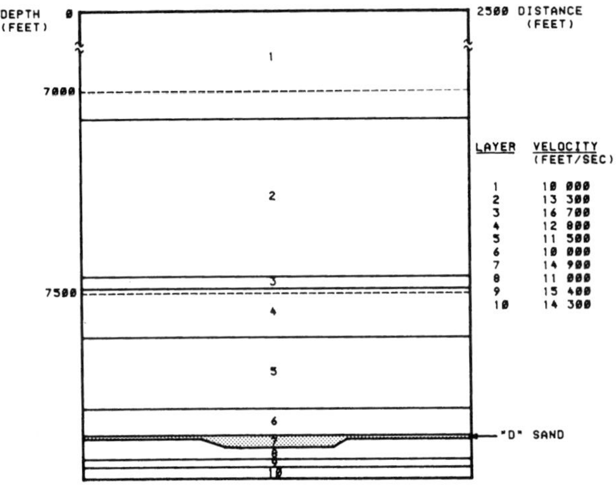

Fig. 5. Geologic model used to study VSP resolution and survey design.

offset distances, geophone level spacings, number of levels, etc.) could be chosen to give a lateral profile length of 2000 ft (600 m). Thus, offset VSPs could be designed to provide data coverage well beyond the next legal drilling location. From these results, the decision was made to proceed with planning and executing an actual survey.

Survey Planning

The first decision to be made was the determination of which well should be used to do the VSP. Well 34-3 was the preferred choice from several points of view. Its location would allow VSP data to be collected over most of the acreage held by Sierra and Berge not already evaluated by drilling. Most important, the fact that Sand D was known to exist in Well 34-3 would allow reliable calibration of the offset VSP data to the well logs and thus increase confidence in the interpretation of the VSP data. On the other hand, Well 34-3 was a producing well. Before the survey was run, the well would have to be shut in and cleaned out. This would result in increased costs, lost production revenue for several days, and the possibility (however slight) of damage to the well and formation. A second concern was the lack of cement behind the casing above a depth of about 6300 ft (1920 m). Without cement to couple the casing to the formation, a path for seismic energy to reach the well geophone (inside the casing) may not exist. This situation will obviously make it impossible to record good VSP data. However, in wells where the casing has been in place for several months or more, such as Well 34-3, natural sloughing may occur to take the place of cement in bonding the casing to the formation. It was hoped that this would be the situation in Well 34-3. After some deliberation, Well 34-3 was selected as the site for the VSP survey.

The original plan called for source offset distances of approximately 5000 ft (1500 m). This distance, when combined with a range of geophone levels from TD to 2500 ft (760 m), would generate a lateral profile approximately 2000 ft (600 m) long. This was more than enough to see to the next legal drilling location and at the same time keep the sources close enough to the well that data quality would not be compromised.

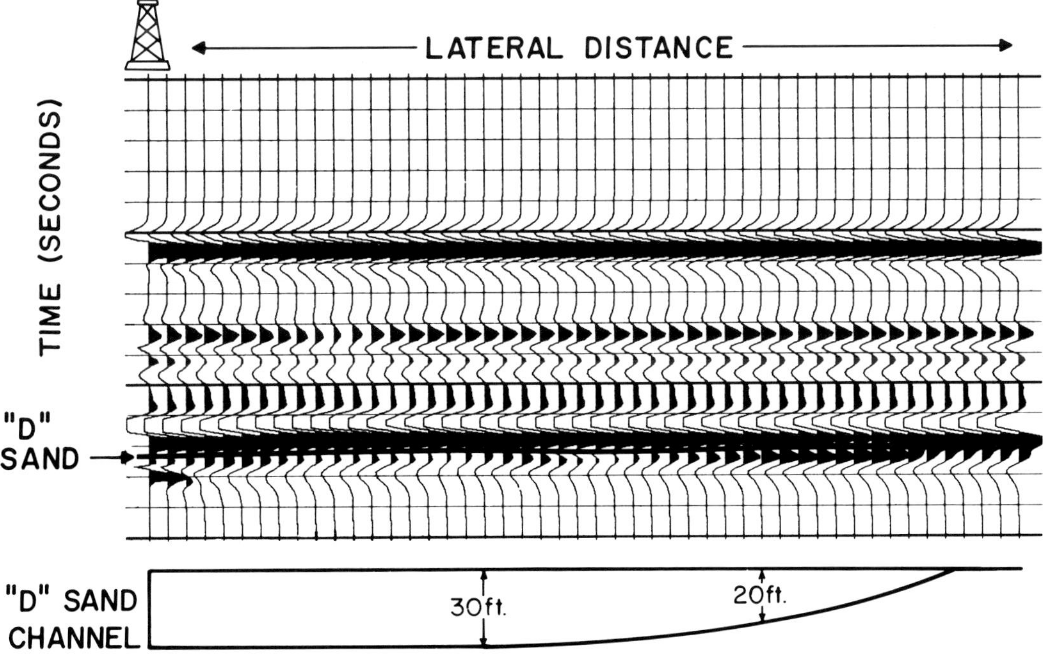

Fig. 6. Results of modeling: synthetic offset VSP data correlated to Sand D thickness.

Figure 7 shows the final survey plan consisting of the five long offsets and one near offset. The actual directions and distances of source locations were chosen to keep the source points on roads to simplify logistics while honoring the original plan as much as possible.

The geophone level spacing was selected to be 75 ft (23 m), which the modeling indicated would generate sufficient lateral resolution (density of reflection points) without unnecessarily increasing the time for the survey by adding additional levels at a closer spacing than needed.

The instrument was a seismic recording unit set up to record both vertical and horizontal channels from the well geophone. The sampling interval was 0.002 s, and the record length was 17 s, consisting of a 12 s vibrator sweep from 10 to 80 cycles/s (10 to 80 Hz) and a 5 s "listening time." The well geophone had a triaxial geophone package and a modified locking arm to allow higher clamping force in the small-diameter (4⅛ in. or 10.5 cm) hole.

Six vibrators were used for this job. Two vibrators were used at each source point, so two runs of the well geophone in the hole were needed, shooting from three offset locations on each run. This put the estimated time for the survey at about 36 hours. Because we did not want to shut in the well for any extended period of time, shooting more than six offsets was considered unfeasible.

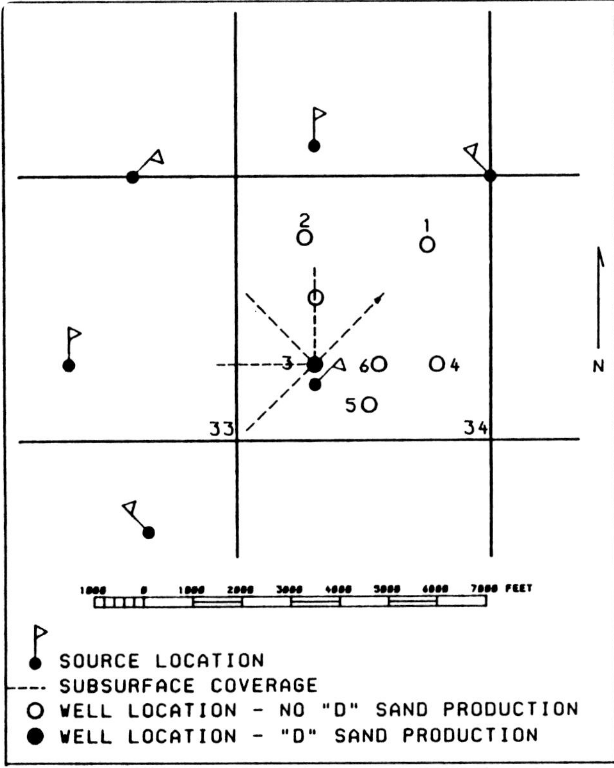

Fig. 7. Original VSP survey plan.

Data Acquisition

When the survey design was finalized, the necessary permits acquired, and the well prepared, the survey was begun (November 29, 1985). The northwest, north, and northeast offsets were the first to be acquired; and as with any well-planned project, problems immediately developed. When the geophone was pulled shallower than 6340 ft (1932 m) in the well, no signal could be observed. The point at which this occurred coincided with the top of cement behind the casing.

The only possible solution to this problem was to adjust the source offset distances so that the desired lateral coverage of 2000 ft (600 m) could be achieved without raising the well geophone higher than the top of cement. Unfortunately, because this problem occurred in the middle of the night, it was not feasible to attempt to move the vibrators to new locations because it would be impossible to survey them in the dark. Therefore, the decision was made to continue and hope that the data could be improved sufficiently in processing to be usable. In the morning, the remaining two far offsets and the near offset were acquired. As Figure 8 shows, the survey plan for the remaining far offsets was redesigned with four source points in each direction. This configuration would allow as much lateral coverage as possible while keeping the geophone in the lower part of the well where the casing was cemented. In addition, the spacing of geophone levels was tightened to 50 ft (15 m) to provide closer spacing of reflection points to attempt to preserve lateral resolution. This was necessary because data collected from the farthest offsets were likely to be of lower quality because of the long travel paths and the probability of refraction and other undesired transmission effects. Fortunately, no further problems were encountered during acquisition, and the survey was completed in 50 hours.

A brief summary of the processing sequence is given in the Appendix.

Interpretation of Data

The final displays consist of the VSP data from each profile transformed to the offset distance, two-way time domain, or CMP seismic domain. A checkshot survey was computed from the zero-offset VSP data to include a time-depth chart, part of which is shown in Figure 9 along with portions of the modeled data, and the zero-offset VSP data. With this chart, the correct two-way traveltime to Sand D is measured, and the seismic response corresponding to Sand D at Well 34-3 can be identified on the zero-offset VSP data. Because the response predicted from the model and that actually observed on the zero-offset VSP were nearly identical, the modeled data were useful in interpreting

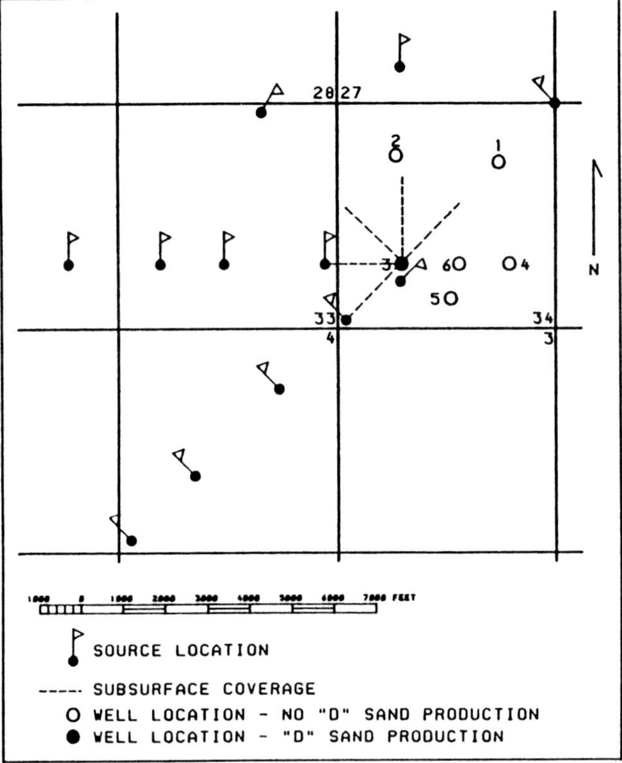

Fig. 8. Revised VSP survey plan.

the offset VSP data. The next step was to identify the Sand D response on the far offset and to map it away from the well by correlating the zero-offset VSP data to the far-offset VSP data. The interpretation of the Sand D channel on the offset data is shown in Figures 10 and 11. Again, agreement of the model data, the near-offset data, and the far-offset data is good.

Two problems complicated the interpretation of these data. The first concerns the poor data recorded on the northwest, north, and northeast offsets in the uncemented casing. A sample of the raw data showing the abrupt decrease in data quality is shown later. After transformation to the CMP seismic domain, this portion of data appears in the area marked on the northeast offset in Figure 10 and similarly on the other two offsets. Because those data are poor, no reliable interpretation can be made through that zone. However, on the north and northeast offsets, data quality appears to improve near the far edge of the data set. These data were recorded with the geophone at the highest levels (shallowest depth) in the well near the surface casing, where there was better coupling to the formation. Therefore, the interpretation was extended on the northeast offset (Figure 10). The north offset also shows the Sand D character at the edge of the data; however, upon comparison with the model data, it was thought that this shape indicated a thinner sand

near the edge of the reservoir. Finally, the northwest offset shows the Sand D character disappearing midway across the data, with no indication of any further buildup. However, this coincides with the location where the data quality deteriorates, as with the north and northeast offsets. Therefore, it cannot be stated reliably that the edge of the buildup is where the Sand D indicator disappears, but only that Sand D extends to at least that point. It may in fact extend farther, but that cannot be ascertained from these data.

The second problem concerns the remaining two offsets to the west and southwest. As described in the Appendix, the processing for these offsets was slightly different than for the other three offsets because multiple source points were used to acquire the data. The result is that where the data collected from each source point overlap with data from the other source points, some smearing occurs, which can distort or disrupt events in the data. The interpretation of the west and southwest offsets is shown in Figure 11. The same procedures described above were again used to identify the Sand D marker.

The preliminary map shown in Figure 12 was constructed, and two potential drilling locations were selected from these interpretations. The first is located 1650 ft (500 m) northeast of Well 34-3 and the second about 1500 ft (460 m) north of Well 34-3 (Figure 12). The northeast drilling location was selected as the primary location because of the superior data quality at that location.

Subsequent Drilling

Well 34-7, located 1650 ft (500 m) northeast of Well 34-3, was spudded December 22, 1985, and completed January 2, 1986. The logs through the Sand D interval are shown in Figure 13; the logs of Well 34-3 are shown for comparison. The Sand D interval in Well 34-7 is 19 ft (6 m), compared with 26 ft (8 m) of Sand D in Well 34-3. Drillstem tests confirmed that the interval was oil-productive.

Continuation of Mapping

To continue delineation of Sand D, another multi-offset VSP was run out of Well 34-7. This survey was conducted open hole immediately after logging operations to avoid problems with uncemented casing as with the first survey. The survey plan is shown in Figure 14. As before, it consists of five far offsets and one near offset. The geophone level spacing, number of geophone stations, and source parameters were the same as before. The survey was started on January 4, 1986, and took 48 hours to complete. The data-pro-

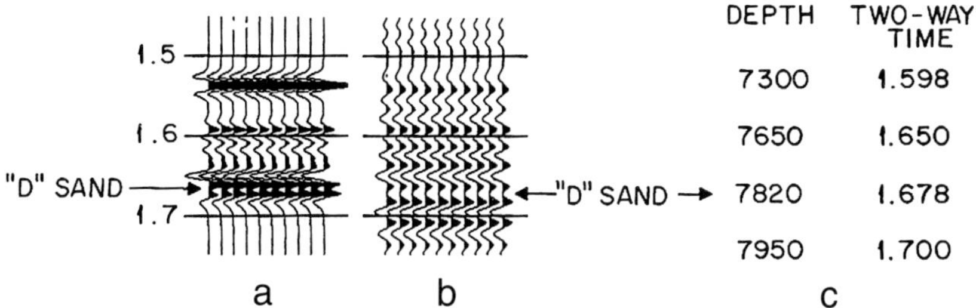

Fig. 9. Correlation of (a) model data and (b) zero-offset VSP data with time/depth data from (c) the check-shot survey.

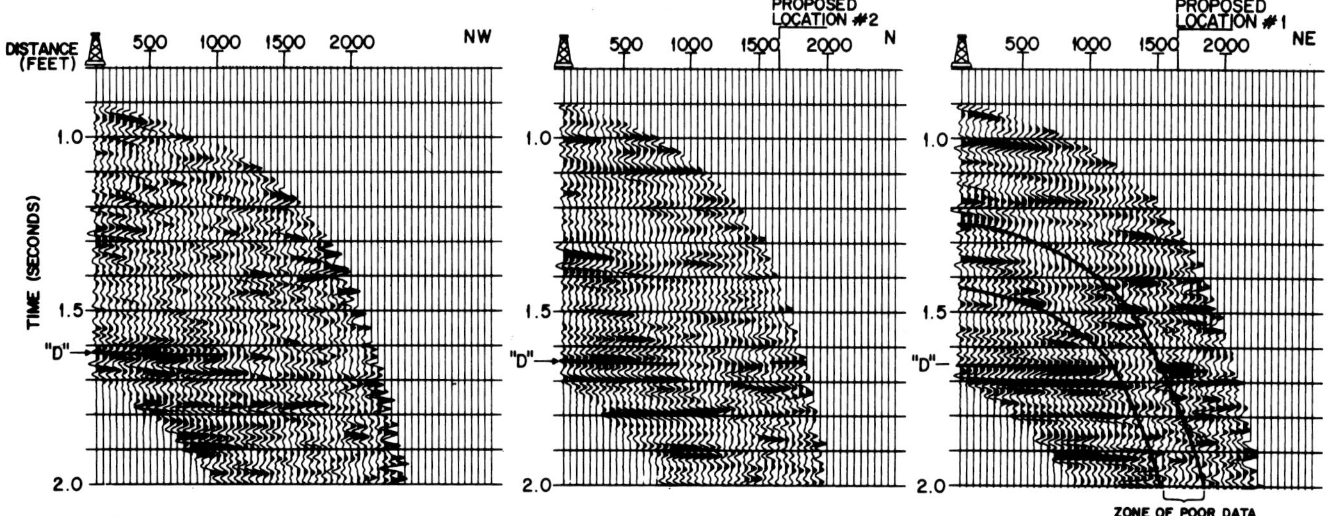

Fig. 10. Final offset VSP data displays from the northwest, north, and northeast profiles.

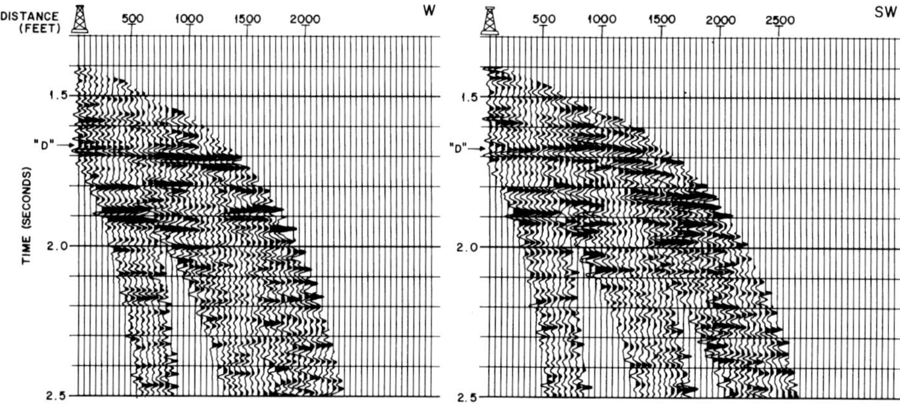

Fig. 11. Final offset VSP data displays from the west and southwest profiles.

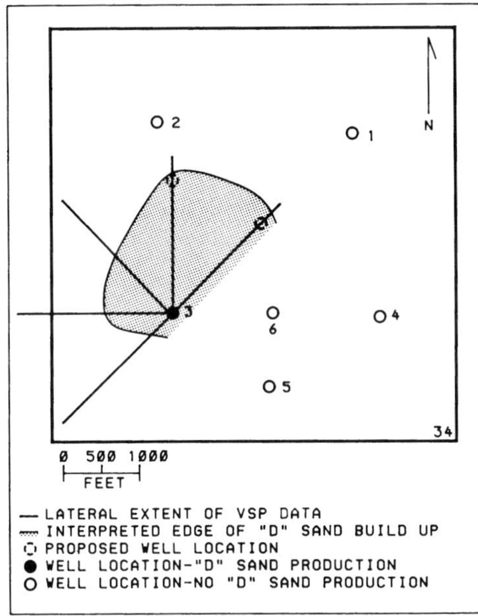

Fig. 12. Results of interpretation of the VSP data set.

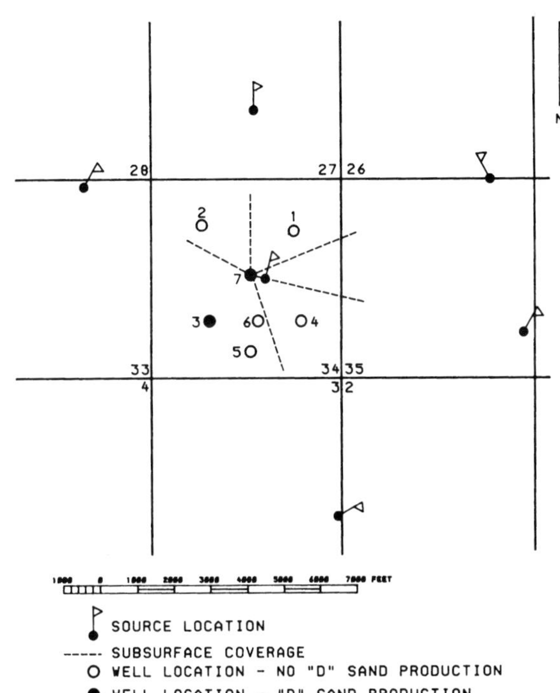

Fig. 14. Well 34-7 VSP survey plan.

cessing sequence was identical to that used for the first survey and is described in the Appendix.

Results of the Second VSP

The interpretation was derived by the same methods used before, with the near-offset VSP data, the time-to-depth conversion from the check-shot table, and the model data to identify and interpret the Sand D event on the offset VSP data after transformation to the CMP seismic domain. The northwest-offset data are shown in Figure 15. The northwest offset was designed to verify the second drilling location north of Well 34-3. The other offsets were designed to complete the delineation of Sand D to the north and east. Figure 16 shows the combined results of both surveys. The second survey confirmed the drilling location to the north of Well 34-3. On the basis of these new data,

however, the location was moved about 200 ft (60 m) south because the Sand D character on both data sets did not show full development at the originally selected site. Moving south should increase the thickness of Sand D, but because of the previously discussed data problems on the north offset of the Well 34-3 survey, the Sand D character (and hence expected Sand D thickness) cannot be determined positively from the data. At the present time, this location has not been drilled to confirm this interpretation; however, plans are pending to drill it in the future.

Recommendations

Offset VSP techniques have been used to delineate a channel-sand field successfully. Drilling based on the

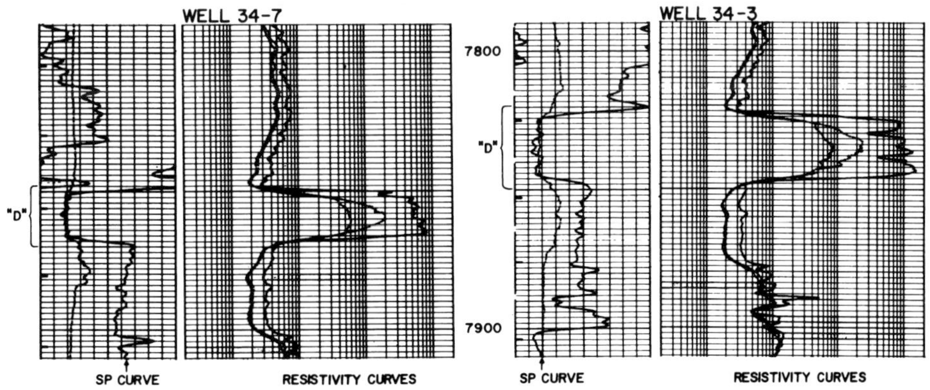

Fig. 13. Comparison of logs over the Sand D interval, Wells 34-7 and 34-3.

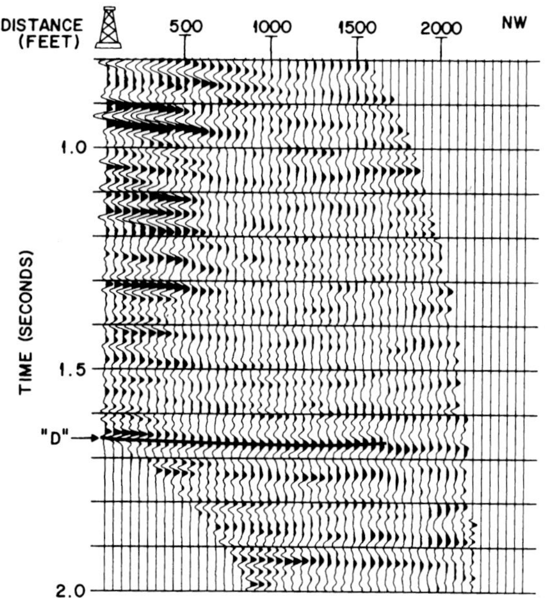

Fig. 15. Final offset VSP data display from the northwest offset well, Well 34-7.

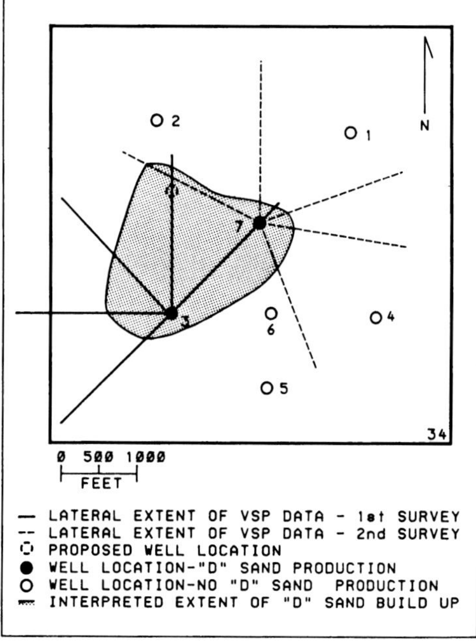

Fig. 16. Results of interpretation of both VSP data sets.

VSP data has proved the validity of the interpretation and demonstrated that VSPs can be used as an aid in locating the best sites for development wells. Also, as a result of this project, some guidelines for successfully using VSPs in field development can be stated.

1. Seismic modeling of the proposed survey should be performed to verify before the survey that resolution necessary to solve the problem at hand is likely to be achieved, to assist in designing the survey parameters to obtain the best possible data, and to assist the interpreter in understanding the recorded data.

2. Surveys should be performed in wells with completely cemented casing or in open hole because poorly cemented casing causes serious degradation of VSP data.

3. Multioffset VSP surveys should be laid out to make as much use as possible of existing well control to confirm modeled results and to lend credence to the interpretation of the VSP data.

4. A near-offset VSP should be run in conjunction with the far offsets to establish velocity control and to aid in correlating the data to the well logs.

It must be emphasized that VSPs are only one of several geophysical methods available for use in developing a field economically.

Acknowledgments

I thank Berge Exploration and Sierra Energy for the use of their data and for their assistance in this project, as well as the many people at Geosource whose time and efforts were invaluable in the preparation of this paper.

Reference

Plybon, S. C., and Oldham, D. W., 1985, Seismic stratigraphic analysis of Lanyard-Lost Creek Field area, Denver Basin, Colorado: Application to Exploration and Field Delineation, Presented at the 1985 AAPG Rocky Mountain Section Meeting, Poster Session, June.

Appendix

Processing VSP Data

The processing of VSP data can be loosely divided into four main steps or groups of steps: (1) preliminary steps, including demultiplexing, gain recovery, editing, stacking, and sorting common level records, picking direct arrival times, and statics corrections; (2) wave-field separation; (3) deconvolution; and (4) transform to offset distance, two-way time domain (so-called VSP/CMP transform). The fourth step is performed only on offset VSP data to present the data in a format that is familiar and useful to the seismic interpreter.

The flow chart in Figure A-1 shows the order of the main steps used to process both of the VSP surveys. As discussed, the Well 34-3 survey encountered problems with data quality in the uncemented portion of the casing, so two of the five far offsets were acquired with multiple source points, which made some slight changes in the processing sequence necessary.

Preliminary Phase

The main objective is to organize and to prepare the data for subsequent steps. After demultiplexing and vibroseis correlation, the data acquired at each level are edited and then summed (stacked) to form a single output trace for each geophone level. Next, the traveltime of the direct arrival is chosen and stored in the trace headers as a reference point for muting and time-shifting the data. After these steps are completed, the data appear as in Figure A-2. The principal wavefields present in the data have been identified.

Wavefield Separation

As stated previously and as shown in Figure A-2, although several different types of energy are recorded during a VSP, only the reflected P-waves are of interest in this case, so the task of separating the reflected P-waves from the rest of the data must be undertaken. A variety of techniques have been devel-

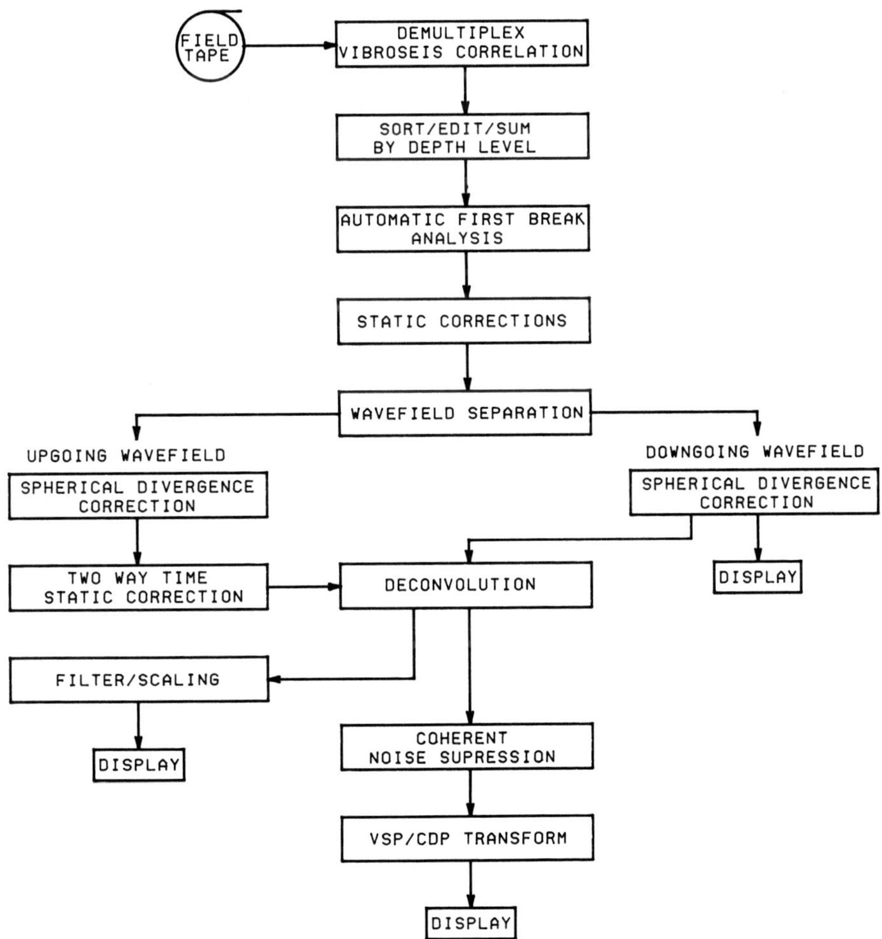

Fig. A-1. Basic VSP processing sequence.

oped for this purpose. In this instance, an averaging technique was used to estimate the direct or "down-going" *P*-wavefield and subtract it from the total data set. This leaves the "upgoing" *P*-waves plus the other wavefields (as identified in Figure A-2). Following deconvolution, a coherency filter was applied to the reflected *P*-wave field to enhance the desired *P*-waves and to suppress these other wave fields, thus completing the wavefield separation process.

Deconvolution

Deconvolution is performed to improve signal characteristics and to suppress undesired events, such as multiples. The deconvolution of VSP data is a signature-type deconvolution that uses the direct *P*-wave field as the signature. Hence, the deconvolution operator is designed on the direct *P*-wave field and then applied to the upgoing *P*-wave field. The operator is updated at every geophone level. The direct *P*-wave field is shown in Figure A-3, while the reflected *P*-wave field following deconvolution and coherency filtering is shown in Figure A-4.

VSP/CMP Transform

The objective is to reorganize the reflected *P*-wave data from the format in Figure A-4 to the offset-distance, two-way-time domain (CMP domain), by assuming that the reflectors have zero dip and no lateral velocity variation. Then, given that the shooting geometry, traveltime to the geophone, and vertical velocity structure are known (from a check-shot survey or zero-offset VSP), one can readily reconstruct all the ray paths to each geophone level and compute the location and two-way traveltime of each reflection

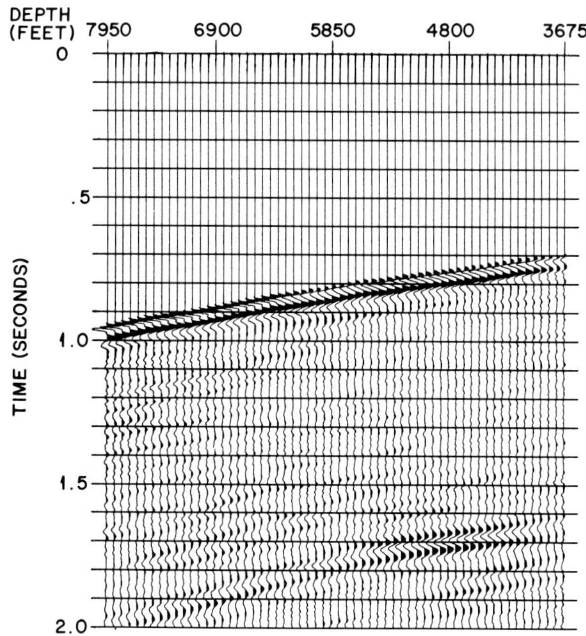

Fig. A-3. Direct (downgoing) waves after wavefield separation.

point. The data are then sorted by reflection point location and output in trace form, where each output trace contains data from common reflection point distances relative to the well-bore. Figure A-5 shows this process schematically.

The two VSP profiles acquired with multiple source points were processed with the same basic steps; however, the data from each source point were proc-

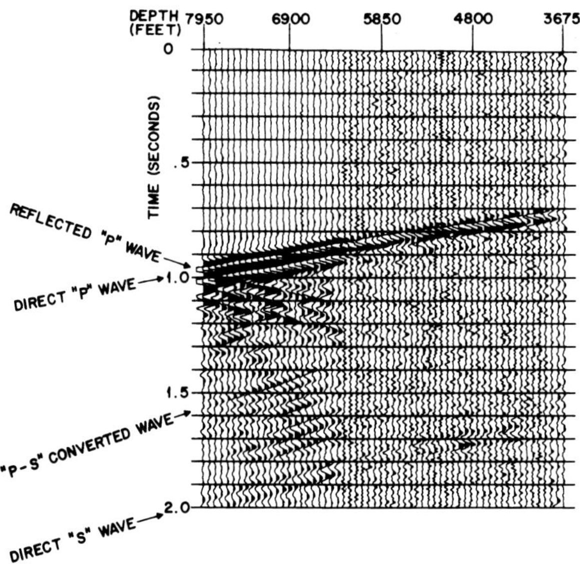

Fig. A-2. Edited and stacked VSP data.

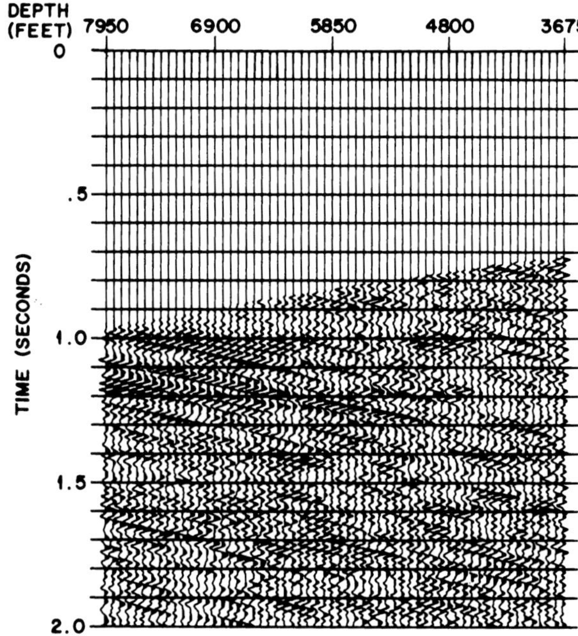

Fig. A-4. Reflected (upgoing) *P*-waves after wavefield separation, deconvolution, and noise suppression.

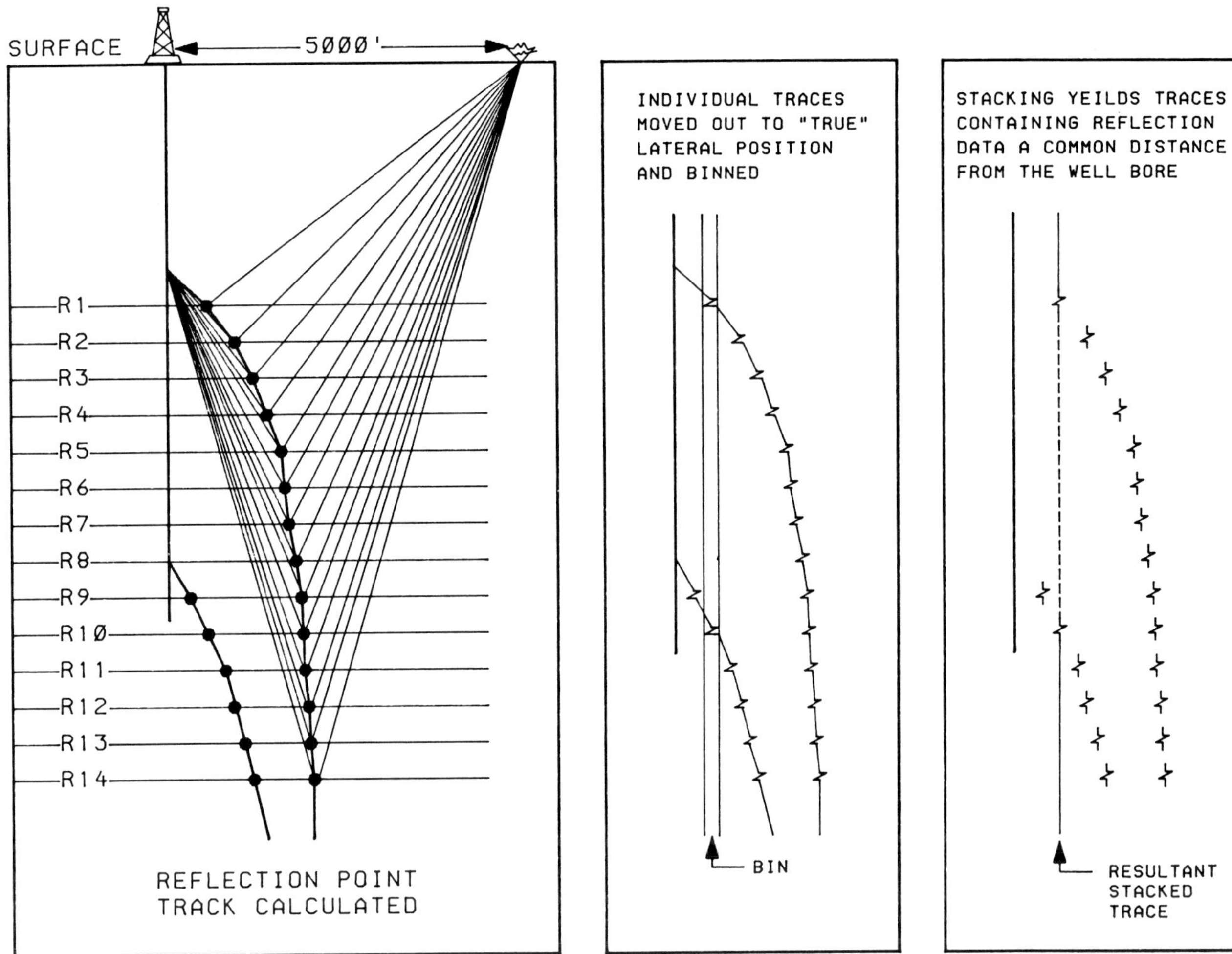

Fig. A-5. Basic concepts of the VSP/CMP transform.

essed independently until the VSP/CMP transform step. At this point, the data sets were combined and transformed together. Some smearing occurs where the lateral coverage from each source point overlaps. Further, the velocity function used in the transform is not exactly correct for each source point; consequently, some discontinuities occur in the final data sets. (See Hardage for more details on VSP processing.)

Reference

Hardage, B. A., 1983, Vertical seismic profiling, Part A: Principles in Handbook of Geophysical Exploration, Geophysical Press. **14A,** Sec. I.

Ekofisk Field Development: Making Images of a Gas Obscured Reservoir

*J. A. Dangerfield**

Introduction

Over the last six years, the geophysical problems associated with the development of the giant Ekofisk Field in the North Sea have provided an interesting case study. The geophysics has been aimed largely at solving the problems of making seismic images through the Tertiary gas which obscures the Chalk reservoir from normal seismic view. The gas abruptly slows the sound and leads to distorted ray paths so that apparent faults are produced. In the most crestal areas attenuation by the gas further results in almost total loss of signal. Although the field was discovered in 1969, surface seismic acquisition had virtually ceased by 1973 with only 14 lines over the field because these problems with gas were considered insuperable. By 1983 the field was considered to be in a mature phase of understanding with 55 wells drilled and little need for seismic images. The field was considered simple, domal, and effectively unfaulted. Waterflood was being planned.

The possibility of being able to undershoot the gas using geophones down a deviated well bore was nevertheless interesting. Acquisition of the first marine walkaway borehole profiles produced excellent seismic data where none had been obtained previously. The profiles indicated the unexpected presence of a major crestal graben, over a kilometer long. The presence of this (or any other faults) was hotly contested and the seismic appearance proposed to be due to gas effects. Subsequent drilling proved the graben's existence and a 45 m fault throw when the well (2/4-K13b) penetrated its western wall. Unfortunately, this was subsequent to sidetracking the well, which had suffered an almost instantaneous loss of 800 barrels of mud into a fracture system that probably fed down into the eastern graben wall.

At this point, it was realized that faults in Ekofisk are a drilling hazard, despite the number of safely drilled wells on the field. A series of borehole profiles were shot in eight wells using various geometries to undershoot the gas and image future drilling locations.

*Phillips Petroleum (Norway), Box 220, Tananger 4056, Norway

All of the profiles were successful in producing clear images, except in the most crestal portion. Various faults were discovered and future well locations moved accordingly. Cost-to-benefits ratios were good: a sidetrack adds some millions of dollars in costs to a well. Nevertheless, the borehole profiles were of the order of twenty times as expensive as normal surface seismic per useful kilometer of line.

Surface seismic methods were therefore reexamined. In 1987 a new series of surface seismic lines was shot with a novel cable orientation. The lines were tangential to the gas cloud so that the cable, and therefore the resulting acoustic ray paths, were kept as clear of the gas as possible. The lines were situated to image the future well locations using 3-D ray trace modeling to predict up-dip migration of the reflection points.

The tangential lines were successful in making clear images, except in the most crestal area. The extra coverage delineated more and larger faults than expected. The extra faults in the Chalk could be seen to be real because the Paleocene immediately above appeared unfaulted, i.e., not severely affected by gas. The shoot was successful in showing the well locations that needed to be moved away from faults. The seismic coverage however, was inadequate to be certain of the major orientations of the faults.

The interpretation suggested a north-northeast fault orientation. The large faults with throws of up to 70 m explained the pressure balance between the upper and lower reservoirs in the local areas of the faults. The two reservoirs normally are separated by a field-wide impermeable zone, the "Tight Zone". The recognition of faulting was instrumental in the later decision to extend the waterflood, from just the lower reservoir, to include both reservoirs.

The size of the faults also suggests the possible presence of strong fracture anisotropy, and therefore strong waterflood flow with that orientation. The possibility of waterflood wells being aligned with producers along this trend, leading to producers watering out, could have serious implications. Data were collected from ongoing core fracture work, formation microscanner (FMS) logs, and anelastic strain measure-

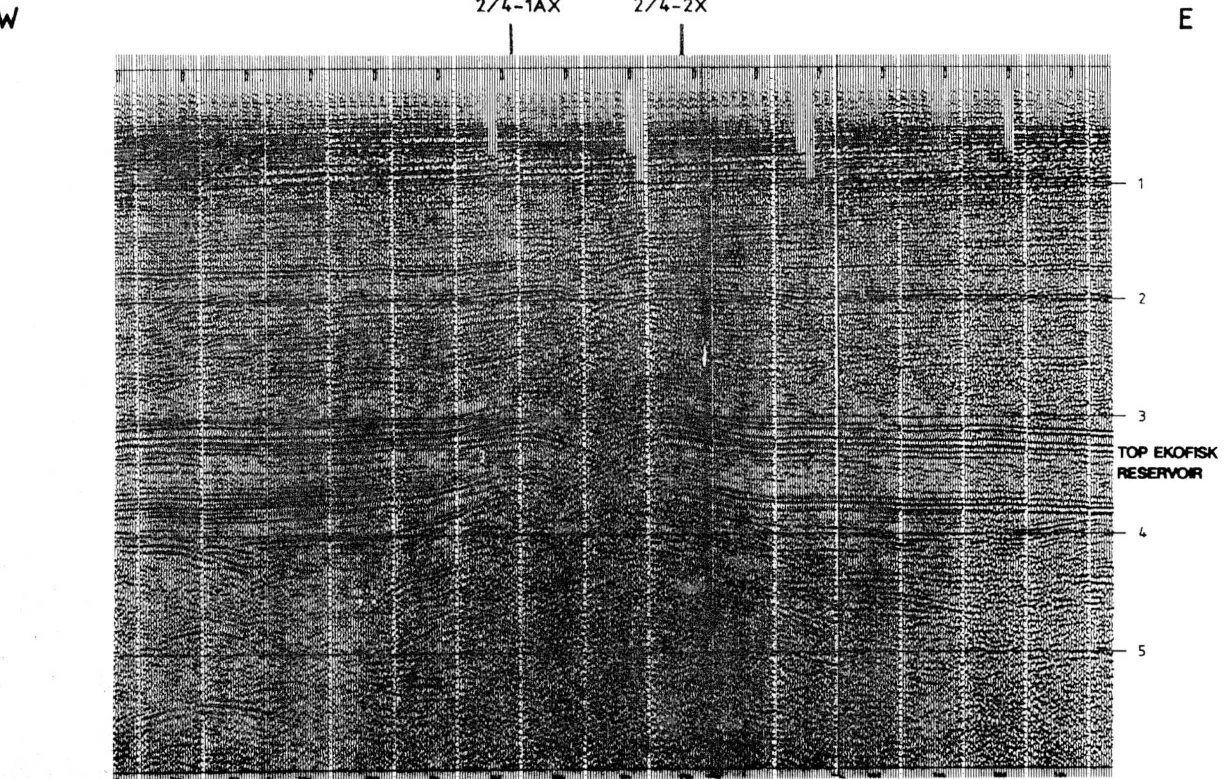

Fig. 1. Seismic section across Ekofisk Field showing data quality problems over field crest.

ments. All of these showed a very dominant north-northeast trend in the northwest half of the field. Waterflood results have been coming in for a year and are being monitored carefully. So far they suggest no problems with early water breakthrough.

Gas Effect

Gas has collected in solution in the Tertiary silty shales above the Chalk reservoir. The gas has leaked from the reservoir over geologic time. The gas dramatically slows sound waves, even to velocities below those of sound in water, so that seismic sections display time depression, multipathing, and apparent faults. Over the crestal area of the field the signal is totally lost due to confused ray paths and absorption.

The field therefore appears as a collapsed anticline in seismic sections (Figure 1). This prevented the field from being a candidate for the earliest exploration wells in the North Sea. Ekofisk was the first oil field discovered in the Norwegian North Sea but, it was only contractual obligation to the drilling rig "Ocean Viking" that led to the field being drilled. The discovery well, 2/4-1 AX, was drilled in 1969 on the apparent high point of the toroidal structure. The field was on production only 18 months later. It was 1973 before a producer was finally drilled into the "collapsed" area,

the giant size of the field realized, and the effect of the Tertiary gas comprehended.

The 3-D extent of the gas is difficult to map. Gas analysis while drilling the crestal 2/4C-11 well showed the gas to extend in irregular concentration from above 3000 ft down to the reservoir at 10 000 ft, with particular concentrations at 4000 and 7000 ft. Mapping on the basis of seismic reflectivity is not useful because the reflectivities are greatest with low concentrations of about 20 percent (Domenico, 1974). A map has been produced of apparent time delay seen on surface seismic which gives a general distribution of the gas (Figure 2), but those delays are also dependent on seismic line orientation with respect to the gas. The seismic data often appear badly faulted but this can be purely an artifact of the gas. In the most crestal portion of the field there is no coherent seismic signal returned at all.

Ray tracing shows the problems that the gas creates in surface seismic: using a model (Figure 3) and considering shots from outside the gas (Figure 4), the problems of ray bending, multipathing, and apparent fault creation are obvious.

Removal of the receiver to depths closer to the reservoir constrains the ray paths (Figure 5). Borehole profiling, therefore, provides significant advantages.

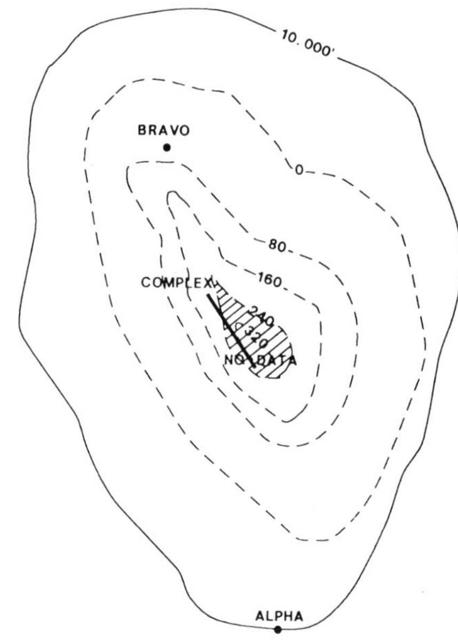

Fig. 2. Apparent time delays (dashed contours) because of effects of gas in Tertiary section.

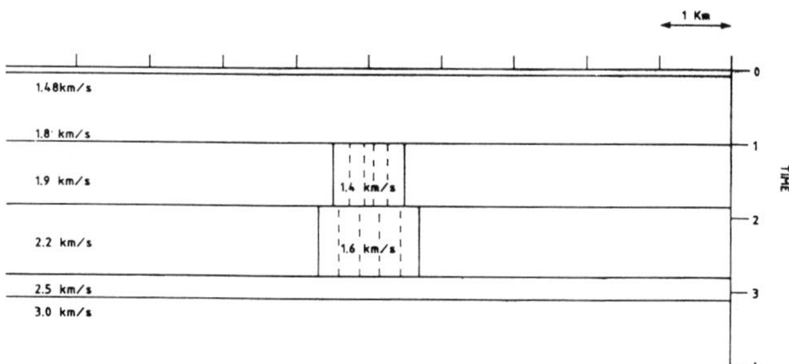

Fig. 3. Velocity model used for ray tracing. Velocities decrease into the center of the gas.

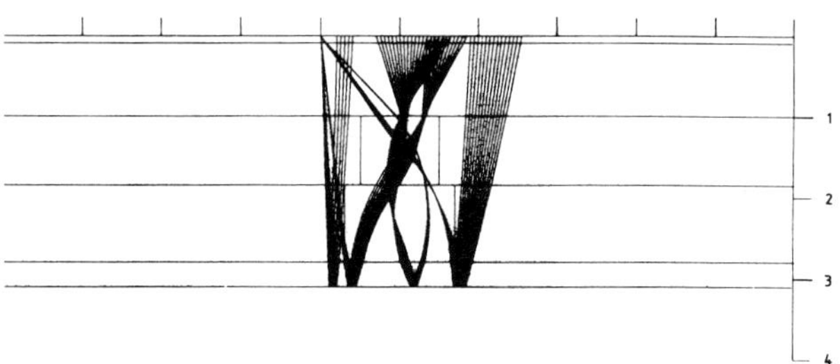

Fig. 4. Seismic raypaths for sources outside the gas region.

Borehole Profiles

Phillips pioneered marine walkaway vertical seismic profiling from deviated wells on the field in 1983. At this time the structure was believed to be a simple unfaulted dome. There was only gentle structural irregularity in the reservoir depth maps which had been made by contouring tops from 55 wells. Figure 6

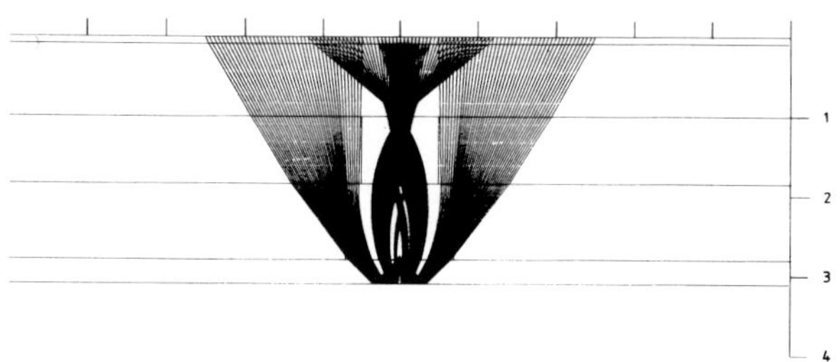

Fig. 5. Seismic raypaths for receiver near the reservoir in a borehole.

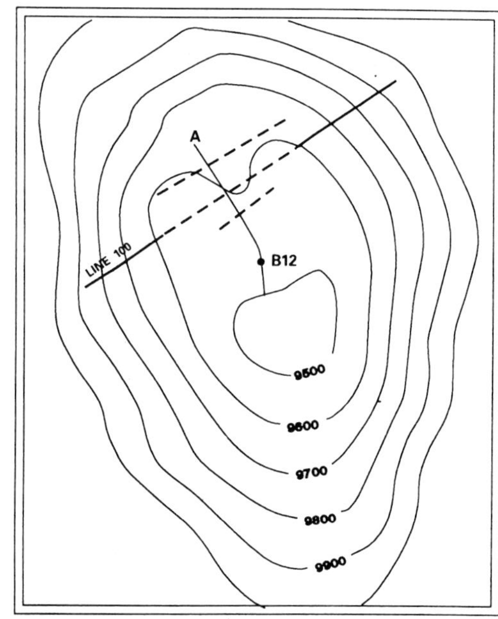

Fig. 6. Map of top Ekofisk Formation showing (dashed lines) source locations for walkaway borehole profiles into deviated B-12 well. Surface line 100 is also shown.

shows the structure and the positions of three walkaway profiles shot from the strongly deviated 2/4 B-12 well. These were the first successful marine walkaway profiles (Christie and Dangerfield, 1984). Surface seismic line 100 shot over the profile locations showed no useful data in this area (Figure 7). The walkaway profiles inserted into the seismic lines at the correct locations showed excellent data (Figure 8). The time depression westward was not structural but due to gas effects. The apparent graben feature mapped linearly across all three walkaways and appeared to show a throw of about 45 m (Figure 9). The interpretation of a graben was hotly contested but did not change the need to waterflood the area. If a graben was present it would be necessary to maintain a positive water pressure in the graben to stop oil being flushed into it from surrounding areas, and to stop the oil then escaping into the layers above and below along the faults.

The waterflood well, 2/4-K13, was lost when 800 barrels of mud disappeared virtually instantaneously, probably into a fracture system associated with the eastern wall of the graben. The subsequent sidetrack penetrated the western wall of the graben and found a 45 m fault exactly at the position indicated by the borehole profiles.

Inspection of the new map (Figure 9) suggested that the contoured structural irregularity in the northwest portion of the field probably indicated other faults.

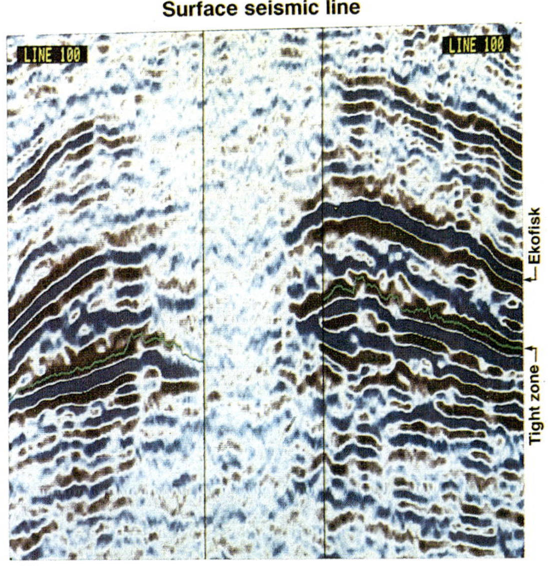

Fig. 7. Portion of surface line 100 showing loss of data over structural crest.

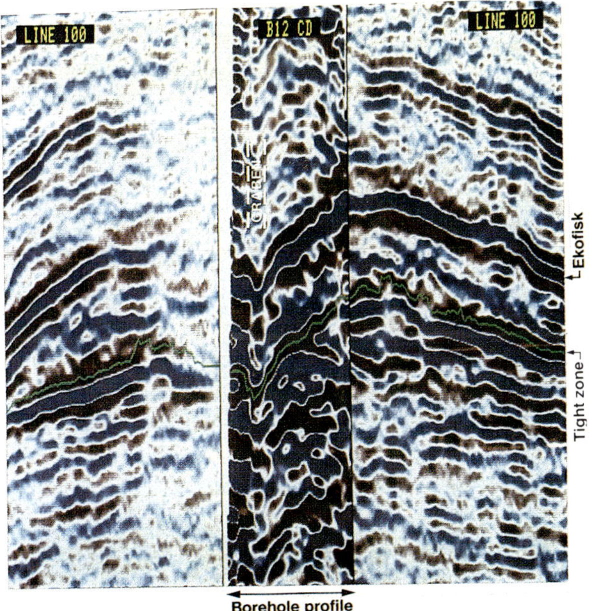

Fig. 8. Portion of surface line 100 with walkaway profile inserted.

It was evident that the new borehole profile methods could provide cost-effective data and that future well locations should be imaged before drilling whenever possible.

A total of eight wells were subsequently used to obtain a variety of stationary long offset, walkaway, and moving-source profiles over the field. In each case the target area imaged was a future reservoir well location and the geometry was set up to undershoot the gas as far as possible. The positions of the resulting 17 profile lines are shown in Figure 10. For precision the positions of the reflections at the reservoir level were mapped using 3-D ray tracing. At Ekofisk Field dips are less than 5° but up-dip migration still displaces reflection positions up to 300 m. The profiles also suffer time shifts one to another because of different travel paths through the gas. The gas does not affect phase. After time shifts, all of the profiles fit each other and the synthetic seismograms with a high degree of confidence. They also fit very well to the seismic data (Figure 8).

Some of these profiles showed large faults, as in the 2/4 K-22 profile in the northeast part of the field (Figure 11). The validity of those faults remained under debate. Almost all the profiles showed clear images despite the fact that all the surface seismic lines over the field have large "holes" with data loss at

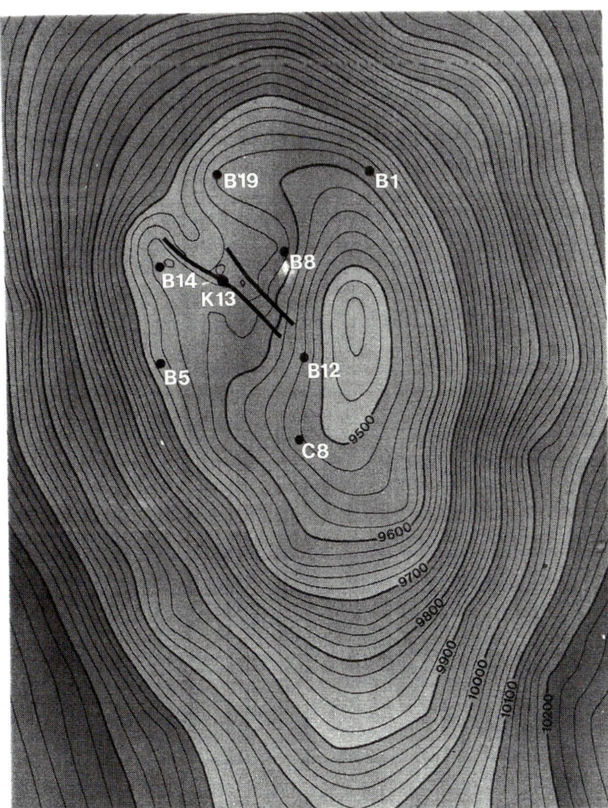

Fig. 9. Contour map showing location of graben seen on all three walkaway profiles into deviated well B-12.

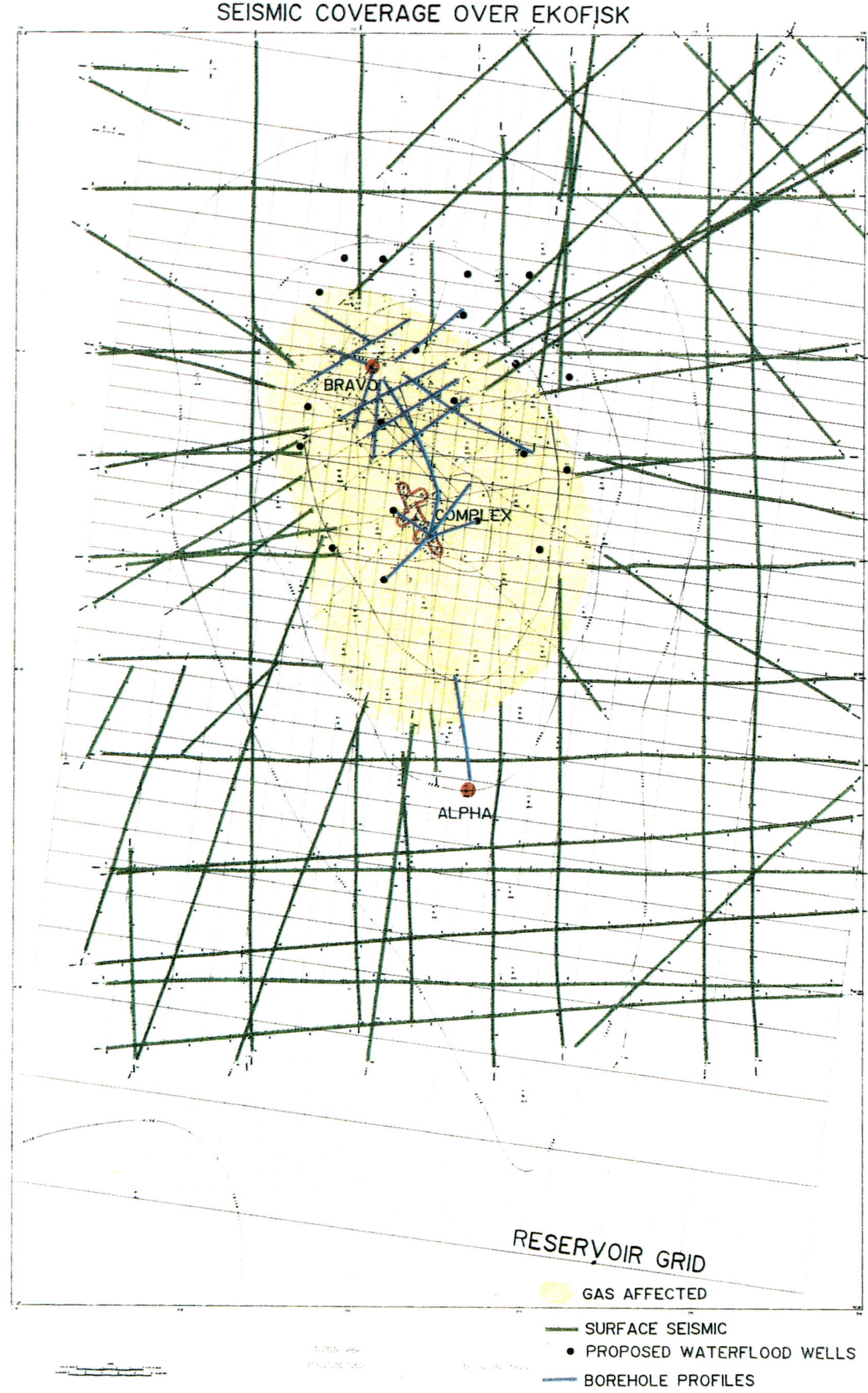

Fig. 10. Profiles used in combined surface/walkaway study.

these locations. Only the profiles obtained from the crestal 2/4C-11 well were severely degraded by gas.

The final maps were made from a combination of well data, the borehole profiles and existing surface sesimic (Figure 12). A number of well locations that were close to potentially hazardous faults were moved accordingly.

Surface Seismic

Surface seismic costs approximately twenty times less per kilometer than borehole profiles at Ekofisk. Various acquisition and processing tests had been carried out but, unfortunately, good data remained elusive. The obvious testing for choice of offsets showed that the signal improved as fold increased, except in those areas where no signal was returned, and there no improvement was seen. Automatic static routines were also tested without worthwhile improvements. Modeling showed undershooting the gas with long cable acquisition to suffer critical angle problems.

We needed both to acquire high-fold seismic data and to keep the cable as clear as possible of the gas affected area. Lines were, therefore, acquired with orientations tangential to the gas (Figures 13 and 14). The raypaths over the whole length of the cable were

thus kept as clear of the gas as possible. Useful fold coverage was increased and there was a small advantage from the updip migration in undershooting the gas. Figure 14 compares the raypaths of a normal star configuration of acquisition over the field with a tangential orientation. The lines were positioned to image the flank waterflood well locations at the reservoir level. 3-D modeling was used to calculate the positions at sea surface from which the reflections would be received from those locations (Figure 15).

At the time of acquisition 19 anchors, each with a tethered buoy, were in place at the field (Figure 13). Three buoy handling vessels were used to drag the buoys to either side as the seismic vessel passed. The hydrophone cable length was also limited to a 1266 m cable because of these obstructions. There were 96 groups of 13.3 m in each cable. The source was a 3240 cu. in. array of air guns.

Weather conditions were poor, necessitating lowering the cable to as low as 11 m with a maximum RMS average noise of 15 microbars and peaks of up to 24 microbars. The predominant swell direction precluded shooting in the desired directions for much of the time and necessitated reorienting some of the lines. With

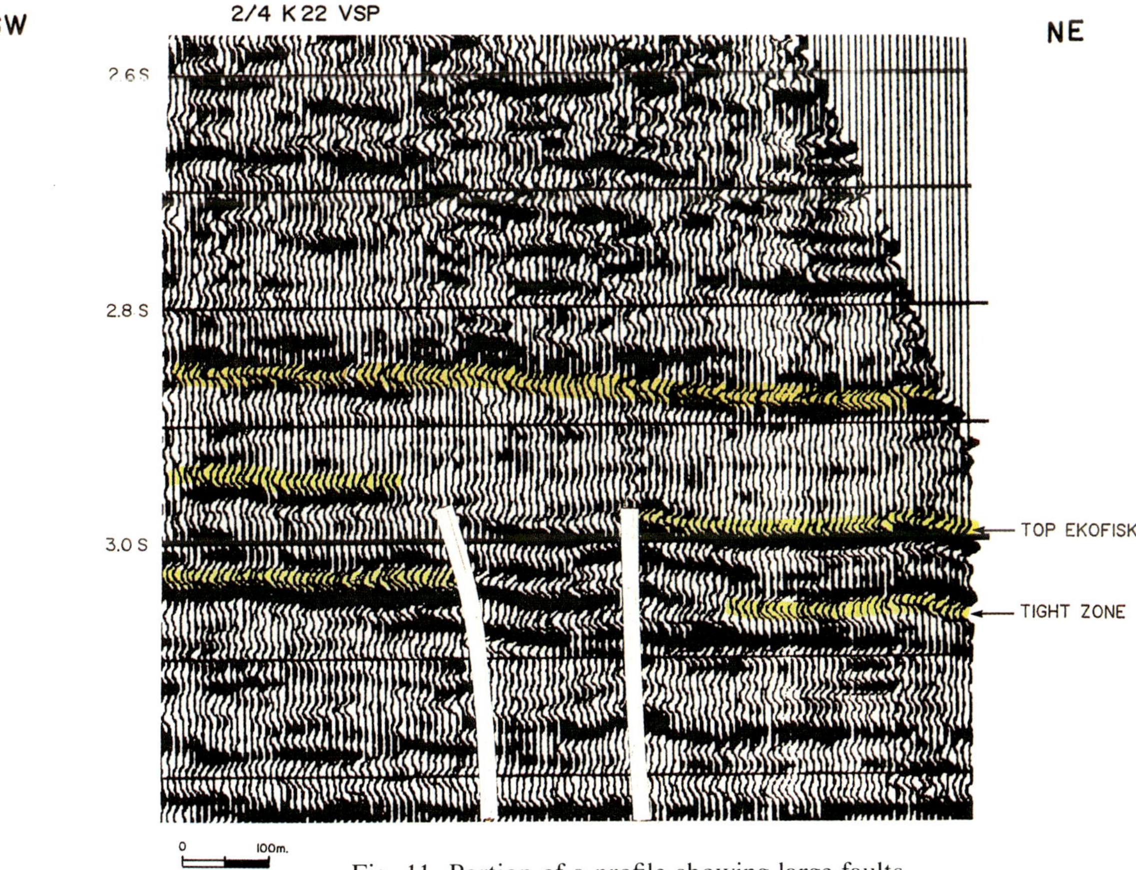

Fig. 11. Portion of a profile showing large faults.

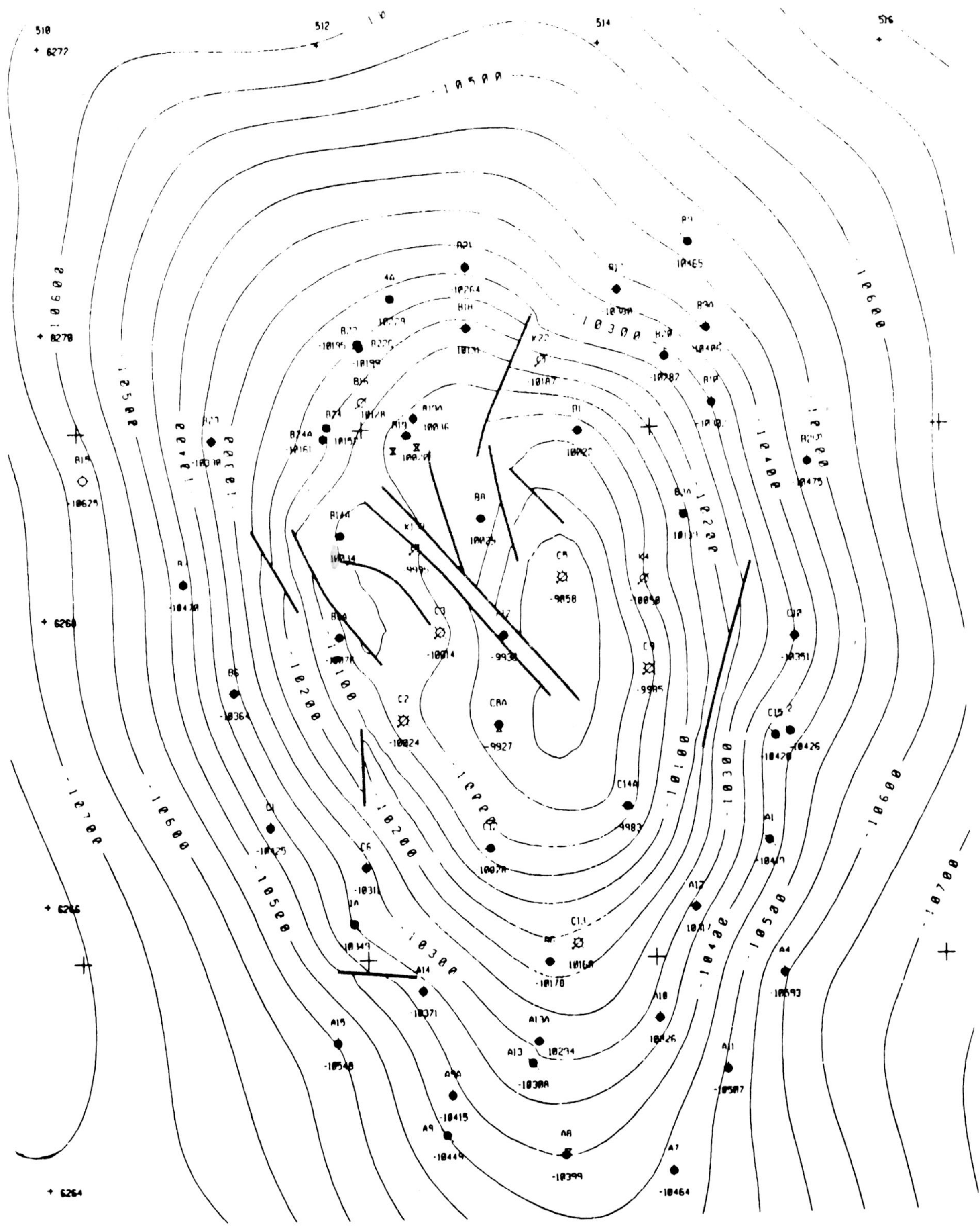

Fig. 12. Interpretation showing faults.

the 3-D ray-trace modeling map it was easy to reorient the lines and still image the targets.

The quality of the final processed data was excellent. Many more and much larger faults were seen than previously expected. Many of these faults could be shown to be real and not due to the gas effect because the top Paleocene above these faults appears unbroken. Figure 16, for example, images the same fault seen in the borehole profile in Figure 11 that was under dispute. This time there was no doubt about the magnitude of the fault.

The new fault map produced (Figure 17) was still contentious because of the extra faults and the poor spatial sampling. The map posited a very strong north-northeast trend with major fault throws of up to 70 m.

There was an immediate benefit to the reservoir model. The faulting explained the equal pressures in upper and lower reservoirs in the vicinity of the faults. Elsewhere the presence of the Tight Zone between the reservoirs keeps them in pressure isolation. Trying to model the pressures without faults has been very difficult.

Within six months, fracture data, FMS log data (using resistivity to show fractures), anelastic strain measurements (using core expansion to show stress-strain relationships), and a shallow seismic 3-D (imaging faults in the first kilometer of section) all showed the same strong trend north-northeast. Current investigations should determine if there is also a conjugate fracture system trending north-northwest and if there is a radial trend due to doming.

Conclusion

This strong north-northeast trend may be very important in the waterflood. Originally a 4-well waterflood pilot study suggested a strong north-south preferential flow of water. The original staggered line drive orientation proposed was designed to keep waterflood wells away from alignments with producers north and south.

The action of the faults themselves is not understood at this point. They might act favorably, if close to a waterflood well, and the fault acts as a conduit for water by providing a large surface area from which water will imbibe into the formation. However, if a producer is too close to a fault then the producer might be watered out early. The early waterflood results are coming in and will be interpreted in light of the new structure maps. The waterflood plans have been extended to include the reservoir horizons both above and below the tight zone.

These maps are expected to show a considerable evolution with our next series of attempts to image the structure with a 3-D surface seismic program and our first 3-D walkaway borehole profile, both shot in the first half of 1989 and currently being processed.

(Text continued on page 109)

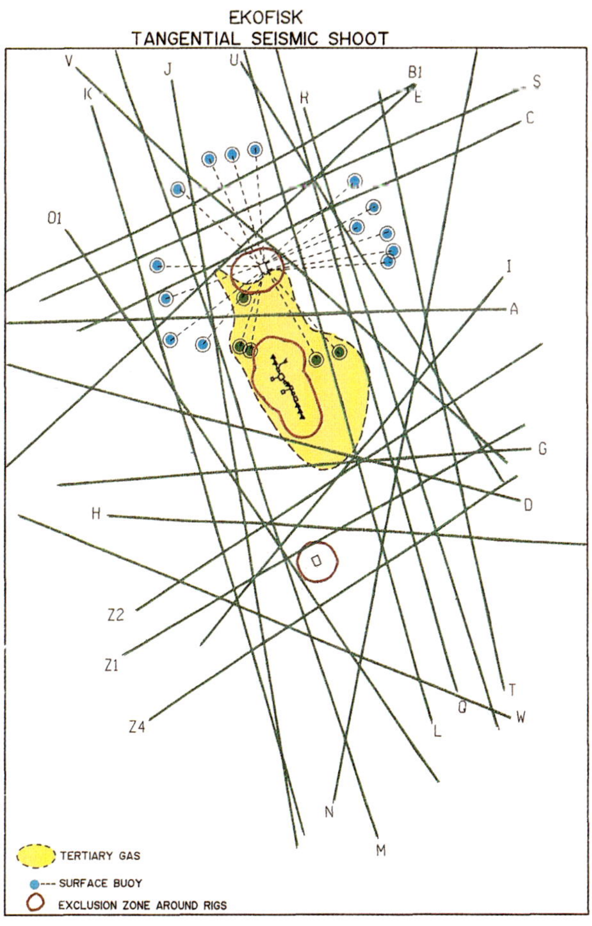

Fig. 13. Locations of tangential surface lines designed to keep raypaths out of the gas-affected region.

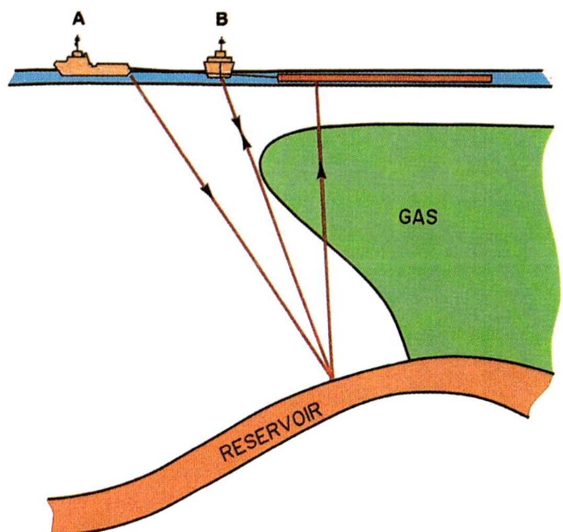

Fig. 14. Diagram comparing raypaths for tangential and radial (star configuration) lines.

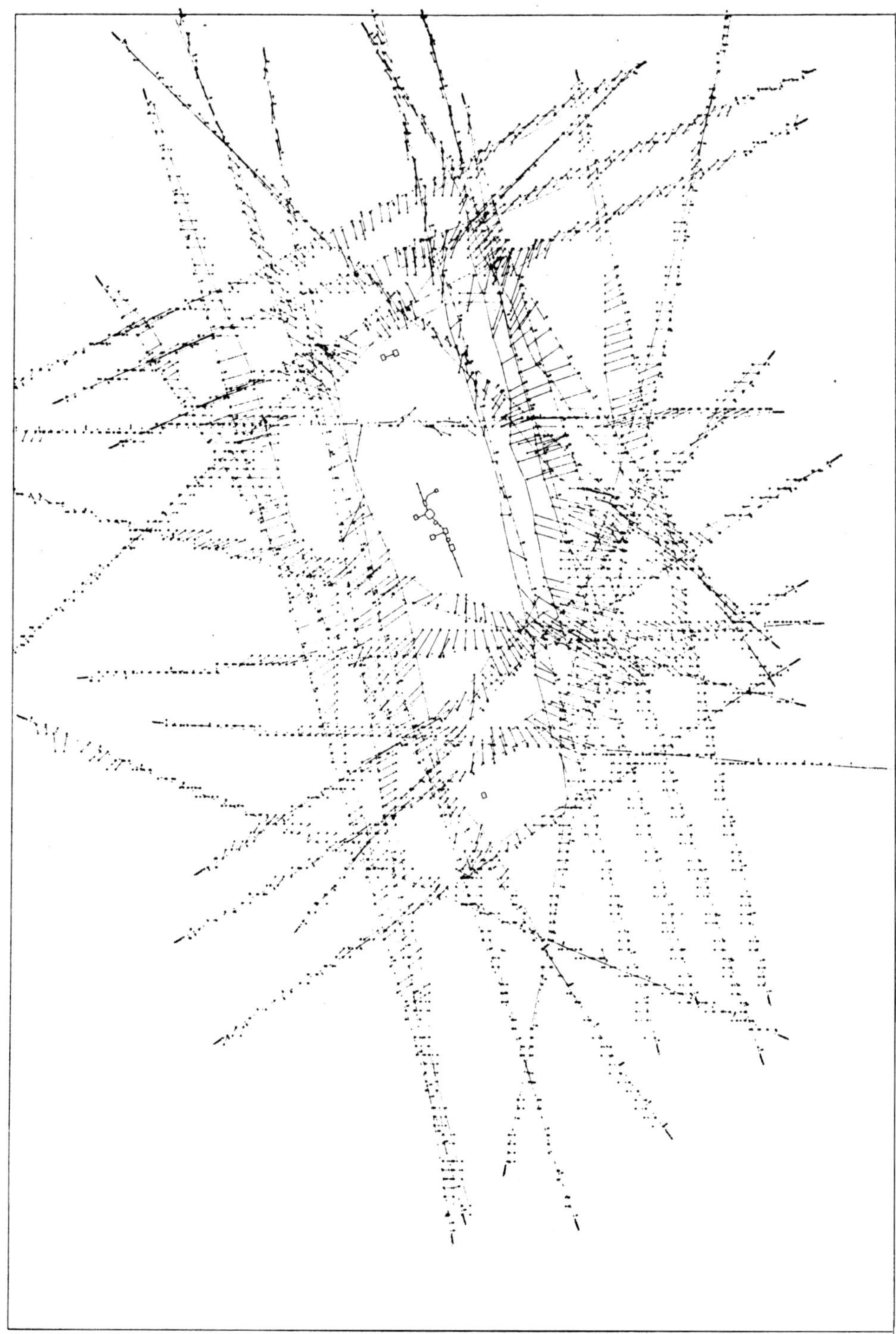

Fig. 15. Up-dip locations of reflecting points on tangential lines.

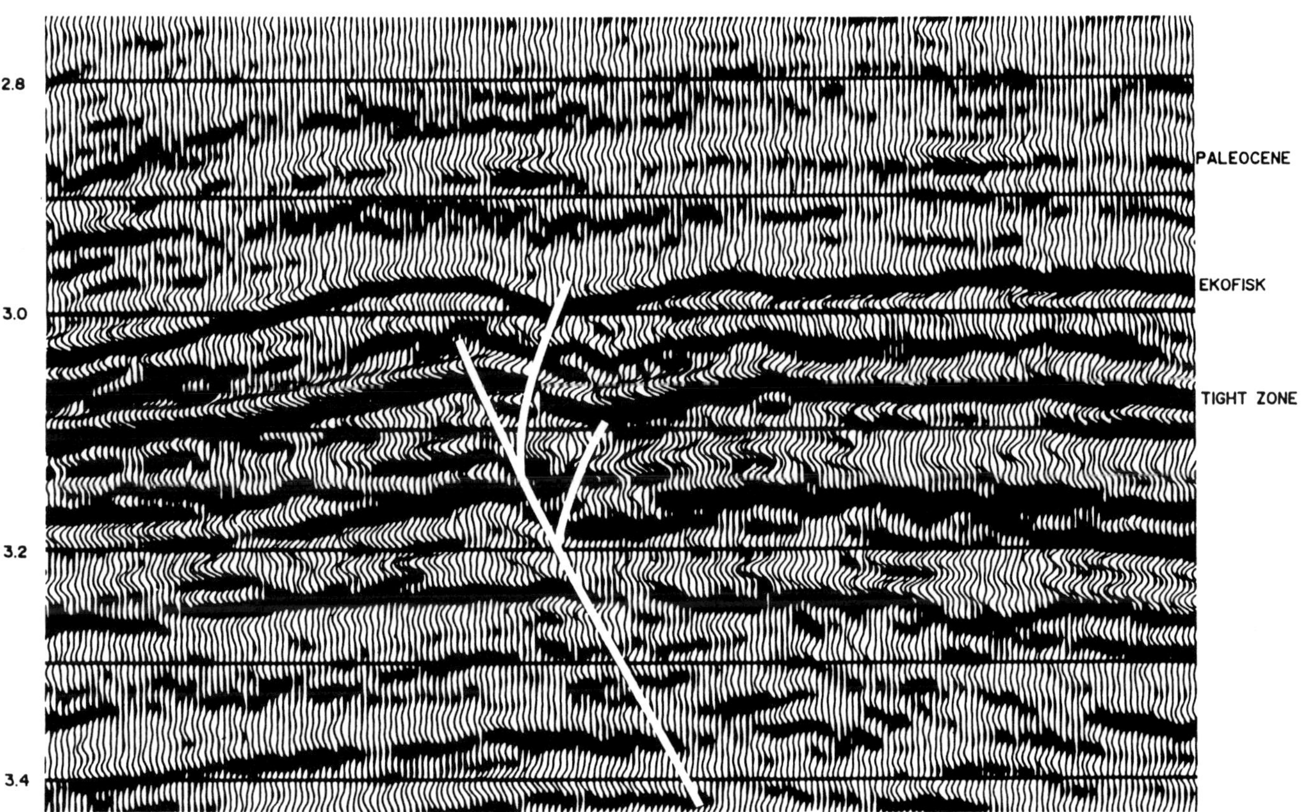

Fig. 16. Portion of seismic line showing definition of faults which do not displace the Paleocene reflection.

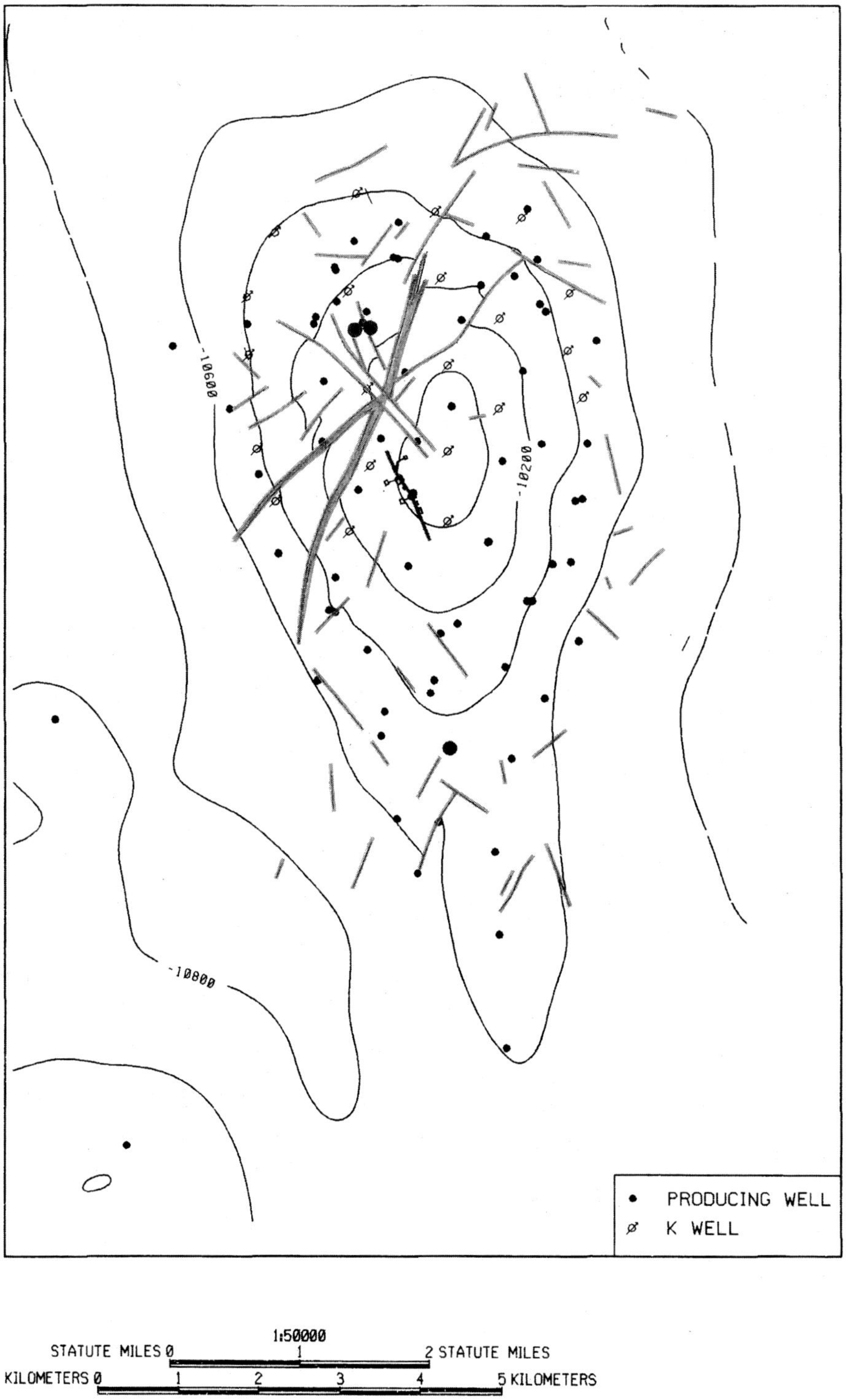

Fig. 17. Final fault map resulting from interpretation of tangential lines.

Acknowledgments

The author thanks the Phillips License 018 group of companies for permission to release this data. The author also thanks Ms. T. Peat for high flying aid during acquisition planning. The paper reflects only the views of the author and not necessarily those of the license 018 group of companies.

References

Brewster, J. and Dangerfield, J.A., 1984, Chalk Fields along the Lindesnes Ridge: Eldfisk: J. of Marine and Petr. Geology, 1, 239–278.

Brewster, J., Dangerfield, J.A., and Farrell, H., 1986, Geology and geophysics of the Ekofisk Field Waterflood: J. of Marine and Petr. Geology, 3, 139–170.

Christie, P. A. F. and Dangerfield, J. A., 1984, Borehole seismic profiles in the Ekofisk Field: Soc. Expl. Geophys. 54th Ann. Internat. Mtg., Expanded Abstracts, 21–23.

Domenico, S., 1974, Effect of water saturation on seismic reflectivity of sand reservoirs encased in shale: Geophysics, 39, 759–769.

Farrell, H. E., 1981, Ekofisk fracture study: Phillips Petr. Co. Internal Report.

Heidbreder, W. L., 1978, European offshore petroleum conference and exhibition: 299–306.

Pekot, L.J. and Gersib, G., 1987, Geology of the Norwegian oil and gas fields: Norwegian Petr. Soc., (Graham and Trotman) 73–87.

Van den Bark, E. and Thomas, O.D., 1981, Ekofisk; first of the giant oil fields in western Europe: Am. Assn. of Petr. Geolog. Bull., 65, 2341–2362.

Wiborg, R. and Jewhurst, J., 1986, Ekofisk subsidence detailed and solutions assessed: Oil and Gas J. (Feb), 47–52.

Ekofisk Field Subsidence Fault Analysis Using 3-D Seismic

*J. A. Dangerfield**

Introduction

At Ekofisk Field the extraction of hydrocarbons has resulted in reservoir compaction of about 8 m. The compaction has been transmitted upward through 3 km of Tertiary shales; the result is a bowl-shaped depression in the sea floor, roughly 4 m deep at its center (Wiborg and Jewhurst, 1985; Dangerfield and Brown, 1985). Comparison with subsidence in the analogous Tertiary shales of the Galveston area suggested subsidence would reactivate any existing faults. Subsidence appears highly unlikely to create new faults.

A three-dimensional (3-D) seismic survey was shot in 1987 with a dense spatial coverage to determine if there was any likelihood of significant Tertiary faults being present. The 3-D survey resolved faults with less than 1 m throw down to a depth of 1 km and proved that no significant faults existed in the area examined. The survey also showed some remarkable glacial and fluvial features, comparable in appearance to those seen in aerial photographs. Their expressions are seen to change with depth on a meter by meter basis. The limits to 3-D resolution in this case are less affected by peak frequency than by spatial density of surface sampling.

The Problem

The subsidence bowl at Ekofisk Field, as measured in 1987, is shown in Figure 1. The bowl appeared very smooth but the level of accuracy of the bathymetric survey was not high enough to resolve any local changes that might be associated with surface faulting. The subsidence bowl at Ekofisk is of a restricted lateral extent not much larger than the area of the reservoir. This information is surprising since the Tertiary is mechanically closer to soil than a rock. The narrow bowl must be explained by near vertical transmission of movement, which raised the possibility of faults controlling the movement. If any faults exist, might they be reactivated and displace the seabottom?

The possibilities of hazards to pipelines and surface structures needed to be investigated. The pipelines

*Phillips Petroleum (Norway), Box 220, Tananger 4056, Norway.

were found to be insensitive to surface faults because they are not rigidly held in place and can accommodate vertical movements by bending. Structures held rigidly in place, like platforms, are more likely to suffer problems, but displacements between legs would have to be at least 15 cm to even begin to cause problems.

A literature survey was conducted to check the likelihood of surface faults being created by compaction and subsidence. The survey revealed 14 oil and gas fields with associated subsidence bowls; their bowls ranged from 0.3 to 8.8 m deep. Faulting only occurred at ground surface in a few cases and then only by reactivation of existing faults.

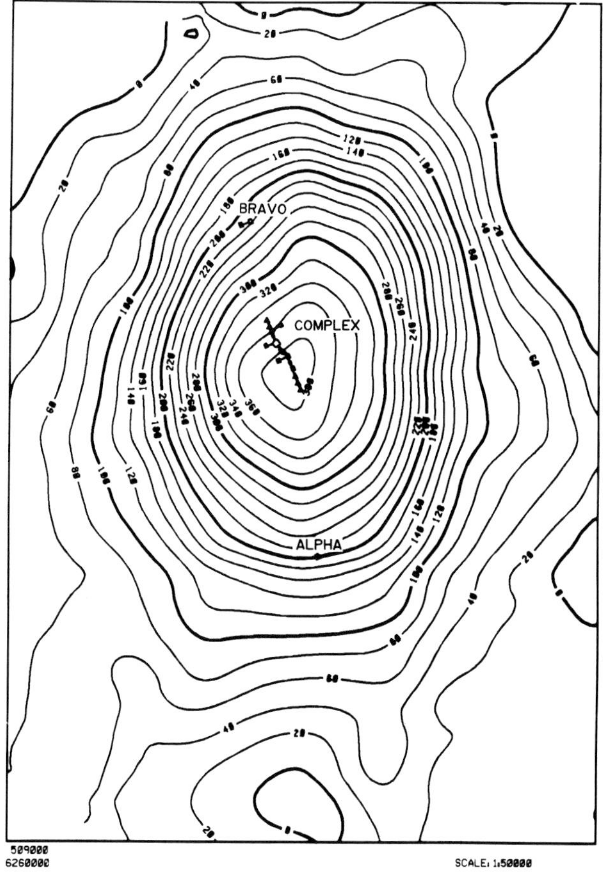

Fig. 1. Subsidence at Ekofisk as of 1987. Contours in centimeters.

Analogy with the Tertiary subsidence bowl around Houston and Galveston suggests that existing faults that are reactivated can break through a thick Tertiary shale sequence to surface, with creep displacements of a couple of centimeters per year or less. It is not considered possible for subsidence to create new faults.

Only fault throws of greater than a few meters, and shallower in depth than 1 km, should have any possibility of reactivation and growth to the seabottom through subsidence. High-resolution sparker, air gun, and water gun seismic lines show a very complex picture in the shallow surface, with ill-defined, discontinuous beds that might, or might not, be faulted on the scale of a meter or so.

Various methods were considered for monitoring surface faulting at Ekofisk, including a microseismic geophone network, a series of tiltmeters on the seabed, and a 3-D seismic survey. The geophone network and tiltmeters were found to be very expensive options, at least ten times the cost of seismic. The 3-D option was much cheaper using the short cable systems originally proposed by Newman in 1984, but might not have sufficient resolution to see small faults with only a meter of throw. It was proposed that very high spatial sampling in the 3-D survey would be required to show any faults. In a uniformly dipping sequence of beds, a fault should be discernible in a time slice as long as the fault displaces a zero-crossing by more than one sample, i.e., 1 ms. This constraint is in contrast to a vertical section where the displacement must be a significant proportion of the wavelet, perhaps 5 ms or so. Figure 2 shows a fault displacement of one sample period and the displacement's expressions in vertical section and horizontal time slice.

Acquisition

Since the two-dimensional lines were so difficult to interpret for small faults, it was uncertain if the 3-D would be successful. We did not, therefore, attempt to examine the whole bowl nor attempt to work beneath the platforms themselves. We were also limited by the presence of the DB102 lifting barge which had 18 anchors out, each with a surface buoy. We examined an area close to the Complex at the center of the field and extending to the Ekofisk Bravo platform in the north. The 3-D volume was 4 x 2 km in areal extent and 1 km deep (Figure 3).

The program was shot using the vessel "Geco Sigma" in the spring of 1987. We used a "flip-flop" system of acquisition, i.e., two separate sources firing alternately, each source being a single P400 water gun. The two guns were spread 20 m apart. A single shot was fired every 10 m along the vessel's track so that there were 20 m between shots for each gun. There were two hydrophone streamers spread 40 m apart,

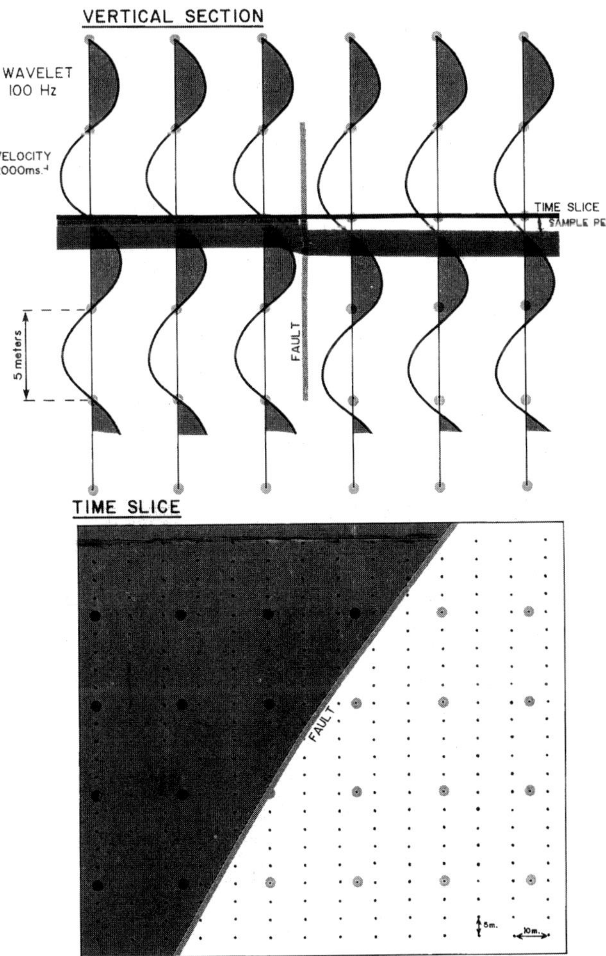

Fig. 2. Schematic of effect of small fault on cross section (above) and time slice (below).

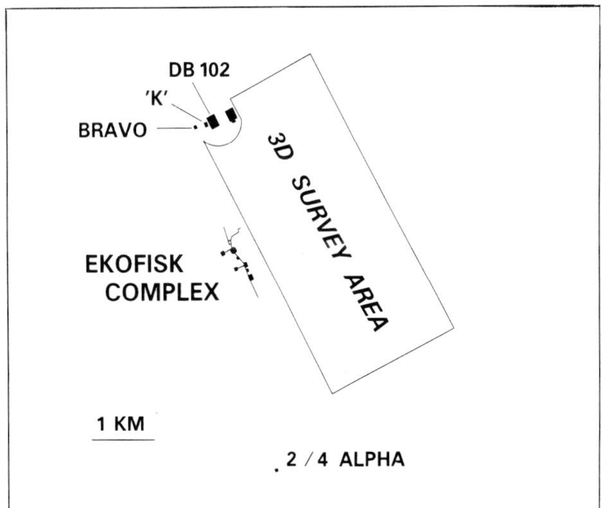

Fig. 3. Location of 3-D subsidence survey oriented along 155° north.

each made up of 48 groups of 5 m each, and towed 2.5 m deep. We used zero in-line offset between the guns and the first hydrophone groups. Four reflection lines were thus collected during each boat pass, with a separation of 10 m between the reflection lines. The bin size was therefore 2.5 x 10 m. The field filter was 27–330 Hz. Navigation was by Microfix, with cable positioning by compass and tailbuoy position checks made by sextant.

Processing

The processing sequence was as follows:

(1) Demultiplex.
(2) Source static correction with individual statics checked by near-trace displays.
(3) Receiver array simulation, 1:2:1 weights after differential normal moveout correction.
(4) Signature deconvolution to minimum phase.
(5) 3-D binning of 6-fold data with 5 x 10 m bins.
(6) 30 dB amplitude compensation to 1 s.
(7) 1500 m/s normal moveout.
(8) Two 120 ms water bottom deconvolution operators with 80 ms gap and 40 ms active operator.
(9) Removal of 1500 m/s NMO.
(10) Residual timing and offset corrections.
(11) Infill missing traces.
(12) Remove 30 dB gain, restore spherical divergence.
(13) Normal moveout correction using one velocity function for the whole survey: 100 ms- 1500 m/s, 550 ms- 1750 m/s, and 1000 ms- 1900 m/s.
(14) Mute.
(15) Stack (6-fold).
(16) Static correction for tides and streamer depths.
(17) 120–80 ms deconvolution.
(18) In-line Kirchhoff summation migration.
(19) Crossline migration.
(20) Zero-phase conversion.

Interpretation

The normal vertical sections taken from the 3-D are not easily interpretable (Figure 4). Small faults may or may not be present on these sections. The time slices, in contrast, show remarkable clarity. Faint events running parallel to the long sides of the rectangle are due to boat tracks but other lineations represent geologic features.

The features in the uppermost 600 ms are most clearly visible because they have suffered only a small amount of compaction and diagenetic change. Several glacial episodes are discernible in the Quaternary section. During pre-Quaternary deposition there was relatively uniform, basinal-shale sedimentation, so few sedimentologic features remain.

The water bottom is approximately 11 000 years old. Assuming the base Quaternary to be 2.5 million years old and 650 ms deep provides an overall average age of 4500 years per millisecond. Allowance for compaction might suggest about 2000 years per millisecond at the seabed and 8000 years per millisecond at the Base Quaternary.

The interval velocities vary with depth from about 1600 m/s at the water bottom to 1750 m/s at the Base Quaternary. The wavelet frequency remains fairly constant at 100 Hz, so a half-wave period (a black peak) is approximately 5 ms in two-way time thickness (Figure 5).

Figure 6 shows both the subsidence bowl in the 3-D area and the positions of pipelines.

The time slice at 99 ms (Figure 7) indicates the curvature of the subsidence bowl by the dark oval area. At the water bottom this area corresponds to a black peak of about 3 m thickness. Subsidence at the center of the bowl was approximately 4 m at the time of data acquisition. The time slice through the water bottom shows pipeline trenches as lineations in the hard sands. The pipeline trenches were originally cut about 1.5 m deep and 4 m wide but would not be visible to a diver since they are filled in. Abandoned trenches are also seen.

The time slices from the water bottom down to 123 ms show clear glacial fluting (i.e., striations) in a roughly east-west orientation; Figure 8 is a slice at 115 ms.

Two small meandering channels can be seen in the northern portion of the time slice at 131 ms (Figure 9) flowing toward the south-southwest, probably produced by thin, decaying ice. In the northern portion of time slice at 133 ms (Figure 10) there is sediment build-up along an old glacier channel edge, which resulted in a distinctly different appearance from the time slice 2 ms higher.

A sharp groove occurs in an "S" shape running northeast-southwest on the time slice from 139 ms (Figure 11). The probability is strong that this groove was ploughed by the basal tip of a wandering iceberg. Shallow submergence of the area about 80 000 years ago is implied.

The narrow channels of fairly constant width on the time slice at 149 ms (Figure 12) are probably glaciation features. They may represent ice margin drainage at the fronts of glaciers at different times. They were truncated by the younger channel in the northeast.

A series of time slices (Figure 13 is the one at 237 ms) shows the development of a river channel running from the northeast to the southwest. The channel itself was probably in a subglacial tunnel and is a couple of hundred meters wide. The ice was probably a few kilometers thick. Within the channel can be seen the course of a river about 20 m wide. The channel is

Fig. 4. Vertical seismic section from 3-D subsidence survey.

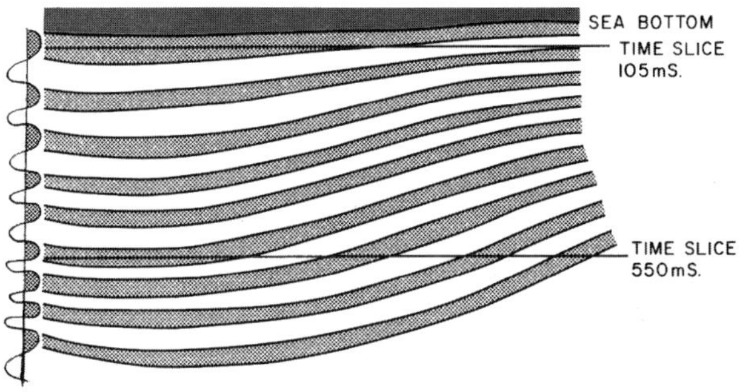

Fig. 5. Sketch of section through subsidence bowl for
100 Hz seismic wavelet.

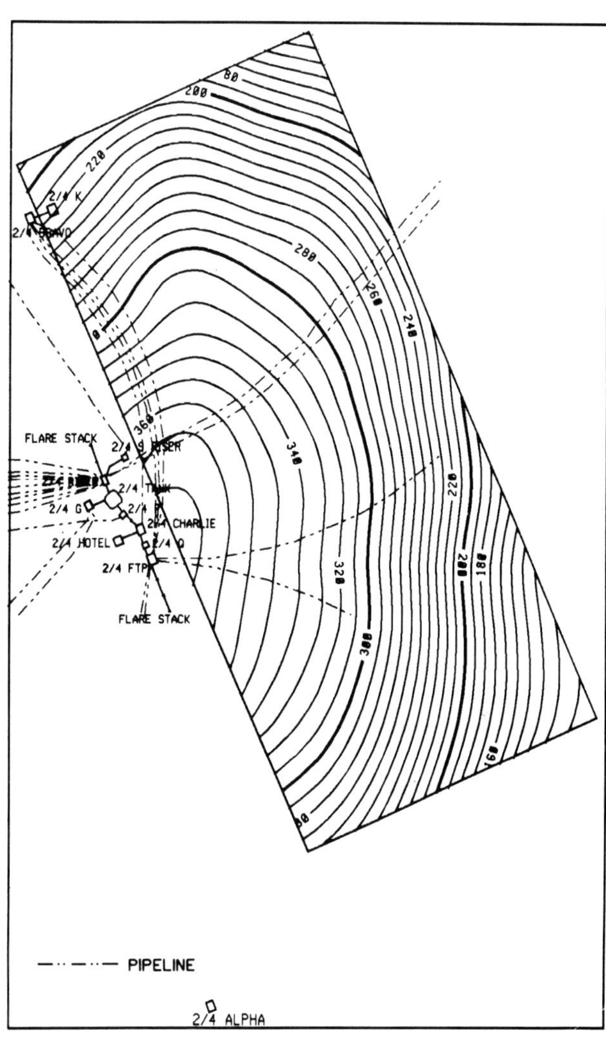

Fig. 6. Water-depth contours over survey area.

roughly 600 000 years old and about 120 m beneath the
current seafloor.

The time slice at 294 ms (Figure 14) is the shallowest
slice in which clearly defined joints appear. These
joints appear as linear features without fault displacements. These, and deeper joints, are made visible by
zones of alteration of acoustic impedance roughly 20 m
wide. The zones might be due to chemical alteration
by ground waters. If this is the case, joints might be
present in the shallower section but invisible to seismic until enough time has elapsed for alteration to
have occurred.

A series of small, subcircular features of high amplitude are seen at 357 ms (Figure 15). These features
appear to occur along lineations that have orientations
north-northeast to south-southwest, similar to the
dominant fracture direction. These might be gas pockets in which gas has collected locally after transmission up through the fractures.

At 555 ms (Figure 16), around 370 m below the
seabed, and below to 1 s there is a depression clearly
expressed by the contours of the time slice. The
depression is about 30 m deep and is centered east
from the Complex. Inspection of the seismic over the
rest of the field outside of the 3-D volume shows the
depression continues deepening toward the southwest.
Small fault dislocations of the bowl are clearly seen.
They represent fault displacements of a few meters.
They are approximately two million years old.

Time slices at 646 ms (Figure 17) and 649 ms (Figure
18) show more small dislocations of the bowl and a
particularly large number of joints. This portion probably represents the base of the Quaternary, roughly
2.5 million years old. Beneath this level the sequence
is of a much more uniform shale deposition in deeper
water, effectively without interesting structure down
to the base of this survey at 1.2 s.

Fault and Joint System

The data set has a resolution limited not by its peak
frequency but by its density of sampling. Fault throws
of less than a meter have been resolved. Joints with
little or no displacement were seen, presumably by
virtue of chemical alteration.

The major faults and joints were mapped as shown
in Figure 19. The general orientations are the same as
those in the reservoir with conjugate systems oriented
in the north-northwest and north-northeast directions,
and dominated by the latter.

Vertical Sections

Copies of two vertical sections from sea surface
down to a depth of 295 ms are included (in-line
sections 50 and 130, Figures 20 and 21). The glacial

(Text continued on page 121)

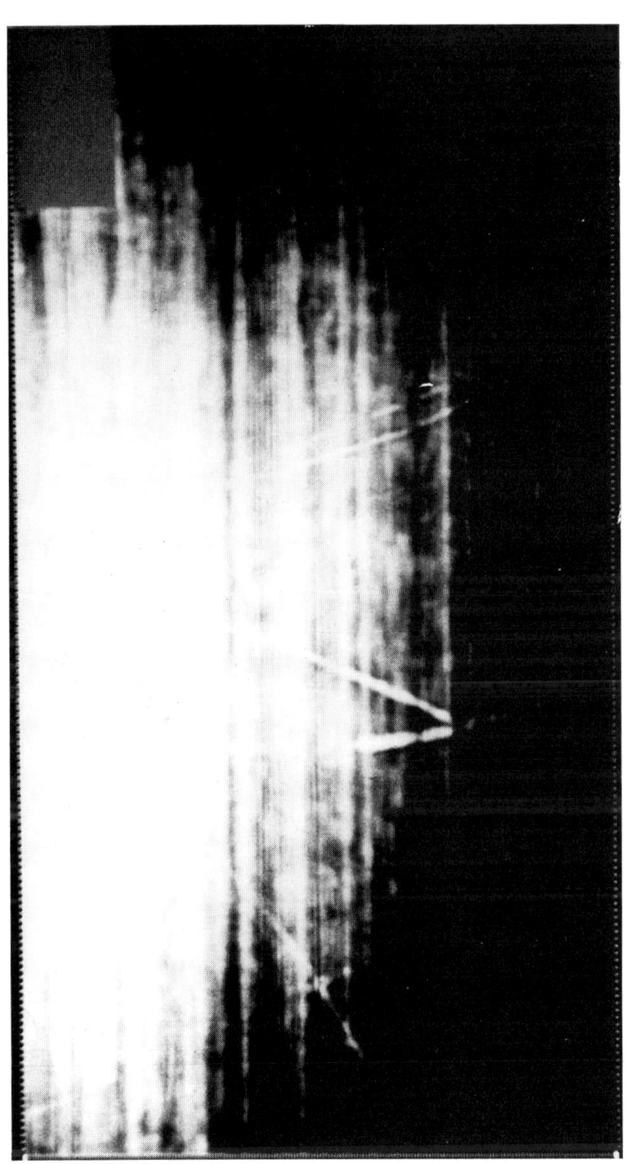

Fig. 7. Time slice at 99 ms (near sea floor).

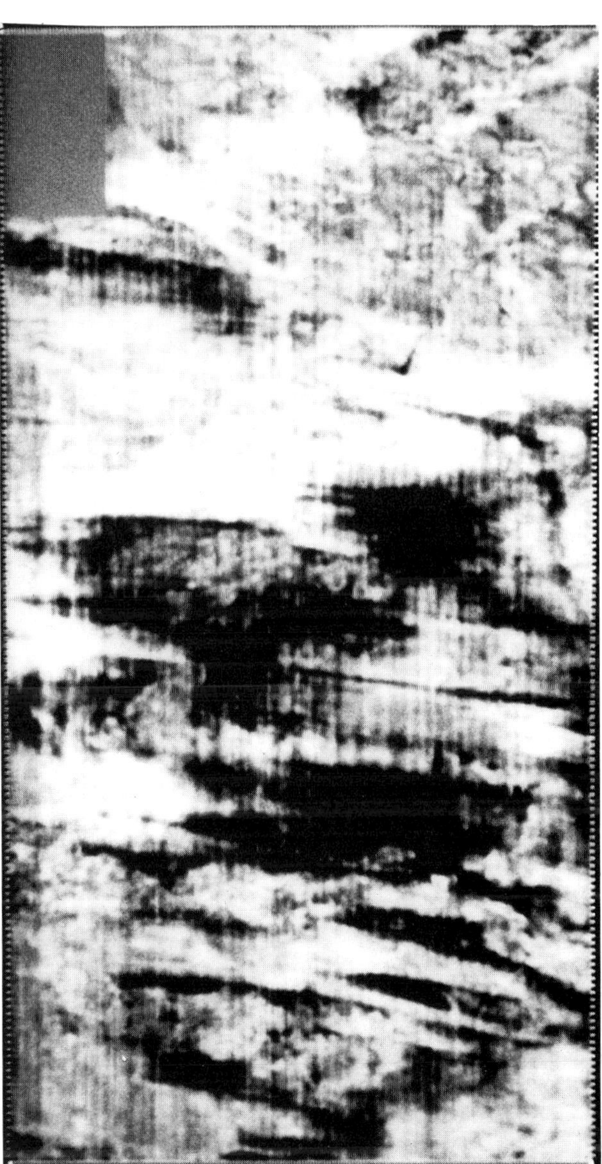

Fig. 8. Time slice at 115 ms showing glacial fluting.

Fig. 9. Time slice at 131 ms.

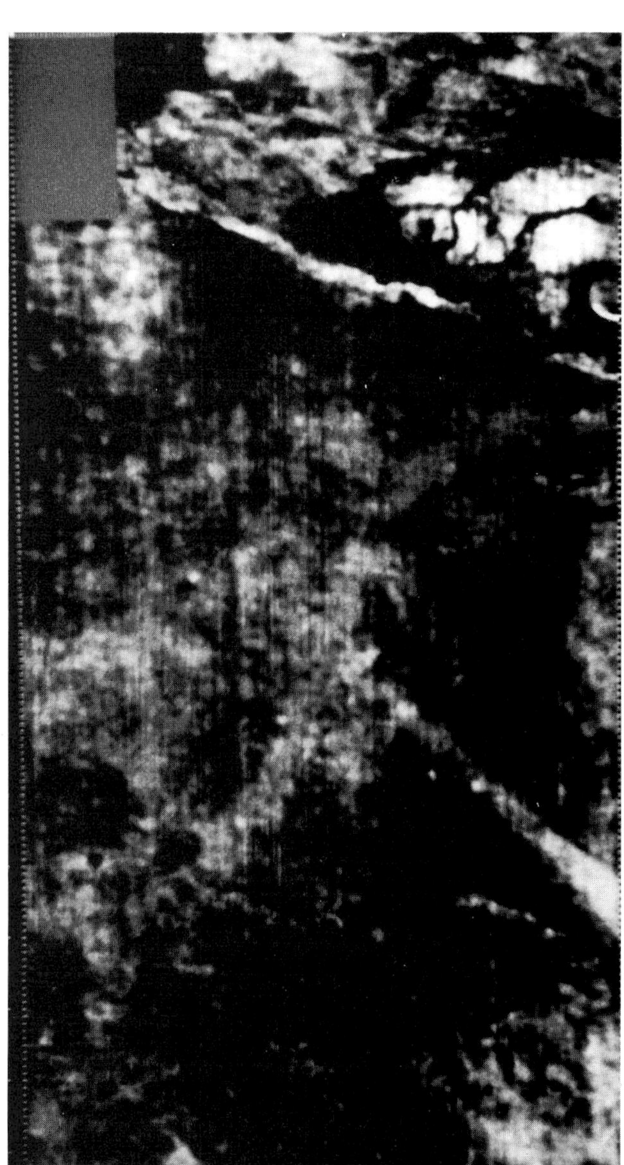

Fig. 10. Time slice at 133 ms.

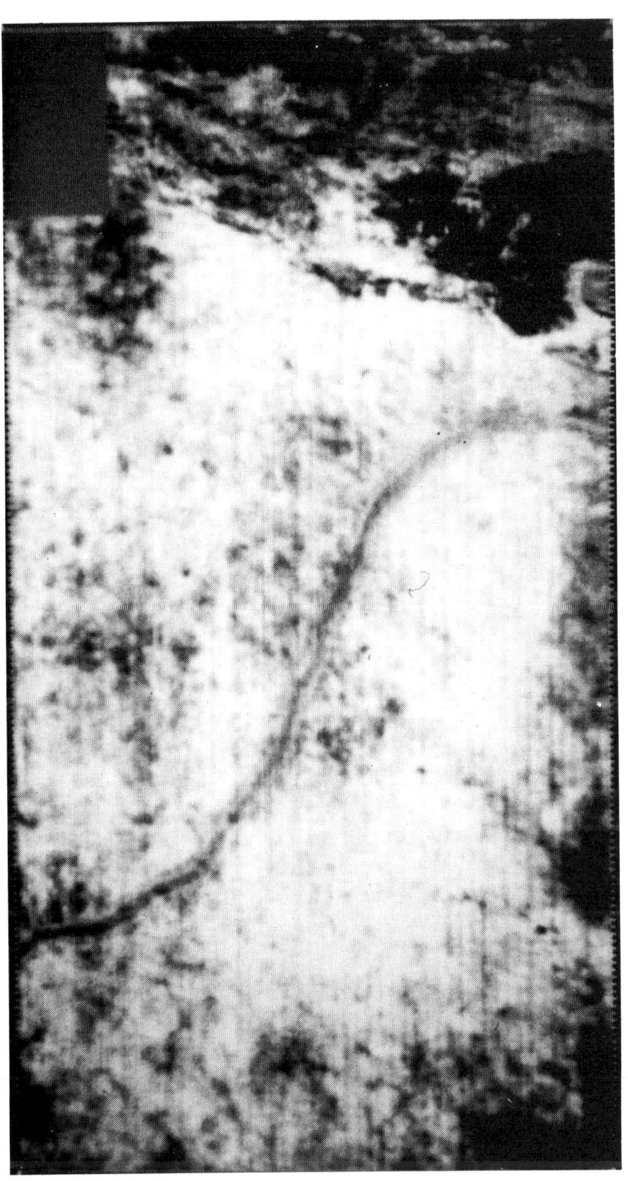

Fig. 11. Time slice at 139 ms showing an S-shaped groove attributed to drag tip of drifting iceberg.

Fig. 12. Time slice at 149 ms.

Fig. 13. Time slice at 237 ms showing a subglacial tunnel with a meandering channel within it.

Fig. 14. Time slice at 294 ms showing jointing without apparent displacement.

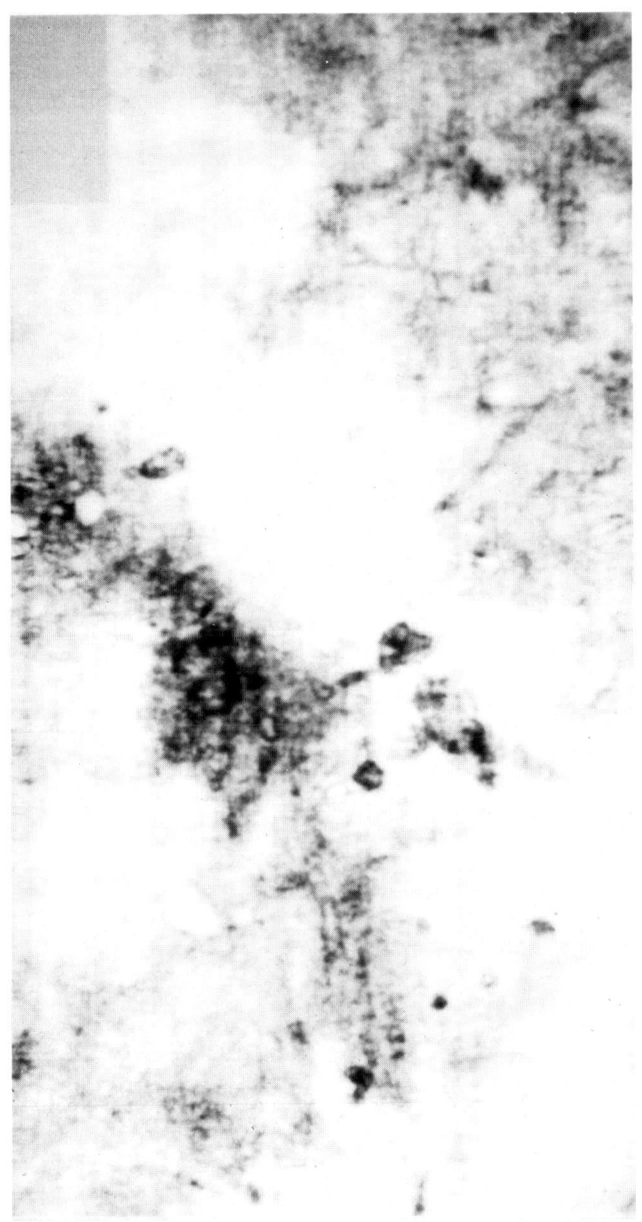

Fig. 15. Time slice at 357 ms showing high-amplitude features, possibly due to gas pockets.

Fig. 16. Time slice at 555 ms showing structural depression and small faulting.

Fig. 17. Time slice at 646 ms.

Fig. 18. Time slice at 649 ms.

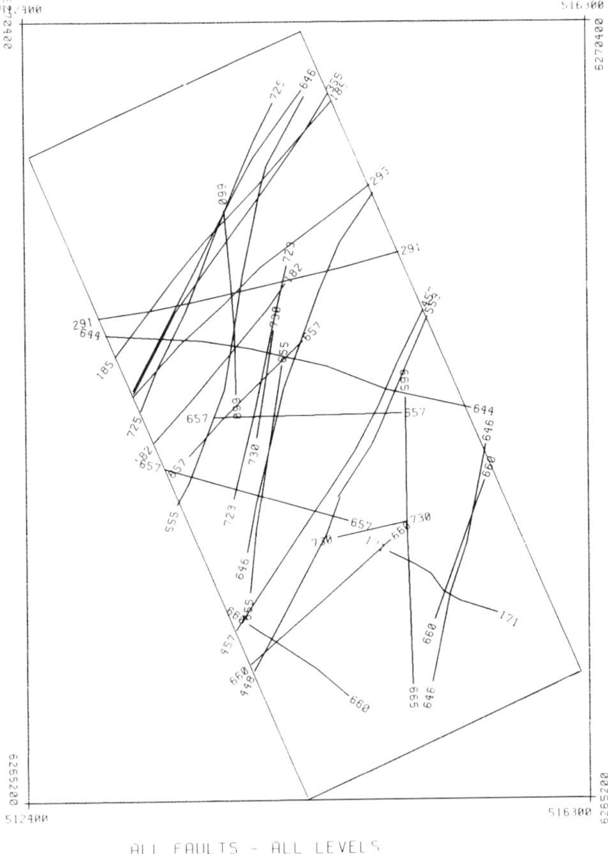

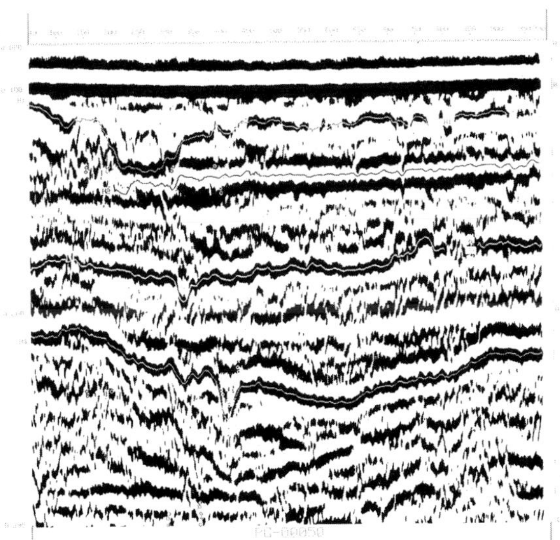

Fig. 19. Map showing major faults and joints.

Fig. 20. Vertical section of line 50 above 295 ms.

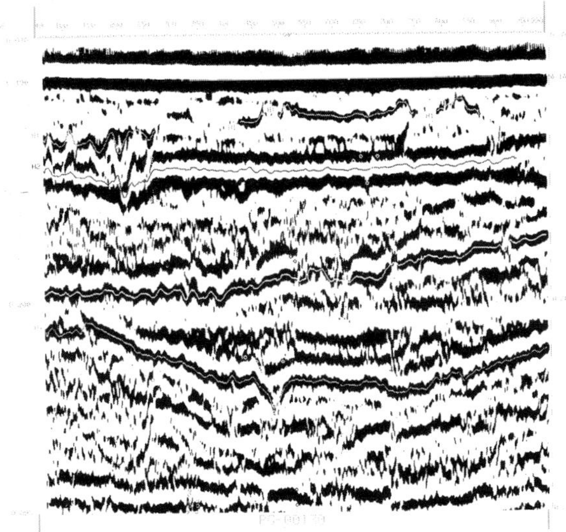

Fig. 21. Vertical section of line 130 above 295 ms.

groove from 133 to 152 ms is partly visible. The subglacial channel from 237 to 250 ms is clear. Most of the other features seen in the time slices are invisible.

Conclusion

Within the area examined there are no significant faults large enough to grow to the sea bottom as subsidence proceeds.

Acknowledgment

The author thanks Adrian Read and Caspar Cronk for their help in identifying the glacial structures depicted. The author also wishes to thank Phillips license 018 group for release of this data set. The views expressed are those of the author alone.

References

Brewster, J., Dangerfield, J. A., and Farrell, H., 1986, The geology and geophysics of the Ekofisk Field waterflood: J. Marine and Petr. Geology, **3**, 139–170.

Dangerfield, K. and Brown, D., 1985, The Ekofisk Field: Proc. N. Sea Oil and Gas Reservoirs: Norwegian Inst. of Technology, 3–22.

Newman, P., 1984, Marine acquisition techniques for precision 3-D surveys: Presented at 46th Ann. Mtg., Euro. Assn. Expl. Geophys.

Newman, P., 1987, 3-D seismic imaging without stack: A case history from the Irish Sea: Presented at the 5th Geophys. Conf. of the Austral. Soc. Expl. Geophys., Expanded Abstracts, 153–157.

Newman, P., 1988, Cost effective 3-D seismic survey during a North Sea winter: 58th Ann. Mtg., Euro. Assn. Expl. Geophys.

Newman, P., 1989, Short-offset 3-D marine seismic applications: 51st Ann. Mtg., Euro. Assn. Expl. Geophys.

Wiborg, R. and Jewhurst, J., 1986, Ekofisk subsidence detailed and solutions assessed: Oil and Gas J., No. 2, 47–52.

Three-Dimensional Data Improve Reservoir Mapping[1]

Piet A. Ruijtenberg, Ray Buchanan,* and Paul Marke**

Introduction

Over the last 25 years, seismic data have increasingly influenced hydrocarbon field development. Analog data before the 1960s had relatively poor resolution and, although often sufficient for exploration and appraisal siting, was generally too coarse and inaccurate to affect development planning significantly. The succeeding digital seismic acquisition allowed for increasingly sophisticated processing and signal enhancement. Further technological improvements in the early 1970s enabled us to recognize a reservoir directly from preserved relative amplitudes and, in favorable cases, to identify reservoir contents. At this time, 2-D seismic data became important in field development. With increased use, however, the limitations of 2-D seismic, caused by inadequate areal coverage and inaccurate reflector positioning, became more apparent. Although seismic wave propagation through the earth is a 3-D phenomenon, all processing and seismic displays assume it to be 2-D, only acting along the axis of the line of acquisition. While this has only a small effect in relatively flat, unfaulted geologic settings, this is not the case in most hydrocarbon fields where out-of-plane energy from adjacent structural elements can cause seismic reflections to be superpositioned and blurred. These weaknesses were recognized in the mid-1970s and led to the first attempts at 3-D surveys. [Dahm and Graebner (1982), Horvath (1985), Ritchie (1986), and Nestvold (1987) provide a complete discussion on 3-D acquisition and processing.]

Today, 3-D seismic surveys are an accepted part of the early data-acquisition process, leading to optimized appraisal sites, refined reserve estimates, and more firmly based development plans. This method provides better control over structural shape (particularly fault orientation and interrelationships) and improves the ability to map stratigraphic variations in detail. The detailed coverage and use of consistent acquisition parameters also make 3-D surveys a powerful tool for investigating seismic amplitude changes. Interactive workstations facilitate rapid display of horizon amplitude maps, and color displays allow subtle amplitude changes to be differentiated. Amplitude changes may be caused by variations in acoustic contrast at top- or intra-reservoir levels that reflect features such as increasing or decreasing porosity and net reservoir development. Changes in pore content may also affect the amplitude and, in some cases, can be used to map hydrocarbon extent or, in the best circumstances, to indicate swept zones from zones of water injection or depletion around producing wells. Minor faults or fractures that are too subtle to note on individual lines often are displayed as linear-amplitude features. Using horizon amplitude maps allows the geologist or seismologist to inject a new level of detail into understanding the field. Future developments in seismic technology will allow an increasing sophistication in our ability to ''see'' at the intrareservoir level.

Growth of 3D Surveys

Figure 1 shows the growth of 3-D surveys for the Shell Group of companies. Initial growth was slowed by technical problems and by the need to justify economically the use of the more expensive 3-D data acquisition and processing methods. Many of the early surveys were acquired in hostile offshore areas, such as the North Sea, where it was easier to demonstrate that significant savings could be made by avoiding

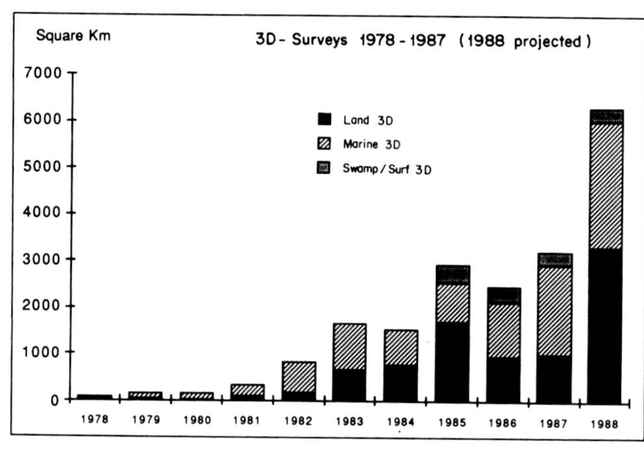

Fig. 1. 3-D surveys by Shell International.

dry-hole locations, thereby reducing (1) the number of appraisal wells needed and (2) the risk in siting development locations that optimized drainage areas. The greatest impact, however, was the ability to match platform size, number of well slots, and production facilities to the more accurately determined field reserves.

These results led to rapid expansion in the use of 3-D surveys from 1981 onward. By 1987, the total number of surveys conducted was about 100. Acquisition of additional areal coverage in 1988 doubled that of 1987. Most of the 1988 surveys were conducted over hydrocarbon fields to provide a better appreciation of reserves and to guide development-well siting. When this can be achieved at a reasonable cost, there is an increasing tendency to include adjacent exploration prospects within the 3-D survey area.

Virtually all these surveys are considered to justify the additional acquisition costs. (Additional cost of 3-D acquisition compared with 2-D acquisition is now very small; typically, a 3-D land survey costs the same as a 2-D survey with 500 m or 1640 ft line spacing (Nestvold, 1987).

The improved definition of 3-D data can change reserve estimates by large amounts. A survey of 10 North Sea fields shows changes in reserves 7 to 81 percent of the amounts predicted from 2-D surveys, with the average change being 36 percent. Obviously, such changes in reserve levels have a major impact on field-development plans and can make the difference between an economic venture and a financial disaster.

A 3-D survey is now almost routine for most new fields and plays an important role in the decision-making process, especially for offshore fields. In a newly discovered field, the benefits of a 3-D survey usually outweigh the cost. This may not be the case in an established field, where the decision depends on the size of the remaining reserves and on the economic impact of potential changes to the development plan; however, many old fields are currently being covered by 3-D seismic surveys. Kluesner (1988) gives an example where a 3-D seismic survey was found to be beneficial, even after more than 200 wells were drilled. The following sections demonstrate how 3-D data can provide better knowledge of structure, reservoir, and the extent of hydrocarbons.

Structure Mapping

The need for 3-D seismic acquisition and processing is more acute where the structure is nonuniformly dipping and extensively faulted. Block IV of the Cormorant field[2] in the U.K. North Sea is an example (Figure 2) (Gaarenstroom, 1984). This large downfaulted block produces from the Middle Jurassic

[2]See also pages 150 to 163.

Brent-sand sequence at about 3000 m (9840 ft), where the reservoir is 80 to 170 m (260 to 560 ft) thick and is eroded on the east. By 1975, three wells had proven the block to be oil-bearing and the interpretation of the close grid of 2-D seismic lines indicated relatively minor internal faulting (Figure 2a). The development plan was for crestal producers with a line of water injection wells at the oil or water contact (OWC) on the west.

When development drilling began in 1982, it became apparent that the block was more complexly faulted than predicted and that these faults acted as barriers, dividing the field into a number of semi-isolated areas. The interpretation of a 1981 3-D seismic survey (Figure 2b) confirmed the complex faulting and showed the isolated blocks to be at right angles to the planned waterdrive direction. The development plan was changed so that producer-injector pairs were located within each fault block. The 3-D interpretation also showed that the fault spacing was, at least locally, of the same order as 100 m (328 ft) line spacing and hence

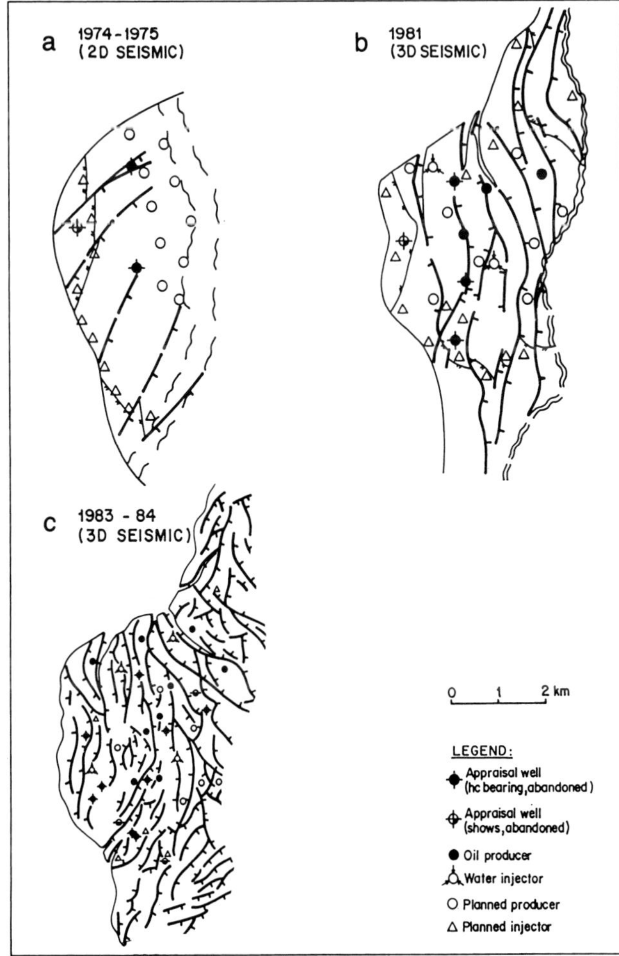

Fig. 2. Comparison of 2-D and 3-D structural maps at Block IV, Cormorant field, U. K. North Sea.

that closer control would be needed to discern the fault pattern fully. A more detailed 3-D survey shot in 1984 with a line spacing of 37.5 m (123 ft) shows Block IV to be even more densely faulted (Figure 2c), and although the dominant structural grain remains the same, the new map has had a considerable effect on the detailed well locations and the reserve-distribution appreciation.

The message from the Cormorant example is simple: plan for 3-D acquisition at an early date so that the interpreted results can be accounted for as early as possible in the development cycle and, for optimal subsurface images, ensure that acquisition parameters use the seismic bandwidth appropriately.

A second example is Thailand's Sirikit field (Figure 3), an onshore field that produces from thin-bedded, Tertiary, fluviolacustrine deposits. The field was discovered in 1981, and it quickly became apparent from both initial appraisal-well results and 2-D seismic interpretation that it was structurally complex. A 3-D survey with a 40 m (130 ft) line spacing was acquired in 1983. Initial interpretation was done from hard copies, and to cope with the huge data set, only every fifth line (the equivalent of a 200 m (660 ft) line spacing) was used (Figure 3a). At a later stage, the survey was reinterpreted on an interactive workstation with all the available data (Figure 3b). Comparison of the two interpretations shows the fault aliasing, which occurs when fault length or frequency is smaller than the line spacing. The "short cut" interpretation shows the faults to be simpler, straighter, and more continuous than indicated by the more detailed examination; hence, the short cut leads to the conclusion that the fault blocks should be isolated, while the detailed interpretation anticipates a degree of communication. The wells would be positioned differently for the two cases.

If 3-D data sets are not fully utilized, especially in complex geologic settings, the result will be a false economy. Because the widespread use of interactive workstations is a recent occurrence, many of the early 3-D surveys were interpreted by hand and only a fraction of the total data sets were used. These older surveys should be reinterpreted. This can be done relatively quickly with automatic-pick software, whereby horizon picks are computer generated by extrapolation from an input grid of interpreted lines. The density of such an input grid is determined by the structural complexity at the objective levels.

Use of Seismic Amplitudes

3-D seismic surveys yield considerably more detail about the reservoir and its contents than conventional interpretation. Seismic reflections are caused by the acoustic impedance (i.e., the product of density and seismic velocity) contrasts between adjacent layers. Assuming a simple case of a thick reservoir overlain by a thick caprock, the signal reflected at the interface is a function of the contrast of acoustic impedance between them. The strength (amplitude) of the seismic reflection can be altered by three principal effects: (1) changes in caprock properties (density, velocity, and lithology); (2) changes in reservoir properties (as above) caused by changes in porosity, mineralogy, or fluid content; and (3) changes in the geometry of the interface such as steep dips, faulting, and fracturing. Because caprocks generally have relatively constant properties over large distances, changes in reflection amplitude are most likely to be caused by reservoir variation and geometry or by both. Use of seismic amplitude data can thus yield valuable information on the reservoir and its contents.

3-D seismic can be used to construct a very detailed amplitude map of a reflector, as shown in Figure 4. Amplitude variations on these maps can be interpreted from calibrations with well data, which, for example, may indicate that low amplitudes always coincide with low porosity in the wells. Alternatively, the shape or patterns on the maps may be interpreted: e.g., the anomaly may have a channel shape, a drainage-pattern shape, etc.

For very thin reservoirs, the amplitude of the corresponding reflection is directly proportional to its thickness, thus complicating the interpretation of amplitude effects. Assuming, however, that reservoir and caprock properties remain constant, amplitude variations can be used to estimate the thickness variations of these thin reservoirs (Brown, 1986; Woock and Kin, 1987).

Minor Faults. In conventional seismic interpretation, faults are recognized as offsets of horizon reflectors. Smaller faults, however, may have such a minute reflector offset that they appear as an inconsequential

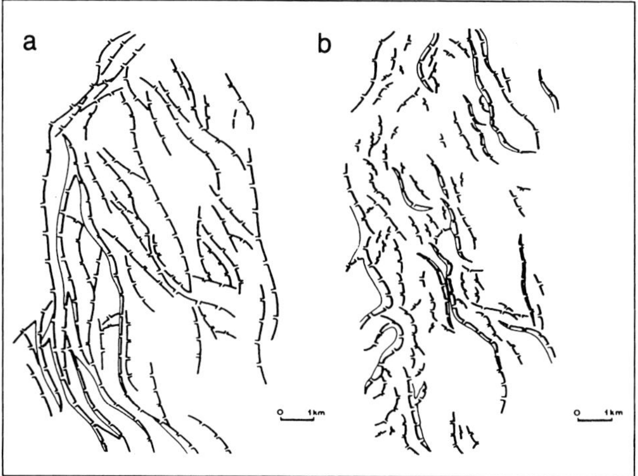

Fig. 3. Comparison of fault interpretation at Sirikit field, Thailand, (a) where 20 percent of the data were used and (b) where all of the data were used.

perturbation. Minor faults, although not directly detected, can be deduced from the local amplitude
reduction that results from imperfect migration, juxtaposition of different lithologies, and interference. A
single line having such an amplitude anomaly is not
significant, but when a number of lines displaying the
anomaly form a linear trend at the top of the reservoir
that is subparallel to the directly mapped fault trends,
the confidence in interpreting the anomaly as a small
fault is greatly increased. Such features can be observed on horizon amplitude maps—i.e., maps of
amplitude variation on a horizon. Figure 5 shows the
modeling principle for amplitude maps like Figure 6.

The solid lines in Figure 6 represent the clearly
defined faults. The offset of the fault causes a break in
the reflection, and the linear-amplitude anomalies subparallel to the faults are considered to be caused by the
minor faults. Confidence in this interpretation is
gained from the observation that such amplitude
anomalies are often seen as the continuation of a
directly mapped fault.

Dip and azimuth (Dalley et al., 1986) are other
attributes useful for mapping small-scale faults. Dip
and azimuth are calculated from a time map (Figure 7)
and then displayed (Figures 8 and 9). Lineaments in
either or both attributes could reveal fault locations.
The time map shows the fault trend; the precise

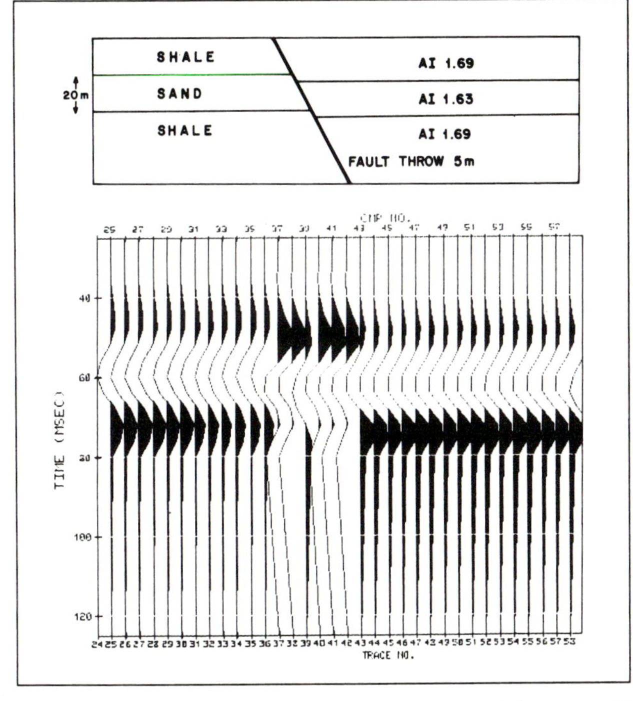

Fig. 5. Modeled amplitude variation caused by a small
fault.

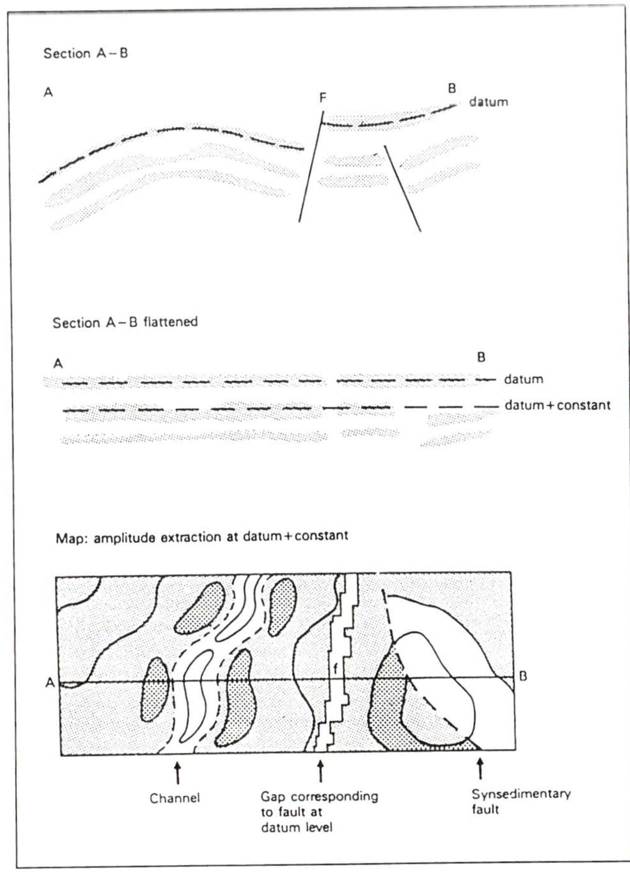

Fig. 4. Principle of the horizon slice or
amplitude map.

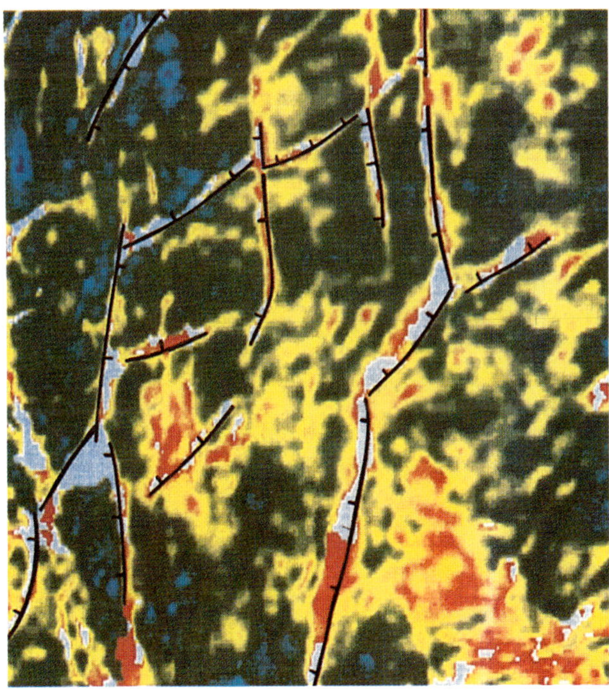

Fig. 6. Amplitude map showing minor faulting expressed as anomalies.

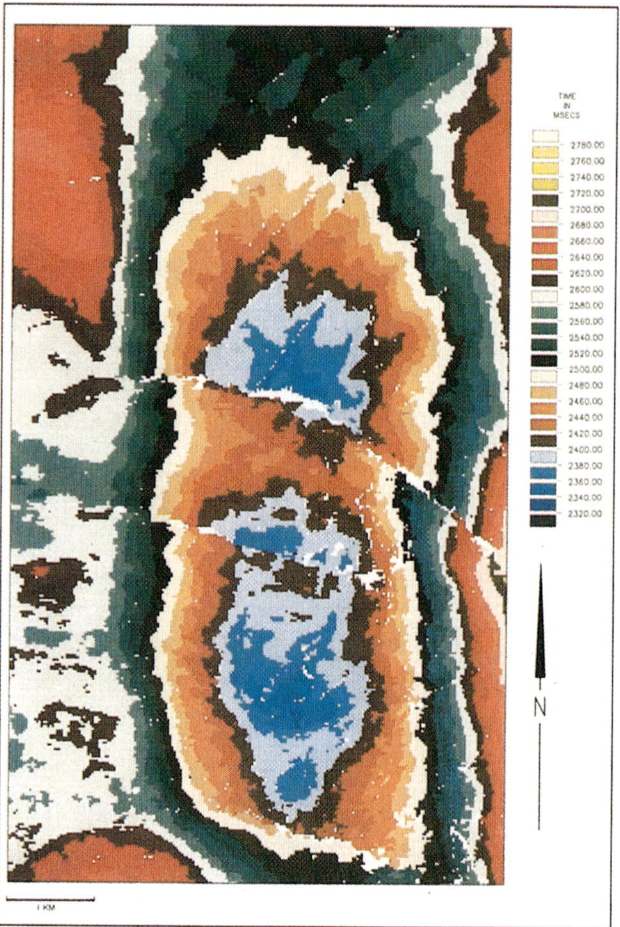

Fig. 7. Time-structure map of faulted anticline.

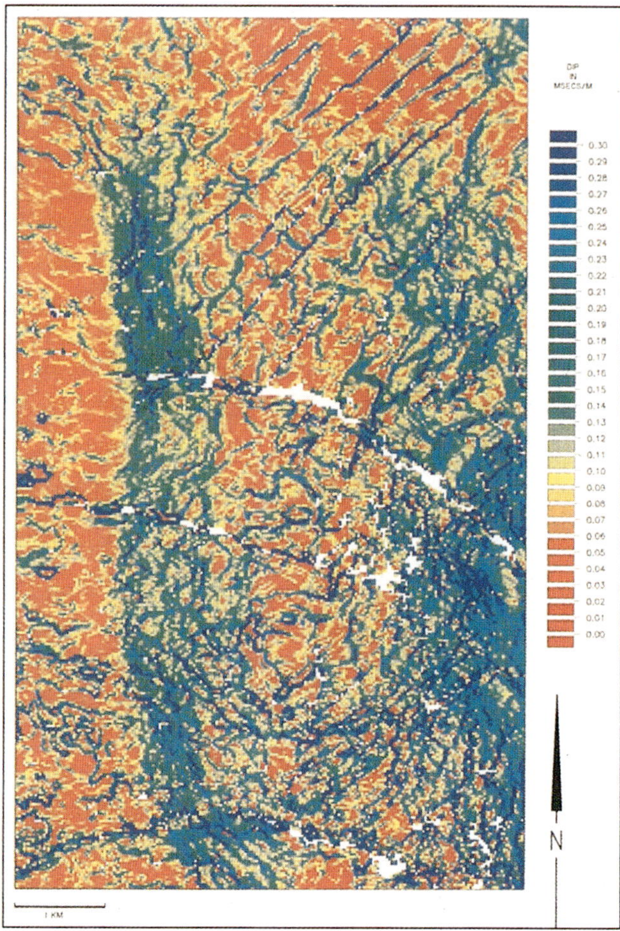

Fig. 8. Dip map calculated from Figure 7.

location and the en-echelon arrangement is clear only on the dip map (Figure 8), especially in the northeast. The azimuth map, displayed as artificially illuminated from the east, gives details that are not apparent on the other maps. Two identifiable fault trends are interpreted to be conjugate fault sets related to transtensional deformation. When combined with well information on fault and fracture orientation, horizon attribute maps (such as amplitude, dip, and azimuth), can provide insight into small-scale reservoir disturbances that can impose a permeability anisotropy or, in extreme cases, that can divide thin-bedded reservoirs into isolated blocks.

Lithological Variation. Use of the horizon slicing technique with an interactive work-station provides a powerful tool for quickly examining amplitudes at different levels within the reservoir sequence (Brown, 1985, 1986). Horizon slices can be constructed for a particular horizon (i.e., top reservoir) and adjusted to the amplitude peak at that level. Deeper levels within the reservoir can be examined by taking new slices at constant time gaps (e.g., 8 ms) below the mapped horizon (see Figure 4). This allows examination of geologically meaningful intervals within the reservoir itself because individual slices can be tied in with well data. Figure 10 shows a modeled case of the decrease in sand-body thickness that arises because

reflections from the top of the sand increasingly interfere with reflections from the unit base. For instance, this situation could occur over a channel sand or, at a larger scale, over a mouth bar, as shown in Figure 11 from the Sirikit field (Flint et al., 1988). The sands of Sirikit are generally very thin, typically less than 5 m [16 ft], and are interbedded with equally thin layers of silt and shale so that the individual reservoirs are below seismic resolution. One of the thicker sands is encased within a 50 m (160 ft) thick lagoonal shale, similar to that shown in Figure 10. The amplitude map in Figure 11 shows that sandstone generates a seismic amplitude that is based on its thickness. Available well controls show the anomaly to be confined to an area of thick (up to 16 m or 50 ft) sand development that decreases (to only 3 m or 10 ft) on the edge of the anomaly.

Channel features can also be discerned from amplitude maps made by horizon slicing. Figure 12, taken from an onshore field, shows a channel-like amplitude trend that appears to be offset by a transform fault. In this case, the interpretation is not confirmed by observation from well data, although it does agree with the geologic setting. An isolated channel reservoir like this would not be drained unless by a dedicated well; however, such studies allow development plans to be adapted to the reservoir distribution.

Fig. 9. Azimuth map calculated from Figure 7.

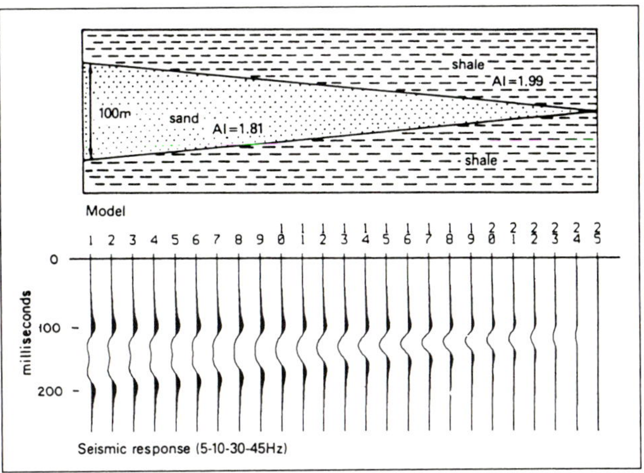

Fig. 10. Modeled seismic response of a sand wedge.

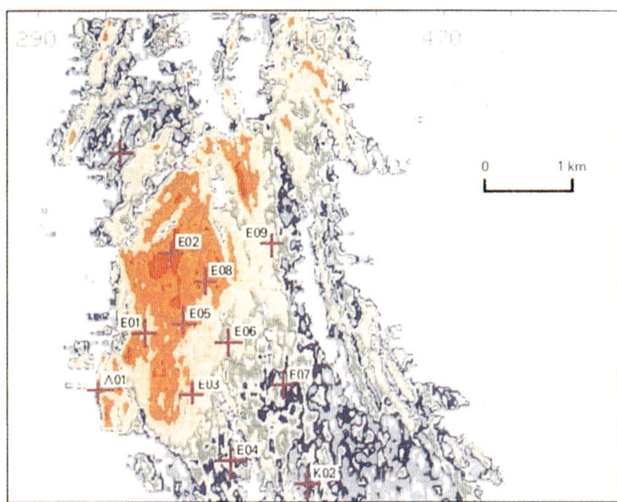

Fig. 11. Amplitude map that is 36 ms below top shale at Sirikit field, Thailand. High amplitude (red) is coincident with regressive deltaic sand body encased in shale. Purple crosses represent well positions.

Fig. 12. Amplitude map showing channel features offset by fault.

Great care is needed with such interpretations because other phenomena can cause similar amplitude variations. Figure 13 shows one possibility, where a very thin coal seam generates a channel-like feature. Figure 14 shows that changes in reservoir porosity can also cause amplitude variation, and hence that diagenetic and depositional changes in the reservoir must be considered. The geologic setting must be accounted for, and it is often necessary to construct theoretical models based on local rock properties to be able to eliminate some of the possibilities and concentrate only on the more probable.

Fluid Content and Porosity. Because changing the fluid content of a reservoir can affect both density and seismic velocity, it follows that changes in fluid content can also affect the acoustic contrast and hence the seismic amplitudes. And, because the porosity of rocks generally decreases with depth, the acoustic impedance of rocks also changes with depth.

Figure 15 shows a typical acoustic-impedance to depth relationship. At shallow depths of about 2000 to 2500 m or 6600 to 8200 ft, water- and oil- and gas-bearing sands have sufficient differences in acoustic impedance to be differentiated on the basis of the seismic amplitude response. Between 2500 m or 8200 ft and depths of 3000 to 4000 m or 9800 to 13 120 ft, the acoustic impedances of oil- and gas-bearing horizons are very similar but are still significantly different from those of water-filled sandstones. In this depth range, hydrocarbon-filled reservoirs can theoretically be separated from water-bearing reservoirs on the basis of the seismic amplitude response, but at deeper levels it is not possible to differentiate between fluid content and amplitude response. [Such curves differ in various areas.–Editor]

In reality, sands are not uniformly thick or constant in properties, and interference from overlying levels may seriously mask or distort the signal so that no contrast remains. However, when fluid-content changes generate amplitude anomalies, they are quickly recognizable by their conformance to structure. An amplitude effect that terminates abruptly at a constant depth is most probably due to a hydrocarbon effect. The strength of 3-D data in this respect lies in the number of these changes observed. A subtle amplitude change observed on a few seismic lines carries little weight; a change seen on tens of lines, however, greatly increases the confidence. The interactive workstation also is invaluable because it enables the most subtle of amplitude features to be enhanced.

Figure 16 is an example from a North Sea field of a gas-bearing, low-porosity-sandstone reservoir at great depth (>4000 m or >13 120 ft)—i.e., below the depth where discernible amplitude effects are usually expected. Yet the horizon amplitude map of the top reservoir clearly shows an amplitude anomaly that coincides with the known gas or water contact. The

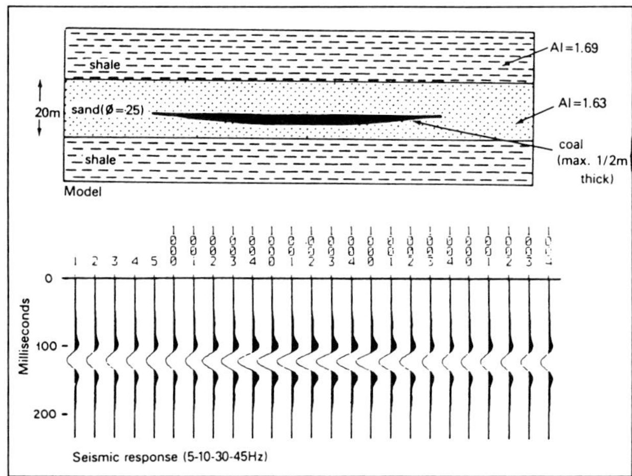

Fig. 13. Modeled seismic response of a thin coal seam.

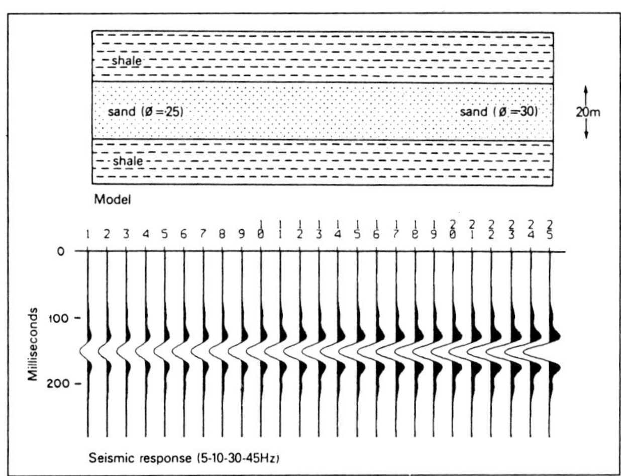

Fig. 14. Modeled seismic response of a reservoir with porosity variation.

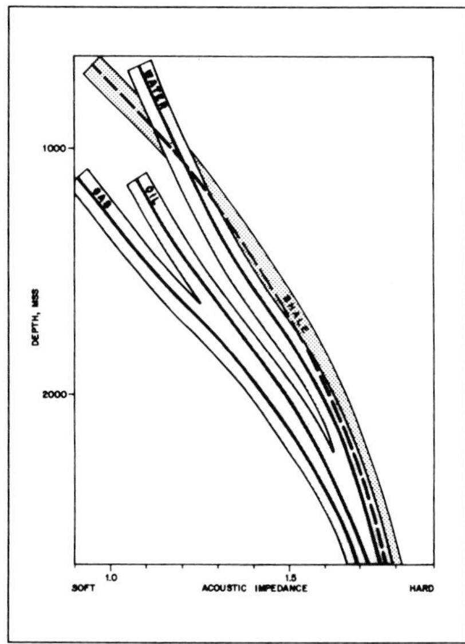

Fig. 15. Acoustic impedance trend curves.

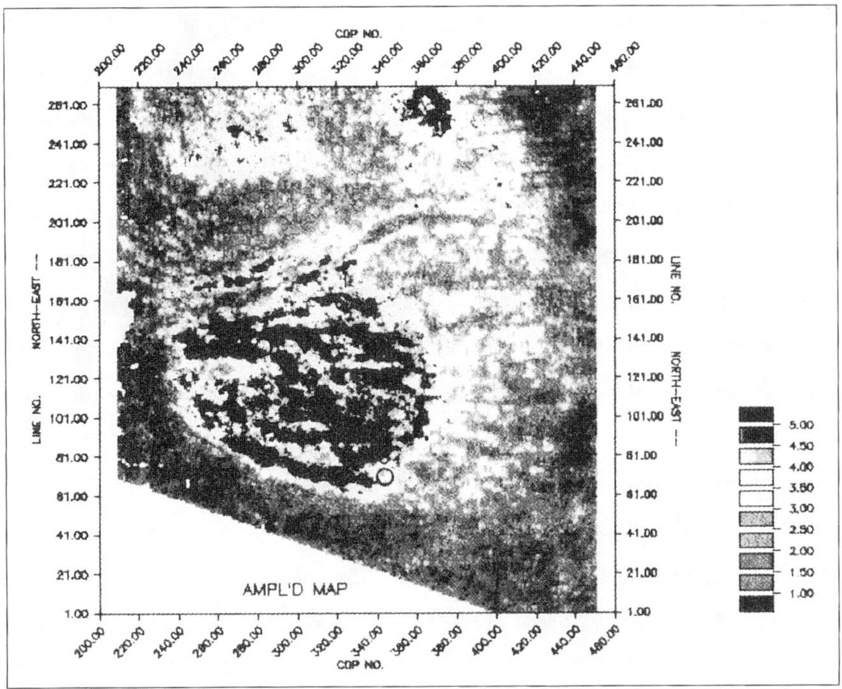

Fig. 16. Amplitude map of a North Sea gas field. Dark circular area coincides with gas-bearing sand.

amplitude pattern is complex and irregular in the east, which is considered to be caused by porosity variation related to the cementation and leaching around faults and not by gas effects. In this example, porosity variations have only a secondary effect on the amplitude variations. It is well known, however, that in many cases porosity is the dominant factor controlling the reservoir seismic response. In such cases, porosity variations can be mapped by studying amplitude variations and calibrating them with well data (Rubbens et al., 1983).

At much shallower depths, where the expected acoustic impedance contrast is even larger because of different fluid contents, one can expect to be able to trace production effects. One may be able to see, for example, the effects of gas-cap expansion or even water injection or gas coning around the wells. In the future, time-lapse 3-D surveys will be more commonly used to monitor the effectiveness of fluid-drainage flood fronts, allowing appropriate adjustments to the development pattern, in line with the observed reservoir response. Although this has been demonstrated to be possible on a very small scale (Greaves and Fulp, 1987), further developments of the technique, in particular shorter turnaround times and cost reduction, are required to allow time-lapse 3-D surveys on a larger scale.

Conclusion

3-D seismic surveys have become a cost-effective tool for mapping hydrocarbon reservoirs and can have a major impact on the volume of reserves estimated.

The greatest benefit, however, is gained from 3-D data that are recorded in a timely manner and completely interpreted, which can be done efficiently with computer-based interpretation systems. In favorable cases, intra-reservoir details (e.g., reservoir sedimentology, porosity, pore fluids, and minor faulting) can be mapped from 3-D seismic data. Knowledge of these details will have a major effect on development strategy and can lead to large cost savings.

Acknowledgments

We thank our colleagues at Shell for their specific comments and help in providing 3-D seismic materials for this paper. Thanks also go to Shell Intl. Petroleum Mij. B. V. for permission to publish this paper and to Koninklijke Shell E&P Laboratorium, Shell U.K. E&P Ltd., Co. Shell de Colombia Inc., Nederlandse Aardolie Mij. B. V., and Thai Shell B. V.

References

Brown, A. R., 1986, Tuning effects, lithological effects and depositional effects in the response of gas reservoirs: Geophys. Prosp., **34**, 623–647.

—— 1985, The role of horizontal seismic sections in stratigraphic interpretations: Am. Assn. Petr. Geol. Memoir 39, paper No. 3.

Dahm, C. G., and Graebner, R. J., 1982, Field development with three-dimensional seismic methods in the Gulf of Thailand: A case history: Geophysics, **50**, 2411–2430.

Dalley, R. M., Gevers, E. C. A., Stampfli, G. M., Davies, D. J., Gastaldi, E. N., Ruijtenberg, P. A., and Vermeer, G. J. O., 1989, Dip and azimuth displays for 3-D seismic interpretations: First Break, **7**, 86–95.

Flint, S., Stewart, D. J., Hyde, T., Gevers, E. C. A., Dubrule, O. R. F., and Van Riessen, E. D., 1988, Aspects of reservoir geology and production behavior of Sirikit Oil Field, Thailand: An integrated study using well and 3-D seismic data: Am. Assn. Petr. Geolog. Bull. 72, 1254–1269.

Gaarenstroom, L., 1984, The value of 3-D seismic in field development: Presented at the 1984 Soc. Petr. Eng. Ann. Tech. Conf. and Exhib., Paper 13049.

Greaves, R. J. and Fulp, T. J., 1987, 3-D seismic monitoring of an enhanced oil recovery process: Geophysics, **52**, 1175–1187.

Horvath, P. S., 1985, The effectiveness of offshore three-dimensional seismic surveys: Case histories: Geophysics, **50**, 2411–2430.

Kluesner, D. F., 1988, Champion field: Role of three-dimensional seismic in development of a complex giant oilfield: Am. Assn. Petr. Geol, Bull., **72**, 207.

Nestvold, E. O., 1987, The use of 3-D seismic in exploration, appraisal and field development: Proc. 12th World Pet. Cong., **2**, 125–132.

Ritchie, W., 1986, Role of the 3-D seismic technique in improving oilfield economics: J. Petr. Tech., 777–786.

Rubbens, I. B. H., Murat, R. C., and van Keulen, J., 1983, Seismic lateral prediction in chalky limestone reservoirs offshore Qatar: Presented at the 1983 Soc. Petr. Eng. Middle East Oil Tech. Conf. and Exhib., Bahrain.

Woock, D. R., and Kin, A. R., 1987, Isopach mapping of gas sands from seismic impedance: Modeled and empirical cases from Ship Shoal Block 134 Field: Am. Assn. Petr. Geol. Bull. **71**, 1143–1151.

Three-Dimensional Seismic Interpretation and Fault Sealing Investigations, Nun River Field, Nigeria

J. D. Bouvier, C. H. Kaars-Sijpesteijn,[‡] D. F. Kluesner,[§] C. C. Onyejekwe,[§]
and R. C. van der Pal***

Introduction

Before 1987, all seismic interpretation in Shell Petroleum Development Company of Nigeria Limited (SPDC) was based on 2-D seismic surveys, which led to considerable structural uncertainty in fields with complex faulting due to inaccurate reflector positioning and aliasing of faults. Planned increases in appraisal and development activity prompted SPDC to acquire 3-D seismic over the major fields. The current plan envisages the acquisition of some 20 3-D seismic surveys by the end of 1990.

Three-dimensional seismic data sets, combined with the use of interactive seismic interpretation workstations, generally provide a superior image of the subsurface compared with conventional 2-D seismic (Brown, 1986; Nestvold, 1987; Buchanan et al., 1988). The denser sampling of the subsurface, typically 25×25 m (82×82 ft), and correct migration of the data result in an improved definition of structure and reservoir geometry.

Many of SPDC's prolific fields are in a semimature stage of development, with current reserve estimates determined from 2-D seismic and well-penetration maps. In addition, recent discoveries have provided considerable impetus for the early appraisal and development of several 8×10^6 to 24×10^6 m^3 (50×10^6 to 150×10^6 bbl) fields. Consequently, 3-D seismic will be acquired to improve hydrocarbon reserve estimates, optimize appraisal drilling, and improve development planning and implementation for both drilling and facilities resources. These surveys, therefore, will provide the foundations for more cost-effective field development.

The fields and structures that will be initially covered by 3-D seismic are those which have (1) large potential for extensions to known accumulations, thus large scope for increased reserves; (2) significant structural uncertainty (i.e., large, complexly faulted fields for which development is expected in the future); and (3) areas of new exploration success for which blanket coverage would greatly aid further exploration, appraisal, and development.

The first 3-D seismic survey acquired by SPDC was in the Nun River field in 1987. This field was discovered in 1960 and is situated in the central part of the Tertiary Niger delta, 100 km (62 mi) west of Port Harcourt and a similar distance southeast of Warri (Figure 1). In this paper, we give an account of the interpretation methods used for mapping and fault-seal studies using 3-D seismic, the possible impact on hydrocarbon reserve estimates, and implications for future appraisal and development.

Geology

The geology of the Tertiary Niger delta hydrocarbon province has been described in several papers (Short and Stäuble, 1967; Weber and Daukoru, 1975; Weber et al., 1978; Weber, 1986; Evamy et al., 1978; Doust, 1988; Knox and Omatsola, 1988). The province is a large, arcuate, wave- and tidal-dominated delta, with deltaic sediments ranging in age from Eocene in the north to Quaternary in the south. The overall regressive sequence of clastic sediments was deposited in a series of offlap cycles that were intermittently interrupted by periods of sea level change. These periods resulted in minor, although sometimes widespread, episodes of erosion or marine transgression. The delta can be divided into three broad lithofacies units. The upper part consists of massive continental sandstones (Benin Formation), which overlie an alternation of paralic sandstones, shales, and clays (Agbada Forma-

©Copyright 1989. The American Association of Petroleum Geologists. Previously published in The American Association of Petroleum Geologists Bulletin, **73**, No. 11, 1397–1414. Reprinted by permission of the American Association of Petroleum Geologists.

*Koninklijke/Shell Exploratie en Produktie Laboratorium, Volmerlaan 6, 2288 GD Rijswijk ZH, The Netherlands.
[‡]Shell Petroleum Development Company of Nigeria Limited, Port Harcourt, Nigeria.
[§]Koninklijke/Shell Exploratie en Produktie Laboratorium, Volmerlaan 6, 2288 GD Rijswijk ZH, The Netherlands. Present address: Shell Petroleum Development Company of Nigeria Limited, Warri, Nigeria.
**Delft University of Technology, Faculty of Mining and Petroleum Engineering, Section of Applied Geophysics, Delft, The Netherlands.

tion). These, in turn, grade downward into predominantly undercompacted, overpressured marine shales, clays, and siltstones with some turbidite sandstones (Akata Formation) (Figure 2).

The progradation of the delta has been dependent on the interaction between rates of subsidence and sediment supply, and modified by synsedimentary faulting. Within the delta, several major growth-fault-bounded sedimentary units are present. These depobelts succeeded one another southward as the delta prograded through time (Figure 2) (Evamy et al.; 1978; Doust, 1988; Knox and Omatsola, 1988). Most of the extensional faulting occurs in the paralic part of the deltaic sequence and has strongly influenced the sedimentation pattern and thickness distribution of sands and shales.

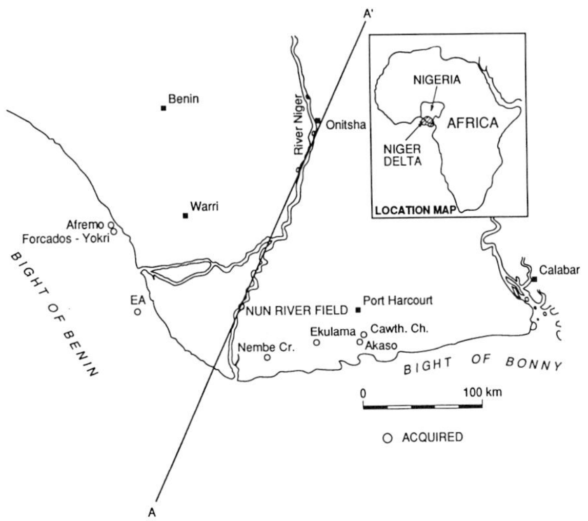

Fig. 1. Niger Delta map showing Nun River field and 3-D seismic surveys of Shell Petroleum Development Company of Nigeria Limited. Names near circles indicate fields.

The Nun River field is situated within a middle to late Miocene depobelt in a large, rollover anticline of a regional southwest-hading growth fault. A large crestal collapse zone, 10 × 5 km (6 × 3 mi) and striking southeast to northwest, is present (Figures 3, 4) and is characterized by numerous syndepositional synthetic and antithetic faults providing fault-dip closures.

The main reservoirs in the Nun River field are fairly simple sedimentologically and lie in the paralic deposits of the Agbada Formation. These deposits are predominantly stacked shoreface sandstones separated by field-wide marine and continental shales (Figure 5). Porosities decrease with depth from about 35 percent at 2288 m (7500 ft) to 18 percent at 4270 m (14 000 ft). Individual sandstones and shales seen in wells are correlatable. However, because the existing wells are confined to the central part of the field, no firm conclusions can be made on the regional paleoenvironmental setting, or on the continuity of sandstone and shale rock units beyond the field boundaries. Within individual depobelts, sandstones and shales are considered to be continuous over a significant area as strong water drives dominate in the nonoverpressured parts of the delta. In Nun River, the shallower D and F sandstones have strong water drives, whereas the deeper G and K sandstones, which are overpressured, have moderate or negligible water drives, respectively.

Historical Development of Nun River Field

The discovery well was drilled in 1960 and found 15 hydrocarbon-bearing sandstone reservoirs (Figure 4). To date, 12 wells have been drilled successfully on the structure with some 40 hydrocarbon-bearing sandstone reservoirs recognized between 2288 and 4423 m (7500 and 14 500 ft). The deepest penetration in the field found oil-bearing sandstones down to 4431 m (14 535 ft), illustrating the deeper prospectiveness of

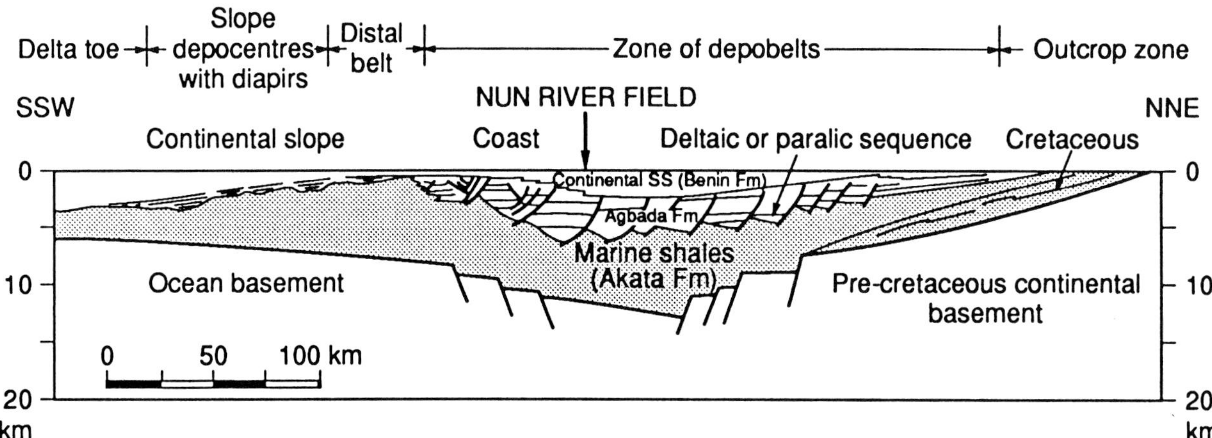

Fig. 2. Schematic cross-section of Niger delta (Doust, 1988). See Figure 1 for location of AA'.

the field. Overpressures are encountered below 3569 m (11 700 ft).

Production was started in 1973 and reached 6400 bbl/day by January 1, 1980. The oil is a sweet, medium to light (31° to 44° API), waxy crude with pour points ranging between 18° and 21°C (65° and 70°F). Current estimates for in-place and ultimately recoverable oil are 39.6×10^6 m^3 (249×10^6 bbl) and 16.2×10^6 m^3 (102×10^6 bbl) respectively, of which 5.1×10^6 m^3 (32×10^6 bbl) have been produced, mainly from the D, F, and deeper overpressured G and K sandstones.

In the 1960s and 1970s, seismic interpretation of the Nun River field was conducted on a widely spaced (2 × 2 km or 1.2 × 1.2 mi) grid of 2-D seismic. In an attempt to improve structural control, infill 2-D seismic was acquired, resulting in an aggregate grid of 0.4 × 1 km (0.2 × 0.6 mi). In 1981, a review of all available

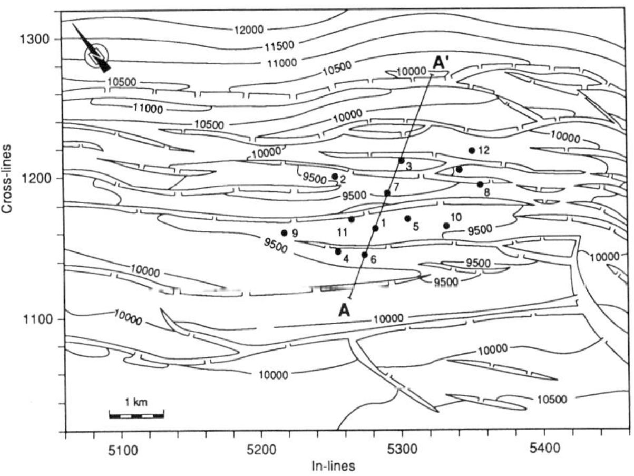

Fig. 3. Schematic E2.0 horizon map. Numbered dots indicate Nun River field wells. See Figure 1 for location of AA'.

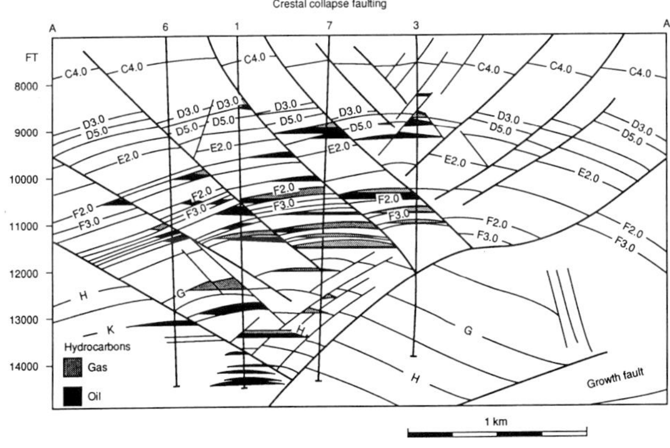

Fig. 4. Schematic geologic cross-section AA' (see Figure 3) through Nun River wells 6, 1, 7, and 3 showing structure and hydrocarbon distribution. Onset of overpressure is between 3569 and 3962 m (or 11 700 and 13 000 ft).

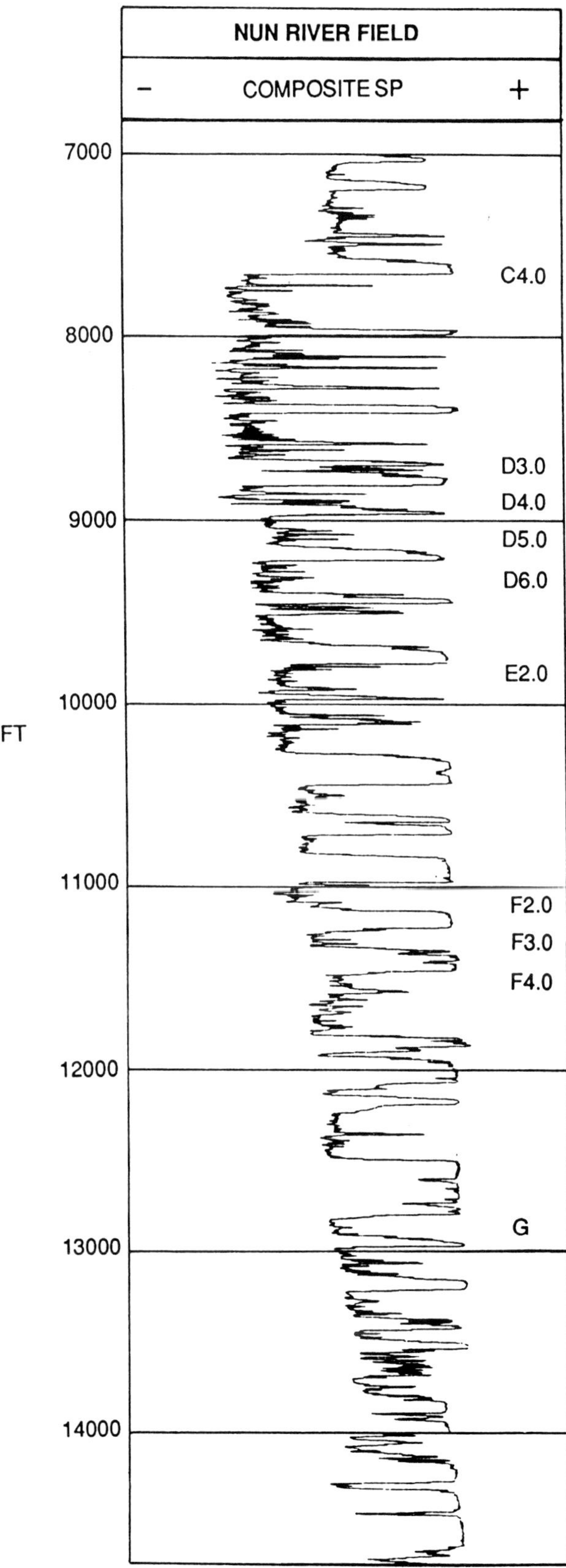

Fig. 5. Nun River field composite SP log. All sandstones are interpreted as stacked shoreface sandstones.

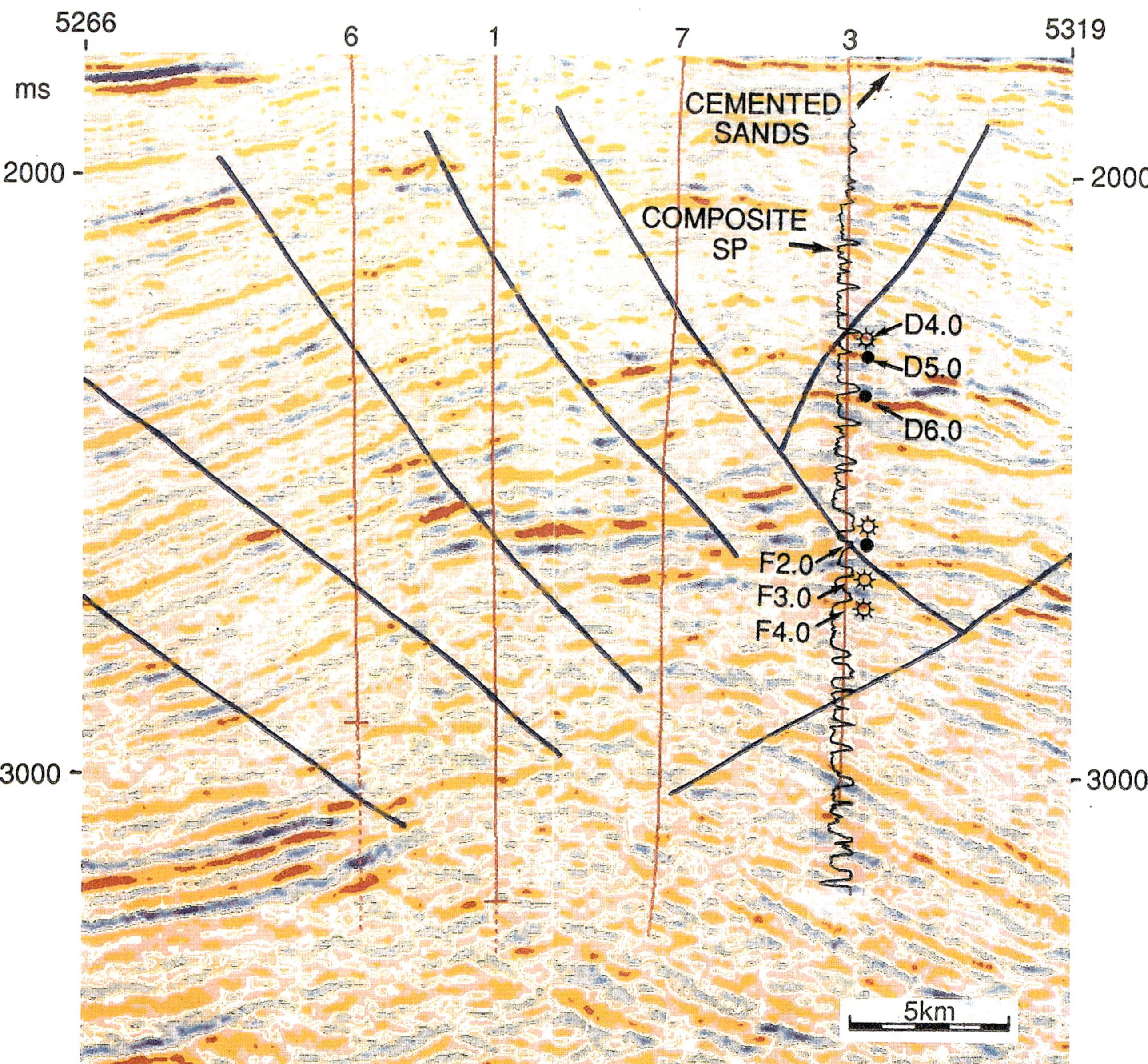

Fig. 6. Seismic section (acoustic impedance) with simplified fault interpretation through Nun River wells 6, 1, 7, and 3 (extension of AA′, Figure 3). Thick sandstones are orange to red (positive amplitude), thick shales are green to blue (negative amplitude). Note seismic to composite SP log correlation. Near well 3, amplitude anomalies (red, indicated by arrows) represent hydrocarbons. High amplitudes near 1800 ms are caused by carbonate-cemented, high-velocity sandstones. Overpressures begin between 2800 and 3000 ms.

2-D seismic lines and the 10 wells drilled up to that time did not provide the level of structural confidence needed to develop the field fully. To gain more information, five appraisal wells, two of which have been drilled, were proposed to delineate the field further. Structurally, these wells showed the field to be unexpectedly complicated. In terms of hydrocarbon appraisal, the wells penetrated significant new hydrocarbon-bearing sandstones. However, in view of the structural uncertainty, it was decided to defer the drilling of the other wells. These structural uncertainties, combined with significant appraisal potential and the production of nearly half of the proven reserves in the field, led to its selection as the first candidate for 3-D seismic.

Quality of Three-Dimensional Seismic Data

The survey was acquired during the fourth quarter of 1986 and the first quarter of 1987. The survey covers an area of 100 km^2 (36 mi^2) centered on the existing wells. Twelvefold coverage was achieved for 75 percent of the surveyed area with a bin spacing of 25 × 25 m (82 × 82 ft). The interpretation was conducted on acoustic impedance data.

Although the seismic is generally good, the quality does deteriorate considerably over the structural crest where the wells are located (Figure 6). This reduction in quality is due to intense faulting and the presence of hard layers, which appear as high-amplitude, discontinuous reflectors, in the overlying continental facies (Figure 6). These layers are thought to be carbonate-cemented sandstones originating from hydrothermal fluids ascending along fault planes. The layers have anomalously fast velocities, 4000 m/s (13 120 ft/s) versus the expected 2600 m/s (8530 ft/s). Their irregular distribution in the collapsed crestal area creates local pull-ups on seismic of up to 76.2 m (250 ft), distorting reflectors underneath. In addition, complex faulting in the crestal area commonly precludes reliable interpretation and mapping below 2.7 s (two-way traveltime).

Seismic-To-Well Match

Seismic-to-well matching was difficult because the scarcity of density logs prevented the generation of reliable synthetic seismograms. Nevertheless, an adequate seismic-to-well match was achieved by correlating time-converted spontaneous potential (SP) logs with acoustic impedance seismic. Gamma-ray logs were not used because some of the sands are radioactive. The tops and bottoms of sandstones on SP logs correspond with zero crossings on acoustic impedance seismic, marking the transition between negative acoustic impedance contrasts (hard shale to soft sandstone) and positive acoustic impedance contrasts (soft sandstone to hard shale). Further calibration is provided by amplitude anomalies corresponding to the main hydrocarbon occurrences, both oil and gas, as seen in the wells (Figure 6).

Interpretation Methods

The Nun River 3-D seismic survey was interpreted on an interactive seismic interpretation workstation. Five key seismic horizons (C3.0, D3.0, D6.0, E2.0, and F2.0 shales) were interpreted and mapped because of their seismic continuity and adequate seismic-to-well correlation. These five key horizons will form the structural framework for further infill horizon mapping. For each horizon, a control grid of northeast-southwest in-lines was defined using every fourth line, i.e., every 100 m (325 ft), except in structurally complex areas, for which every other line, or every 50 m (160 ft), was interpreted (Figure 7). Only the horizons were interpreted, fault cut-outs were left as blanks. Line-to-line interpretation was performed semiautomatically by transferring horizon picks from line to line and editing interactively. Crosslines and arbitrarily oriented lines were used to check and improve the consistency of the interpretation.

The interactively interpreted control grid was used as input to an automatic tracking program. The program first interpolated the control grid, extended the interpretation to each bin point, and, in turn, centered the horizon pick on an amplitude extremum. Fault cutouts were interpreted as part of the horizon. In addition to horizon picking, the automatic tracking procedure was used to extract seismic attributes (e.g., amplitude, instantaneous phase, etc.) along x and y coordinates of relevant seismic loops for future analysis.

Dip and Azimuth Displays Applied to Fault-Pattern Mapping

To enhance the structural interpretation, instantaneous dip and azimuth displays (Dalley et al., 1989) were created for each mapped horizon. These displays reveal the magnitude of local dip (inclination of a horizon in the subsurface expressed in ms/m) and azimuth (measured as an angle in degrees from a local reference direction) for each horizon pick (Figures 8, 9). The displays are used to map subtle trace-to-trace vertical time shifts, which can be less than the sample rate of the data. These features may include minor faults, flexures, or stratigraphic features such as channels. Subtle trace-to-trace shifts might be overlooked or inconsistently picked in interpretations using manual digitization. This chance for error justifies using sophisticated automatic-tracking programs. However, the full benefit of this technique can be exploited only with 3-D seismic. Similar shifts might be created on 2-D seismic by static, noise, or migration effects. By studying the distribution of these features, the resolu-

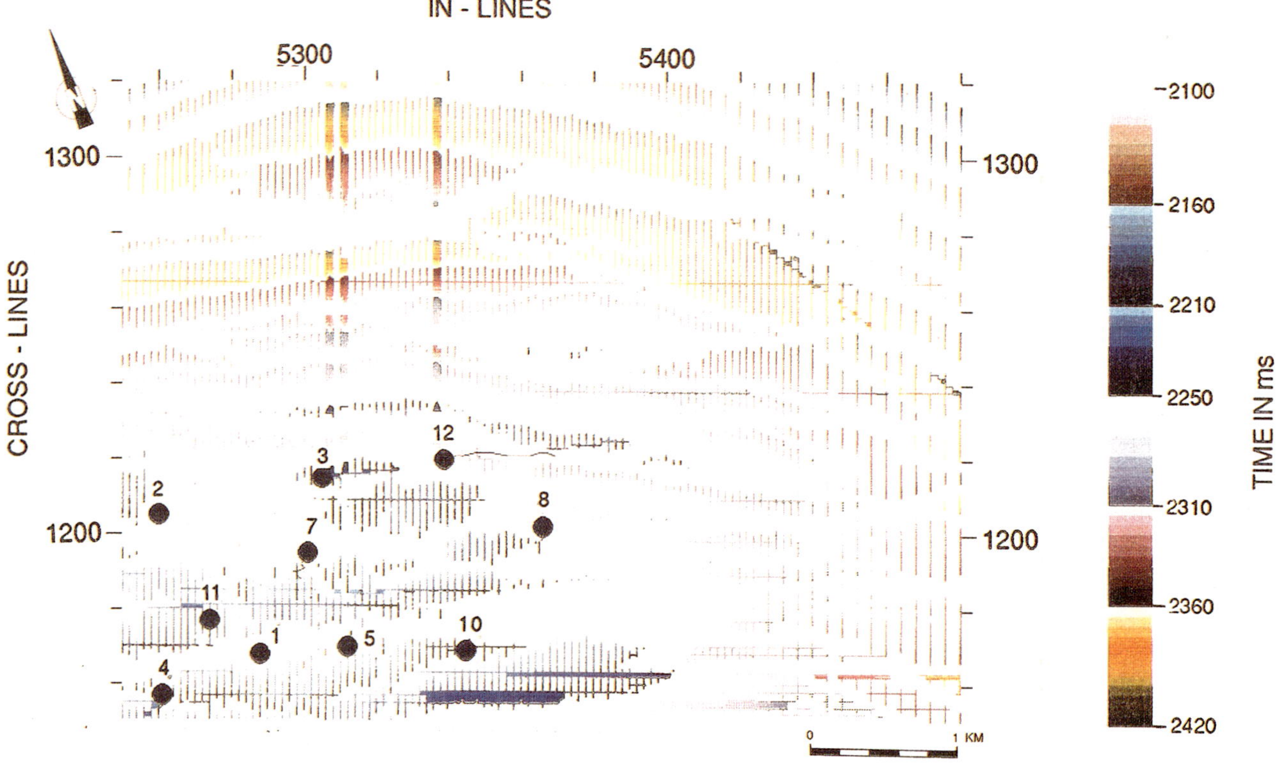

Fig. 7. D3.0 shale map, interpreted control grid of 3-D seismic.

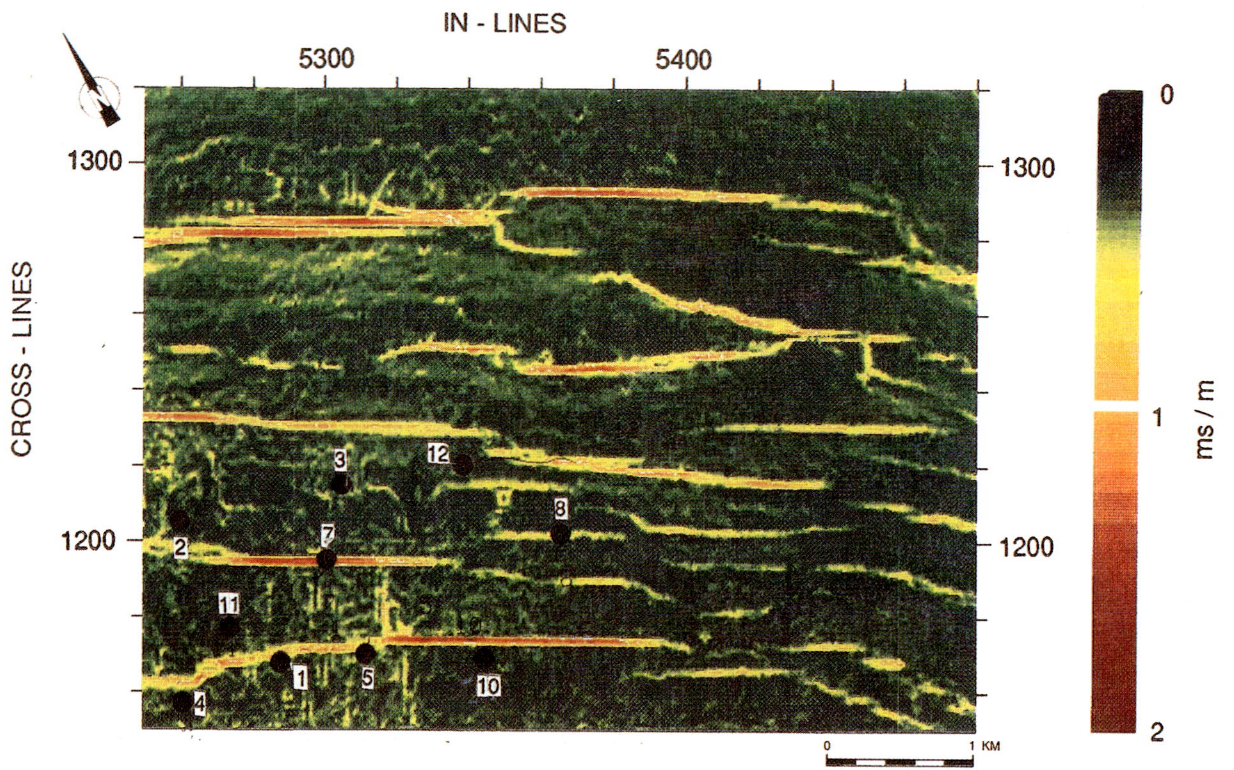

Fig. 8. D3.0 shale map, 3-D seismic dip display. Fault dips are approximately 0.5 to 1.5 ms/m (2-way traveltime) or 35° to 65°.

tion of 3-D seismic allows us to distinguish geologic lineaments from geophysical artifacts.

Dip and azimuth displays were used extensively in the structural interpretation. On an azimuth display (Figure 9), a triangular domain of northeastern dip protrudes into the southwestern flank of the Nun River field anticline. In this area, the continuity of faulting and direction of fault had changed rapidly over a very short distance. This discontinuity of faulting easily can be missed on a grid of interpreted in-lines and cross-lines. This complex fault system is believed to have developed contemporaneously with, and to accommo-date the stress field created by, a major north-south oriented fault whose tip can be seen in the southeast-ern corner of Figure 3.

Comparisons among the control grid of interpreted seismic lines (Figure 7) and the dip and azimuth displays (Figures 8, 9) show that each map contains specific information not available on the other two. These three sources of information were graphically integrated to delineate fault patterns. Figure 10 shows the increased structural resolution from dip processing combined with line interpretation, and Figure 11 shows the azimuth display combined with the dip display. The fault pattern was mapped from a combi-nation of these displays and integrated with an auto-matically picked horizon map (Figure 12). This proce-dure was conducted for each of the five key horizons. Finally, the vertical consistency of the mapped fault pattern was checked and interactively corrected.

Dip and azimuth computations have limitations. If the difference in dip between the fault scarp and the horizon is minimal, possibly because of drag into the fault, the fault is not visible on a dip display (Figure 10). If fault throw and heave are very small, the dip of a fault scarp may be aliased (i.e., appear smaller than it really is) and be indistinguishable from the dip of the horizon. Likewise, if a fault scarp has the same azimuth as the horizon it intersects, the fault will not be visible on an azimuth display (Figure 11).

Infill-Horizon Mapping

Infill-horizon maps, or maps located between key seismic horizons, are generally produced by manually adding isochore thicknesses to the key seismic hori-zons. This method is time consuming and is commonly inaccurate, particularly when sedimentary growth, fault dips, and patterns change rapidly with depth. The flexibility provided by interactive interpretation work-stations is such that infill horizon maps can be derived quickly by automatically adding isochore thicknesses on the workstation. The newly created infill horizon is still an approximation because it is a surface parallel to the key seismic horizon from which it was derived. Automatic tracking is used to focus the new horizon onto the correct loop and to fine-tune the interpreta-

tion. Dip and azimuth displays, and displays of other seismic attributes, such as amplitude and area, are then applied to map reservoir quality and/or pore fill (Figures 13, 14).

Amplitude Mapping

In Nigeria, sandstones have a comparatively lower acoustic impedance than shales. The seismic acoustic impedance data were processed such that sandstones have positive amplitudes (orange, Figure 6), and shales have negative amplitudes (green to blue, Figure 6). Hydrocarbon-bearing sandstones have an even lower acoustic impedance than water-bearing sand-stones; likewise, they have an increased, positive amplitude (red, Figure 6). Generally, in the Nun River field, seismic anomalies (red) are related to the pres-ence of hydrocarbons. The absence of a seismic anom-aly in a closure does not preclude the presence of oil.

Amplitude anomalies broadly conform to structure with their base congruent to an isotime in blocks 2 and 3 of the F2.0 horizon map (Figure 13). In this case, the base of the amplitude anomalies correlates with known oil/water contacts from wells (Figure 14). The gas/oil contact cannot be observed on seismic. Indeed, acous-tic impedance trends versus depth computed for Nun River water, oil- and gas-bearing reservoirs indicate that the acoustic impedance contrast between the light F2.0 oil (39° API) and water is significantly larger than between oil and gas. This effect is further accentuated in this case by the high gas/oil ratio of the F2.0 oil (32.5 m^3/STB or 1148 ft^3/STB). The anomaly encompasses discontinuous faults in blocks 3a, 3b, and 3c, indicat-ing a common oil/water contact in these sub-blocks. This area was previously mapped on 2-D seismic as three separate fault blocks, with separate hydrocarbon accumulations.

Amplitude anomalies that broadly conform to struc-ture can be used to fine-tune structural depth contours locally (Enachescu and Demers, 1988). For example in block 3c, a structural time contour parallel to the base of an anomaly in the east diverges away from the anomaly toward the western part of the block (Figure 13). This divergence from the structural contour is the result of a local pull-up created by the anomalously high-velocity carbonate-cemented sands in the upper part of the paralic sequence. Because well data indi-cate that the oil/water contact is flat, the edge of the amplitude anomaly follows an isodepth contour. This information was used to correct locally the contours derived from a time-to-depth conversion (Figure 14).

Conversely, a strong divergence of an apparent amplitude anomaly from structure may indicate an-other cause for the anomaly other than hydrocarbons. The positive amplitude anomaly in block 1 (Figure 13) strongly diverges from structural contours and is likely

(Text continued on page 141)

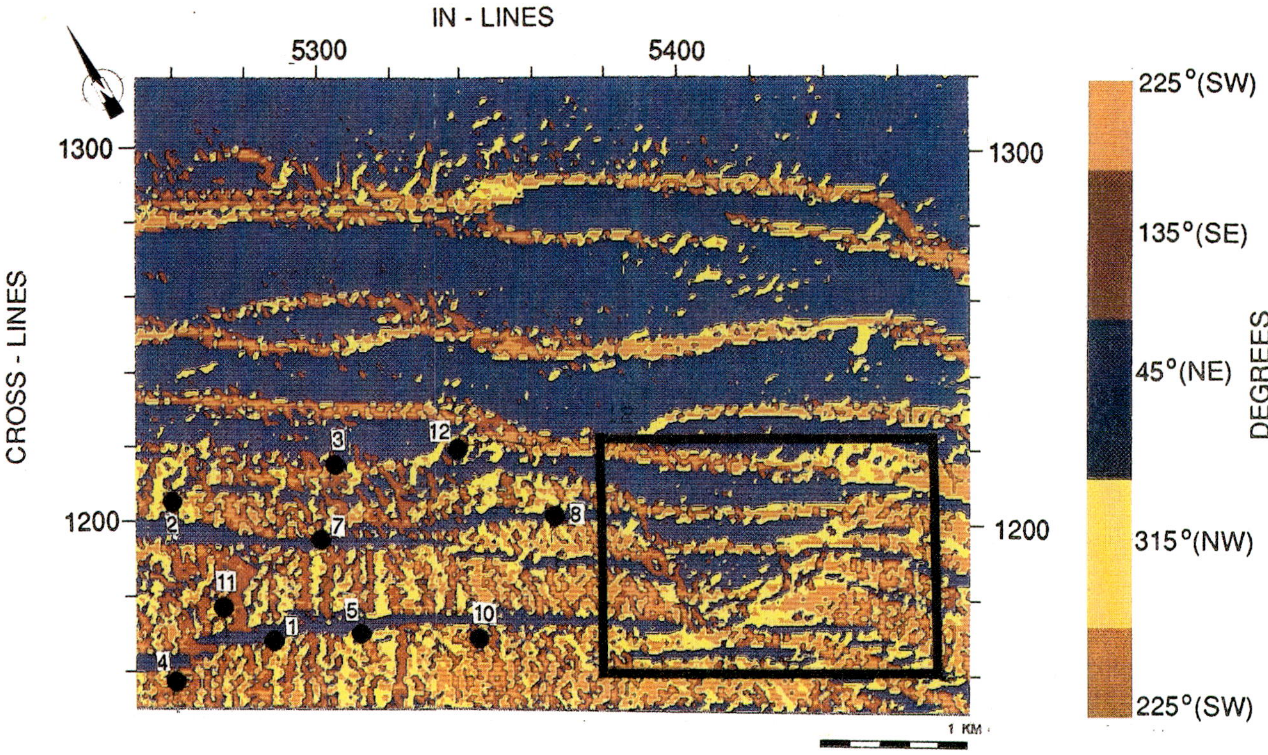

Fig. 9. D3.0 shale map, 3-D seismic azimuth display. Black box highlights triangular protrusion of northeastern dips into southwestern flank of Nun River anticline and rapid change in fault dip.

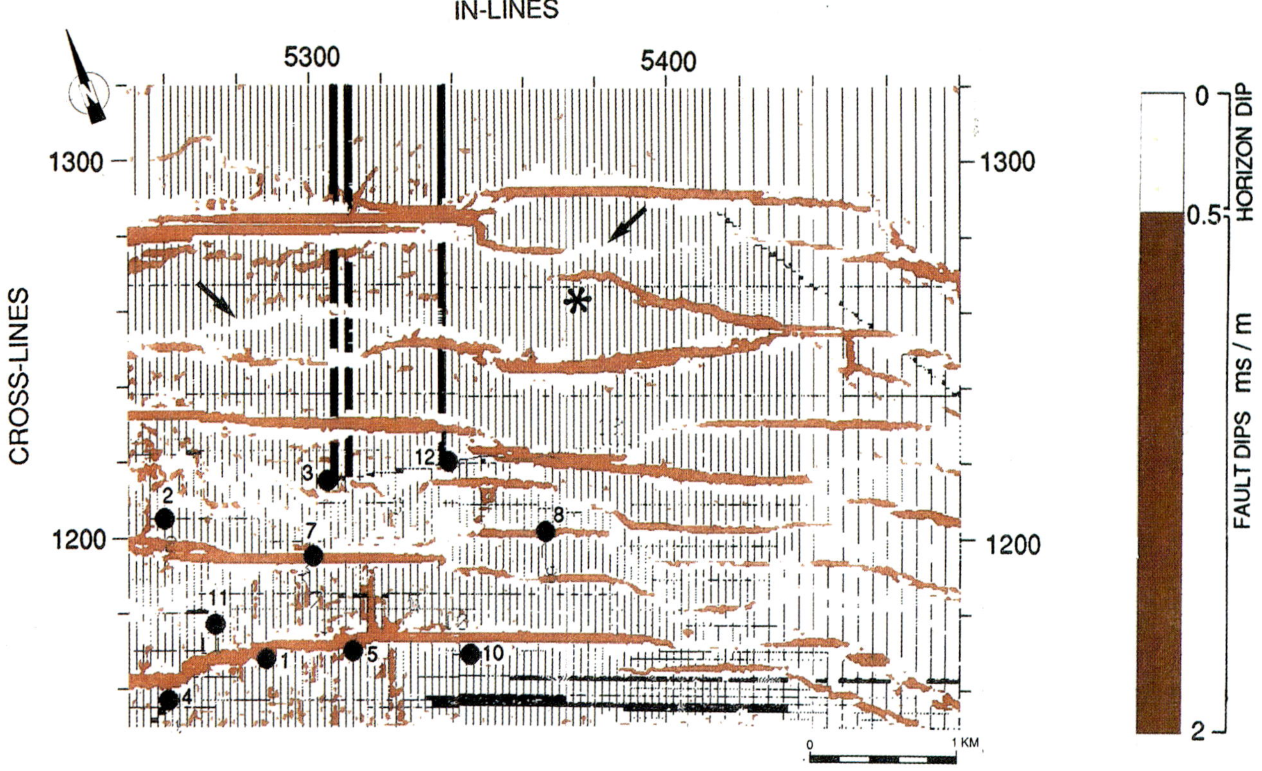

Fig. 10. Fault dips (red) and interpreted control grid (black) map. Arrows indicate two faults not expressed on dip display (fault dip equals horizon dip) but seen on azimuth display (stars on Figure 11). Asterisk indicates continuation of fault (flexure) some 70 m (230 ft) west of last interpreted fault control.

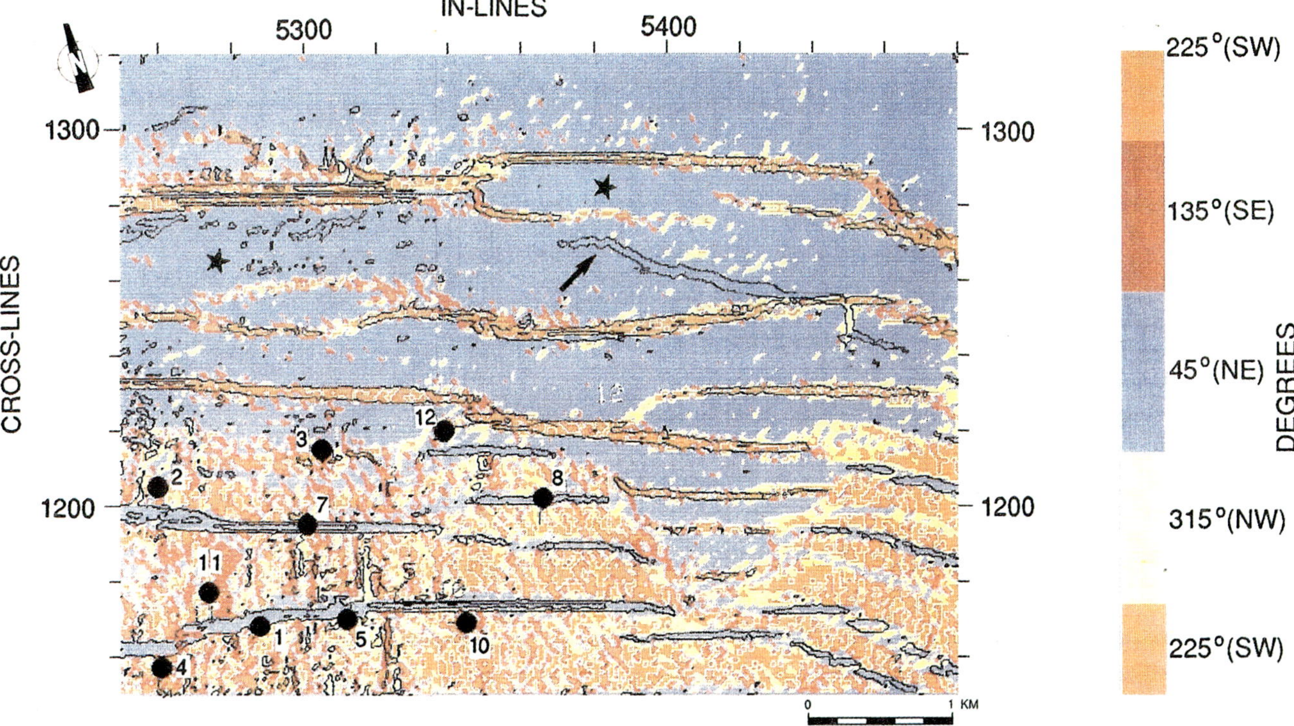

Fig. 11. Azimuth data map (in color) combined with contoured dip data (black). Arrow indicates fault not expressed on azimuth display (fault plane azimuth equals horizon azimuth) but shown in dip display. Stars refer to faults not expressed by dip computations (Figure 10).

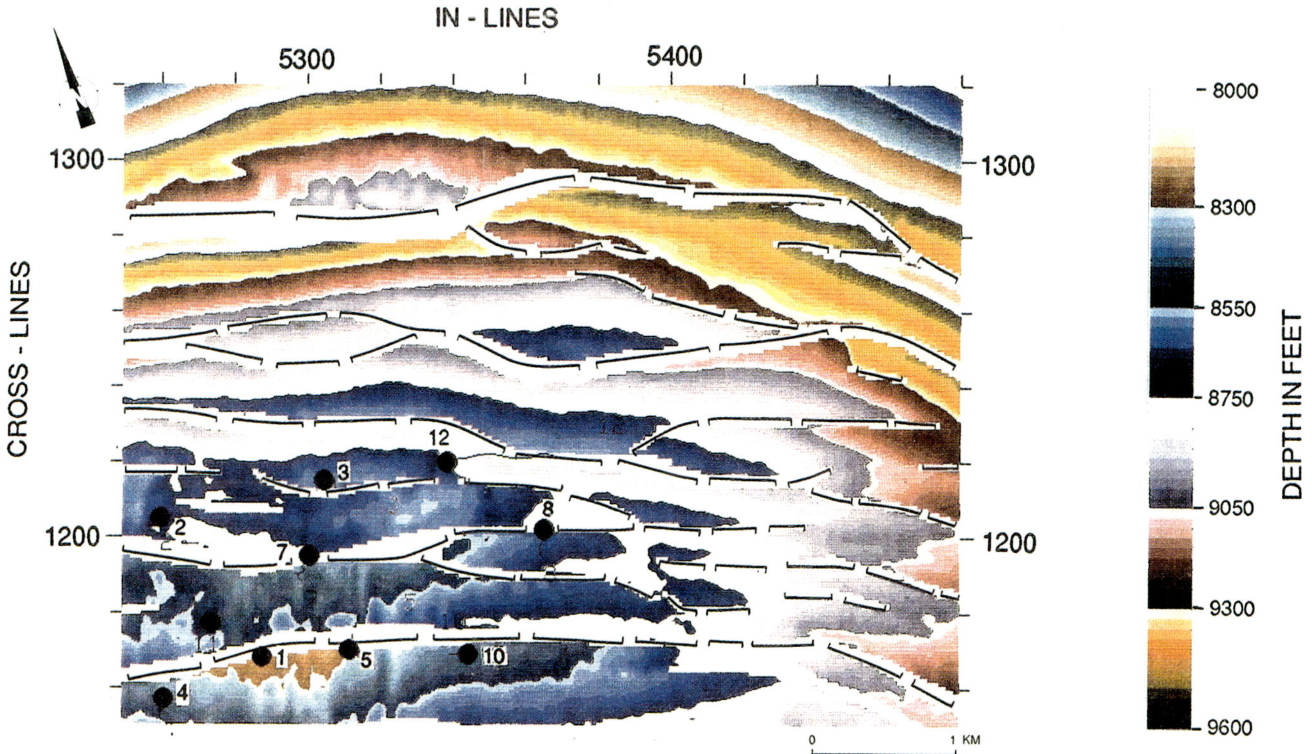

Fig. 12. D3.0 shale horizon map.

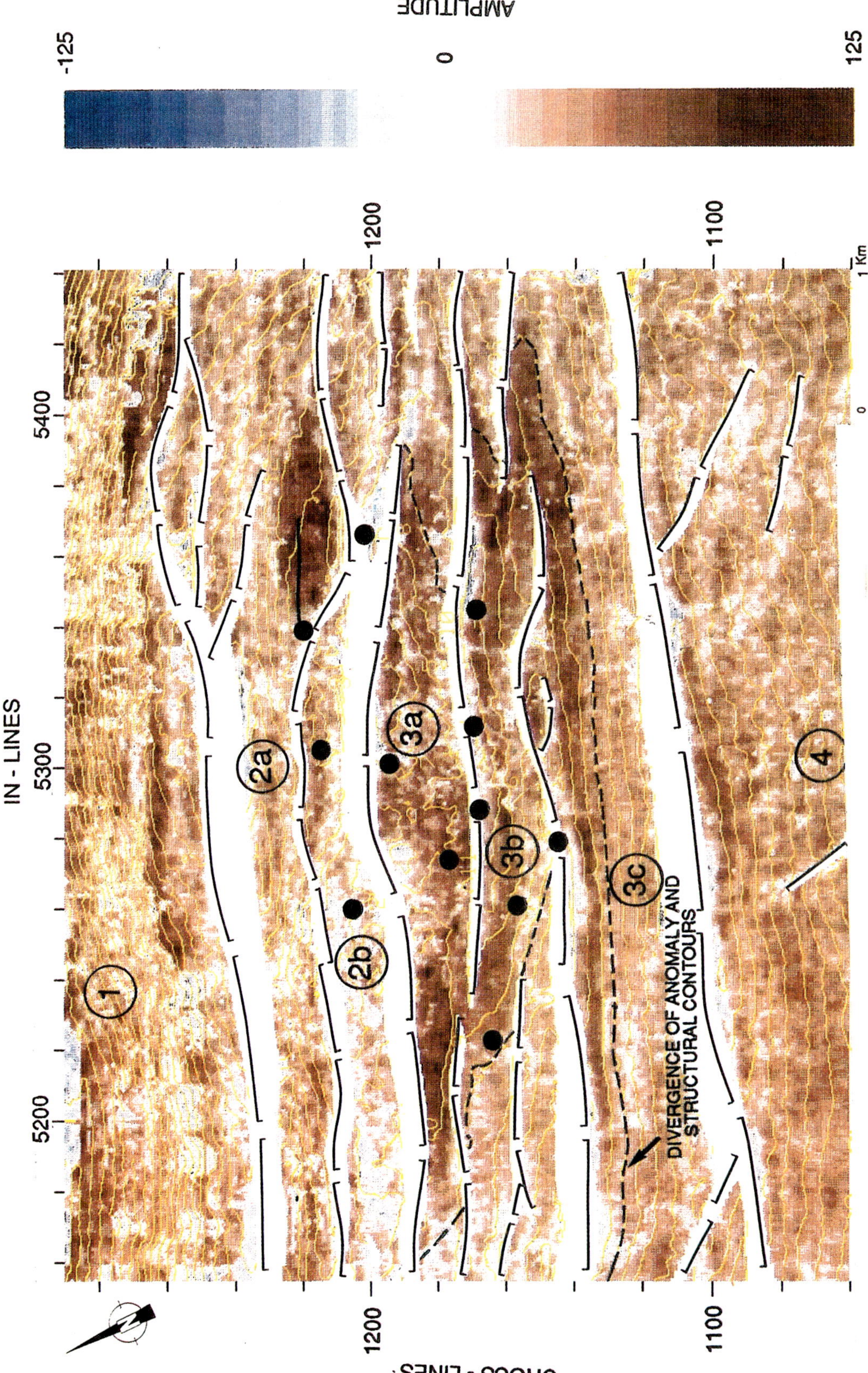

Fig. 13. F2.0 seismic horizon map showing amplitude and structural data. This map was derived from the F2.0 shale-horizon by subtracting 24 ms to approximate seismic loop corresponding to top of sandstone reservoir and automatic tracking. Seismic amplitudes are blue (negative amplitude) to red (positive amplitude). Structural contours (in 2-way traveltime) at top reservoir level are indicated by yellow lines. Dashed black line outlines divergence of structural contours and base of anomaly as result of velocity pullups occurring mainly in the western part of the field. Numbers in circles label fault blocks. Black dots are well locations.

to be a geophysical artifact rather than a direct indication of hydrocarbons. An additional prominent amplitude anomaly is present on the flank of the field in block 4 (Figure 13). This anomaly is conformable to structure and is a promising appraisal target.

Comparisons of Two 2-D and 3-D Seismic Maps

The Nun River field structural configuration based on 3-D seismic has changed considerably compared to the previous 2-D seismic interpretation (Figures 15, 16). Generally, only the strike of the major faults has remained the same, but the complexity and discontinuity of the faulting have increased, resulting in fewer isolated blocks and a more open structure. Interpreted fault configuration changed significantly in the central part of the crestal collapse, where faults were previously mapped as continuous, unbroken features (Figure 15). The faults are now mapped as a discontinuous series of faults, rapidly varying in throw as well as hade (Figure 16). Also, numerous minor faults have been identified and mapped for the first time.

The previously mapped twin culminations (Figure 15) have been effectively reduced to a single culmination. The combination of a single culmination with the discontinuous nature of the faulting means that more reservoirs are in communication than previously envisaged. This will have important implications on connected original oil in place, primary sweep efficiency, and access to aquifer and pressure support. Material-balance calculations based on production histories, combined with this improved knowledge of reservoir communication, result in better estimates for recoverable hydrocarbons.

Numerous blocks on the northern and southern flanks, particularly blocks 1, 2a, and 4, offer considerable appraisal potential (Figure 16). In these blocks, numerous stacked reservoirs either have been partially tested by a vertical well or not penetrated at all. The blocks will be appraised with wells deviated behind faults to ensure the appraisal of the full prospective sequence in an optimum position.

Fault-Sealing Investigations

The Nun River reservoir maps show numerous fault-dip closures that are either hydrocarbon bearing, water bearing, or untested. The absence of hydrocarbons in a potential reservoir trap may be the result of nonsealing faults or the lack of charge. Various theories and discussions have been put forward on source rock location and migration paths in the Niger delta (Frankl and Cordry, 1967; Short and Stäuble, 1967; Weber and Daukoru, 1975; Evamy et al., 1978; Ejedawe et al., 1984; Ekweozor and Daukoru, 1984; Lambert-Aikhionbare and Ibe, 1984; Doust, 1988; Knox and Omatsola, 1988). The abundance of hydrocarbons in the Nun River field makes a lack of charge

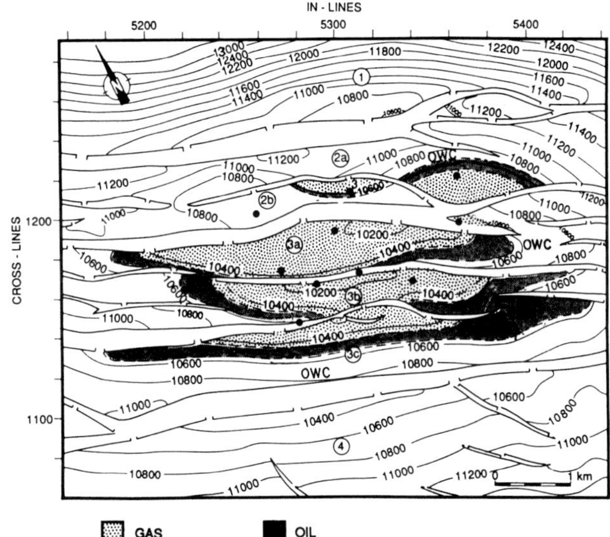

Fig. 14. F2.0 reservoir map with hydrocarbons. Depths are in feet below mean sea level. OWC = oil/water contact. Contour interval 200 ft.

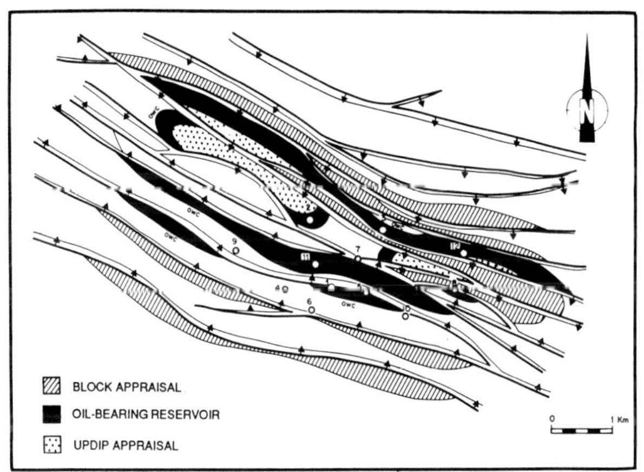

Fig. 15. D5.0 reservoir map based on 2-D seismic data. OWC = oil/water contact.

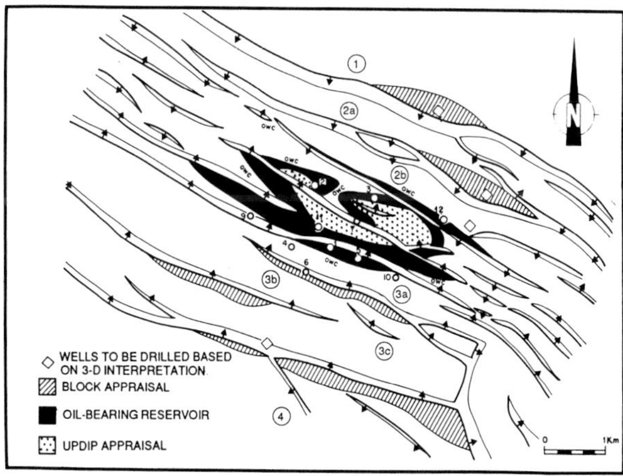

Fig. 16. D5.0 reservoir map based on 3-D seismic data. Numbers in circles indicate fault blocks. OWC = oil/water contact.

unlikely. The major factors that control hydrocarbon distribution within the field are lateral spillpoints at the termination of discontinuous faults, and seals or lack of seals along fault planes.

Fault closures require an impermeable lithology over the trap area (i.e., top seal) and lateral juxtaposition of the entire hydrocarbon-bearing reservoir against a sealing lithology along or within the fault plane (Smith, 1966, 1980; Downey, 1984; Watts, 1987). In the Nun River field, top seals are provided by field-wide marine and continental clays and shales (Figures 4, 5). Lateral seals are provided by the juxtaposition of these impermeable clays and shales against hydrocarbon-bearing sandstones along fault planes. In addition, clay smears from individual ductile clay layers are emplaced along fault planes during faulting and provide a seal to migrating gas and fluids in the Niger delta, as in other deltaic hydrocarbon provinces [e.g., the Gulf Coast of the United States and the Baram delta in Northwest Borneo (Weber et al., 1978; Smith, 1980; Mandl, 1988)]. Thus, two variables need to be assessed when evaluating the seal along fault planes in the Niger delta: (1) juxtaposition of lithology and (2) clay smear (Figure 17).

Fault Slicing of 3-D Seismic for Evaluation of Juxtaposition and Seal Along Fault Plane

A fault slice is an extraction of seismic amplitudes from 3-D seismic along a surface parallel and adjacent to a fault plane, either in the footwall or hanging wall, and is displayed as a projection onto a vertical surface (Figure 18). Fault slices provide continuous, unfaulted, strike sections within narrow fault blocks of complexly faulted fields. The fault slices are useful for deriving fault throw, juxtaposition of sandstones and shales along a fault plane, and for planning well trajectories (Figure 19). Fault slicing of 3-D seismic was introduced by Brown et al. (1986, 1987) to illustrate juxtaposition of lithology, and was further developed by van der Pal (1988, 1989). Other uses of fault-slice projections include mapping amplitude anomalies and spillpoints, and mapping events in seismostratigraphy.

Fault slices from 3-D seismic have been used to evaluate the juxtaposition of reservoirs and shale seals along the "K" fault plane in the Nun River field. The "K" fault plane was studied because of adequate 3-D seismic coverage, well penetrations, fault-dip closure traps with known hydrocarbons, and related seismic amplitude anomalies to calibrate the fault-seal results. Fault-slice projections were generated for the "K" fault outside the zone of fault disturbance at 75 m (250 ft) into the footwall and hanging wall (Figure 20). Obviously, the closer the slices are to the fault plane, the more accurate the juxtaposition. However, if the slices are generated too close to the fault, the deteri-

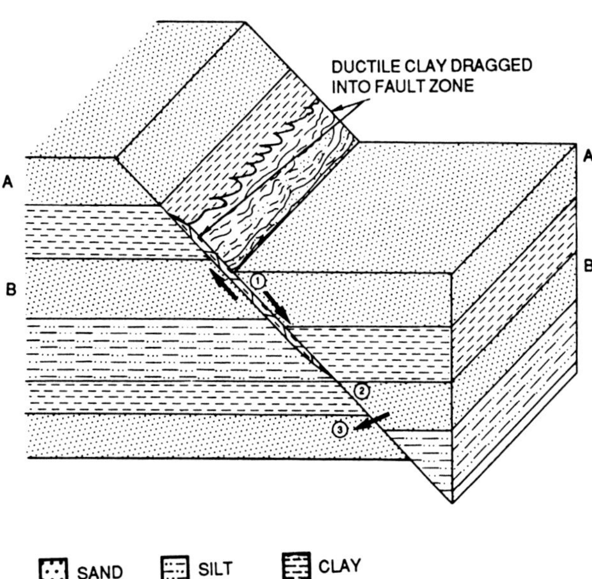

Fig. 17. Fault seal and nonseal: (1) dragging of ductile clays into fault plane during faulting creates clay seal between two sandstones (A and B); (2) juxtaposition of reservoir to impermeable clay bed; and (3) sandstone-to-sandstone window, or leak in fault plane creating possible spillpoint to migrating hydrocarbons. (Modified from Smith, 1980, and Downey, 1984).

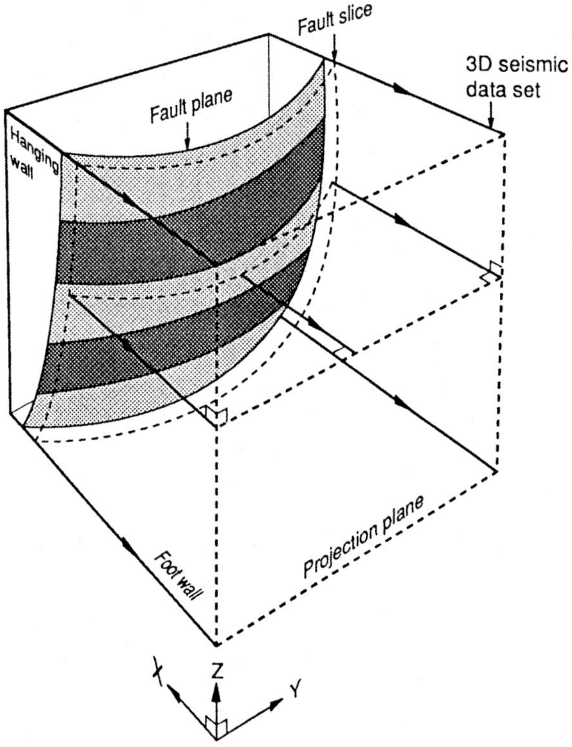

Fig. 18. Position of fault slice and projection plane in relation to footwall and hanging wall of dipping normal fault.

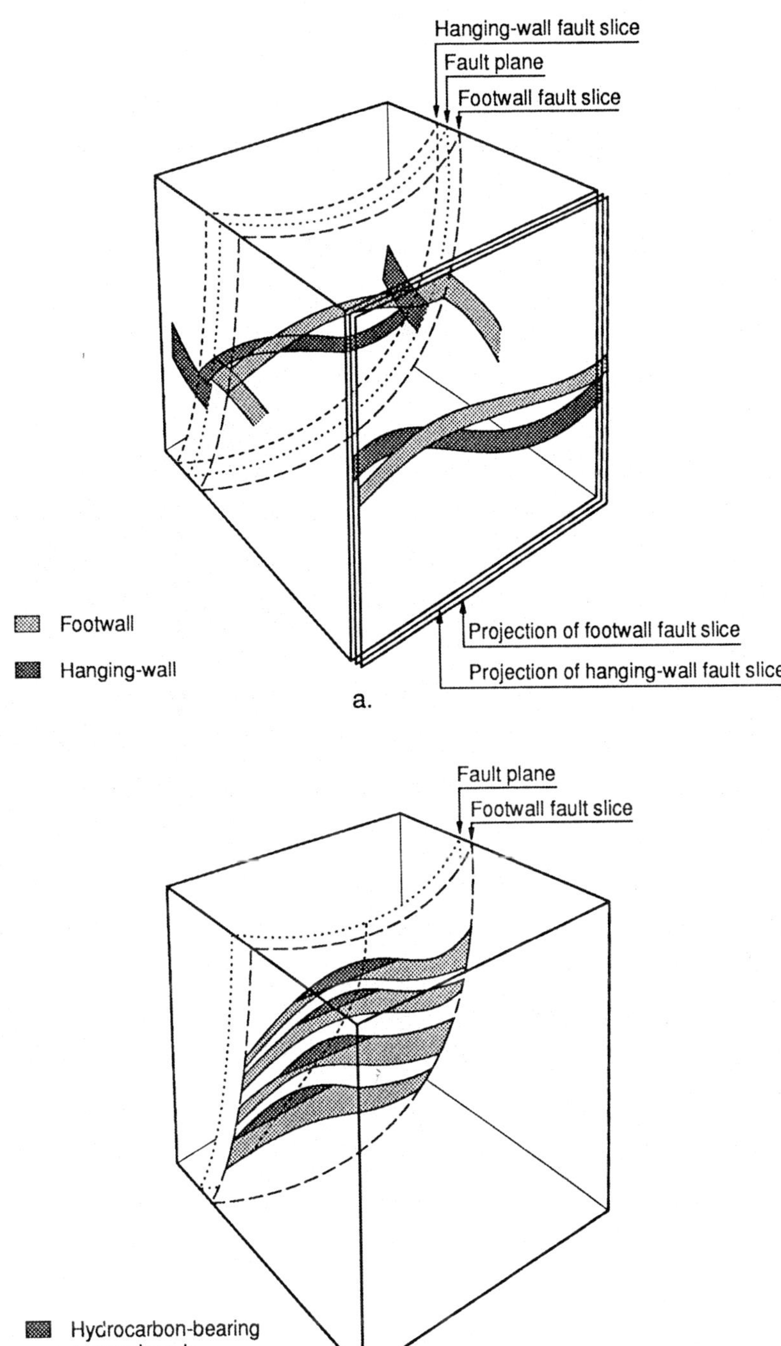

Fig. 19. Two examples of application of fault slices: (a) use of hanging-wall and footwall fault slices to determine reservoir seal juxtaposition; and (b) use of footwall fault slice to determine optimum well trajectory.

oration of the seismic commonly produces poor-quality slices, which are difficult to interpret.

For reservoir/seal definition, a composite, stratigraphically complete, SP log was derived from wells that have penetrated the "K" fault (Figure 5). The composite SP log was rescaled and correlated with the key seismic horizons on the acoustic impedance fault-slice projection to interpret the main sandstone and shale strata (Figures 21, 22). Hydrocarbon zones in wells fit with amplitude anomalies seen on seismic.

Fault-throw contours and juxtaposition of lithology were obtained for the "K" fault plane by superimposing the interpreted footwall and hanging-wall fault slices (Figures 23, 24). The displacement variation of the fault, the juxtaposition of shale to shale and

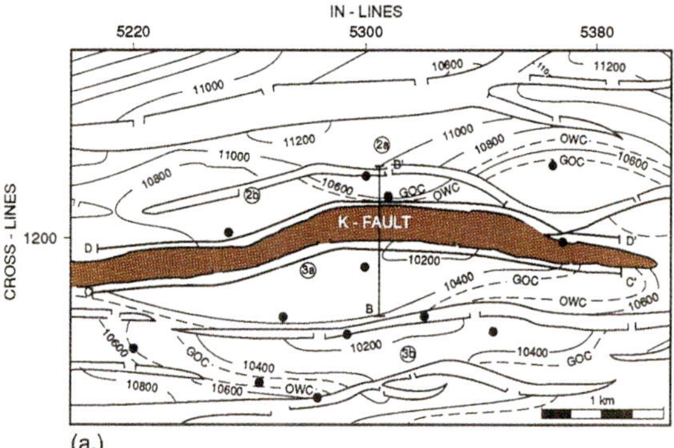

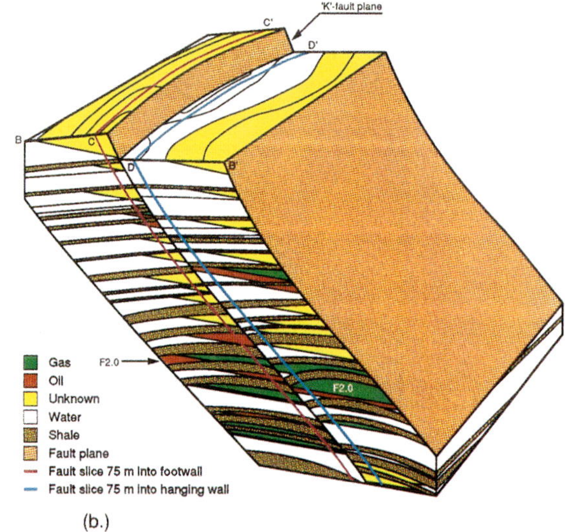

Fig. 20. Nun River field fault slicing: (a) F2.0 horizon location map (contour interval 200 ft) showing location of "K" fault (brown), in-line section (BB′), and fault slice sections at 75 m (250 ft) into footwall (CC′) and hanging wall (DD′); and (b) "K" fault depth section (BB′), hydrocarbon distribution, and fault slice traces at 75 m (250 ft) into footwall (CC′) and hanging wall (DD′). GOC = gas/oil contact, OWC = oil/water contact.

sandstone to shale, and potential sandstone-to-sandstone leaks or windows are well illustrated. Because the fault slices are 75 m (250 ft) away from the fault plane (Figure 20), corrections for bed dip and fault drag need to be applied to find the true fault throw and juxtaposition at the fault plane. In this example, the required correction for fault drag and dip is smaller than that warranted by the resolution of the seismic interpretation and has not been applied in deriving Figures 23 and 24.

Clay Smear Seal Evaluation Along Fault Plane

The presence of hydrocarbons trapped against sandstone-to-sandstone windows along the "K" fault plane, juxtaposed to other hydrocarbons or water (Figure 24) highlights the importance of clay smear as a sealing element in the Nun River field. This sealing mechanism has been the subject of earlier laboratory experiments and field observations.

Lehner and Pilaar (1974, personal communication; and in Weber et al., 1978) performed a series of ring-shear tests of sand/clay systems to simulate faulting and clay smear emplacement in a deltaic sandstone-shale sequence. The ring-shear apparatus consisted of a ring-shaped container, the upper half of which was fixed while the lower half rotated (Mandl et al., 1977). In the experiments, "the apparatus was filled with zones of sand alternating laterally with vertical bands of clay. These bands were sheared off under various simulated overburden pressures along a median slip plane. The clay bands sheared off in this fashion formed one continuous, multilayered, clay gouge along the median slip plane of the shear zone (Figure 25) and proved to be an effective seal to vertical water flow when tested" (Weber et al., 1978).

Lehner and Pilaar (1974, personal communication; and in Weber et al., 1978) compared the results of the ring-shear apparatus tests with field observations of syndepositional normal faulting of a clastic Tertiary deltaic sequence in the open-cast lignite mines of Frechen, West Germany (Figure 26). The actual mechanism of clay-gouge emplacement is probably more effective and involves a larger amount of clay squeezed into the fault zone (F. Lehner, 1989 personal communication) than in the laboratory model. They observed clay gouges in the shear zones of faults, and these originated from both the upthrown and downthrown sides of shale source beds. The clay gouges within the shear zone decreased in thickness away from the source beds, and individual clay gouges from separate source beds coalesced to form one multilayered clay gouge. Thicker shale source beds provided longer, more continuous clay gouges than thinner source beds.

A relationship for indicating clay smear potentials (CSP) was devised by Speksnijder (1987; 1987, per-

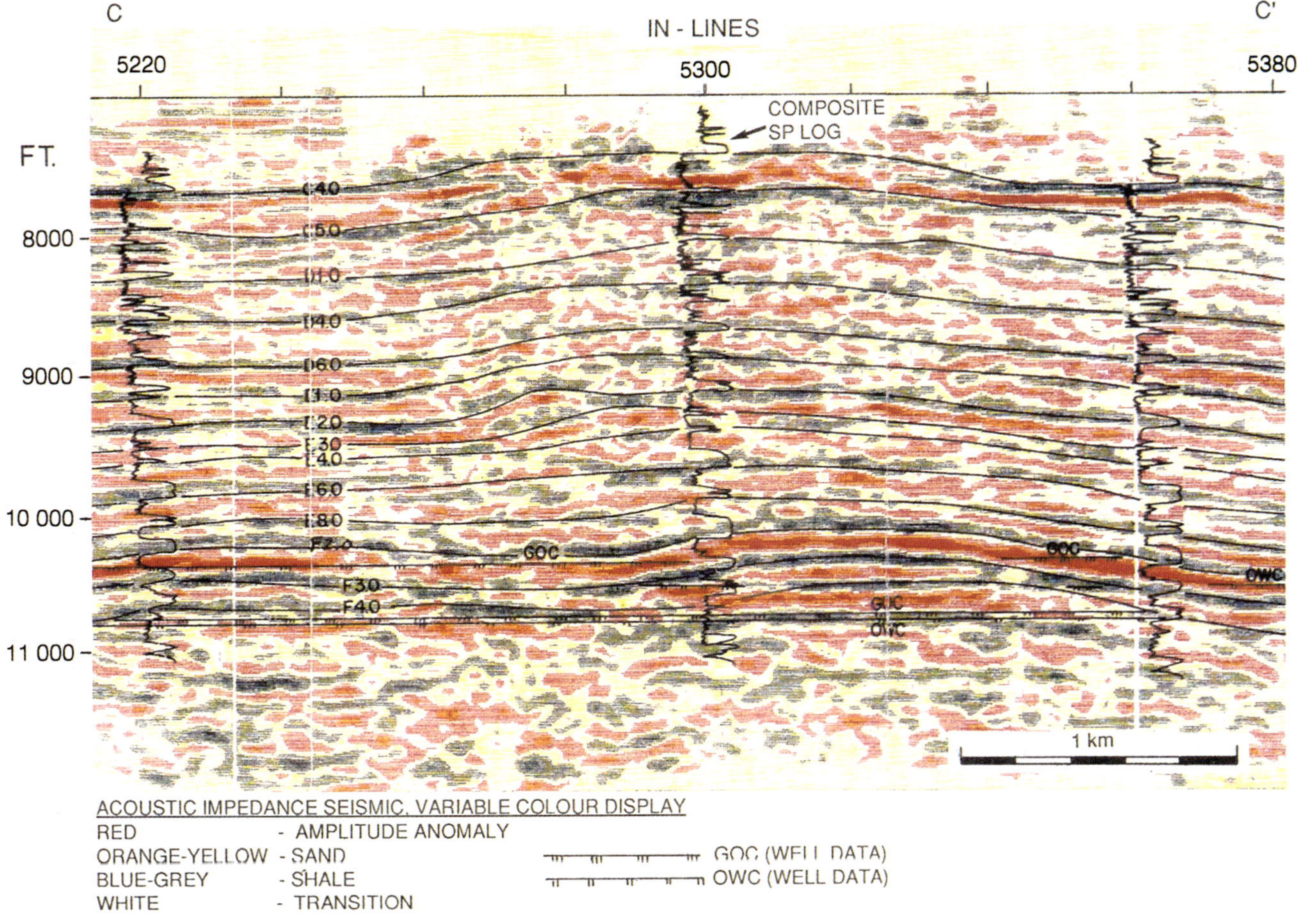

Fig. 21. "K" fault slice from 3-D seismic (see Figure 20, CC') at 75 m (250 ft) into footwall with superimposed composite SP log. Note qualitative correlation between seismic events, lithology, and hydrocarbons. GOC = gas/oil contact, OWC = oil/water contact, WUT = water up to.

sonal communication) using the observations of Lehner and Pilaar (1974, personal communication; and in Weber et al., 1978) as a guide. In this relationship, the CSP represents the relative amount of clay that has been smeared from individual shale source beds at a certain point along a fault plane. The CSP increases with shale source bed thickness and with the number of source beds displaced past a particular point along a fault plane. The CSP decreases with increased fault throw or fault displacement. The effectiveness of a clay smear in providing a seal to migrating fluids or gas depends primarily on the number and thickness of shale source beds contributing to the clay-smear seal. To determine local clay-smear sealing versus non-sealing cutoff values, individual CSP values need to be calibrated against known trapped hydrocarbons in sandstone-to-sandstone windows along a fault plane(s).

To evaluate seals along the "K" fault plane further, the mapping of juxtaposition and fault throw from fault slices was combined with the calculation of CSPs. CSPs were calculated using clay thickness and fault throw derived from the composite SP log and fault-slice projections, respectively. The resultant high, medium, and low CSP values were plotted in the sandstone-to-sandstone windows (Figure 27).

Comparisons between CSP values in hydrocarbon-bearing and water-bearing windows along the fault plane were used to calibrate sealing versus non-sealing CSP values (Figure 27). Hydrocarbons are trapped in windows that have high to medium CSP values. The oil/water contacts seen in the D4.0 and D5.0 sandstones in the footwall may be the result of the loss of critical clay-smear seals. Low CSP values represent little chance for the presence of continuous clay smear seals that can trap hydrocarbons. The increase in CSP values below 2926 m (9600 ft) reflects an increase in thicker shales, thus an increase in fault seal from smearable clay as well as from juxtaposition against impermeable strata.

The combination of fault slicing of 3-D seismic to obtain reservoir-to-seal juxtaposition and fault throw with the calculation of CSP values across windows in the fault plane provides a useful tool for ranking

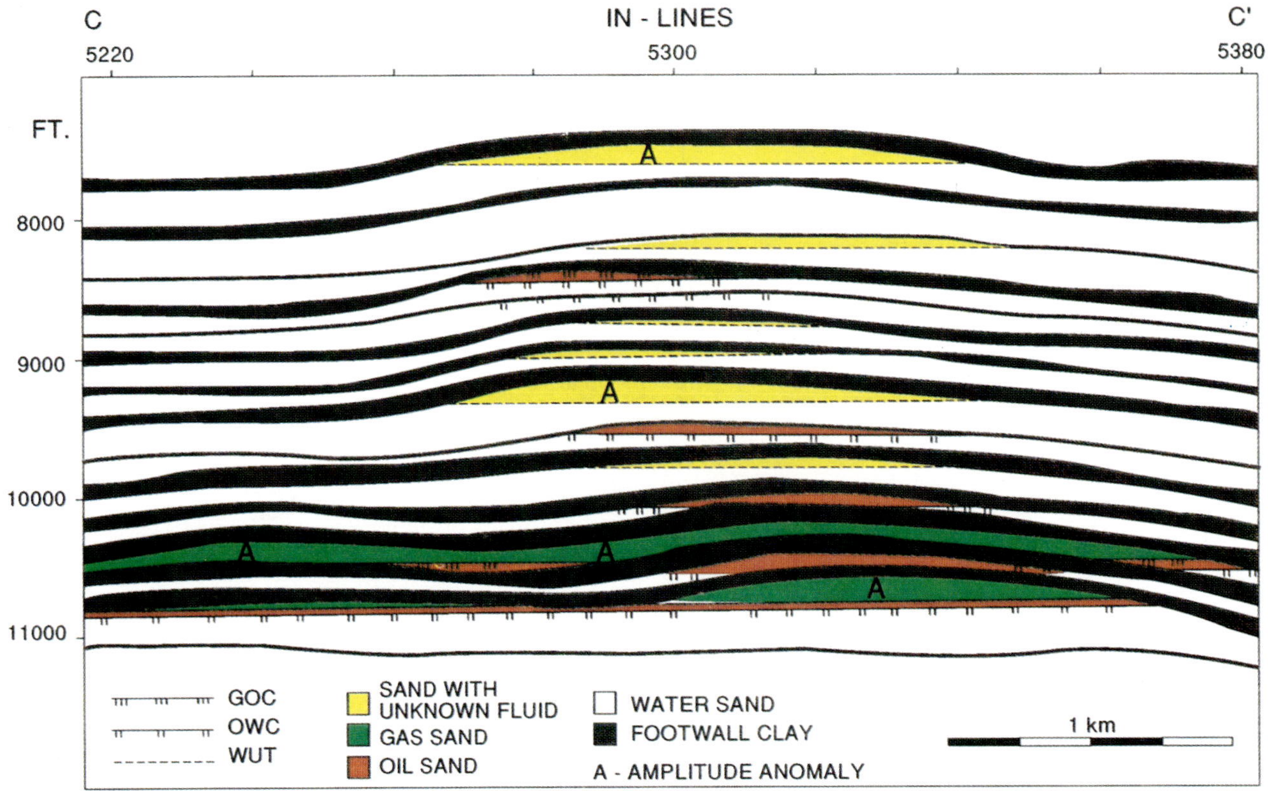

Fig. 22. Interpreted fault slice at 75 m (250 ft) into footwall, derived from Figure 21 by correlating composite SP log with key seismic horizons and seismic amplitudes. Hydrocarbon contacts and water levels are from well data. GOC = gas/oil contact, OWC = oil/water contact.

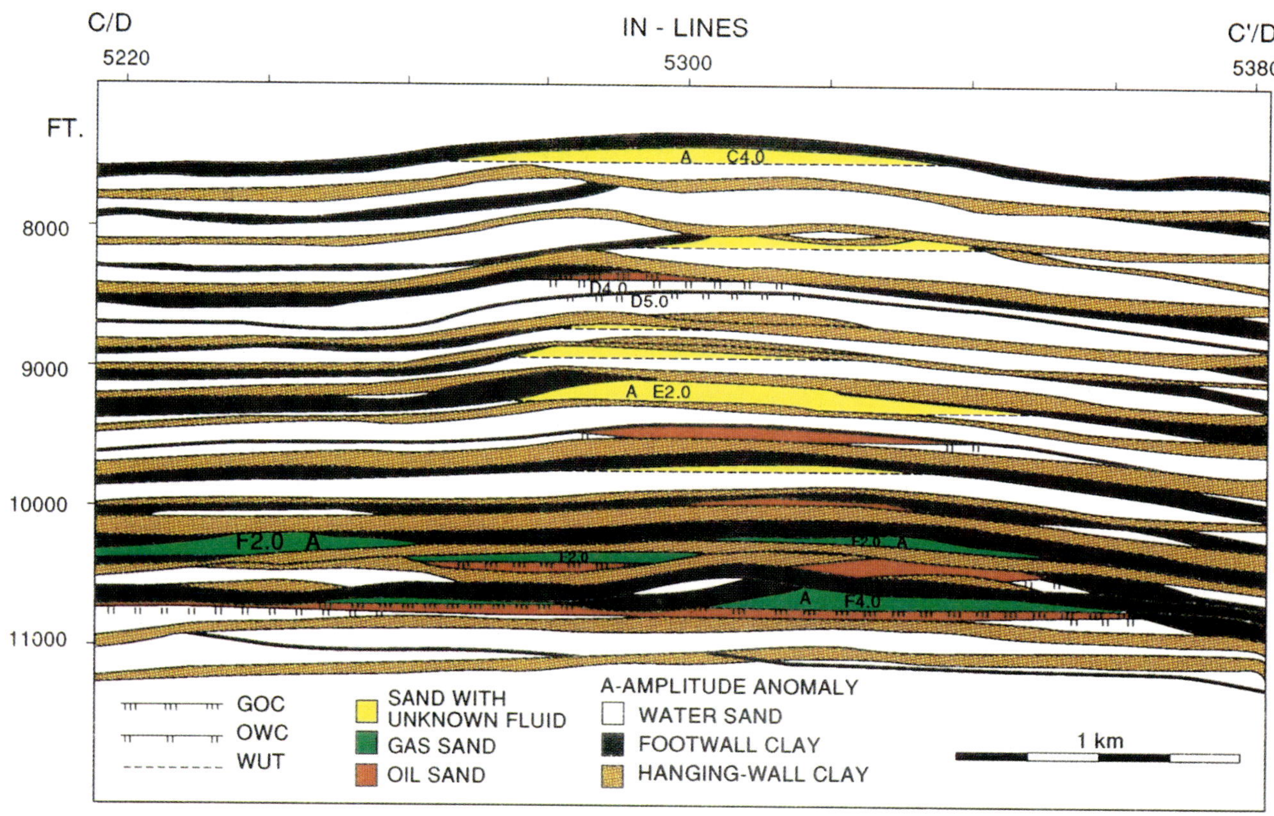

Fig. 23. True vertical fault throws in feet of "K" fault derived from superimposing footwall (CC') and hanging-wall (DD') fault slices and measuring horizon offset. See Figure 20 for locations.

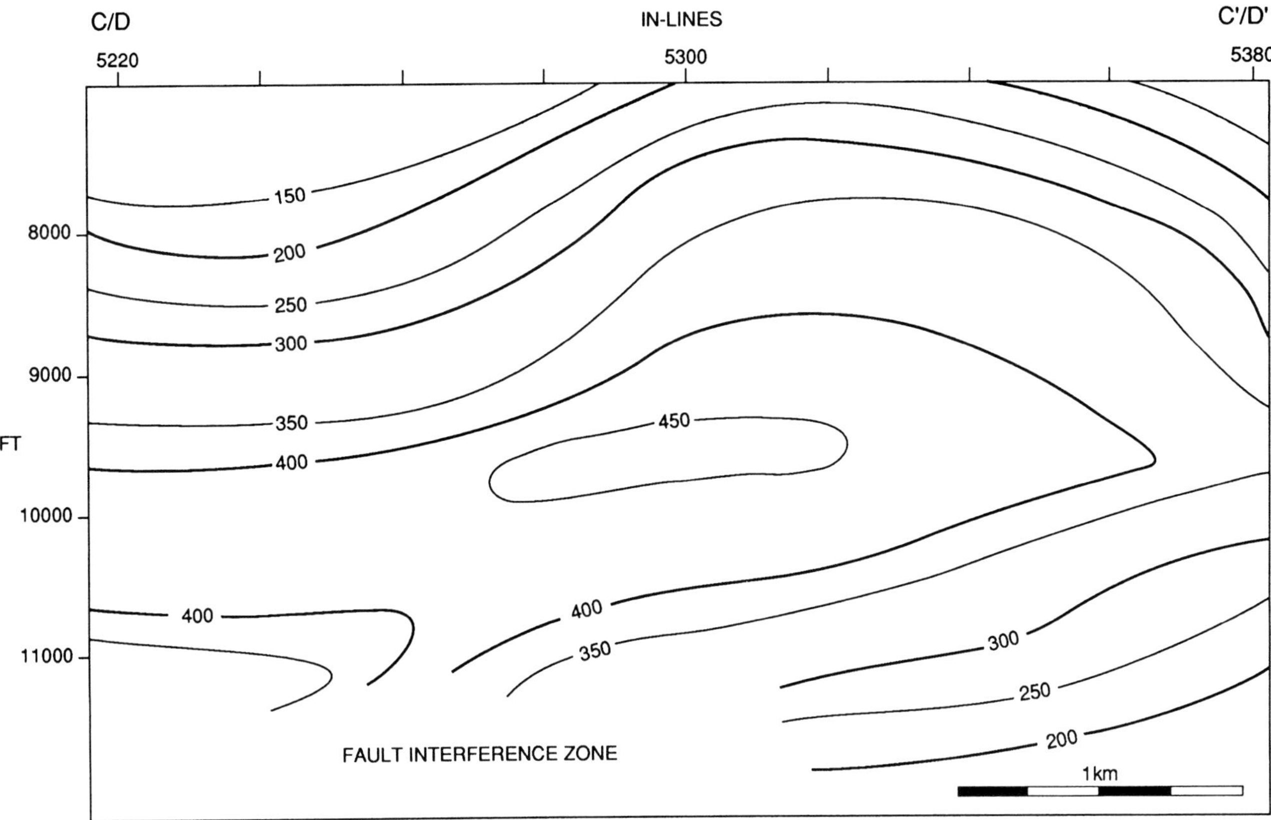

Fig. 24. "K" fault plane juxtaposition cross-section created by superimposing interpreted footwall (CC') and hanging-wall (DD') fault slices to derive juxtaposition of shale-to-shale, shale-to-sandstone, and sandstone-to-sandstone windows. Footwall hydrocarbon contacts and water levels are plotted within sandstone-to-sandstone windows. Obviously, some sealing mechanism (clay smear along "K" fault plane), as well as juxtaposition of shale-to-sandstone, is creating seal to hydrocarbons. See Figure 20 for locations. GOC = gas/oil contact, OWC = oil/water contact, WUT = water up to.

exploration and appraisal prospects. In addition, the identification of potentially weak clay-smear seals may also be used in the management of producing reservoirs. For example, fault breakdown may occur along weak clay-smear seals as the result of increased pressure differentials from production on one side of a fault. This fault breakdown may lead to either commingled production or gas or water ingress across the fault plane from an adjacent block. If potentially weak clay-smear seals are identified in a producing reservoir, pressure maintenance programs may be initiated to prevent fault breakdown and safeguard isolated production.

Conclusions

The interpretation of the Nun River 3-D seismic survey has provided a detailed structural interpretation of the reservoir previously unattainable with 2-D seismic and well data. Although formerly elongate accumulations have been reduced in size, numerous partially tested and untested closures, commonly with associated seismic amplitude anomalies, are apparent

on new maps. These closures offer considerable scope for increasing the estimated hydrocarbon reserves.

The application of dip and azimuth displays and of fault-slicing techniques on interactive seismic interpretation workstations have been used to map fault pat-

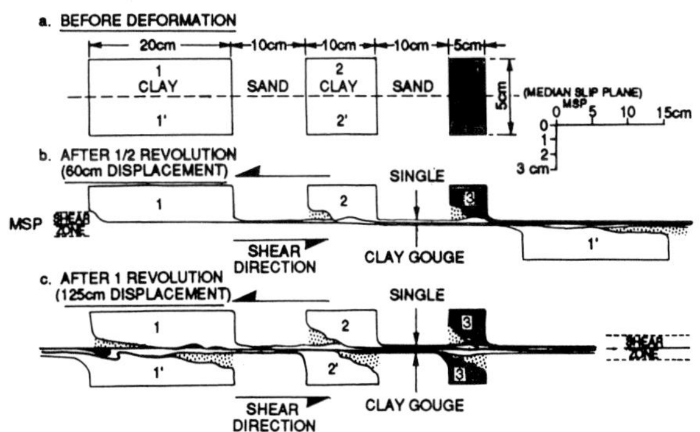

Fig. 25. Development of multilayered, single clay gouge in alternating sand/shale sequence sheared in ring-shear apparatus (by Lehner and Pilaar in Weber et al., 1978).

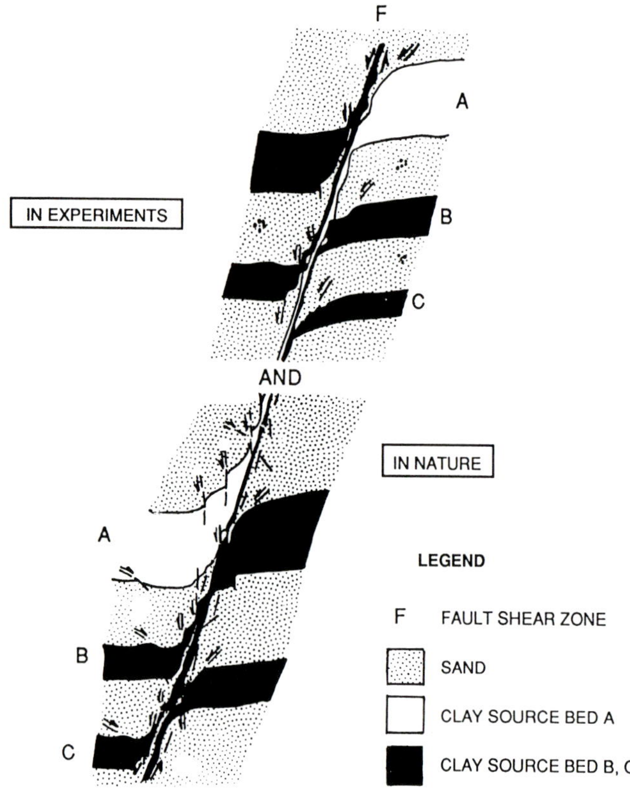

Fig. 26. Comparison of clay smears in fault shear zone as produced in ring-shear apparatus with field observations in Frechen brown-coal pits in West Germany. Note multilayered clay gouge (by Lehner and Pilaar in Weber et al., 1978).

terns and juxtaposition, respectively. Reservoir-to-seal juxtaposition from fault slices, combined with clay-smear seal analysis along sandstone-to-sandstone windows or potential leaks along the fault plane, has been used to derive a complete analysis of fault seal along the fault plane. This increased knowledge of the lateral extent of faults, reservoir compartmentalization, fault seals, and reservoir communication will result in a more optimum development of the field.

Acknowledgments

This paper is based on data acquired and interpreted by Shell Petroleum Development Company of Nigeria Limited (SPDC). In support of this work, Koninklijke/Shell Exploratie en Produktie Laboratorium (KSEPL) has developed specific techniques and methods, which are presented in this paper.

The writers thank reviewers N. L. Watts of SPDC, Lagos, Nigeria; M. F. S. Petch, W. T. Horsfield, J. C. Mondt, I. van der Molen, and S. H. Kromoidjojo of KSEPL, Rijswijk, The Netherlands, for their comments, critique, and support. In addition, the writers thank the many Shell geologists and geophysicists who were closely associated with the project, in particular, S. O. Odumodu, P. L. E. Boegner, and D. J. Guilj of SPDC, Port Harcourt, Nigeria, and C. F. W. Höcker, and M. Peters of KSEPL, Rijswijk, The Netherlands.

The writers are indebted to Shell Internationale Petroleum Maatschappij, Shell Petroleum Develop-

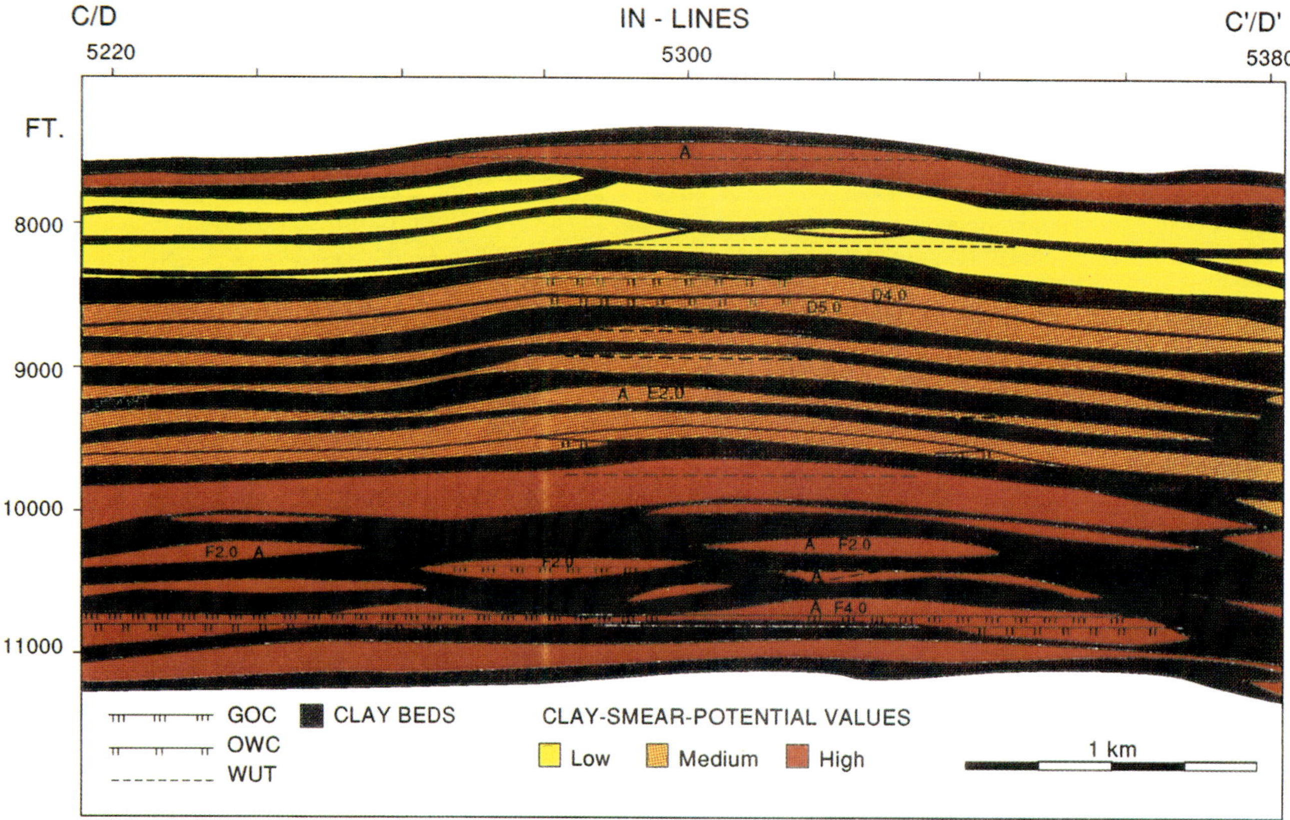

Fig. 27. Clay-smear potential (CSP) values along sandstone-to-sandstone windows of "K" fault plane.

ment Company of Nigeria Limited, Shell Internationale Research Maatschappij B. V., Nigerian National Petroleum Corporation, and Nigerian Department of Petroleum Resources for permission to publish this paper.

References

Brown, A. R., 1986, Interpretation of three-dimensional seismic data: AAPG Memoir **42**, 194.

Brown, A. R., Dahm, C. G., and Graebner, R. J., 1981, A stratigraphic case history using three-dimensional seismic data in the Gulf of Thailand: Geophysical Prospecting, **29**, 327–349.

Brown, A. R., Edwards, G. S., and Howard, R. E., 1986, Fault slicing—the study of faults from a production perspective (abs.): Society of Exploration Geophysicists Annual Meeting Program, 499–500.

——— 1987, Fault slicing—a new approach to the interpretation of fault detail: Geophysics, **52**, 1319–1327.

Buchanan, R., Marke, P. A. B., and Ruijtenberg, P. A., 1988, Application of 3-D seismic to detailed reservoir delineation: Society of Petroleum Engineers International Meeting on Petroleum Engineering Proceedings, 91–97.

Dalley, R. M., Gevers, E. C. A., Stampfli, G. M., Davies, D. J., Gastaldi, C. N., Ruijtenberg, P. R., and Vermeer, G. J. O., 1989, DAZIL (dip, azimuth, illumination): First Break, **7**, no. 3, 86–100.

Doust, H., 1988, The Niger delta: hydrocarbon potential of a major Tertiary delta province: Koninklijk Nederlands Geologisch en Mijnbouwkundig Gezelschap Symposium on Coastal Lowlands, Geology, and Geotechnology Proceedings (1987): Dordrecht, Netherlands, Kluwer Academic Publishers, 203–212.

Downey, M. W., 1984, Evaluating seals for hydrocarbon accumulations: AAPG Bulletin, **68**, 1752–1763.

Ejedawe, J. E., Coker, S. J. L., Lambert-Aikhionbare, D. O., Alofe, K. B., and Adoh, F. O., 1984, Evolution of oil-generative window and oil and gas occurrence in Tertiary Niger delta basin: AAPG Bulletin, **68**, 1744–1751.

Ekweozor, C. M., and Daukoru, E. M., 1984, Petroleum source-bed evaluation of Tertiary Niger delta: reply: AAPG Bulletin, **68**, 390–394.

Enachescu, M. E., and Demers, J. H. L., 1988, DHIS validation for the Amauligak oil field, Canadian Beaufort Sea: Society of Exploration Geophysicists Annual Meeting, Expanded Abstracts, 765–768.

Evamy, B. D., Haremboure, J., Kamerling, P., Knaap, W. A., Molloy, F. A., and Rowlands, P., 1978, Hydrocarbon habitat of the Tertiary Niger delta: AAPG Bulletin, **62**, 1–39.

Frankl, E. J., and Cordry, E. A., 1967, The Niger delta oil province—recent developments onshore and offshore: 7th World Petroleum Congress Proceedings, **2**, 125–209.

Knox, G. J., and Omatsola, E. M., 1988, Development of the Cenozoic Niger delta in terms of the "escalator regression" model and impact on hydrocarbon distribution: Koninklijk Nederlands Geologisch en Mijnbouwkundig Gezelschap Symposium on Coastal Lowlands, Geology and Geotechnology, The Hague (1987): Dordrecht, Netherlands, Kluwer Academic Publishers, 181–202.

Lambert-Aikhionbare, D. O., and Ibe, A. C., 1984, Petroleum source-bed evaluation of Tertiary Niger delta: discussion: AAPG Bulletin, **68**, 387–389.

Mandl, G., 1988, Mechanics of tectonic faulting, models and basic concepts: developments in structural geology: Amsterdam, Elsevier, 4, 44, 107.

Mandl, G., de Jong, L. N. J., and Maltha, A., 1977, Shear zones in granular material, an experimental study of their structure and mechanical genesis: Rock Mechanics, **9**, 95–144.

Nestvold, E. O., 1987, The use of 3-D seismic in exploration, appraisal and field development: 12th World Petroleum Congress Proceedings, **2**, 125–132.

Schlumberger Technical Services, Inc., Dubai, U. A. E., 1989, The three forces of faulting, faulty communication: Middle East Well Evaluation Review, no. **6**, 16–17.

Shell Internationale Petroleum, Maatschappij, B. V., 1988, Horizon processing techniques for recognition of structural geology on 3-D seismic: Research Disclosure 29473, 792.

Short, K. C., and Stäuble, A. J., 1967, Outline of geology of Niger Delta: AAPG Bulletin, **51**, 761–779.

Smith, D. A., 1966, Theoretical considerations of sealing and non-sealing faults: AAPG Bulletin, **50**, 363–374.

——— 1980, Sealing and non-sealing faults in Louisiana Gulf Coast salt basin: AAPG Bulletin, **64**, 145–172.

Speksnijder, A., 1987, The structural configuration of Cormorant Block IV in context of the northern Viking Graben structural framework: Geologie en Mijnbouw, **65**, 357–379.

van der Pal, R. C., 1988, Fault slicing: a use of three-dimensional seismic data for fault analysis, practical aspects: Research Report, Koninklijke/Shell Exploratie en Produktie Laboratorium, Rijswijk, and Delft University of Technology, Faculty of Mining and Petroleum Engineering, Section of Applied Geophysics, Delft, The Netherlands, 1–28.

——— 1989, A new method for construction of fault slices from three-dimensional seismic data (abs.): European Association of Exploration Geophysicists Conference Technical Programme and Abstracts of Papers, 46.

Watts, N. L., 1987, Theoretical aspects of cap-rock and fault seals for single- and two-phase hydrocarbon columns: Marine and Petroleum Geology, **4**, 274–307.

Weber, K. J., 1971, Sedimentological aspects of oil fields in the Niger delta complex: Geologie en Mijnbouw, **50**, 559–576.

——— 1986, Hydrocarbon distribution patterns in Nigerian growth fault structures controlled by structural style and stratigraphy: Journal of Petroleum Science and Engineering, **1**, 91–104.

Weber, K. J., Daukoru, E., 1975, Petroleum geology of the Niger delta: 9th World Petroleum Congress Proceedings, **2**, 209–221.

Weber, K. J., Mandl, G., Pilaar, W. F., Lehner, F., and Precious, R. G., 1978, The role of faults in hydrocarbon migration and trapping in Nigerian growth fault structures: 10th Annual Offshore Technology Conference Proceedings, **4**, 2643–2653.

Cormorant: Development of a Complex Field

J. H. Stiles, Jr. and J. W. McKee[‡]*

Introduction

The Cormorant Field provides an example of development plan changes which are required as the nature of a large, complex field becomes better known. The number and locations of wells and the completion method have changed as the structural, stratigraphic, and diagenetic nature of the field was refined. A better understanding of a key development constraint, the drilling reach, has allowed development plan changes with significant cost savings. Close surveillance of geologic results and field performance has been necessary to allow almost continuous updating. Studies and evaluations undertaken by Esso to evaluate such changes for each of the field's four fault blocks are discussed. These studies complement work done by the operator, Shell U.K. Exploration and Production.

Cormorant is located in 500 ft of water in the U.K. North Sea about 95 miles northeast of the Shetland Islands (Figure 1). The field (Figure 2) was discovered in 1972 and was delineated with 11 additional wells. The field is composed of four major fault blocks, covers an area of over 10 000 acres, and contains over 1.5 billion barrels of oil. Estimated ultimate recovery is about 600 million barrels.

Development started in 1978 with installation of the South Cormorant platform followed by a platform on North Cormorant in 1982. The area between the two platforms is developed with subsea technology. Single satellite well P-1 started production in 1981 and was followed by installation of an Underwater Manifold Centre (UMC) in 1982.

Geology and Reservoir Fluid Properties

The field, located along the west of the North Viking Graben, is composed of four westerly dipping fault blocks. The structural style is the result of two tectonic phases. The earlier Caledonian phase resulted in a northeast-southwest structural grain, while the Late Kimmerian phase created the north-south faulting and associated westerly dipping fault blocks. The latter occurred during and after deposition of the Brent Group. As a result, the Brent thickens to the west and is truncated or down faulted to the east. Dip on the flanks varies from 8 to 12 degrees.

The oil is contained in the Middle Jurassic Brent Group which is divided into seven members representing a regressive-transgressive sequence within a northerly prograding delta system (Figure 3). Overall Brent thickness ranges from 200 ft to more than 500 ft. The basal Broom Formation is an offshore lag deposit composed of fine to coarse grained sandstone of highly variable quality. The formation is overlain by the Rannoch which was deposited in a lower to upper shoreface environment and is composed of micaceous sands and shales, with quality generally improving upward. The overlying Etive is a beach and barrier bar complex which forms the most important oil producing member. The succeeding lower, mid, and upper Ness

*Exxon Company, International 820 Gessner, Houston, TX 77024. Formerly Esso Exploration and Production, UK Ltd., Exploration House, 21 Darmouth Street, London, SWIM9BE, U.K.
[‡]Retired from Esso Exploration and Production: 2 Churchfields Ave., Weybridge, Surrey, England.

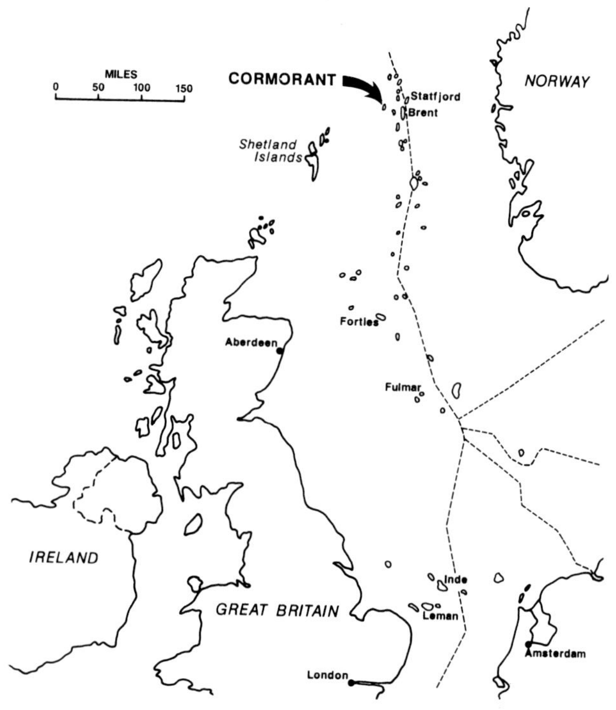

Fig. 1. Map of North Sea.

Members are fluvial and lagoonal sediments deposited in back-barrier and delta-top environments. The Tarbert Formation, which represents a transgressive phase, is absent in Block I and most of Block II. Upper Jurassic shales form the top seal.

Diagenesis appears as kaolinite and mica and as pore-throat reduction due to compaction and quartz overgrowth. In general, Tarbert and Ness sandstones are quartzose while those of the Etive, Rannoch, and Broom are feldspathic and contain more kaolinite. Illite exists in the aquifer. A more detailed description of Cormorant geology can be found in Budding and Inglin (1981) and Gaarenstroom (1984).

The reservoir contains highly undersaturated crude with areal variations in fluid properties (Table 1). Water viscosity is 0.3 centipoise, and the mobility ratio varies from favorable in the south (Block II) to neutral or slightly unfavorable in the north (Block III).

Initial Development Plan

Although the original plan (Figure 2) varied from block to block, the principles were similar: (1) one or two rows of producers located at the crest just to the west of the truncation surface or bounding fault, (2) a single row of injectors drilled just beyond the oil-water contact, (3) producers and injectors fully perforated to waterflood all layers concurrently. Development has been implemented much as planned in some areas

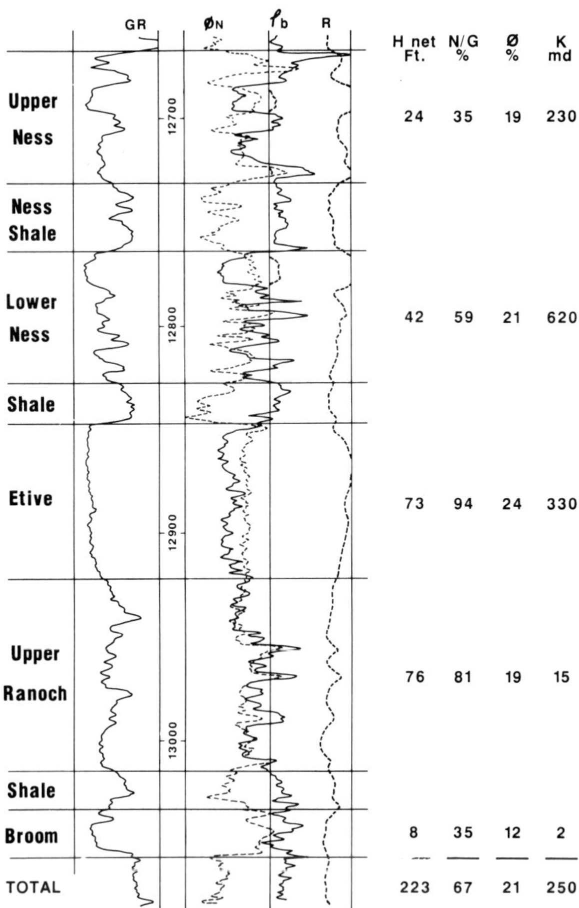

Fig. 3. Type log—UMC Well W-3.

while significant changes have been necessary elsewhere.

A discussion of the individual blocks follows, starting with the least complicated (Block III) and proceeding to the most complex (Block IV).

Block III

Development of Block III (Figure 4) has gone much as planned although some changes have been required. Two rows of producers were planned as a contingency against lack of sand continuity, but this was altered to a single row when early well results showed excellent continuity. The injectors were to be located in the aquifer beyond a fault zone but this was altered after

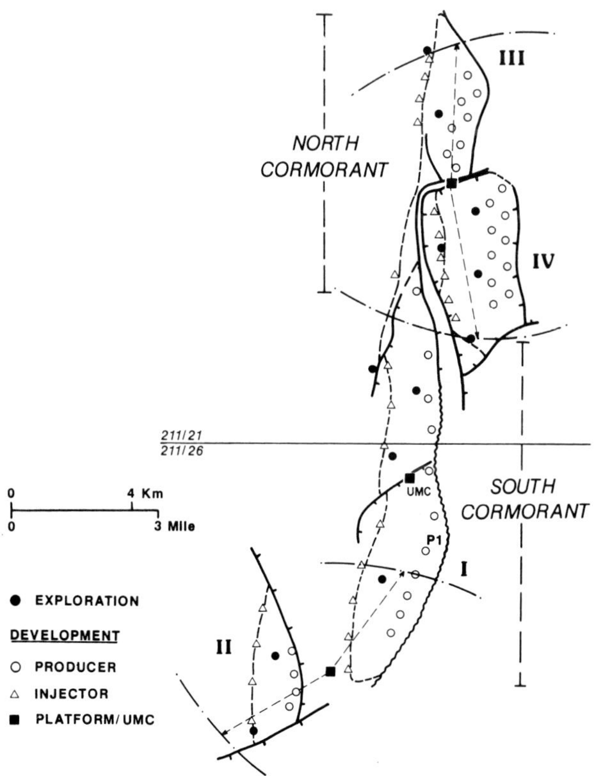

Fig. 2. Map of Cormorant Field.

Table 1. Reservoir Fluids Properties

Block	Pb psi	Rs scf/stb	μo cp	Boi rb/stb
III	865	210	1.32	1.16
IV	1480	380	1.07	1.23
I	1830	580	0.71	1.41
II	2955	800	0.55	1.44

the first injector, CN-19, found poor rock properties and pressures some 400 psi higher than in the oil column, indicating significant flow restriction across the faults. Three subsequent injectors were drilled to the east of the fault zone. Each found good communication with the oil column, and reservoir pressure and rates have been maintained at high levels.

Other Block III uncertainties were the platform drilling reach and the need for subsea injection in the north of the block. After drilling a highly deviated injector (CN-03), an evaluation was undertaken. A geologic study showed the major formations to exhibit excellent sand continuity. A simulation study of the Etive included 2-D cross-sections (Figure 4) to investigate displacement process and generate pseudo-relative permeabilities, and a 2-D areal model to investigate recovery in the north of the block. The results from the northern section are shown in Figure 5. Permeabilities were assigned based on foot-by-foot core data from well 211/214, and the cross-section was flooded with a fully perforated injector located below the oil-water contact, representing CN-03, and an upstructure producer. Water saturations are shown at water breakthrough and indicate a tendency for water to underrun the oil column with the higher permeability lower section being flooded preferentially.

The areal grid for the Etive (Figure 6) covered the entire block so that pressure gradients created by existing and planned wells would be included. The grid was refined in the area of interest in the north of the block.

While published correlations, based on pattern flooding, indicate improved recovery with an injector to the north, the saturation contours after ten years of production show good displacement to the north of CN-03 with existing injectors. The results showed lack of justification for an expensive subsea injection well.

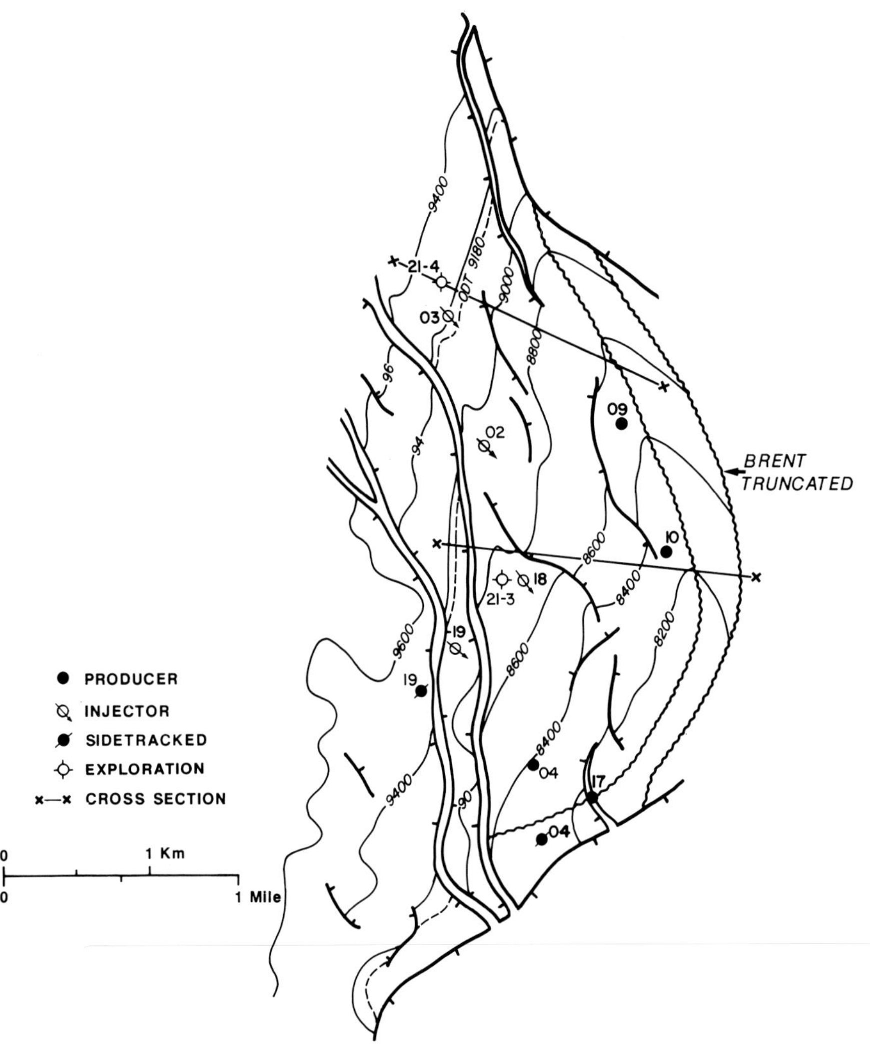

Fig. 4. Block III Top-Brent Structure map.

Block I

Before proceeding with subsea development for the major part of Block I, a study of the block was undertaken. Key questions concerned the recovery achievable with the limited number of wells available with the Underwater Manifold Centre (UMC), and the performance of such a development with fully perforated wells. The study included a detailed geologic interpretation, three simulation cross-sections and a 2-D areal simulation using pseudo-relative permeabilities generated from the cross-sections. The areal grid and the locations of the cross-sections are shown in Figure 7. The results of a thermal simulation of cold water injection into a hot reservoir were also incorporated into the areal study as were the complicated wellbore and subsea flowline hydraulics. The results of the study indicated (1) that recovery was relatively insensitive to well spacing assuming good sand continuity, (2) there was no significant benefit from dual zone completions, (3) injection rates would be moderate due to injection of cold water, and (4) producers should be located to encounter a full reservoir section

near the truncation feature. The last item created a significant seismic challenge in field development.

The nature of the problem can be seen in the structural section shown in Figure 8. While desirable to locate producers as high as possible above the oil-water contact, thereby delaying water breakthrough and minimizing attic oil, a well with the upper formations truncated was undesirable in that the truncated layers could not be waterflooded concurrently. While truncated wells could be sidetracked later, this would delay the start of the waterflood in these layers, many of which are of poorer quality and need the maximum production and injection from the start for good recovery. The challenge was to get as near the truncation as possible without encountering it. This challenge was met with P-2, the first UMC well, but P-3 encountered a severely truncated section. Analysis of the seismic line (Figure 9) suggested an alternative seismic interpretation.

The initial interpretation assumed that the dip seen on the flank of the structure at the truncation (Kimmerian unconformity) level continued over the struc-

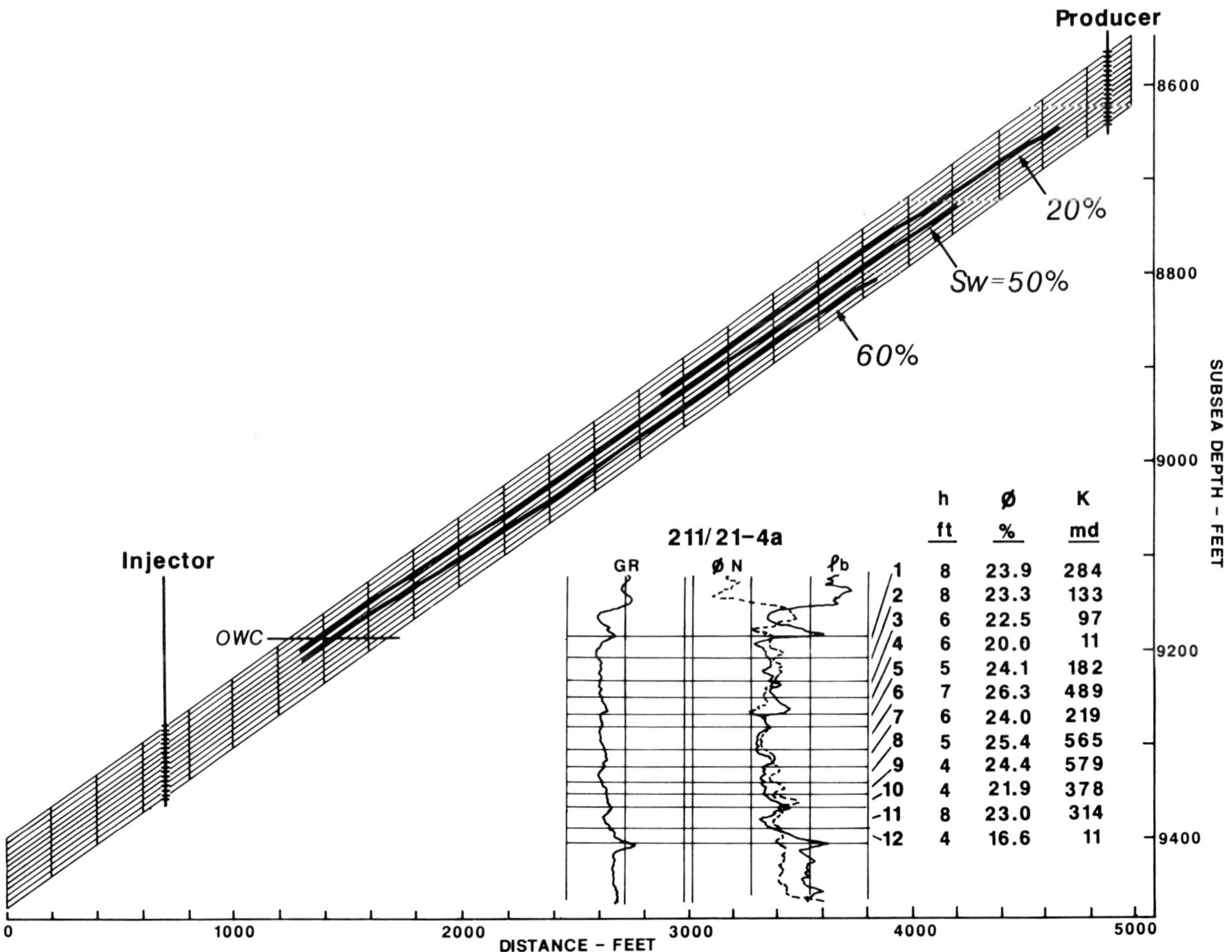

Fig. 5. Simulation cross-section of Etive Formation—North of Block III.

tural crest with continuation of the underlying shale across the crest. This interpretation suggested absence of Brent erosion until reaching the point where the seismic event at the Top Brent level began to flatten (near P-3). However, the P-3 well drilling results at this location found the Brent deeply eroded. Reevaluation suggested that at the P-3 location the shale interval between the Kimmerian unconformity and the top of the Brent thins in such a way that destructive interference occurs with loss of signal at the truncation level. A revised interpretation locates the onset of Brent erosion where loss of signal occurs at the truncation level (see time slice Figure 10). Accordingly, the well was sidetracked a short distance to the west (P-3A) where a full section was encountered. This move yielded a superior producer having a higher initial Productive Index and in communication with more oil in place.

Subsequent confirmation of the revised interpretation led to a simplification of Block I development and a major cost saving. The location for P-6, the most northern UMC well, was found to be at the limit of

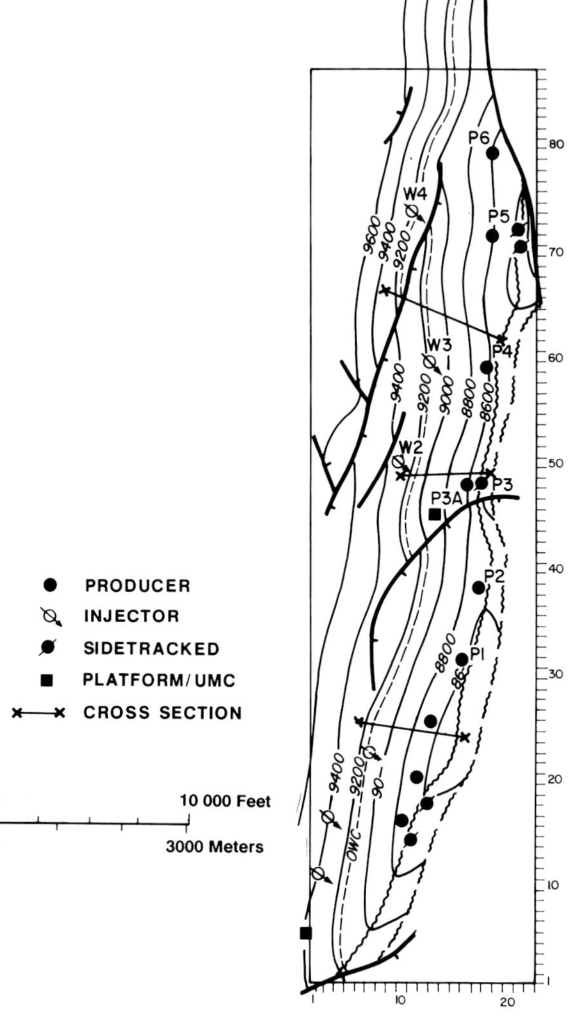

Fig. 7. Top-Brent structure map of Block I with areal simulation grid.

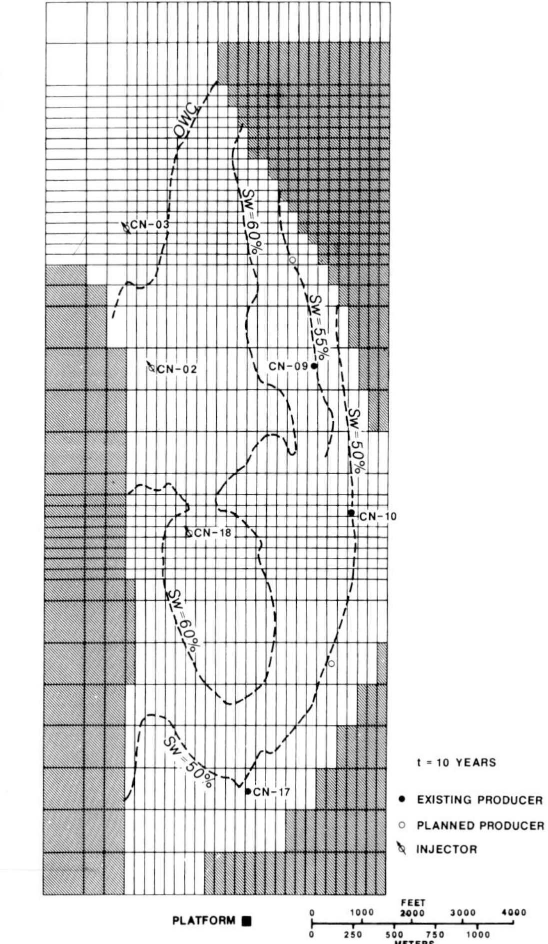

Fig. 6. Block III (Etive) Areal simulation grid and results.

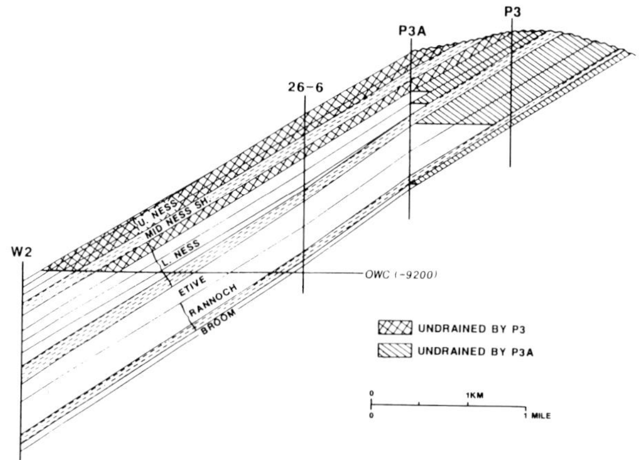

Fig. 8. Structural cross-section through wells P-3 and P-3A.

North Cormorant's revised drilling reach. An evaluation of the pros and cons of drilling the well from the platform versus the UMC showed the risk of sidetracking as the critical issue, with the lower angle UMC well having the advantage if the likelihood of sidetracking was high.

Following review of the seismic sections through each producer in the block, the risk of severe truncation was judged to be small, and the well was drilled from the North Cormorant platform with a saving of £15 million (or $25 million). The saving was large owing to elimination of expensive insulated subsea flow lines. The well encountered a thick Brent group section with only slight truncation.

Figure 10 is a time slice at 2360 ms, just above the Kimmerian unconformity, which illustrates the onset of major Brent truncation between locations P-3 and P-3A.

Block II

Block II (Figure 11) is the smallest block with much of the 100 million barrels of oil-in-place concentrated in the Upper Ness and Etive sand. The original development plan (Figure 2) included four crestal producers and four flank water injectors. Early production and pressure history (Figure 12) from the first producers (CA-03 and CA-33) indicated only about one-half of the mapped oil-in-place was supporting production. Pressure recovery during shut-in periods suggested repressuring due to: (1) wellbore cross-flow from pressure-supported layers, or (2) flow around or through a fault partially segmenting the block. Segmentation was supported by the relatively high pressures found in the first injector, CA-34. Injection into this well at rates of 4000–5000 barrels of water per day was much lower

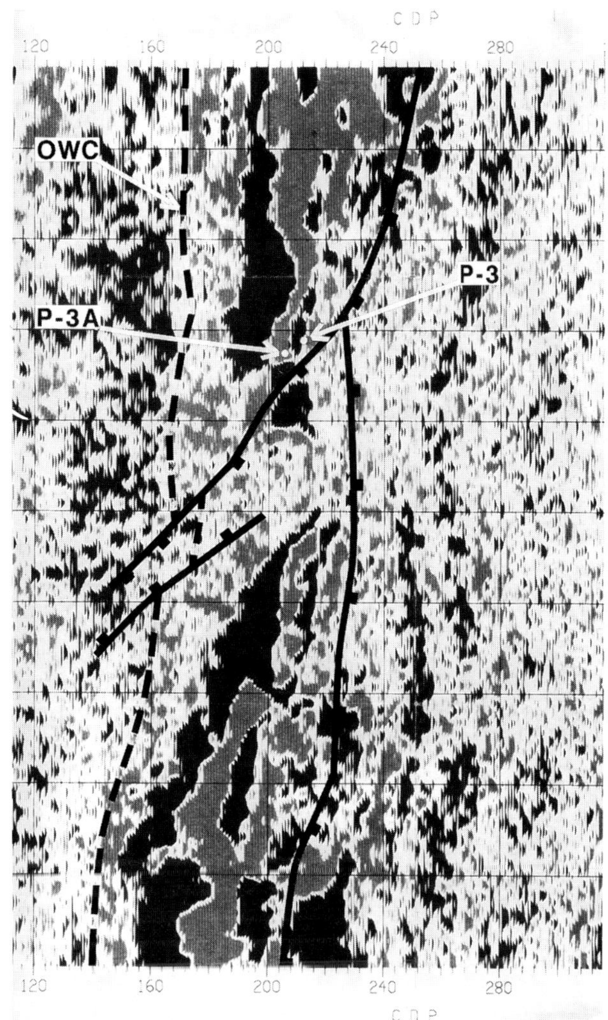

Fig. 10. Seismic time slice at 2360 ms—Block I.

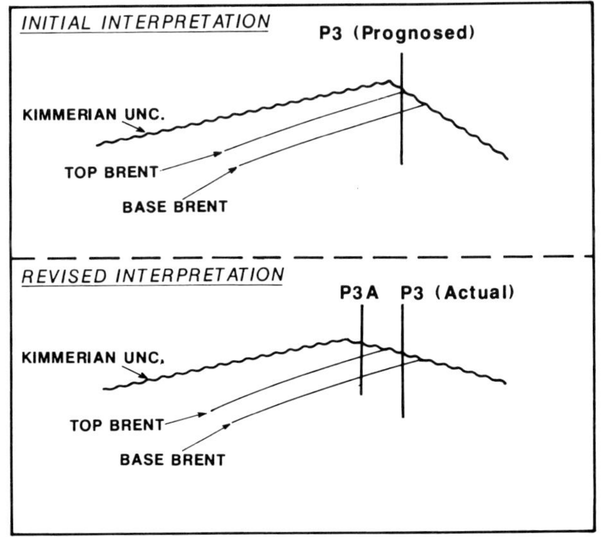

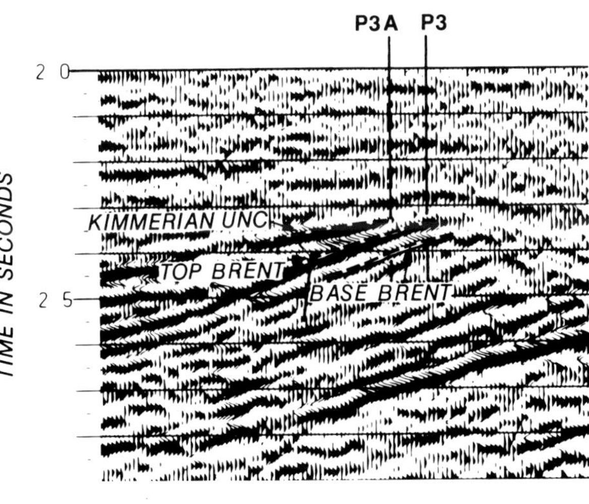

Fig. 9. Seismic section through wells P-3 and P-3A.

than expected, with very little water entering the usually good quality Etive sands, which yielded most of the 6000–8000 barrels of oil per day per well from the two crestal producers. RFT pressures in injector CA-30 and in suspended well 211/26-2, taken after considerable water was injected into CA-34, showed an Upper Ness pressure 1400 psi higher than in the Etive. Little pressure response was noted in the crest-

al producers which continued to decline as flowing pressures approached the bubble point. Understanding was hampered by lack of core material downstructure and the inability to measure Upper Ness and Etive pressures separately in the fully perforated crestal wells.

An integrated geologic and reservoir engineering study of the block was undertaken, and a detailed seismic study revealed a fault just to the west of the crestal producers. Although the fault had offset of only about 50 ft, and could not be traced over the full length of the block, it could create a significant flow restriction in the thin Upper Ness sands. This possible flow restriction led to a strategy involving the drilling and possible sidetracking of an additional producer to resolve one aspect of the problem.

CA-35 was drilled upstructure and found the formations uniformly pressure depleted, thereby confirming that in the Upper Ness there was a significant flow restriction which would prevent depletion of the downstructure oil with existing crestal producers. The well was accordingly sidetracked to the west to provide a drainage point for the downstructure Upper Ness oil. The well found the Upper Ness above initial pressure and the Etive pressure depleted. The well was placed on production, being completed as a dual to allow pressures in the two formations to be monitored separately.

Results from petrographic analysis of core material from injector CA-30 and from upstructure producers offered explanations for the poor Etive injectivity. First, the mineralogy of the Upper Ness, where injectivity was good was found to differ significantly from that in the Etive, where injectivity was poor. As shown in Table 2, the fraction of feldspar and kaolinite in the Etive is higher than in the Upper Ness. In addition samples from the Etive in injector CA-30 were found to contain fibrous illite. Studies by Shell and others (Heavyside et al., 1983 and de Wall et al., 1986) had indicated that these elongated clay structures can have a dramatic effect on permeability. An example of the illite structure which we found in CA-30 is shown in Figure 13.

Tests comparing liquid permeabilities measured on preserved core plugs with routine air permeabilities following extraction and drying indicate that routine core results significantly overstated permeabilities. The delicate clay structures which result in low in-situ

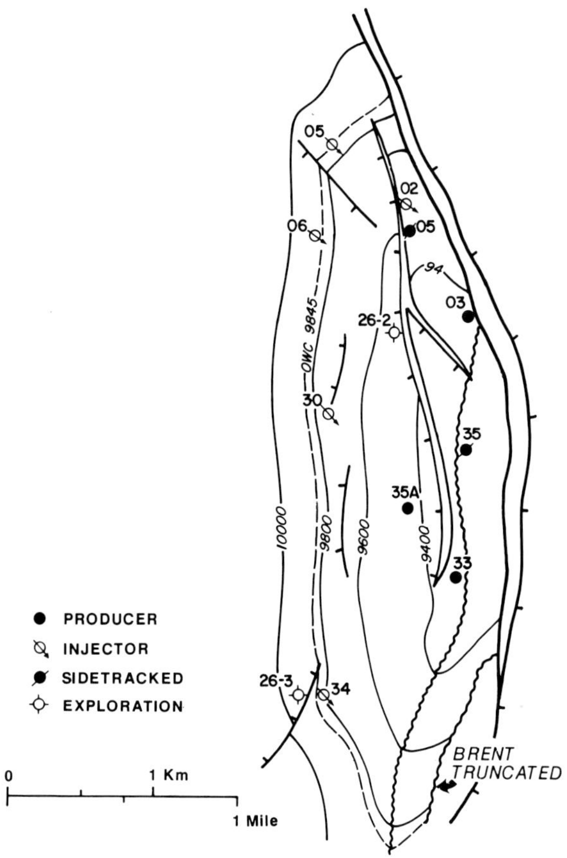

Fig. 11. Top-Brent structure map—Block II.

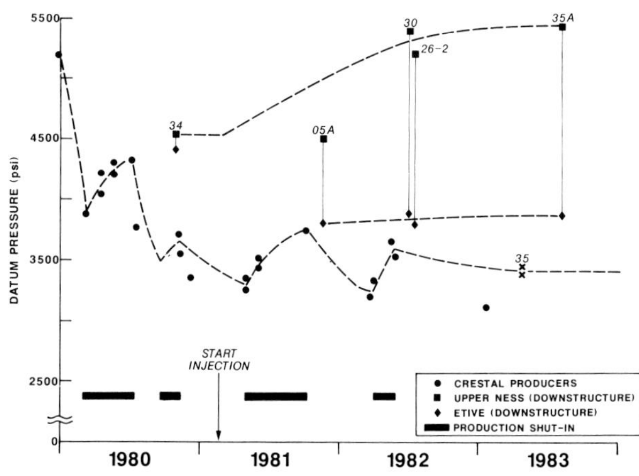

Fig. 12. Early Block II pressure history.

Table 2. Block II Petrographic Results

	Upper Ness	Etive
No. of samples	24	50
Feldspar (%)	3	10
Kaolinite (%)	4	7

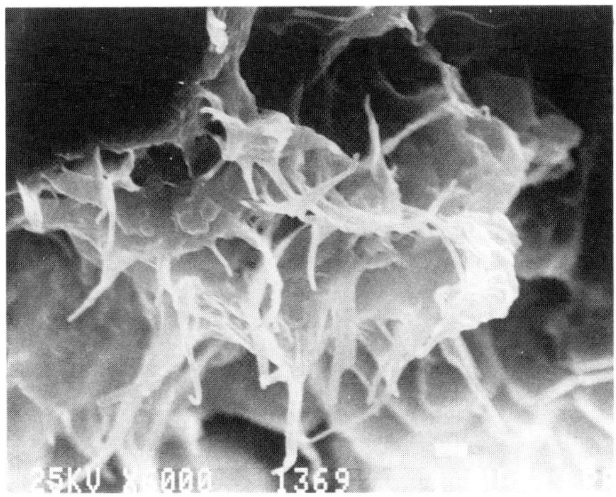

Fig. 13. Scanning electron microscope photograph of illite structure—well CA-30 (Etive).

permeability are broken during extraction and drying, thereby increasing the size of pore openings and the resulting air permeabilities. Comparisons of permeability measurements from producer CA-35A and injector CA-30, shown in Table 3, illustrate the adjustments to routine core analysis results. This comparison shows that routine air permeabilities on Etive cores from the aquifer are overstated by a factor of 13. Our analyses showed that poor Etive injectivity was due to low permeability rock and not to wellbore damage. Failure to apply the correction factors to air permeability measurements from the aquifer would have led to continued attempts to stimulate the Etive to achieve rates consistent with the routine air permeability results.

A 3-D simulation study of the block was the final step in the evaluation. The objectives were to reconcile the revised reservoir description with the complex pressure-production history of the block and to evaluate alternate development plans. The Etive grid from the model is shown in Figure 14. A series of cross-sections were constructed through the block, and flooded to depletion to derive pseudo-relative permeability functions for the 3-D model. The mid-structure

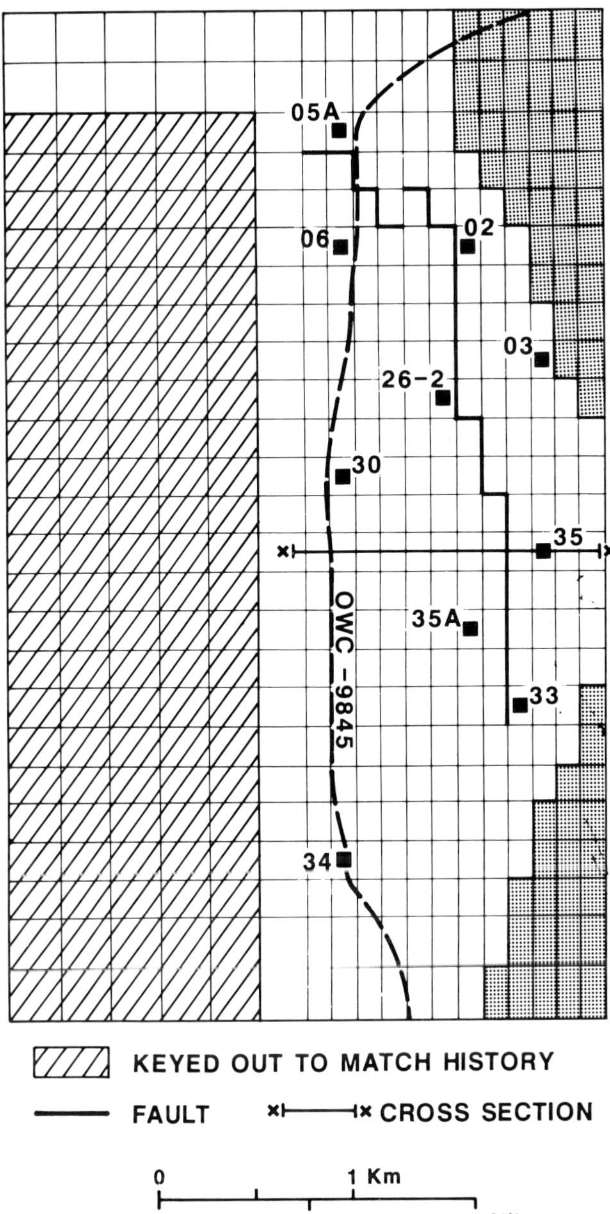

Fig. 14. Etive simulation grid—Block II.

Table 3. Block II Special Core Analysis Results

	Upstructure (CA-35A)		Downstructure (CA-30)	
	Upper Ness	Etive	Upper Ness	Etive
No. of samples	4	10	10	13
K liquid (md)	174	73	219	2.4
K air (md)	310	163	521	31
K air ÷ K liquid	1.8	2.2	2.4	12.9

seismic fault and a minor fault to the northwest were assumed to be sealing over their defined lengths. RFT pressures were the primary matching parameter but with note made of routine pressure data and flowmeter results. The key variable in the history match was the length of the sealing fault across the reservoir. A special feature of the model was modification of the rate routine to allow repressuring through wellbore cross-flow during extended shut-in periods. The RFT pressure match achieved for the Upper Ness and the Etive are shown in Figure 15.

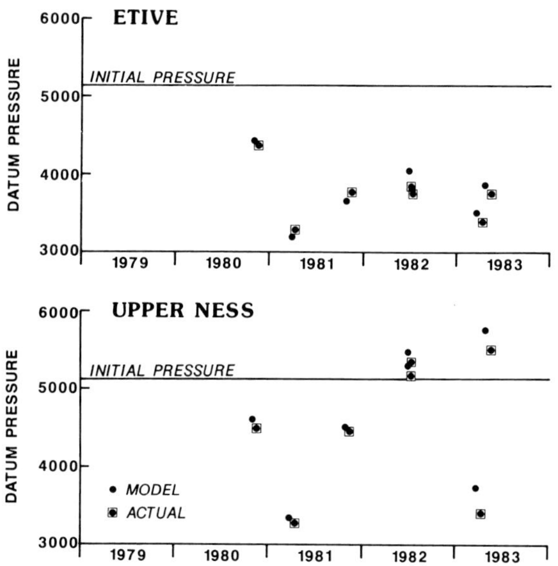

Fig. 15. Upper Ness and Etive pressure history match—Block II.

Prediction runs were made for two key cases. The first case assumed that the original development plan, with crestal producers and downstructure injectors, was continued for the Etive. The second case assumed a revised plan with the upstructure and downstructure areas being waterflooded separately. The forecasts for the two cases are compared in Figure 16; they show that the revised plan not only recovers slightly more oil but does so far more efficiently.

The restricting nature of the fault within the Etive was confirmed through an extended interference test conducted by Shell, and as a result the development plan for Block II was altered such that the upstructure and downstructure areas were developed separately. Injector CA-02 was drilled upstructure to waterflood the crestal accumulation, and an additional injector was also drilled in the aquifer (CA-06) to increase downstructure Etive injection. To improve monitoring

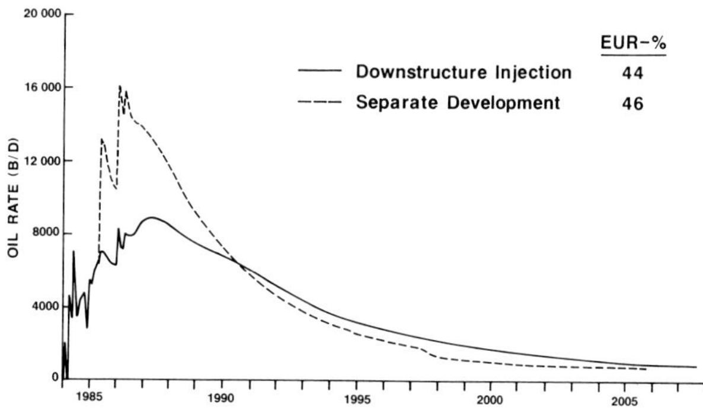

Fig. 16. Predictions of oil rate versus time—alternative Block II developments.

of individual formations and provide greater reservoir control, these new wells were completed as duals.

Block IV

The initial development plan for Block IV included two rows of upstructure producers and a single row of injectors located just below the oil-water contact. Both producers and injectors were to be single completions with all sands perforated.

The original seismic survey was shot in 1976 on a 500 m spacing. While severe structural complexity was indicated, it was unclear how faults should be connected from line to line. The mapping of likely fault orientations was possible following a 3-D seismic survey shot in 1979. The throws of the faults varied, however, and their effect on field development was uncertain. This uncertainty was clarified by early well performance.

Well CN-29 (Figure 17), the first well drilled in the block, encountered a complete Brent Group section at

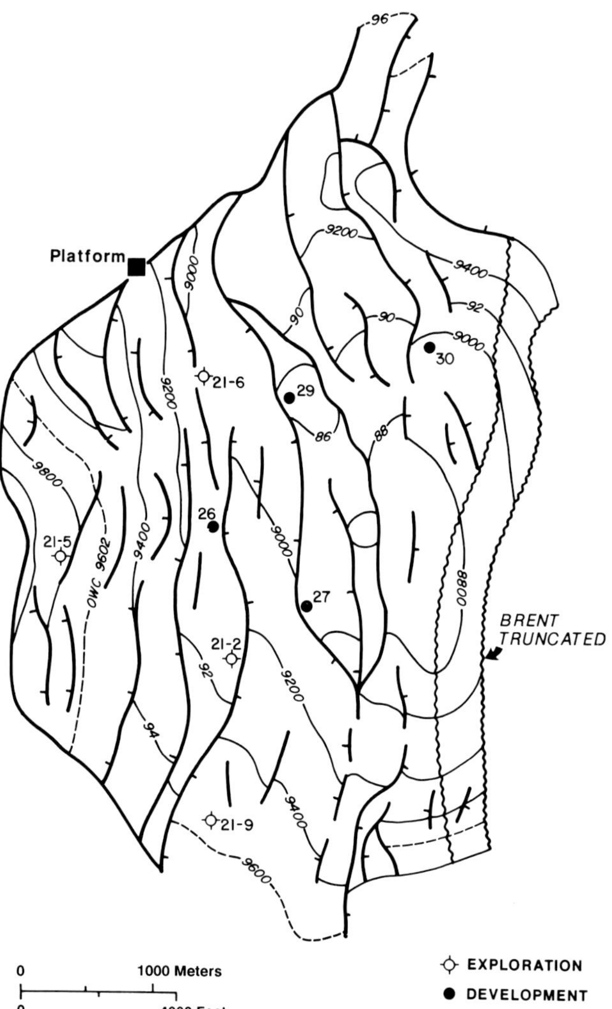

Fig. 17. Early Top-Brent structure map—Block IV.

initial pressure of 5250 psi. Early pressure decline (Figure 18) was much more rapid than expected. Material balance calculations indicated an oil-in-place value of less than 100 million barrels, less than one-sixth of the mapped volume. The second well, CN-27, found severe pressure depletion in line with CN-29, and the segmented nature of the block was confirmed by subsequent wells CN-26 and CN-30. Their pressures showed CN-29 and CN-27 to be essentially isolated from the surrounding areas.

The original plan was clearly inappropriate and a revised delineation and development strategy was formulated. Key components of the strategy were: (1) identify likely sub-blocks based on seismic interpretation, (2) achieve significant pressure depletion prior to injection to improve oil-in-place calculations, (3) identify the injector location for a sub-block at the time the producer is selected, (4) locate wells and drill in a sequence to determine the sealing nature of faults, and (5) reduce delineation risk by drilling wells which can be either a producer or an injector.

Integration of geophysical interpretation into the strategy is critical in: (a) avoiding faulted or truncated sections, (b) avoiding segments with small oil-in-place, and (c) being unable to find an acceptable injector location to support a good producer. This integration is done by conducting a detailed seismic interpretation of the entire area of interest before selecting a well location and identifying the sidetrack alternatives.

The selection of early well locations in the west of the block (Figure 19) provides an example of the application of the strategy. With early field performance indicating that the centrally located wells were largely isolated, an early objective was to initiate production to the west. The producer location is shown as CN-26 but the location for a supporting injector was needed as well. The seismic interpretation indicated that it was unlikely that the well could be supported by drilling an injector into the aquifer to the west due to the numerous north-south trending faults

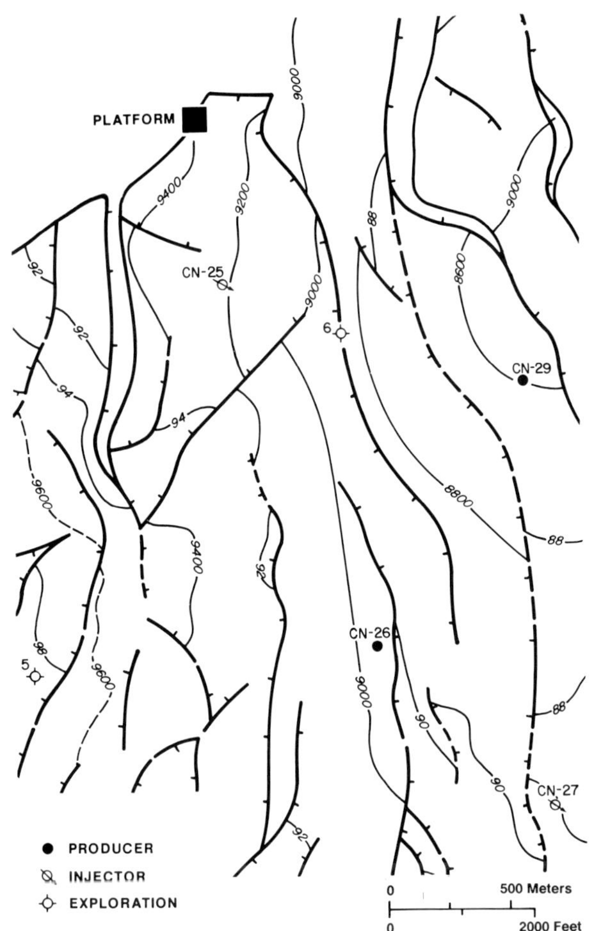

Fig. 19. Top-Brent structure map—West of Block IV.

which separated this area from CN-26. Displacement in a north-south direction was judged more appropriate. It was agreed that the well should be north of CN-26, but there was concern that a well too far north would be separated from CN-26 by the fault running southwest-northeast and a north-south fault which CN-26 nearly intersected. The agreed strategy was to locate the injector, CN-25, as near the large boundary fault as possible to minimize the volume of oil to the north. The well would determine conclusively whether the faults between CN-26 and CN-25 sealed. If pressures were near original, indicating sealing faults, the well would be completed as a producer, thereby developing a separate sub-block. The well would provide an immediate benefit from either outcome.

The CN-25 well encountered a good reservoir section with pressures depleted by over 2000 psi indicating good communication with CN-26. The well, completed as an injector at a rate of over 30 kilobarrels of water per day, reversed the production decline from CN-26 and increased its rate from 10 to over 20 kilobarrels of oil per day.

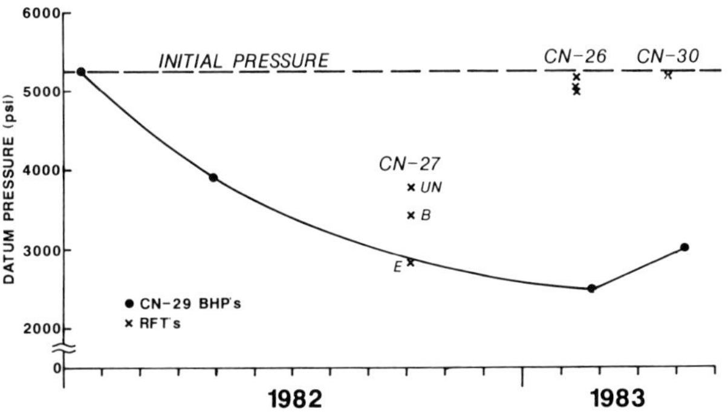

Fig. 18. Early pressure history—Block IV.

With formation tops in early wells differing significantly from predictions, efforts were made to improve the seismic data through reprocessing. While this reprocessing met with some success, analysis indicated that conditions during data acquisition had not been optimum. Improvements in seismic techniques, both in acquisition and processing, since the original survey led to a decision to shoot a second 3-D seismic survey in 1984 on closer (37.5 m) grid spacing. Esso's interpretation based on that survey is shown in Figure 20. Major objectives of future drilling were to: (1) provide injection support and displacement for existing and future producers, (2) delineate the northeast and southeast parts of the block, (3) determine the sealing nature of faults, and (4) further define variations in rock and fluid properties.

Figure 21 compares seismic data for an east-west section through the block for: (a) the original 1979 3-D survey, (b) the second 3-D survey shot in 1984, and (c) reprocessing of the 1984 survey. The 1984 survey improved fault definition over the 1979 survey and also improved reflector resolution. Reprocessing of the 1984 data set resulted in further improvement in reflector resolution and continuity as well as surpressing multiples. Each of these improved the use of seismic in guiding the development plan and in locating wells.

Permeability Contrast

Work throughout Cormorant has indicated that accurate reservoir description is critical, not only in correctly modeling the water-oil displacement mechanism within a single formation, but also in managing the distribution of injection and production among the various formations. Permeability assignment is particularly important in obtaining valid history matches at Cormorant due to the ambiguous nature of pressure and production data. In the fully perforated, stratified section, routine bottom hole pressure measurements do not indicate individual sand pressures, and production and injection has to be allocated among sands based on periodic flowmeter results.

Initial studies showed considerable scatter in plots of porosity versus permeability and early attempts to correlate core permeability to log response were not entirely successful. A thorough coring program was clearly needed, and the core coverage which has been achieved is shown in Figure 22. Over 12 000 feet of core have been recovered and analyzed and the results are valuable not only in reservoir simulation studies but also in planning perforating, testing, and stimulation programs.

The data from the Etive in well CN-29 is an example of the use of core data in reservoir development and modeling. As shown in Figure 23, a plot of porosity versus permeability using all data points indicates a reasonable fit. However, a detailed analysis shows the inadequacy of this approach as separate populations were identified within the data. Specifically the section at the base of the sand exhibits higher permeabilities for a given porosity value than the upper part of the section. Use of a single trend, in conjunction with log calculated porosities, would not correctly define the stratification, and thus the displacement mechanism in this area of the field.

Examination of log responses in Figure 23 shows that the higher permeabilities correspond to the interval with higher log resistivity readings and with a larger difference between the readings of the density and neutron logs. Similar relationships have been

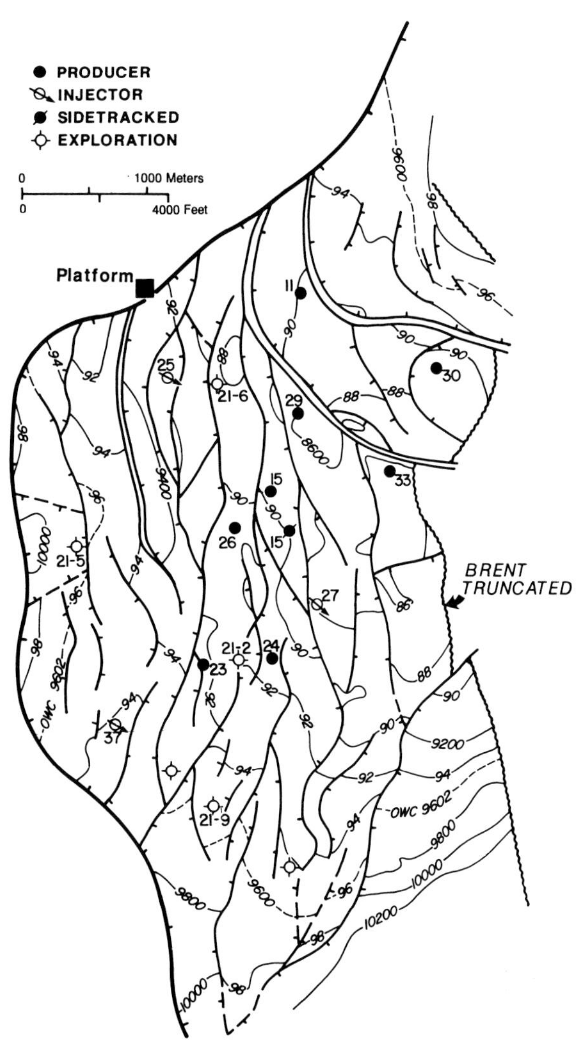

Fig. 20. Current Top-Brent structure map—Block IV.

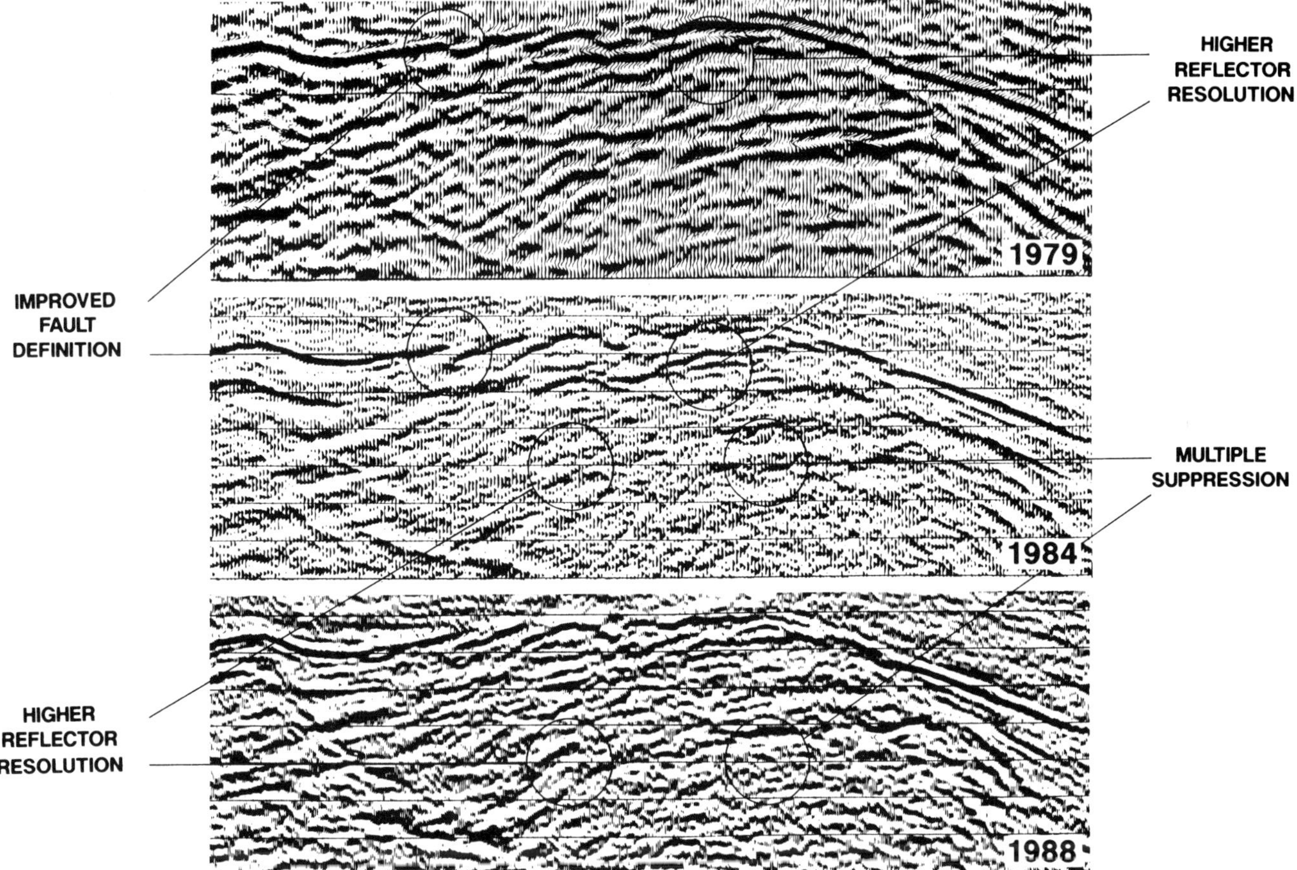

Fig. 21. Comparison of seismic survey quality—Block IV. Upper: 1979 survey; center: 1984; lower: 1988 reprocessing of 1984 survey.

observed elsewhere in Cormorant and indicate the potential for improved permeability estimates in uncored wells through detailed reservoir description work.

Completion Method and Perforation Strategy

The potential for improving field performance through dual or selective completions has been evaluated throughout field life. Early studies showed little or no benefit, except in Block II, for other than single, fully perforated completions. Subsequent well results have shown, however, that permeability contrast within the section is often more severe than originally estimated. Additionally the high and low permeability sands are distributed throughout the section rather than being conveniently grouped for use in dual completions. This problem is addressed by comparing a layer's fraction of oil-in-place with its fraction of permeability-thickness (kh), the latter being a guide to the distribution of production or injection if the well is uniformly perforated. Some zones have excess kh, i.e., they have a much larger fraction of kh than oil-in-place, and these layers will be flooded much faster than the overall section. This can be undesirable where all sands are open in producer wellbores.

The solution to this situation has been to partially perforate sands with excess kh, in effect relying on a mechanical skin to adjust production and injection distribution. Partial perforation of thick layers has been done by shooting only a few feet at the top of the section. Thin layers have been dealt with using a reduced perforating density (one shot per foot) and smaller diameter, less efficient perforating guns. An example of the use of partial perforation to manage injection distribution in Cormorant is discussed in Stiles and Valenti, 1987.

Conclusions

Close surveillance and integration of geologic results with field performance has resulted in numerous changes to the development plan. A number of specific lessons have been learned from our development experience with Cormorant.

1. While 3-D seismic is proving a valuable guide, 3-D should not be allowed to totally control field development, and there is a need for a well thought-out strategy which recognizes seismic limitations.

2. 3-D seismic has been particularly useful in selecting individual well locations. As an example, integration of detailed seismic analysis with well results has made it possible to locate producers in Block I in near optimum fashion and to replace an expensive subsea well with a much less expensive well from a nearby platform.

3. Development wells have encountered sections with permeability contrast significantly different from that envisaged in initial development studies and these differences have led to changes in completion and perforating strategy. A procedure to accurately assess permeability contrast and the resulting effect on completion strategy is needed throughout field development.

4. Partial perforation of intervals has been used to improve production and injection distribution in wells where several sands are perforated.

5. It is important at Cormorant to accurately define permeability distribution within a sand to model water-oil displacement correctly, and identification of different rock types within the same formation is required to do this. There is scope to use log response for such identification in uncored intervals or wells.

6. The integration of seismic, petrographic studies, special core analysis and well test analysis into a simulation study allowed a change to the development plan for a complex area of the field (Block II) to be made with greater confidence.

7. Routine core analysis greatly overstates permeabilities where delicate clay structures are damaged during core handling, resulting in enhanced permeability. A portion of the core from early wells should be preserved at the well site to allow adjustment factors for routine air permeabilities to be developed. Failure to recognize that permeabilities are overstated can lead to incorrect stimulation of poor injectors and to inappropriate initial development plans.

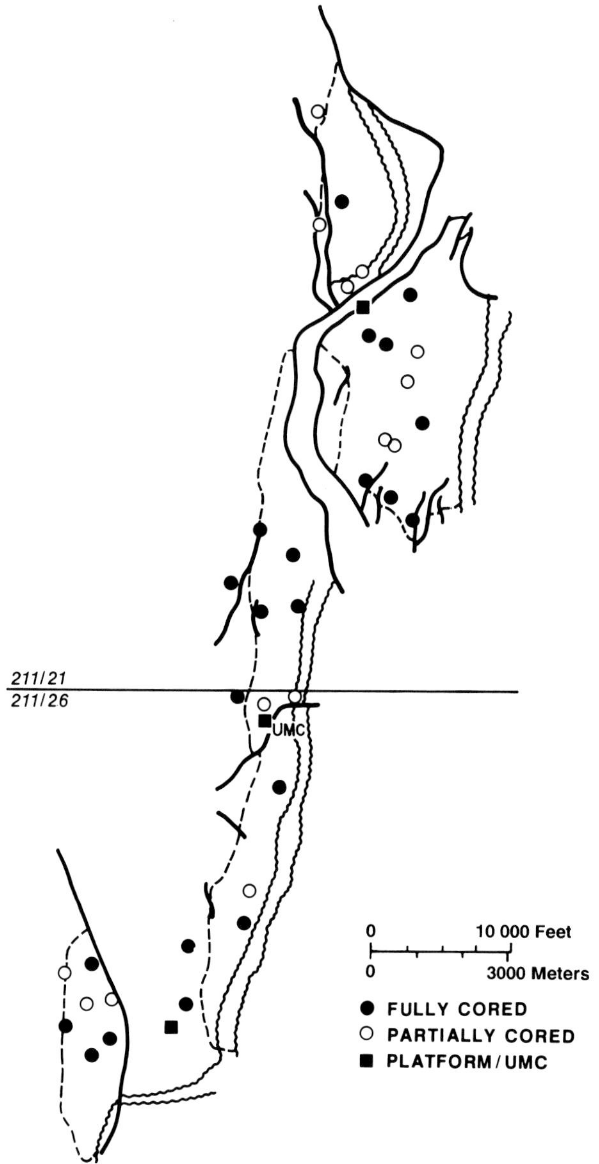

Fig. 22. Map of Cormorant with core coverage.

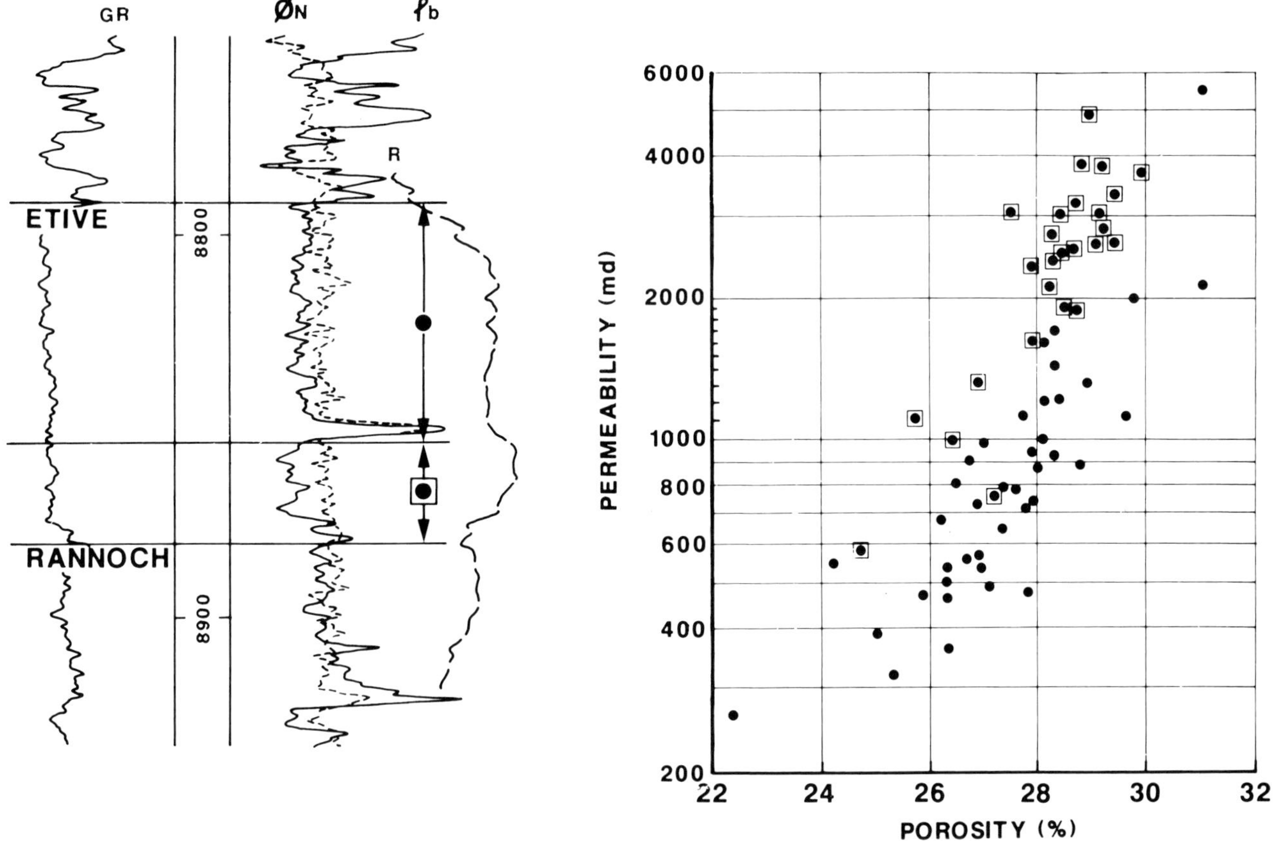

Fig. 23. Portion of CN-29 log with porosity versus permeability for Etive. Gamma ray log (GR), neutron porosity (φN), density (ρb), resistivity (R).

Acknowledgments

The authors express appreciation to Shell U.K. Exploration and Production for giving permission for publication of this paper.

References

Budding, M.C., Inglin, H.F.,1981, Reservoir geological model of the Brent Sands in southern Cormorant in Petroleum Geology of the Continental Shelf of Northwest Europe, Heyden & Son Ltd.

de Wall, J.A., Bil, K. J., Klantorowicz, J. D., and Dicker, A.I.M., 1986, Petrolphysical core analysis of sandstones contailning delicate illite: Presented at the 10th European Formation Evaluation Symposium, Paper Z.

Gaarenstroom, L., 1984, The value of 3D (3-dimensional) seismic in field development: Presented at the 59th Ann. Soc. Petr. Eng., Am. Inst. Min.,Metall. Eng. Technical Conference, SPE Paper 13049.

Heavyside, J., Langley, G.O., Pallatt, N.,1983, Permeability characteristics of Magnus Reservoir Rock, 8th European Formation Evaluation Symposium.

Stiles, J. H., Valenti, N. P., 1987, The use of detailed reservoir description and simulation studies in investigating completion strategies, Cormorant-UK North Sea: Presented at Offshore Europe 87, Soc. Petr. Eng. Paper 16553.

Application of Three-Component VSP Data on the Interpretation of the Vulcan Gas Field and Its Impact on Field Development[1]

M. D. Noble, R. A. Lambert,* H. Ahmed[‡] and J. Lyons[‡]*

Introduction

Results of three-component Vertical Seismic Profile (VSP) surveys acquired during the initial phase of development drilling of the Vulcan Gas Field, situated in the UK sector of the Southern North Sea, are described. A main objective of the survey was to acquire high-resolution VSP seismic data along lines of varying azimuth, radiating from the wellbore. These lines would supplement the existing seismic coverage of the gas-bearing Rotliegendes Sandstone reservoir and possibly highlight any potential drilling hazards or structural complexities not previously mapped, which could have an impact on the location of later development wells.

The three-component data were acquired in a near-vertical well using four orthogonally arranged, fixed-offset (2 km), energy sources. A three-component borehole geophone was used in conjunction with a gyro tool for measurement of horizontal-component geophone azimuth.

The results of both conventional vertical component only and of three-component VSP data processing are compared. The impact of these data on the current interpretation of the Vulcan Field is discussed. In addition, the fixed-offset-source VSP data are directly compared with conventional (normal-incidence) VSP data subsequently acquired along deviated wells with trajectories similar to the original VSP seismic lines.

It was concluded that, in this case, three-component VSP processing of the data was necessary to provide a useful sub-surface image. The fixed-offset VSP technique utilized in this case produced useful seismic data up to some distance from the wellbore and should therefore be recognized as a very useful location verification tool for future development drilling.

Recent advances in both VSP acquisition and processing of offset-source data now enable the interpretation geophysicist to view high-quality lateral seismic profiles around the locality of the borehole. For the interpretation geophysicist involved in detailed geophysical mapping of an oil or gas field, particularly during the initial development drilling stages, the acquisition of such data has advantages over conventionally acquired surface seismic data. The VSP's unique geometry ensures better resolution than for surface seismic data due to the shorter energy travel path and the relatively better coupling of the geophone to the formation away from near-surface noise. Also, multiple-generated noise can be directly identified and suppressed by deconvolution operators deterministically derived from the measured downwaves in the VSP. In addition, VSP acquisition and processing time is relatively short compared to that of acquiring additional two-dimensional (2-D) or three-dimensional (3-D) surface seismic data.

By locating the source radially away from the borehole, a series of lateral VSP profiles can be obtained which may yield critical structural and stratigraphic information about the reservoir or zone of interest. Such data may be used to advantage in the development of a field by arranging for the lateral profiles to illuminate future development well locations, particularly if these locations are in areas of sparse or poor surface seismic and well data.

In addition, three-component VSP borehole-geophone recording and processing is now available to the industry, enabling the complete wavefield to be investigated in the subsurface (e.g., DiSiena et al., 1984; Ahmed et al., 1986). This can be used to enhance the compressional (P) wave reflected image as well as revealing useful shear (S) wave information to assist in geologic and reservoir engineering studies.

A case history of the Vulcan Gas Field is presented which illustrates the use of fixed-offset VSP data in improving the existing structural interpretation of the Rotliegendes Reservoir over the East Vulcan fault block and its impact on development drilling, present and future.

[1]Paper presented at the 49th Meeting of the European Association of Exploration Geophysicists, Belgrade, June, 1987. ©1988 EAEG. Published in First Break, **6**, 131–149. Reprinted with permission.

*Conoco (UK) Limited, Park House, 116 Park Street, London W1Y 4NN, UK.

[‡]Seismograph Service (England) Limited, Holwood, Keston, Kent BR2 6HD, UK.

The Vulcan Gas Field

The Vulcan Gas Field is situated in the UK Sector of the Southern Gas Basin of the North Sea, lying approximately 60 km offshore and in an area of 30 m water depth (Figure 1). The field straddles the 49/21 and 48/25b block boundary and is operated by Conoco (UK) Ltd. Gas-bearing reservoir sands are contained within the Permian Rotliegendes Leman Sandstone Formation. The Rotliegendes Sandstone in this area is composed predominantly of stacked aeolian sand-dune facies and is sealed by the overlying Zechstein Carbonate and Evaporite sequences. The structure consists of two large en-echelon tilted fault blocks cut by numerous small faults of variable strike (Figure 2a). The source of gas is believed to be the underlying Carboniferous Coal Measures. A generalized geologic cross-section and stratigraphic column are shown in Figures 2b and 3.

The 49/21-6 discovery well was drilled in March, 1983, and tested gas at a combined rate of 35 MMSCF GPD, followed by appraisal wells 48/25b-2 and 48/25b-3 the following year which tested at 25 and 11 MMSCFGPD, respectively (see Figure 2a for well locations). The field is currently under development.

Geologic Setting

The Vulcan field lies within the axial region of the inverted Sole Pit Basin which underwent deepest burial in late Jurassic–Early Cretaceous time and which was subsequently uplifted by as much as 1830 m in Middle and Late Cretaceous time (Walker and Cooper, 1987). As a result, most of the Cretaceous and Jurassic section has been eroded. Although the post-Permian section is relatively unfaulted, there are numerous faults which cut the Rotliegendes Sandstone and which have probably been active in various times during the burial and inversion phases. They may also

have been active during the deposition of the Rotliegendes Sandstone itself, having implications for development drilling.

The deep burial appears to have had an important diagenetic effect on the porosity and permeability of the Rotliegendes Sandstone such that reservoir quality appears to vary across the Vulcan field. However, the observed variations in reservoir quality do not appear to be solely the result of depth of burial, but also appear to be related to depositional facies. The most favorable reservoir section is known as Zone-3 (after Conway, 1986) occurring approximately 100 m below the top of the Rotliegendes. Well logs and velocity analysis show that the acoustic impedance at the top of Zone-3 is such that a seismic event is just detectable on surface seismic data (Figure 3). In addition, evidence from wells in adjacent gas fields suggest that porosity and permeability degradation is likely in the immediate vicinity of faults which cut the Rotliegendes Sandstone. No direct evidence is seen of this in the Vulcan field.

Field Development Plan

The initial development plan consists of a maximum of 20 development wells, 10 on each of the East and West Vulcan fault blocks with the first well (R01) to be drilled in a crestal position on East Vulcan.

During the latter half of 1985, 620 km of 40-fold seismic data were acquired (using an airgun array) over the field in the form of a 500 × 1000 m 2-D grid in preparation for development drilling. Resulting seismic data quality was considered to be good, although the Top Rotliegendes seismic event and the Top Basalanhydrit seismic event (key events for depth conversion) appeared poorly resolved in places (Figure 4a). Correlation with the synthetic seismogram produced from the 49/21-6 well suggests that this may be due to the effect of intra-Zechstein short-period multiples (Figure 3). A top Rotliegendes depth map of the East Vulcan fault block, based on 1985 surface seismic data is shown in Figure 4b.

The Rotliegendes Sandstone Reservoir, being tight, dictated that a number of the development wells would have to be hydraulically fractured prior to production. The development philosophy was therefore based on the following criteria:

(1) Drill the Zone-3 structural highs to ensure maximum productivity. The Top Rotliegendes Reservoir depth map shown in Figure 4b provides an indication of the Top Zone-3 structure. However, if intra-Rotliegendes faults had been active during sand deposition, a more detailed seismic stratigraphic interpretation of Zone-3 would be necessary requiring higher-resolution data than the surface seismic data.

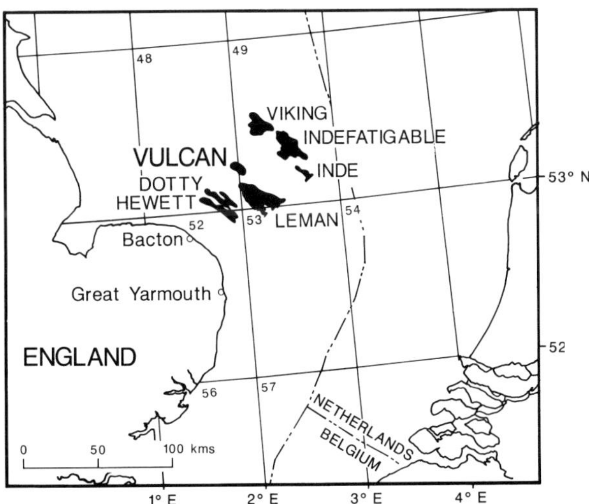

Fig. 1. Location map of the Vulcan field.

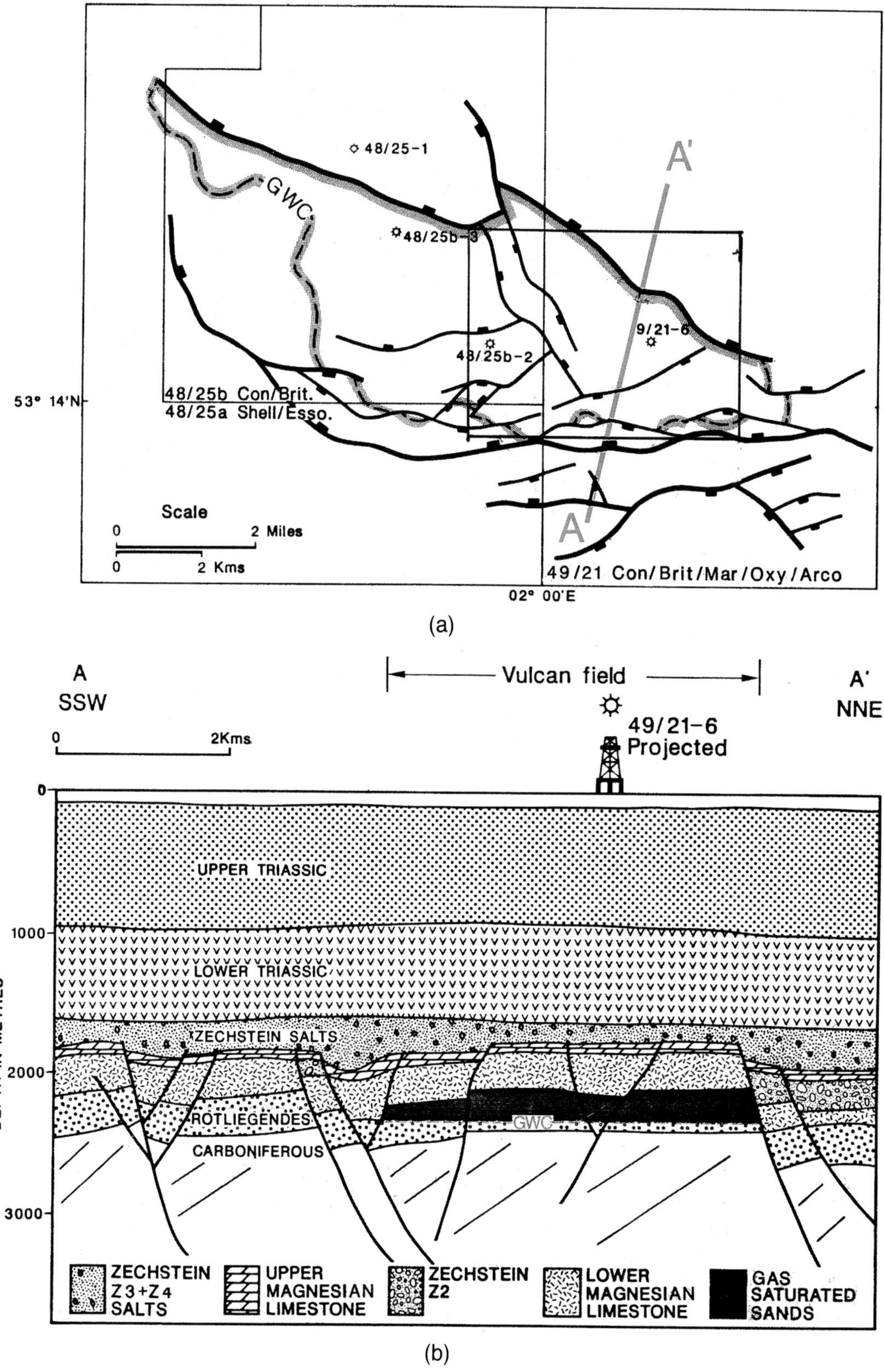

Fig. 2. (a) Fault pattern at Top Rotliegendes reservoir showing the location of exploration wells. The box indicates the area of Figures 4b and 5. (b) Geologic cross-section AA' (see location on Figure 2a).

(2) Preferentially drill those areas having the best predicted reservoir properties.

The prediction of reservoir properties over the East Vulcan fault block from only one existing well would be subjective for reasons explained above. Consequently, a better understanding of the structural–stratigraphic reservoir model would be required to assist with such predictions. This would also necessitate the use of high-resolution seismic data.

(3) Avoid faults within the reservoir because of the likely reduction of porosity and permeability, the possibility of a reduced drainage area and potential drilling problems.

Only the location of major faults are known with any certainty from the surface seismic data. Even these, however, are subject to some uncertainty because of the smoothing effect of intra-Zechstein multiples on the top Rotliegendes seismic event, thus reducing the detection of minor faults at this level.

(4) Maintain optimum well spacing for maximum possible drainage of the reservoir. At the same time, sample as many individual fault blocks as possible for the investigation of sealing faults.

An accurate fault pattern map at Top Rotliegendes and Zone 3 level would therefore be required.

(5) For highly deviated wells drilled through the reservoir, ensure the borehole azimuth is coincident with the least principal stress direction (σ_3) for efficient hydraulic fracturing.

If the wellbore azimuth is significantly different from the least principal stress direction (σ_3) then vertical boreholes through the reservoir are necessary for efficient hydraulic fracturing. For any deviated well this will require an S-shaped borehole which will severely limit the possible reach from the platform.

This last criterion is particularly important in tight gas reservoirs such as the Vulcan field. The direction of σ_3 has been traditionally inferred from the mapped fault trends obtained from the surface seismic interpretation.

How could development geophysics contribute to assisting the development drilling program further? To improve the quality of the interpretation, particularly the fault pattern, additional 2-D or even 3-D seismic data could have been acquired and processed. How-

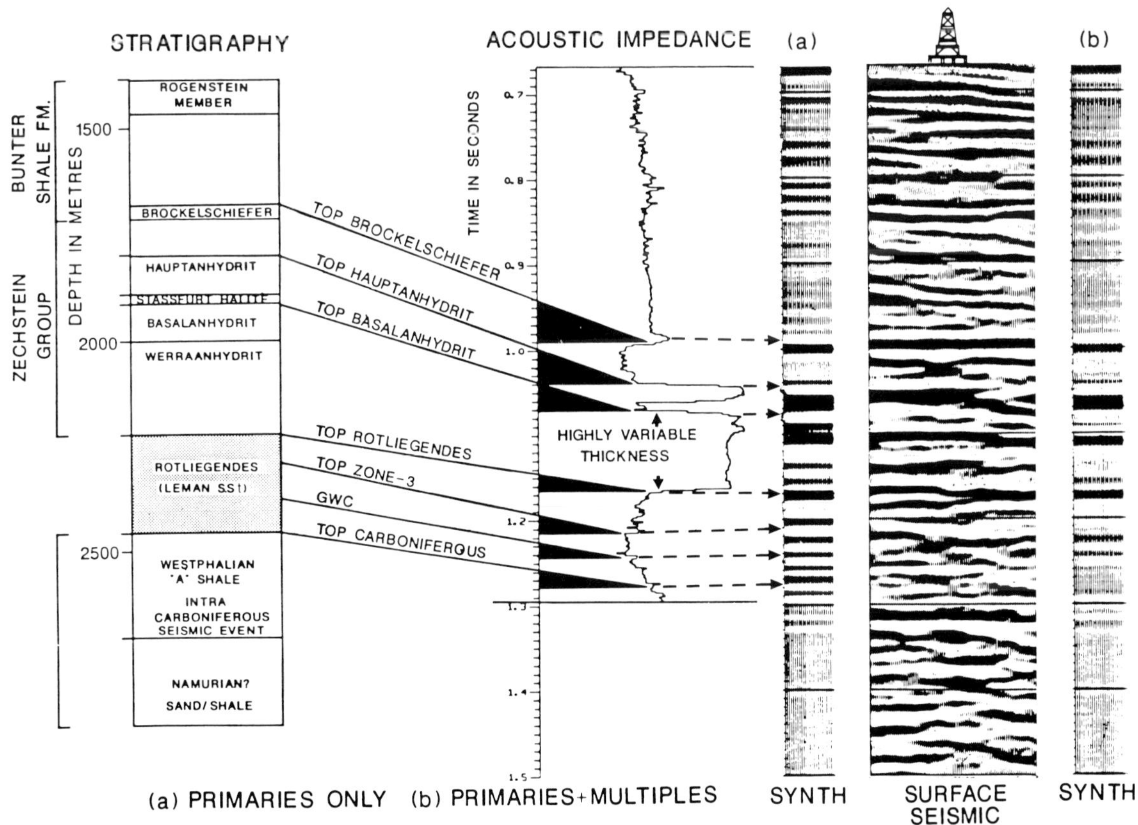

Fig. 3. Correlation of surface seismic data with the 49/21-6 well-log data; the synthetic seismogram was derived from the acoustic impedance measured in the well and a wavelet extracted from the surface seismic at the well location.

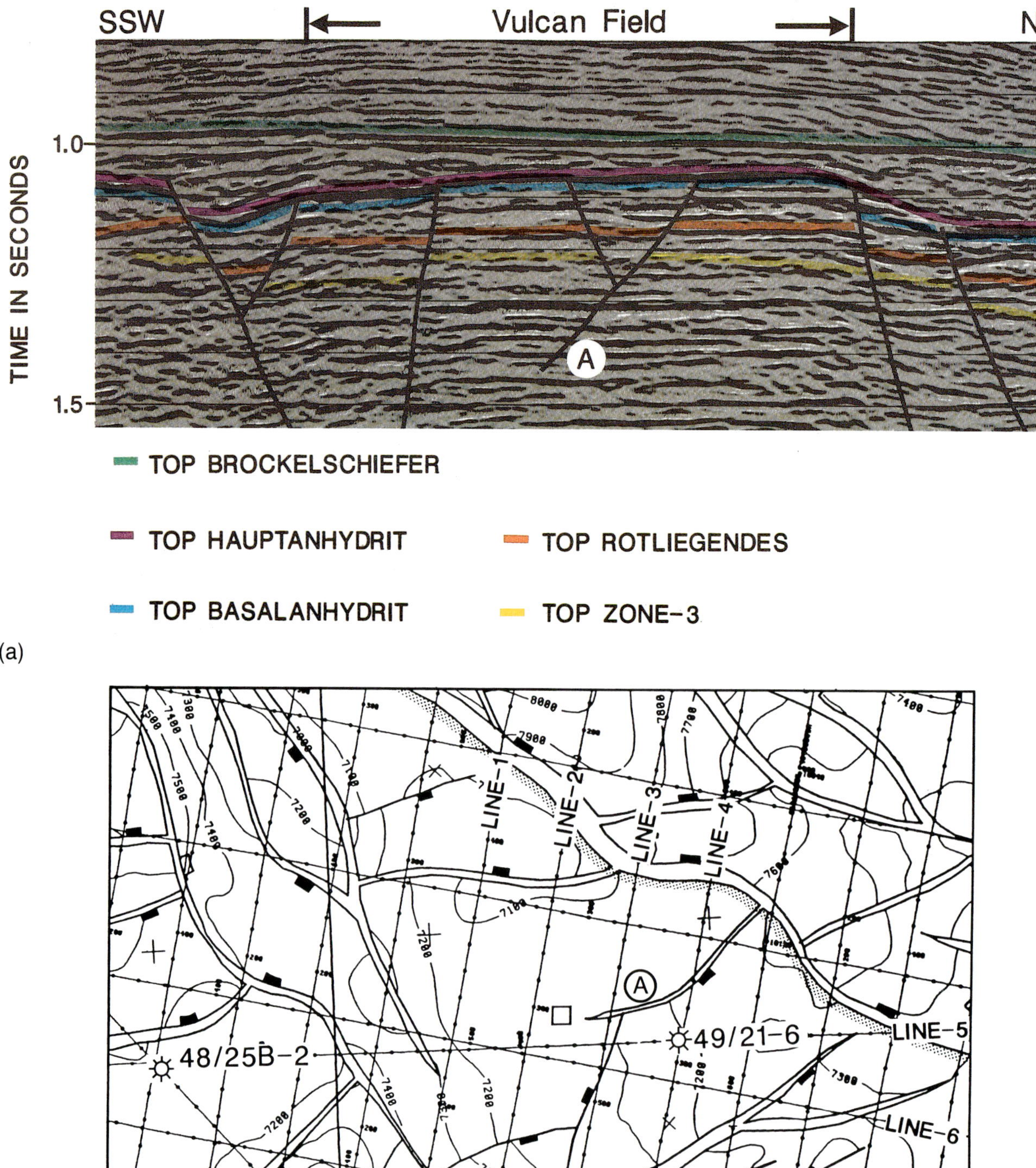

Fig. 4. 1985 2-D seismic survey. (a) Seismic Line 3 (see location Figure 4b) showing interpretation of the Top Rotliegendes and other key seismic events. (b) Depth map of the Top Rotliegendes for East Vulcan.

ever, this was probably outside the time frame of the project and there was no guarantee of achieving better data quality and resolution than the recently acquired surface seismic data. Instead, it was decided to run a high-resolution multioffset VSP program in the first development well (R01), the exact details of which would be determined by presurvey modeling.

Ideally, the offset VSP survey would yield high-resolution seismic data which would address above criteria (1), (2) and (4). Recent work has shown this to be the case (Cramer, paper presented at the 61st Annual Meeting of the SPE, New Orleans, 1986). In addition, recording the VSP with a three-component borehole geophone would enable the investigation of the complete seismic wavefield (*P*- and *S*-waves) around the well. It should then be possible to determine rock parameters and the preferred fracture orientation within the reservoir from a direct combination of seismic velocities of *P*- and *S*-waves (Ahmed et al., 1986), so addressing criteria (2) and (5) above.

Presurvey Planning

The first Vulcan development well (R01) was to be a near-vertical well and would be located in the center of the East Vulcan fault block. As such, and being the first of a total of 10 wells, R01 was an ideal candidate for the acquisition of the multi-offset VSP program. The VSP profiles are shown in Figure 5. The near-vertical borehole would enable relatively easy acquisition and processing while at the same time the central location of the well would enable the critical evaluation of impending development-well locations through the production of lateral VSP profiles. The VSP was expected to yield high-resolution data which could be used to enhance the structural and stratigraphic interpretation of the East Vulcan fault block. In addition, it was planned to acquire three-component VSP data if possible to record the complete wavefield. This enables the VSP processing to be enhanced and thus improve the *P*-wave reflection image as well as providing *S*-wave information for quantitative rock property calculations and hydraulic fracture applications.

The VSP survey objectives could be summarized as follows:

(1) Acquire two high-resolution VSP seismic profiles intersecting the R01 borehole location, de-

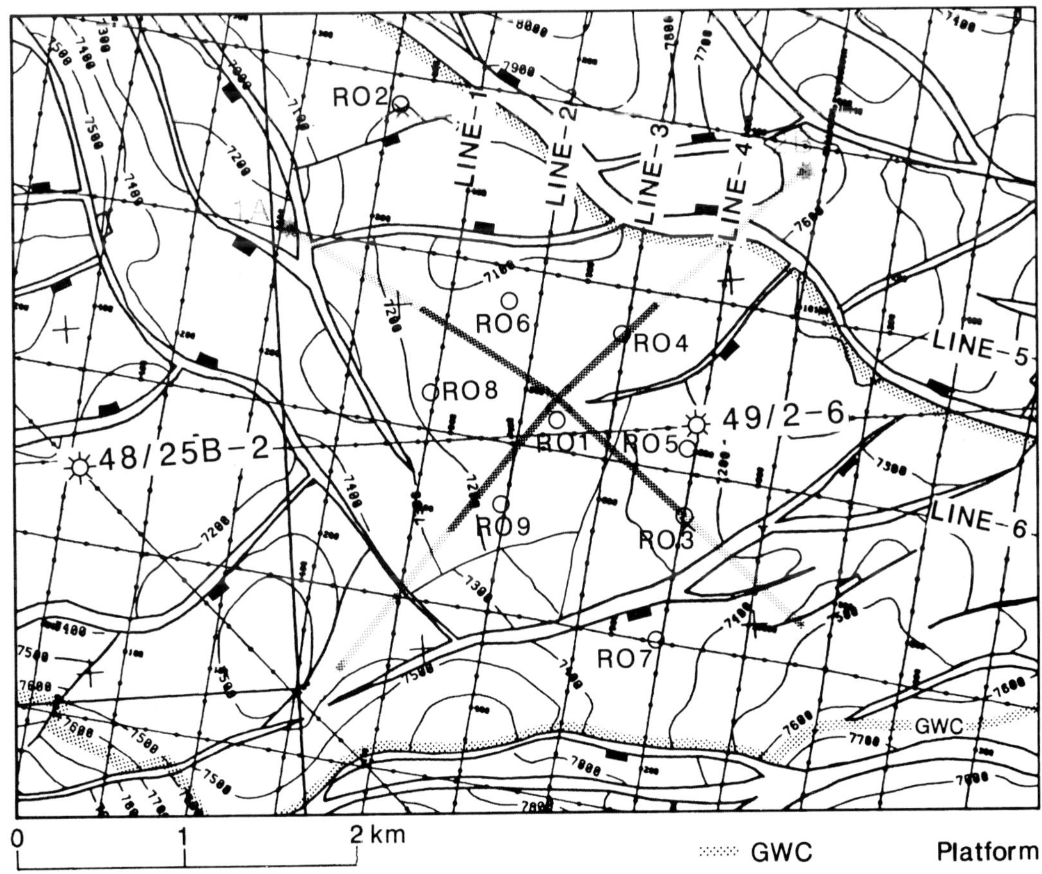

Fig. 5. Actual R01 VSP source locations and the illuminated area at Top Rotliegendes.

signed to sample as many planned well locations as possible; in particular the R03, R04, R08 and R09 locations (see Figure 5). These were among the closest wells to the R01 location and lay in relatively undrilled areas.

(2) Learn as much as possible about the distribution of rock properties within the reservoir formation (the Rotliegendes Sandstone).

(3) Accurately determine the interval velocity dis-

tribution with depth to assist in time–depth conversion for structural mapping.

(4) Identify the main multiple generators within the surface seismic data for enhancing subsurface structural interpretation and, furthermore, with a view to seismic reprocessing.

(5) Determine the azimuth of the least principal stress (σ_3) for efficient directional drilling and hydraulic fracturing.

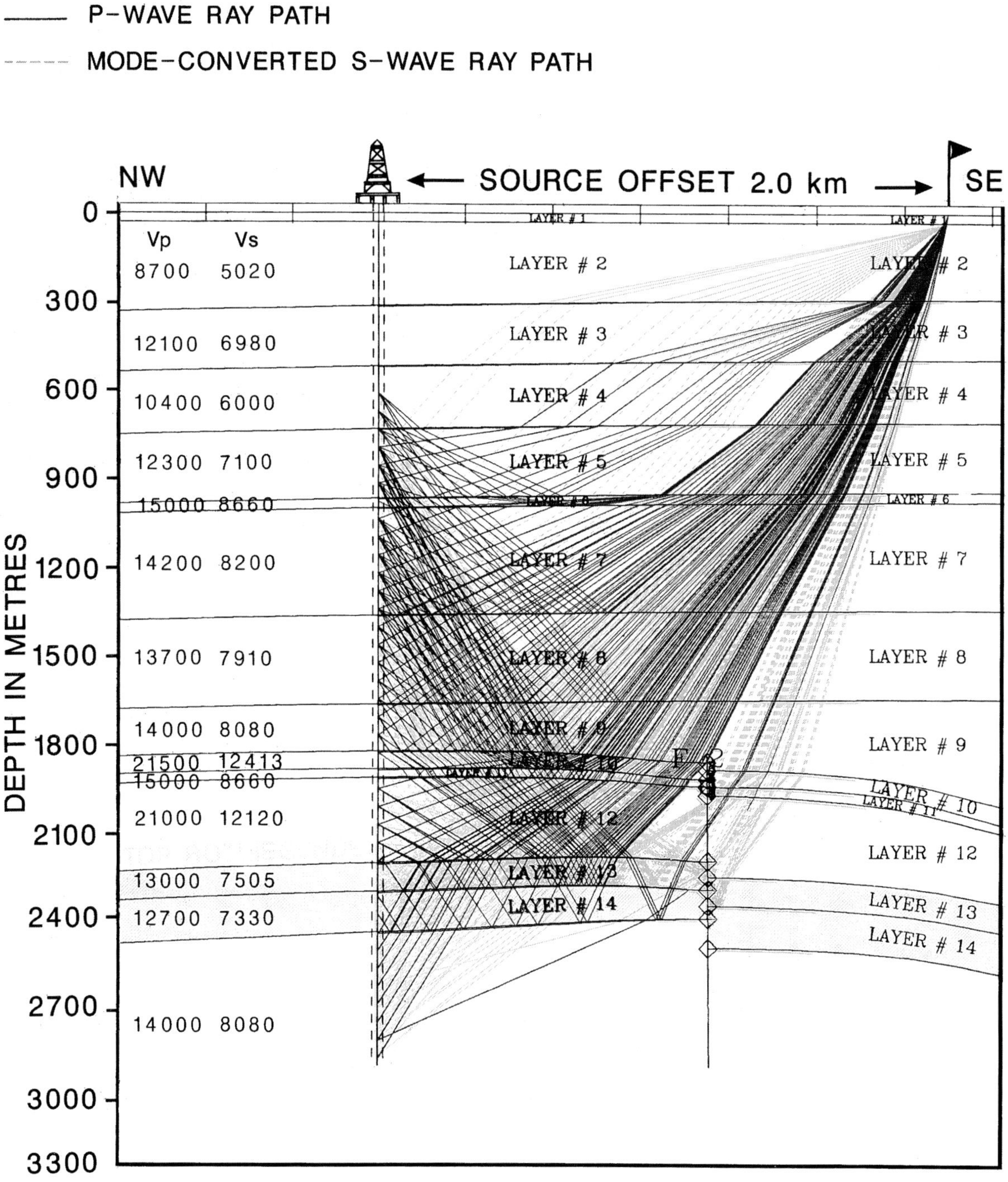

Fig. 6. Ray-trace modeling of an initial geologic model to investigate and determine the R01 VSP survey parameters. Here a source offset 2 km southeast of the platform has been modeled. Layers 13 and 14 comprise the Rotliegendes.

Presurvey Modeling

To decide the optimum parameters for the VSP survey in the proposed R01 well a number of horizontal offsets for the location of the energy source were selected and the results of ray-trace modeling were investigated to consider the following aspects:

(1) The maximum source offset from the wellhead before critically refracted arrivals reached the geophone ahead of the direct arrival, causing problems in VSP data processing.
(2) The number of geophone levels and maximum geophone spacing that defines the highest usable frequency below spatial aliasing.
(3) Lateral coverage of the area of interest (top and base of the Rotliegendes) which is controlled by the source location and the total span of the geophone array in the well.
(4) The effect of wave-mode conversion at particular interfaces in the subsurface.

Figure 6 shows a ray-tracing plot for a source located southeast of the well. Solid lines represent raypaths of the P-waves and dashed lines represent raypaths of the mode-converted S-waves. Such a display is useful in determination of the subsurface illumination/coverage which can be obtained, prior to acquiring the VSP survey. It should be noted that increased illumination of the reservoir is achieved by making use of mode-converted SV-waves.

A synthetic VSP was generated to provide the seismic response of both vertically and horizontally polarized geophones for the source-receiver geometry given in Figure 6 and is displayed in Figure 7. This shows the time–depth curve defined by the first arrival at each receiver position in the well and also the different expected arrivals of the wavefields (P- and mode-converted SV-waves, up and downgoing) with their true amplitudes. Multiples have not been included in this investigation and therefore are not shown in Figure 7. Similar displays were obtained by modeling for energy sources in other directions around the well but they are not presented here.

From these results and to achieve all the objectives described above, the following VSP surveys were recommended:

Fixed-source offset VSPs with source locations at a distance of about 2 km and in different directions radially from the wellhead, with the following recording parameters:

Recording at 15 m geophone spacing for the interval between total depth (TD) and 1500 m in the well to obtain the required lateral illumination of the Rotliegendes with high resolution. This also allows recording of the amplitude variations which is required for a determination of rock parameters away from the drilled zone.

Recording at 30 m geophone spacing for the interval between 1500 and 900 m in the well to extend the

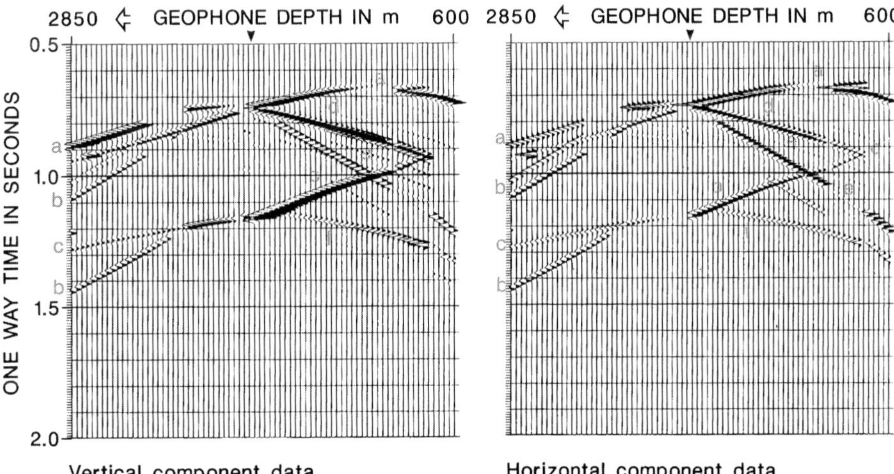

Fig. 7. Synthetic VSP obtained for the R01 VSP geometry shown in Figure 6 as vertical-component data (left) and horizontal-component data (right). Recorded seismic events are identified as direct P-wave (a), SV-wave converted from transmitted P-waves (b), P-wave converted from transmitted SV-waves (c), SV-waves at Top Hauptanhydrit (e) due to conversion of incident P-wave, and reflected P-waves at Top Hauptanhydrit (f) due to conversion of incident SV-waves. The arrow on the depth scale indicates the intersection of Top Hauptanhydrit by the well.

cover and information provided by the above further away from the borehole.

Use of a three-component borehole geophone to record the VSPs, which will enable better wavefield separation in the data processing stage, the enhancement of the *P*-wave reflected image, and the extraction of *P*- and *S*-wave velocities.

A zero-offset VSP (rig source) should also be recorded with the previous geophone spacing along the well from near-surface to TD in order to derive a velocity–depth function of the *P*-waves. This will also help in identification of multiple patterns in the *P*-downwaves and in image reconstruction of the offset VSP data.

VSP Data Acquisition

Between June 30 and July 1, 1986, VSP data were acquired with a three-component borehole geophone assembly (SSL-GCH 300-single arm, one vertical and two horizontal geophones fixed orthogonally) for four fixed-source offsets: 1A, 4A, 2B and 3B (Figure 5). To acquire the four fixed-offset-source VSPs and in order to keep the rig downtime to a minimum, two boats were used to deploy the energy source (a single Bolt airgun 160 in^3 operating at constant chamber pressure of 1600 psi) at two different locations simultaneously. Hence, two runs were needed to acquire the VSP for the four source locations.

To measure the azimuth of the horizontal-component geophones at each recording level during the VSP survey the tool arm (which is inline to one horizontal component), was taken as a reference to a gyro tool (Gyrodata Wellbore Surveyor) fixed on the bottom of the geophone assembly. There was no evidence of cross-feeding or any other effects on the VSP recorded data related to this combination of the two tools because seismic recording followed the gyro readings, with the gyro spin motor turned off.

The advantage of using a gyro in combination with a three-component VSP survey are as follows:

To provide accurate azimuth of the horizontal components needed for data processing.

A comprehensive borehole deviation survey can be acquired at the same time as the VSP survey; this has commercial implications in reducing rig downtime and wireline charges.

VSP Data Processing

Data for the four fixed-offset VSPs were processed using both the vertical component only as well as all

three components. The processing route used (Figure 8) can be simplified into the following steps:

Data preparation,
Wavefield separation,
Multiple suppression, and
Reflection image reconstruction.

Only in the second stage do the vertical component only and three-component processing approaches differ markedly.

Data Preparation

Field tapes were first demultiplexed, then the data for each shot examined to remove any poorly recorded or noisy traces. Gun delay correction has then been carried out for each recording, by reference to the arrival times monitored by a near-field hydrophone (3 m below gun). Then multiple shots to each geophone level were stacked to improve the signal-to-noise ratio. The final stage of data preparation was to compensate for the effect of source output variation and the bubble energy produced by using a single Bolt airgun. This was done by designing an operator from the source signature recorded by the near-field hydro-

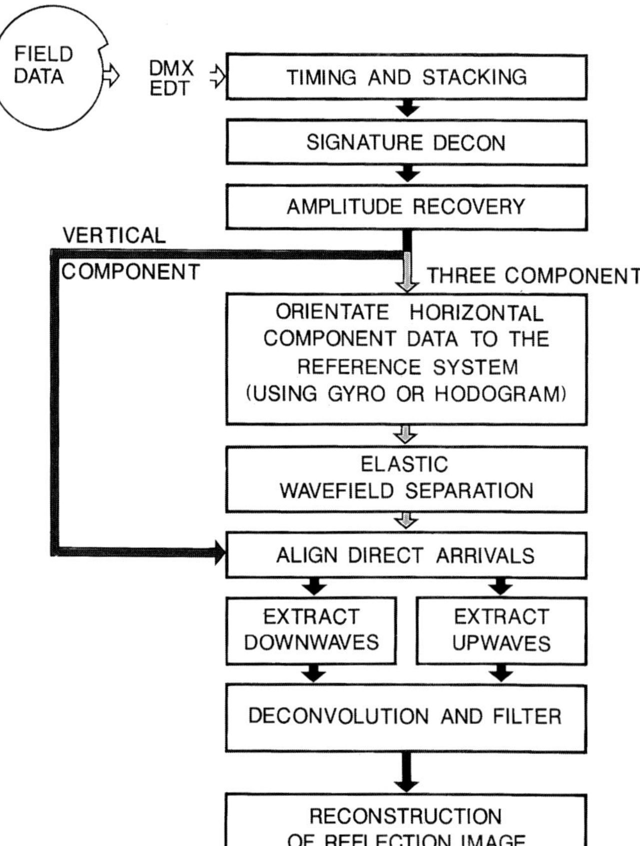

Fig. 8. Data processing flow chart for both a vertical-only and three-component VSP.

phone to deterministically deconvolve the geophone data.

Wavefield Separation

The elastic wavefield recorded by a downhole geophone consists principally of body waves, compressional (P) and shear (S). Shear-waves can be further divided into SV-waves, with vertically polarized particle motion, and SH-waves with horizontal particle motion. SV-waves and to a lesser extent SH-waves are generated in a marine environment by the mode conversion of incident P-waves at high acoustic impedance boundaries. In this part of the southern North Sea, the principal interfaces at which P and SV mode conversions take place can be identified from the raw R01 VSP data as the Top Muschelkalk Halite and Top Hauptanhydrit.

In the 2-D plane the seismic wavefield recorded by a vertical (Z) and a horizontal (H) geophone for incident P- and S-raypaths can be described by the equations shown in Figure 9a. It is clear from these equations that each geophone records both P- and S-wave components. This is further complicated by P- and S-waves arriving from below the geophone which can be demonstrated by a similar analysis.

Separation of the P- and S-wavefields is handled in two fundamentally different ways when processing the vertical component only and the three-component data sets.

Vertical Component Data

A geophone polarized to respond only to the vector components of the elastic wavefield along the vertical axis records both upgoing and downgoing P and SV arrivals. The apparent velocity of these arrivals between adjacent geophones will vary with depth as a result of the changes in their incident angles. The relationship between incident angle (θ) and apparent velocity (V_a) measured at two successive geophone positions (G1 and G2) is given in Figure 9b. It illustrates that for an incident plane wave to a geophone array, the difference in travel path distance (ΔL) is much shorter than the geophone interval (ΔZ). Hence, the apparent velocity will increase with increasing angles of incidence to the vertical.

Variations of apparent velocity with angle of incidence for arrivals to a group of geophones can result in similar apparent velocities being recorded for arrivals of different wave types. Hence, aliasing between wavefields can be a problem when separating wavefields using conventional velocity filters resulting in distorted upwave amplitudes. However, velocity-filtering techniques remain the only available tools for wavefield separation when processing a single geophone component. Acceptable results can be obtained by estimating each downgoing wavefield using median filters, and then removing the estimate from the data. To obtain the reflected P-wavefield, each of the downgoing P- and SV-waves and the upgoing SV-waves

(a) Downgoing wavefield contributions to the recorded components

(b) Apparent velocity incident angle relationship

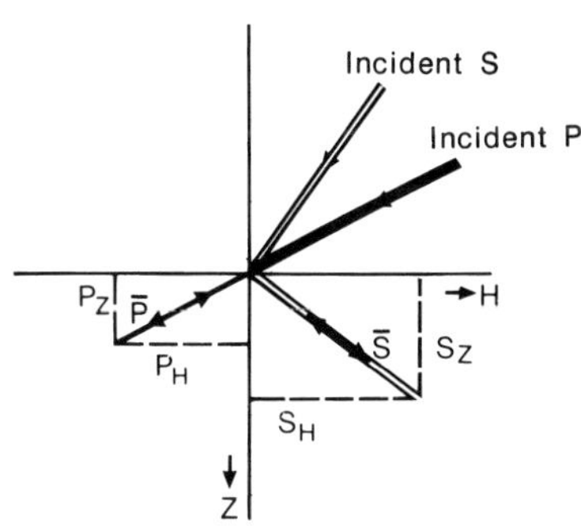

$\bar{P}$ = particle motion to incident P

$\bar{S}$ = particle motion to incident S

$Z = P_Z + S_Z$

$H = S_H - P_H$

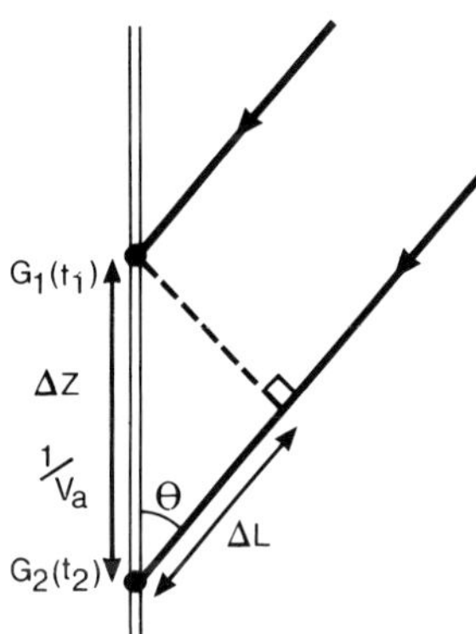

Actual velocity (V) $= \dfrac{\Delta L}{t_2 - t_1}$

Apparent velocity (Va) $= \dfrac{\Delta Z}{t_2 - t_1}$

Fig. 9. Wavefield analysis demonstrating the fundamentals of the elastic mode-separation filter used to resolve the recorded components for the complete wavefield.

must be estimated and subtracted in turn. This was carried out for the vertical component of the R01 offset VSP data.

Three-Component Data

Many of the problems described above, associated with recording and processing of a single vertically polarized geophone, are relieved by processing the data recorded by a three-component borehole geophone assembly. Firstly, the amplitude variations with angle of incidence will be recovered from the components representing the vectors of each arrival in the vertical plane. Secondly, the wave modes of an arrival can be identified from the particle motion and velocity recorded by each geophone component, which enables separation of the P- and SV-waves by polarization techniques.

In general, the first stage in wavefield separation for data recorded using a three-component borehole geophone is to remove the random orientation of the horizontal components between recording levels. This is achieved by transforming the recorded three-component data into a fixed reference coordinate system. Conventionally, the vertical plane containing the source and geophone defines the vertical axis (Z) and one horizontal axis (HS) of the fixed coordinate system. The other horizontal component (HT) is perpendicular to the source azimuth (HS). For the R01 VSP data, the geophone azimuth at each level was measured by a gyro (as described above) so that the horizontally recorded data could be transformed to HS and HT components. As a result of this transformation, any horizontally polarized shear waves will be maximized in the HT directions, while P- and SV-arrivals will be recorded by both Z and HS. Separation of P- and SV-wavefields from the Z and HS components is done either by using the polarization technique (Galperin, 1983; Ahmed et al., 1986) or polarization filters (e.g., Dankbaar, 1985).

For the four fixed-offset VSPs recorded in R01, a polarization filter was used to separate the different components of the observed wavefields. A detailed description of this technique is beyond the scope of the paper. However, in short, the filter coefficients are determined from the ratio of apparent and actual formation velocities of P- and S-waves and the response of the Z and HS components to the particle motion around the borehole (see Figure 9). In this way the elastic wavefields can be separated either as a scalar, or as a vector along the original Z and H directions. The most significant benefit from using this scheme is that the amplitudes are preserved, which in turn allows reliable stratigraphic interpretation and rock parameter determination. It should be noted that separation of the elastic wavefields using such a polarization filter is limited by the first alias frequency of the

S-waves. As a consequence, careful consideration of shear velocities and the geophone depth interval must be made during presurvey modeling to ensure that an acceptable bandwidth is obtained from the VSP data. Separation of the upgoing and downgoing energy to the geophone is now simplified to estimating and subtracting the corresponding downgoing arrivals from the separated P- and S-wavefields, using a median filter.

Multiple Suppression

With the P-wavefield now decomposed into its upgoing and downgoing constituents using a median filter, an operator is designed from the downwaves to deterministically deconvolve the upwaves for multiple reflection attenuation. However, unlike a zero-offset VSP, for a fixed-offset VSP the travel paths will differ between upgoing and downgoing arrivals to the geophone (Figure 6). Hence, the multiple tail of the upwaves from each reflector will not entirely be a scaled version of the downgoing wavefield. Only a short operator can be used that will account for short-period multiples generated in the near-surface or close to the geophone location, where the travel paths can be considered the same. Long-period multiples will usually have normal move-out (NMO) and therefore cannot be correctly handled by a constant deconvolution operator for all traces.

Reflection Image Reconstruction

Ray trace image reconstruction has been applied to the deconvolved upwaves from each of the four offset sources using the scheme described by Dillon and Thompson (1984). The reflection images obtained have been combined at the well location and displayed as two northwest–southeast and southwest–northeast sections to give more easily interpretable seismic lines.

It should be noted that in using such imaging schemes where laterally continuous horizontal layers are assumed, the data are not forced to agree with a preconceived geologic model. This assumption will, however, have the effect of decreasing dip toward the source and increasing dip away from the source. Similarly, reflection points will appear further downdip than they actually are.

Interpretation of VSP Data

The interpretation and conclusions sections of this paper deal only with the results of the vertical component and three-component VSP data processing. The studies into rock parameter calculations and hydraulic fracturing applications are continuing and may be presented in a future paper. The drilling results of the R01 well were rather a surprise, with the Top Rotliegendes being encountered about 48 m shallower than prognosed at a location approximately 80 m north of the intended target. A thinner Upper Magnesian Limestone section was encountered than had been prognosed, but this only accounted for less than 16 m depth error at Top Rotliegendes.

The correlation of well-log data with the rig-source VSP data and the identification of primary seismic events is shown in Figure 10, and correlation of the synthetic seismogram and VSP-transposed traces for this well are shown in Figure 11 together with the nearest seismic (strike) line. The synthetic seismogram has been constructed using a wavelet extracted from the surface seismic data which generally, for the existing exploration wells, gives a good tie at Top Rotliegendes level. However, there is a time discrepancy at Top Rotliegendes level between the surface seismic and the synthetic/VSP of about 10 ms, which is not seen on the overlying seismic horizons (the surface-seismic time being the longer). It appears likely that the Top Rotliegendes has been upfaulted in this area which would explain the drilling results and the time mistie seen above. The nearest fault (fault A), however, is seen only on seismic line 3 (Figure 4); no continuation of the fault is seen on seismic lines located adjacent to the borehole.

The R01 zero-offset VSP was used to correlate from the well logs to the offset-VSP profiles enabling these to be interpreted.

Northwest–Southeast VSP Profile

The interpretation of the vertical component reflection image along the northwest–southeast direction is shown in Figure 12. Generally, data quality is disap-

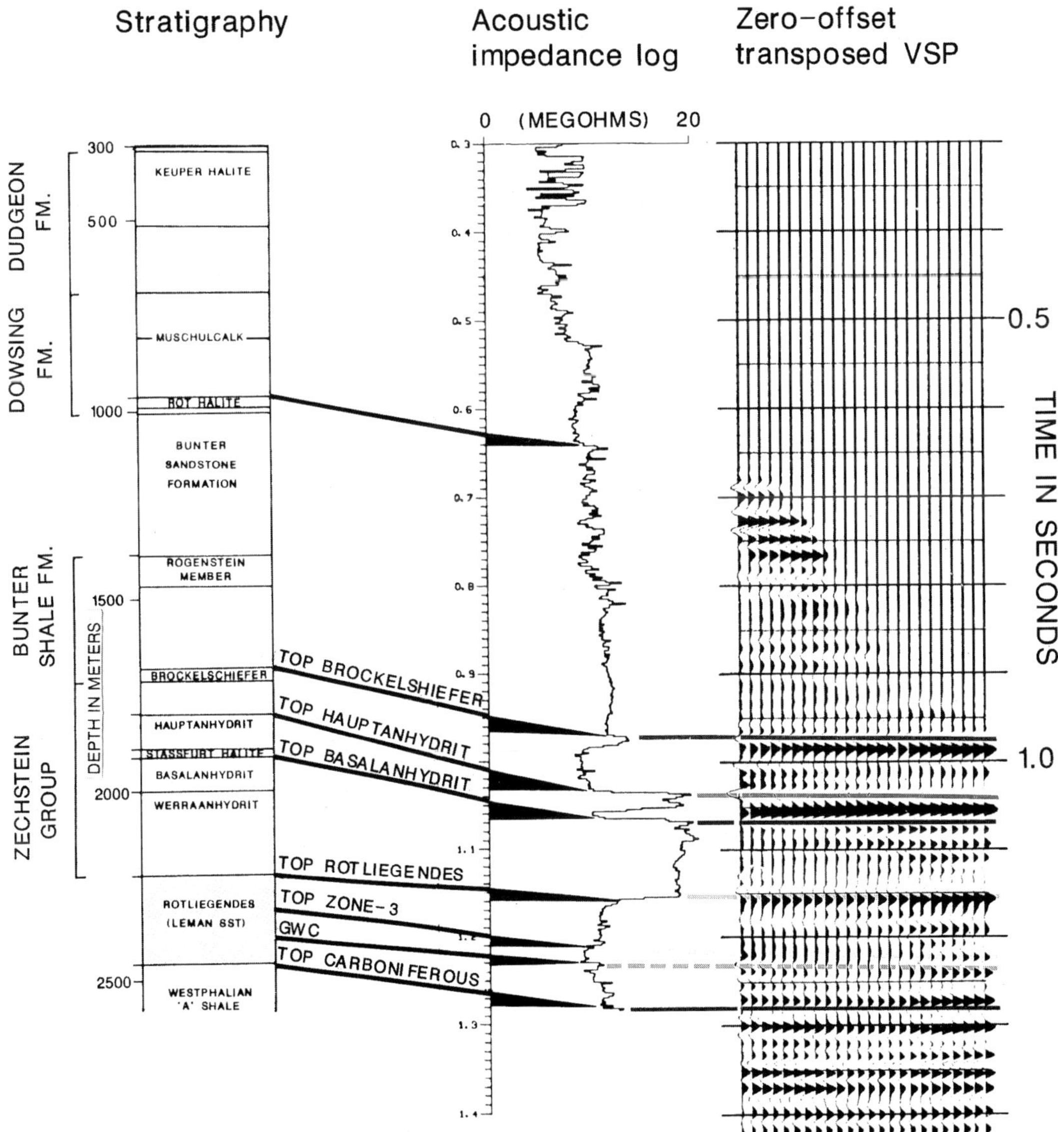

Fig. 10. Correlation of zero-offset VSP results with well-log data from the R01 well.

pointing in terms of reflection continuity and character, making correlation with the surface seismic data difficult. The blank data zone at about 1.075 s (just below Top Hauptanhydrit) around the borehole was predicted by the pre-survey modeling (Figure 6) and is a result of the refraction of raypaths within the high interval-velocity Upper Magnesian Limestone. A number of faults are observed at Top Rotliegendes level which were not interpreted on the surface seismic data, including a fault just to the southeast of the R01 location. This fault may be a continuation of Fault A (Figure 4b).

The corresponding image obtained from processing of the three-component data is shown in Figure 13. Data quality has been significantly improved over the vertical component only processing. More continuous Top Hauptanhydrit and Top Rotliegendes seismic events are observed and the imaging of faults appears more reliable. Note the Top Zone-3 seismic event which appears to deepen into the fault just southeast of the borehole. This event appears on the deconvolved upgoing wavefield display for the zero-offset VSP (Figure 10) as a primary event. Note also the addition of the shallower data on this display at around 0.5 s which has resulted from the contribution of the horizontal geophones.

Southwest–Northeast VSP Profile

Figure 14 shows the interpretation of the vertical component VSP image for the southwest–northeast profile. Again, data quality is rather disappointing for the same reasons as discussed above. Figure 15 shows the corresponding three-component processed image. Significant data quality improvement is observed particularly in the northeast offset which has been made more continuous at Top Rotliegendes level; note the disappearance of fault B on this offset (Figure 14).

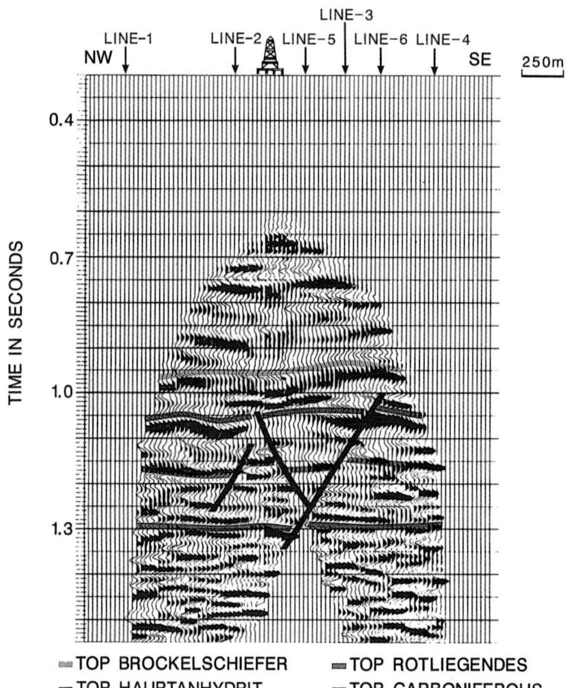

Fig. 12. Structural detail on northwest–southeast VSP profile obtained from processing of vertical-component data only.

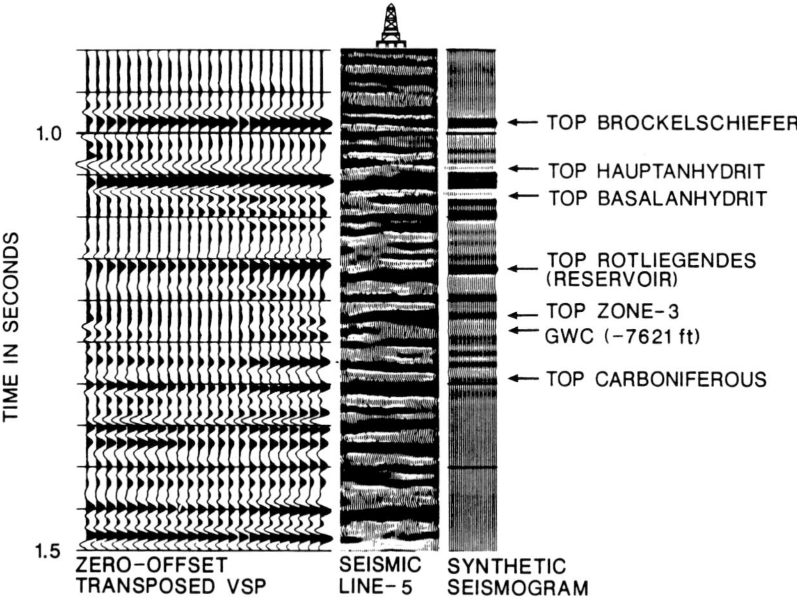

Fig. 11. Correlation of zero-offset VSP results with surface seismic data of Line 5 (near to R01 well location) and the synthetic seismogram derived from the well logs.

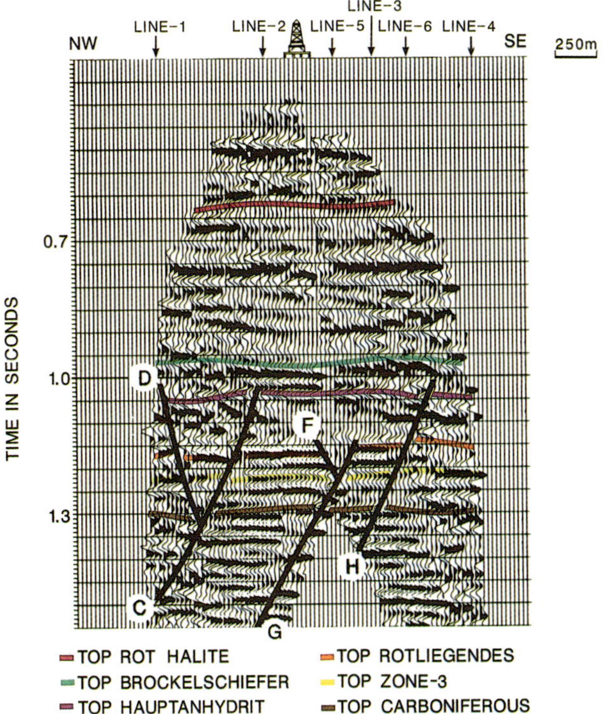

Fig. 13. Structural detail on northwest–southeast VSP profile obtained from processing of three-component data.

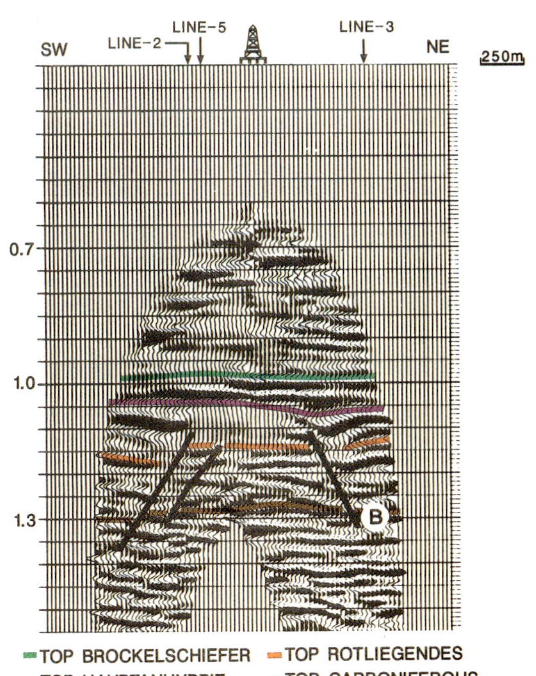

Fig. 14. Structural detail on southwest–northeast VSP profile obtained from processing of vertical-component data only.

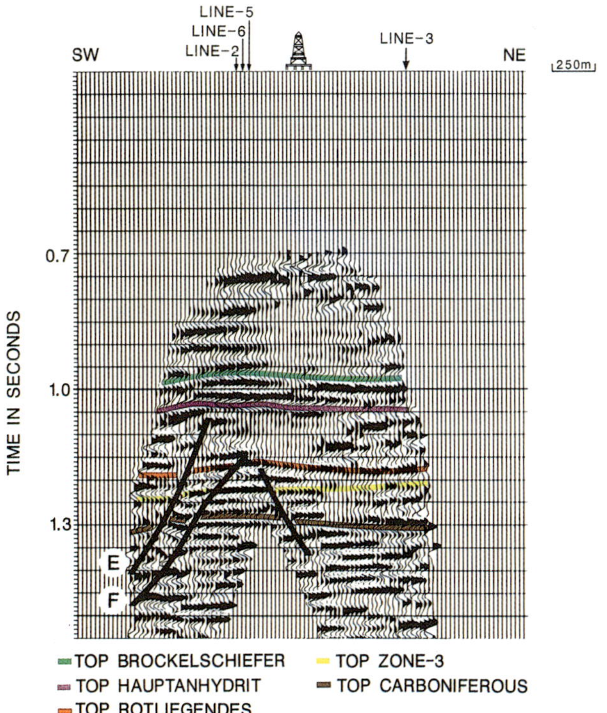

Fig. 15. Structural detail on southwest–northeast VSP profile obtained from processing of three-component data.

Note also the Top Zone-3 reflection which appears clearly just southwest of the borehole and the apparent thinning of the Top Rotliegendes down to the top of the Zone-3 interval to the northeast end of the profile.

In this case it can be seen that the three-component VSP processing has produced a much more useful subsurface image than the vertical component only processing. This is due to recording the complete wavefield in the subsurface and the robust nature of the applied wavefield separation technique which preserves the true amplitude of each P-wave arrival.

The fault interpretation taken from the three-component VSP profiles has been transferred to the Top Rotliegendes depth map as shown in Figure 16. Note the presence of a graben on the northwest profile, bounded by faults C and D, and a possible extension of fault A to the southwest profile (fault F).

Returning to the surface seismic data, Figure 17 shows the interpretation of seismic line 1 in the region of these features made after the acquisition of the R01 VSP. Figure 18 shows a similar comparison for line 2. For seismic line locations see Figure 4b.

Impact on Development Drilling

The interpretation of the three-component VSP data described above became available at the beginning of September 1986, by which time the drilling of R03 was underway (Figure 16). Thus the VSP interpretation

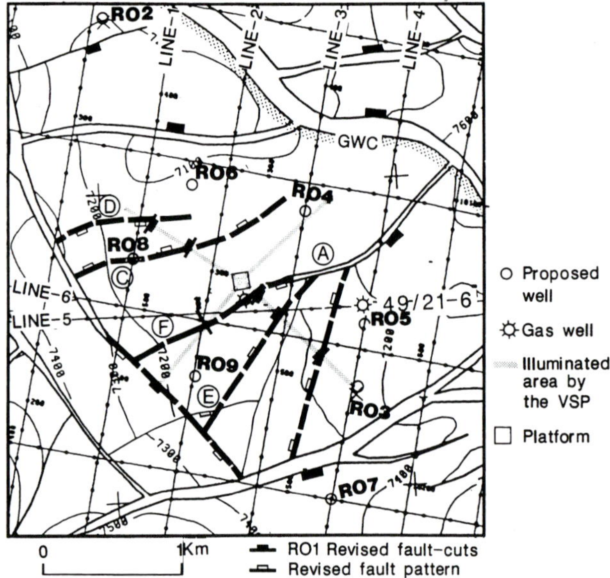

Fig. 16. East Vulcan Top Rotliegendes structure map with revised fault pattern superimposed after interpretation of R01 VSP (three-component processed data).

was able to impact R04 onward. As shown on the northeast R01 VSP profile, the R04 location appeared to be structurally sound and since there was a possibility of shallow Zone-3 at this location, this appeared to be a good production well location. The R08 location, however, now appeared to be situated close to the southerly bounding fault of the graben (fault C) and this location was moved slightly south to avoid the fault (see Figure 17). The R09 location now appeared to be situated in a downfaulted terrace from the R01 location (between faults E and F), but was safely distant from any fault to be acceptable (see Figure 18).

In short, the acquisition, processing and interpretation of the R01 three-component VSP survey has been carried out in a time frame enabling the results of this survey to impact the field development program by the critical evaluation of proposed development well locations.

If the validity of the VSP interpretation could be checked directly by drilling and then running further VSP surveys, the use of the three-component offset-VSP technique as a tool for field development could be qualitatively ascertained. This was to be the next stage.

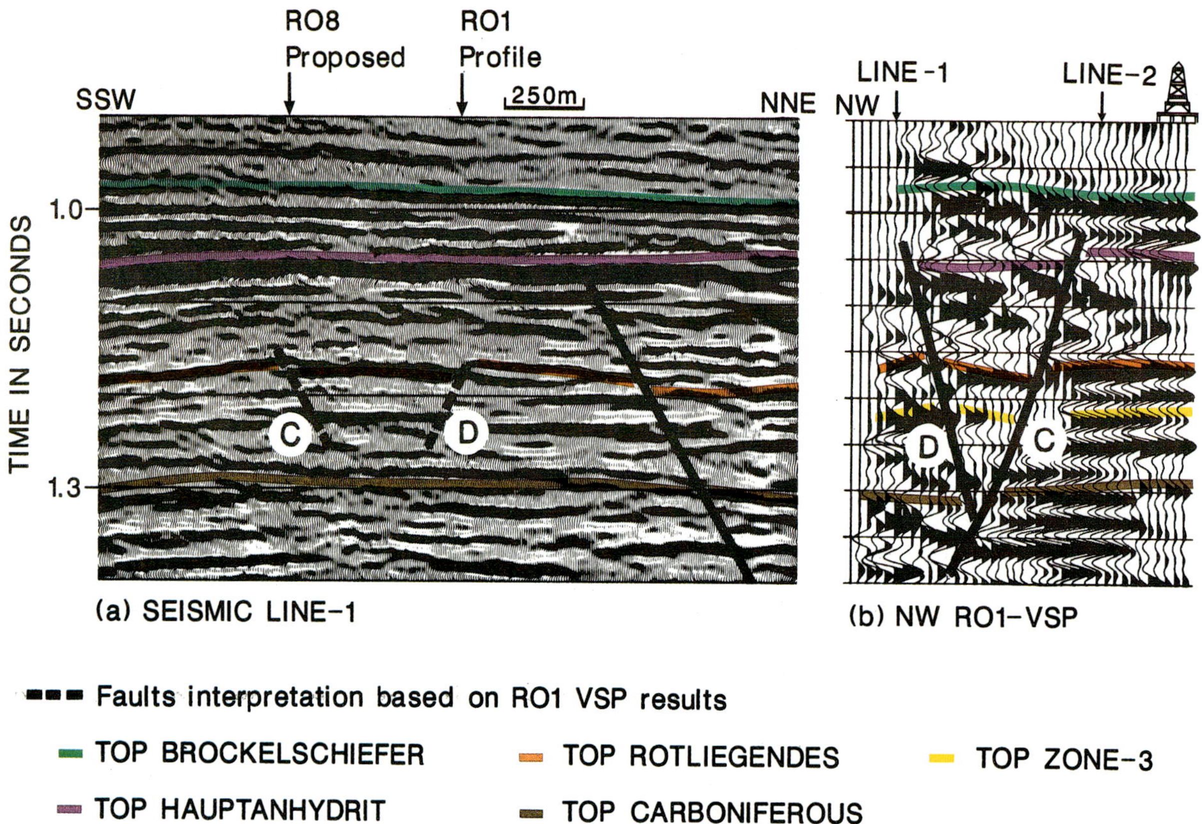

Fig. 17. Comparison of data quality and interpretation between surface seismic (Line 1) and three-component offset-VSP data in the region of a shallow graben northwest of the platform.

Development Drilling Results

At the time of writing this paper, the R08 well was being completed. All the wells including R08 had been successful and, except for R01 and R02, had encountered Top Rotliegendes to within 12 m of prognosis at an average depth of 2195 m (subsea). For each of the deviated wells R02–R07 (Figure 16 for locations), normal-incidence (moving source—moving geophone) VSP surveys were acquired to further enhance the structural mapping of the field. These VSPs were acquired using a gimballed three-component borehole geophone and processed as vertical component only since the contribution of the horizontally polarized geophones for the *P*-wave is minimal in this case. They commonly exhibit high-resolution data. Since these VSP profiles run along the track of each borehole, they provide a very thorough check of the R01 fixed offset VSP data, particularly in the case of the R03 and R04 VSPs which overlie directly the southeast and northeast offset-VSP profiles, respectively (Figure 16). Figures 19 and 20 show a direct comparison between the R01 VSP profiles and the R03 and R04 VSP profiles, respectively, all displayed with the same parameters

(50 Hz high-cut filter and 200 ms equalization window). In addition, Figure 21 shows the interpretation of the R02 VSP upwaves, which substantiates the fault interpretation from the northwest R01 VSP profile (Figure 13).

It is observed that fault interpretations are common to both sets of VSP data even where surface seismic data cannot fully resolve the same detail. However, it is also observed that the fault-cuts derived from the two VSP data sets are displaced from each other horizontally by about 100 m. The explanation for this is probably that the reflection image obtained from the R01 VSP has been incorrectly mapped during data processing, whereas the normal incidence VSP data indicates the more realistic fault positions.

The fault cuts from the R02–R06 VSPs have been superimposed on to the R01 VSP fault interpretation as shown in Figure 16. There appears to be a general agreement between the two VSP data sets, albeit with some fault positioning differences as described above.

It is interesting to note that, as predicted by the northeast R01 VSP profile, the top of Zone-3 in well R04 was encountered about 27 m shallower than that originally prognosed from the R01 drilling results.

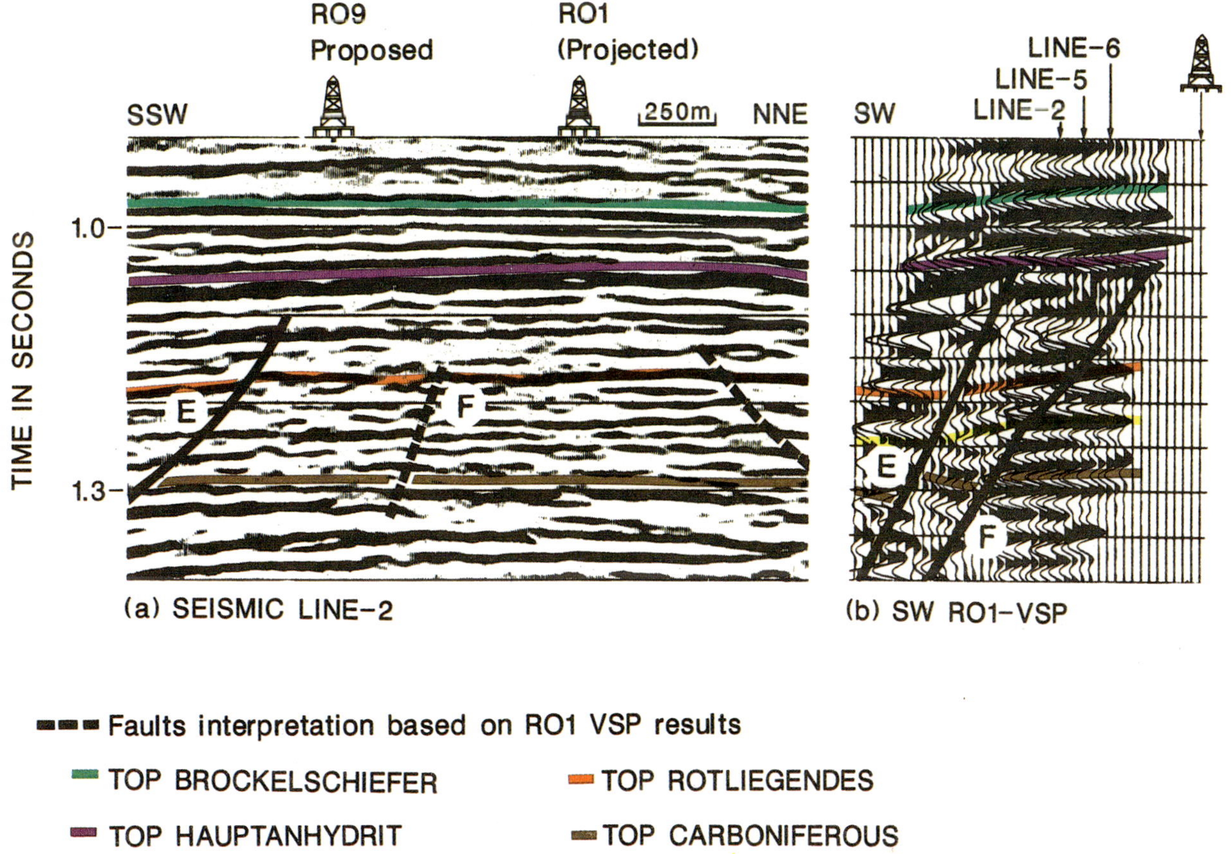

Fig. 18. Comparison of data quality southwest of the R01 well location between surface seismic (Line 2) and three-component offset-VSP data.

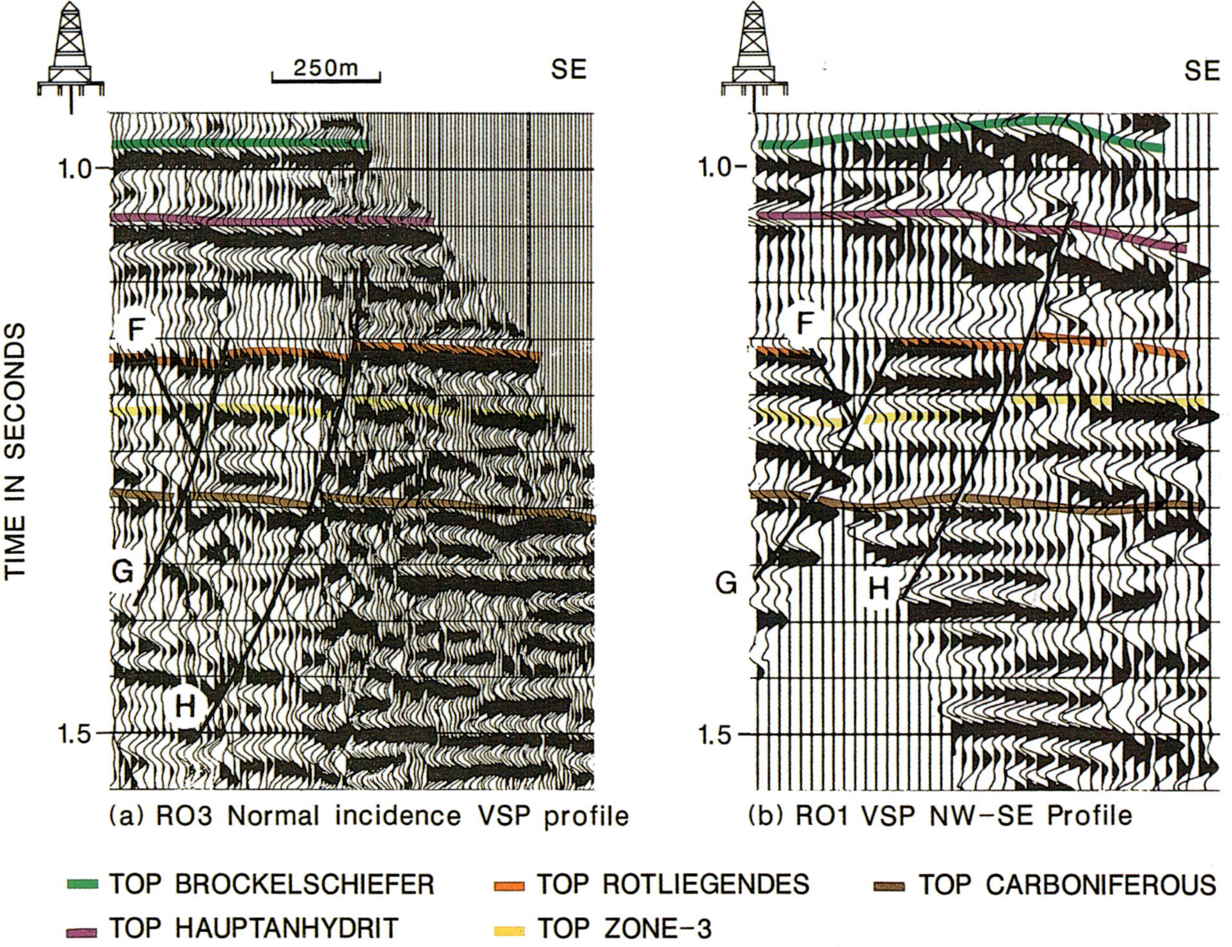

Fig. 19. Comparison of data quality and interpretation between R03 normal-incidence VSP data and R01 offset VSP profile.

Discussion

Full three-component processing of the acquired VSP data has contributed significantly to improve the structural understanding of the East Vulcan field. The necessity of recording and processing the VSP data as three-components is demonstrated by the improved reflection image which has led to a better structural interpretation of the subsurface around the R01 well. This, in turn, has been confirmed by the subsequent VSP surveys in the development wells drilled along the fixed-source offset trajectories of the R01 VSP. In fact, the improvement in the quality of the reflection image achieved by three-component VSP data processing is a result of two factors:

First, the total vector of the wavefield (i.e., incident and reflected) is obtained from recording the VSP data in three dimensions. In this way lateral continuity of the seismic events results, in particular, for formations with high acoustic impedance and for the VSP records at shallow depths where the raypaths arrive at the geophone at high incident angles (i.e., more or less horizontally). For the reflected energy in high dip and complex structural areas, the same argument applies.

Second, the applied wavefield separation technique is based on the polarization of the particle motion associated with the different components of the elastic wavefield at the geophone. Consequently, true amplitudes are preserved which has implications for stratigraphic interpretation. In contrast, spectral notches and amplitude distortion will result from using velocity filters for wavefield separation in *vertical component only* processing, due to spatial aliasing between the wave-modes.

The R01 VSP fault interpretation (Figure 22) indicates a much more detailed fault pattern than that derived from interpreting the surface seismic data (Figure 4b). In retrospect, it could be stated that some of these faults should have been interpreted on the surface

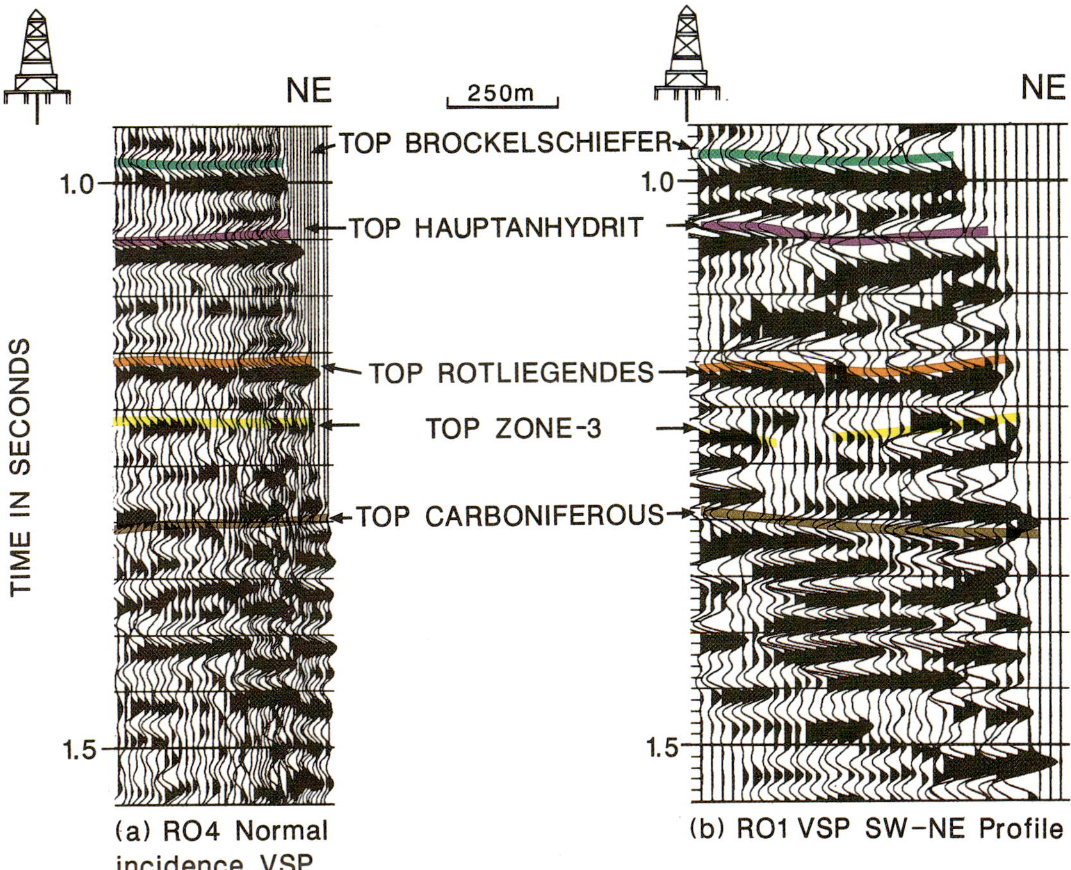

Fig. 20. Comparison of data quality and interpretation of the R04 normal-incidence VSP data and R01 offset VSP profile.

seismic data before the R01 VSP was acquired and processed. This is possible, but these features were originally discounted on the basis that they were probably multiple-interference and/or noise-induced discontinuities. The R01 VSP profiles provided data which significantly revised the fault interpretation and which has been substantiated by subsequent drilling and additional VSP interpretation.

The fault trends on the East Vulcan fault block now appear to be dominantly northwest–southeast and southwest–northeast. The relationship of the East Vulcan and West Vulcan fault blocks now becomes more apparent from the observed fault pattern (Figure 22). Both fault blocks are bounded by northwest–southeast trending faults, but internally dissected by numerous southwest–northeast trending faults, dividing each into a series of southwest–northeast trending horsts and grabens. The East and West Vulcan fault blocks appear to have been divided along the northwest–southeast trending shallow graben separating the 48/25B-2 and 49/21-6 wells, as suggested by the similar shape of the relatively younger, graben-bounding faults and the offsetting of the southwest–northeast trending faults within the graben. The southwest–

northeast trending faults may be the result of reactivation of intra-Carboniferous, Caledonian-trending faults, evidence for which has been seen on the surface seismic data.

The impact of this interpretation on future drilling is twofold: first, the structural understanding of the area has been enhanced and consequently the prediction of porosity and permeability variations and reservoir zonation can be improved. Second, there are more accurately mapped structural fault blocks for which the reservoir engineer can now aim in order to ensure optimal gas recovery.

In addition to the interpretations discussed above, the investigation into rock properties (using a combination of P- and S-wave data), not described in this paper, may yield information concerning the anisotropy and stratigraphy of the reservoir formations. This can be used for planning further development drilling and hydraulic fracturing programs, ultimately enhancing gas recovery.

Conclusions

It has been demonstrated that the acquisition, processing, and interpretation of three-component off-

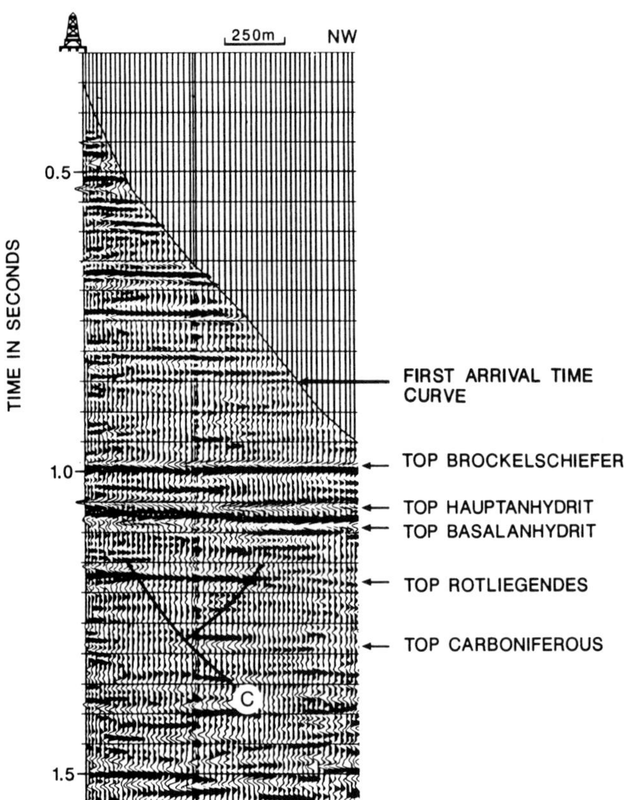

Fig. 21. Interpretation of normal-incidence VSP data from the R02 well.

set-VSP data on the first development well of the Vulcan Gas Field have led to a much improved structural interpretation of the Top Rotliegendes Sandstone reservoir. In this particular case, three-component VSP processing was required to provide a useful subsurface image rather than conventional vertical-component processing. This is due to recording the complete wavefield in the subsurface and the robust nature of the applied wavefield separation technique which preserves the true amplitude of the reflected energy.

The VSP interpretation was able to impact reservoir development by providing a critical evaluation of proposed well locations. Subsequent drilling and acquisition of normal incidence VSPs in deviated wells along the same azimuth as the three-component offset VSP profiles substantiated the interpretation of the latter. As a result, further geologic and reservoir engineering studies may be enhanced.

This case history illustrates the importance of considering the three-component fixed-source-offset VSP as an intermediate step toward improving structural interpretation between the exploration and development phases of a hydrocarbon field.

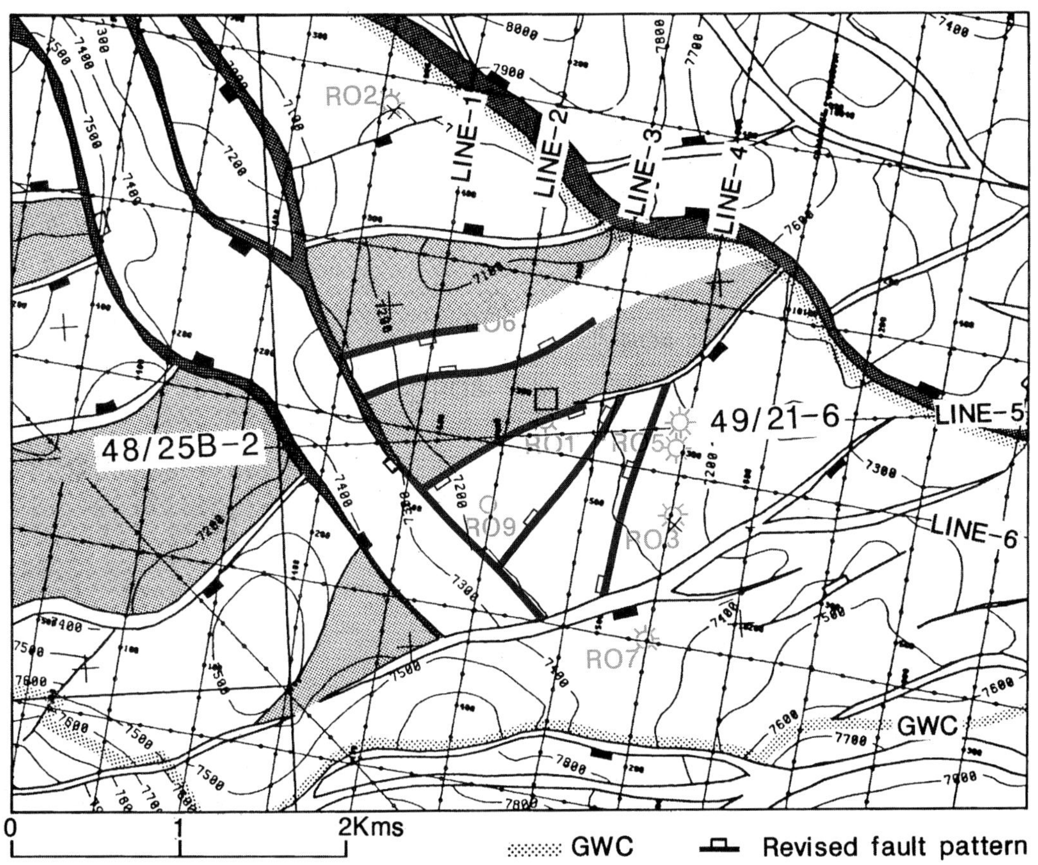

Fig. 22. The structural relationship between East and West Vulcan at Top Rotliegendes in the light of revised fault interpretation from VSP profiles.

Acknowledgment

This paper is published by permission of Conoco (UK) Ltd, Seismograph Service (England) Ltd, Britoil plc, Arco British Limited, Marathon International Petroleum, and Occidental Petroleum (UK) Ltd. The authors acknowledge that the interpretations presented in this paper are provisional and that the contents of this paper do not necessarily represent the opinions of all the licensees of Blocks 49/21 and 48/25b.

References

Ahmed, H., Dillon, P. B., Jonstad, S. E., and Johnston, C. D., 1986, Northern Viking Graben multi-level three-component walkaway VSPs—a case history: First Break, **4**(10), 9–27.

Conway, A. M., 1986, Geology and Petrophysics of the Victor Field, *in* J. Brooks, J. C. Geoff and B. van Hoorn (Eds), Habitat of Palaeozoic Gas in NW Europe: Geological Society Special Publication **23**, 237–249. Blackwell Scientific Publications, Oxford.

Dankbaar, J. W. M., 1985, Separation of P- and S-waves: Geophysical Prospecting, **33**, 970–986.

Dillon, P. B., and Thompson, R. C., 1984, Offset source VSP surveys and their image reconstruction: Geophysical Prospecting, **32**, 790–811.

Disiena, J. P., Gaiser, J. E., and Corrigan, D., 1984, Three-component vertical seismic profiles: orientation of horizontal components for shear-wave analysis, *in* M. N. Toksoz and R. R. Stewart (Eds), Vertical seismic profiling Part B: Advanced concepts, 171–188. Geophysical Press.

Galperin, E. I., 1983, The polarization method of seismic exploration: Reidel, Dardrecht, Holland.

Walker, I. M., and Cooper, W. G., 1987, The structural and stratigraphic evolution of the north-east margin of the Sole Pit Basin, *in* J. Brooks and K. Glennie (Eds). Petroleum Geology of North-West Europe: Graham and Trotman.

Chapter 5: Reservoir Description
Defining Reservoir Properties

Marc H.F. DeBuyl

Definition of Terms and Concepts

Soon after an appropriate number of delineation wells are drilled, development geologists, petroleum engineers, and financial analysts need to generate a reservoir model for a field that is to be commercially produced. This model will be updated periodically for technical, financial, and tax purposes as new data become available. This modeling activity is defined as "reservoir description" and generally consists of predicting the areal distribution of those reservoir properties that affect development drilling decisions, reserves estimates, production forecast, and field management plans.

Given the diversity of these objectives, the reservoir description generated by a development geologist is generally different than that created by a petroleum engineer or financial analyst. Their respective purposes being different, each person involved in reservoir management may require a model of the reservoir that provides different information. For instance, important properties such as net reservoir thickness, porosity, lithology, and fluid saturation are necessary if a geologist is to produce a model for estimating reserves in place. However, a petroleum engineer must add permeability and fluid flow properties to the database to generate a model that can be used for predicting recoverable reserves and production rates.

In both cases, the reservoir models are derived from a smooth interpolation of discrete geological and petrophysical measurements and well tests obtained at sparse well locations. The experienced development geologist will guide the interpolation process with a depositional model based on sedimentary and paleontologic data judiciously interpreted in the context of the local and regional stratigraphy. In concert with this effort, the petroleum engineer must estimate the recoverable reserves of the field, and determine how to optimize production while minimizing expenditures. Thus, the engineer is primarily concerned with hydraulic units that control the fluid flow behavior of the reservoir and is less concerned about the depositional processes that created those stratigraphic units. It is, of course, desirable that all reservoir descriptions for a field be consistent regardless of which discipline—geology, engineering, or geophysics—produces the model, and that apparent discrepancies be reconciled or accounted for systematically in the data analysis.

The Role of Seismic Data in Reservoir Description

Often the role of 2-D and 3-D seismic data is limited to providing only a structural model for the reservoir. The interpretation of seismic data, however, has evolved over the last twenty years so that seismic stratigraphy models can now be produced. These models are particularly helpful in understanding the local and regional depositional history of the basin. This additional seismic information is valuable not only in hydrocarbon exploration, but also in defining reservoir models by ensuring that the depositional mechanisms that created the reservoir facies are compatible with the paleogeographic and sedimentary history of the area.

In heterogeneous rock units, conventional reservoir models interpolated from well data alone can result in inaccurate resource appraisals and production forecasts and adversely affect the cost-effectiveness of field development programs. Based on Department of Energy (DOE) report EIA 0216–(87), approximately three-quarters of the U.S. domestic reserve addition results from redefining existing reservoirs and from enhanced recoverability. These facts emphasize the need for improved description methods which use data that can define reservoir properties between wells. Seismic data are ideal for this purpose, so this discussion is primarily concerned with illustrating how seismic data can contribute to inter-well reservoir descriptions.

Chapter 5 describes and illustrates seismic methodologies that provide spatially dense control on reservoir parameters such as net thickness, gross thickness, net porosity, porosity-thickness, lithologic variations, V_p/V_s, and Poisson's ratio. Recently published or updated papers illustrating different aspects of seismic

*Bureau of Economic Geology, Austin, Texas. Formerly with Atlas Wireline Services.

reservoir description are presented first. Some case histories of successful applications of these methodologies are then reviewed. Case histories allow readers to determine whether seismic data can help solve specific reservoir modeling problems and convey a sense of how accurate the resulting reservoir models are.

Basic Issues To Be Confronted

First, we discuss how seismic data contain information about the petrophysical properties of hydrocarbon reservoirs. Reservoir boundaries generally are characterized by interfaces where acoustic impedence properties change rather abruptly. These properties include compressional- (P-) and shear- (S-) wave velocities, P- and S-wavefield attenuations, and bulk density. As rock parameters such as lithology, porosity, pore pressure, fluid saturation, cementation, and temperature change, these acoustic properties change and produce variation in the reflectivity contrasts at reservoir interfaces. Such contrasts affect the amplitudes, phases, and frequencies of the reflected seismic signals, and these seismic effects can be analyzed to provide estimates of the petrophysical properties of the reservoir.

This type of seismic analysis can raise many questions about seismic data processing as well as seismic interpretation procedures. In addition to questions related to processing and interpretation, the following fundamental issues must be kept in mind:

Seismic data provide only averaged information about the acoustic properties within an appreciable volume of rock and do not have the vertical resolution of wireline logging. On the other hand, seismic data provide spatially dense information along each seismic profile, and this continuous spatial imaging of reservoir conditions provides important control needed in a reservoir description.

Layered earth models inverted from seismic data are inherently nonunique.

Seismic attributes extracted from reflected seismic wavefields are ambiguously related to reservoir properties of interest.

These problem areas will be reviewed and some seismic processing and interpretation techniques will be discussed that help mitigate some of the difficulties that they introduce.

Resolution Problems

Chapter 2 presented the convolutional model of the seismic trace. Based on this model, the seismic response of a sequence of spikes, corresponding to a sequence of reflection coefficients at layer interfaces, can be constructed by summing the wavelet responses of each interface. When the interfaces are closely spaced, say by less than a quarter of the signal wavelength, the composite seismic response is an interfering complex, and the wavelet amplitude and waveshape are no longer linearly related to the reservoir properties. However, analysis techniques can approximate the wavelet interference effects produced by thin layers and estimate the layer thicknesses and petrophysical properties that produced these effects. For example, an interpretation of thin bed wavelet interference and its effect on composite reflection waveforms has been used to predict layer thickness, porosity, and net pay in surprisingly thin beds (Brown et al., 1984).

Other successful thin bed interpretation techniques have focused on the stratigraphic inversion of the seismic signal. Iterative forward modeling (Gelfand and Larner, 1984) and L-1 norm techniques (Oldenberg et al., 1983) often provide an improved resolution of the thin bed structure of the subsurface when recursive methods (Lindseth, 1979) are not acceptable.

However, the layers produced by all of these inversion techniques are "thick" relative to the scale of measurements used in core analyses and wireline logs. Hence seismic reservoir models average petrophysical properties over a vertical interval of appreciable thickness. This averaging must be taken into account when calibrating seismically estimated parameters to well-derived reservoir properties.

Nonuniqueness Issues

In striving for greater vertical resolution in the processing of seismic data, we cannot avoid the fact that the resulting reservoir model is not unique. To reduce the degree of nonuniqueness, while at the same time "broadening" the seismic bandwidth, we must introduce external constraints. These external constraints are generally derived from well data, and because they are imposed differently in various inversion processes, they produce multiple reservoir models that satisfy the seismic imaging requirements. Specific geologic knowledge of the field is invaluable in selecting the most plausible of these several solutions to the inversion problem, and hence, in reducing the uncertainties and the nonuniqueness of the interpretation.

Ambiguity Questions

The amplitudes of reflected seismic signals depend primarily on variations in acoustic impedance, which is the product of velocity and density. Hence changes in either rock velocity or density will contribute to variations in the seismic response of the reservoir. A number of petrophysical properties, such as lithology, porosity, pore pressure, and fluid saturation, affect both rock velocity and density. Therefore, to estimate reservoir properties using seismic data we must quantify the respective contribution of each petrophysical parameter to the acoustic measurements. Two methods, deterministic modeling and stochastic modeling, have been used to establish the numerical relationships needed for these quantifications and to minimize ambiguities between seismic measurements and reservoir descriptions.

Deterministic Reservoir Modeling

In the deterministic modeling approach, we typically use empirical relationships to relate the reservoir parameters to seismically derived attributes. Unfortunately, these parameters are known only at well locations and must be interpolated between the control points. The sensitivity of a selected reservoir property to changes of several petrophysical parameters must, therefore, be evaluated. This is an essential step in ensuring that the cumulative uncertainty resulting from the interpolation of several control parameters between well locations does not exceed the expected maximum variation of the reservoir property that is being predicted.

This concept is illustrated by Figure 1 which graphs porosity as a function of lithologic composition for different seismic impedance and fluid saturation values. From this diagram, we can see that a wide range in lithology, porosity, or saturation results in a constant acoustic impedence. Therefore, if we want to predict reservoir porosity from a seismically determined estimate of impedance, we would first have to estimate rock composition and saturation. The magni-

tude of the errors involved in extrapolating these two petrophysical parameters (rock composition and saturation) will directly affect the accuracy of the porosity prediction. A critical issue in reservoir description is to determine whether the total porosity error resulting from all of these extrapolations is within acceptable bounds.

To illustrate the use of deterministic calibration curves, assume that a reservoir unit at a control well is known to have 10 percent porosity, a water saturation of 100 percent, and an acoustic impedance of 10 000 g.m/cm^{3}s, which is point C in Figure 1. If seismic data imply that the reservoir impedance decreases to 6 000 g.m/cm^{3}s at some lateral distance from this well, then the curves in Figure 1 indicate that the following petrophysical changes could be possible:

The water saturation and mineralogy remain constant, but the porosity increases to 32 percent (point E).

The mineralogy does not change, but the water saturation reduces to a very low value and the porosity increases to 24 percent (point D).

The porosity remains at 10 percent, but the water saturation becomes low and the mineralogy changes from predominantly sand (coordinate C′) to predominantly shale (coordinate B′). The new reservoir condition is point B.

The porosity is 10 percent, the water saturation remains approximately 100 percent, but the mineralogy changes from C′ to A′, so that the reservoir is described by point A.

A reservoir modeler will have to decide if one of these solutions best fits the boundary conditions imposed by all other available data, or whether some other condition is more likely.

In general, in deterministic modeling we try to characterize the variability of those reservoir parameters that are the greatest contributors to changes in acoustic properties. Graphic or numeric relationships such as shown in Figure 1 are ideal for these purposes. Changes of lithology and porosity usually have a first-order effect on seismic impedances (Angeleri and Capri, 1982; de Buyl et al,, 1988; de Buyl et al., 1987). However, an even greater contribution may result from the presence of gases (Ostrander, 1982; Greaves and Fulp, 1987; Backus, 1987) or from temperature gradients induced, for example, during a fire flood (Tosaya and Nur, 1984; Pullin et al.,1986). By contrast, variations of permeability may not be directly derivable from seismic data. However, Chapter 6 shows that an indirect mapping of permeability excur-

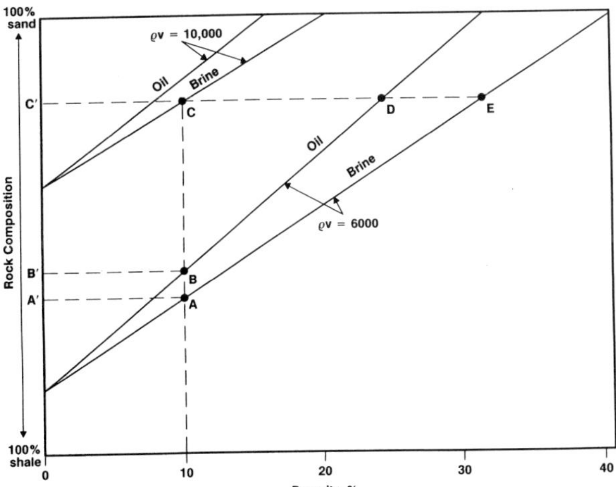

Fig. 1. Typical deterministic calibration curves relating reservoir properties (lithology, porosity, porefluid) and acoustic impedance (ρV).

sions above certain threshold levels is potentially feasible with seismic data and may provide a way to detect preferential flowpaths in reservoirs.

Subsequent papers included in Chapter 5 present deterministic models based on the direct interpretation of seismic reflection amplitudes as a means for estimating reservoir properties, and others illustrate additional deterministic approaches.

Stochastic Modeling

It is essential to quantify the imprecision involved in the spatial description of a reservoir and to evaluate how reservoir parameter errors affect the predictions of reserves and production. It is also important to model the relationships between seismic measurements and the reservoir properties that have the greatest effect on the acoustic response. In deterministic approaches, these relational models are created by crossplotting seismic impedance data at well locations against log-derived petrophysical properties of the reservoir (e.g., Figure 1). Such deterministic approaches, however, treat the data as spatially independent observations and do not capitalize on the existence of spatial patterns in the variations of rock properties. In addition, the reliability of deterministically derived reservoir parameters are sometimes not easy to assess, either quantitatively or qualitatively. For these reasons, some reservoir modelers prefer a stochastic, or geostatistical, approach because:

Geostatistics uses spatial auto-correlation functions to analyze patterns in modeling data and to determine preferential directions of variability;

Spatial cross-correlation functions also measure the sensitivity of seismic data to changes in reservoir parameters;

Geostatistical reservoir models honor well data within error bounds that are consistent with measurement accuracies;

Expert information, such as geologic knowledge about the reservoir, can be incorporated objectively and quantitatively; and

Geostatistical approaches provide measures of the reliability of the reservoir models that are useful for risk assessment.

Some important geostatistical approaches to reservoir modeling are exemplified in the papers by Doyen and Guidish and by Brac et al. which conclude Chapter 5.

References

Angeleri, G. and Capri,. R., 1982, Porosity prediction from seismic data: Geophys. Prosp., **30**, 580–607.

Backus, M., 1987, Amplitude versus offset, A review: 49th Ann. Internat. Mtg., Soc. Expl. Geophys., Expanded Abstracts, 359–364.

Brown, A., Wright, R., Burkhart, K., Abriel, W., and McBeath, R., 1984, Tuning effects, lithologic effects, and depositional effects in the seismic response of gas reservoir: Presented at 46th Ann. Mtg., Eur. Assn. Expl. Geophys., London.

DOE/Energy Information Administration, 1988, U.S Crude Oil, Natural Gas Liquids Reserves, 1987 Ann. Rept.: DOE/EIA-0216(87).

de Buyl, M., Guidish, T., and Bell, F., 1988, Reservoir description from seismic lithologic modeling: Soc. Petr. Eng., Paper 15505, J. Petr. Tech.

de Buyl, M., and Ullah, S., and Guidish, T., 1987, Seismic reservoir description, substantiation by reservoir simulation: Soc. Petr. Eng. Paper 16781. Presented at the 62nd Ann. Soc. Petr. Eng. Conf., Dallas.

Gelfand, V. and Larner K., 1984, Seismic lilthologic modeling: The Leading Edge, **8**, No. 11, 30–35.

Greaves, R. and Fulp, T., 1987, Three dimensional seismic monitoring of an enhanced oil recovery process: Geophysics, **52**, 1175–1187.

Lindseth, R.O., 1979, Synthetic sonic logs—a process for stratigraphic interpretation: Geophysics, **44**, 3–26.

Oldenburg, D. W., Scheuer, T., and Levy, S., 1983, Recovery of acoustic impedance from reflection seismograms: Geophysics, **48**, 1318–1337.

Ostrander, W. J., 1982, Plane-wave reflection coefficients for gas sands at non-normal angles of incidence: 52nd Ann. Internat. Mtg., Soc. Expl. Geophys., Expanded Abstracts, 216–218.

Pullin, N.P., Matthews, L.W., and Hirsche, W. K., 1986, Techniques applied to obtain very high resolution 3-D imaging at an Athabasca tar sands thermal pilot: The Leading Edge, **6**, No. 12, 10–15.

Tosaya, C.A. and Nur, A.M., 1984, Monitoring of thermal EOR fronts by seismic methods: Soc. Petr. Eng., Paper 12744, Presented at the 59th Ann. Soc. Petr. Eng. Conf.

Net Producible Gas Sand Mapping by Interactive Analysis of Seismic Time and Amplitude[1]

Alistair R. Brown[*]

Introduction

Garden Banks is a region offshore Louisiana and Texas outside the 200 m isobath (Figure 1) where the hydrocarbon prospects are principally in Pleistocene sands. In 1977 and 1978, Chevron and their partners made several gas discoveries in this area. In particular they penetrated sands at a depth of about 4500 ft (1400 m) which indicated good production. These sands are variable in thickness, porosity, and the number of lobes present, and the principal task facing the production geologist and geophysicist is to map net producible sand.

A three-dimensional (3-D) seismic survey was recorded over the gas sands in 1981. Because of the known and estimated structural and stratigraphic dips, close subsurface spacings were used: the line spacing was 50 m and the common-mid-point (CMP) spacing was 12.5 m. Figure 2 shows a north-south vertical section from the 3-D survey. The black rectangular box encloses the bright reflections associated with the gas sands. The north structural dip is assumed to be generated by a salt swell, and the irregularity of the sands is considered to result from slumping. The reservoirs are clearly stratigraphically controlled up-dip and the reflections from the sand interfaces are variable in amplitude. The objective of the interactive interpretive project was to utilize the seismic amplitude information to estimate the effective sand thickness of the gas sands.

Direct detection of gas sands on the basis of anomalous amplitude has been popular for many years. A thorough treatise on the criteria for detection was provided in Backus and Chen (1975). Most practical uses of seismic amplitude for hydrocarbon detection have been qualitative, and the difficulties of doing otherwise were explained in Domenico (1976).

It is now realized that wavelet processing to zero phase significantly aids in identifying reservoir reflections and also helps in relating individual amplitudes to individual reservoir interfaces. Careful wavelet processing was thus an important element in the processing of this 3-D data. The quantitative use of amplitude to estimate the aggregate thickness of sand units encased in shale is discussed in Meckel and Nath (1977), Neidell and Poggiagliolmi (1977), and Schramm et al. (1977). 3-D migration images subsurface reflection data more accurately than two-dimensional (2-D) migration (French, 1974), and hence the resulting amplitudes should provide more detailed and quantitative information. Brown et al. (1981) showed the ability of 3-D data to reveal subtle stratigraphic features on the basis of amplitude.

A small portion of the Garden Banks 3-D survey was identified as the study area (Figure 3). The area measures 2.0 km north-south by 2.2 km east-west. The drilling platform to the east caused a gap in 3-D data coverage immediately to the east of the study area. Four deviated wells from the platform and one vertical well provided evaluation points for the sands under study. Figure 4 excerpts the log information for two of these wells. The interpreted producible sand zones and the resistivity logs clearly demonstrate the problem: both Upper and Lower Sand members contain many lobes of producible sand and there is considerable lateral variability in the sands over a short distance.

Interactive Interpretation

Interactive interpretation of seismic data is now in widespread use (Gerhardstein and Brown, 1984). As more and more interpreters complete interactive projects, the resulting benefits have become increasingly clear.

The success of the Garden Banks project can, to a significant extent, be attributed to the interactive system. The benefits of the interactive system are reviewed before the results of the projects are described.

[1]This paper is adapted and updated from two published papers: Brown, A.R., Wright R.M., Burkart, K.D., and Abriel, W.L., 1984, Interactive seismic mapping of net producible gas sand in the Gulf of Mexico: Geophysics, **49**, 686–714, and Brown, A.R., Wright, R.M., Burkart, K.D., Abriel, W.L., and McBeath, R.G, 1986, Tuning effects, lithological effects and depositional effects in the seismic response of gas reservoirs; Geophys. Prosp., **34**, 623–647.
[*]Consulting Reservoir Geophysicist.

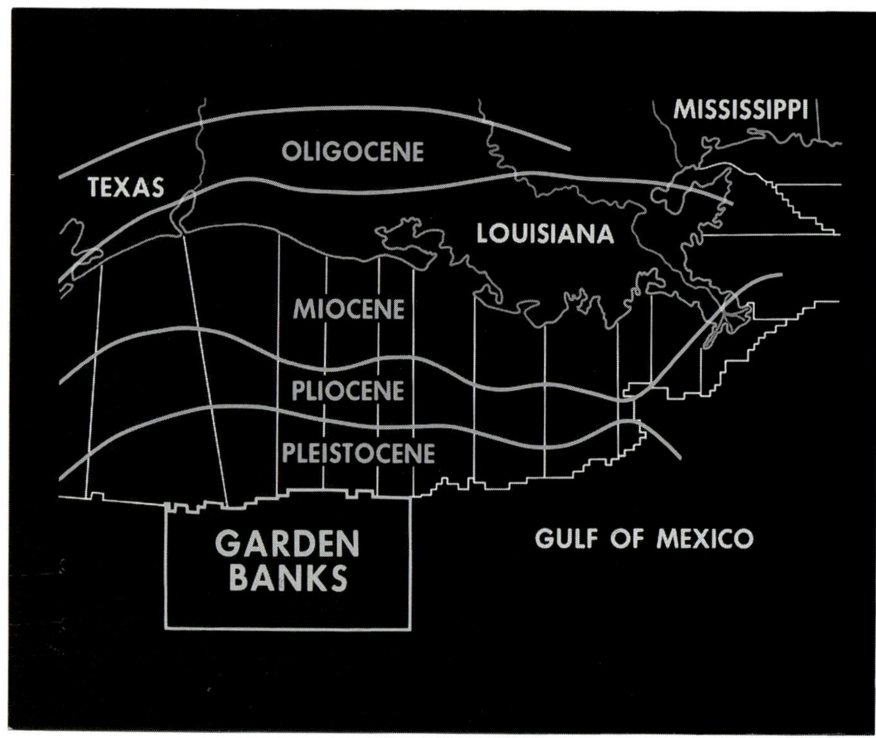

Fig. 1. Location of Garden Banks area offshore Louisiana and Texas.

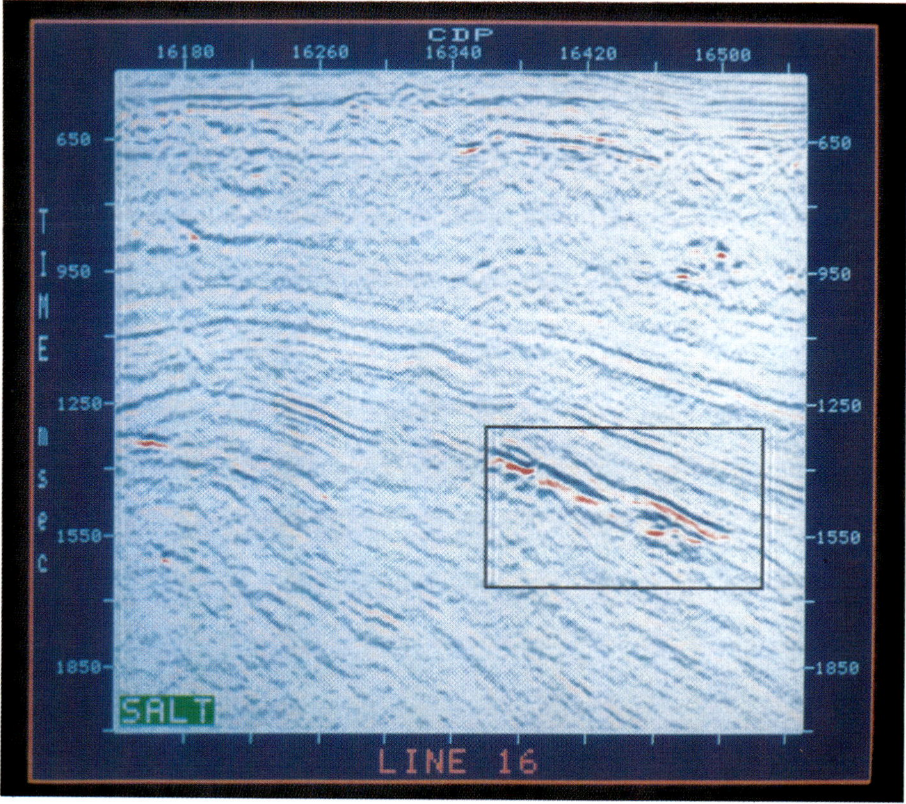

Fig. 2. Line 16 from the 3-D survey showing the setting of the bright reservoir
events on the north flank of a structure probably caused by a salt swell.

Data management. The interpreter needs little or no paper; the selected seismic data display is presented on the screen of a color monitor and the progressive results of interpretation are returned to the digital database.

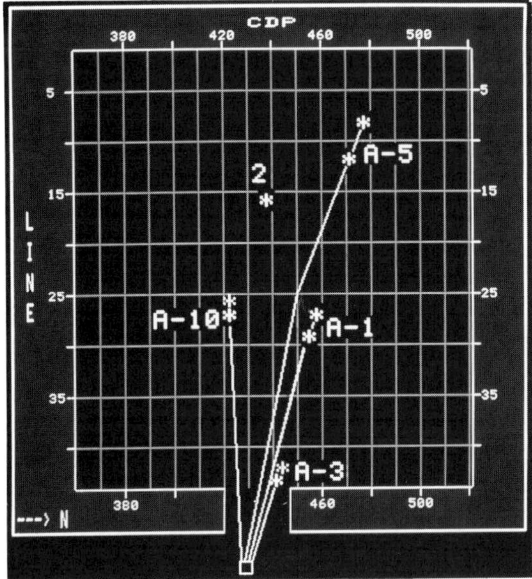

Fig. 3. Location of the five wells within the study area showing the points of intersection with the Upper and Lower Sands.

Color. Flexible color display provides the interpreter with maximum optical dynamic range adapted to the particular problem under study.

Image composition. Data images can be composed on the screen so that the interpreter views what is needed, no more and no less, for the study of one particular issue. Slices through the data volume are designed by the user in order to customize the perspective to the problem.

Idea flow. The rapid response of the system makes trying new ideas easy. The interpreter can rapidly generate innovative map or section products in pursuit of a better interpretation.

Interpretation consistency. The capability to review large quantities of data in different forms means that the resulting interpretation should be more consistent with all available evidence. This is normally considered the best measure of interpretation quality.

More information. Traditional interpretive tasks performed interactively will save time; however, the extraction of more detailed subsurface information is more persuasive and far-reaching.

Manipulative Sequence

Figures 5 through 8 illustrate the data in the north-south vertical section from every fourth line. The vertical scale is enlarged, so the north structural dip is exaggerated. Each data portion has the same CMP and

(Text continued on page 197)

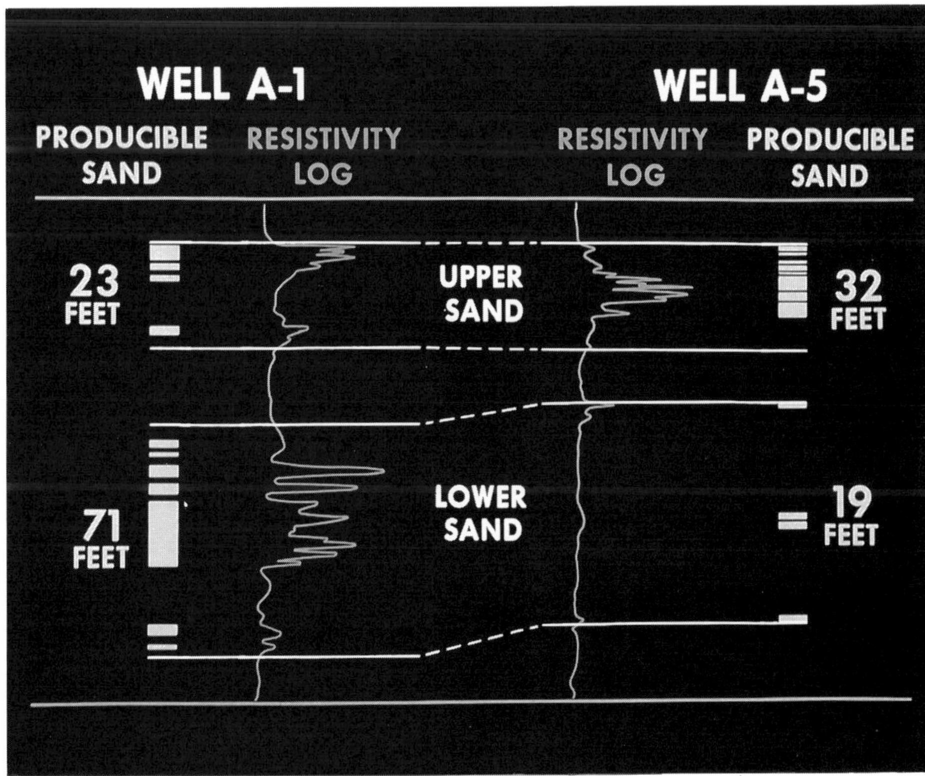

Fig. 4. The interpreted lobes of producible gas sand penetrated by wells A-1 and A-5.

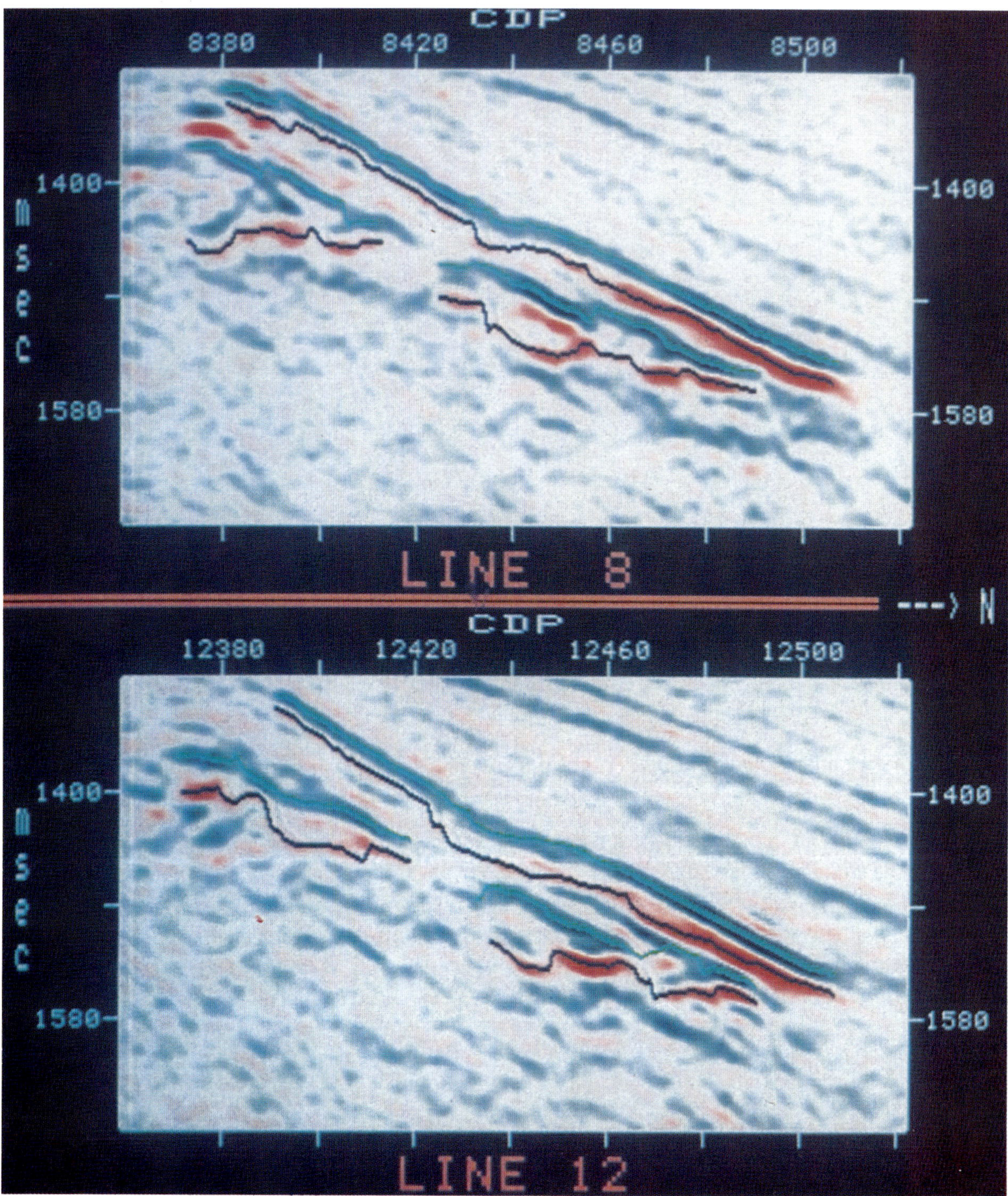

Fig. 5. Lines 8 and 12 showing tracked horizons for top and base of Upper and Lower Sands.

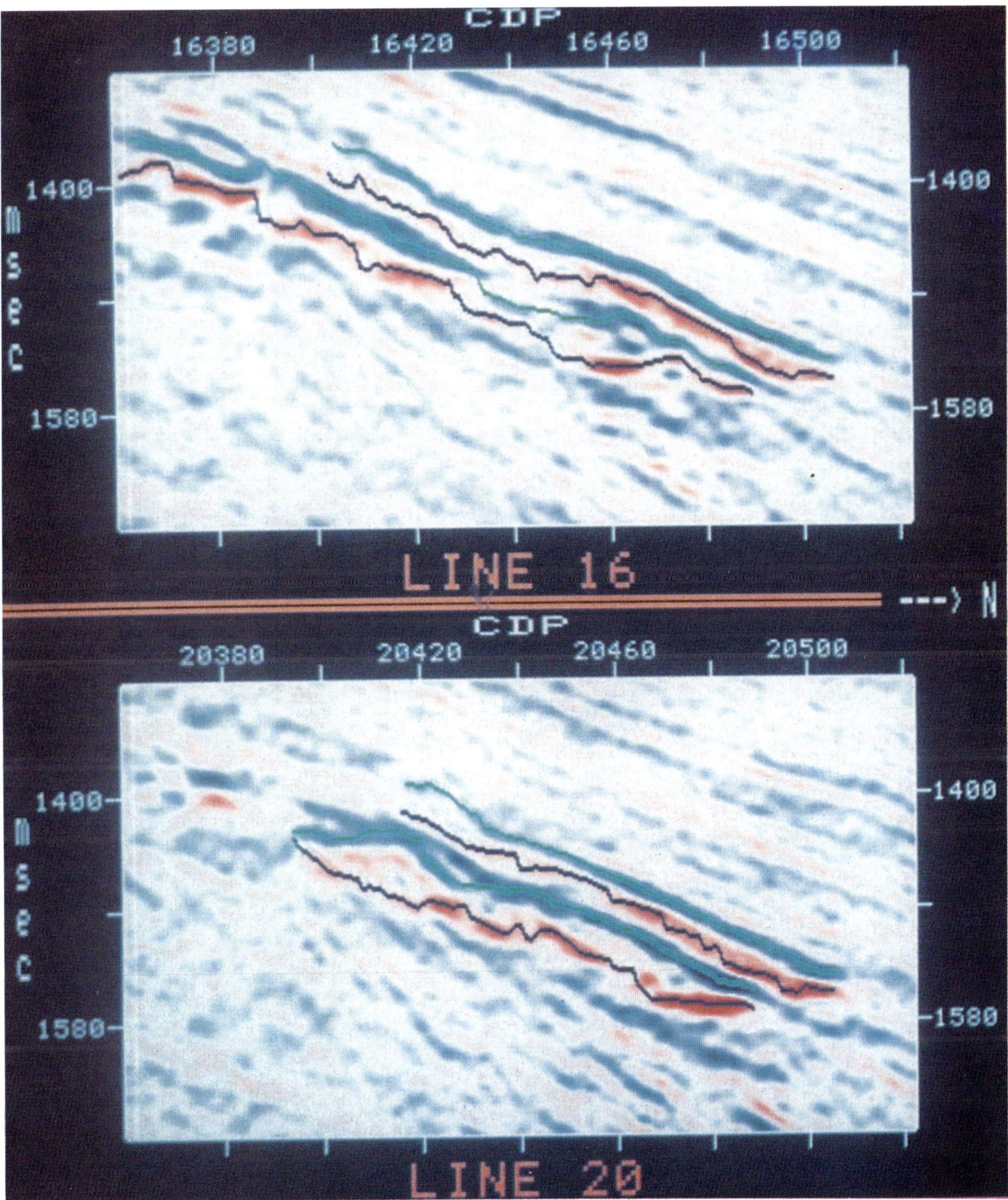

Fig. 6. Lines 16 and 20 showing tracked horizons for top and base of Upper and Lower Sands.

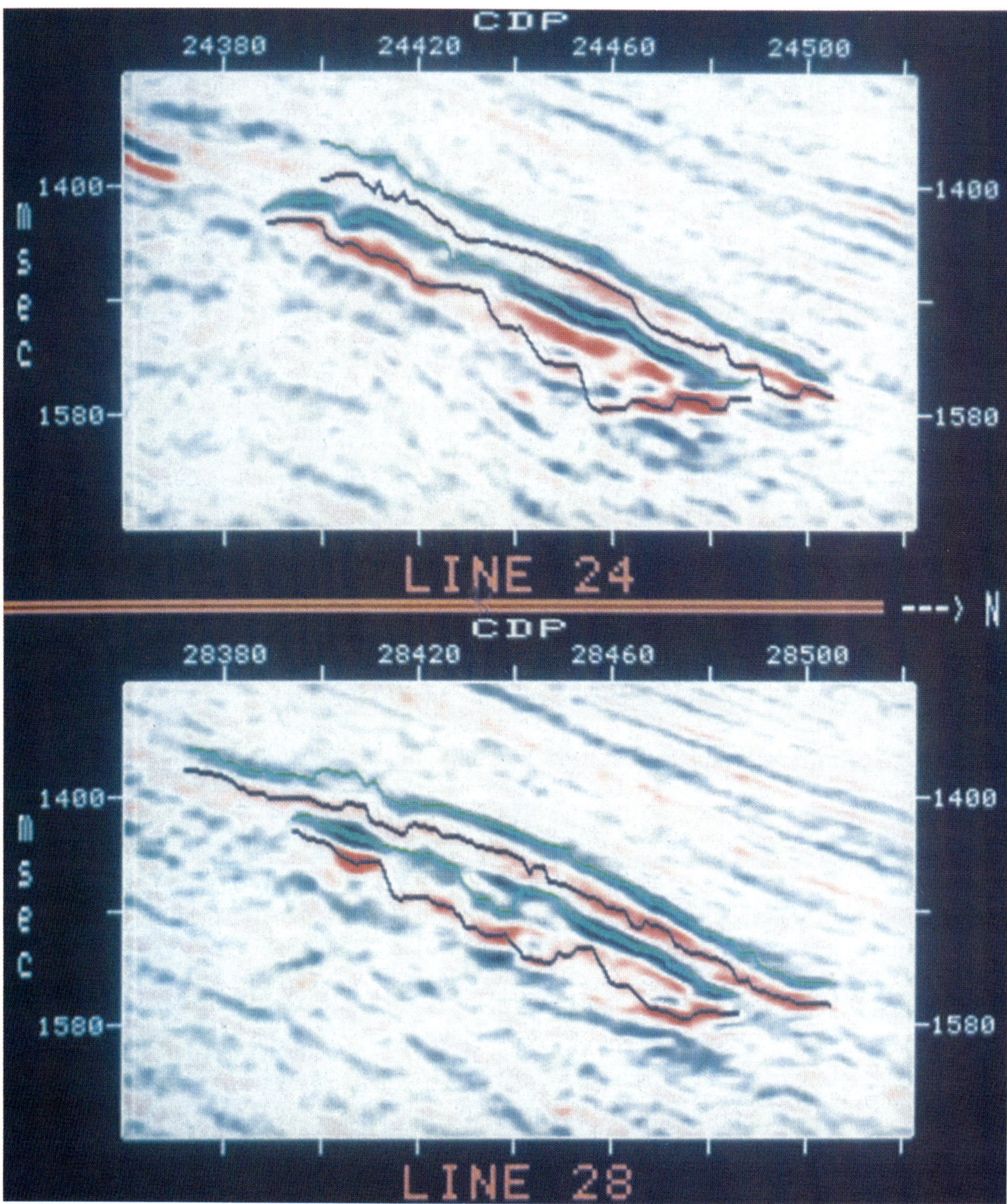

Fig. 7. Lines 24 and 28 showing tracked horizons for top and base of Upper and Lower Sands.

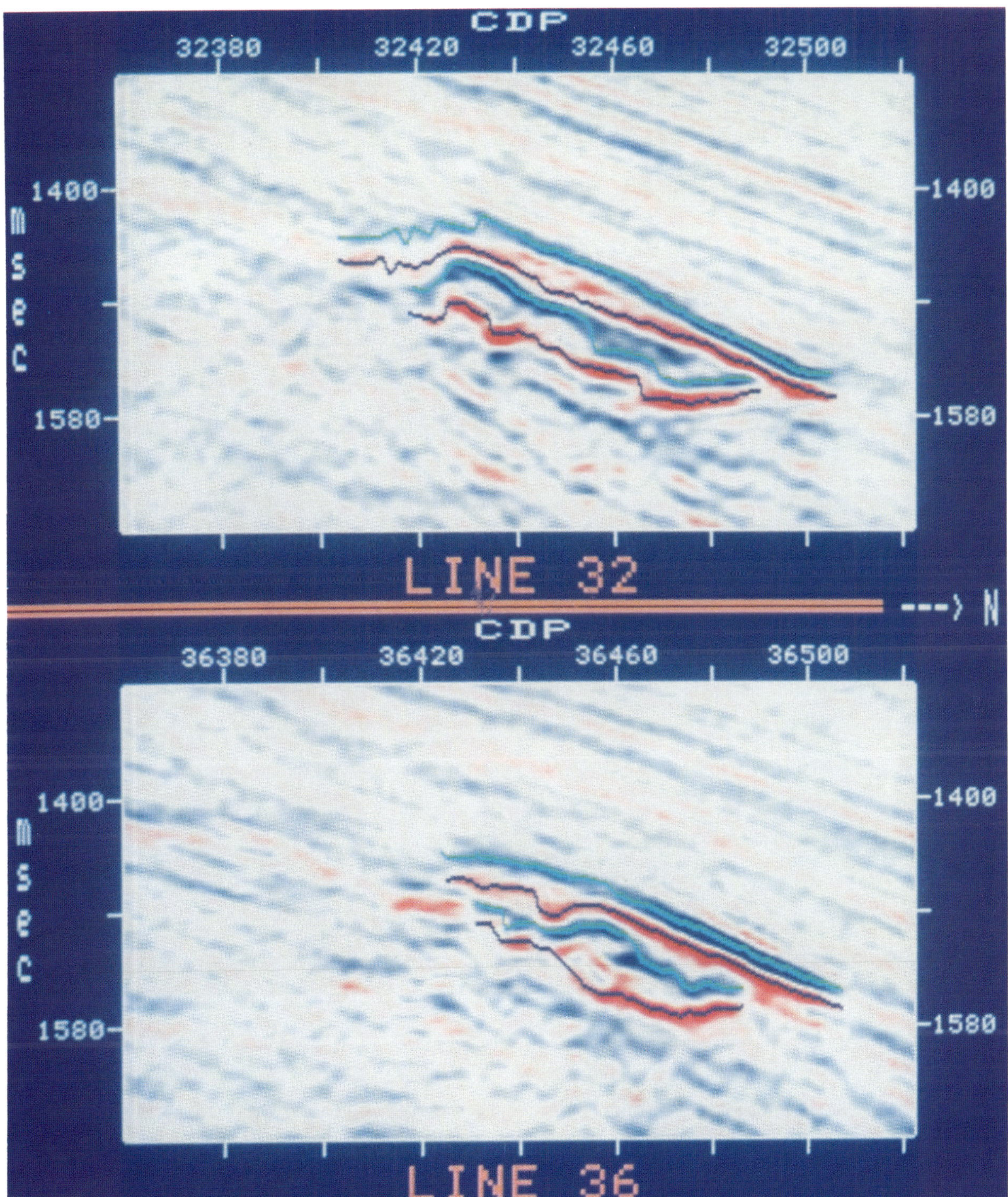

Fig. 8. Lines 32 and 36 showing tracked horizons for top and base of Upper and Lower Sands.

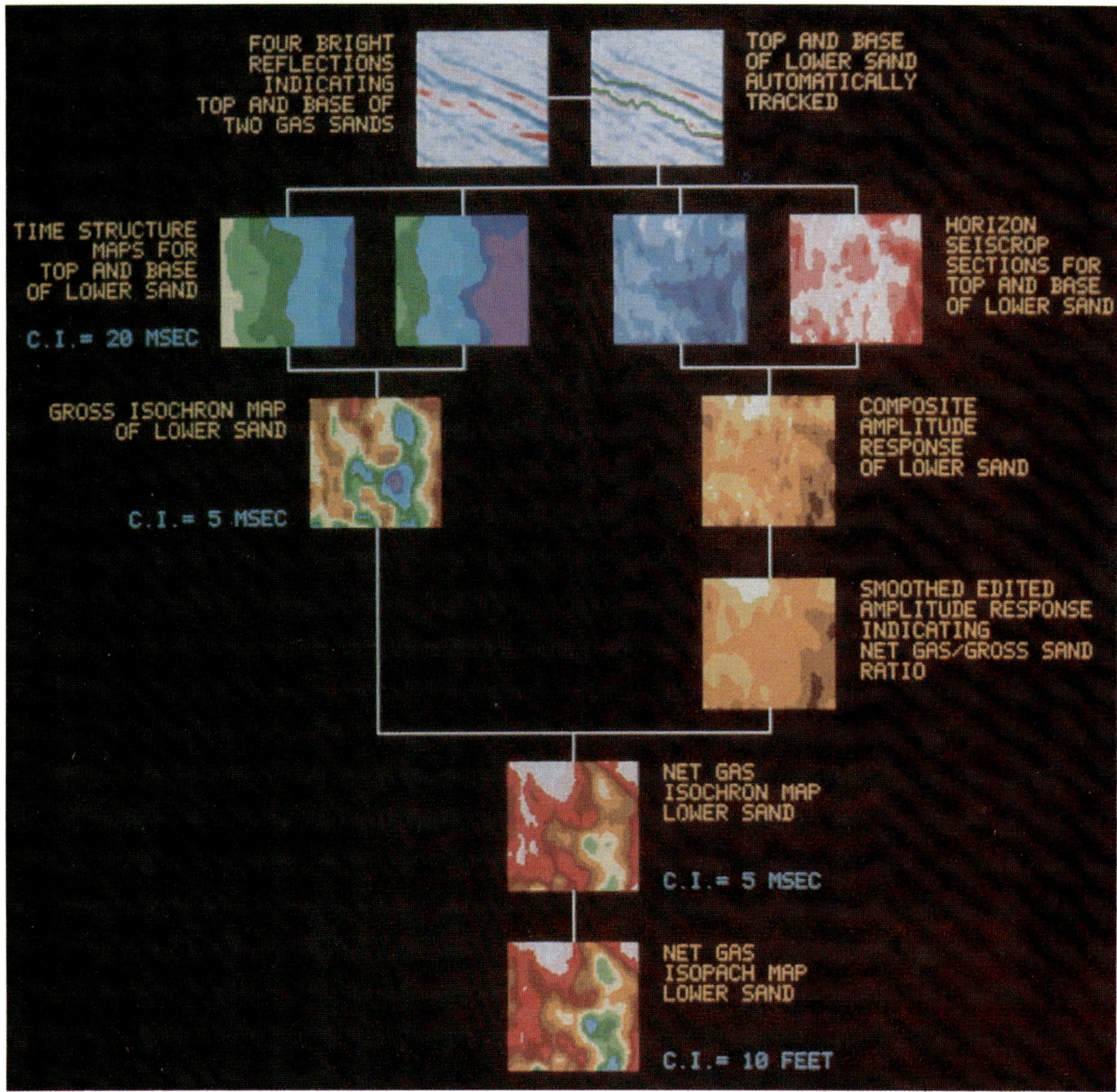

Fig. 9. Interpretive sequence used and intermediate products generated in the course of deriving net-producible-gas sand maps. C.I. equals contour interval. Time-structure maps show dip down to the right; the purple area is the flat spot at the base of the Lower Sand. The isochron and isopach maps have greens and blues indicating the thicker zones. All the four amplitude products have darker colors indicating the higher amplitudes. Horizon Seiscrop section is another name for horizon slice, namely the spatial amplitude distribution over an interpreted seismic horizon.

time range as that outlined by the black box in Figure 2.

In general there are four bright (high amplitude) events, and these events were identified as the top and base of the Upper and Lower Sands. At the downdip limits in several places are prominent flat spots. Automatic tracking was used to follow the events wherever possible so that the track followed the true maximum amplitude of the peak or trough. Manual editing was required over less than 1 percent of the total horizon distance tracked. However, this is not an indication that the horizon interpretation was easy; at times the selection of the event segment for the automatic tracker was quite difficult. Using a two-window screen display as illustrated in each of Figures 5 through 8, we were able to study the tracked horizons from the last line and the data for the next line while tracking the present line. The four horizons were only tracked where they were at least somewhat bright and the flat spots, where present, were tracked as the bases of the gas accumulations. The final tracks used are superimposed on the sample vertical sections in Figures 5 through 8.

The act of interactively tracking each data point caused a time and amplitude value to be stored in the prospect data base. The sequence of manipulations then applied to these times and amplitudes is charted in Figure 9. The tracked times directly yielded a structural contour map for each of the four horizons. These contour maps were then gently smoothed to remove the roughness of the automatic tracking and subtracted to yield a gross ischron map for each of the two sands.

The display of the stored horizon amplitudes directly yielded a horizon slice, also known as a horizon Seiscrop™ section, for each horizon; this type of section is defined in Brown et al. (1981). For each sand the horizon slice from the top and base reflections were added together (in absolute values) to yield the composite amplitude response for that sand. This summation constituted a selective vertical integration, analogous to an inversion process, whereby the response of the sand was accentuated relative to that of encasing material.

Because of constructive interference between the wavelets from the top and bottom of each sand interval, the reflection amplitudes were boosted when their time separation was around one-half seismic period; this effect is known as tuning (Widess, 1973) and the tuning thickness was here determined to be between 15 and 16 ms. Proper removal of tuning effects was a critical part of the procedure. The magnitude of the tuning effects was estimated both statistically and deterministically. Figure 10 shows a crossplot in which the composite horizon amplitude is plotted against measured gross isochron thickness for each data point

tracked. The upper envelope of this crossplot resembles a tuning curve and the multiple maxima indicate multiple side lobes in the effective wavelet. Figure 11 shows several wavelets and their corresponding tuning curves, and the number of tuning curve maxima obvi-

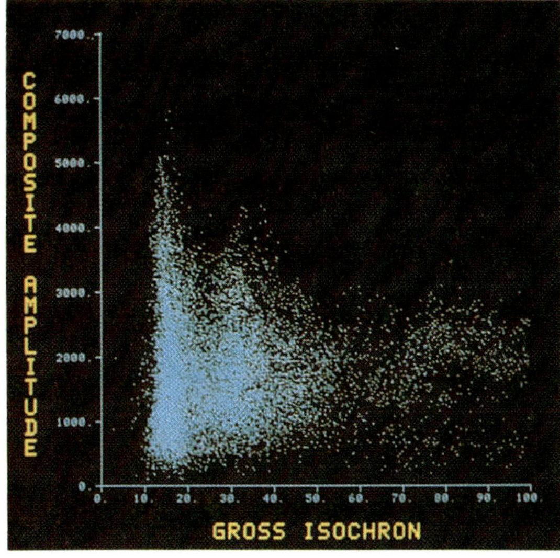

Fig. 10. Interactive crossplot of composite amplitude against gross isochron for the whole prospect area.

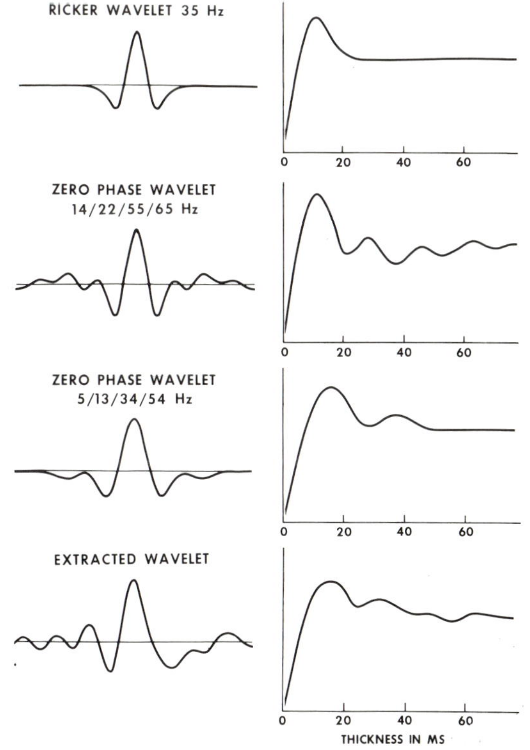

Fig. 11. Various wavelets and their corresponding deterministic tuning curves.

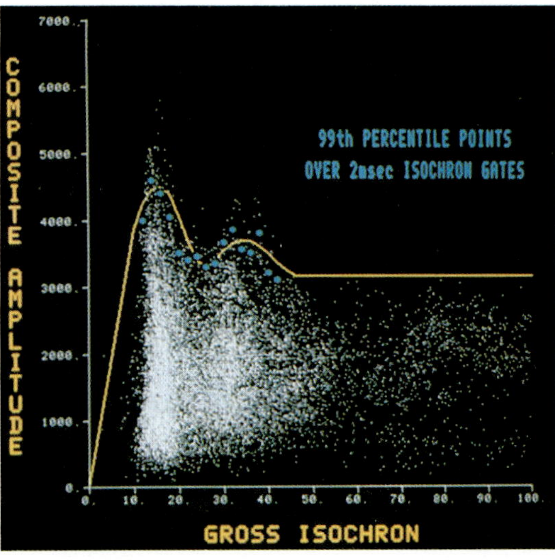

Fig. 12. Same crossplot as in Figure 10 analyzed to yield 99th percentile points and showing these points in relation to interpreted tuning curve.

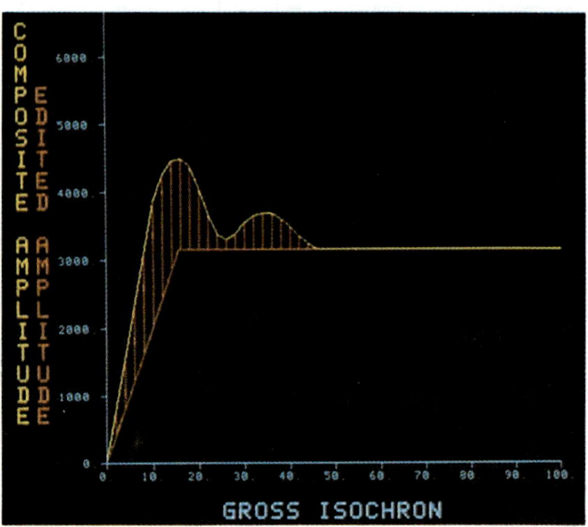

Fig. 13. The editing of tuning effects.

gas sand isochron map was converted to a net gas sand isopach map for each sand.

ously derives directly from the number of wavelet side lobes. The crossplot of Figure 10 was statistically analyzed over 2 ms isochron gates to yield 99th percentile points plotted in blue in Figure 12. In this way overtuned and excessively high amplitude points were eliminated. The yellow tuning curve in Figure 12 was then adopted as the correct tuning curve for this reservoir data, as it fits the statistical points and also conforms to the deterministic tuning curve derived from the zero-phase wavelet in the data (third curve, Figure 11). The tuning effects were then edited from the amplitudes according to Figure 13; each amplitude was reduced by a multiplicative factor according to its gross isochron so that the response shown in yellow was reduced to that shown in orange.

The resultant edited composite amplitude response was then used as an indicator of the proportion of the sand interval capable of producing gas, the higher amplitudes indicating higher proportions and the lower amplitudes lower proportions. In order to turn this amplitude response into a map of the ratio of net gas sand to gross seismic interval, we simply divided by the maximum amplitude, having established from a nearby well that this maximum amplitude corresponded to 100 percent effectiveness. The multiplication of this ratio map by the *gross* isochron map then gave the *net* gas sand isochron map for each sand. Producible gas sand was assumed to have a constant velocity, and this velocity was measured from sonic logs to be 5000 ft/s. Using this measurement, the net

Results for Upper Sand

The time structure maps for the top and base of the Upper Sand are shown in Figures 14 and 15 where the maps have already been gently smoothed. A three-point by three-point spatial filter was used with weights selected so that the amount of smoothing was approximately equal in the two orthogonal directions, considering that the subsurface spacing was four times greater in one direction than the other. The black area surrounding the color within the blue border of the study area is where the horizons were not tracked because they were judged insufficiently bright.

The gross ischron map for the Upper Sand appears in Figure 16. Note a distinct thickness trend running northwest-southeast.

The similarity between the horizon slices from the top and the base of the Upper Sand (Figures 17 and 18) reinforces that the major cause of the lateral changes in amplitude is variations within the sand rather than its encasing material and hence supports the summation of the two to yield the composite amplitude response (Figure 19). Removal of tuning distortions yielded the edited amplitude response of Figure 20. Note that the high amplitude lineation parallel to the northern boundary of the area tracked and visible in each of Figures 17, 18, and 19 is absent in Figure 20. This lineation was caused by constructive interference between the reflection from the top of the reservoir and the one from the fluid contact, and clearly this needed to be removed by the editing procedure. Figure 20 is, after calibration to a nearby well, also a map of net gas to gross sand ratio for the Upper Sand.

(Text continued on page 204)

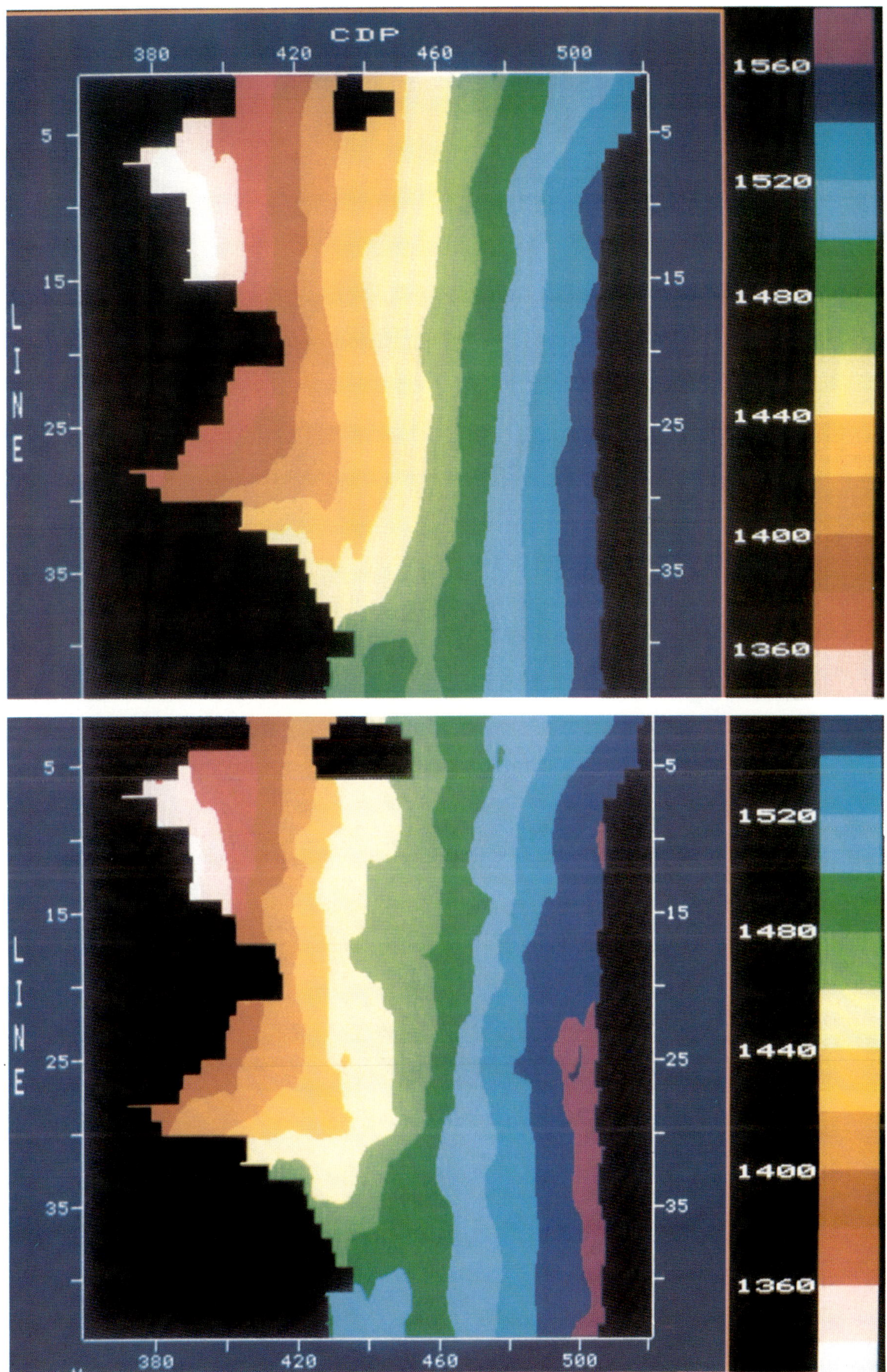

Fig. 14–15. Time-structure map, Upper Sand. Top of sand (above), base (below).

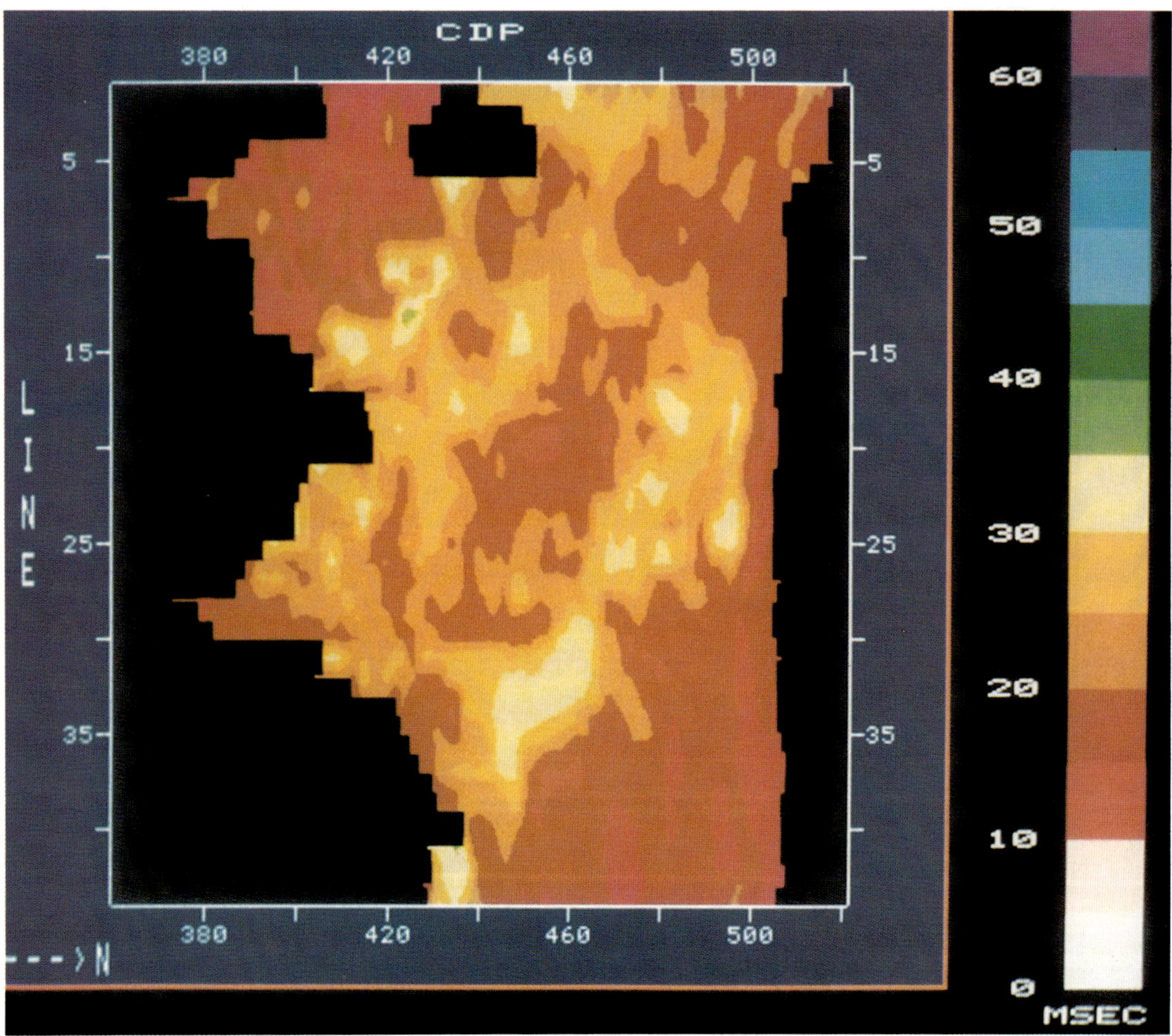

Fig. 16. Gross isochron map, Upper Sand.

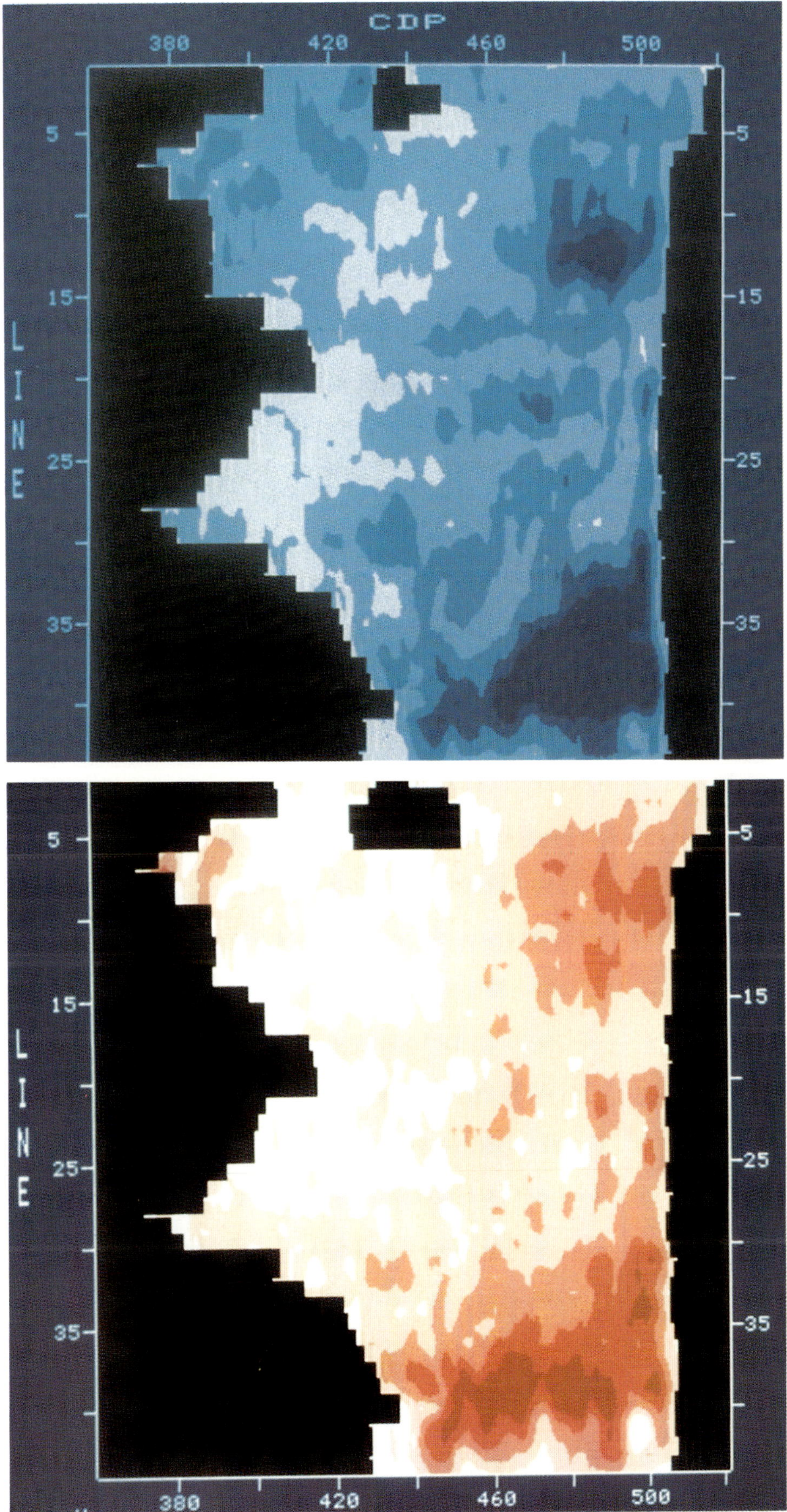

Fig. 17–18. Horizon slice, Upper Sand. Top of sand (above), base (below).

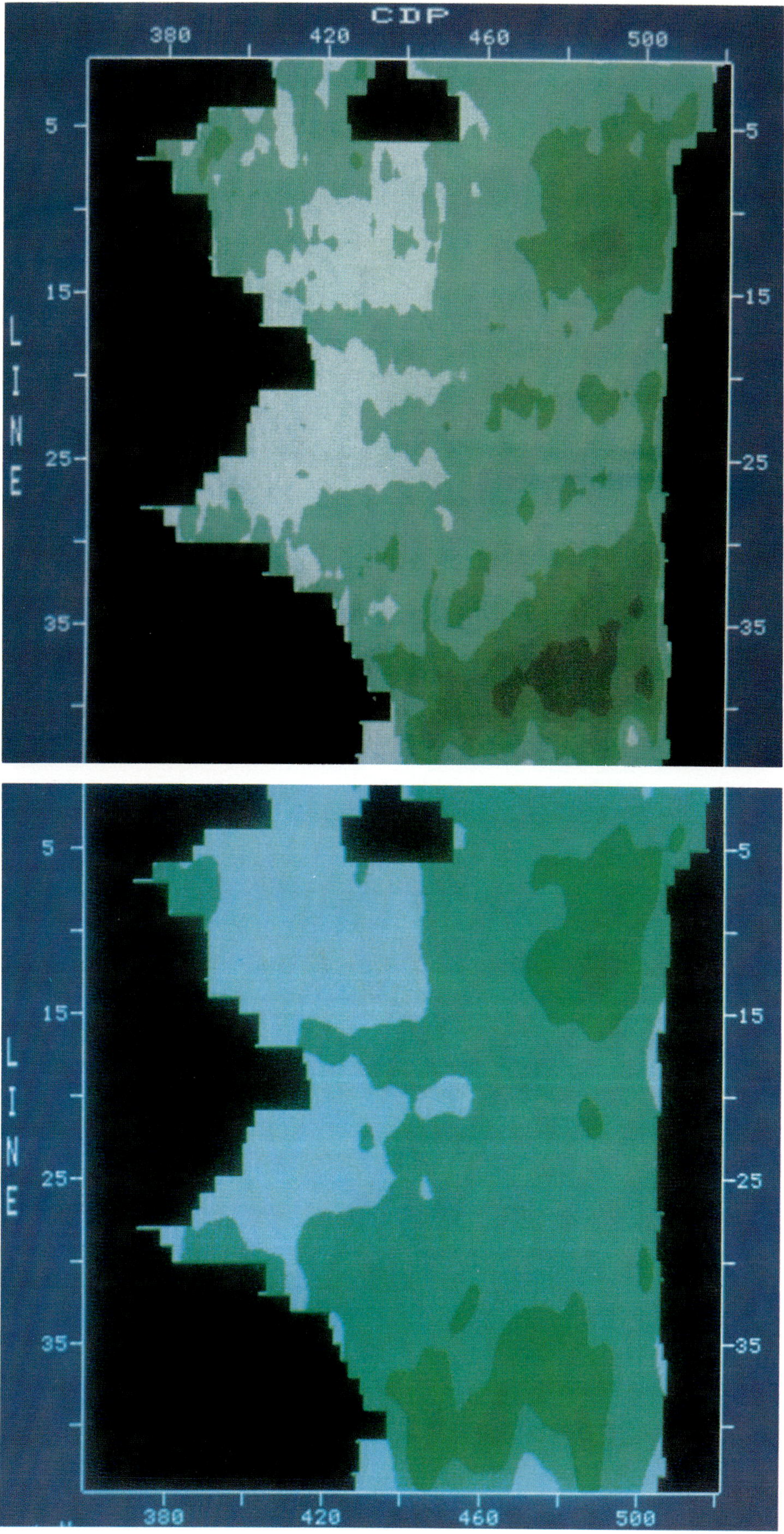

Fig. 19–20. Composite amplitude response, Upper Sand (above) and after editing of turning effect (below).

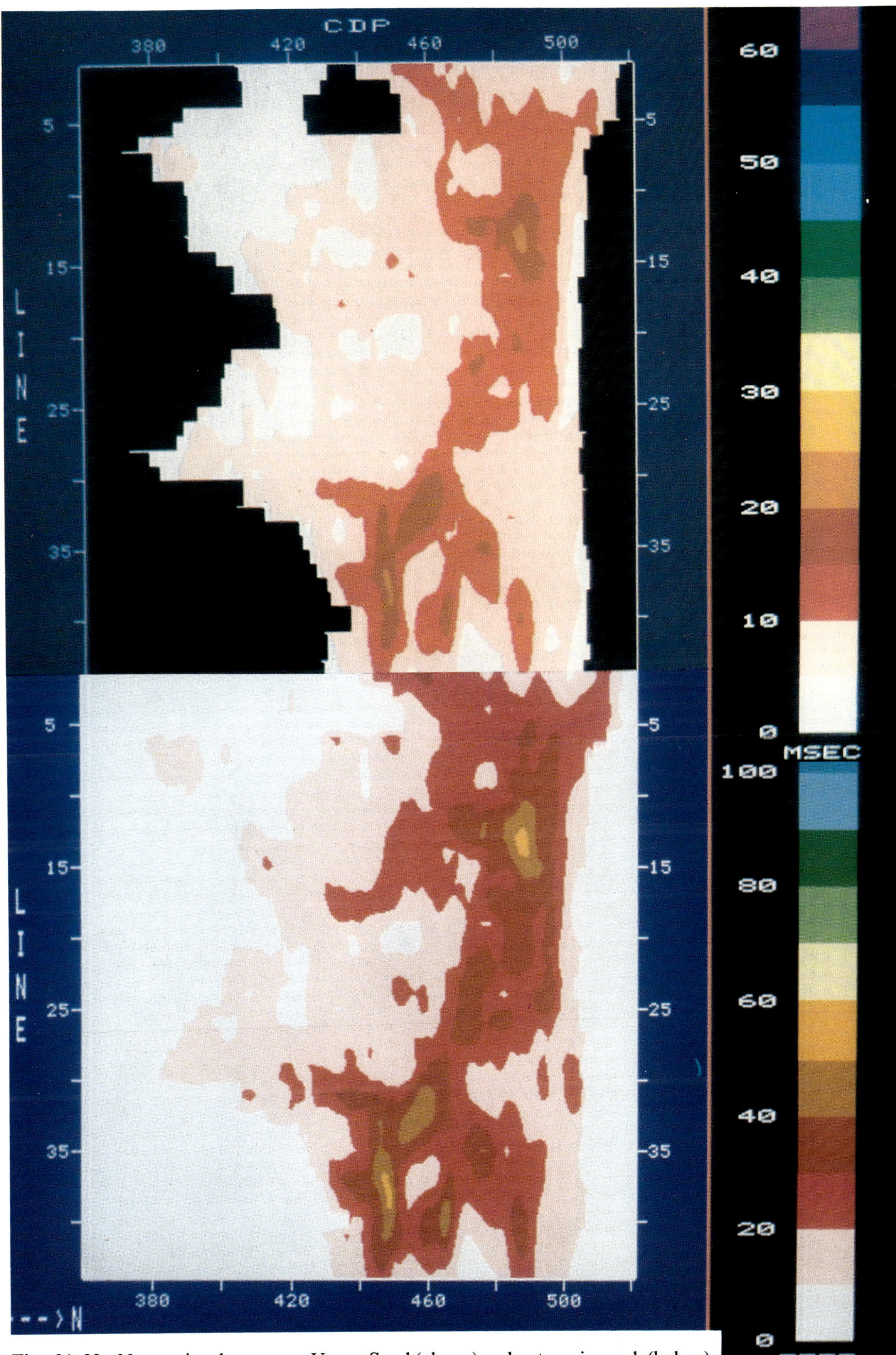

Fig. 21–22. Net gas isochron map, Upper Sand (above) and net gas isopach (below).

The product of this net gas to gross sand ratio map (Figure 20 scaled) and the gross isochron map (Figure 16) directly yielded the net gas sand isochron map (Figure 21). By comparing Figure 21 with the gross isochron map (Figure 16), note that the southwesterly gross thickness trend has disappeared indicating that it contains little gas, whereas the northeasterly gross trend is strongly preserved, indicating that it is an effective reservoir. Application of the constant 5000 ft/s gas sand velocity then converted Figure 21 to the net gas sand isopach map of Figure 22. Here we have removed the distinction between where the sand was not tracked, and hence assumed to be thin or absent, and where it was computed to be effectively thin; both are shown here in white.

By integration of the net gas sand isopach map of Figure 22, the total reservoir volume of the Upper Sand within the study area is computed to be 13 000 acre ft (17 million m^3).

Results for Lower Sand

Figures 5 through 8 demonstrate that the Lower Sand is more variable in thickness and also reaches much greater thicknesses than the Upper Sand. The Lower Sand reflections were also more difficult to track. However, the manipulative sequence used was exactly the same, namely that charted in Figure 9. The resultant net gas sand isopach map is shown in Figure 23. The same color legend applies to the two isopach maps of Figures 22 and 23 so that they can be readily compared. It is clear that the Lower Sand is more variable in net producible gas thickness and also contains larger values than the Upper Sand.

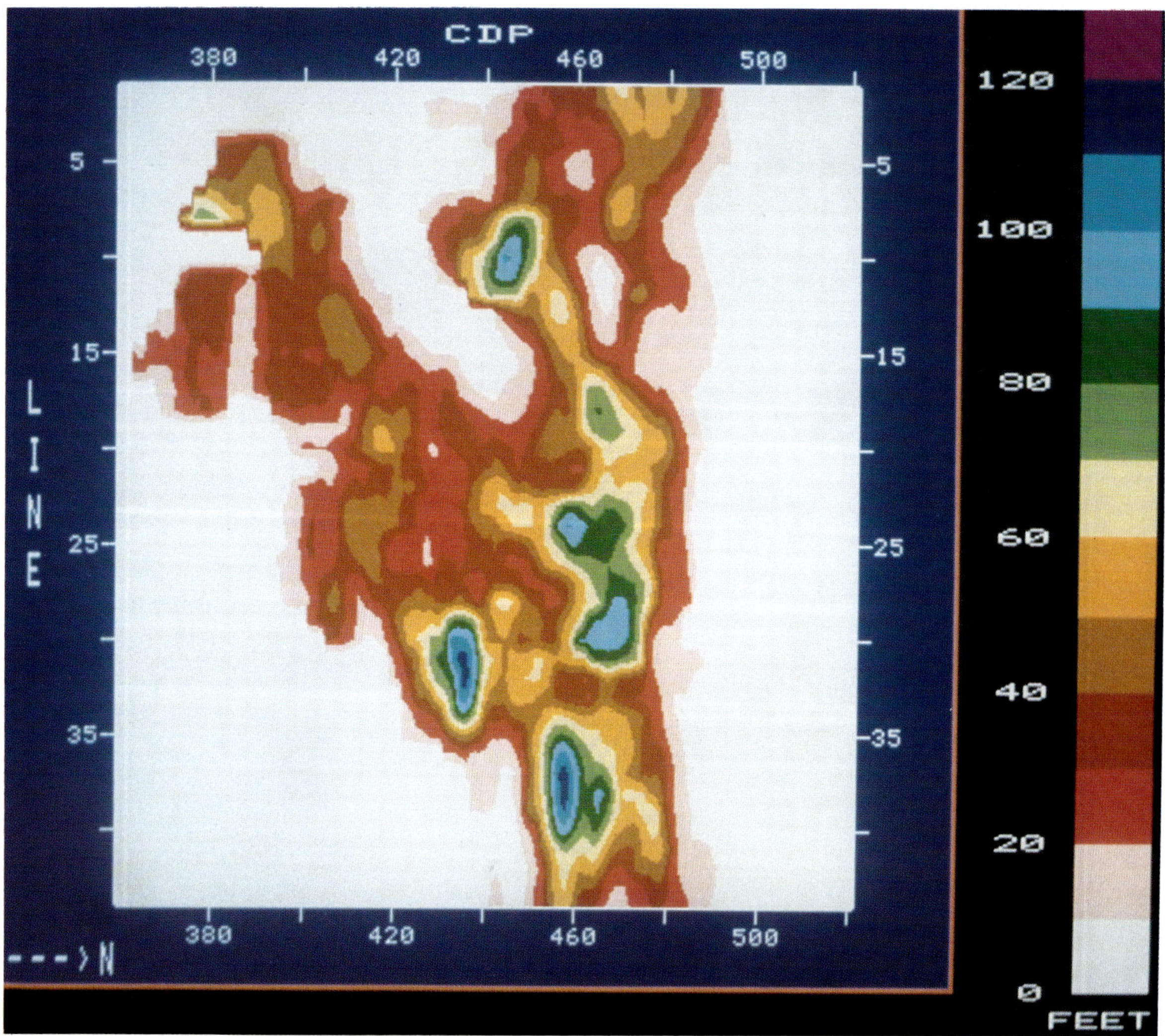

Fig. 23. Net gas isopach map, Lower Sand.

By integration of the net gas sand isopach map of Figure 23, the total reservoir volume of the Lower Sand within the study area was computed to be 20 000 acre ft (26 million m^3).

The Well Ties

The five wells in the study area (Figure 3) each provided a control point for the Upper and Lower Sands against which the accuracy of the net gas sand isopach maps could be assessed. Figure 3 shows the location at which each well penetrated the top of each sand; these locations were considered accurate to within plus or minus 100 ft (30 m). Using a version of Figure 3 as an overlay to the isopach maps of Figures 22 and 23, the seismic net gas sand values at the wells were read. These readings are tabulated in Table 1 and have as a superscript an estimate of the uncertainty of the reading. This number should be added to and subtracted from the prime value in each case to give the range of thicknesses read within the 100 ft (30 m) radius location error circles.

Interpretation of the well logs provided the vertical net gas sand values at the wells, which are also tabulated in Table 1, the sand lobes considered producible in wells A-1 and A-5 are illustrated in Figure 4. This interpretation is also subject to some uncertainty; the superscripts in Table 1 should again be both added to and subtracted from the prime values in order to indicate the range of net gas sand thicknesses interpretable from the well logs.

The initial seismic interpretation was performed independently of the well information. The first net gas sand isopach maps indicated the need to re-examine the horizon tracks for the Lower Sand at wells A-3 and A-10 and for the Upper Sand at well 2. The first two of these were simply adjusted and the isopach maps remade. However, a local diminution in amplitude on line 26 at the location of well A-10 suggests a minor fault or other cause for further very local thinning. We made no adjustment to the tracks for the Upper Sand at well 2 because some very abrupt line-to-line changes in horizon tracks would have been required. However, line 16 suggests that a local deepening of the base Upper Sand reflection at CMP 16437 (Figure 6), the location of well 2, is quite reasonable.

Our accumulated experience from the horizon tracking in general and the well tying in particular is that the thickness of the sand bodies varies very abruptly indeed. On the crosslines, where trace spacing is 50 m, there are occasional indications of aliasing. We therefore conclude that the subsurface sampling interval, while being adequate for all structural dips, was too coarse in the crossline direction for proper imaging of some of the depositional dips.

Comparison of the final net gas sand values from the seismic isopach maps and from the well logs (Table 1) reveals that five of the ten sand thickness measurements agree within 2 ft and all but two agree within their estimated uncertainty ranges. The tie at well 2 is greatly improved if the values for the Upper and Lower Sands are added together.

Conclusions

In Pleistocene sands of the Gulf of Mexico, seismic amplitude has been used as a calibration factor to determine the net thickness of the producible gas sands. This determination led to accurate mapping of

Table 1. Comparison of net gas sand thicknesses read from seismic isopach maps and those interpreted from well logs. Superscripts are estimates of the uncertainty, plus or minus, of the sand thickness values.

	WELL				
	A-1	A-3	A-5	A-10	2
NET GAS SAND UPPER MEMBER					
FROM WELL LOGS	23^{10}	42^{10}	32^{8}	14^{8}	34^{18}
FROM SEISMIC ISOPACH MAPS	25^{6}	44^{6}	32^{7}	13^{2}	18^{3}
NET GAS SAND LOWER MEMBER					
FROM WELL LOGS	71^{20}	19^{10}	19^{4}	12^{4}	0
FROM SEISMIC ISOPACH MAPS	58^{15}	12^{6}	20^{7}	25^{7}	8^{5}

ALL VALUES IN VERTICAL FEET

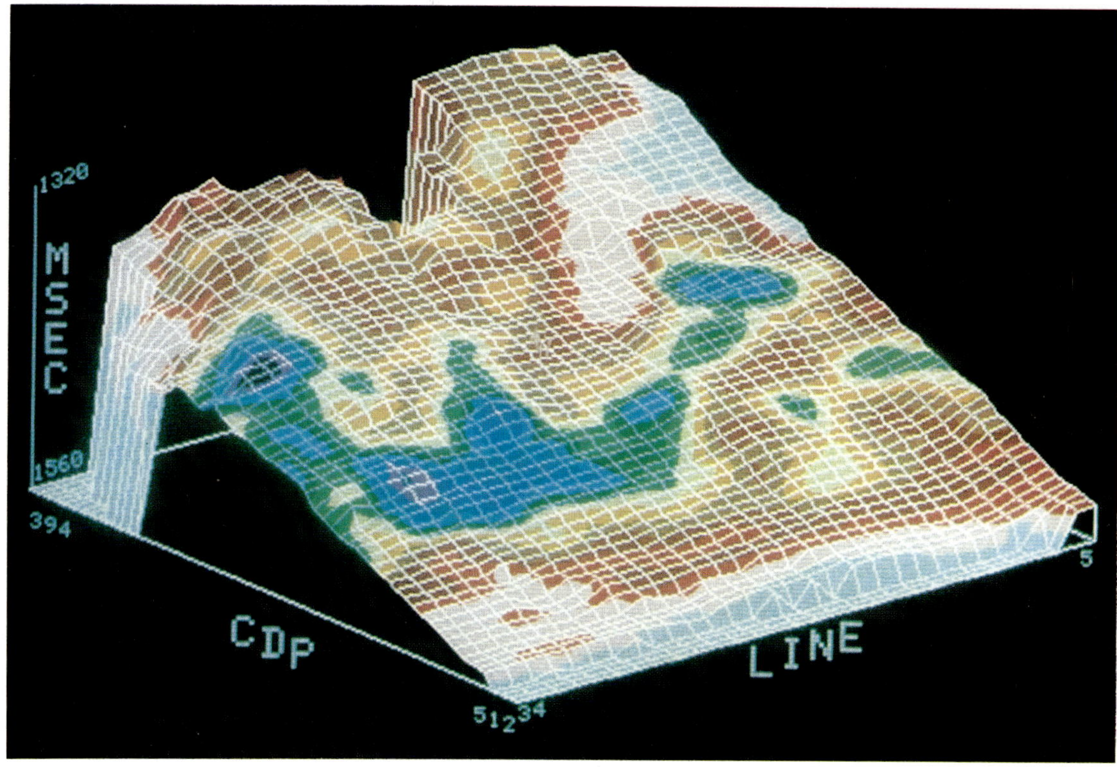

Fig. 24. Total net gas sand isopach map superimposed on the structure of the top of the reservoir. The greens and blues indicate the thicker net gas zones.

highly varying sands over two intervals. Figure 24 shows a final product of total net gas sand, the summation of the results from the two reservoirs, superimposed on the structure of the top of the Upper Sand.

Critical to the success of the project were amplitude focusing provided by 3-D migration; unambiguous relationship between amplitudes and reservoir interfaces provided by zero-phaseness, and automatic horizon tracking, color display, data manipulation, and innovative idea flow provided by an interactive interpretation system. The data point spacing of 50 m in the crossline direction was marginal; a closer spacing would have been preferred.

The procedure used here assumes a regular distribution of gas sand lobes within the gross sand interval, where the gross thickness is above the tuning thickness. The procedure also assumes porosity to be uniform. An approach which addresses nonuniform distribution and variable porosity is presented in the following paper.

Acknowledgments

We thank the joint participants in the 3-D survey for permission to publish results of the project. The participants were Chevron U.S.A., Union Oil Company of California, CNG Producing Company, Phillips Petroleum Company, and ICI Delaware.

References

Backus, M.M., and Chen, R.L., 1975, Flat spot exploration: Geophys. Prosp., **23**, 533–577.

Brown, A.R., Dahm, C.G., and Graebner, R.J., 1981, A stratigraphic case history using three-dimensional seismic data in the Gulf of Thailand: Geophys. Prosp., **29**, 327–349.

Domenico, S.N., 1976, Effect of brine-gas mixture on velocity in an unconsolidated sand reservoir: Geophysics, **41**, 882–894.

French, W.S., 1974, Two-dimensional and three-dimensional migration of model-experiment reflection profiles: Geophysics, **39**, 265–277.

Gerhardstein, A.C., and Brown, A.R., 1984, Interactive interpretation of seismic data: Geophysics, **49**, 353–363.

Meckel, L.D., Jr., and Nath, A.K., 1977, Geologic considerations for stratigraphic modeling and interpretation, *in* Seismic stratigraphy-Applications to hydrocarbon exploration: Am. Assn. of Petr. Geol. Memoir **26**, 417–438.

Neidell, N.S., and Poggiagliolmi, R., 1977, Stratigraphic modeling and interpretation-geophysical principles and techniques, *in* Seismic stratigraphy-Applications to hydrocarbon interpretation: Am. Assn. of Petr. Geol. Memoir **26**, 389–416.

Schramm, M.W., Jr., Dedman, E.V., and Lindsey, J.P., 1977, Practical stratigraphic modeling and interpretation, *in* Seismic stratigraphy-Applications to hydrocarbon exploration: Am. Assn. of Petr. Geol. Memoir **26**, 477–502.

Widess, M.B., 1973, How thin is a thin bed?: Geophysics, **38**, 1176–1180.

Estimated Pay Mapping Using Three-Dimensional Seismic Data and Incremental Pay Thickness Modeling[1]

*Dennis B. Neff**

Introduction

Modern interpretation workstations offer convenient extraction of amplitude and time thickness (isochron) information from 3-D seismic data. Incorporating this information into the task of reservoir pay mapping is problematic, in part due to nonuniqueness in the seismic waveform expression of hydrocarbon reservoirs. By combining a simple forward modeling technique with crossplot cluster analysis, the seismic amplitude and isochron data can be converted more accurately to estimates of reservoir pay thickness and distribution. The methodology for integrating the petrophysical and 3-D seismic data is detailed here and illustrated with a case history from an Offshore Gulf of Mexico, Pleistocene gas sand reservoir.

Methods

The methodology as portrayed in the flow chart of Figure 1 uses well established and extensively publicized theories and practices including (A) rock velocity and density estimation via petrophysical equations; (B) vertical incidence, convolutional modeling; (C) seismic amplitude and isochron extraction and analysis; (D) three-variable crossplot analysis; and (E) computer mapping. Steps A and B, as illustrated in this paper, closely follow Crain and Boyd (1979) and Neff (1990) who use petrophysical equations to modify velocity and density curves for hydrocarbon and non-hydrocarbon bearing rocks. These derived curves are then used as input to convolutional models for further empirical analysis of seismic waveforms.

Steps C and E of Figure 1 were illustrated for a 3-D data volume by Brown et al. (1984) and Brown et al. (1986) and in the previous paper where they detailed net producible gas mapping from 3-D seismic data and further crossplot analysis of the amplitude and isochron data. This paper parallels Brown's work by using 3-D amplitude and isochron maps as input data,

but extends his work by including forward model analysis as input to three-variable crossplots (Figure 1, steps A and D). The crossplot analysis and polygon design steps then ultimately define how the seismic amplitude and isochron data are transformed into an estimated pay map.

The forward modeling technique called Incremental Pay Thickness (IPT) modeling is used to establish an empirical relationship between pay thickness versus seismic amplitude and seismic time thickness. As

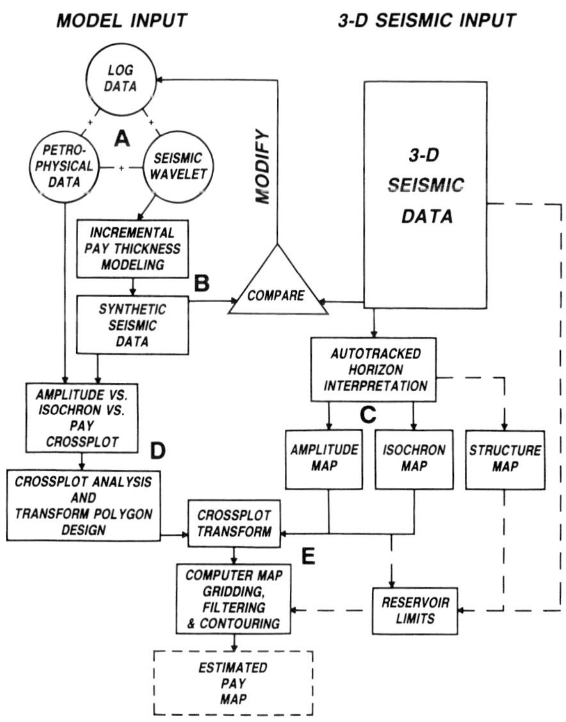

Fig. 1. Flow chart of the model-assisted approach for determining an estimated pay map from 3-D seismic data. Log data, petrophysical data, and seismic wavelet information are input to convolutional models which ultimately define a crossplot transform. Seismic data are analyzed for amplitude and isochron information which are transformed into units of pay via the model-derived crossplot transform. Computer mapping algorithms and external determinations of reservoir limits then define the final estimated pay map.

*Director, Interpretation Technology, Phillips Petroleum Company, 590G Plaza Office Building, Bartlesville, OK 74004.
[1]Originally published in Geophysics, **55**, 567–575.

detailed by Neff (1990), this modeling approach allows analysis of gross interval thickness, net pay thickness, net porosity feet of pay, and hydrocarbons in place because additional petrophysical data including porosity and water saturation are used as input log data. Figure 2 illustrates the key elements of IPT modeling that are essential to this reservoir estimation technique. The original input model values of shale volume (V_{sh}), porosity (POR), sonic and density (DEN) are shown in Figure 2a. The interval that has potential for

pay is also shown (Pay) along with the revalued sonic and density (HC) curves through that potential pay interval. The original "wet" curves and each additional 1 ms of "pay" thickness have a unique synthetic seismogram shown in Figure 2b. Vertically aligned below each wiggle trace are the accompanying measured or calculated values of trough amplitude, peak amplitude (AMP), trough-to-peak (apparent) isochron, actual time thickness of pay (ISO), net feet of pay (NET), net porosity feet of pay (POR), and hydrocar-

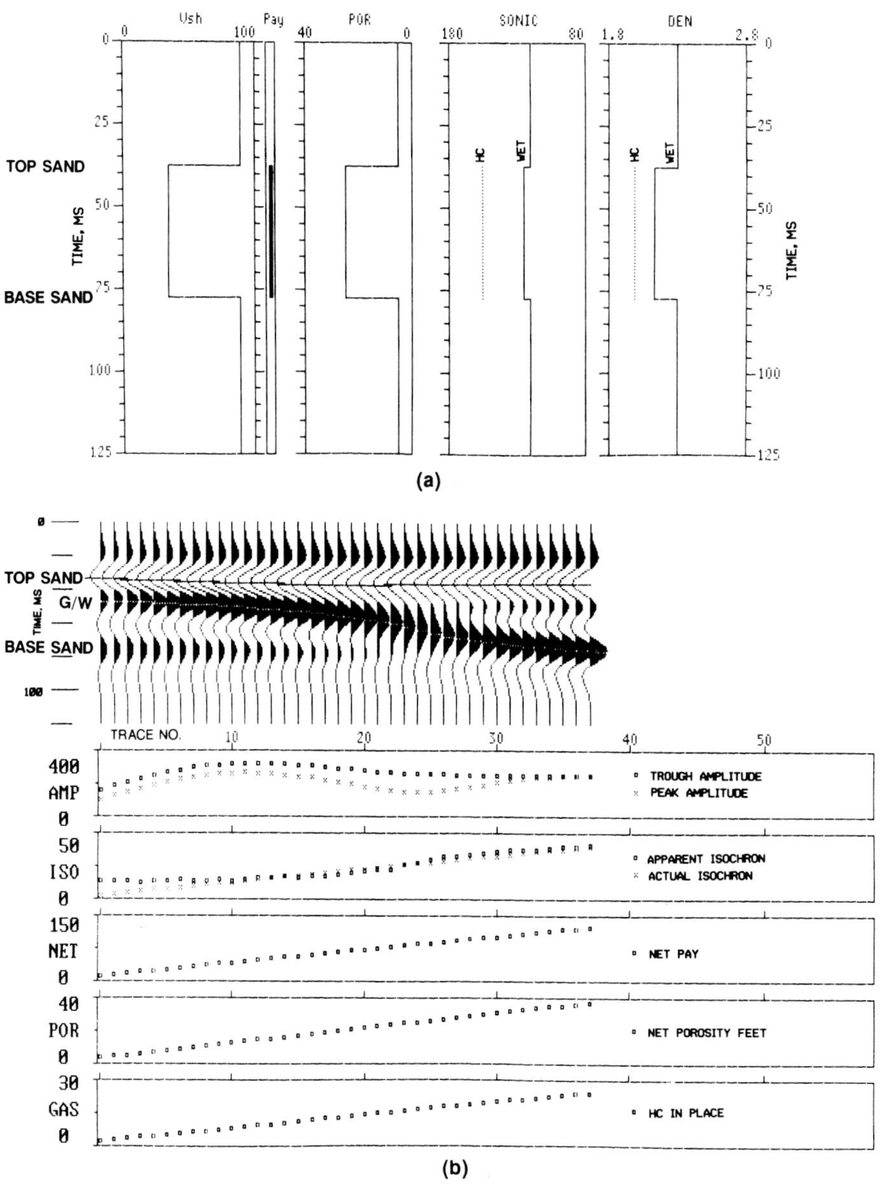

Fig. 2. (a) Input model curves for volume of shale, pay zones, porosity, sonic and density values. Initial sonic and density curves are solid lines (WET) and recalculated pay values are dotted lines (HC). (b) Model output including vertically aligned wiggle trace with highlighted amplitude minima and maxima, graphic plot of extracted amplitudes, graph of actual and apparent isochron time thicknesses, and graphs of net feet of pay, net porosity feet of pay, and hydrocarbons in place.

bons in place (GAS). These data serve as the input to the three-variable, crossplot analysis (Figure 1, step D).

Examples of three-variable crossplots are given in Figure 3. Trough and peak amplitudes are displayed separately so that any significant symmetry or variances are readily noted. The integers 0 to 9 are used to denote the partitioned Z-axis variable net feet (Figure 3a) or net porosity feet (Figure 3b). For example, 0 in Figure 3a represents 0 to 15 net feet, 1 represents 15 to 30 net feet, the integer 2 represents 30 to 45 net feet, etc. Also illustrated in the upper graph of Figures 3a and 3b are least-squares-fit lines through the integer 5 which help define areas of common pay thickness and the trend of boundary lines that separate different data clusters. The integers 0 to 9 and least-squares-fit lines aid in selecting the final relative pay parameter (i.e., net feet, net porosity feet, or hydrocarbons in place) that can most reliably be mapped from the 3-D amplitude and isochron data.

Transforms

Simple models such as shown in Figure 2a were analyzed to define an appropriate mathematical function(s) to transform the seismic data into values of estimated pay. One set of models included porosity variations of 20 percent, 25 percent, and 30 percent with pay thickness ranging from 0 ft to 140 ft (43 m) as illustrated in Figure 2a. A second series of models modified the curves of Figure 2a to include interbedded shale such that sand-to-shale ratios ranged from 60 percent to 100 percent. Porosity was held constant at 25 percent for the interbedded shale models. A third group of models had shale-to-silt-to-sand transitional boundaries. The top-of-pay was always midway in the transition zone and these zones ranged from a few feet to 150 ft (46 m). Porosity was again held constant at 25 percent.

Figure 3 exemplifies some of the results from this simple model analysis:

(1) Fundamental, three-layer model, amplitude-isochron tuning curves (Neidell and Poggiagliolmi, 1977; Neff, 1990) are not always present as shown by the *peak* amplitude versus isochron graphs of Figure 3.

(2) Thin bed (13–15 ms) amplitude-isochron trends, which move through the plotted points in the direction shown by vector $\mathbf{P}_b$, are different from resolved bed responses, which follow trends like the dashed line as shown by vector $\mathbf{P}_a$.

(3) Amplitude is significantly affected by porosity changes. The amplitude contribution to vector $\mathbf{P}_a$ is the vector a, and the isochron contribution is vector i.

(4) Net feet trends are significantly different from net porosity feet trends (Figure 3a, $\mathbf{P}_a$ versus Figure 3b, $\mathbf{P}_a$).

Additional significant observations from the simple model runs which are not shown include:

(5) Interbedded shale models dominantly affect the isochron measurements of the amplitude-isochron graphs in contrast to porosity models which dominate amplitude.

(6) Transitional bed boundaries cause significant amplitude *and* isochron variations on the amplitude-isochron graphs.

These results which are supported by crossplot analysis of several hundred blocked models emphasize that no single mathematical equation or function can universally transform data with multiple varying parameters such as porosity, shaliness, transitioning, and water saturation. The simpler approaches of the

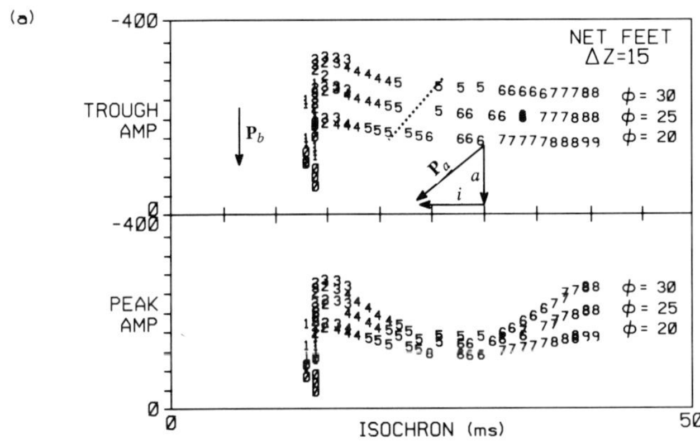

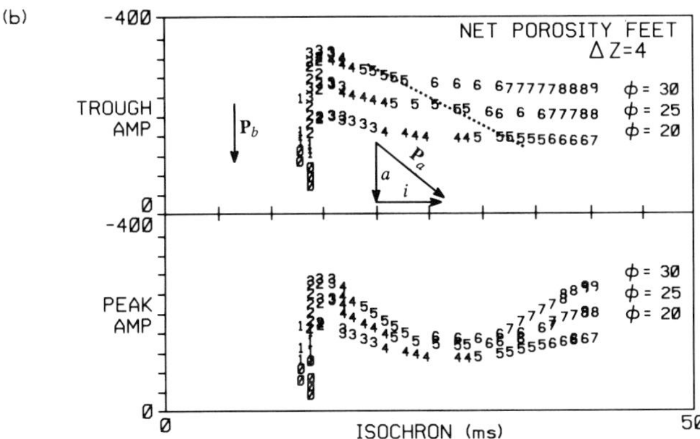

Fig. 3. (a) Three-variable crossplot display based on the model of Figure 2. The y-axis (horizontal) is trough or peak amplitude, the x-axis (vertical) is the isochron time between minimum and maximum amplitudes, and the z-axis (out of plane) is the partitioned net feet of pay calculated from the input model and displayed as the integers 0 to 9. Each graph shows three data trends which relate to models with 30 percent, 25 percent and 20 percent porosity. (b) Three-variable crossplot with the z-axis integers defined by net porosity feet of pay.

last paper thus need extension. An iterative, interpretive approach is a more appropriate and robust means to investigate probable map trends of estimated pay given one or more assumptions of dominant parameters such as porosity, shaliness, etc.

Several crossplot displays and statistical analyses (mean, mode, standard deviation, and least-squares fit) are needed to achieve the final estimated pay map. Figure 4a shows one composite crossplot display of over 450 traces for some of the blocked models of porosity, interbedded shale and varied water saturation. By isolating and plotting only limited ranges of net feet, net porosity feet, or hydrocarbons in place, the amount of overlap between successive clusters is observed always to be at least 50 percent and often exceed 100 percent overlap. Figure 4b shows clusters for 1–3 net porosity feet (integer 0) and 15–17 net porosity (integer 7). While the overlap of integers

representing common pay thickness is evident in Figure 4a, statistical analysis of mean values and least squares fit lines show a distinct progression of the centers of clusters as relative pay values increase.

Smoothly varying estimates of pay thickness are graphically created as illustrated in Figure 4c. Polygons are defined as most inclusive of the clusters of a small range of net porosity feet values (i.e., 2 net porosity feet). The shape and spacing between each polygon is influenced by the shape and separation between the statistically defined centers of each cluster. The transformation is completed by assigning the specified net porosity feet value of the polygon to all points within that polygon. Data points that lie outside the polygons are temporarily given null values and reassigned valid values during the mapping.

Inspection of the polygons of Figure 4c shows that values up through 4 net porosity feet are strictly amplitude dependent and include only the tuned isochron times of 13 and 14 ms. Other net porosity feet polygons are both amplitude- and isochron-dependent. The more vertical boundary lines such as 12 and 14 net porosity feet indicate a stronger dependence on iso-

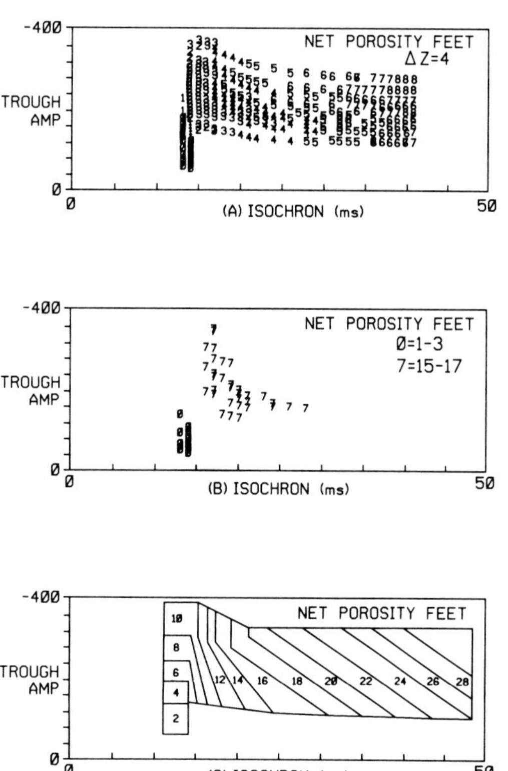

Fig. 4. (a) Three-variable crossplot for models which tested porosity, interbedded shale, and water saturation changes. Over 450 model data points are shown. (b) Three-variable crossplot of data from (a) but with limited *z*-axis displays. 0 represents 1 to 3 net porosity feet and the 7 represents 15 to 17 net porosity feet. The polygon outlines are the same as those depicted in (c) for 2 and 16 net porosity feet. (c) Three-variable crossplot with only the polygon outlines that define the relationship between a trough amplitude-isochron data value and an estimated pay value of net porosity feet. The polygon increment is ±2 net porosity feet.

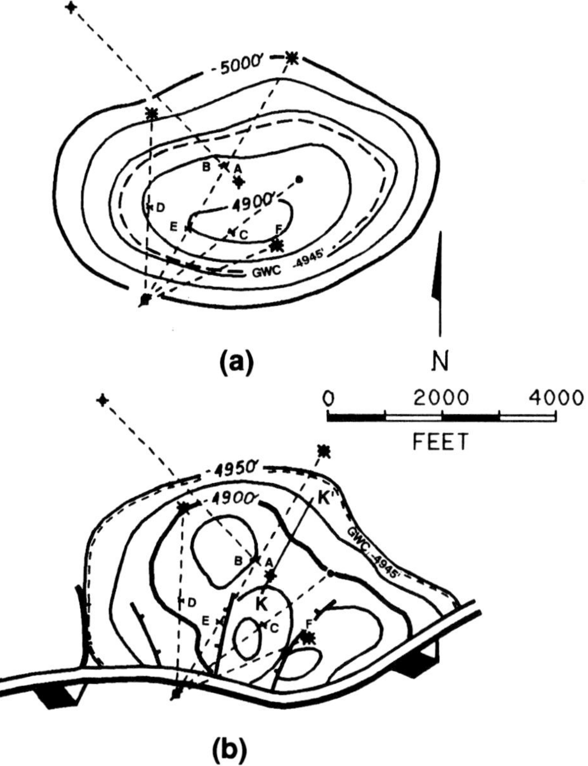

Fig. 5. (a) Original structure map of lower B-2 reservoir sand that was based upon six points of subsea well control and 2-D seismic data. (b) Upper B-1 reservoir sand map based upon 3-D seismic data and well control. The structural style of the 3-D interpretation shows more fault detail and north-south structural trends.

chron measurements. The lower angle lines such as bound 22 through 28 net porosity feet indicate a more uniform contribution of both the amplitude and the isochron values. In general, boundary lines above tuning are more diagonal if porosity is the dominant variation in the reservoir models and boundary lines become even lower angled if interbedded shale is a major component in the sand body geometry. These generalizations were substantiated during the modeling of several Gulf of Mexico seismic anomalies. Likewise, significant variations to these generalizations were also noted, the details of which follow.

Data Case Test

One analysis of the methodology involved mapping a depleted reservoir in an area of active exploration, offshore Louisiana. The Pleistocene, Trimosina A reservoir had been penetrated by six wells at a depth of approximately 5000 ft subsea and yielded 13.8 BCFG from two sands before abandonment. A 3-D seismic survey processed to zero phase became available after the field was depleted. An original reservoir structure map is shown in Figure 5a. This map was based upon six points of well control and 2-D seismic

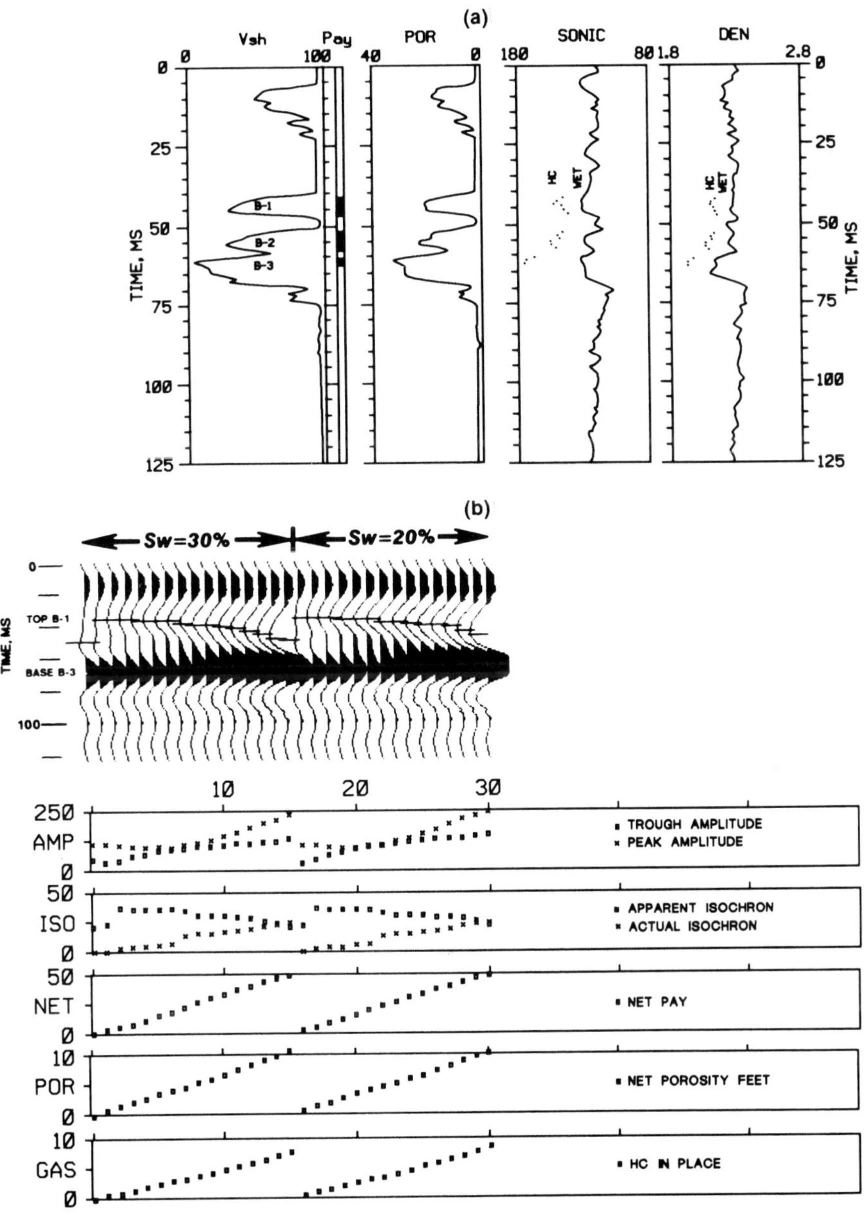

Fig. 6. (a) Input model curves for well A of Figure 5. Dots to the left of the sonic and density curves represent recalculated pay values with an S_w of 0.3. (b) Wiggle trace and graphic displays for the input models of (a). The amplitude and isochron responses are anomalous compared to the simple model of Figure 2.

data. The 3-D seismic data significantly altered the structural interpretation, as presented in Figure 5b. Not only does the 3-D seismic map show the reservoir to be more extensive than originally mapped, the presence of minor north-south trending faults are verified by 20 ft (6 m) and 30 ft (9 m) missing sections in wells E and F, respectively.

As a test of the methodology previously detailed, only the original discovery well A is used for input modeling. The resulting estimated pay map is then compared to the known values for the remaining five wells in order to judge the effectiveness of the modeling and crossplot transformation process. Figure 6 shows the original IPT model for well A. A critical element of this reservoir's waveform response is shown in Figure 6b where the apparent isochron thickness decreases in contrast to continually increasing pay thickness (ISO graph versus NET graph). Figure 7 illustrates the strong similarity of the synthetic waveform responses to the actual 3-D seismic data. The location of the seismic profile (K'-K) is shown in Figure 5b. The anomalous waveform character associated with the 47 net feet (14 m) of pay sand was explicitly detailed by Neff (1990) and further discussions here concentrate on the crossplot analysis of that IPT model data.

Prior to crossplot analysis, the input model of Figure 6a is further modified to reduce B-1 and B-2 sand thicknesses by 5 to 10 ft (2 to 3 m) and to increase porosities by 3 percent. Ultimately 105 traces are used as input to the three-variable crossplots of Figure 8. These data points show an anomalous trend as compared to the more basic models portrayed in Figure 3. The lower values of pay (integers 1, 2, and 3) plot to the lower right of the graph while high values of pay (integers 7, 8, and 9) plot to the upper left of the graph.

Graphs of trough and peak amplitudes versus net feet, net porosity feet, and hydrocarbons in place all show very similar trends to the graphs in Figures 8a and 8b. The average amplitude graph (peak amplitude minus trough amplitude divided by two) of Figure 8c has slightly tighter clustering of pay variables, and is chosen as the most appropriate data from which to design a polygonal graphic transform (Figure 10, overlay).

Creation of the final amplitude and isochron maps is the next step in the process. There is ambiguity during the autotracking stage of the interpretation as to where to pick the top-of-pay trough event near the edges of the reservoir. It is further complicated because time structure maps on the autotracked trough show less structural dip and a greater extent to the reservoir than do autotracked maps based on the underlying peak event. Figures 6 and 7 provide insight as to how the waveforms should respond near the reservoir edge. With this understanding, the zero edge of the reservoir is defined on each seismic line through an interactive process of comparing a trough amplitude map, a peak amplitude map, a time structure map on the peak, and seismic wiggle trace screen plots enhanced by a color amplitude background display. These reservoir limits are then used in conjunction with fault boundaries in subsequent computer mapping to exclude all points outside these boundary lines during the gridding and

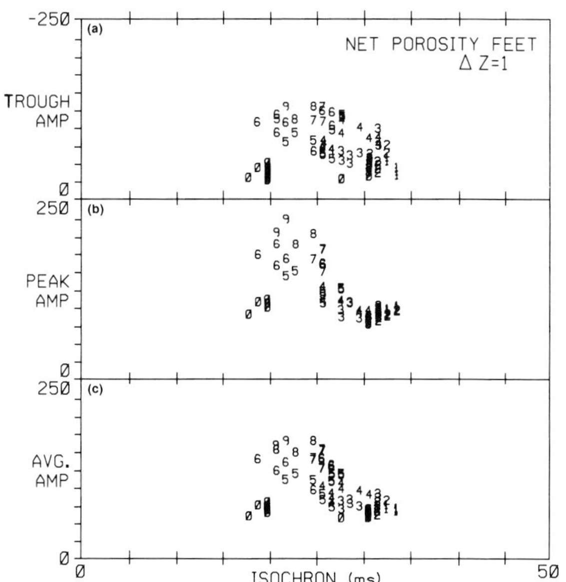

Fig. 8. (a) Three-variable crossplot showing trough amplitude, apparent isochron and net porosity feet values for well A models including those of Figure 6. (b) Three-variable crossplot of peak amplitude, apparent isochron and net porosity feet values for well A models. (c) Three-variable crossplot of average amplitude, apparent isochron, and net porosity feet values for well A model.

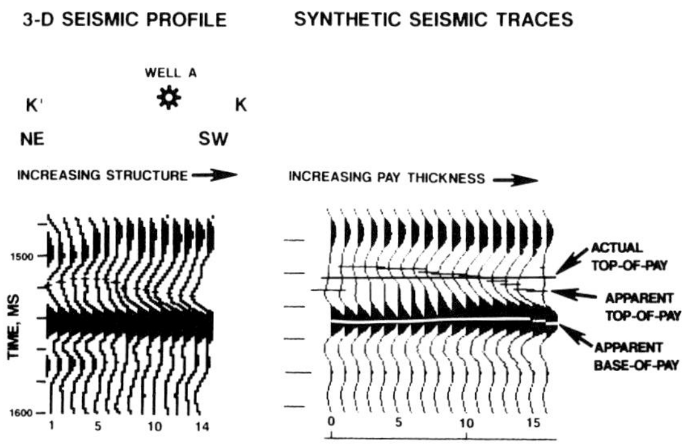

Fig. 7. Actual seismic profile through well A compared to synthetic traces of Figure 6b. Common waveform responses include increasing trough and peak amplitudes and delayed trough times with increasing structure, broadening peak waveforms with increasing structure, and isochron thinning with increasing pay thickness.

filtering of amplitude, isochron, and estimated pay maps.

Figure 9a shows a final computer-generated isochron map. Typical computer mapping runs include grid spacings of 100 ft and two passes of a least-squares fit, twenty-one point circular filter. Small interior faults are not defined for the computer runs but the southern and western faults are honored by mapping algorithms. Minor smoothing and editing is later done along the fault boundaries. Because there is only 10 ms of isochron variance across the reservoir, the map of Figure 9a shows meandering contours frequently associated with low-relief maps. The trend of decreasing isochrons as previously discussed is obvious along profile K-K'.

The average amplitude map of Figure 9b also is computer contoured with only reservoir bounding faults. This map shows three north-south trending, average amplitude closures of magnitude 90 or 100. In each case, the closures are bounded on the east and/or west by one of the minor north-south trending faults.

A crossplot of the data input into the amplitude and isochron maps is shown in Figure 10. Superimposed on this graph are the transform polygons that are designed by isolating the ±1 net porosity feet values of Figure 8c. The polygon design process is identical to that detailed for Figure 4. The polygons are rescaled to compensate for the amplitude differences between the model data graph of Figure 8c and the real data graph of Figure 10. The shape of the polygons for net porosity feet values of 1 to 6 shows that amplitude is the dominant-net-porosity indicator. For values of 7 to 13, the isochron becomes more critical, as shown by the more diagonal polygon boundary lines. Plotted data points outside the polygons are assigned null values and those within boundary lines are assigned the mean value of the polygon (i.e., 1.5, 2.5, 3.5, etc.).

The final computer map of net porosity feet is shown in Figure 11. Again, minor interior faults are not used in the computer mapping, but locations of these north-south trending faults are vividly expressed as contour lows between the high-valued closures. The dominance of the amplitude-biased polygons is observed by comparing the amplitude map of Figure 9b with the pay map of Figure 11. While the northwest-southeast trends of the isochron map of Figure 9a are not reflected in the north-south trends of the net porosity feet map, it is consistent that the maximum valued closures (i.e., 9 and 10 contours) are located within the areas of lower isochron values (i.e., 20 ms).

The net porosity feet values, as calculated from logs of the six wells, are listed in Table 1 along with the estimated pay as taken from the net porosity feet map

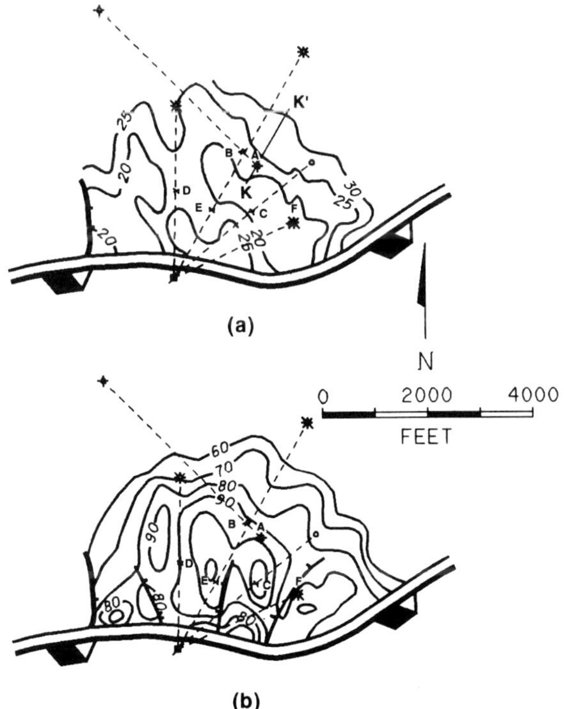

(a)

(b)

Fig. 9. (a) Seismic trough-to-peak isochron map for the B sand reservoir. The apparent isochron values decrease toward the structural crest near wells C and F. (b) Seismic average amplitude map for B sand reservoir. The amplitude values increase toward the structural crest. The minor north-south faults follow trends of reduced amplitude.

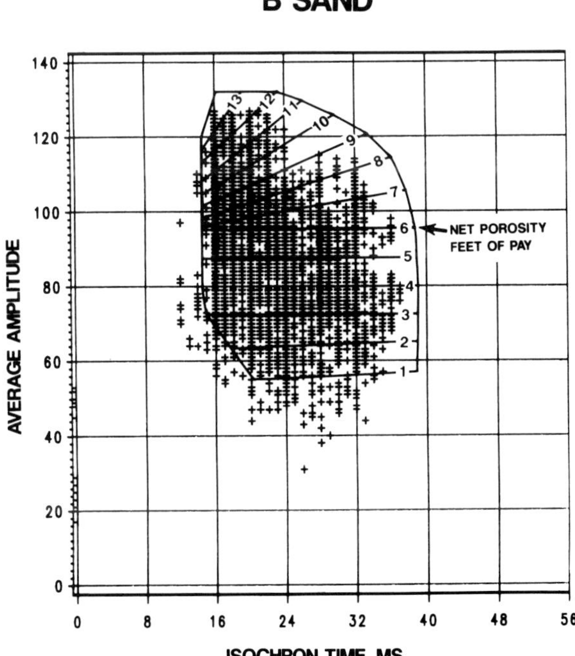

Fig. 10. Crossplot of average amplitude versus apparent isochron values extracted from the 3-D seismic data. The polygons based upon the models and crossplots of Figures 6 and 8 are superimposed. The values of net porosity feet are defined as 1.5, 2.5, 3.5, etc. for each progressively higher valued polygon.

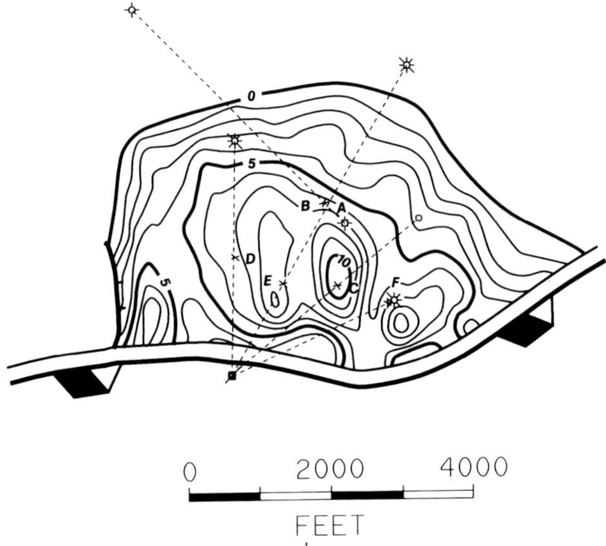

Fig. 11. Estimated pay map of net porosity feet for the B sand reservoir. The seismic based net porosity feet values as interpolated from the map are compared to log values in Table 1.

of Figure 11. Five of the six wells show less than 10 percent error when the numbers of Table 1 are compared. Well B had 15 percent error, which is attributed primarily to anomalously high water saturation values in the lower 15 ft of the B-2 sand. This lowers the total net feet of pay and in turn gives a lower value for net porosity feet. It also is noted that well A does not match the map values exactly, even though this well is used for the type of logs in input seismic modeling. This mismatch is attributed to the facts that (1) the model-based crossplot of Figure 8 has over 100 model responses besides the synthetic response for well A; (2) the polygon transform of Figure 10 honors a trend of data values and is not biased to match just well A; and (3) the map contours of Figure 11 include grid interpolation, areal filtering, and contour smoothing. Therefore, the overall trends and relative pay values are intentionally honored by this methodology and detailed revisions to compensate for the remaining 1 percent to 15 percent error are left to other methods of contour editing.

Table 1. Comparison of well log measured values of net porosity feet versus those predicted by the seismic-crossplot-mapping technique portrayed on the map of Figure 11.

Well	Log measured	Seismic predicted
A	7.6	7.0
B	5.4	6.2
C	10.3	10.4
D	6.2	6.8
E	7.6	8.1
F	7.1	7.0

Conclusions

The use of model-derived crossplot transforms has significantly improved the ability to extend 3-D seismic amplitude and isochron data into estimated pay maps, as illustrated by this case history. The complex waveforms observed in this Pleistocene reservoir demonstrate the need for a versatile and robust technique which the methodology described here appears to meet. Other reservoir studies also have shown the need for both simple and complex transforms. Nonuniqueness in the seismic expression of petrophysical parameters continues to be a difficult problem which requires individual assessment for each reservoir. By testing multiple seismic attributes versus pay thickness parameters in the crossplot analysis procedure, better solutions to the nonuniqueness problem are possible.

References and Further Reading

Angeleri, G. P., and Carpi, R., 1982, Porosity prediction from seismic data: Geophys. Prosp. **30**, 580–607.

Brown A. R., Wright, R. M., Burkart, K. D., and Abriel, W. L., 1984, Interactive seismic mapping of net producible gas sand in the Gulf of Mexico: Geophysics, **49**, 686–714.

Brown, A. R., Wright, R. M., Burkart, K. D., Abriel, W. L., and McBeath, R. G., 1986, Tuning effects, lithological effects and depositional effects in the seismic response of gas reservoirs: Geophys. Prosp., **34**, 623–647.

Brown, A. R., 1987, The value of seismic amplitude: The Leading Edge, **7**, no. 10, 30–33.

de Buyl, M., Guidish, T., and Bell, F., 1988, Reservoir description from seismic lithologic parameter estimation: J. Petrol. Tech. **40**, 475–482.

Crain, E. R., and Boyd, J. P., 1979, Determination of seismic response using edited well log data: 7th Formation Evaluation Sympos., Canadian Well Logging Society.

Domenico, S. N., 1974, Effect of water saturation on seismic reflectivity of sand reservoirs encased in shale: Geophysics, **39**, 759–769.

———, 1976, Effect of brine-gas mixture on velocity in an unconsolidated sand reservoir: Geophysics, **41**, 882–894.

Gelfand, V. A., and Larner, K. L., 1984, Seismic lithologic modeling: The Leading Edge, **3**, no. 11, 30–34.

Hardage, Bob A., 1987, Seismic stratigraphy: Handbook of geophysical exploration: Seismic exploration, **9**, 195–197.

Maureau, G. T. F. R., and Van Wijhe, D. H., 1979, The prediction of porosity in the Permian carbonate of eastern Netherlands using seismic data: Geophysics, **44**, 1502–1517.

Meckel, L. D., Jr., and Nath, A. K., 1977, Geologic considerations for stratigraphic modeling and interpretation, *in* Seismic stratigraphy—Applications to hydrocarbon exploration: AAPG Memoir **26**, 417–438.

Neff, D. B., 1990, Incremental pay thickness modeling of hydrocarbon reservoirs: Geophysics, **55**, 556–566.

Neidell, N. S., and Poggiagliolmi, E., 1977, Stratigraphic modeling and interpretation—Geophysical principles and techniques, *in* Seismic stratigraphy—Applications to hydrocarbon exploration: AAPG Memoir **26**, 389–416.

Widess, M. B., 1973, How thin is a thin bed?: Geophysics, **38**, 1176–1180.

Wyllie, M. R. J., Gregory, A. R., and Gardner, G. H. F., 1958, An experimental investigation of factors affecting elastic wave velocities in porous media: Geophysics, **23**, 459–493.

AVO Analysis of 3-D Seismic Data Identifies Untested Reservoirs in Old Gas Field, California[1]

K. H. Van Sickle, W. L. Soroka,* and J. E. Valusek[‡]*

Introduction

The seismic technique amplitude-versus-offset (AVO) has gained widespread attention among explorationists because AVO promises to help them recognize subtle stratigraphic targets and to identify hydrocarbons with more confidence. AVO effects exist because the amplitude along a seismic reflection varies with offset distance (or angle of incidence) between wave source and receiver. The magnitude of these variations depends on changes in velocity, density, and other rock and fluid properties from one subsurface layer to another.

In the Sacramento Basin of California, the use of AVO as a hydrocarbon indicator has been limited to two-dimensional (2-D) seismic data for nearly ten years. Mobil Exploration and Producing U.S. decided to test a proprietary AVO algorithm developed by Mobil's Dallas Research Lab (DRL) for use with three-dimensional (3-D) data to identify and evaluate potential untested turbidite reservoirs in the basin's Grimes Gas Field. Interpretation of a standard 3-D data set and a specially processed AVO data set was performed on a Landmark interactive workstation. As a result, Mobil drilled three successful gas wells in 1989, proving the viability of this technique to reduce drilling risk.

Background

The Grimes Gas Field is located approximately 60 miles north of Sacramento, California. The Grimes properties, outlined in green (Figure 1) encompass about 13 square miles of rich farm country. The 160-acre well spacing was established 25 years ago. The dots in Figure 1 indicate well locations—four per square-mile section. Cumulative production in the field to date is over 250 billion cubic feet (BCF). The most prolific reservoir in the Grimes Field is the Armstrong Sand (called Forbes Zone 1 in this paper) of the Upper Cretaceous Forbes formation. Forbes is

made up of erratic turbidite reservoirs of submarine fan channel-levee complexes and lobe deposits at depths of 5000 to 10 000 ft (Figure 2).

As old wells watered out, several Mobil leases were in jeopardy so the decision was made to acquire a 3-D seismic survey (indicated by the white rectangle in Figure 1) of the southern portion of the field, in which Mobil holds a 75 percent interest. Existing 2-D data were too sparse to properly image rapid changes in stratigraphy associated with turbidite sands. A six-square-mile Vibroseis 3-D survey (with 173 two-mile lines) was acquired in two separate passes by Mobil and Western Geophysical in November 1987 and April 1988, respectively. These two data sets were merged in processing, completed in December 1988, and 3-D AVO processing was completed in January 1989. Final interpretation of both 3-D seismic and 3-D AVO was completed in April 1989 using the Landmark workstation. High data quality is necessary to effectively evaluate amplitude variations with offset. In the Grimes area, most of the survey ranged from 18- to 36-fold, except at the outer boundaries and data quality was relatively high. The survey was processed in 110 foot bins.

The key objectives of the Grimes Gas Field project were to:

identify untested turbidite gas reservoirs that occur as discontinuous stratigraphic traps ranging in size from ten to several hundred acres, and up to 100 ft thick,

evaluate the effectiveness of the 25-year old, 160-acre well spacing to efficiently produce remaining reserves,

test the AVO technique in a 3-D, rather than a 2-D, mode, and

drill economic well locations and increase production.

[1] Published in The Leading Edge, July 1990, **9**, No. 7, 18–22.
*Mobil Exploration and Producing Services, Box 650232, Dallas, Texas, 75265-0232.
[‡] Landmark Graphics Corporation, Houston, Texas.

Interactive Interpretation

Conventional Data

Interpretation of conventional 3-D data on the workstation spanned several months. Fifteen gas-prone stratigraphic layers—each with a number of distinct traps—were mapped using interactive techniques such as horizon flattening, time slicing, and creation of arbitrary seismic lines. First, a regional marker was mapped and flattened to approximate the original depositional surface. Then, each stratigraphic interval below it was mapped using the regional trend as a guideline. Amplitude extractions associated with gas were far more important than structure maps, because there is only a slight southerly dip to the area. Extractions were displayed with color graphics. By interactively "turning off" all but the highest amplitudes, the seismic displays eliminated unnecessary clutter and highlighted the zones most likely to contain gas. Besides data management, this was one of the main advantages of using the workstation. Unfortunately, seismic amplitudes can indicate lithology changes as well as gas. To narrow the possibilities for error, an AVO algorithm was employed.

AVO Data

Mobil's patented AVO process is called Lithology-Hydrocarbon-Indicator (LHI). LHI processing was carried out on a Cray supercomputer, the results were output to tape in standard SEG-Y format, and then were loaded into the workstation in the same manner as conventional 3-D seismic. Interactive graphic displays highlighted positive AVO responses (with bright

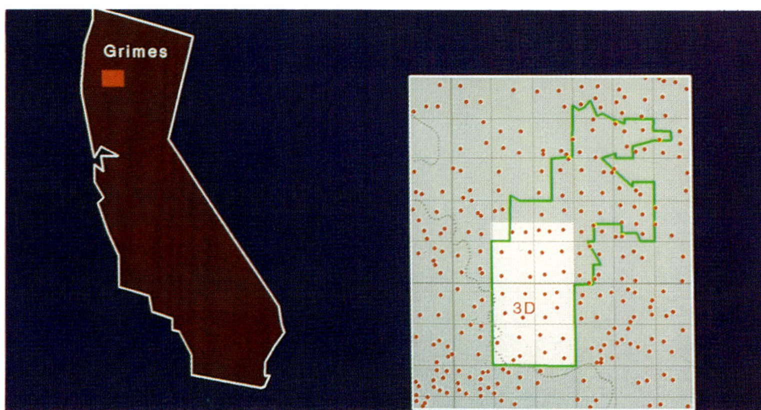

Fig. 1. Grimes Gas Field, showing lease boundaries, well locations, and 3-D seismic survey. Each square represents one square mile.

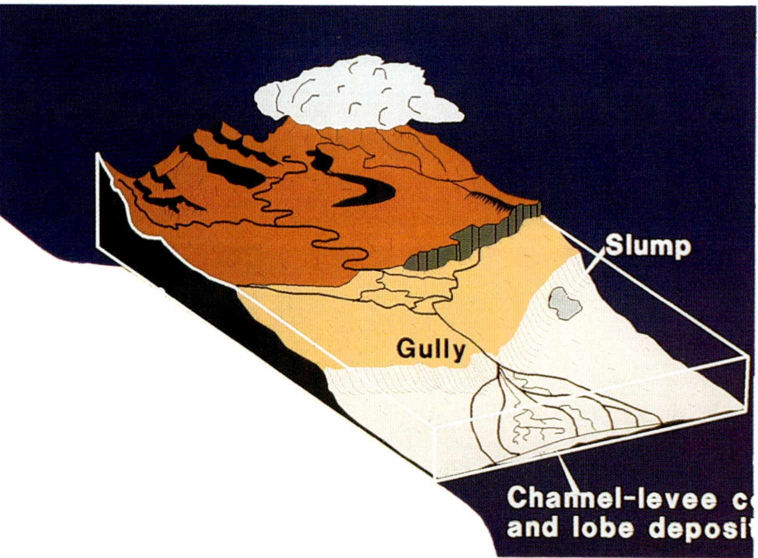

Fig. 2. Geologic model of Upper Cretaceous Forbes Formation.

colors) and eliminated negative responses. The brighter the colors, the higher the increase of amplitude with offset and, thus, the greater likelihood of gas. Of course, because AVO is highly dependent on data quality, there is not a perfect match with the high amplitude zones.

The goal of AVO analysis is to distinguish between gas and lithologic features such as salt, basement, or other rock types that cause sudden changes in reflectivity. When P-waves encounter an acoustic impedence contrast in the subsurface, some P-wave energy is reflected at the interface, some transmitted, and some converted to reflected and transmitted S-wave energy. The reflected P-wave amplitude varies with offset, according to how the P-wave and S-wave reflection coefficients (R_p and R_s) match. If R_p and R_s are nearly equal, amplitude decreases as offset increases, but if R_p is significantly larger than R_s, amplitude increases with offset. This increase is a consequence of the conservation of energy, and is formally described by the Zoeppritz equations.

A consideration of simple rock physics makes amplitude variations easy to understand. P-waves are sensitive to both pore fluids and rock material, but S-waves are sensitive primarily to the rock matrix. Thus, P- and S-wave reflectivity is approximately equal from an interface between two nonporous rocks. In contrast, a gas-water interface within a porous reservoir produces a large R_p (because of the fluid change) and a small R_s (because rock type does not change, Soroka, W., et al., 1990). Thus, even a small amount of gas causes both a sharp decrease in P-wave velocity and a corresponding increase in amplitude with offset, indicating the presence of the gas.

One limitation of this technique, however, is that it cannot discriminate between commercial and noncommercial hydrocarbon accumulations. Also, with a 15-ft limit of resolution, thinner gas sands are invisible. Nevertheless, AVO analysis can help seismic interpreters adjust their estimates of the probability of success for the thicker reservoirs, which are more likely to contain commercial gas.

Ten known reservoirs above the limit of resolution were mapped in the Grimes Field to test the reliability of AVO responses before attempting to map unknown reservoirs, and a positive AVO response occurred on 90 percent of the known gas zones. Figure 3 shows the results from Forbes Zone 1, the most prolific reservoir in the field. Figure 3a is the conventional amplitude extraction map, showing only the highest positive amplitudes. Four wells have produced over 16 BCF from this sand to date. Figure 3b, the corresponding AVO map, also shows a positive response. While it is not as continuous as the amplitude data, the positive AVO response generally matches the highest amplitudes.

Also, the AVO showed a mostly negative response (white) when tested on the basement reflector (Figure 4). While these results increase confidence in the application of 3-D AVO, obviously there is room for further refinement. Higher quality data would considerably improve AVO evaluation.

Predictions

Cross-section A-A' (Figure 5) is a seismic line created to pass through, or near, several wells and an area of potential interest (marked by a solid triangle). It shows high amplitude events due to both gas and lithology. The basement is a relatively continuous amplitude, as shown also in Figure 4. Gas symbols indicate known pay zones. The second pay interval in the well at left has no associated amplitudes because the sand thickness is below the limit of resolution.

Several unknown high amplitude events (marked Zones C and 14 in Figure 5) were analyzed next using AVO. Figure 6 illustrates the results in map view. Forbes Zone C, a 35 to 50 acre amplitude anomaly (Figure 6a) along line A-A', had a correspondingly high positive AVO response (Figure 6b) at the proposed location marked with the solid triangle. This strongly implied a new gas reservoir, stratigraphically discontinuous with zones in the four surrounding wells. One advantage of 3-D over 2-D seismic is the ability to image the small reservoirs. Unless a 2-D line just happened to have crossed this area, Zone C probably would not have been detected. Because of the relatively good correlation between conventional and AVO amplitudes, Forbes Zone C was assigned a high probability of success (Ps) as a gas-producing reservoir.

Interpretation of Zone 14, however, was slightly more ambiguous. On the one hand, very strong amplitudes over a large portion of the survey suggested a possible lithologic event (Figure 7a). On the other hand, unevenly distributed positive AVO responses (Figure 7b)—strongest in the area of interest—suggested the possibility of gas. As a result, Zone 14 was assigned a much lower Ps. A third, possibly depleted, pay zone (below Zone C and above Zone 14) was also predicted.

Drilling Results

The Grimes Operating Unit (GOU) 32/33-5 was drilled in early 1989 to test these amplitudes identified from the combined analysis of conventional 3-D seismic and 3-D AVO data sets. Calculation of reserves and the location of the well were based primarily on interpretation of the standard 3-D, whereas the probability of success was determined from the AVO data. Forbes Zone C was the primary target.

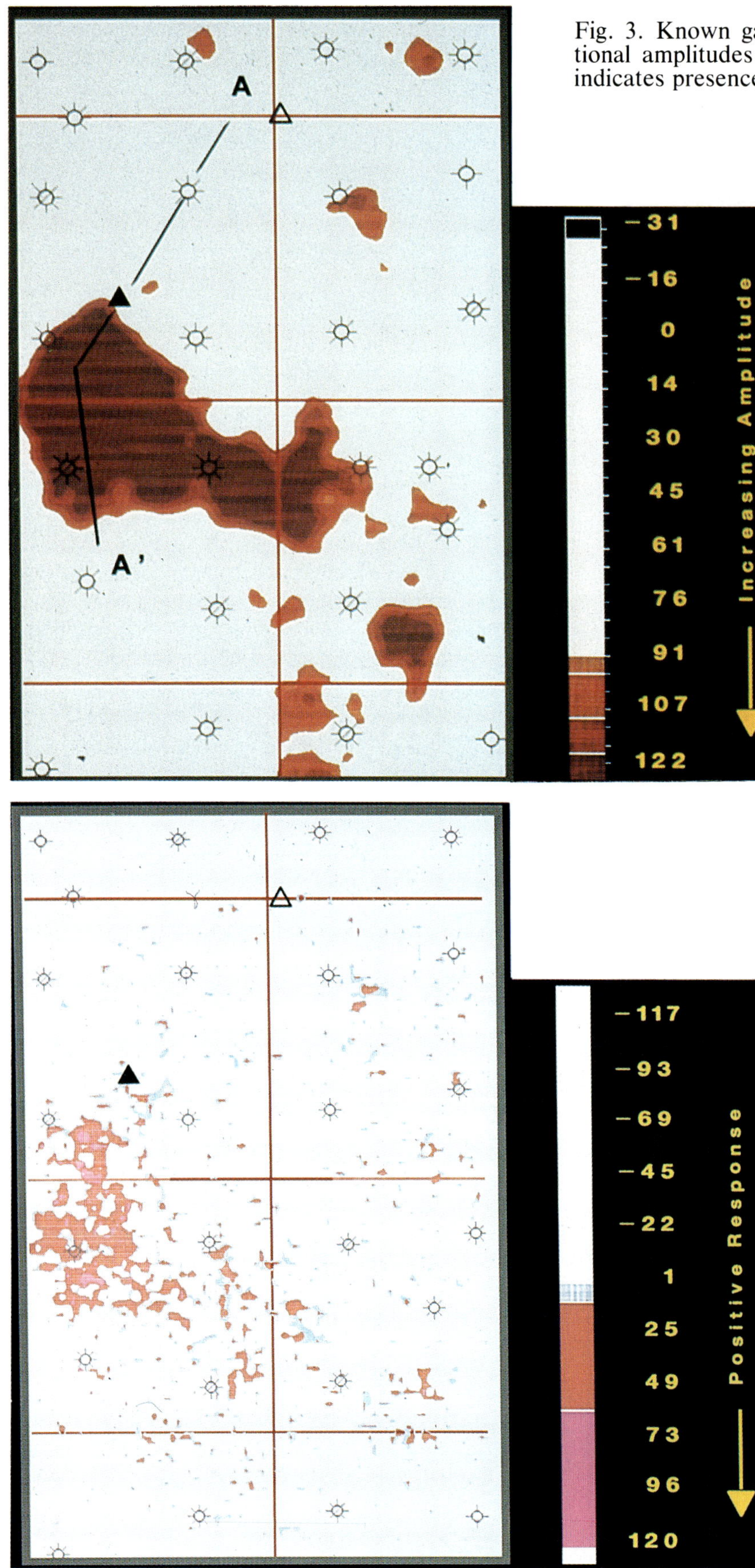

Fig. 3. Known gas reservoir: Comparison of (a) conventional amplitudes and (b) AVO responses. Positive AVO indicates presence of gas.

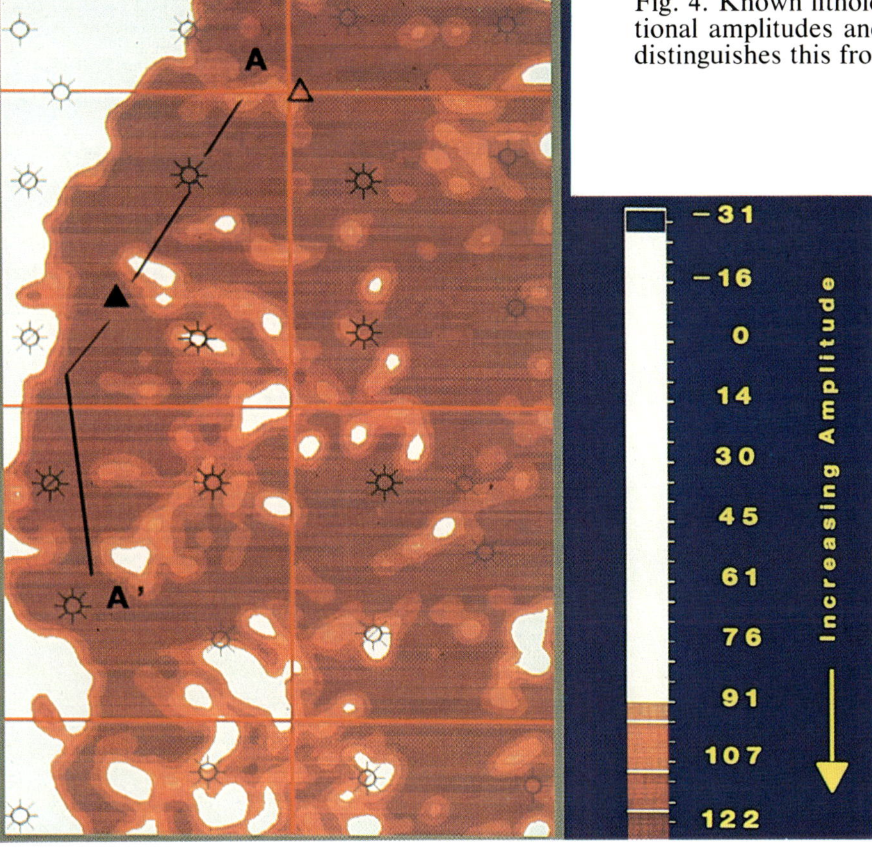

Fig. 4. Known lithologic event: Comparison of (a) conventional amplitudes and (b) AVO responses. Negative AVO distinguishes this from gas-related amplitudes.

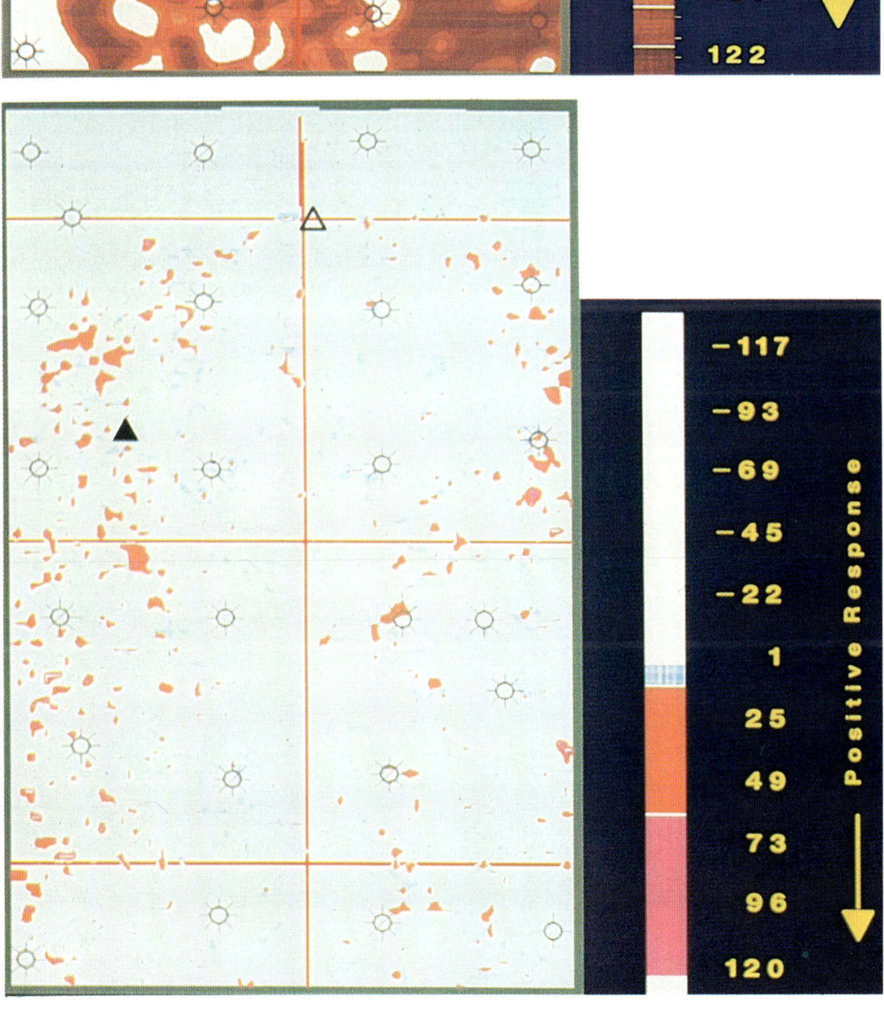

Drilled between four existing wells, GOU 32/33-5 confirmed the presence of this new reservoir, which had a 45 ft pay zone and tested initially at 4.5 million cubic feet per day (MMcfd). Production from Zone C at the initial flow rate increased Mobil's total production from the 28-year-old Grimes Field by more than 50 percent. Four shows below Zone C, two possibly depleted, had 62 ft of sand. The deeper target, Forbes Zone 14, turned out to be a low velocity lithologic event. While unfortunate, it was not entirely a surprise—as was reflected in the low probability of success that had been assigned from the AVO data.

After validating the AVO technique with one drilling success, two additional wells were drilled in 1989 using the same data. Combined, they tested at nearly 3 MMcfd.

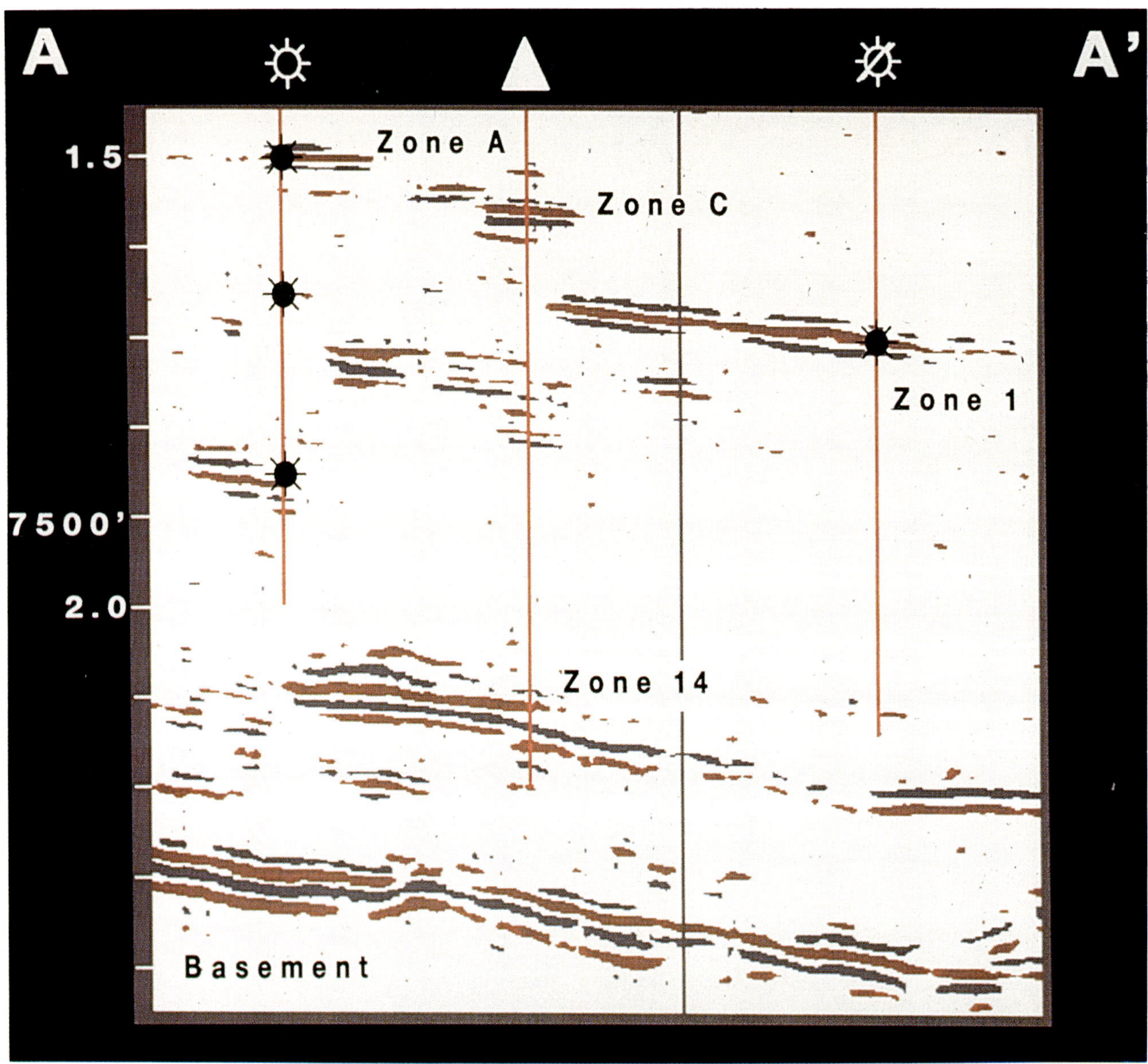

Fig. 5. Arbitrary seismic cross-section showing untested high amplitude events at Zone C and Zone 14, between and discontinuous with known producing zones (indicated by gas symbols).

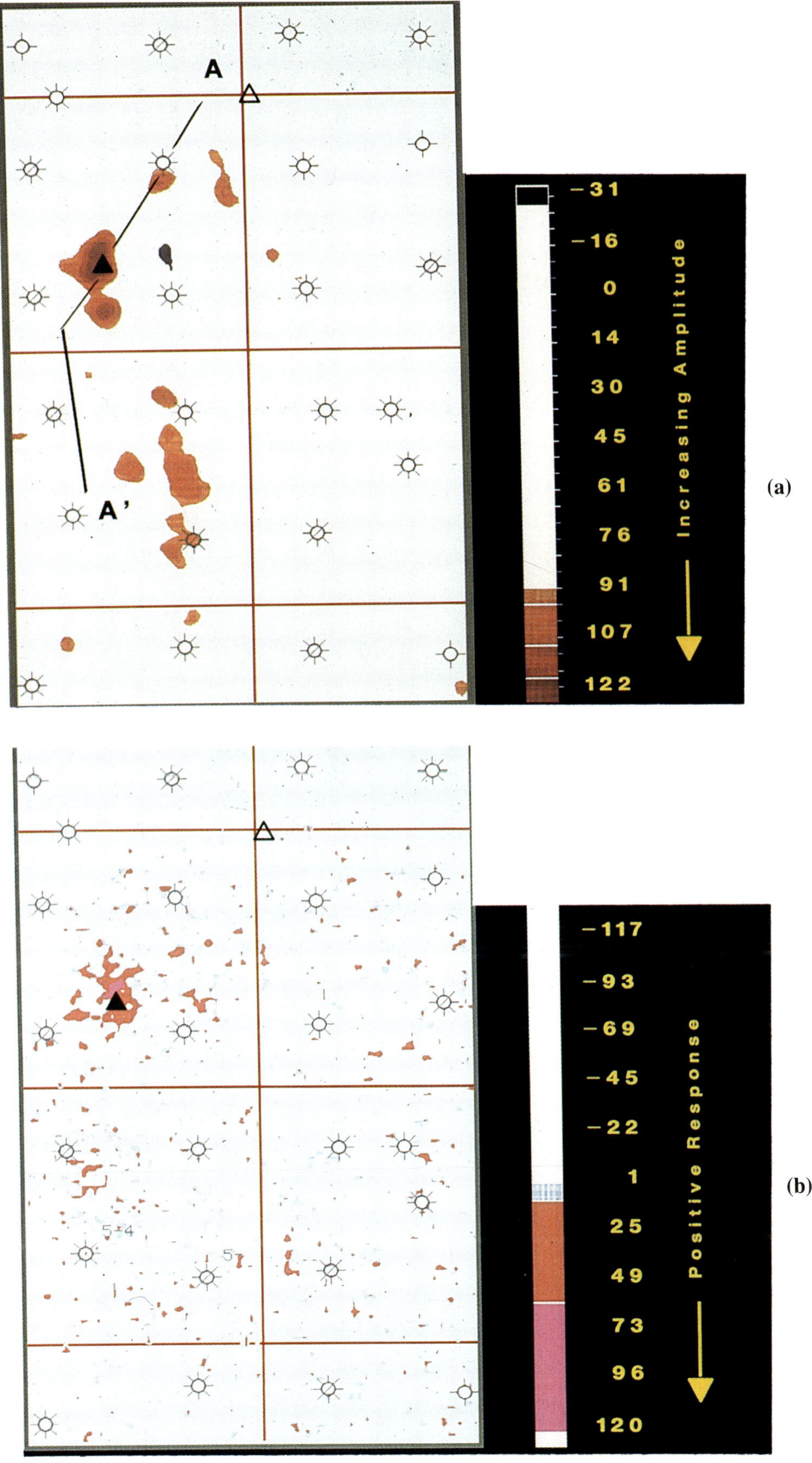

Fig. 6. Untested Zone C anomaly (at solid triangle): Comparison of (a) conventional amplitudes and (b) AVO responses. Positive AVO suggested gas, which drilling confirmed.

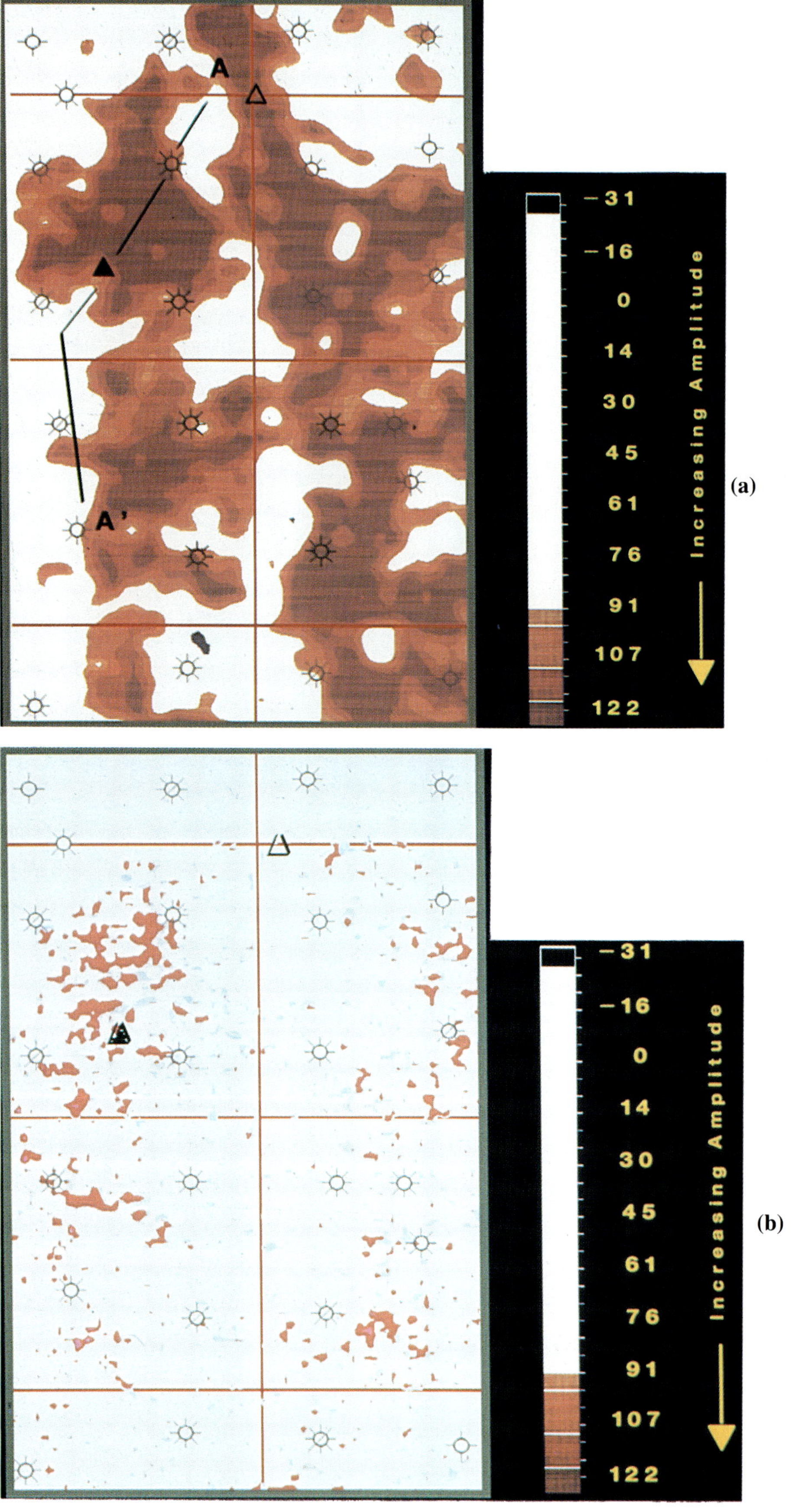

Fig. 7. Untested Zone 14 anomaly: Comparison of (a) conventional amplitudes and (b) AVO responses. Mostly negative AVO suggested a lithologic event, but irregular positive AVO in area of interest suggested possibility of gas. Drilling revealed no gas.

The main results of 3-D geophysical evaluation of the Grimes Gas Field may be summarized as follows:

other potential untested reservoirs and possible well locations were identified,

the historic 160 acre well spacing proved inappropriate for the most efficient development of gas in these turbidite sands,

the 3-D AVO technique proved successful in reducing the risk of predicting and drilling untested reserves,

more accurate parameters for economic calculations were acquired,

three wells, with combined initial flow rates of almost 7.5 MMcfd, increased Mobil's total field production rate by 100 percent in 1989,

the entire project cost was less than half the cost of drilling a single well in this area.

Conclusions

Because turbidite reservoirs change so rapidly, imaging the 3-D seismic data on a workstation proved critical to determining the areal extent of positive AVO responses. The 3-D AVO interpretation technique helped Mobil evaluate untested gas amplitudes with more confidence, and reduced development drilling risk by more accurately determining the probability of success for individual target zones. While 3-D AVO should not be taken as a perfect solution to prospect or field evaluation, when used in conjunction with other interactive tools and geophysical techniques, interpretations will benefit.

Acknowledgments

The invaluable assistance of the following Mobil geoscientists made the Grimes Gas Field project a success: Kathy Stollenwerk, who provided geologic input; Kim Knollenberg, who provided reservoir engineering input; Phil North, who processed the 3-D seismic and 3-D AVO data sets; and Tom Fitch, who contributed his AVO expertise and support.

References and Further Reading

Burnett, R. C., 1990, Seismic amplitude anomalies and AVO analyses at Mestena Grande Field, Jim Hoggs Co., Texas: Geophysics, **55**, 1015–1025.

Hilterman, F., 1990, Is AVO the seismic signature of lithology? A case history of Ship Shoal-South Addition: The Leading Edge, **6**, No. 6 15–22.

Rutherford, S., and Williams, R., 1989, Amplitude-versus-offset variations in gas sands: Geophysics, **54**, 680–688.

Soroka, W., Fitch, T., Van Sickle, K., and North, P., 1990, Three-dimensional amplitude versus offset analysis: 60th Ann. Internat. Mtg., Soc. Expl. Geophys., Expanded Abstracts, 1530–1532.

Yu, G., 1985, Offset-amplitude variation and controlled-amplitude processing: Geophysics, **50**, 2697–2708.

Complex Reservoir Characterization By Multiparameter Constrained Inversion[1]

Ruben D. Martinez, Bruce E. Cornish,* Anthony J. Rebec,* and Michael P. Curtis[‡]*

Introduction

Detailed reservoir characterization is a requirement for the most effective recovery of in-place hydrocarbons.

The characterization of reservoirs requires the integration of different data types to define a reservoir model. Traditionally, geologic, well log, and core data have been the most commonly used data sets for reservoir model building (Raymer and Burgess, 1980). Well log and core data provide detailed information about the vertical variation of many reservoir properties but they are restricted to regions adjacent to the borehole.

More recently, seismic data have played an increasingly important role in describing reservoirs away from the wellbore (Curtis et al., 1983; Angeleri and Carpi, 1982; Martinez, 1985). With its superior lateral resolution, seismic data can contribute to a well-defined geometric description of structural and stratigraphic aspects of the reservoir (Graebner et al., 1981). Also, it is well known that the seismic amplitude variations are linked to changes in acoustic impedance that can be related to reservoir properties which include lithology and porosity.

A method that combines the high vertical and lateral resolution of the well log and seismic data, respectively, appears to be the most effective approach to the characterization of reservoirs in two or three dimensions (Cornish and King, 1988; King, 1988).

Reservoir Characterization Approach

A complete system for reservoir characterization using multiparameter constrained inversion is described in Figure 1. This approach consists of three inversion steps that deal with model optimization using the least squares criterion.

First, the seismic data are inverted using broadband constrained inversion (BCI). BCI simultaneously makes use of an initial interpretation, seismic data, and well log data to obtain a high resolution acoustic impedance model.

Well log inversion (WLI) processing is done in parallel with BCI to estimate lithologic volume fractions and porosity from well log data. The combined acoustic impedance, lithologic, and porosity models derived from BCI and WLI are used to perform the final inversion process: lithology constrained inversion (LCI) (see Figure 1).

LCI provides final models in which the lithology and porosity parameters are allowed to vary spatially. These models are consistent with the broadband acoustic impedance and the lithologic and porosity models obtained in BCI and WLI, respectively.

The field data example used to illustrate the complete approach is from a basin characterized by sand and shale sequences (Figure 2). The target formation consists of two sand members; an upper marine member and a lower continental member. Both are oil-bearing formations but the marine member is the main oil producer in the area.

The sedimentary sequence that characterizes the target formation is transgressive in nature. Above the marine sands, massive marine shales are present. Below the target sands, tuffs (volcanics) are present. The sand and shale content of the tuffs is highly variable. (Figure 3). The target sands (marine and continental) have a gross thickness of 30 m at depths

[1]Presented at SEG/EAEG Research Workshop on Reservoir Geophysics, July 31–Aug. 3, 1988, Dallas, TX.
*Halliburton Geophysical Services, 1 Fluor Daniel Ave., P.O. Box 5019, Sugar Land, TX. 77487-5019.
[‡]Halliburton Reservoir Services, 5950 N. Course, Houston, TX 77072.

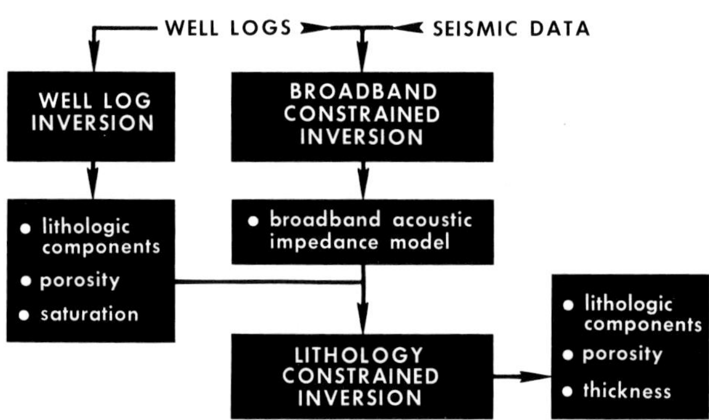

Fig. 1. The complex reservoir characterization sequence is comprised of three inversion steps which integrate seismic, borehole, and geologic data.

that vary from 1600 to 1750 m. These sand bodies are spatially distributed in a complicated manner due to structural growth and complex depositional processes.

Figure 2 shows the location of the profile extracted from a 3-D seismic data volume used to illustrate the techniques presented here. The five wells which exist along this profile were used for calibration purposes. The raw migrated seismic data corresponding to the profile is shown in Figure 4. The target sands are located between 1.43 and 1.50 s at the S-1 well location.

Broadband Constrained Inversion

The objective of broadband constrained inversion (BCI) is to combine seismic, geologic, and borehole information to obtain an optimized broadband acoustic impedance model suitable for reservoir delineation (Cornish and King, 1988).

In order to achieve this objective, a diversity of data types must be brought together (Figure 5). This may be easily accomplished in an interactive workstation environment. Some of this information is quantitative, like the recorded seismic and well log data, while some of it is interpretive, such as unit boundaries, fault locations, and sequence or facies changes. Furthermore, certain types of information may be finely resolved vertically but irregularly distributed in space (at well locations), in contrast to the uniformly distributed but grossly sampled (in time or depth) seismic data and its horizon interpretation.

The effective integration of data exhibiting such extreme variations in scale requires the establishment of a vertical to spatial linkage; this linkage is provided by the surface seismic data amplitudes. The seismic data provide lateral continuity and contain much of the information that constrains the areal distribution of the reservoir model.

The BCI process begins with the definition of an initial acoustic impedance model (Figure 6), built using all the geologic and borehole information as well as a structural and stratigraphic seismic interpretation. The model is parameterized in terms of event reflectivity and delay time. An estimate of the seismic trace is formed and compared with the actual seismic data to

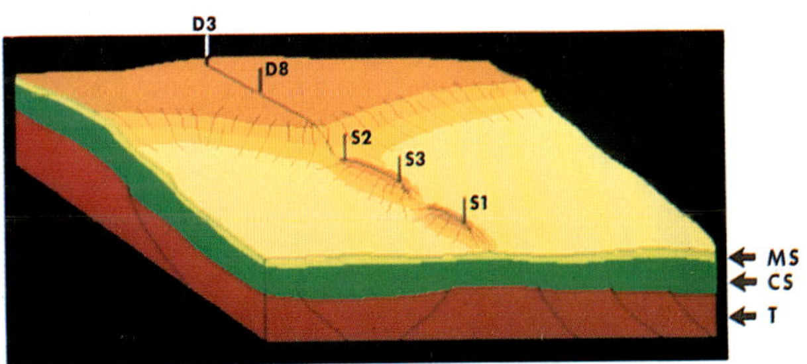

Fig. 2. Schematic view of the distribution of marine sands (MS), continental sands (CS), and tuffs (T), with well locations indicated.

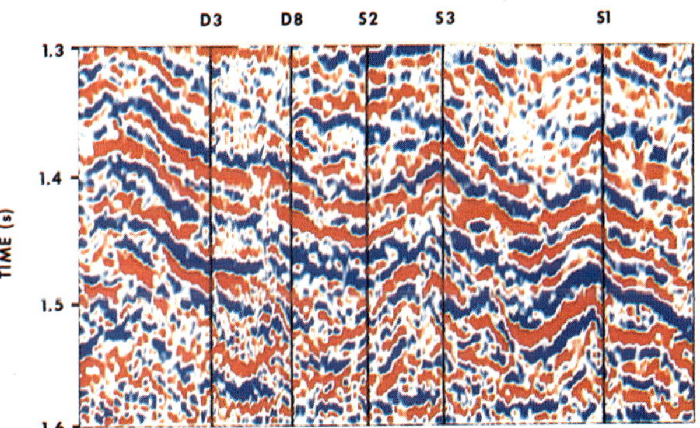

Fig. 4. Raw 3-D migrated seismic data set. The blue peaks and the red troughs correspond to positive and negative amplitudes, respectively.

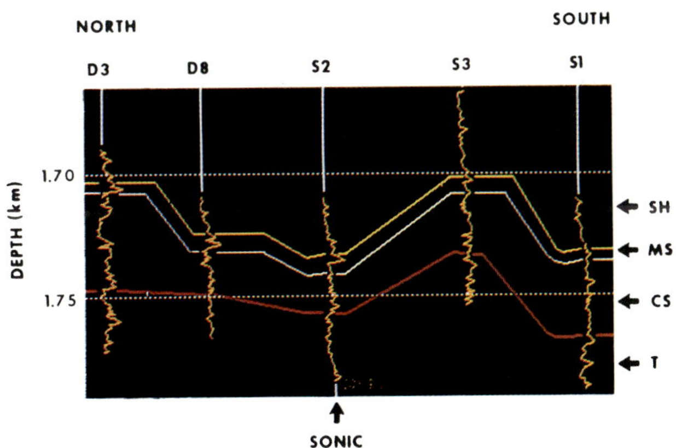

Fig. 3. Sonic log correlation section showing the paleostructure of the marine sands (MS), continental sands (CS), and tuffs (T). The symbol SH represents shale.

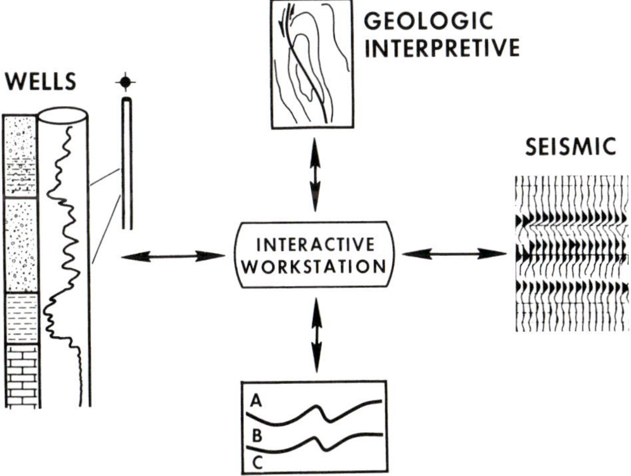

Fig. 5. Data integration concept in an interactive workstation environment.

yield a residual error trace. The error is used to update the model parameters through sensitivity estimates. Constraints provided by the model and data variances are used to control the stability and resolution of the inversion process. Since the relationship between the delay time parameter and the seismic data is nonlinear, several iterations may be required to converge to a final model.

The stochastic algorithm used to compute the model updates is given by

$$m = mo + [\underline{\mathbf{G}}^T\underline{\mathbf{G}} + \mathbf{Cn}\ \mathbf{Cm}^{-1}]^{-1}\ \underline{\mathbf{G}}^T\ (s - d)\,,$$

(1)

where m is the updated model, mo is the initial model, $\underline{\mathbf{G}}$ is a sensitivity matrix, $\mathbf{Cn}$ is the noise covariance matrix, $\mathbf{Cm}$ is the model covariance matrix, d is the seismic data, and s is the estimated seismic data. The perturbations $(m - mo)$ are computed based on the residual error $(s - d)$ but constrained by $\mathbf{Cn}$ and $\mathbf{Cm}$ (Cornish and King, 1988).

The optimized acoustic impedance shows vertical resolution beyond that provided by the bandlimited seismic data, thus reducing the classical resolution problem associated with bandlimited seismic logs.

The initial and final acoustic impedance models obtained from the a priori and seismic data are shown in Figure 7. Also, Figure 8 shows the raw migrated data (top) and the synthetic traces derived from the final acoustic impedance model filtered with the same bandwidth of the seismic data (bottom). The synthetic data panel clearly illustrates the effectiveness of the BCI algorithm in noise rejection.

Comparisons between seismically derived acoustic impedance and the well log acoustic impedance for the five wells reveal that the impedance anomalies associated with the target sands (1.4 to 1.52 s) have been recovered by the BCI process with significant accuracy (Figure 9).

Well Log Inversion

Well log inversion (WLI) is carried out in parallel with BCI to estimate lithologic volume fractions and

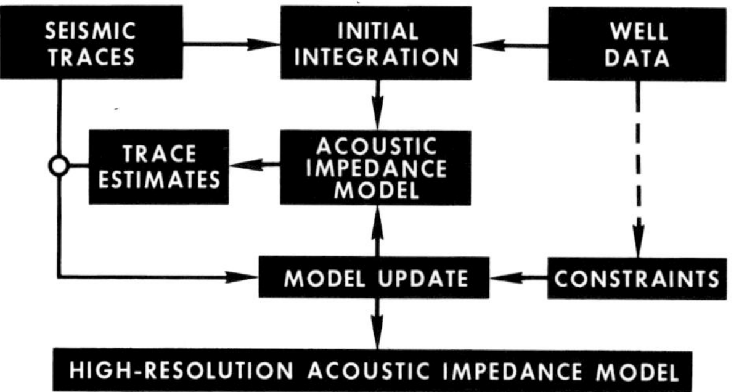

Fig. 6. Broadband constrained inversion (BCI) system.

porosity using well logs from the five wells (see Figure 10). The first step in WLI requires the use of all the available geologic data to parameterize the lithologies known to exist in the prospect. Well log analysis is performed to define rock matrix and fluid parameters. Based on the rock matrix and fluid models already defined, the volume fractions of lithology and porosity are calculated at every sample (in depth or two-way time) by solving a set of linear equations of the form:

$$\mathbf{v} = \begin{bmatrix} \mathbf{R} \\ \underline{\mathbf{G}} \end{bmatrix}^{-1} \begin{bmatrix} b \\ 0 \end{bmatrix},$$

(2)

where $\mathbf{v}$ is a vector containing the volume fractions, b is a vector containing the well log observations, $\mathbf{R}$ is the matrix containing the response equations, and $\underline{\mathbf{G}}$ is a matrix containing the constraints which constitute the a priori information included in the system.

WLI was accomplished using sonic, density, and gamma ray logs from all five wells. In this case, the lithologies have been parameterized as sand and shale. The results of the inversion process are presented in Figure 11. The target sands are at approximately 1.450 s and are well defined by the presence of some interbedded shales. Porosities tend to increase in proportion to the sand volume fraction, as expected.

The sonic, density, and gamma ray logs were calculated using the final lithologic and porosity models derived in the WLI process (Figure 13). For comparison, Figure 12 shows the measured sonic, density, and gamma ray logs. Only minor differences exist between the two sets of logs, showing that the lithologic and porosity models successfully reproduce the well log measurements.

Lithology Constrained Inversion

At this point, all of the elements required to perform lithology constrained inversion (LCI) are available; the broadband acoustic impedance model and the volume fractions of lithology and porosity at every well location (see Figure 1).

The seismic interpretation and the a priori information previously used in BCI are combined with the volume fractions of lithology and porosity derived at the well locations by the WLI process. This integration produces an initial model for each lithology and porosity for all traces across the traverse being analyzed.

The initial model is subsequently used to compute acoustic impedance estimates (see Figure 14). A comparison between synthetic acoustic impedance and broadband acoustic impedance (derived by BCI) produces a residual error. Since the relationship between acoustic impedance and volume fractions of lithology and porosity is nonlinear, it is necessary to iterate until the residual error is minimized in the least squares

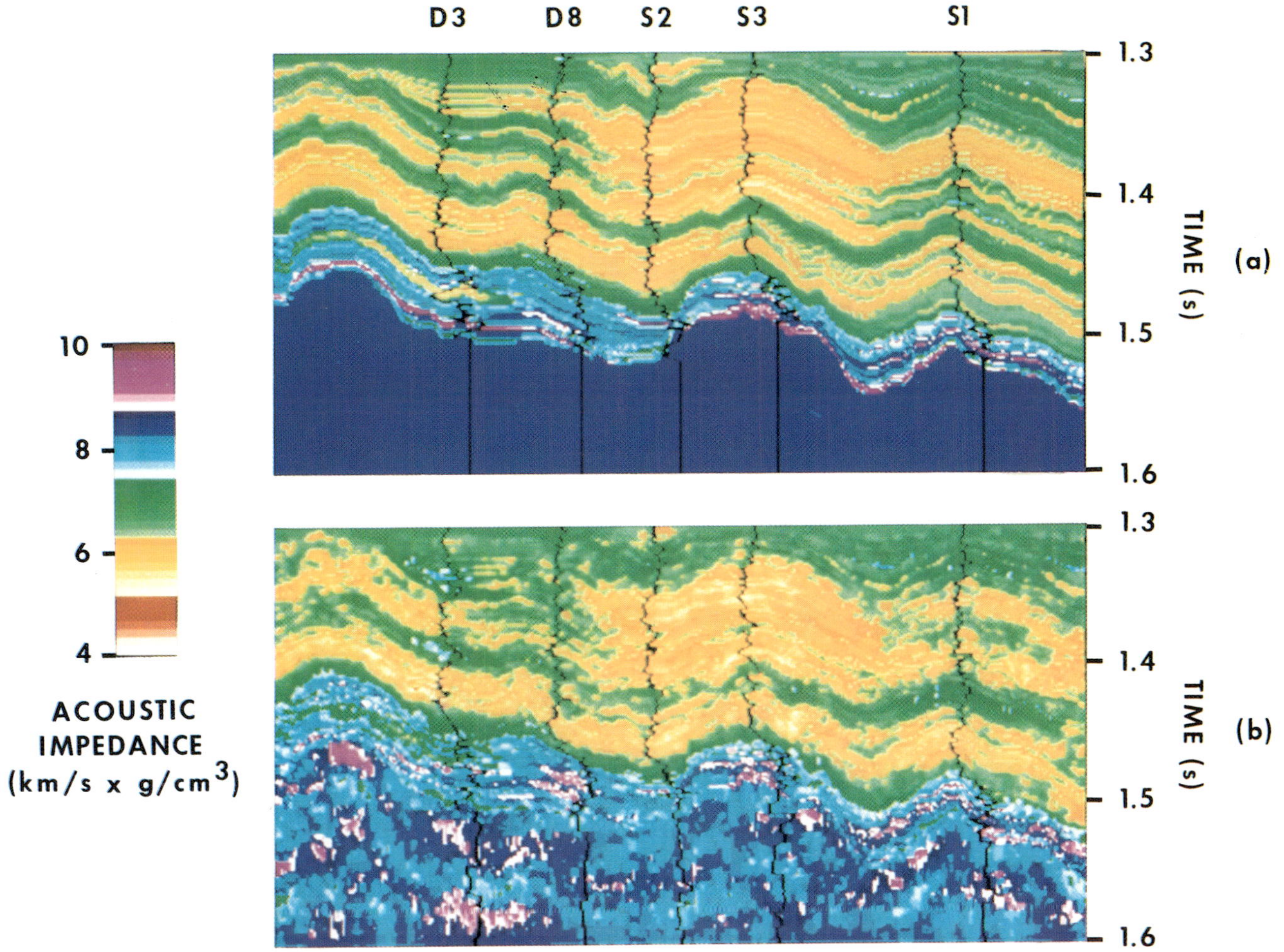

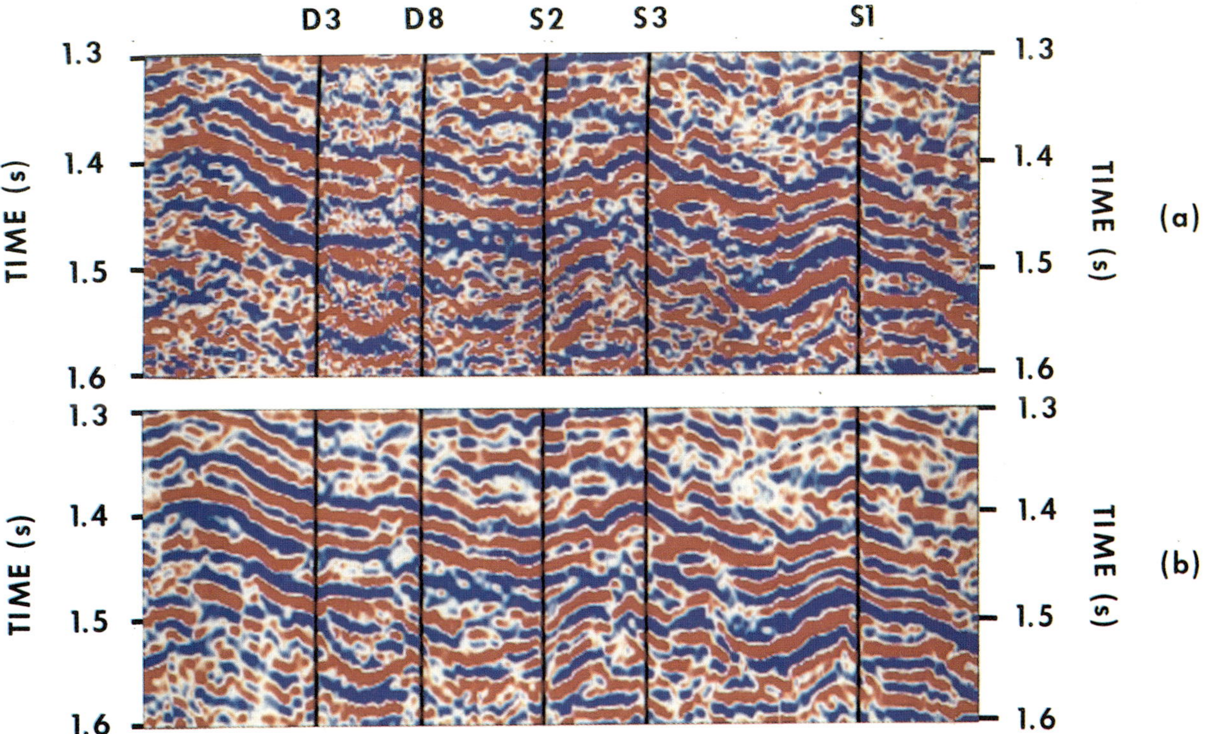

Fig. 7. Initial acoustic impedance model (a) as created with a priori information and final acoustic impedance model (b) after broadband constrained inversion.

Fig. 8. Raw 3-D migrated seismic data (a) and bandlimited synthetic seismic data (b) as derived from the final acoustic impedance model. The blue peaks and the red troughs correspond to positive and negative amplitudes, respectively.

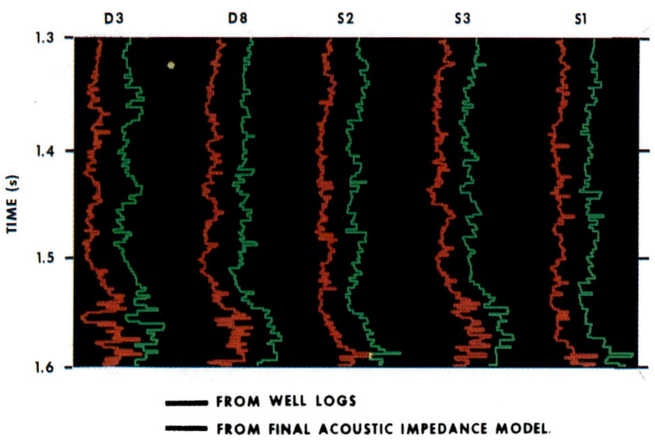

Fig. 9. Comparison of acoustic impedance logs derived from well log data and broadband constrained inversion at the five well locations.

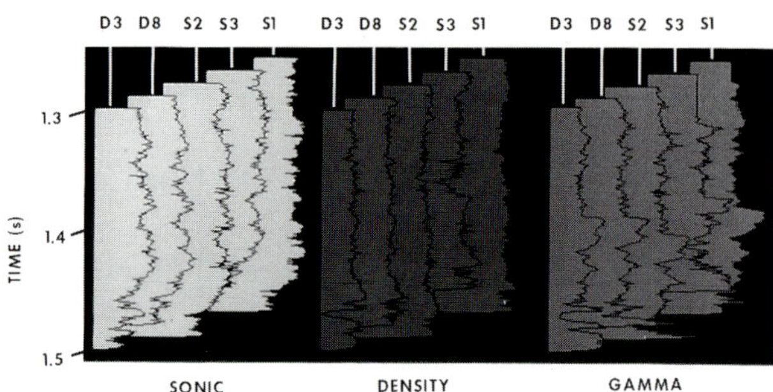

Fig. 12. Measured sonic, density, and gamma ray logs at wells D3, D8, S2, S3, and S1.

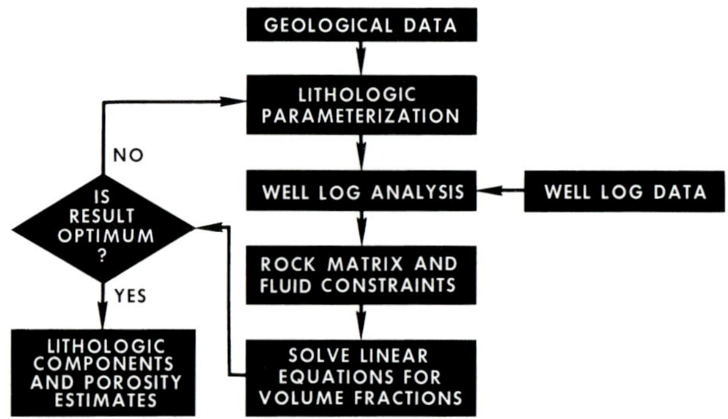

Fig. 10. Well log inversion (volumetric analysis) system (WLI).

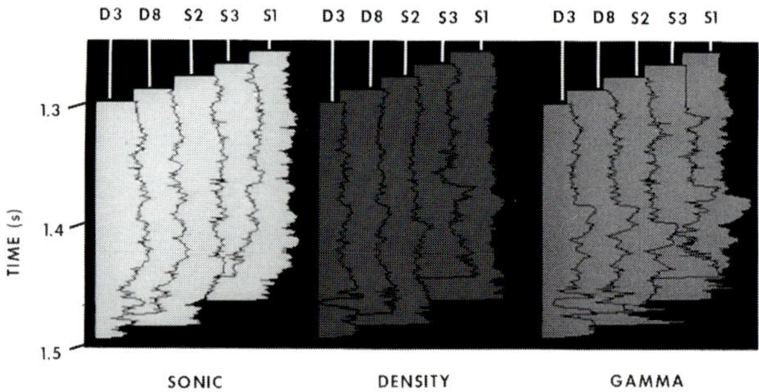

Fig. 13. Sonic, density, and gamma ray logs computed from the final lithologic and porosity models at wells D3, D8, S2, S3, and S1 (to be compared with Figure 12).

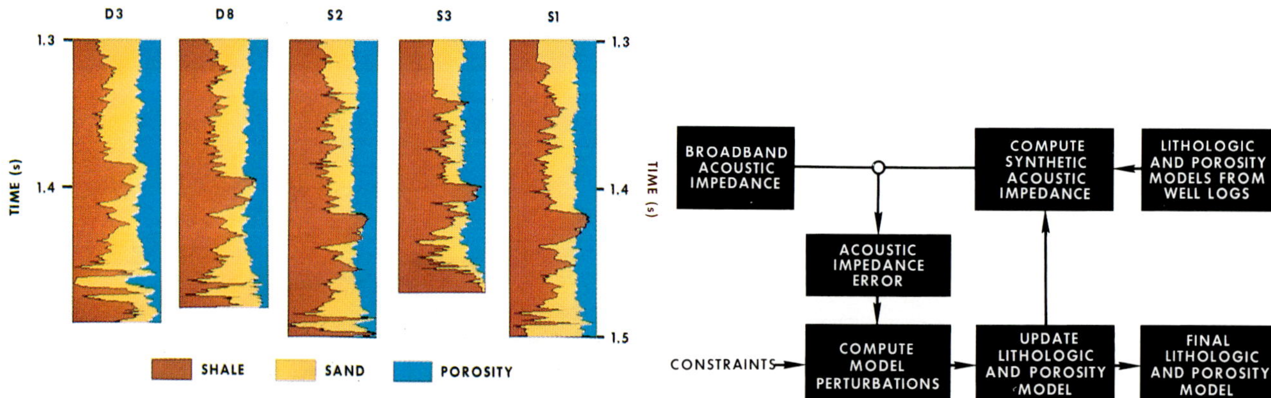

Fig. 11. Lithologic and porosity models derived from well log data following well log inversion.

Fig. 14. Lithology constrained inversion (LCI) system.

sense. Updates for the volume fractions of lithology and porosity models are estimated using the residual error and the sensitivity calculated with the current model.

The optimization problem just described is underdetermined because acoustic impedance from seismic and constraints from well logs are the only available data. In the presence of mixed lithologies, the number of unknowns becomes greater than the information available. In these instances (as in this example) a minimum length solution for a set of linearized equations is given by:

$$(v - vo) = \begin{bmatrix} \underline{A} \\ \underline{K} \end{bmatrix}^{-1} \begin{bmatrix} (z - zo) \\ 0 \end{bmatrix}, \qquad (3)$$

where $(v - vo)$ are the volume fraction of lithology and porosity updates, $(z - zo)$ is the residual error, $\underline{A}$ is the sensitivity matrix, and $\underline{K}$ is the matrix containing the constraints (which are related to geologic conditions).

The forward model used by LCI is based on the definition of acoustic impedance given by:

$$Z = \frac{\rho}{\Delta t}\, C\,, \qquad (4)$$

where

Z = acoustic impedance in (ft/s)(g/cm^3) or (m/s)(g/cm^3)
ρ = bulk density in g/cm^3
Δt = transit time in μs/ft or μs/m, and
$C = 10^6$.

A variety of ways may be used to relate density and transit time to volume fractions of lithology and porosity. However, it is well accepted that density is approximately a linear function of the volume fractions of lithology and porosity (Doveton, 1986). Transit time is not completely linear for either very low or high porosities, which normally correspond to carbonates and undercompacted sediments. For porosities in the range of 20–30 percent, the linearity assumption between transit time and volume fractions of lithology and porosity is acceptable. When the porosities are either too low or too high, other models are invoked; for example, the porosity-transit time transformation of Raymer-Hunt-Gardner (Leslie and Mons, 1982) or the *P*-wave velocity relationships between porosity and clay content proposed by Han et al. (1986) for shaly sands. In the field data example presented here, the porosities oscillate within the range in which transit time may be assumed to vary linearly with volume fractions of sand, shale, and porosity (based on a modified time-average equation to include shale content).

The acoustic impedance model (bottom of Figure 7) and the initial lithologic and porosity models are combined in LCI to obtain final volume fractions of lithology and porosity (Figures 15, 16, and 17). Sand, shale, and porosity profiles have been calculated as percentages. At a particular trace and time sample the summation of the volume fractions of lithology is unity, or 100 percent. Figure 18 shows the calculated

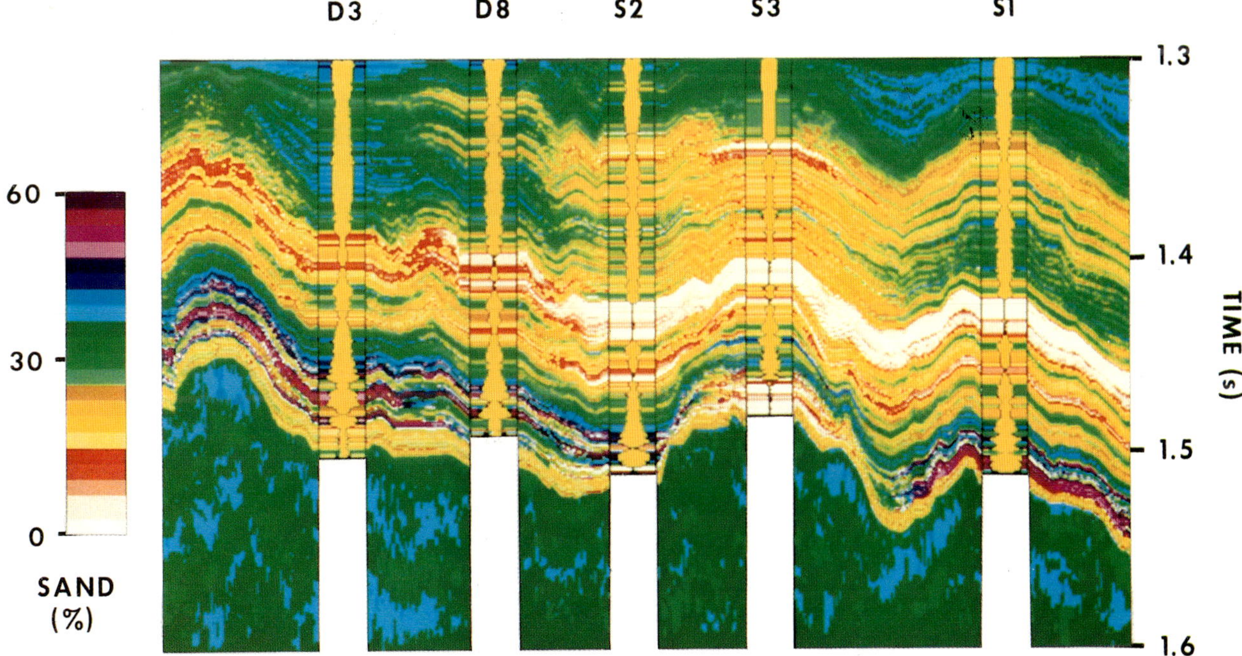

Fig. 15. Sand volume fraction profile. The sand models derived from well log data are inserted at the five well locations for comparison.

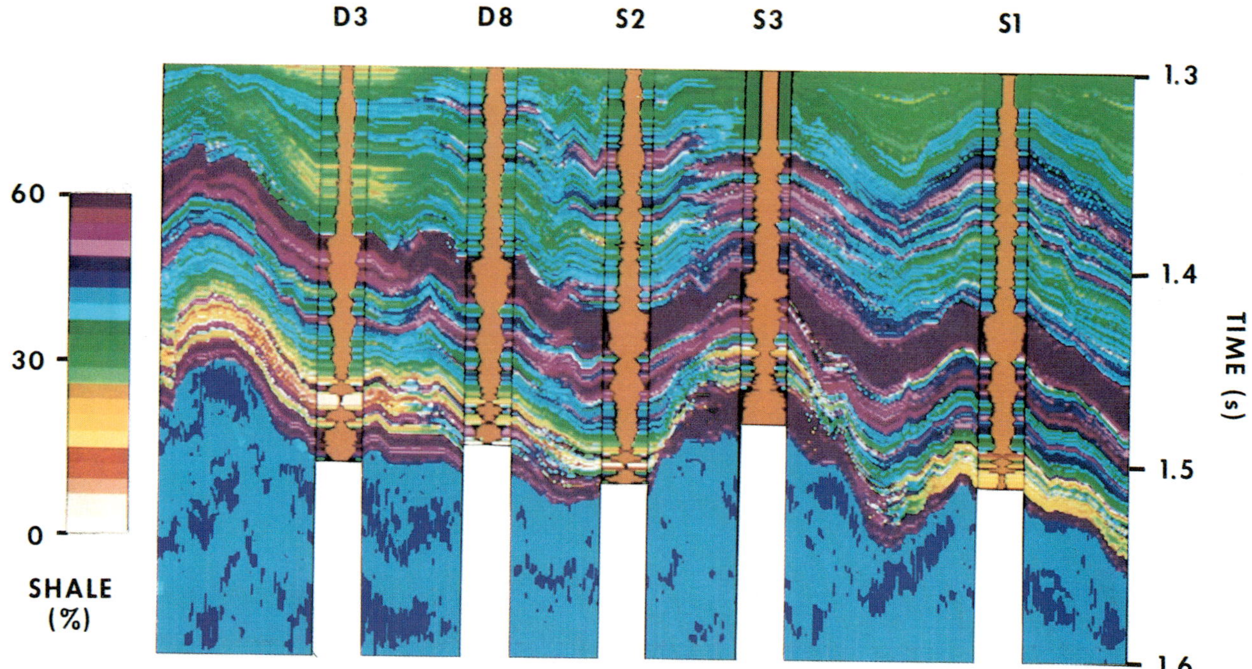

Fig. 16. Shale volume fraction profile. The shale models derived from well log data are inserted at the five well locations for comparison.

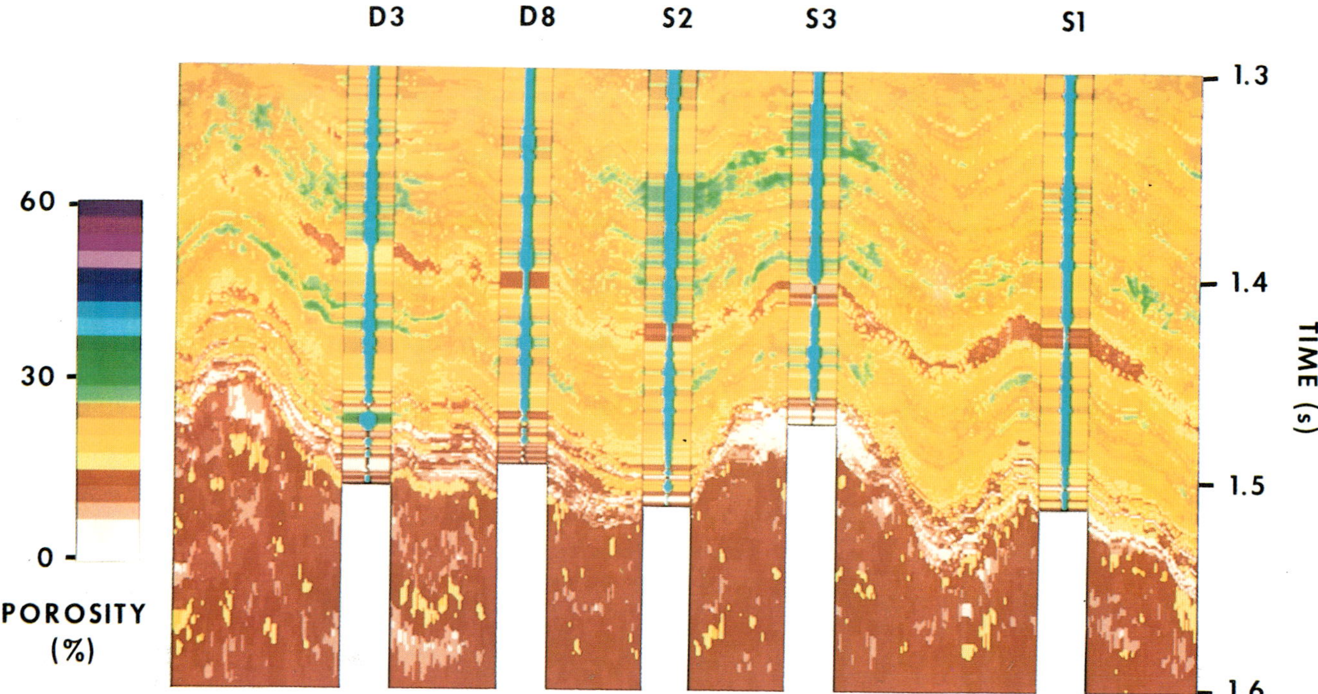

Fig. 17. Porosity profile. The porosity models derived from well log data are inserted at the five well locations for comparison.

acoustic impedances from the initial (top) and final (bottom) models of lithology and porosity. Note the similarity of the acoustic impedance model derived by BCI (bottom of Figure 7) and the calculated impedance model from the LCI final results (bottom of Figure 18). This demonstrates that the lithologic and porosity models fit the data (BCI impedance model).

To illustrate, acoustic impedance, sonic, and density logs computed from the volume fractions of lithology and porosity derived by LCI are compared with the well log acoustic impedance, measured sonic, and measured density from well D-3 (Figure 19). The correlation between the two sets of logs is good. The presence of differences is attributable to noise in the seismic and well log data.

Figure 20 shows a comparison of the lithologic and porosity models derived using WLI with those computed by LCI. The large variation of the velocity and density of the fluids with respect to the velocity and density variations of the sands and shales is explained by the larger sensitivity of the acoustic impedance to porosity variations.

Transit time and density sections are estimated using the estimated volume fractions of lithology and porosity (Figures 21 and 22). This method avoids the use of any predefined density-velocity relationship to separate velocity and density from acoustic impedance. These two sections are useful in the calculation of thickness (estimated transit time) and in engineering applications such as estimating density.

Finally, Figure 23 shows a profile of potential reservoir zones. This section is constructed by limiting the volume fractions of lithology and porosity profiles according to conditions considered optimum for development and production. In this example, the good reservoir zones are considered to have sand content of 55 to 65 percent, shale content of 0 to 15 percent, and porosity from 20 to 35 percent. Within these zones the values of the product of porosity and thickness are color coded. At the top of Figure 23, the total porosity-thickness (meters of porosity at each trace location) is plotted against horizontal distance along the traverse. Clearly, where wells D-3 and S-1 are located the

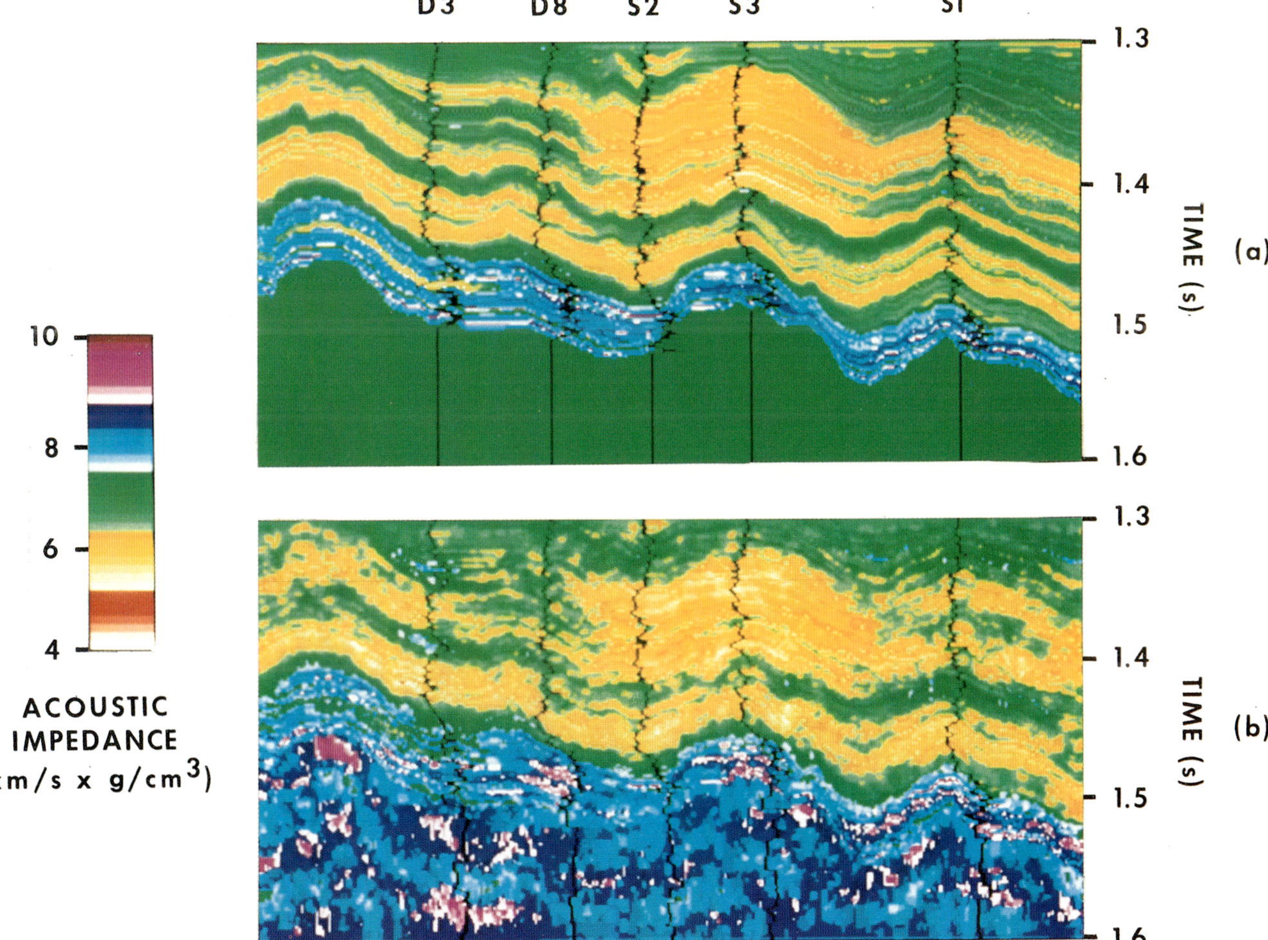

Fig. 18. (a) Initial acoustic impedance model calculated from the initial lithologic and porosity models, and (b) the final acoustic impedance model calculated from the lithologic and porosity models shown in Figures 15–17.

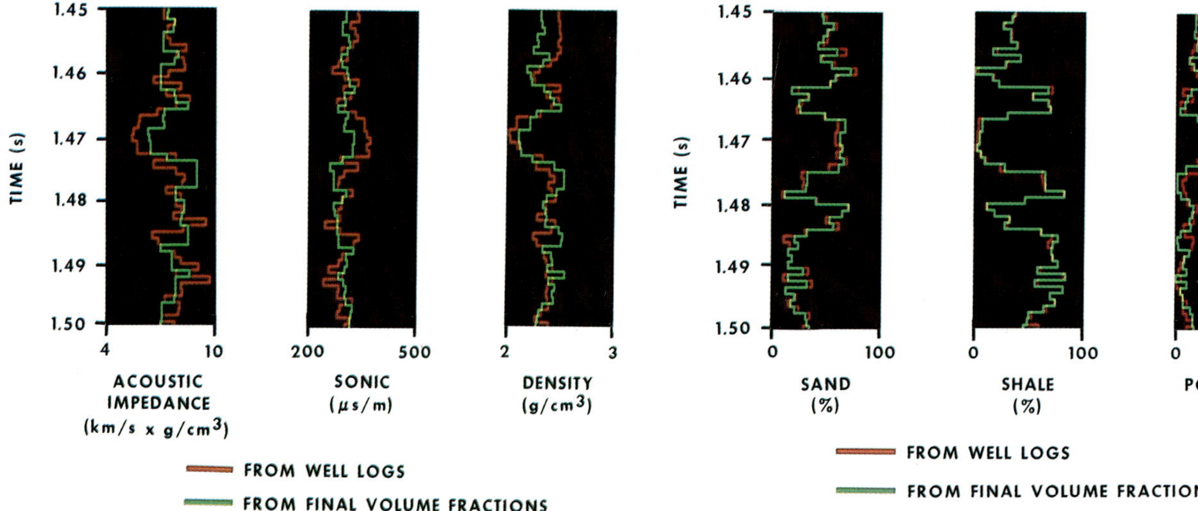

Fig. 19. Measured acoustic impedance, sonic, and density compared with the acoustic impedance, sonic, and density calculated from the final lithologic and porosity models at the D3 well.

Fig. 20. Comparison of sand, shale, and porosity logs derived from well logs and the optimized acoustic impedance model using WLI and LCI, respectively.

reservoir sands are thicker and more porous, a result which is consistent with measured well information.

components, porosity, and thickness varying laterally and vertically.

The optimized lithologic components and porosity results are consistent with seismic and borehole data and provide a detailed view of the spatial distribution of the reservoir rocks.

Summary and Conclusions

An approach has been presented for the estimation of important reservoir parameters such as lithologic

The advantage of this approach is that it applies a systematic integration of all data sets resulting in an improved reservoir model. This more accurate model

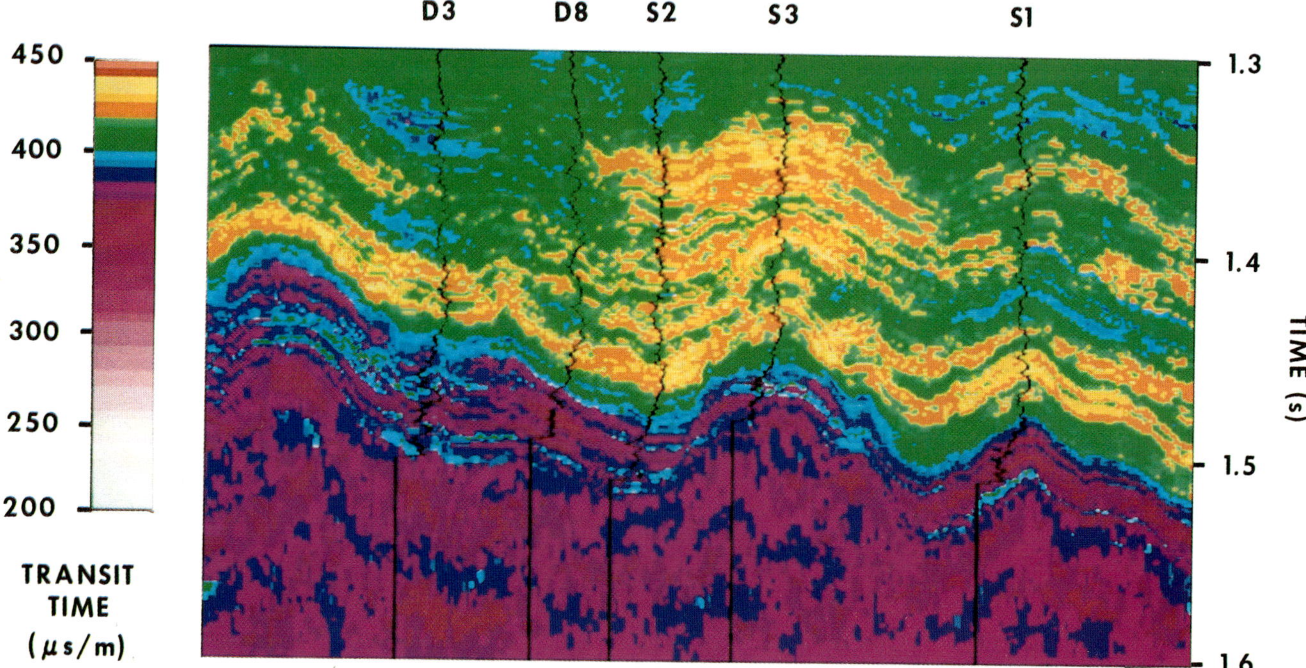

Fig. 21. Estimated transit time profile. The traces displayed at the well locations are the corresponding measured sonic logs.

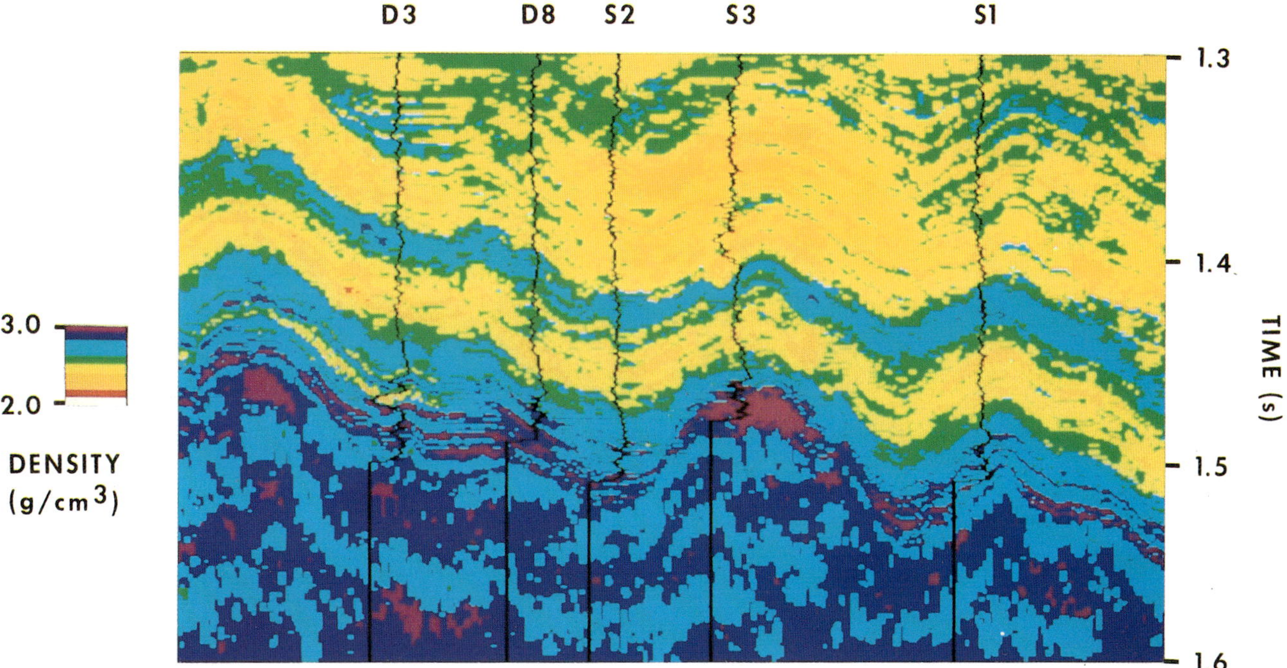

Fig. 22. Estimated density profile. The traces displayed at the well locations are the corresponding measured density logs.

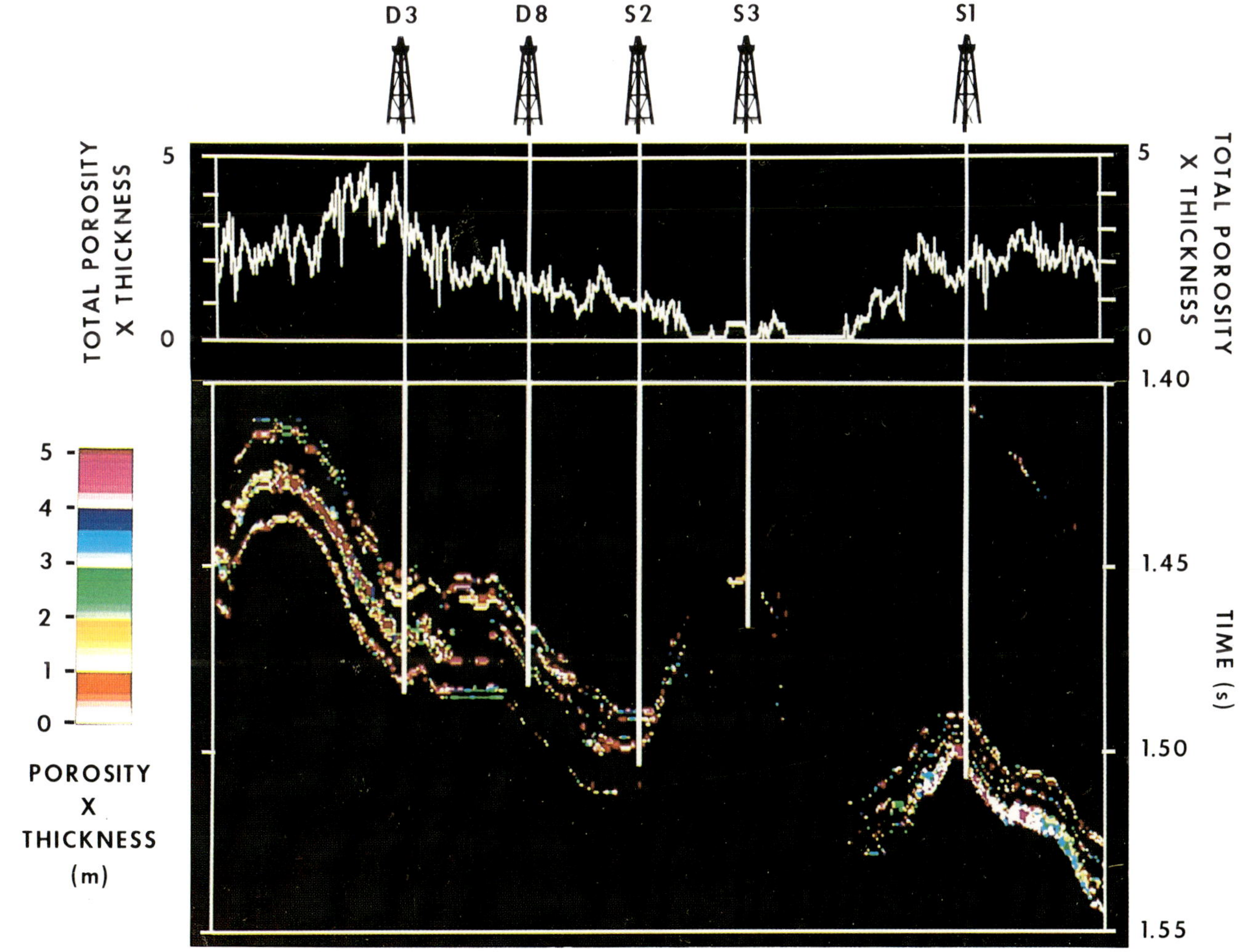

Fig. 23. Profile of potential reservoir zones. The plot at the top shows the lateral variation of the total porosity thickness.

can be used for simulation purposes and also can be periodically updated with production data.

Acknowledgments

The authors wish to thank Empresa Nacional del Petroleo (ENAP), Chile, for kindly providing the data sets that helped illustrate the techniques presented. We are grateful to Mr. Mladen Vrsalovic of ENAP for authorizing the use of the seismic and borehole data. Special thanks go to Mr. Roberto Acosta of ENAP for providing interpretation support to initialize this project.

References

Angeleri, G. P., and Carpi, R., 1982, Porosity prediction from seismic data: Geophys. Prosp., **30**, 580–607.

Cornish, B. E., and King, G. A., 1988, Combined interactive analysis and stochastic inversion for high-resolution reservoir modeling: Presented at the 50th Mtg. European Assn. Expl. Geophys.

Curtis, M. P., Martinez, R. D., Possato, S., Saito, M., 1983, 3-dimensional seismic attributes contribute to the stratigraphic interpretation of the Pampo Oil Field, Brazil: Presented at the 53rd Ann. Internat. Mtg., Soc. Expl. Geophys.

Doveton, J. H., 1986, Log analysis of subsurface geology: Concepts and computer methods: John Wiley & Sons.

Graebner, R., Wason, C., and Meinardus, H., 1981, Three-dimensional methods in seismic exploration: Science, **211**, 535–540.

Han, D. H., Nur, A., and Morgan, D., 1986, Effects of porosity and clay content on wave velocities on sandstones: Geophysics, **51**, 2093–2107.

King, G. A., 1988, The application of seismic methods for reservoir description and monitoring: Presented at the SEG/China Petr. Soc. Production Geophysics Meeting.

Leslie, H. D., and Mons, F., 1982, Sonic waveform analysis: Applications: Presented at the 33th Ann. Logging Symposium, Soc. Prof. Well Log. Assn.

Martinez, R. D., 1985, Deterministic estimation of porosity and formation pressure from seismic data: Presented at the 55th Ann. Internat. Mtg., Soc. Expl. Geophys.

Raymer, L. L., and Burgess, K. A., 1980, The role of well logs in reservoir modeling: Presented at the 55th Ann. Mtg., Soc. Petr. Eng., Am. Inst. Min., Metall.

Reservoir Petrophysics of Bima Field, Northwest Java Sea[1]

E. Poggiagliolmi, V. R. Checka,[‡] R. C. Roe,[§] and R. Purantoro[§]*

Introduction

In reservoir evaluation and development, the assessment of petrophysical properties, such as porosity, mineralogy, and type of pore fluid, normally are supplied by the available borehole information (well logs, well tests, and core data). These measurements generally are accurate in the depth direction but their lateral penetration is shallow, hence large errors can be introduced by interpolation between and extrapolation from well locations. In contrast, surface seismic measurements are laterally continuous but have a low vertical resolution and provide acoustic rather than petrophysical information. However, recent advances in seismic inversion techniques permit petrophysical parameters to be calculated from high-resolution seismic acoustic impedance (Weathers and Helm, 1984). Therefore, quantitative petrophysical measurements now can be obtained from borehole calibrated seismic data. Furthermore, measurement accuracy can be assessed by statistical analysis of errors inherent within the seismic data and the petrophysical and acoustic relationships.

The purpose of this case study is to demonstrate an integrated multidisciplinary method used to map the reservoir porosity, fluid, and thickness distribution of the Bima Field from calibrated seismic data (EnTec, 1984, 1986, 1987). The Bima Field covers an area of approximately 2500 hectares and is situated 100 km north of Jakarta (Figure 1). A total of 67 seismic lines (600 km) of 1979 to 1986 vintage were processed concurrently with information taken from 30 wells (Figure 2). Due to the large amount of seismic and well data, reference is made to one line running east-to-west across the field (4390-SD-86, Figure 2) and one on-line well (ZU-11, Figure 2).

The seismic processing techniques, inversion of seismic data to high-resolution acoustic impedance, and subsequent petrophysical calibration are discussed. The inversion method used in this study is

[1] Reprinted with permission of Indonesian Petroleum Association.
*Elnusa—Entec Energy Consultants Ltd., St. Anne House, 20/26 Wellsley Rd., Croydon CRD 96B, England.
[‡] ARCO International Oil and Gas Company
[§] Atlantic Richfield Indonesia Inc.

termed Micromodelling and is a proprietary technique developed by EnTec Energy Consultants Ltd.

Geological Setting and Stratigraphy

The Bima Field is controlled geologically by metamorphic crystalline schist basement rocks which form a north-to-south backbone to the Bima Field. The western and northern limits of the field are delineated by major faults. The southern and eastern limits are ill-defined and the basement rocks dip toward the east.

Thick Oligocene deltaic clastic deposits (Deltaic Talang Akar) occur on the flanks of the basement ridge and infill faulted basinal structures and lows. The thickest deltaic accumulations occur in the eastern part of the field. Coal-rich shales overlie the clastics and are in turn overlain by widespread reefal limestones of Miocene age which are characterized by large buildups along the western edge of the basement highs. The carbonate rocks are divided into the Marine Talang Akar and the Batu Raja formations. Carbonate deposition was followed by a regressive phase during which shales were deposited with occasional laterally persistent limestone stringers (ZU-24 formation).

Both oil and gas occur in separate reservoirs within the clastic (Deltaic Talang Akar) and carbonate (Marine Talang Akar and Batu Raja) formations. For the purpose of this petrophysical study, the Batu Raja limestone is divided into upper and lower formations and the lower Batu Raja includes the Marine Talang Akar.

The principal objectives of the study were to delineate the gas cap and to map the porosity, fluid, and thickness distribution of the upper Batu Raja carbon-

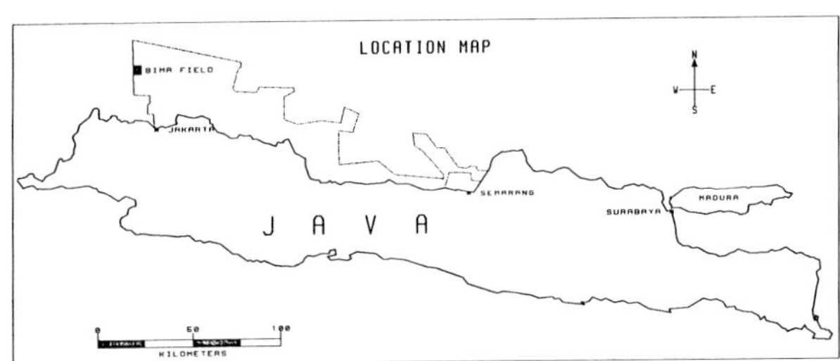

Fig. 1. Bima Field location map.

"

ates using borehole-calibrated seismic data (EnTec, 1987).

A typical stratigraphic sequence is displayed in Figure 3 for well ZU-11. Selected logs are shown in the time domain at a 2 ms sample interval. All the well logs were edited where necessary and the sonic log was calibrated to the surface seismic using available borehole seismic data.

Seismic Data

The seismic processing sequence made full use of well data to optimize processing parameters and to maintain calibration of the seismic data. Emphasis was placed on achieving a broad bandwidth to maximize the vertical resolution, attenuation of multiple reflections through the zone of interest, and a high signal-to-noise ratio.

A multiline, multiwell approach was used throughout all seismic processing stages. Interactive wavelet extraction was carried out on the migrated data at the well locations to obtain the best possible estimate of the seismic wavelet. The wavelet extraction was followed by wavelet processing of the seismic traces, during which the wavelet was converted to its zero-phase equivalent. The wavelet processing significantly enhanced the resolution through the zone of interest.

The final wavelet-processed section for line 4390-SD-86 is shown in Figure 4. Several reflections are identifiable through the zone of interest (0.86 to 0.98 s). These reflections are related to stratigraphic features within the Batu Raja and Talang Akar formations. The coal-rich shales particularly are well defined through the region of 0.94 s.

Seismic acoustic impedance was obtained by Micromodelling the wavelet-processed seismic data. The Micromodelling technique is based upon the most effective use of information content in the seismic trace to obtain a reflection series (consistent with the number of degrees of freedom allowed by the bandwidth). The reflection series is subsequently inverted to obtain acoustic impedance. This technique is an iterative process which successively approximates the reflectivity to generate a solution compatible with the wavelet supplied. Micromodelling fully utilizes bandwidth down to the point where the signal-to-noise ratio equals unity.

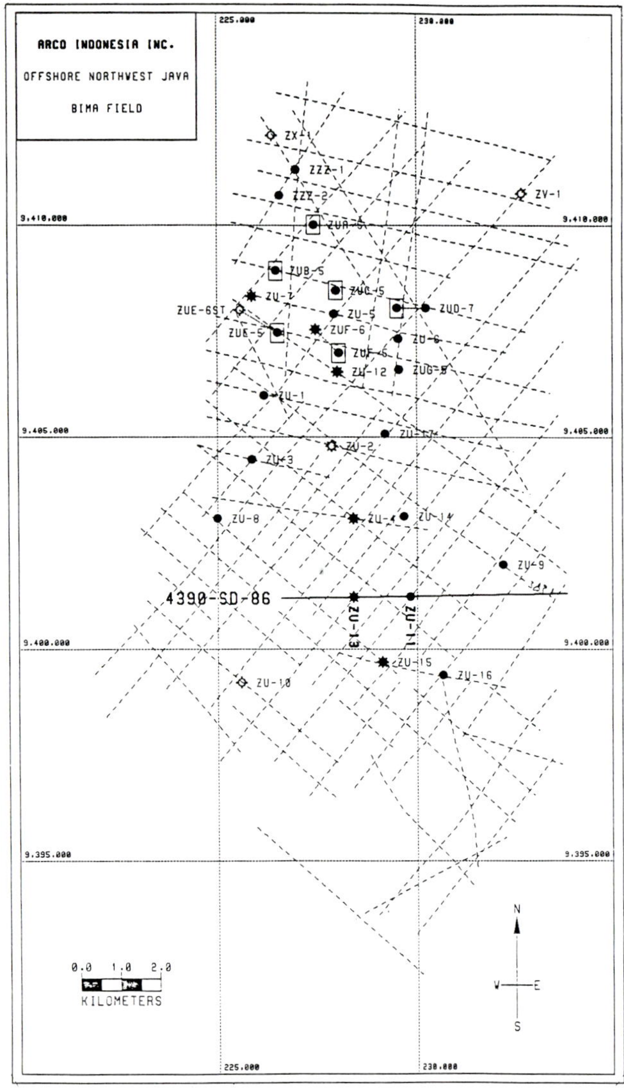

Fig. 2. Bima Field study area showing seismic lines and well locations.

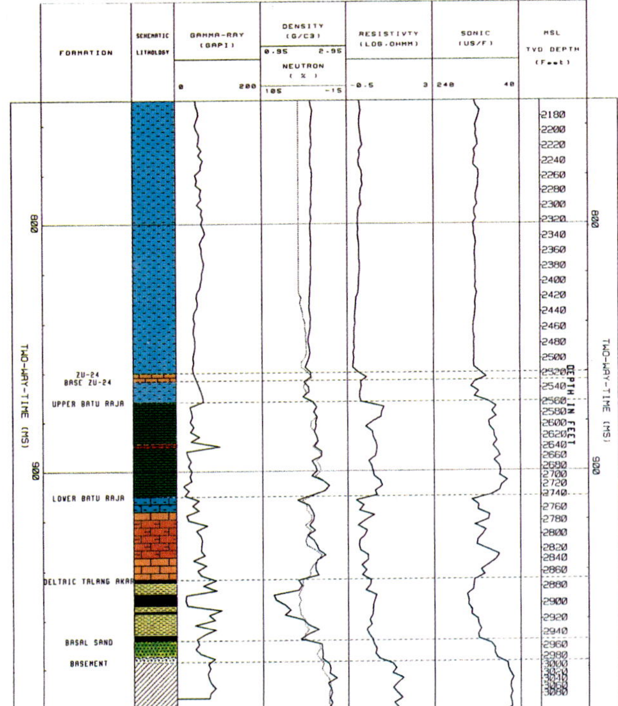

Fig. 3. General stratigraphy for well ZU-11 with wireline logs.

The Micromodelling of the wavelet-processed data was carried out and subsequently color-contoured to aid lithostratigraphic interpretation. The minimum resolution obtained by Micromodelling was 4 ms, corresponding to a thickness of approximately 6 m. A color contoured Micromodelling section is shown for a portion of line 4390-SD-86 around well ZU-11 (Figure 5). The top of the Batu Raja is readily apparent and occurs as a dark blue interval at approximately 0.86 s.

The drop in acoustic impedance at 0.91 s corresponds to an increase in shaliness that is typical of the lower Batu Raja. The dark brown coal-rich Talang Akar shales dominate the section by their low acoustic response at 0.95 to 0.97 s. Below the shales, the Talang Akar sandstones vary widely in thickness and acoustic response while the basement appears as a strong event in the region of 0.98 s.

The three main formations were delineated on all the Micromodelling lines for the entire field. Reference was made to on-line wells for stratigraphic character correlation. The gas-oil contact (GOC) within the upper Batu Raja clearly was visible by its relatively flat positive acoustic response and also was delineated on the Micromodelling. In contrast, the oil-water contact

(OWC) could not be identified due to the close similarity in acoustic impedance between oil-saturated and brine-saturated rocks. As a result, the OWC was obtained by extrapolation from the wells. This approach was considered acceptable since the OWC varied little from well to well across the field.

Micromodelling Calibration

Color-contoured Micromodelling profiles are well suited for delineating reservoir intervals. They are characterized by high resolution, and color aids the interpretation by enhancing continuity of stratigraphic intervals. In order that quantitative measurements can be obtained from the Micromodelling, the acoustic amplitudes through the zone of interest are calibrated to petrophysical parameters determined from well data. The principles of Micromodelling calibration are based upon the assumption that the acoustic properties of reservoir rocks depend primarily upon their mineralogical composition, porosity, and type of pore fluid.

The approach used for the Bima Field was to develop a multivariate petrophysical model that was tailor made to the petrophysical properties of each

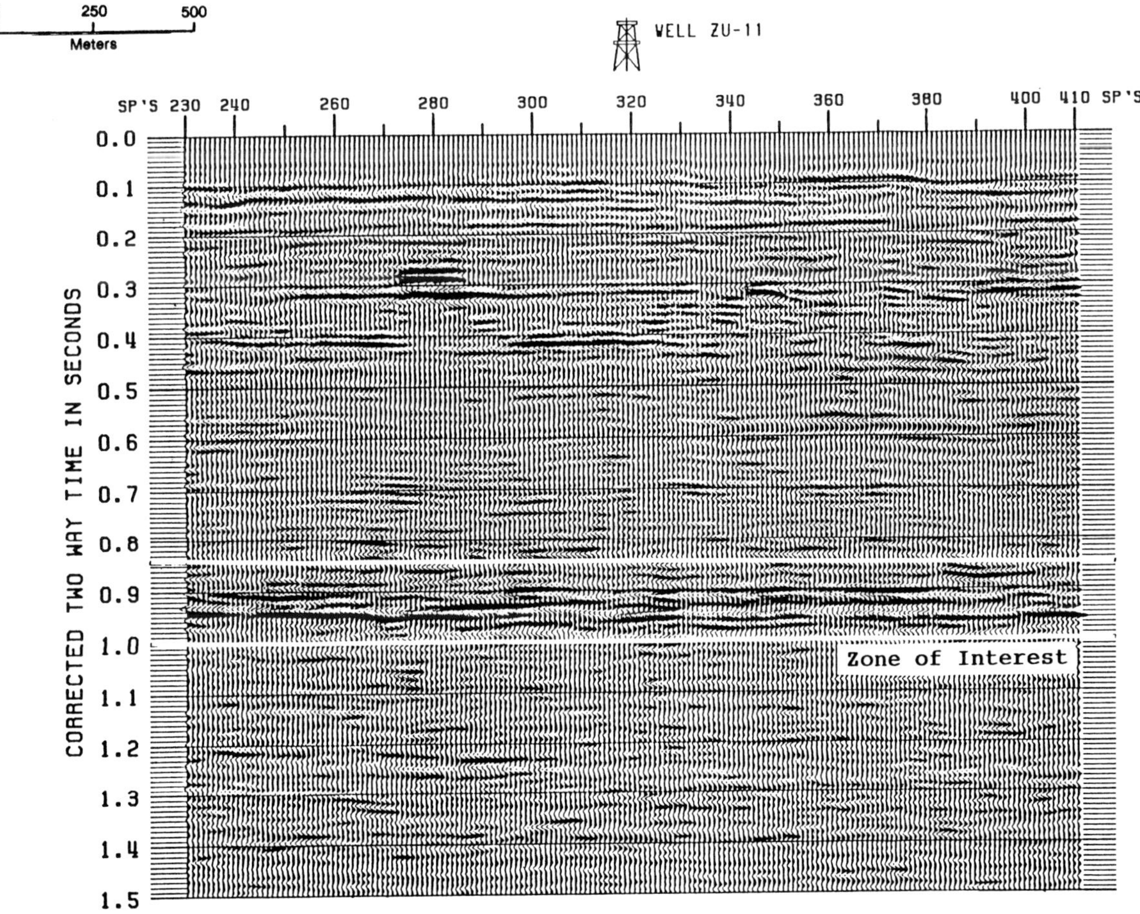

Fig. 4. Wavelet processed seismic section, line 4390-SD-86.

formation to be calibrated. This model was achieved by using all the available borehole information in a multivariate and multiwell statistical approach to estimate the effective porosity and associated error. Core data played an important role in the construction of the model, both in the selection of input parameters and in the validity tests of the results. An interactive approach was used to obtain the necessary parameters. Reference was made continually to the Micromodelling and borehole data throughout the construction of the petrophysical model. A number of control wells not employed in model construction also were used to test the validity of the calibration.

The calibration model was designed to transform the Micromodelling to effective porosity for the upper and lower Batu Raja formations and the Deltaic Talang Akar. The calibration model varied considerably for the three formations due to their different mineralogies. Compensation was built into the model for the Upper Batu Raja to account for the presence of gas.

Further modeling was also performed to convert the calibrated Micromodelling data from the time to the depth domains. An analysis of errors in the models was made to determine the confidence in the predicted porosity for the three formations. Further variance inherent within the seismic data was combined with the petrophysical error to obtain the total level of confidence. The error components of the seismic data included residual multiples, random noise, and amplitude changes due to overburden effects. Seismic noise levels were determined on the common-midpoint (CMP) gathers. The effect of post-stack processes, such as deconvolution, wavelet processing, etc., were accounted for in the analysis. The total level of confidence was determined for an error distribution around the mean effective porosity of ± 3 percent.

The calibration model was applied to the respective formations and the effective porosity (and variance) was computed. The results of the calibration for well ZU-11 and the Micromodelling at the well location are

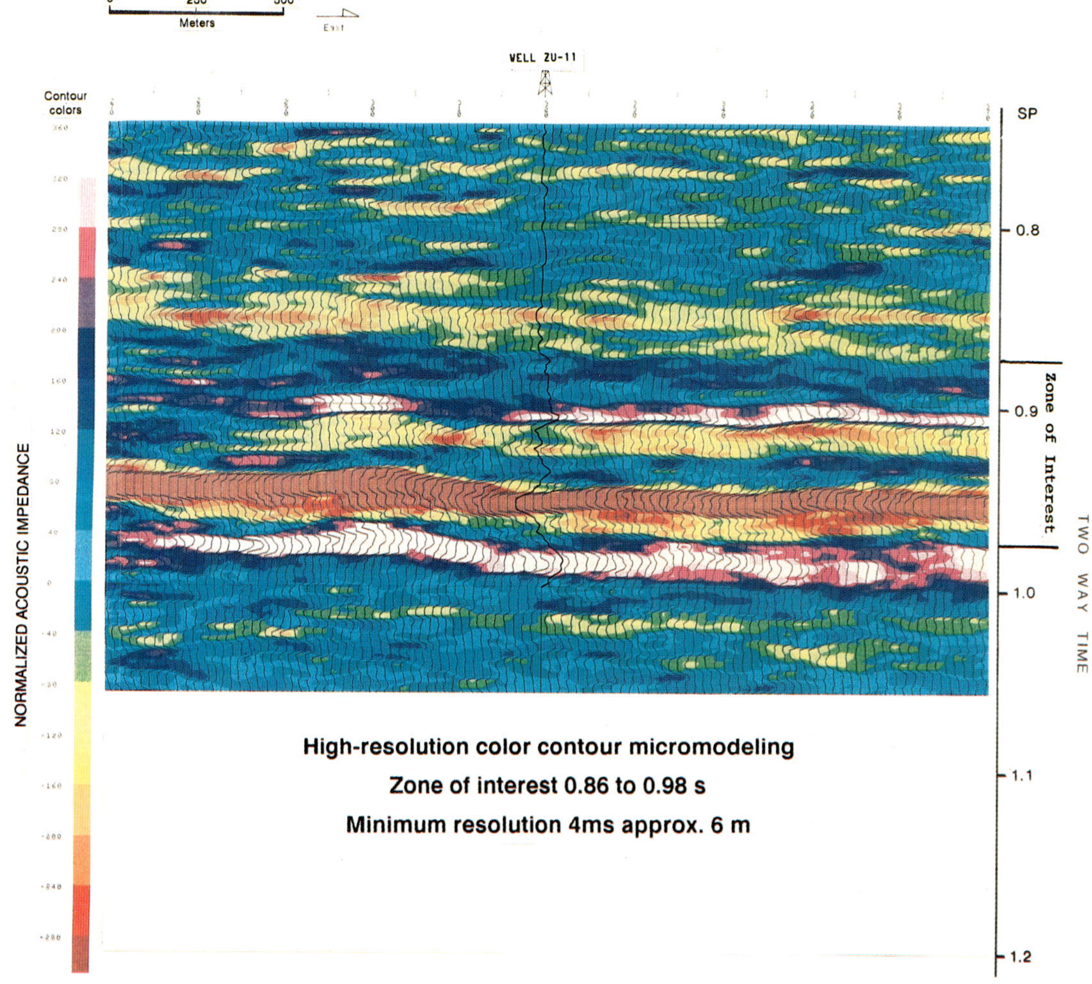

Fig. 5. Color countured Micromodelling of line 4390-SD-86 with inserted acoustic impedance log for ZU-11 well.

shown in Figure 6. The upper curve (A) shows the porosity in the time domain computed from the well acoustic impedance with the core porosities over-printed as asterisks. The calibration model was applied to the Talang Akar irrespective of the presence of coal; the effect of coal was to produce very high apparent porosity values. An analysis of well porosity frequency distributions indicated that porosities in excess of 36 percent were unlikely to occur. Consequently, whenever the porosity exceeded this value, C was used to indicate potential coal.

The central curve (Figure 6b) shows the porosity predicted from the Micromodelling trace at the well location. Similar to the upper curve, good correlation occurs between the Micromodelling porosity and core porosity, and again the presence of coal is indicated by excessively high calculated porosity. The level of confidence (Figure 6c) in the predicted Micromodelling is in the region of 80 percent through the carbonate formations and is extremely low (less than 20 percent) in the region of the coal-rich Talang Akar shales.

The Micromodelling porosity is displayed for the region of line adjacent to the well ZU-11 in Figure 7. This display is in the form of a color plot, termed a ''Reservoir parameter distribution'' (RPD) display. Porosity within the upper Batu Raja varies widely from less than 5 to 36 percent. A number of values in excess of 36 percent have been predicted by the calibration. These values may be realistic where they occur within a definable porosity interval (SP 345–895 ms) or unrealistic where they occur at the top of the

Batu Raja. In the latter case, it is likely that smectite rich caprock may have been included in the calculation. The effective porosity derived from well data has been inserted at the well location in Figure 7.

The porosity of the lower Batu Raja is generally low and does not exceed 20 percent. The coals of the Talang Akar clearly are identifiable by their high apparent porosity, while the underlying sands vary in porosity from 5 to 36 percent.

The associated level of confidence in the RPD section shows the predicted porosity to be reliable for the Batu Raja and low for the coal-rich Talang Akar (Figure 8). In general, the level of confidence in the predicted porosity of the upper Batu Raja is greater than 70 percent, based upon a margin of error of ± 3 percent. Such high levels of confidence can be achieved only when acoustic impedance and effective porosity are closely related and the quality of the seismic data is good.

The Micromodelling porosity was converted to the thickness domain so that the cumulative pore space (integrated porosity) within the gas, oil, or brine intervals could be calculated for the Upper Batu Raja. An example of the cumulative pore space for the three fluid intervals is shown for the entire line 4390-SD-86 (Figure 9).

Maps were produced showing the integrated porosity (volume/unit area) through the gas and oil intervals for the Upper Batu Raja (Figures 10 and 11). These maps provide an assessment of the potential reservoir resources with extremely good lateral (spatial) resolu-

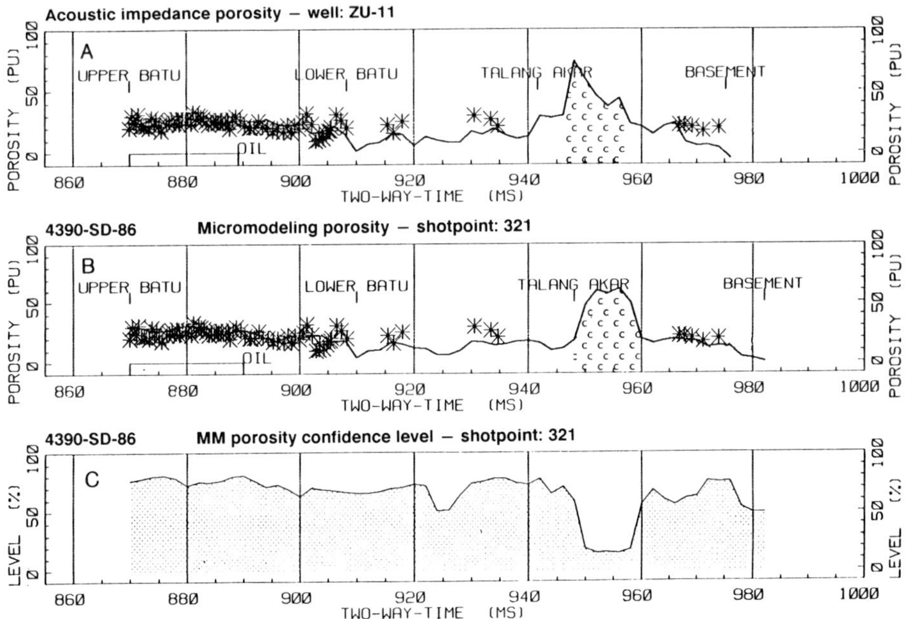

Fig. 6. Well porosity A, Micromodelling porosity B, and Micromodelling porosity confidence level C. Asterisks indicate core porosities. The C's in the Talang Akar formation indicate probable coal.

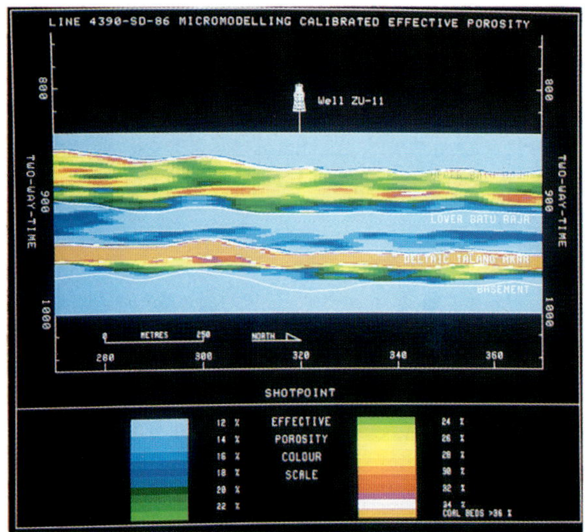

Fig. 7. Reservoir porosity distribution with inserted well porosity.

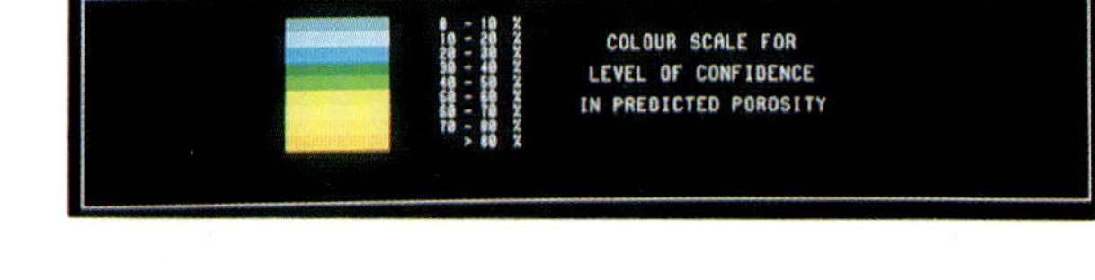

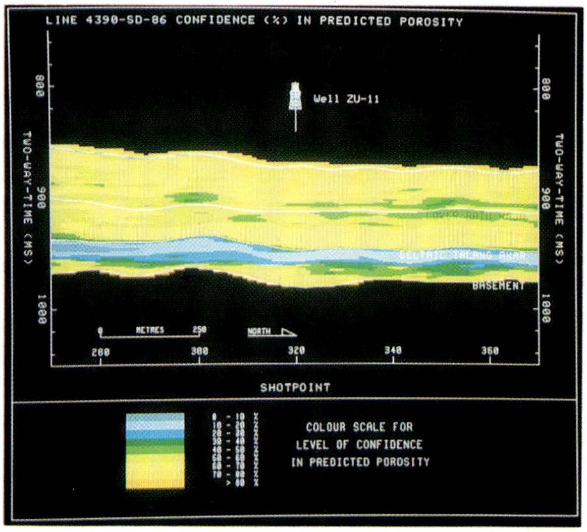

Fig. 8. Level of confidence in reservoir porosity prediction.

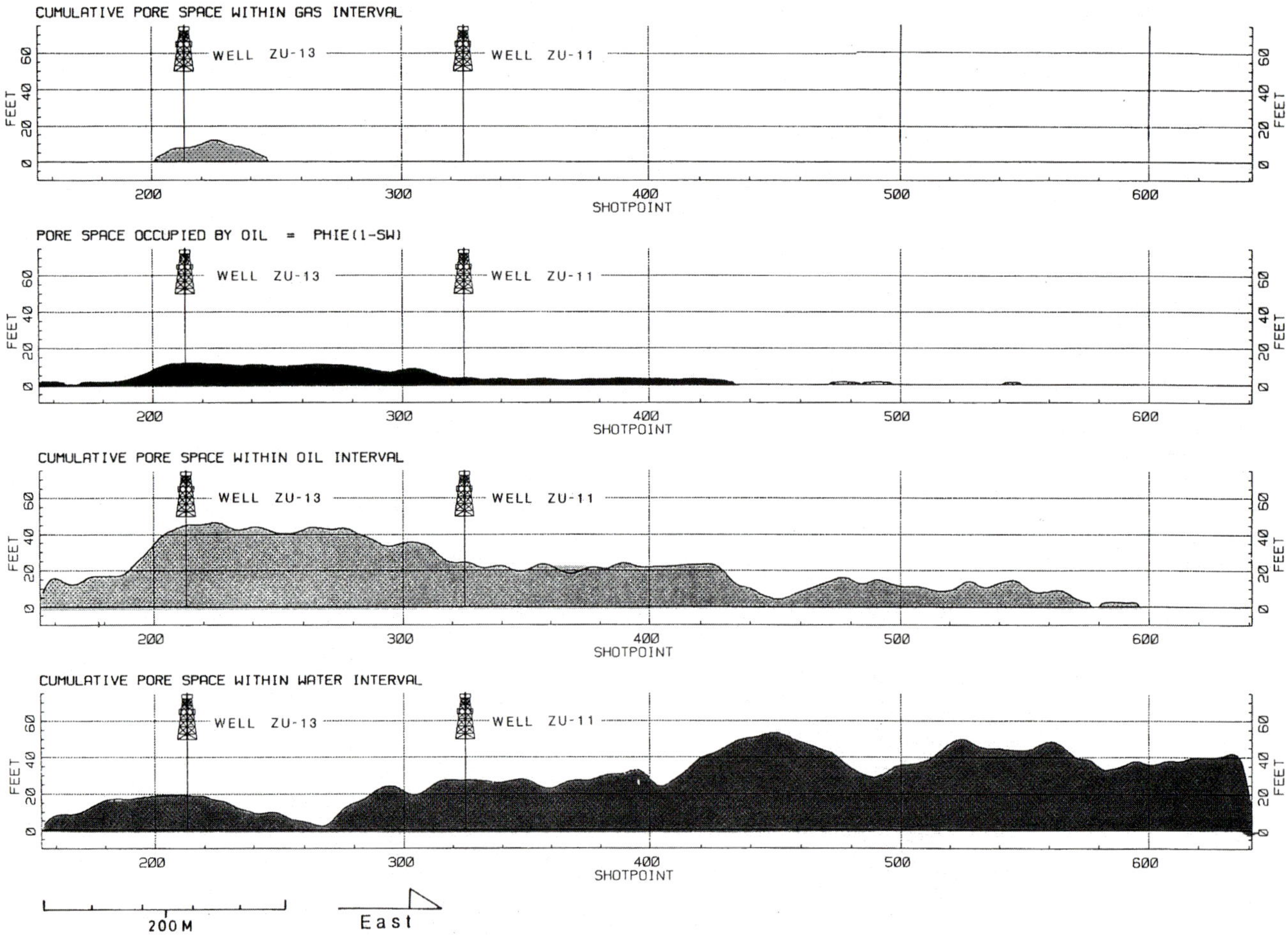

Fig. 9. Cumulative pore space within Upper Batu Raja reservoir for gas, oil, and brine intervals.

tion. In comparison a map showing the pore volume in the Upper Batu Raja generated from the available wells is insufficient to characterize reservoir heterogeneities (Figure 12). The maps were generated using the porosity derived from the Micromodelling of all the lines shown in Figure 2. The map showing the distribution of cumulative pore space within the oil column clearly is related to the structure (Figure 11). The distinct northwest trend shown corresponds to the structural high of the limestone build-up. The dark blue defines the edge of the OWC. The highest values occur within the central and southern portion of the field where values of up to 50 ft are obtained. It should be emphasized, however, that this map does not show the net oil distribution, but rather the pore space within the oil interval. Net oil is difficult to obtain because the water saturation (SW) can only be estimated or inferred and cannot be measured from seismic data alone.

The cumulative pore space within the gas column is considerably lower than that for the oil (Figure 10). This is not due to a lower porosity but rather to a much thinner reservoir interval. Again, the dark blue defines the edge of the gas cap and the maximum pore space achieved is in the central part of the field where the structure is at its shallowest depth.

Conclusions

Mapping petrophysically calibrated seismic data provides detailed quantitative reservoir information that cannot be obtained from well data alone. This method depends strongly upon an integrated multidisciplinary approach with careful attention to detail throughout all stages in the calibrated processing sequence. The results of the calibration are quite valuable in reservoir evaluation, not only because they provide porosity and fluid and thickness distribution maps but also because the statistical reliability of the method is obtained. As a result, cumulative pore space maps of the oil and gas columns are ideally suited for estimating in-place reserves and can be used as an integral part of development planning.

Postscript

This article is based on a paper given at the Indonesian Petroleum Association in October 1988. Since

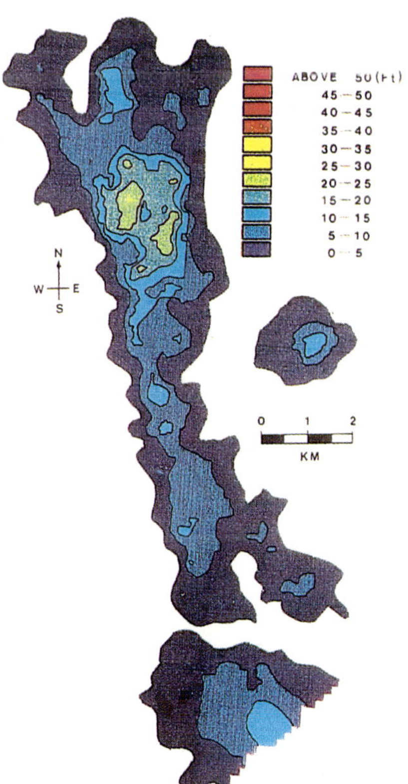

Fig. 10. Volume of pore space for gas column within the Upper Batu Raja, Bima Field.

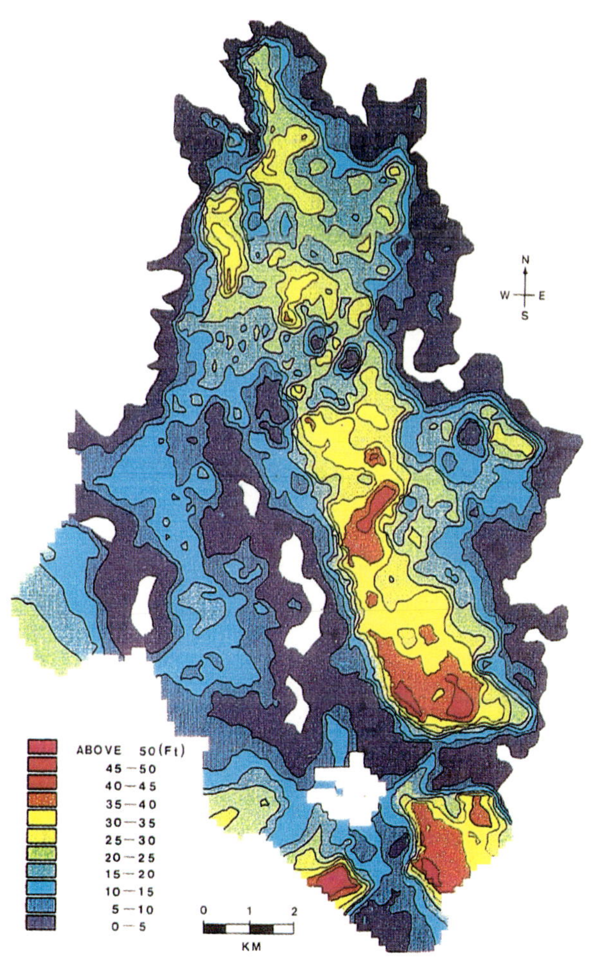

Fig. 11. Volume of pore space for oil column within the Upper Batu Raja, Bima Field.

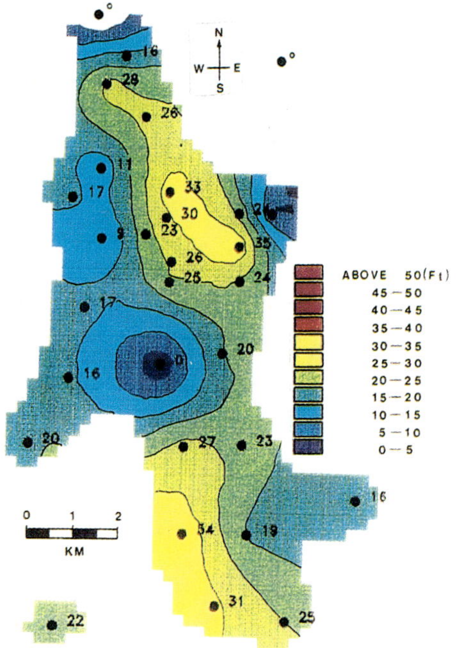

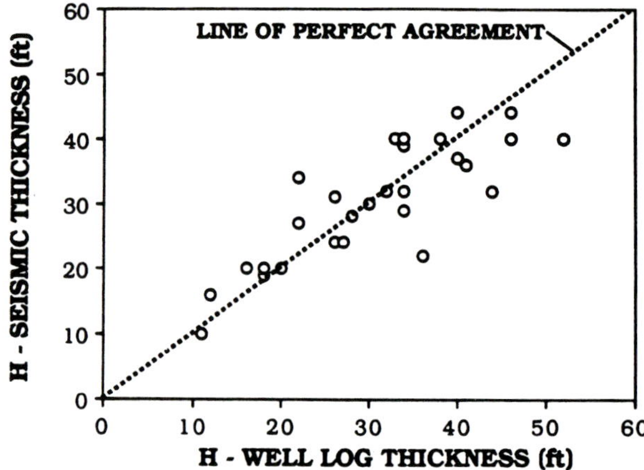

Fig. 14. Seismic-derived thickness for the hydrocarbon bearing interval in the Upper Batu Raja compared to well values.

Fig. 12. Oil volume per unit area, Bima Field.

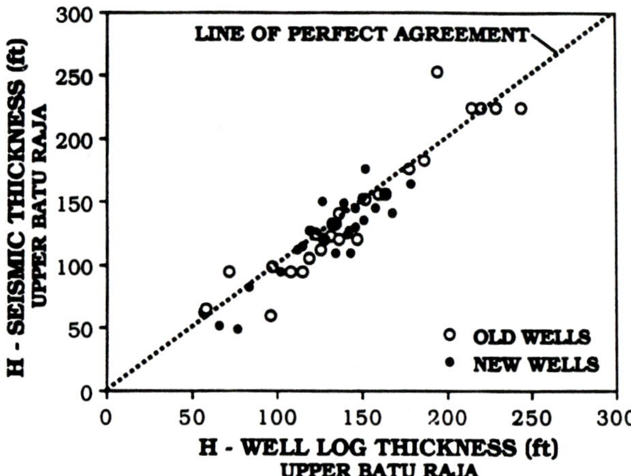

Fig. 13. Seismic-derived thickness for the Upper Batu Raja compared to well values.

was excellent as shown by the crossplot of Figure 13. Excellent agreement was also obtained (better than 12 ft) from a comparison of seismic and log thickness of the hydrocarbon-saturated layer in the Upper Batu Raja (Figure 14). The results described have now been successfully used by ARCO to determine the buildup edge/reservoir limits, basement topography, lithology distribution, hydrocarbon saturated layer thickness, and aquifer thickness associated with the Bima Field.

Acknowledgments

The authors are grateful to the management of Atlantic Richfield Indonesia Inc., Pertamina, and P.T.Elnusa for permission to publish this paper. Special thanks are also expressed to all those who participated in the preparation of the manuscript and to the technical staff at EnTec for their valuable contribution.

References

EnTec Energy Consultants Ltd., 1984, Bima Field reservoir delineation study, Line 3386-SD83, Vol. 1.

EnTec Energy Consultants Ltd., 1986, Bima Field integrated reservoir delineation study, Part 2, Vols. 1 and 2.

EnTec Energy Consultants Ltd., 1987, Bima Field reservoir delineation study, Vols. 1 to 4.

Weathers, L. R. and Helm, R. B., 1984, Stratigraphic delineation of the Baong Sandstone by high-resolution acoustic inversion in the North Sumatra Basin: Proceedings Indonesian Petr. Assn. 13th Annual Convention.

Woodling, G. S., Kaldi, J. G., Oentarisih, K. T., and Roe, R. C., 1990, Integrated reservoir simulation study of the Bima Field, Offshore NW Java: Proceedings Indonesian Petr. Assn. 19th Annual Convention.

then a detailed study has been performed by ARCO to optimize development and production strategies for the Bima Field (Woodling et al., 1990). As part of this study the accuracy of the thickness and pore volume maps for the Upper Batu Raja were tested by comparing them with the results of new wells. The agreement

Seismic Discrimination of Lithology and Porosity, A Monte Carlo Approach

P. M. Doyen and T. M. Guidish[‡]*

Introduction

In heterogeneous formations, spatial variations of petrophysical properties cannot be inferred reliably from sparsely distributed well data. After careful data processing, lateral variations of seismic amplitudes can be used to help predict interwell changes in reservoir properties, such as porosity and lithology.

One of the main challenges in this seismically based prediction process is the averaged and relatively imprecise nature of the seismic information. First, the acoustic parameters that can be derived from band-limited and noise-contaminated amplitude data are inherently nonunique. Second, changes of acoustic parameters, such as seismic impedances, generally cannot be related to a unique physical cause. Instead, variations of acoustic properties integrate the effects of a number of different geologic variables, such as porosity, lithology, fluid type and saturation, pressure, temperature, and rock microgeometry. In practice, the seismic prediction is restricted to rock properties, such as porosity, lithology, and gas occurrence, that contribute the most to changes in acoustic properties. Even so, in view of the limited information content of seismic data, it is important to quantify the reliability of seismically derived reservoir models and evaluate how modeling errors affect predictions of reserves and production.

Conventional methods for estimating reservoir properties, such as porosity, from seismic data rely on empirical or regression formulas that are constructed, for example, by crossplotting seismically derived impedances against porosity measurements in wells. Such approaches treat the data as spatially independent observations and ignore the existence of spatial patterns in the variations of rock properties. Moreover, the reliability of the resulting reservoir models is rarely assessed.

Here we present a two-step Monte Carlo procedure for predicting the spatial variations of lithology and porosity from seismic and log data. This technique,

based on the work presented in Journel and Alabert (1990) and Doyen et al. (1988) accounts for the ambiguous nature of the seismic information. Rather than calculating a unique reservoir model, the Monte Carlo method provides multiple, equally probable models, each of which is consistent with the seismic and log data available. The variations between models reflect the uncertainty in the seismic prediction process and can be used to assess risk in reservoir development.

In the next section, the Monte Carlo procedure for the simulation of lithology is briefly described. A similar method applies for the simulation of porosity. In the third section, the Monte Carlo method is applied in two steps to generate cross-sectional lithology and porosity models of an oil-producing formation along a high-resolution seismic line.

Monte Carlo Simulation of Lithology

Figure 1 illustrates the lithology modeling process, which combines seismic and well data. The figure depicts two wells intersecting a seismic section, which is represented by a regular grid of acoustic impedance (Z) observations. In practice, the acoustic impedances are obtained from the inversion of an amplitude stack section, as explained in the next section. At each pixel $\bar{x} = (x, z)$ of the seismic grid, a lithologic indicator variable B is defined by

$$B(\bar{x}) \begin{cases} = 1 \text{ if } \bar{x} \text{ is in sand} \\ = 0 \text{ if } \bar{x} \text{ is in shale.} \end{cases} \qquad (1)$$

This unknown lithology indicator must be inferred at each point $\bar{x}$ from observations of the variable B in nearby wells and from the knowledge of the local seismic impedance value, $Z(\bar{x})$, at that point. The advantage of using acoustic impedances in inferring lithology is that the seismic data provide spatially dense information between boreholes. However, the seismic information is limited in that, in general, acoustic impedance is not uniquely related to lithology. That is, the impedance ranges for sands and shales often overlap. For instance, depending on the presence of gas, pore pressure, age, mineralogy, and porosity, reservoir sands can exhibit higher or lower impedances than those of adjacent shales. Here, we do

*Western Geophysical, a Division of Western Atlas International, 455 London Road, Isleworth, Middlesex TW7 5AB.
‡Western Atlas Software, a Division of Atlas International, 10205 Westheimer Road, Houston, TX 77042.

not assume that there is a unique correspondence between B and Z. Instead, we model their dependence statistically using a spatial crosscorrelation function.

Monte Carlo lithology simulation is based on estimating at each pixel of the model the probability p that the pixel corresponds to a sand. This probability, calculated from the well and seismic information, is called a conditional probability. It is defined at each pixel $\bar{x}$ of the cross-section as:

$$p(\bar{x}) = \text{Prob } \{B(\bar{x}) = 1 | B(\bar{x}_1), \cdots, B(\bar{x}_n), Z(\bar{x})\}, \tag{2}$$

where, in the conditioning data set, $B(\bar{x}_1),\ldots,B(\bar{x}_n)$ represent n binary-valued observations of the lithology at spatial locations $\bar{x}_1,\ldots,\bar{x}_n$ in the wells, and $Z(\bar{x})$ is the seismic impedance at $\bar{x}$ (see Figure 1). The probability is calculated at each pixel using a Linear-Mean-Square (LMS) estimation procedure. The LMS estimate, $p_{\text{est}}(\bar{x})$, is given by

$$p_{\text{est}}(\bar{x}) = \sum_{i=1}^{n} \omega_i(\bar{x})B(\bar{x}_i) + \alpha(\bar{x})Z(\bar{x}) + c(\bar{x}), \tag{3}$$

where weights $\omega_1,\ldots,\omega_n$ and α assigned to the data are determined at each pixel $\bar{x}$ by minimizing a mean square prediction error criterion. This minimization only requires knowledge of the spatial autocorrelation and crosscorrelation structures of the variables B and Z. It is performed by solving a system of normal equations analogous to a cokriging system (Journel and Huijbregts, 1978; Doyen, 1988). In equation (3), the constant, $c(\bar{x})$, is determined from the condition that the average proportions of sand and shale in the simulated models be equal to those observed in the wells. Note that the relative magnitude of weight α in

equation (3) depends on the degree of local crosscorrelation existing between seismic impedance and lithology; i.e., the stronger the correlation, the larger the weight value. In practice, this crosscorrelation is determined by comparing the lithology interpreted in wells with impedance data derived from seismic traces that are in the direct vicinity of the wells. The probability estimate p_{est} mixes binary-valued observations with impedance data, corresponding to values outside the interval [0,1]. In practice, before performing the LMS estimation of p_{est}, the Z data are transformed so that the new values lie between 0 and 1.

The simulation method involves the sequential estimation of the conditional probability p_{est} at all pixels of the model, or equivalently, at all sample points of the seismic impedance profile. The algorithm for generating one sand/shale model can be summarized as follows.

For all pixels $\bar{x}$ in the cross-section:

1. Obtain the LMS estimate
   $p_{\text{est}}(\bar{x}) = \text{Prob}_{\text{est}} \{B(\bar{x}) = 1 | \text{data}\}$.
2. At $\bar{x}$, draw a simulated value B_s that is equal to 1 or 0 with probability $p_{\text{est}}(\bar{x})$ and $1 - p_{\text{est}}(\bar{x})$, respectively, and
3. Add the simulated value $B_s(\bar{x})$ to the conditioning data set of the probability in Step 1 above.

Note that, when the first location $\bar{x}$ is considered in the simulation grid, the probability p_{est} is calculated only from the observations of B in wells. However, when the ith pixel is selected, the probability estimate is calculated from the $(i - 1)$ previously simulated values, in addition to the well observations. The simulation process stops when all pixels of the model have been assigned a sand or shale value. The lithologic model constructed using this process is not unique. In fact, repeated application of the algorithm yields different models by varying the order in which the model pixels are visited and the sampling of the sand and shale probability distribution. It can be shown that the models simulated using the above algorithm all have the following properties:

The models reproduce the vertical sand and shale sequences observed in wells.

The models reproduce the average sand and shale proportions, or equivalently the average net-to-gross ratio inferred from the well data.

The areal extent and thickness of the simulated sand and shale units are constrained by the spatial autocorrelation structure of the lithologic data.

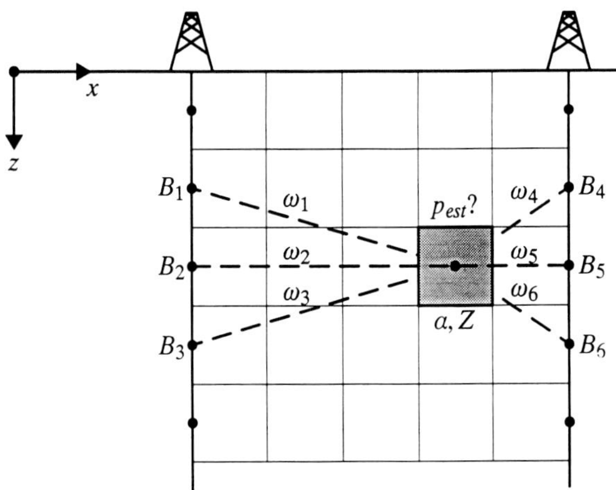

Fig. 1. The sand probability, p_{est}, is calculated from the seismic impedance, Z, and from the lithologic data in wells. ω_i and a are statistically determined weights.

The models are consistent with the seismic data in that they agree with the crosscorrelation between lithology and acoustic impedance.

Finally, note that a similar Monte Carlo procedure can also be applied to generate porosity simulations.

Case Study of a Canadian Reservoir

The Monte Carlo technique was applied to obtain cross-sectional lithology and porosity models of the Mannville Formation along the high-resolution seismic line displayed in Figure 2. The line intersects four closely spaced oil-producing wells numbered 1 to 4. A full suite of interpreted logs was available at each of the wells. The target formation is located between approximately 450 and 550 ms two-way traveltime. This time interval corresponds to depths ranging from 480 to 650 m. Note the high temporal resolution of the seismic data: the signal bandwidth extends from about 20 Hz to more than 200 Hz. This broad bandwidth was achieved by using specialized acquisition and processing techniques similar to those discussed in Pullin et al. (1986). Note also the complex lateral changes of seismic amplitude in the target zone. These changes reflect the variability of the lithology and porosity in the Mannville Formation. The purpose of the Monte Carlo modeling was to delineate lateral lithofacies and porosity changes that could affect the performance of oil recovery operations. In view of the high frequency content of the data, we were interested in obtaining detailed reservoir models with a vertical resolution on the order of 5 m, or approximately one-quarter of the dominant seismic wavelength.

Figure 3 shows a flowchart of the reservoir modeling process, starting with the processed seismic amplitude data and the interpreted well logs. First, the seismic amplitude data were converted to acoustic impedance estimates by applying a recursive inversion process similar to that described in Lindseth (1979). Figure 4 shows the estimated impedance section for the target seismic window after conversion from traveltime to depth. A low-frequency trend calculated from the sonic and density logs was added to the seismically derived impedances. Note the fine layering of the impedance model which reflects the high-frequency content of the data and the complex interfingering of the sands and shales in the Mannville Formation. In order to check the quality of the inversion process, we compared inverted seismic impedances and log-derived impedances as a function of depth at the four wells along the seismic line. For well 1, this comparison is shown in Figure 5 with the two impedance curves sampled at a 2 m spacing. Though lacking the fine details apparent on the impedance log, the seismic impedance function closely approximates the log across the depth interval corresponding to the Mannville Formation.

Figure 5 also displays the sand and shale intervals which were interpreted from the gamma ray log at well 1. This well penetrated five sand members, labeled A through E. The five sands are oil saturated with the lower sand being the main producer. Comparing lithology and impedance at well 1, it is apparent that the

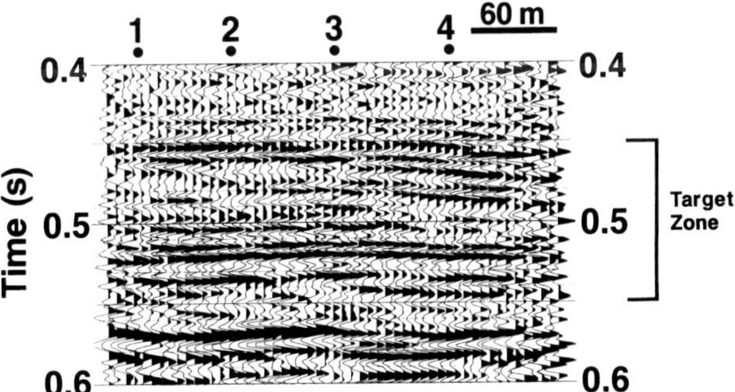

Fig. 2. Relative-amplitude-processed stack section intersected by four oil-producing wells.

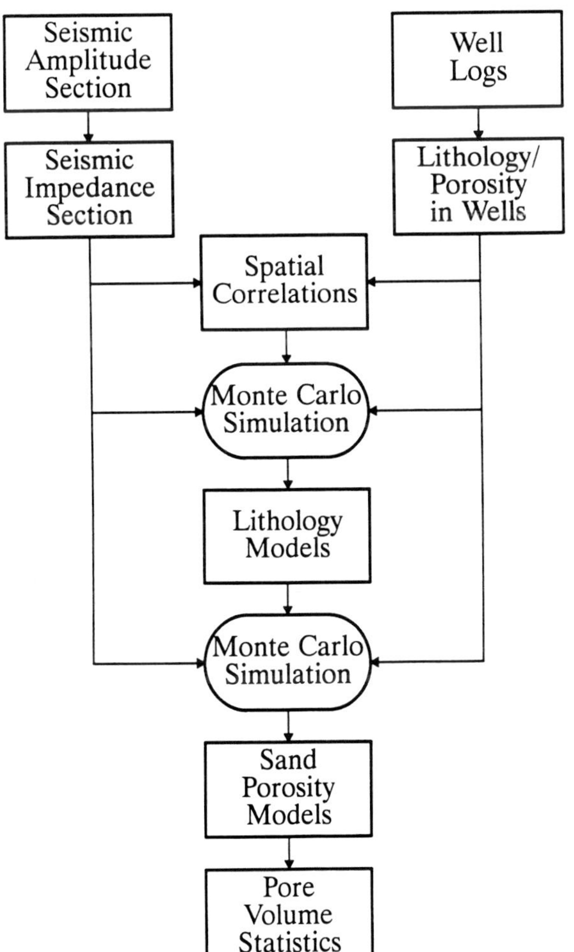

Fig. 3. Flowchart of the reservoir modeling process.

major sands, such as C and E, tend to have lower acoustic impedances than adjacent shales. However, due to the trend of increasing impedance with depth, a deep sand, such as E, may have an impedance as high as a shallower shale, such as the shale between sands B and C. To remove this feature, which would adversely affect the seismic discrimination of lithology, the linear trend depicted in Figure 5 was subtracted from the impedance depth function at each CDP along the seismic line. Figure 6 displays seismic impedance frequency histograms for sands and shales after removal of the linear depth trend and normalization to a reference impedance of 5000 m/s g/cm^3. These histograms were calculated from the four wells along the seismic line. The two histograms show that sands exhibit lower impedance, on average, than shales. However, there is considerable overlap between the impedance ranges of the two lithologies. The histograms were used to estimate the coefficient of crosscorrelation between the variables B and Z. This coefficient is needed in the calculation of the sand probability, p_{est}.

Figure 7 shows four sand and shale cross-sectional models generated during the first simulation step (see Figure 3) from the seismic impedance section and from the binary lithologic data at wells 2 to 4. Well 1 was used as a hidden well where the accuracy of the lithology prediction can be evaluated. Note that the black and yellow bands overlaid at the wells correspond to the log-interpreted sand and shale intervals. The simulations were performed at all sample points of the impedance profile, which is composed of 53 traces, each containing 86 depth samples. The spacings between samples in the horizontal and vertical directions are 6 m and 2 m, respectively. In the Monte Carlo modeling, geometrically anisotropic, exponential cor-

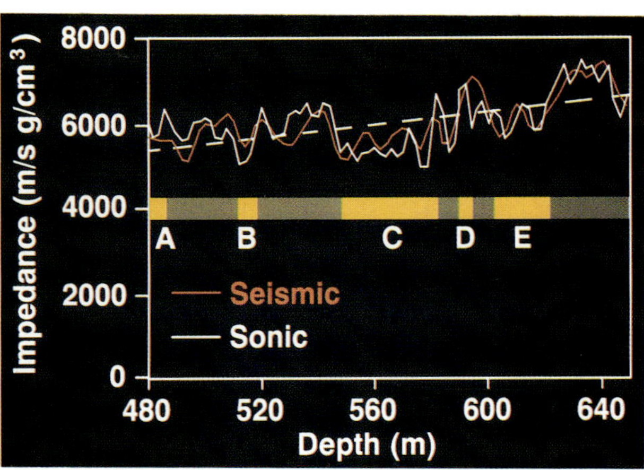

Fig. 5. Comparison of sonic impedance and inverted seismic impedance at well 1. The five sand members penetrated by the well are labeled A through E.

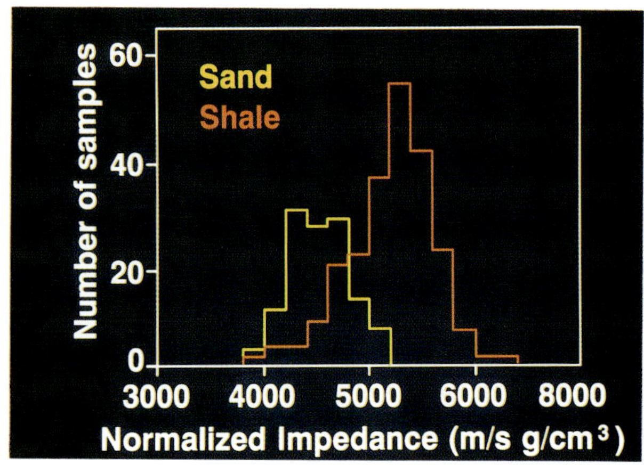

Fig. 6. Normalized impedance histograms for sands and shales.

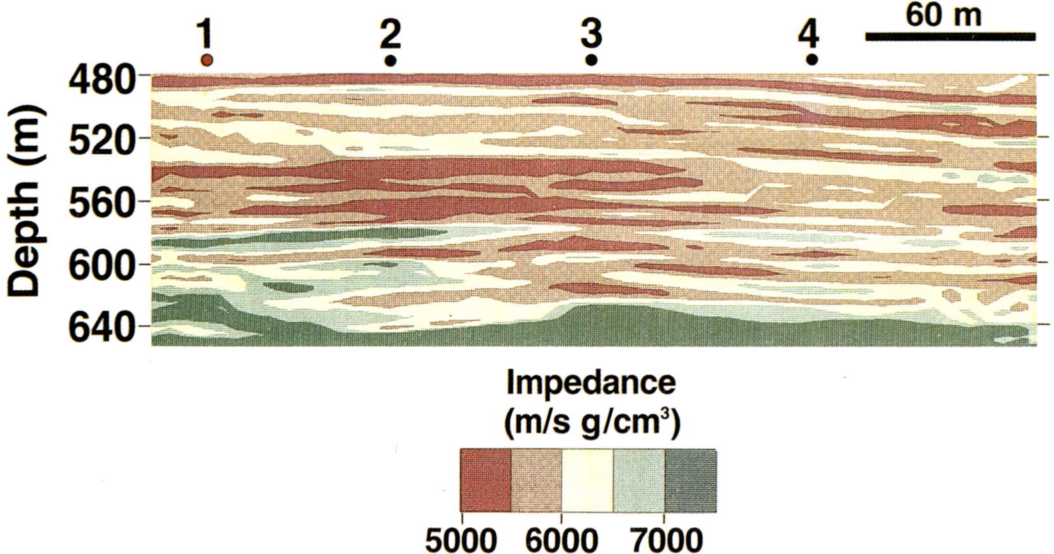

Fig. 4. Depth-converted seismic impedance section of the target zone.

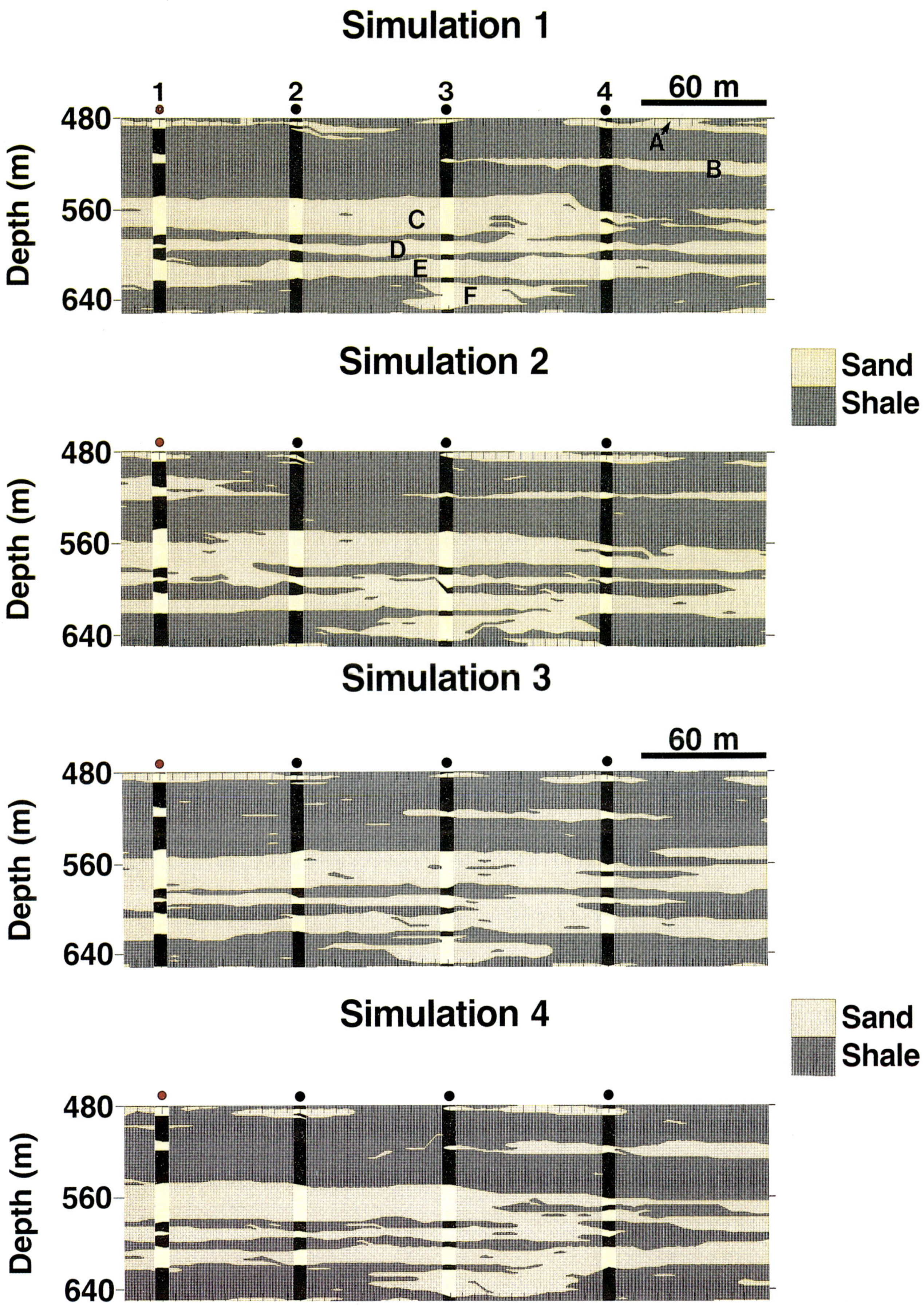

Fig. 7. Four lithologic models. The black and yellow vertical bands at the wells indicate the log-derived lithology. Well 1 is a hidden well in the simulation process.

relation models (Journel and Huijbregts, 1978) were selected for variables B and Z.

As indicated in the flowchart, the modeling of lithology was followed by the simulation of porosity within the sand units. Figure 8 shows four sand and shale cross-sections with overlays of the sand porosity distributions generated during the second simulation step. The sand porosity simulations were also derived by combining the impedance data and porosity observations in wells 2, 3, and 4. The simulated images represent alternative lithology and porosity models that are consistent with the seismic and well-log information available. Note in particular that all the models tie exactly at wells 2, 3, and 4. Differences between simulations reflect the uncertainty in describing interwell reservoir heterogeneities.

Examination of Figures 7 and 8 shows that certain large-scale simulated features consistently appear in all models, while other higher-frequency patterns are more variable from one simulation to another. The variability in the high frequency details of the models reflects the limited model resolution that can be achieved from the data. For example, comparison of the lithology models in Figure 7 shows that the gross morphology of the sand members labeled C through F in simulation 1 is well constrained. In addition, the position of sands C, D, and E are predicted accurately at test well 1. By contrast, variations from simulation to simulation in the geometry of sands A and B indicate that the lateral extent of these thin sands is poorly defined. Similarly, for porosity, comparison of the models displayed in Figure 8 shows that the gross contours of the high and low porosity zones are resolved, while the exact position and details of the porosity highs and lows are not well defined. In addition to dependence on spatial frequency, the simulation variability is controlled by the quality and proximity of the observations that are used in the modeling. For example, differences between lithology models are greatest in areas of the seismic profile where the impedance values lie in the overlap interval between the impedance ranges of the sand and shale classes. One such area corresponds to the shallow part of well 1 where there is no acoustic contrast between sands A and B and the encasing shales (see Figures 5 and 7). Note that, as expected, the differences between models also increase away from the wells.

In summary, the Monte Carlo approach enables us to explore the range of possible lithology and porosity models that agree with the information at hand, and therefore assess the uncertainty in the reservoir description process. In addition, compared to smoothly interpolated reservoir models, simulated models, with their high-frequency details, better mimic the variability existing in real reservoirs. This feature is particularly important in reservoir flow simulation studies

since the flow behavior may be controlled by the detailed interconnectivity of the sand compartments and by small scale heterogeneities within the sands. To contrast the properties of simulation and smooth estimation, Figure 9 compares two simulations of porosity as a function of depth in hidden well 1 with a smooth porosity curve estimated from the same log and seismic data using the cokriging technique (Journel and Huijbregts, 1978; Doyen, 1988). Both simulated and cokriged predictions are displayed with the porosity curve derived from the logs. Note that, in this example, the porosity calculation was performed in both the sands and shales. Although the cokriging prediction is the most accurate in a mean square sense, it tends to overestimate the lows and underestimate the highs of the porosity log. In contrast, the simulated curves are less accurate, on average, but they better emulate the vertical variability, in particular the sharp transitions, observed on the log curve. This comparison between simulation and cokriging illustrates the trade-off that exists between accuracy and resolution in reservoir modeling.

Monte Carlo reservoir models can also be used to calculate risk-qualified estimates of hydrocarbon reserves. In this study, we performed two runs of 100 simulations, conditioned by seismic and well data. From each joint simulation of lithology and porosity, we calculated an estimate of the total pore volume—or more precisely pore area, since the models are 2-D—within the sands by integrating the porosity values across the 2-D model cross-section. Figure 10 shows histograms of pore area predictions obtained by pooling together the predictions from each set of 100 simulations. In the first set, wells 1 and 2 were not used in conditioning the simulations. In the second set, only well 1 was a hidden well. Both sets of simulations used the seismic impedance profile to constrain the modeling. Comparison of the histograms shows that the pore area prediction uncertainty, as measured by the spread of the distributions, decreases, as expected, when the number of conditioning wells is increased.

Conclusion

We have introduced a two-step Monte Carlo procedure for modeling reservoir geometry and porosity by rigorously combining seismic impedance and log data. In the first step, the spatial arrangement of the producing and nonproducing facies, sand and shale in this case, is simulated. This step is followed by the simulation of the lateral distribution of porosity within the sand units. The simulation procedure yields a family of equally probable reservoir models, which are conditioned by both the seismic and well data, and by the autocorrelation and crosscorrelation structures of the data. Compared with well-log-derived models, the seismically conditioned simulations are better constrained spatially and provide more detailed and reliable images of interwell reservoir

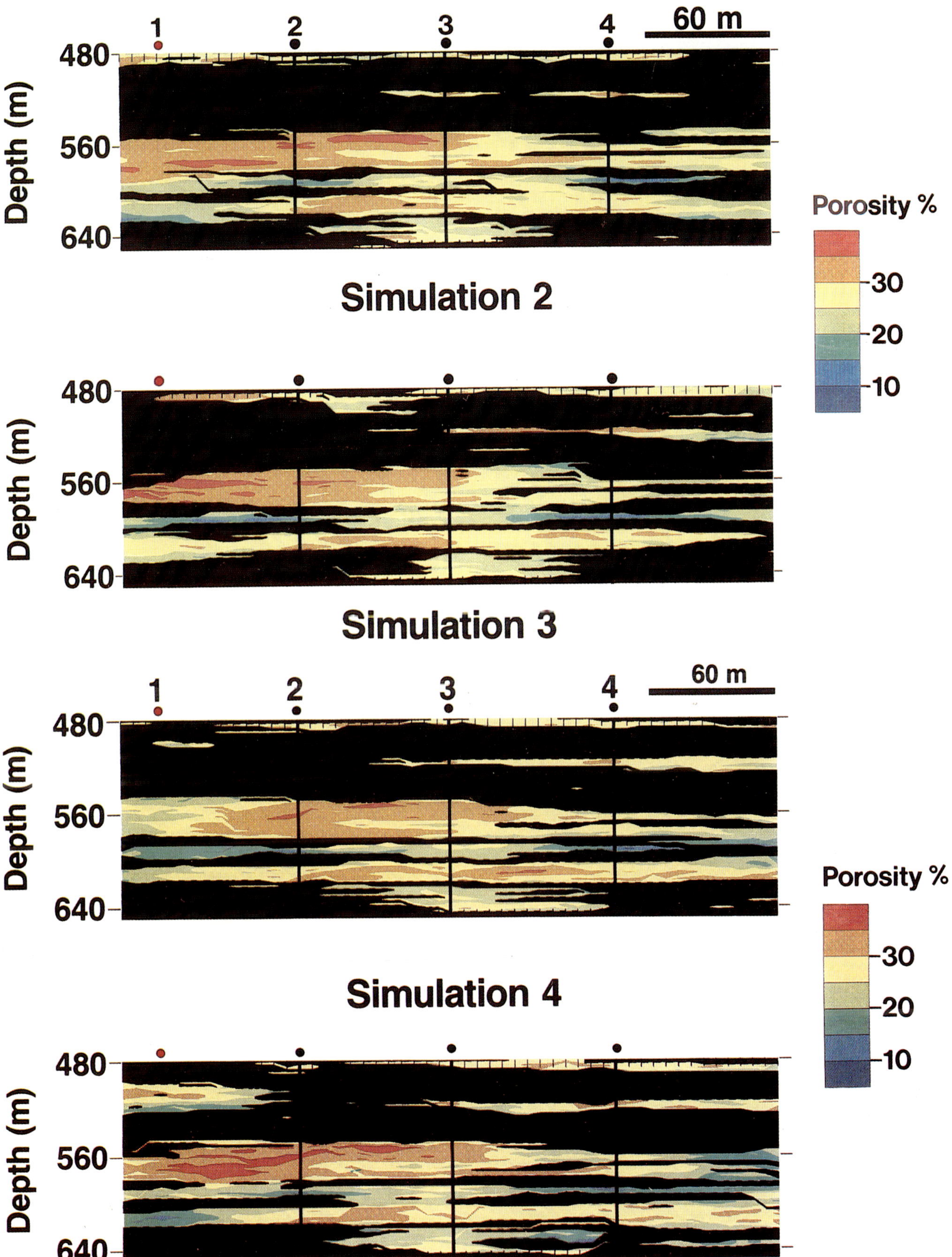

Fig. 8. Four lithology and sand porosity models. Well 1 is a hidden well in the simulation process.

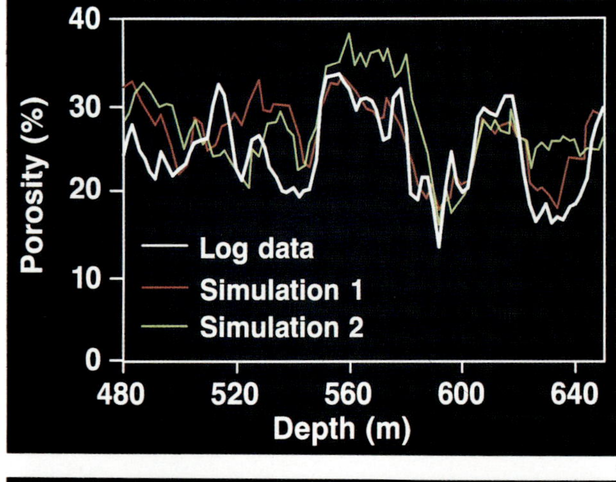

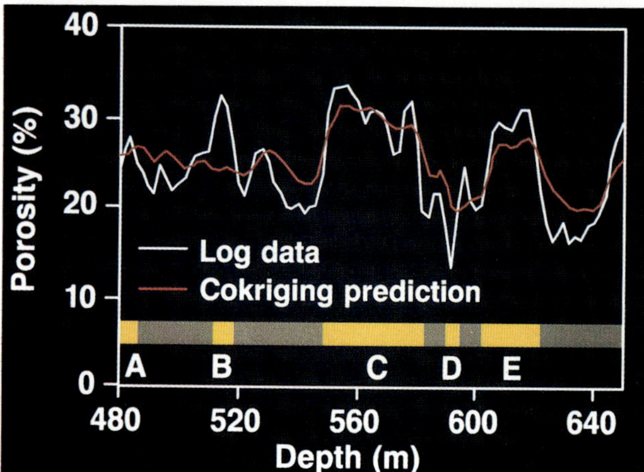

Fig. 9. Cokriged and simulated porosity predictions compared with porosity log data at hidden well 1.

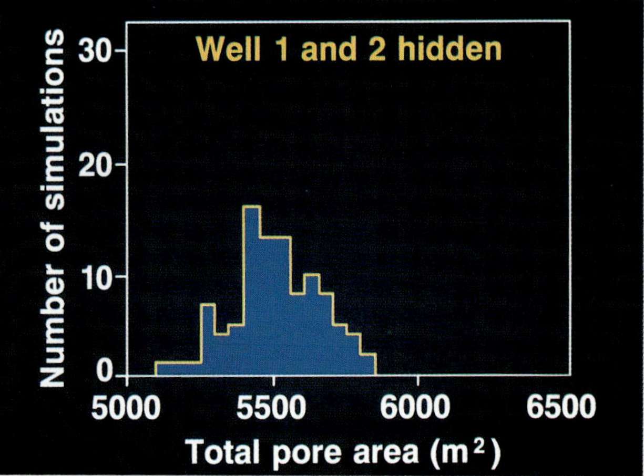

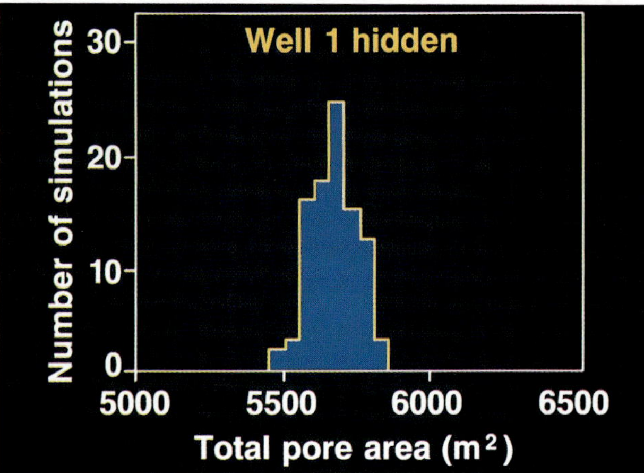

Fig. 10. Histograms of sand pore area predictions calculated from two sets of 100 simulations.

heterogeneities. However, in keeping with the band-limited and ambiguous nature of the impedance data, the seismically consistent simulations still differ in the connectivity of the sand units and in the details of the porosity distribution. Analyzing the variability among successive simulations is useful for assessing the uncertainty associated with the subsurface models. In particular, error-qualified estimates of the reservoir pore volume are obtained from frequency histograms derived by combining many simulations.

Acknowledgments

We are grateful to Amoco Canada Petroleum Company Ltd. for releasing the seismic and interpreted log data for this publication.

References

Doyen, P. M., Guidish, T. M., and de Buyl, M. H., 1988, Lithology prediction from seismic data, a Monte Carlo approach: Presented at the 58th Ann. Internat. Mtg., Soc. Expl. Geophys.

Doyen, P. M., 1988, Porosity from seismic data: A geostatistical approach: Geophysics, **53**, 1263–1295.

Journel, A. G., and Alabert, F. G., 1990, New method for reservoir mapping: J. Petr. Tech., **42**, No. 2, 212–218.

Journel, A. G., and Huijbregts, C. J., 1978, Mining geostatistics, Academic Press.

Lindseth, R. O., 1979, Synthetic sonic logs—a process for stratigraphic interpretation: Geophysics, **44**, 3–26.

Pullin, N. E., Matthews, L. W., and Hirsche, W. K., 1986, Techniques applied to obtain very high resolution 3-D seismic imaging at an Athabasca Tar Sands thermal pilot: Presented at the 56th Ann. Internat. Mtg., Soc. Expl. Geophys.

Inversion With a priori Information: An Approach to Integrated Stratigraphic Interpretation

J. Brac,* P. Y. Déquirez,* F. Hervé,* C. Jacques,* P. Lailly,* V. Richard,* and D. T. van Nhieu*

Introduction

A detailed interpretation involving lithologic modeling and porosity mapping is necessary for an accurate assessment of reservoirs. In a detailed reservoir study, the interpreter uses a large data set comprised of seismic sections, well logs, and geologic information, with the seismic sections being used to extrapolate information from wells. Stratigraphic deconvolution commonly is used to improve the seismic resolution, which is of paramount importance. This processing technique requires interactive preprocessing to establish a consistency between acoustic impedance logs and the seismic information.

Direct lithologic or stratigraphic modeling is also popular. In this approach, a lithologic model, i.e., an acoustic impedance or porosity cross-section, is obtained by extrapolating and interpolating log information between wells, with the stratigraphy being defined from an interpretation of the seismic section.

These two approaches are complementary, but the interpreter still must answer the following questions:

How should the lateral variations of events on the seismic section be interpreted in terms of lithology or porosity change?

How should the stratigraphic model be corrected so that it matches the seismic section?

The automatic procedure proposed to answer these questions is to construct an acoustic impedance cross-section that fulfills three criteria:

(1) Matches the acoustic impedance logs at different well locations.
(2) Agrees with our knowledge about reservoir stratigraphy, and
(3) Generates a synthetic seismic section that matches the observed seismic section.

*Institut Francais du Pétrole, Box 311, 92506 Rueil-Malmaison, Cedex, France.

The joint fulfillment of these three criteria guarantees the consistency of the interpretation. A similar approach proposed in Gelfand and Larner (1984) uses a lithologic model, guessed by the interpreter, which is improved until that model produces a best fit between synthetic and observed seismic sections. Our method uses a different strategy and has advantages that are emphasized in a case study. The main operations performed in our approach are depicted in Figure 1.

A Classic Stratigraphic Interpretation Problem

We want to determine the geometry of a carbonate reservoir and to access the distribution of porosity within that reservoir. Some wells, 2, 3, and 4 in Figure 2, produce oil, whereas well 1 is dry. Figure 2 shows the stacked seismic section with preserved amplitudes. The reservoir is located at the level defined by the arrow. $S^{obs}(x, t)$ will denote this seismic section [obs stands for observed, t is time, and x defines the common midpoint (CMP) location]. The time sampling interval is $\Delta t = 2$ ms and the CMP spacing is $\Delta x = 20$ m.

Borehole information is available for well 1. Great care has been taken to correct some measurement errors and to make the log information consistent with the seismic information. It is especially important that discontinuities in the impedance log be located at the correct two-way traveltime τ. The acoustic impedance distribution $Z_1{}^{\log}(\tau)$ (Figure 3), obtained in this manner, was used to compute the wavelet $w_1(t)$ from the seismic traces in the vicinity of well 1 [see Richard and Brac (1988) for the method]. Assumptions on the lateral changes of the wavelet allow the construction of a sequence of wavelets $w(x, t)$. In this study we chose $w(x, t) = B(x) w_1(t)$, where $B(x)$ is an estimated scaling factor for each CMP having a magnitude of $0.7 \le B(x) \le 1.2$. This sequence of wavelets is used for the stratigraphic deconvolution of the seismic section Figure 4).

Direct Stratigraphic Modeling: The Stratigraphic Model

We first construct an acoustic impedance cross-section, the "stratigraphic model" denoted $Z^{\text{strat}}(x, t)$, that integrates the logging and stratigraphic information. To fit the log information we set

$$Z^{\text{strat}}(x_i, \tau) = Z_i^{\log}(\tau), \quad i = 1,\ldots,I \qquad (1)$$

where $Z_i^{\log}(\tau)$ is the impedance log versus the two-way traveltime τ for well number i located at $x = x_i$ and I is the total number of wells.

We then formalize the stratigraphic information. The substratum is layered, and the layer geometry is

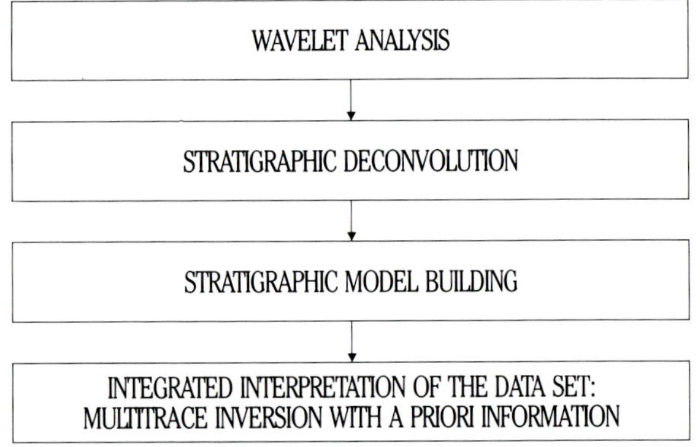

Fig. 1. Flow chart showing the main operations performed during the suggested stratigraphic interpretation procedure.

controlled by the local sedimentation. The substratum is heterogeneous when observed in the direction orthogonal to the layers, but if we follow an event (isochron line), the substratum appears much smoother but is still inhomogeneous. Therefore, isoch-

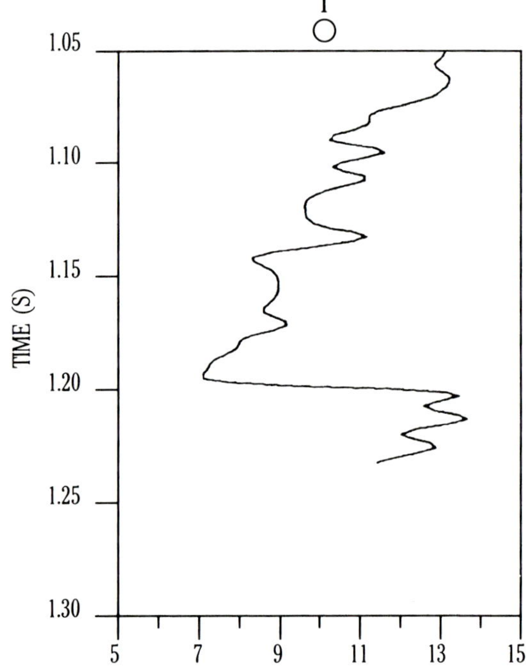

Fig. 3. Acoustic impedance log from borehole 1.

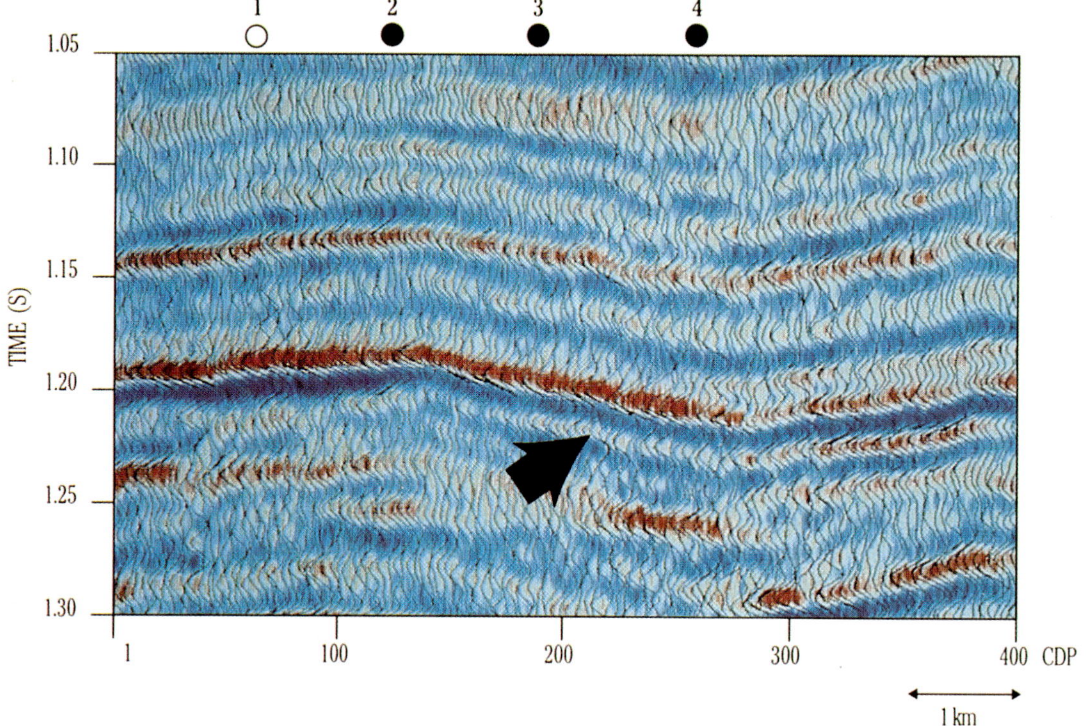

Fig. 2. CMP stacked seismic data. The arrow indicates the carbonate reservoir.

rons define correlation lines. These correlation lines may stop (unconformities) or may be discontinuous (faults). The geometry of any correlation line can be obtained from horizons picked on zero-phase seismic section (Figure 4) if we have defined the types of deposition (onlap, offlap, concordant) between the horizons. In the field data example shown, we used the concordant mode.

The direct stratigraphic modeling technique relies on these stated considerations. The technique enables construction of the stratigraphic model Z^{strat} by linearly varying the values $Z_i^{\log}[\tau(x_i)]$ along the correlation lines $\tau(x)$, respecting equation (1). Note that the stratigraphic model minimizes the objective function,

$$J(Z) = \int_0^X \int_0^T (\nabla Z \cdot \mathbf{1})^2 \, dx \, d\tau \qquad (2)$$

where $\mathbf{1}(x, \tau)$ is the unit vector tangent to the correlation line at (x, τ) and the intervals $[0, X]$ and $[0, T]$ define the domain of interest. In the example under examination, we used just one well (extrapolation mode) and therefore the stratigraphic model is constant along the correlation lines (Figure 5). If we define a modeling operator F, which is a one-dimensional (1-D) seismic trace model $S(x, t)$ without multiples, by

$$F(Z) = S(x, t) - \int R(x, \tau)w(x, t - \tau)d\tau,$$
$$\qquad (3)$$

with

$$R(x, \tau) = \frac{d}{d\tau} [\log Z(x, \tau)], \qquad (4)$$

then a residual section $S^{\mathrm{obs}} - F(Z^{\mathrm{strat}})$ (Figure 6) shows that the stratigraphic model is inconsistent with many seismic events. More precisely, the residual traces have low amplitude in the vicinity of well 1, which validates the consistency of the log, $Z_1^{\log}(\tau)$, and the computed wavelet $w_1(t)$, but the residuals increase substantially away from the well. More importantly, the stratigraphic model deviates from the unknown actual medium within the reservoir.

Integrated Interpretation of the Data Set: The Optimal Model

The interpreter naturally wants the most likely model given the available information. Therefore, we compare the likelihood among all the possible models. We define for the set of all possible impedance cross-sections an a priori probability function that is assumed to be Gaussian with an expectation Z^{strat}, i.e., the stratigraphic model becomes the a priori model. The construction of the covariance operator relies on several assumptions. First, the logs, $Z_i^{\log}$, are assumed to be exact since the interpreter usually puts confidence in the log information after appropriate preprocessing and, therefore, assumes that the errors in the stratigraphic model arise only from errors in the

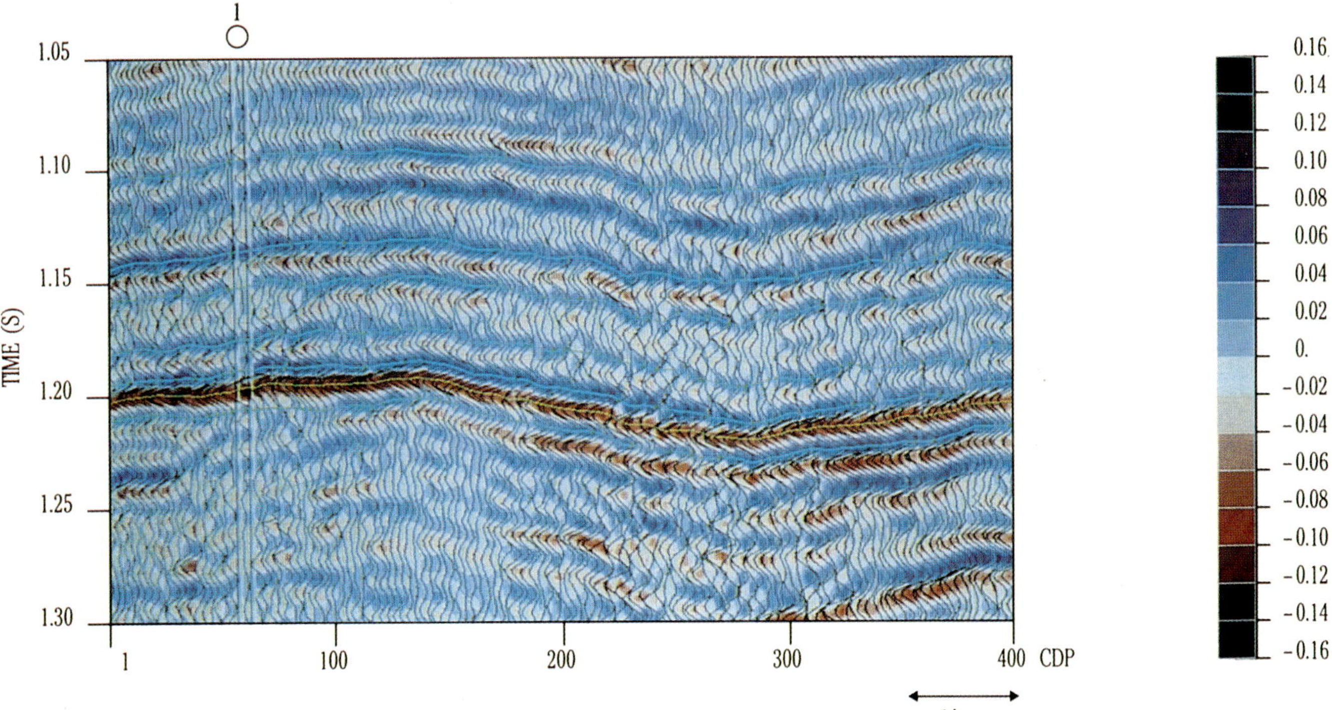

Fig. 4. Filtered reflectivity log from borehole 1 (high cut 110 Hz) inserted into the zero-phase section. The picked seismic time horizons tied to well 1 are used in the direct stratigraphic modeling.

geometry of the correlation lines and the heterogeneous nature of the actual geological formations along these lines. Using formalized stratigraphic rules, the errors in the stratigraphic model are strongly correlated along the stratigraphic isochron lines and are not correlated in directions orthogonal to these isochrons.

Second, the covariance operator is chosen to be exponential (Frankel and Clayton, 1986) along the correlation lines and diagonal in the orthogonal direction. Moreover, the covariance operator must model the conditional probability given condition (1). Therefore, the covariance operator depends on two parameters, a standard deviation $\sigma(x, \tau)$ and a correlation length $\lambda(x, \tau)$, both of which may vary spatially. It is required that λ must go to 0 in the vicinity of a discordance. Taking into account the correlations of the medium on both sides of a fault requires extensions to these simple concepts.

Third, seismic noise is modeled by a Gaussian random function with zero expectation that is not correlated in space but is in time, with the amplitude spectra of the signal and noise being identical. Consequently the covariance operator C_s describing the seismic uncertainties is diagonal in x, and the magnitude of C_s is defined by the noise level which may vary spatially in the (x, t) plane.

Assuming that the interpreter has estimated the seismic noise level as well as the parameters σ and λ, then we can look for the one model, which will be called the optimal model, that maximizes the a posteriori probability function given the observation S^{obs} (maximum likelihood criterion). If we assume that the seismic observation is independent of the a priori knowledge of the impedance model, the optimal model is that single impedance cross-section, among all those that fulfill requirement (1), that minimizes the objective function (Tarantola and Valette, 1982):

$$J(Z) = \|Z - Z^{\text{prior}}\|_Z^2$$
$$+ \{F(Z) - S^{\text{obs}}, C_S^{-1}[F(Z) - S^{\text{obs}}]\} \quad (5)$$

where the comma denotes the L_2 scalar product and

$$\|Z\|_Z^2 = \int_0^X \int_0^T \left[\frac{1}{2\lambda\sigma^2} Z^2 + \frac{\lambda}{2\sigma^2} (\nabla Z \cdot \mathbf{1})^2 \right] dx d\tau.$$
$$(6)$$

Some Constraints of the Formulation

Note that the future model will be (1) the stratigraphic model when we have no confidence in the seismic section, or (2) the classic pseudo-log section (Becquey et al., 1979) when we have no confidence in the stratigraphic model. Therefore, the optimal model is bounded by these two extreme models, which guarantees the robustness of the method.

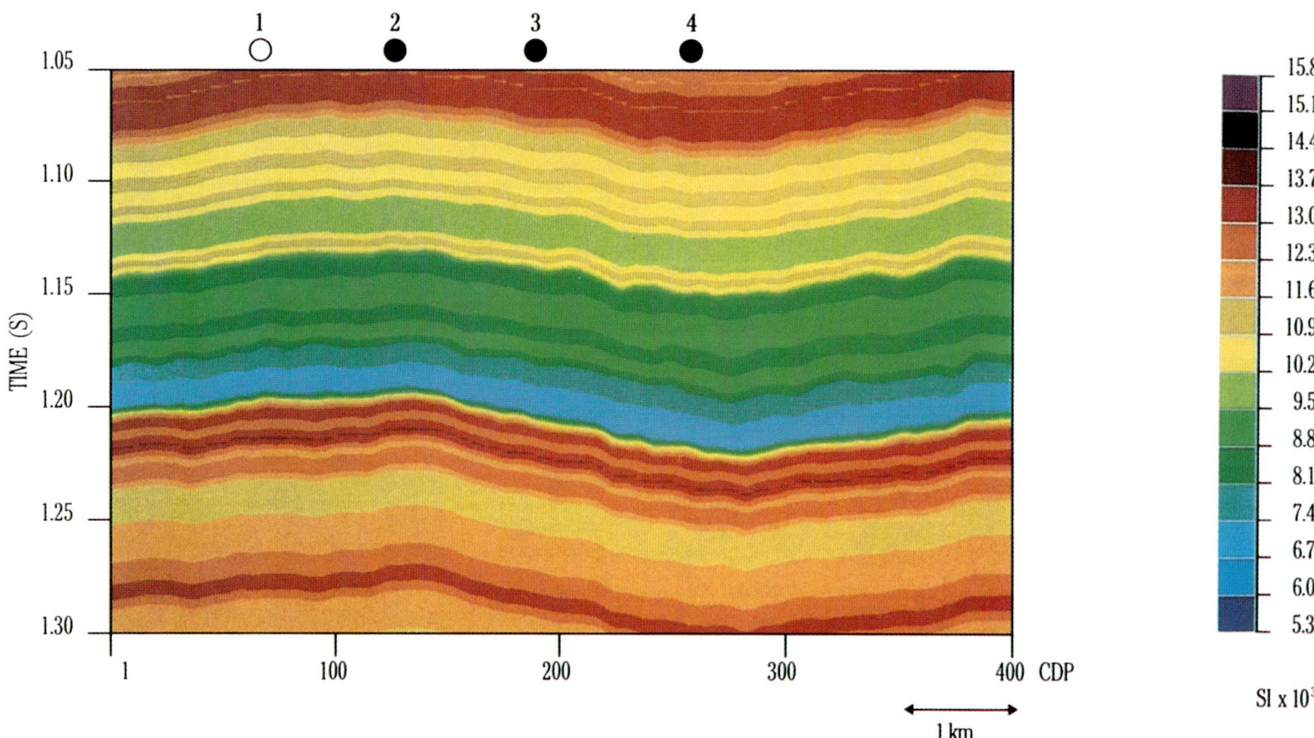

Fig. 5. Impedance cross-section obtained by stratigraphic modeling on the basis of borehole information from well 1 only.

The seismic misfit function [second part of $J(Z)$ in equations] aims at correcting the geometry of the stratigraphic model and at introducing heterogeneities in order to fit the seismic information.

The additional geologic term [first part of $J(Z)$ in equations] allows a stabilization of the ill-posed deconvolution problem. In particular, this term allows low frequencies to be present in the solution. A reliable result cannot be obtained at frequencies beyond the upper seismic frequencies since the seismic misfit function cannot correct the stratigraphic model. Therefore, these high frequencies have been filtered during the construction of impedance logs from borehole data (Figure 5). Moreover this geologic term provides, for sufficient correlation lengths, a laterally structured optimal model which will prevent the synthetic seismic section from fitting the incoherent noise in S^{obs}. In other words, a partial elimination of the noise will be possible with important consequences on the vertical resolution: specifically, reliable information can be extracted from the edges of the seismic frequency band. This improved resolution should in turn improve the lithologic interpretation of the optimal model.

The geologic term couples the impedance traces from one CMP to another. The minimization of $J(Z)$ given by equations is then a two-dimensional (2-D) problem. Thus, an overall inversion of the entire seismic section is achieved rather than a standard trace-by-trace inversion (Oldenburg et al., 1986, for instance). This multitrace approach forces a lateral structuring of the solution model and prevents the solution from following undesired deviations. Moreover, the interpretation of logs between several wells [the constraint of equation (1)] can be handled without difficulty.

Assessment of the Carbonate Reservoir

Figure 7a shows the optimal model obtained from the a priori model and the seismic section previously presented (Figures 5 and 2), with covariance matrices defined by constant parameters, namely $\lambda = 250$ m, a standard deviation σ corresponding to a relative error of 1 percent, and an estimated value of 30 percent for the signal-to-noise (S/N) ratio in the seismic section (Richard and Brac, 1988). Of course the optimal model matches the log information at well 1. Moreover, as a consequence of the strong correlation required ($\lambda = 250$ m), the result is layer structured. The optimal model accounts for the coherent information of the seismic section, and the associated residual section (Figure 8) mainly reveals noise. This improvement of the S/N ratio is the reason for the high resolution obtained in the optimal model. However, we have no information beyond the upper seismic frequencies.

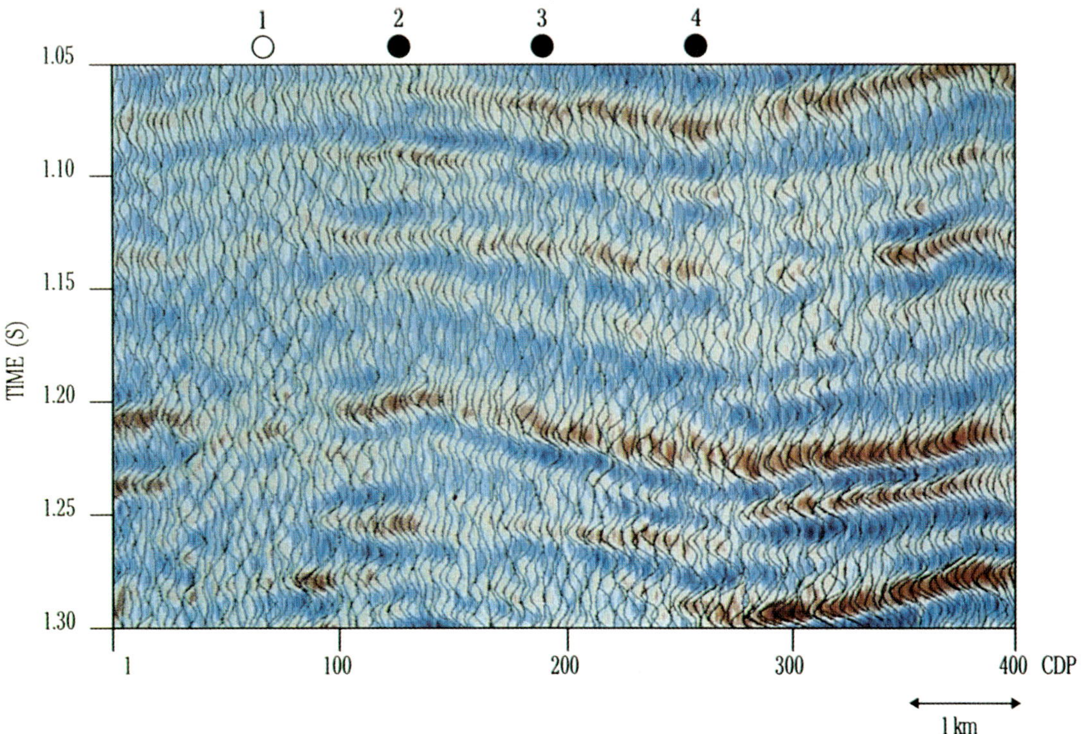

Fig. 6. Residual seismic section associated with the stratigraphic model shown in Figure 5 (observed section—synthetic section).

Figure 9 illustrates the reflection coefficient section derived from the optimal impedance cross-section. There are many more details directly visible than in the trace by trace deconvolved data in Figure 4.

The optimal model can be used to predict logs. Figure 10 shows a satisfactory correlation between the impedance log from borehole 4, after the log was filtered to the seismic frequency band, and the pseudo-log of the optimal model at this location. The local low acoustic impedance zone located at about 1.22 s (arrow in Figure 7a) in well 4 is oil productive with high porosity. From dry well 1 to productive well 4, the optimal model clearly displays impedance variations between 1.21 and 1.23 s, which indicate changes in reservoir thickness and relative quality. The change from nonporous to porous carbonates takes place at about trace number 130. The determination of the geometry of the hydrocarbon reservoirs now becomes easier as a result of better resolution and closer log correlation. Using the impedance distribution given by the optimal model and an experimental relationship between impedance and porosity, delineation of the distribution of porosity in the producing interval (Figure 7b) and estimation of reserves in situ are possible.

Choice of the Parameters of the Covariance Operators

Choosing the covariance parameters λ and σ is the responsibility of the interpreter. Indeed, the interpreter is the only one who can specify the confidence associated with the different data used to describe a reservoir. For instance, the optimal model might be the stratigraphic model in the case of poor seismic data and of a substratum with weak lateral variations. Computer analyses can assist the interpreter in this difficult task, for instance, by displaying random impedance sections associated with a given covariance matrix.

Experiments with synthetic data can often confirm foreseeable effects of the covariance operator parameters on the final solution. Let us just say that we need only rough estimates of these parameters. Moreover, to carry out an a posteriori analysis of the contradic-

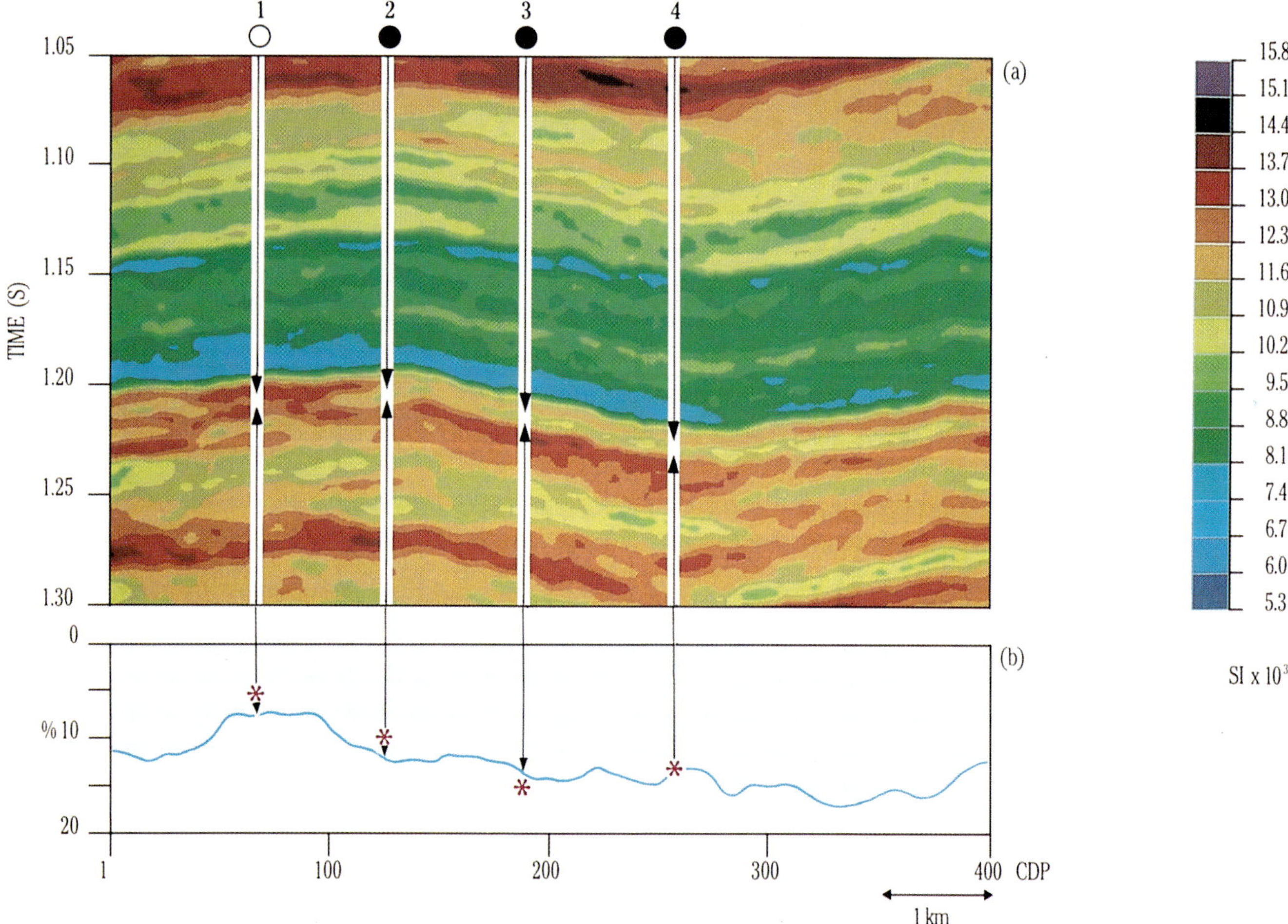

Fig. 7. (a) Optimal impedance cross-section obtained from the integrated interpretation of the data set. (b) Porosities predicted in the reservoir compared to porosities calculated from borehole data (*).

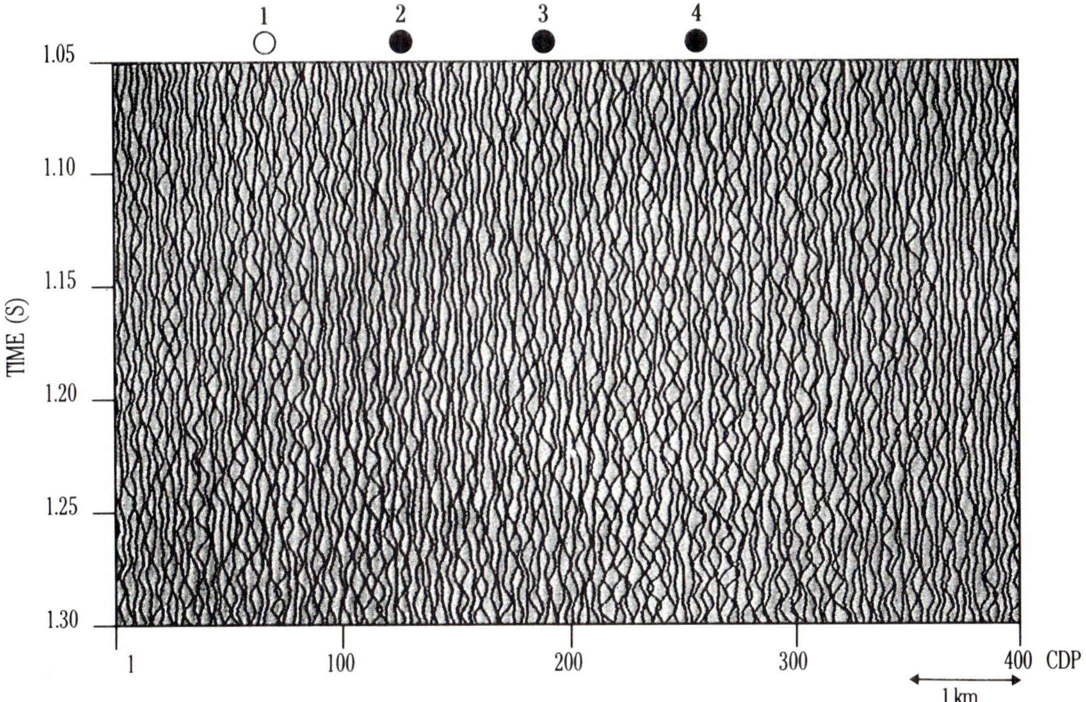

Fig. 8. Residual seismic section associated with the optimal model shown in Figure 7.

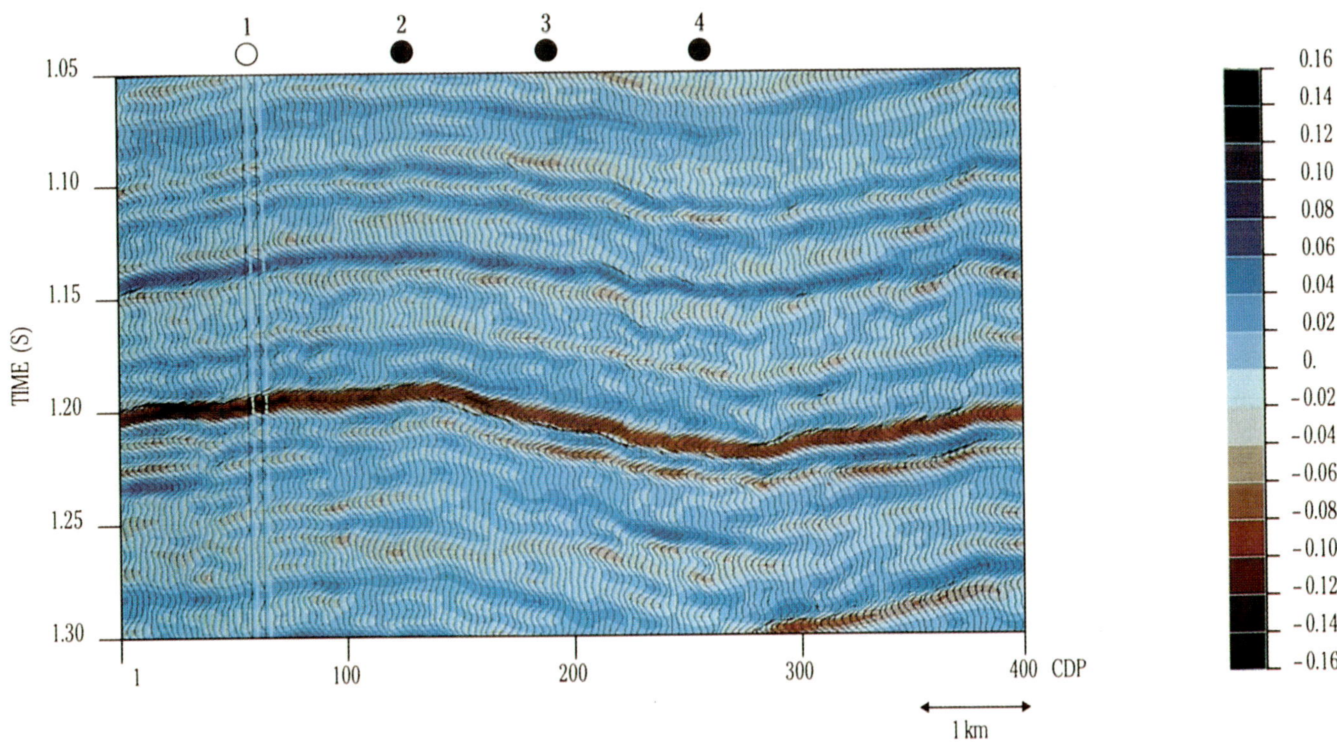

Fig. 9. Reflectivity section associated with the optimal model shown in Figure 7. Such a conversion of the seismic section to a broad-band zero-phase section presents the advantage of improving the resolution without introducing noise.

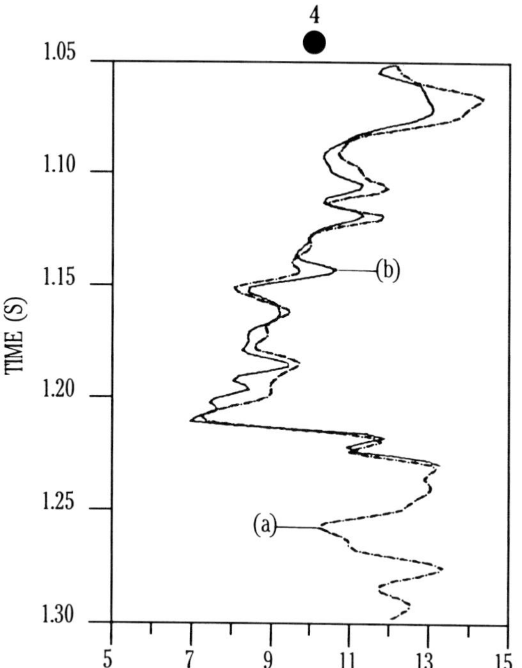

Fig. 10. Comparison between (a) pseudo-log obtained on the basis of the optimal impedance model shown in Figure 7 and (b) acoustic impedance log from borehole 4.

tions that may exist within the reservoir data set and to review the possible reservoir uncertainties that may result from these contradictions is possible. This flexibility enables several reservoir assumptions to be tested. Finally, different choices of these parameters may be equally likely. Thus our approach yields a set of possible optimal models that allow the interpreter to choose from a set of possible reservoir interpretations.

Conclusions

The seismic inversion method proposed in Gelfand and Larner (1984) tests or improves a lithologic inter-

pretation and generally produces several equally plausible lithologic distributions that satisfy the available data constraints. Our approach differs in that we first determine an acoustic impedance cross-section that synthesizes the coherent information within the total data set (well logs, stratigraphy, and seismic sections). This approach allows a consistent and automatic interpretation of the complete seismic section. We next interpret the optimal stratigraphic impedance model in terms of lithology, porosity, etc. The high vertical resolution and the strong data structuring of the optimal model make constructing lithologic models easier. However, since the optimal model does not contain information beyond the upper seismic frequencies, several lithologic models are possible, just as with the Gelfand and Larner approach.

Acknowledgment

We thank Total Compagnie Francaise des Pétroles and especially S. Nicoletis and P. Montaud for providing the data and for permission to publish this paper.

References

Becquey, M., Lavergne, M., and Willm, C., 1979, Acoustic impedance logs computed from seismic traces: Geophysics, **44**, 1485–1501.

Frankel, A., and Clayton, R. W., 1986, Finite difference simulations of seismic scattering: Implications for the propagation of short-period seismic waves in the crust and models of crustal heterogeneity: J. Geophys. Res., **91**, 6465–6489.

Gelfand, V., and Larner, K., 1984, Seismic lithologic modeling: The Leading Edge, **3**, No. 11, 30–35.

Oldenburg, D. W., Levy, S., and Stinson, K., 1986, Inversion of band-limited reflection seismograms: theory and practice: Proc. of Inst. Elect. Electronic Eng., **74**, 487–497.

Richard, V., and Brac, J., 1988, Wavelet analysis using well log information: 58th Ann. Mtg., Soc. Expl. Geophys., Expanded Abstracts, 946–949.

Tarantola, A., and Valette, B., 1982, Inverse problems = Quest for information: J. Geophys., **50**, 159–170.

3-D Seismic and Well Logging Study of a West Siberian Lowland Oil and Gas Field: Application to Petrophysical Characterization[1]

P. Renoux and C. Etienne**

Introduction

The western Siberian basin (Figures 1 and 2) was caused by a Hercynian rift with formation of grabens, connected to the Ural orogeny. During an emergence period at the end of the Triassic, erosion created a hydrographic network. Several transgressions introduced the source rock deposits during the Jurassic. Basin conditions changed to a shelf depositional environment in a regression during the early Cretaceous and during this period, the main reservoir series were deposited. The area studied is located directly above a basement high.

In cooperation with the Ministry of Petroleum in the former Union of Soviet Socialist Republics (USSR), the Compagnie Génerale de Géophysique (CGG) recorded, in 1985, a three-dimensional (3-D) grid of 60 km^2 north of the town of Surgut in Western Siberia. Using information from available well logs, a calibrated stratigraphic interpretation was made to define the petrophysical characteristics of the different reservoirs; this interpretation involved the mapping of structure, porosity, and reservoir quality.

Seismic Acquisition and Processing

The zone studied in the southeastern part of the 3-D grid was 4 km east-to-west and 2.5 km north-to-south. The recording spread (Figures 3 and 4) had five parallel east-west lines 2400 m long, spaced 200 m apart, with 48 geophones at 50 m intervals on each line. A series of 20 shots along a north-to-south line was recorded with individual shots spaced 50 m apart, resulting in a 25 $\times$ 25 m bin. After each shot sequence, the spread and shot line moved 300 m westward and the sequence was repeated. The data set was 20 fold.

Processing was carried out in two steps (Figure 5). The first step resulted in a 3-D x, y, t block and the second in a depth and acoustic impedance block. Two runs of automatic statics were implemented, the first for medium frequencies (medium wavelengths) and the second for high frequencies (short wavelengths). The choice of stacking velocities and mute functions was made with the help of a vertical seismic profile (VSP) as a model.

Well data were evaluated and tied with the seismic data. About 15 wells were drilled in the survey area. The logs recorded in these wells mainly involved resistivity and spontaneous potential (SP) curves. Acoustic and density logs were recorded in only three wells (3500, 530, and 3364), which serve as references for the entire survey. VSPs were recorded in wells 3500 and 3364; quality was excellent, showing high-frequency content, and correlation with the surface seismic data was straightforward. The 3-D block was inverted using one VSP as a model and the second as a control (Figure 6). The bandwidth of the seismic data is practically identical to the VSP bandwidth and zero phase; increasing impedance corresponds to the maximum amplitude of the black peak. Conversion to impedance used complementary data from the sonic logs for the low-frequency range which is not present in the seismic bandpass. Spatial interpolation of this low-frequency data was performed for application to the 3-D grid.

In order to integrate well data and lithologic information, the 3-D block in x, y, t must be converted to x, y, z (depth) as accurately as possible, and an initial conversion was made using the VSP data. The 3-D impedance block was tied to the producing horizons using the vertical wells in the area. The top of the sandstone producing layers, recognizable as an increase in acoustic impedance, was retied precisely with the impedance breaks. Depth correction maps were then produced and a new depth conversion was made, tying the vertical wells. Deviated wells were used to cross-check the accuracy of the depth conversion.

Identification of Horizons

The fact that the structures are continuous was verified using a seismic workstation to reexamine the entire 3-D block. A large number of logs were recorded at well 3500 and these logs were used to tie the

[1]Condensed and edited by R. E. Sheriff.

*Innovation, Research, and Industrialization, Compagnie Génerale de Géophysique, 1 rue Leon Migaux, Massy, Cedex, 91341 France.

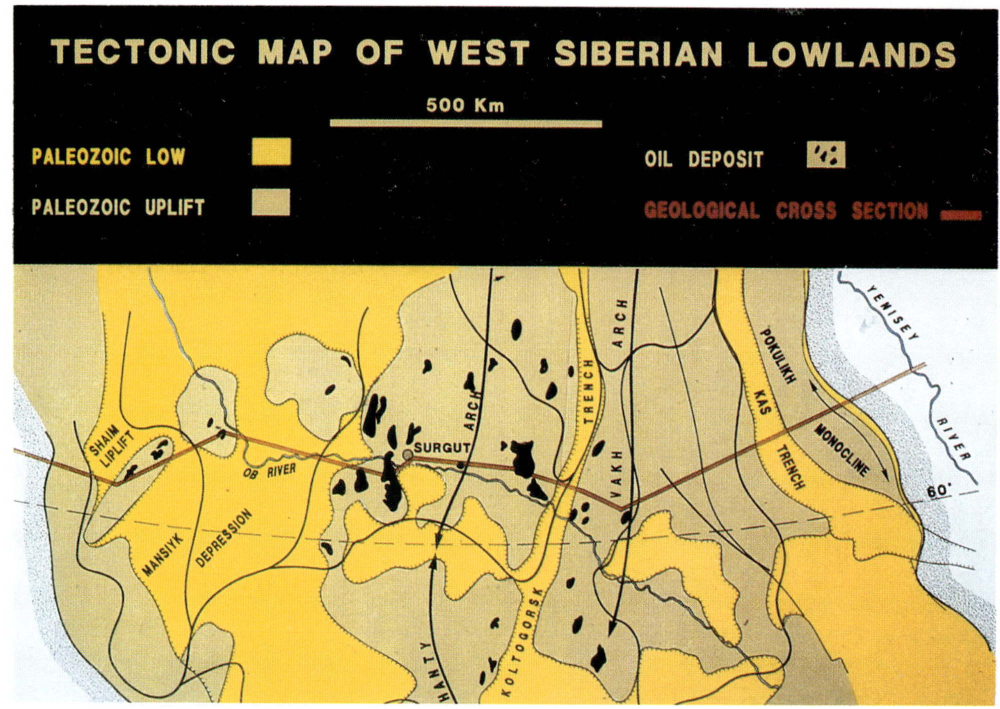

Fig. 1. Location of the west-to-east geologic cross-section of the west Siberian lowlands. (Bazanov et al.)

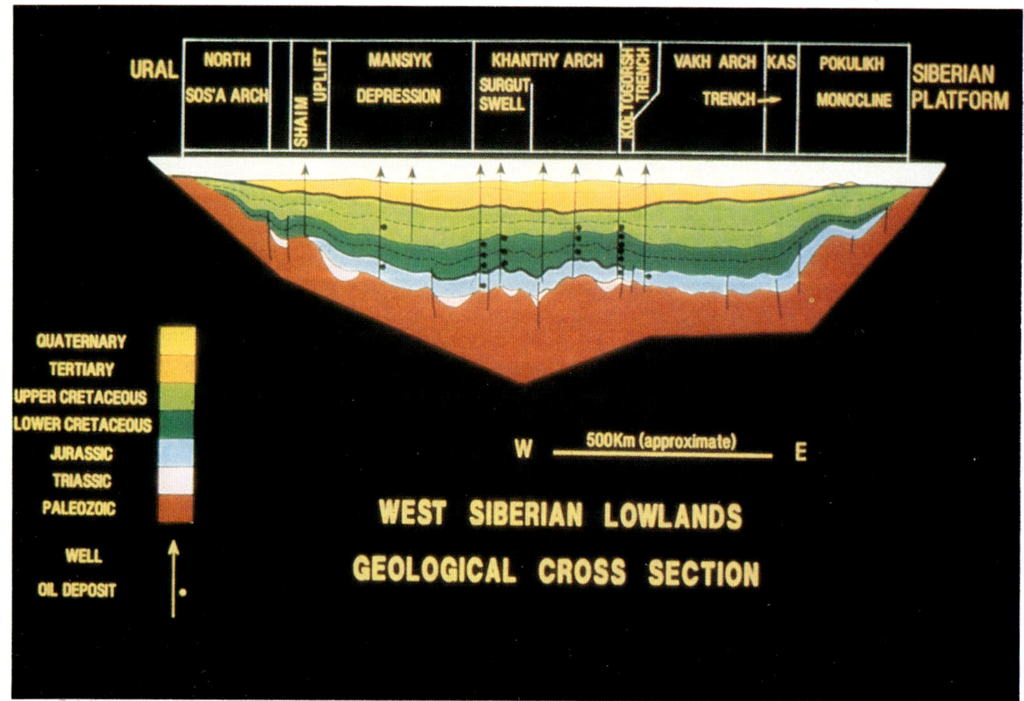

Fig. 2. Geologic cross-section of the west Siberian lowlands. The Hercynian graben is mainly filled by Tertiary, Cretaceous, and Jurassic beds lying unconformably on Triassic. The source rocks are Jurassic. The reservoirs belong to the Neocomian layers. (Bazanov et al.)

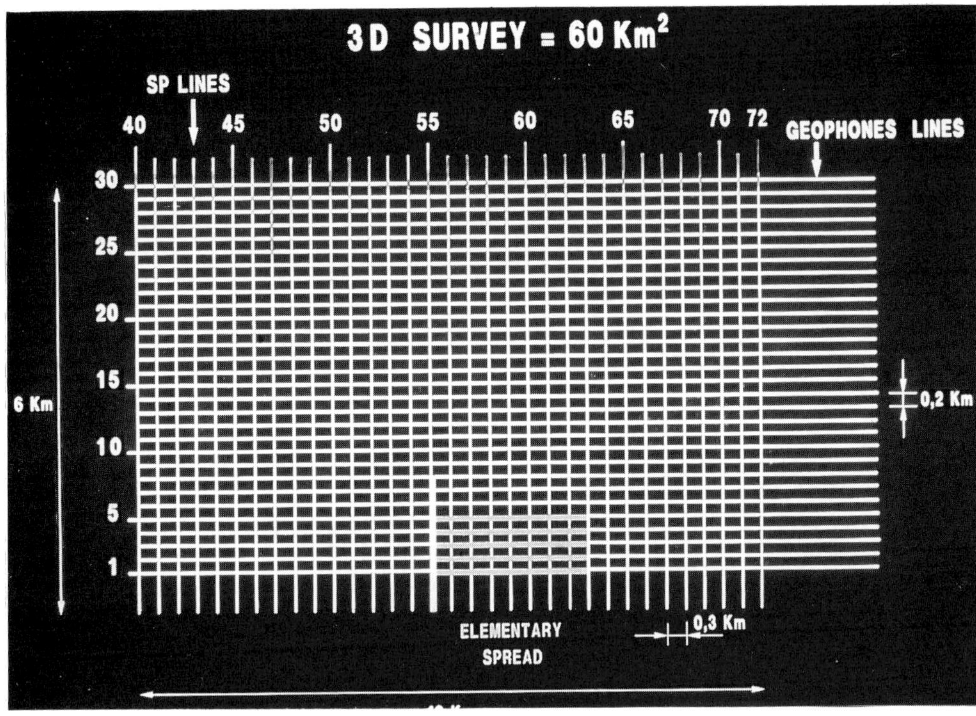

Fig. 3. Total 3-D survey grid; the present study was taken from the southeastern part.

Fig. 4. The recording spread has five parallel lines 2400 m long, spaced 200 m apart. There are 48 geophone groups on each line at 50 m intervals. A series of 20 shots was recorded on each set of five lines, located along a north-to-south line at 50 m intervals. The result is a 25 m square investigation unit.

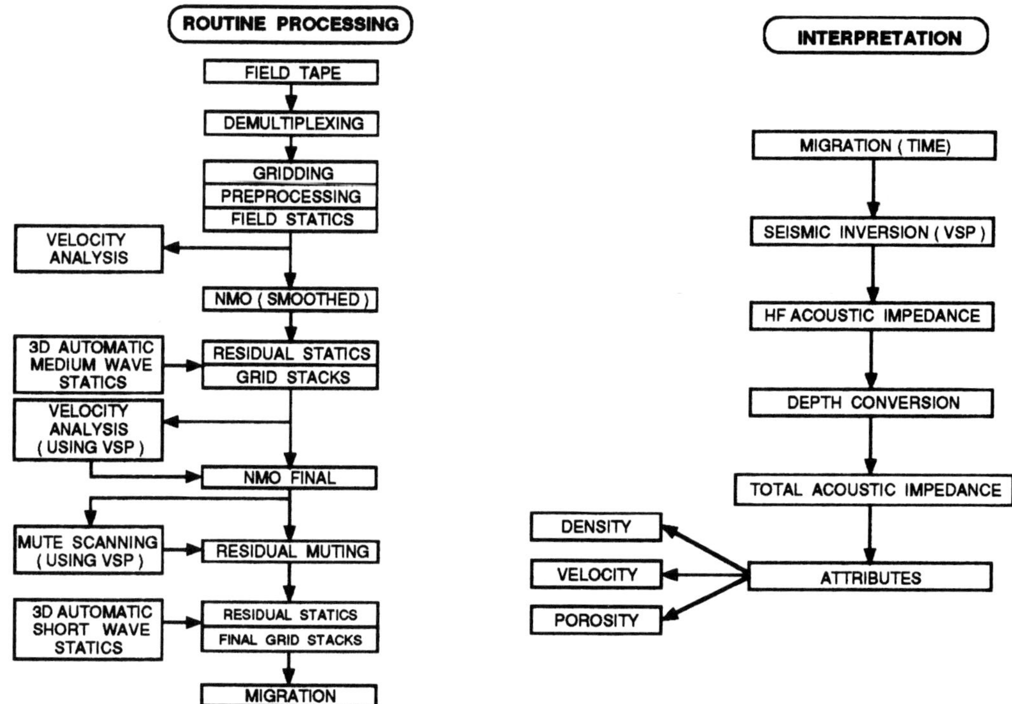

Fig. 5. The processing sequence for the 3-D Siberia project which consists of two main steps. The conventional process on the left results in a 3-D time block, while the more interpretive approach on the right results in a 3-D depth block and total acoustic impedance, allowing a study of density, velocity, and porosity parameters.

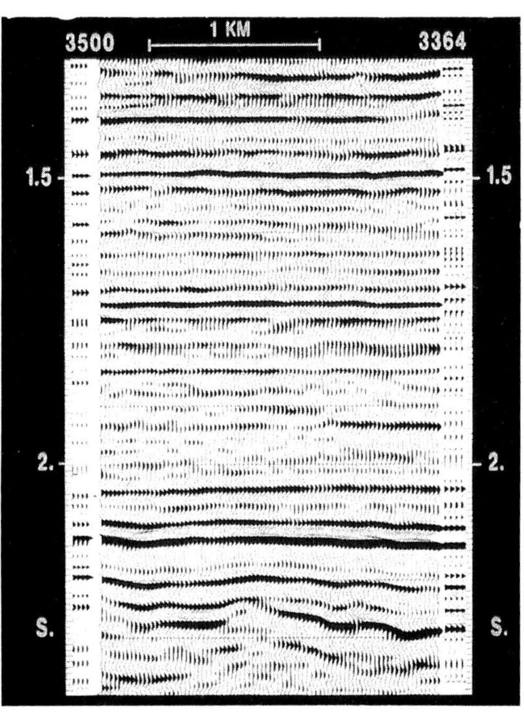

Fig. 6. One-dimensional zero phase inversion using VSP as model and control.

geologic horizons with the seismic characteristics of the VSPs. It is easy to recognize these horizons on the seismic sections. Four families of horizons correspond to the major series (Figure 7):

(1) The erosional surface of the Triassic;
(2) Three sandstone layers in the Jurassic, which are in part reservoirs inserted between source series of coal and bituminous shale. The Jurassic series finish with the Bagenov-B, which is a 20 m deposit of bituminous shale and an excellent seismic marker.
(3) The basin conditions in the lower Cretaceous generated clinoforms and are topped by horizon B-10. Reflectors represent low energy. Only B-18 is a reservoir.
(4) Above this, the shelf conditions in the lower Cretaceous series are characterized by north-to-south channels. Horizons AC-7 and B-2, 3 are pay zones.

The lithologic logs reveal only two main sediment components, sand and shales. Assuming that acoustic impedance and depositional energy are high for sands and low for shales, impedance values can be associated with the lithology as a function of increasing sand content (Figure 8), according to the following classification:

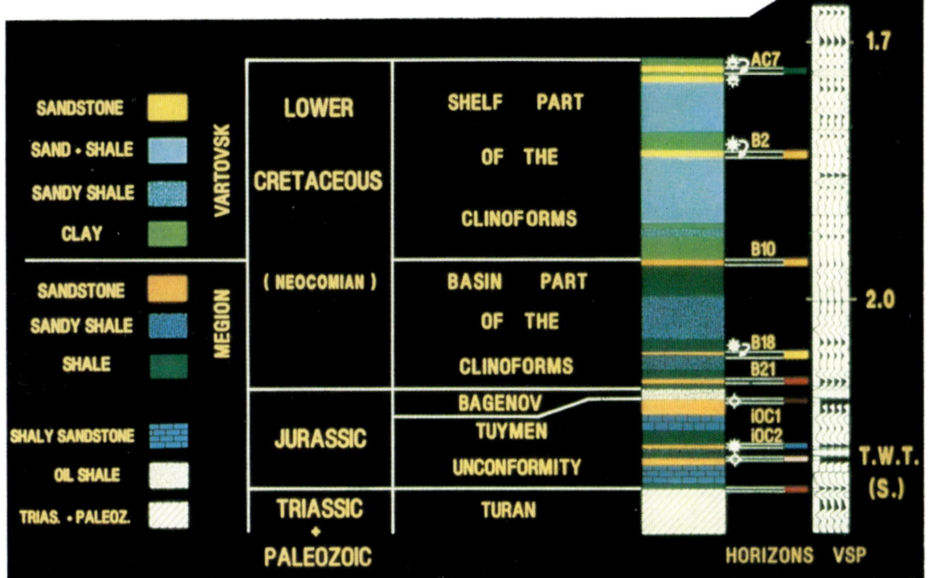

Fig. 7. Correlation between geologic series and seismic horizons, allowing identification of the reservoirs. The lithology consists only of sandy shale sediments.

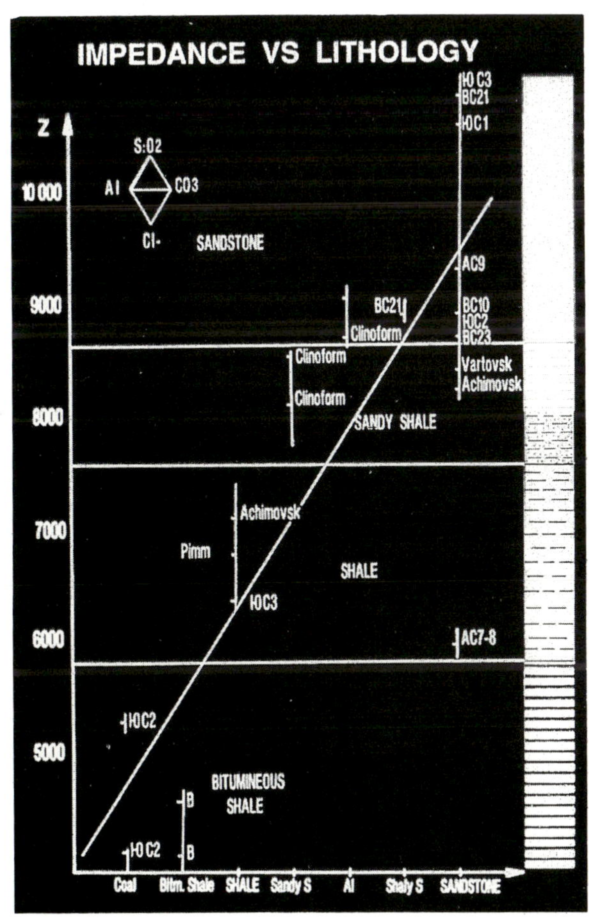

Fig. 8. Relationship between impedance and sandstone percentage. Energy of deposition increases from coal to sandstone.

1-Coal

2-Bituminous Shale

3-Shale

4-Sandy shale

5-Shaly sand

6-Sandstone.

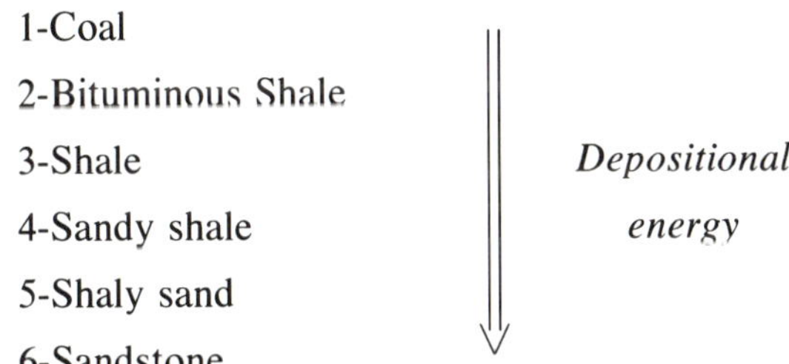

These relationships are established statistically for the wells of the survey and Table 1 summarizes this classification.

For the Jurassic, values are clearly divided: 4000–7000 for shales, 9000–11 500 for sandstones. For the

Table 1. Acoustic impedance—lithology identification.

Impedance $(g/cm^3)(m/s)$	Lithologic nature
4000–6000	Coals and bituminous shales
6000–7000	Jurassic shales; gas reservoirs AC-7 and 8
7000–7500	Sandy shales; Pimm·shales; Achimovsk shales
7500–8700	Sandy shales; shaly sands (B-10, Cretaceous clinoforms)
8700–10 000	Reservoir sandstones of IOC-2, B-2,3, AC-9
10 000–11 000	Reservoir sandstones of the Jurassic (IOC-1)
11 000–12 000	Sandstones of IOC-3

Note: Basement values vary between 10 000 and 11 000.

Lower Cretaceous, values are continuous and relatively high (7000–9000 for the Megion sandy shales). Colors were associated with this lithologic study so as to reveal the various series and lateral variations. This color selection was also cross-checked by a log at each well.

The quality of the SP logs is good, with responses correlating accurately with porous sand beds. Deflections appear in line with sandstone reservoirs, while in shaly sediments the curve corresponds to the shale baseline. Figure 9 indicates two SP responses in the Neocomian: shale baseline values in prograding sequences and repeated deflections in porous shelf deposits. Figure 10 shows correlation of a resistivity log with the impedance section. Jurassic coal and bituminous shale source beds are characterized by high resistivity values, while in the sandy sediments resistivity values decrease in inverse proportion to the shale content.

Interpretation

Different levels from the Triassic to the Cretaceous will be studied in order of decreasing depth. The major episodes in the history of the basin can be defined on the two lines shown in Figures 11 and 12.

Basement

The depth contour map of the basement (Figure 13) clearly shows aspects of the erosion surface corresponding to the immersion phase at the end of the Triassic. The north-to-south Hercynian direction is clearly visible. A southwest-to-northeast paleo-valley network crosses the surface. Two structural highs at the north and west are visible; these affect subsequent sedimentation. There is about 400 m of relief.

Jurassic Series

During the Jurassic this epi-continental basin underwent generalized immersion interrupted by partial regressions. Jurassic layers show coastal onlap on the Triassic unconformity. This resulted in recurrent sequences with frequent lateral variations and progressive modification of the depositional environment (lacustrine, palustrine) during the Bajocian to the

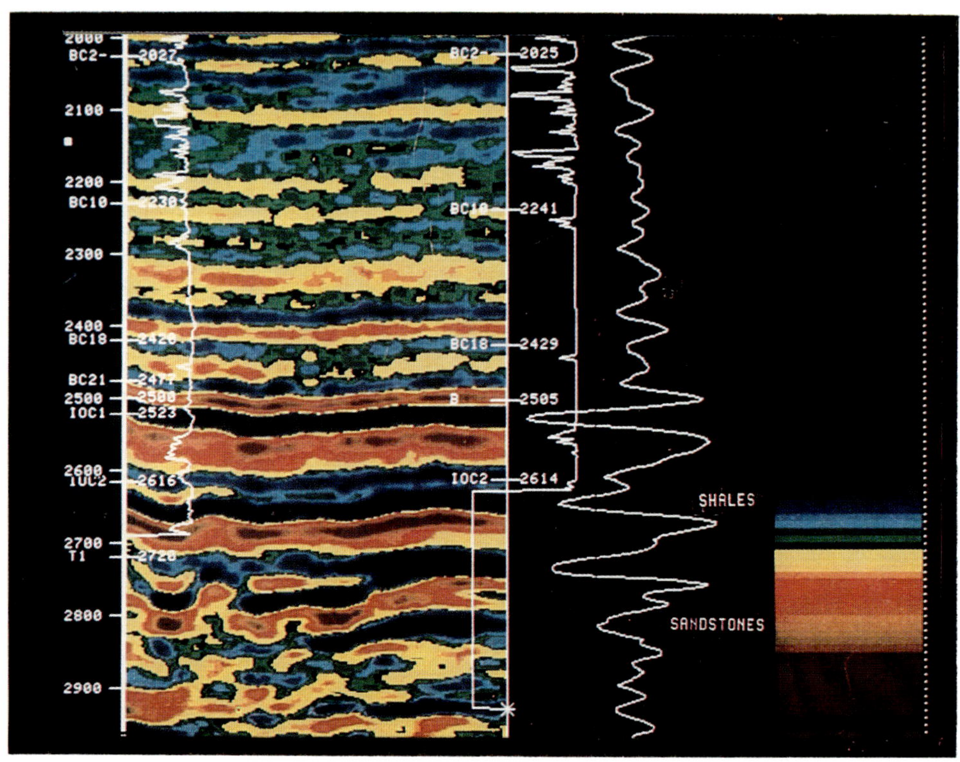

Fig. 9. Correlation between the calibrated impedance section and SP logs. Sandstones are shown in red, sandy shales in yellow and green, shales in blue, and coal in black. Note the large deflections in the porous shelf Neocomian deposits.

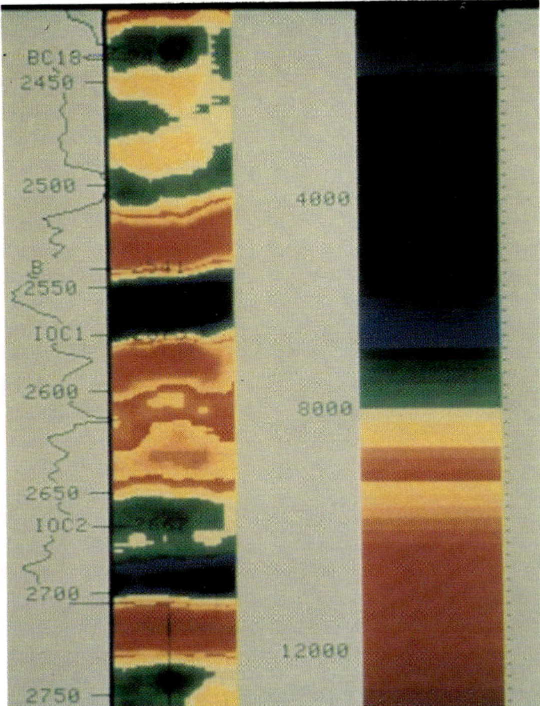

Fig. 10. Impedance (color coded) versus resistivity log; resistivity increases to the left. High resistivity correlates with source beds (2550 m, 2680 m), consisting mainly of bituminous shale and coal deposits. Note the low resistivity of the shale layer at 2500 m. Data from well 3355.

Bathonian, then marine up to the Volgian, with maximum transgression during the Callovian period. Modeling (Figure 14) illustrates the invasion of paleovalleys by transgression with correlative deposits of source rocks. This invasion can be verified by a constant depth impedance section showing coal deposits (in black on Figure 15).

The east-to-west section (Figure 16) shows alternation of sandstone layers interbedded with coal-bearing series. Lateral variations are visible with shaly sandstone in depressions and coal-bearing shale at the top, featuring a coal dome. The position of source rock seals changed as the water level rose. Some sandstones are oil-bearing reservoirs, such as IOC-2. Figure 17 shows that, flattened on the pyritic shales above IOC-2, there are correlations between three wells spaced at 1 km intervals; correlation between sandstone beds is difficult, suggesting lateral facies variations.

The terminal Jurassic (Volgian) consists of an euxinic phase with deposition of sealing Bagenov bituminous shales. This key bed, 20 to 25 m thick, extends over one million square kilometers. The bituminous pyritic shales are characterized by low density and low seismic velocity. This bed, which corresponds to the maximum negative amplitude (B in Figure 16), also contrasts with surrounding layers of sandstone and is a good indicator of the seismic polarity.

(Text continued on page 269)

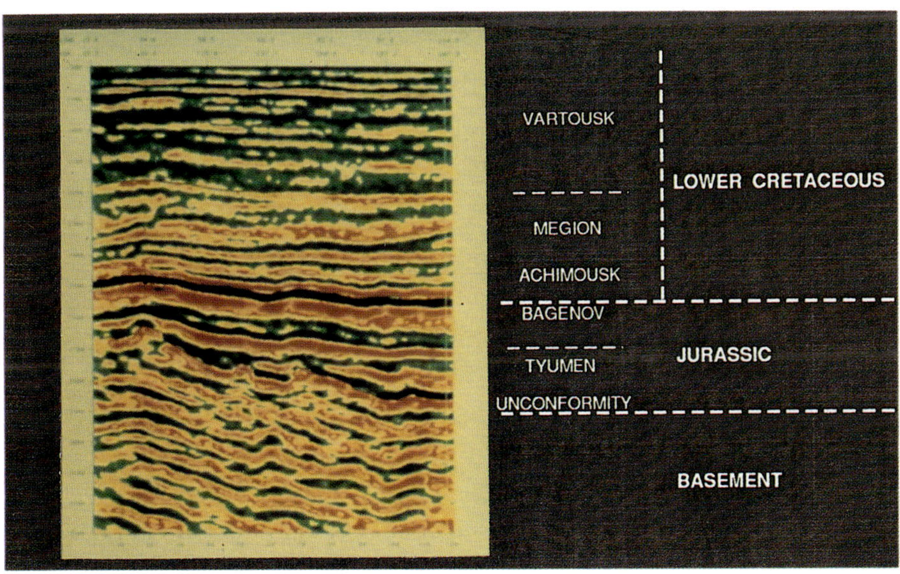

Fig. 11. Identification of main series of the 3-D block on an arbitrary line. The seismic character of the reflectors from each geologic period facilitates understanding of sedimentation: onlap for the Jurassic, offlap for the Megion, and parallel for Vartovsk deposits (Neocomian age).

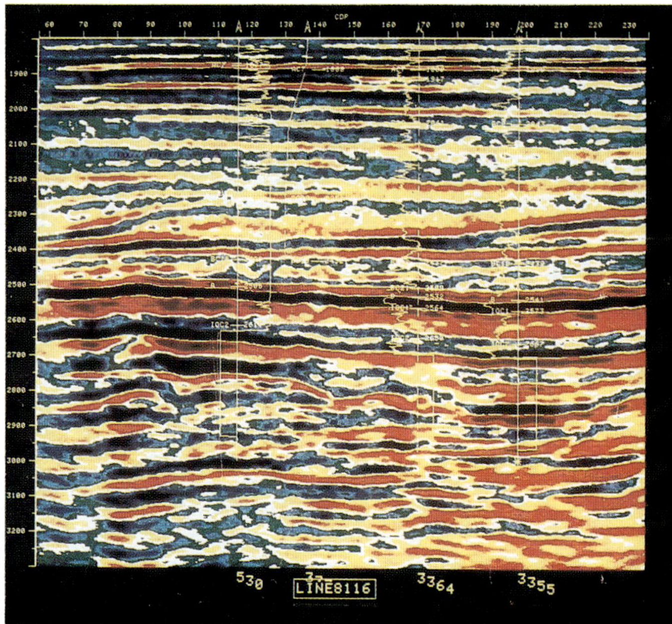

Fig. 12. East-to-west line showing ties between logs and seismic. Note the two SP curve facies in the Neocomian. The accuracy of correlation is confirmed by the high resistivities of the major source rock beds (2500 m depth).

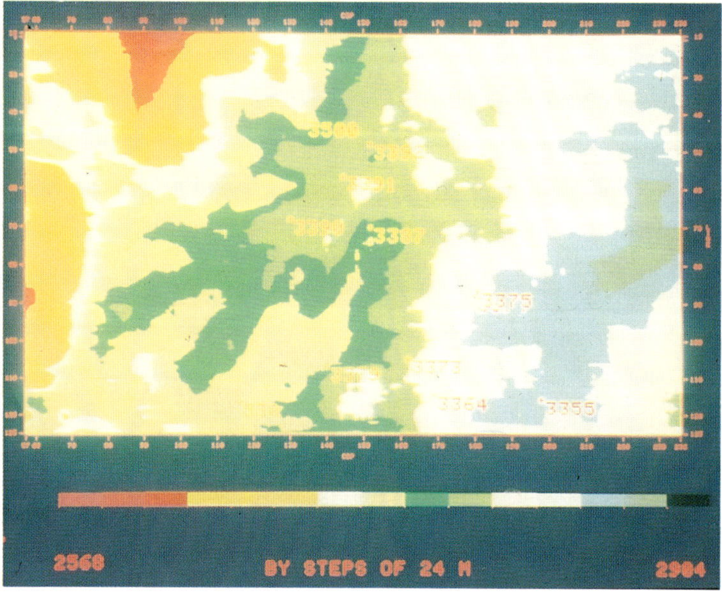

Fig. 13. Depth map of the Triassic unconformity. The Hercynian north-to-south trend and the southwest-to-northeast paleo-valley network are clearly visible.

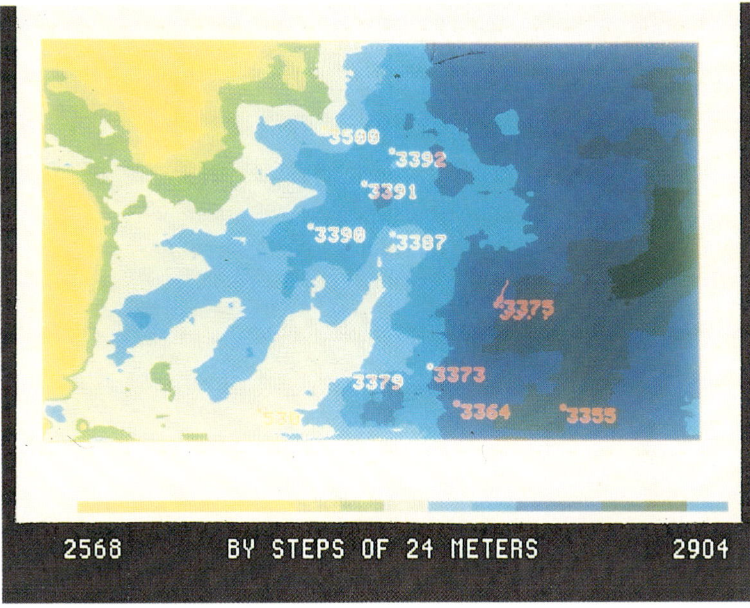

Fig. 14. Modeling of the Jurassic transgression (Bajocian Batho-nian age). Lacustrine conditions in the paleo-valleys induced propitious conditions for source rock deposits.

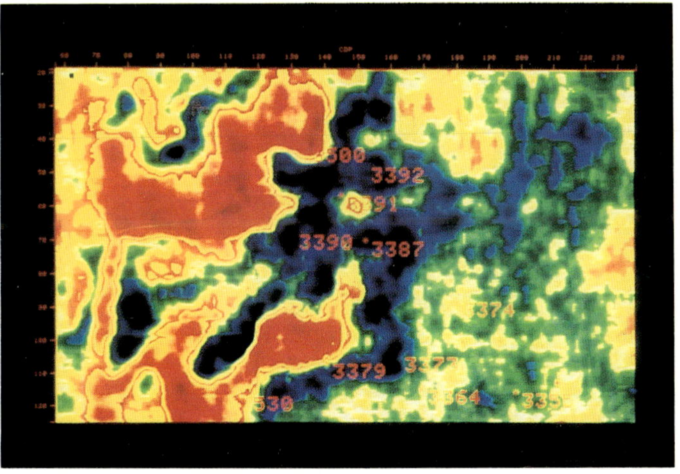

Fig. 15. Depth slice at 2664 m showing coal deposits in the paleo-rivers, confirming the hypothesis formulated in Figure 14. See also Figure 16.

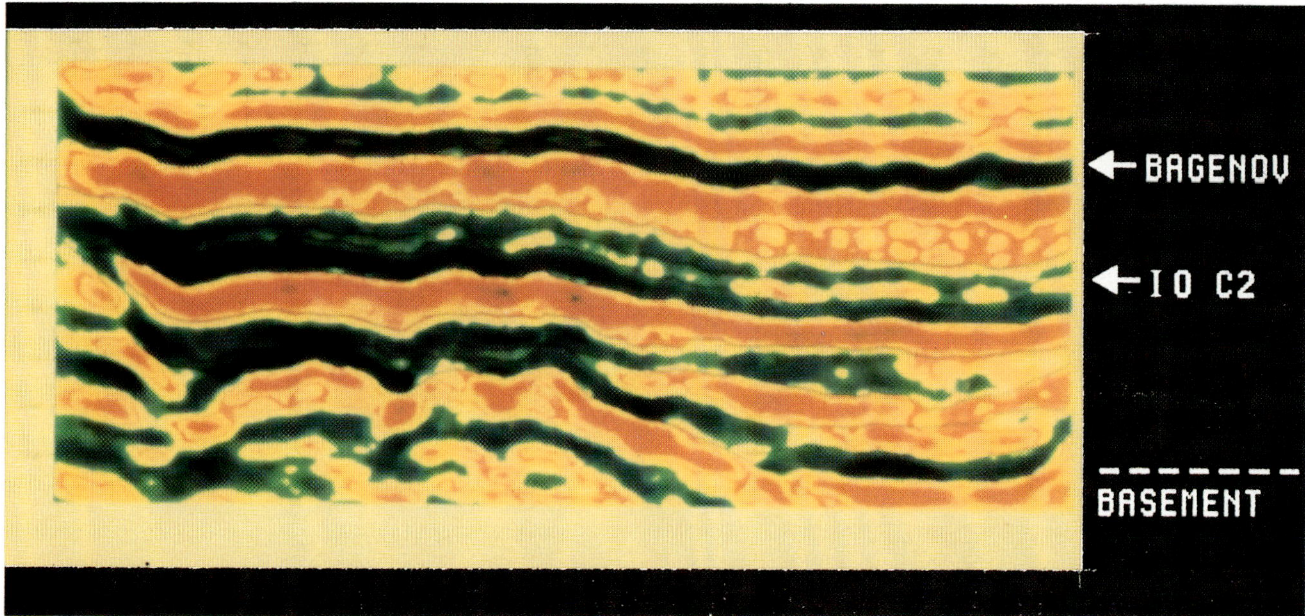

Fig. 16. Jurassic transgression depositional system. Coastal onlapping layers feature a coal dome (lateral variations of facies), and transgression is maximum at the Callovian age. Euxinic conditions during the late Jurassic (Volgian) induce a 20 m thick deposit of Bagenov bituminous shales. Sandstone beds such as IOC2 are partially productive. Paleo-river cross-sections are visible, filled by carboniferous sediments.

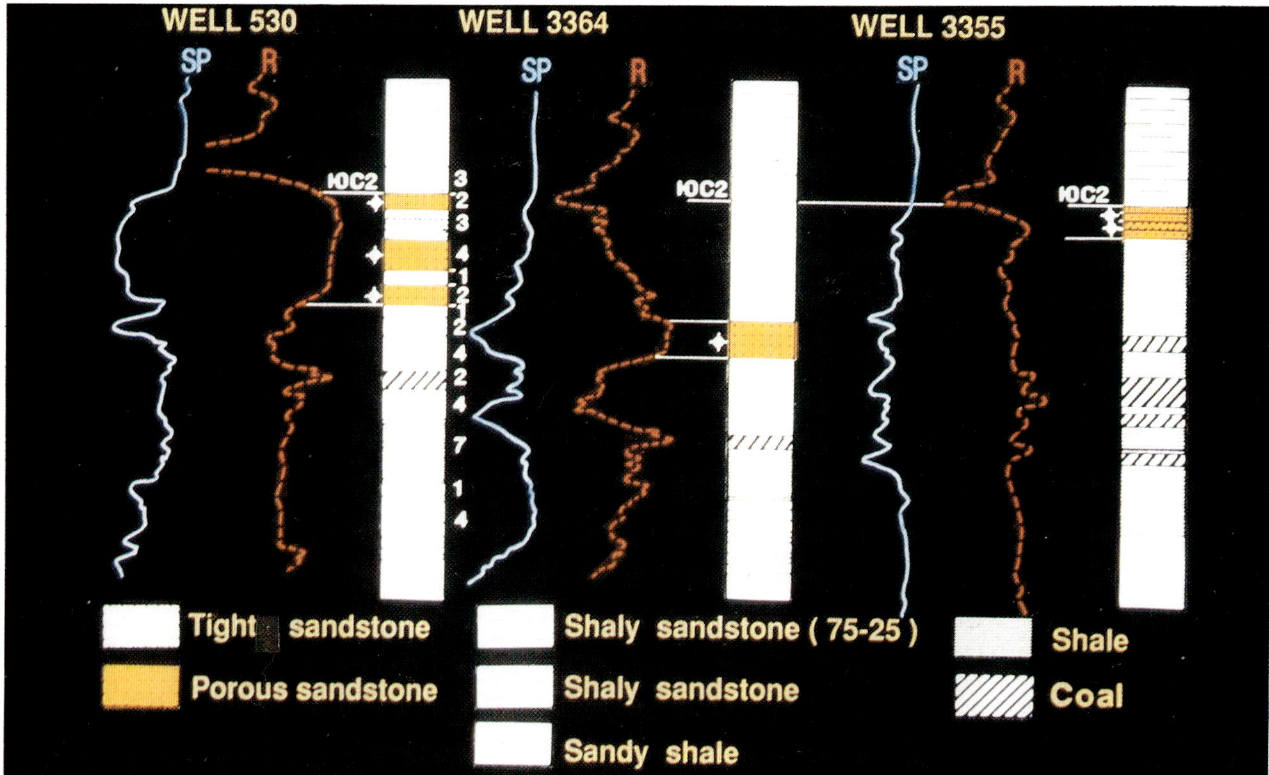

Fig. 17. Flattening of Jurassic logs at the seal of the reservoir (pyritic shales) showing that correlation between sandstone beds is difficult, suggesting considerable lateral variations in facies (IOC2).

Cretaceous Series

The Cretaceous horizons are Neocomian. They can be broken into two series corresponding to different paleogeographic environments.

The Megion (Valanginian) period corresponds to deposition of shale-sandstone layers that form east-to-west progradation (Figure 18); these clinoforms display typical offlap slope facies. Deposition is caused by Valanginian rivers from the raised Siberian shelf associated with renewed tectonic activity at the beginning of the Cretaceous. The base of the clinoforms (Achimovsk formation) shows good oil-bearing potential. Reflections compose a wedge of lensoid-shaped units that pinch out updip and downdip. These progradations which extend over several kilometers with an average thickness of around 100 m are not clearly differentiated on the logs. Level B-10 at the top of the Megion beds characterizes a general flattening of the area. B-10 is a discontinuous reflector formed by sandstones of varying thickness (up to 10 m).

The Vartovsk (Hauterivian-Barremian) consists of alternating shale and sandstone beds. These deposits result from repeated withdrawal of the sea toward the northwest, with progressive replacement by wide paleo-rivers (Figure 19) and reservoir deposition. Reservoirs are noted from reservoir BC-2,3 upward. This reservoir sandstone is continuous in the area studied; the sandstone is capped by the key marker of Pimm shales around 20 m thick. Above the Pimm shales, sandstone and shale series alternate for 80 m, interrupted by a southeast-to-northwest channel, which is easily distinguished on the east-to-west sections in constant time (details follow). The next layer is the sandstone series AC-7,8 consisting of intercalated oil and gas reservoirs. The presence of gas is characterized by very low acoustic-impedance values.

Study of Reservoir BC-2,3

Three reservoirs belonging to the Vartovsk series were studied. The BC-2,3 reservoir sandstone and the surrounding series present sufficient homogeneity (Figure 20) to allow a detailed study of their characteristics. Various sedimentary levels around the reservoir show remarkable homogeneity, for example, between the most distant wells, as indicated by Table 2:

The BC-2,3 reservoir at its base can have shale veins less than 1 m thick. Level BC-2,3 is a good continuous seismic marker over the survey area. In order to obtain as much information as possible, three BC-2,3 picks were made: (1) maximum impedance value (crossover amplitude), (2) average impedance value of the layer top (positive amplitude peak), and (3) minimum impedance value of the Pimm shales (crossover amplitude). These provided data to map the depth (Figure 21), thickness (Figure 22), and maximum impedance (Figure 23) of the reservoir, and the impedance of the sealing Pimm shale (Figure 24).

The depth map shows the raised western area with a slight depression in its center. The impedance map shows high values oriented southwest-to-northeast. The Pimm shale impedance map values are 6000–7000 for the major part of the survey but 7500–8000 in the eastern part, the higher values possibly representing a higher proportion of sandstone.

Examination of logs (Figure 25) shows that in certain wells the reservoir layer includes thin shale layers at the base. There is no relationship between SP or

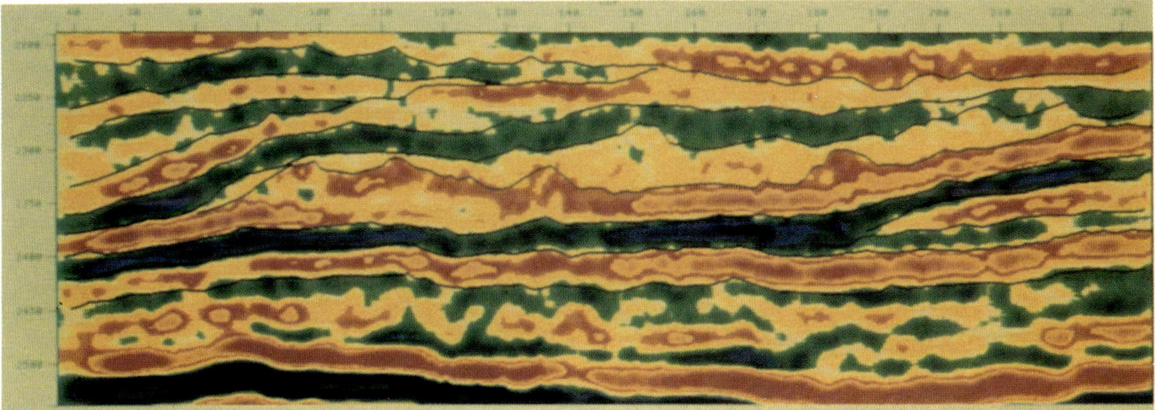

Fig. 18. Lower Neocomian Megion clinoforms: east-to-west progradation showing typical offlap slope facies. Due to Valanginian river deposits, they are about 100 m thick. They induced filling of the basin and platform environment of the Vartovsk series.

resistivity and acoustic impedance. For example, there is high impedance at well 3387 despite intercalated shale layers, and low impedance at well 530 which has only sandstone. The shale layers at the base of the reservoir appear to have no influence on impedance.

Porosity modifies the acoustic parameters. Using all the wells in the survey area, an impedance and porosity chart can be produced (Figure 26), revealing a fairly linear relationship. This chart provides the basis for displays graduated in porosity.

Logs recorded in 1982 indicated 6–8 m of oil only around wells 3500, 3390, and 530; other exploration wells indicate oil and water (3391, 3379) with oil thickness of 3–6 m. Permeability values range around 200 mD. Two rows of injection wells surround the field in the concerned block. Production indicates two particularly productive areas: the area of well 536 and along a north-to-south line in the vicinity of wells 534 and 1481. Lower production is located in the center and around the injection wells. Two areas can be delineated (Figure 27):

Area A, the structural top outside the high impedance zone, with high porosity and initially high production of oil only; and

Area B, not a structural top, outside the high impedance area, with high porosity and high production.

Superimposing the impedance map on the depth map helps localize the better quality part of the reservoir. Water circulates in the zones of high porosity, while the flow of injected water is much slower in areas of low porosity. Thus, the southwest-to-northeast line of lowest production may correspond to an area of lower porosity. The impedance map could be used to position injection wells.

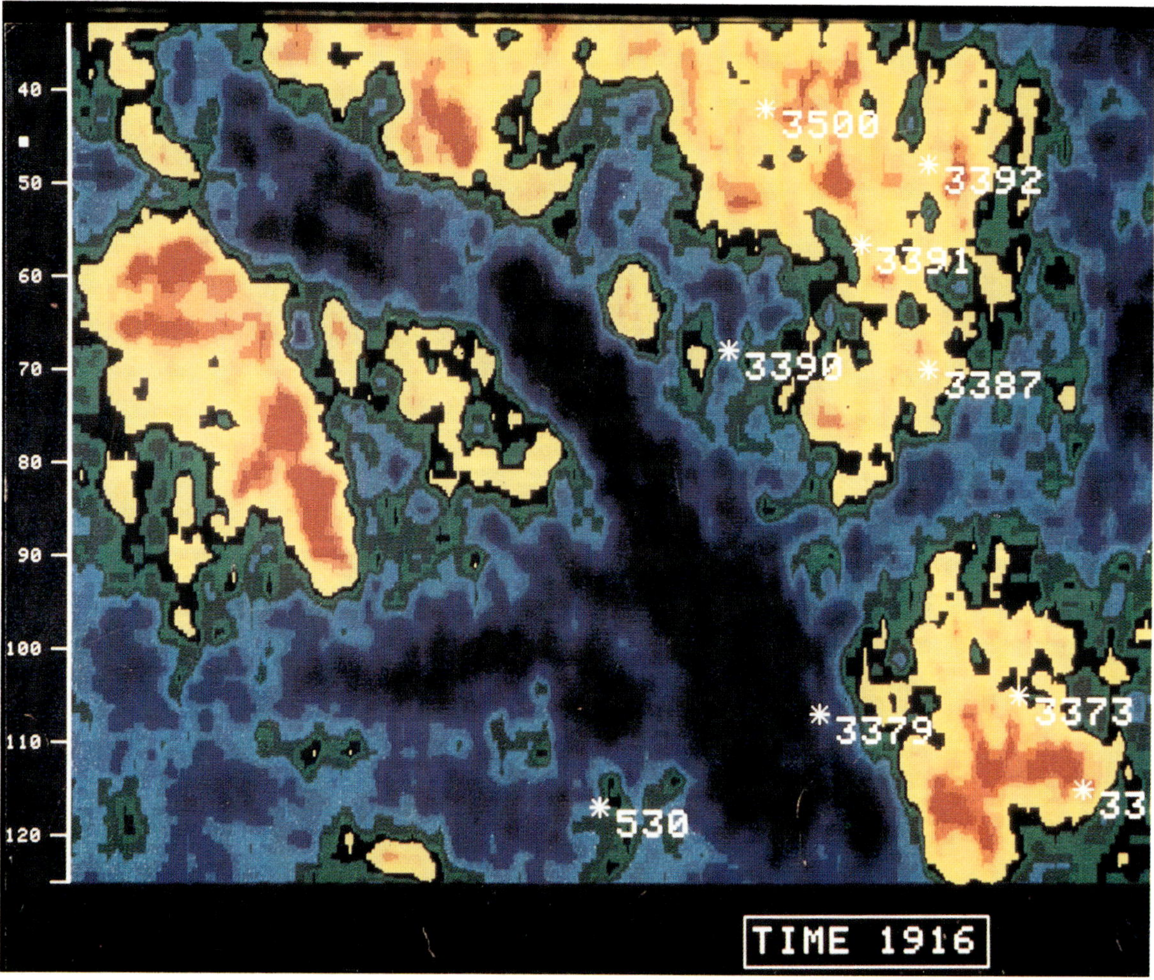

Fig. 19. Vartovsk paleo-river network. During the Hauterivian Barremian period, withdrawal of the sea toward the west was accompanied by the development of the paleo-river network. Width is 250 m. Tributaries are visible southwestward. Note the northwest curvature forming the trough between old basement paleo-tops.

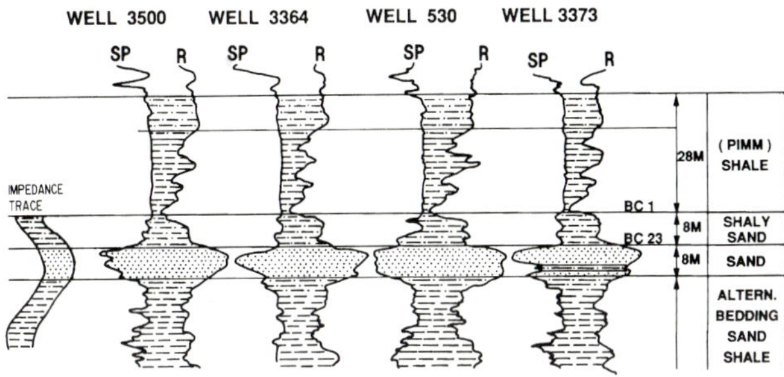

Fig. 20. SP and resistivity characteristics of reservoir BC-2,3, flattened at the top of BC-1. Shaly sands and oil-bearing sandstones are 8 m thick. The isopachous characteristics of these layers are propitious for an impedance study. One trace has been drawn on the left side, showing seismic locations of the events.

Table 2. Thickness of BC members

Lithology	Well 3500	Well 3355
Pimm shales	22 m	22 m
Shaly sand BC-1	6 m	7 m
Reservoir sandstone BC-2,3	8 m	8 m
Alternating shale-sandstone	30 m	30 m

Study of Reservoir AC-9

Reservoir AC-9 belongs to the Vartovsk series. The geometry is dictated by the path of a paleo-river. This series is about 15 m thick and composed of alternating sandstone and shale layers. The reservoir thickness varies between 3 and 13 m. The formation seal consists of shales 8–18 m thick, which constitute the transition to reservoir AC-8. This reservoir contains oil and, locally, gas.

This reservoir could be identified on the amplitude section by its peculiar channel shape (Figures 28 and 29). A time slice clearly shows the low amplitude area elongated from south to northwest between the two Triassic paleo-tops. Comparison between the impedance map and the depth contour map (Figures 30 to 33) shows that the area of minimum impedance values corresponds to a structural top. One major low impedance axis can be distinguished, from southeast to northwest, crossed in the south by a southwest-to-northeast axis. The low impedance axis corresponds to the channel. Reservoir closure is about 10 m.

The hydrocarbon content of the reservoir correlates to the impedance map. Areas containing oil and gas correspond to low values: oil and gas impedance values are below 7700; water zone impedance values are over 8000. Calculated porosity values are larger in

(Text continued on page 277)

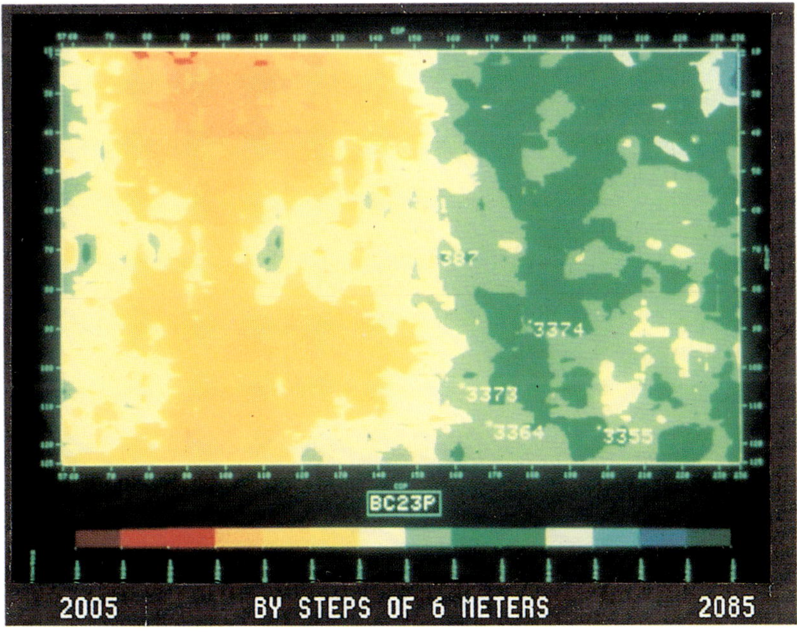

Fig. 21. BC-2,3 reservoir depth map. On the studied area, the depth variation is around 15 m. The Hercynian north-to-south structural trend is still visible.

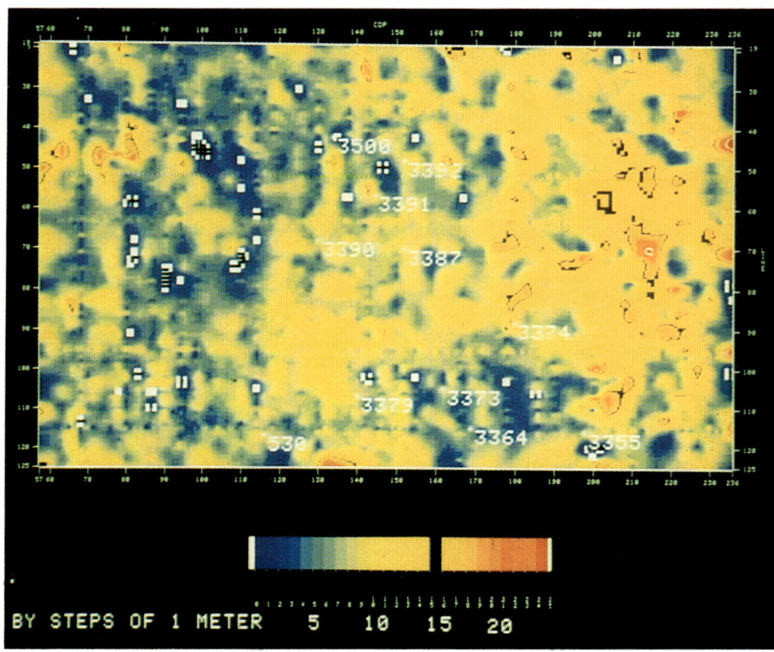

Fig. 22. Isopachous map of BC-2,3. Results are similar to those of the logs. No major thickness variations which could affect impedance values were induced.

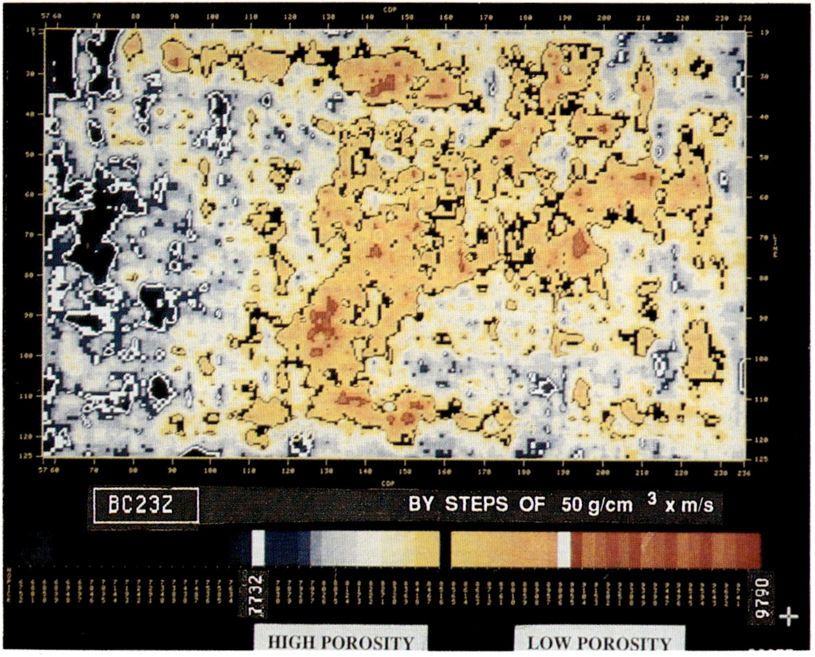

Fig. 23. BC-2,3 reservoir impedance map indicates values ranging between 8000 and 9000. A southwest-to-northeast high axis appears, which can be studied to explain the relationship with porosity.

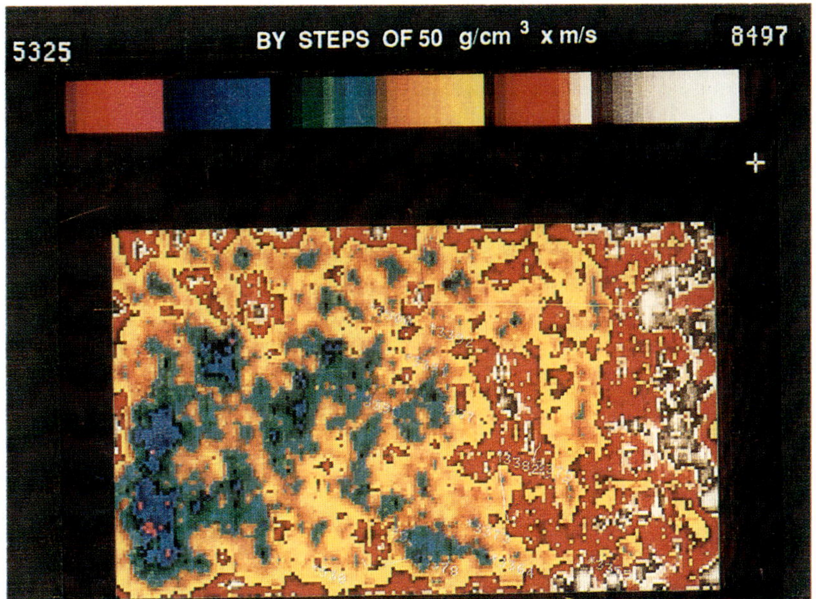

Fig. 24. BC-2,3 seal-impedance map. In order to check the specificity to the reservoir of the impedance anomaly, the Pimm shale impedance map has been calculated. The southwest-to-northeast high-impedance axis does not appear in the shale layer. Values range from 6000 to 7500. See Figure 8.

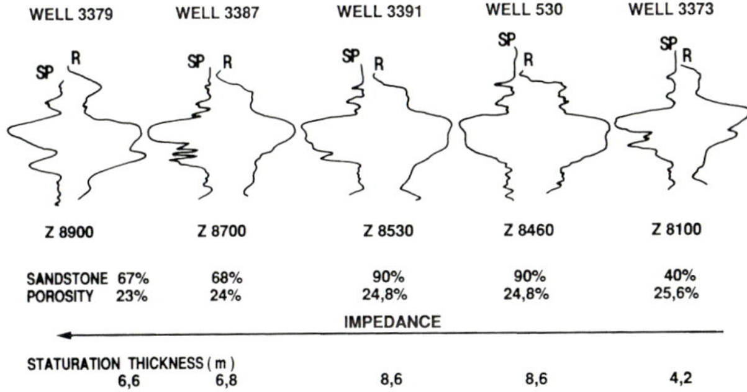

Fig. 25. Shale factor in BC 2,3 reservoirs. Decreasing impedance values can be explained by two parameters: increasing shale content or porosity. Logs are classified according to increasing impedance. The presence of shale layers does not affect impedance values. In fact, because they are located at the base of the reservoir, they do not interfere with the reflection coefficient. One parameter that varies in the same direction is the porosity value, inducing decreases in both density and velocity.

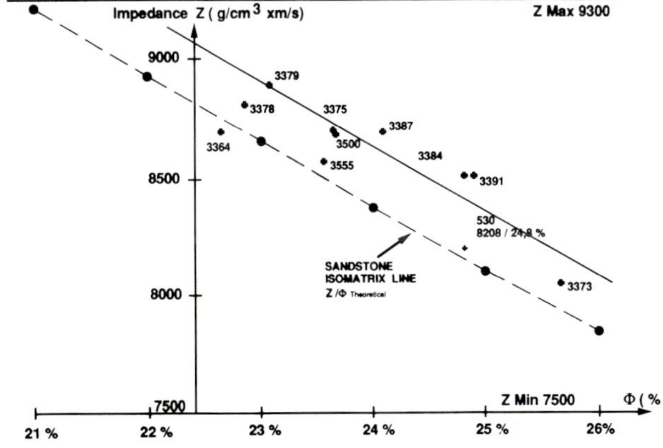

Fig. 26. Impedance versus porosity. The crossplot of 12 well values features a linear regression, proving the interconnection. An impedance map can sometimes be considered as a porosity map of considerable value for reservoir engineering.

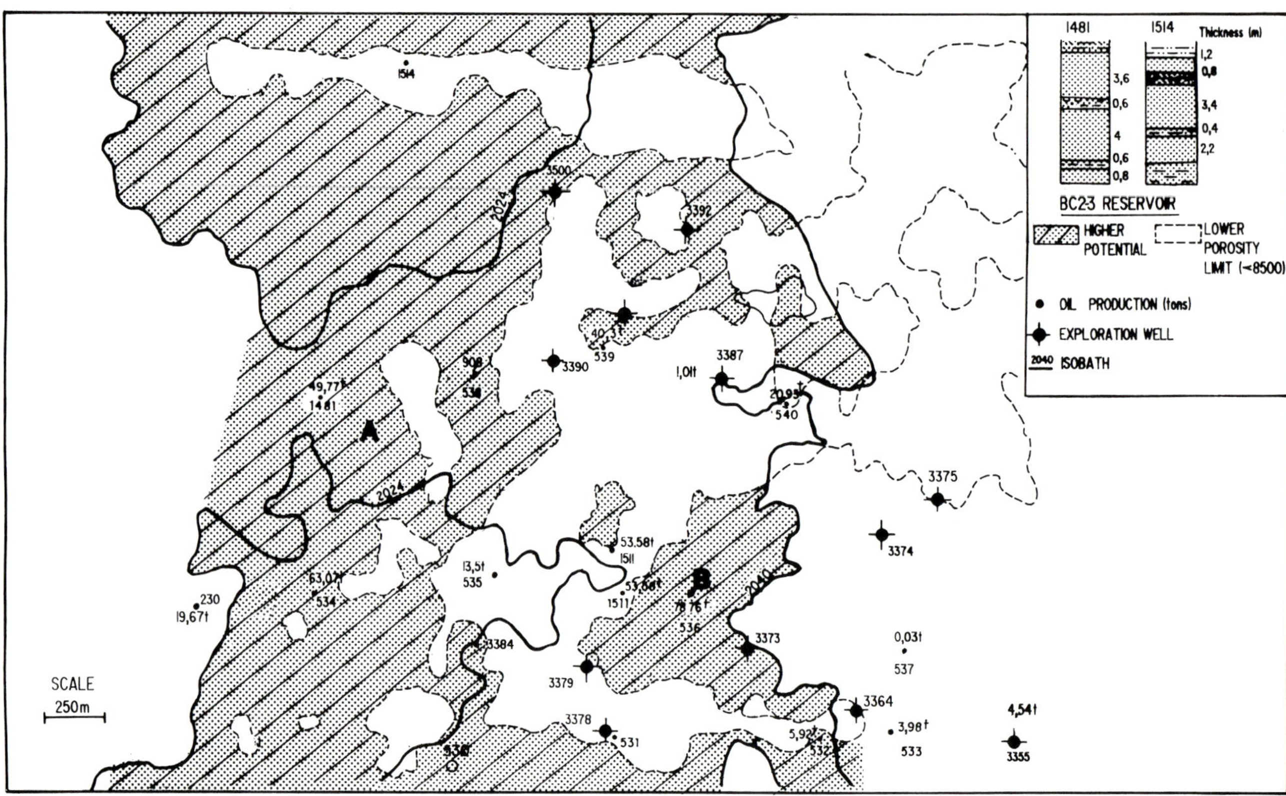

Fig. 27. Correlation between depth (heavy contours), impedance, and production wells. The central high impedance area shows a low production rate. Low closure values indicate that porosity is a more determining factor than depth (in this case) of a pay zone.

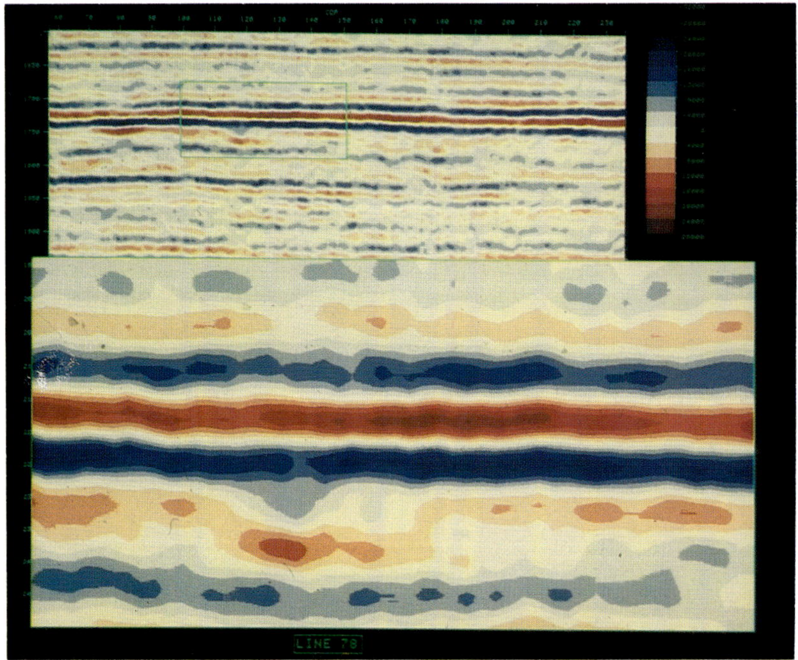

Fig. 28. The channel appears on the amplitude section and characterizes Vartovsk shelf deposits. Negative amplitudes are shown in blue, positive amplitudes in red.

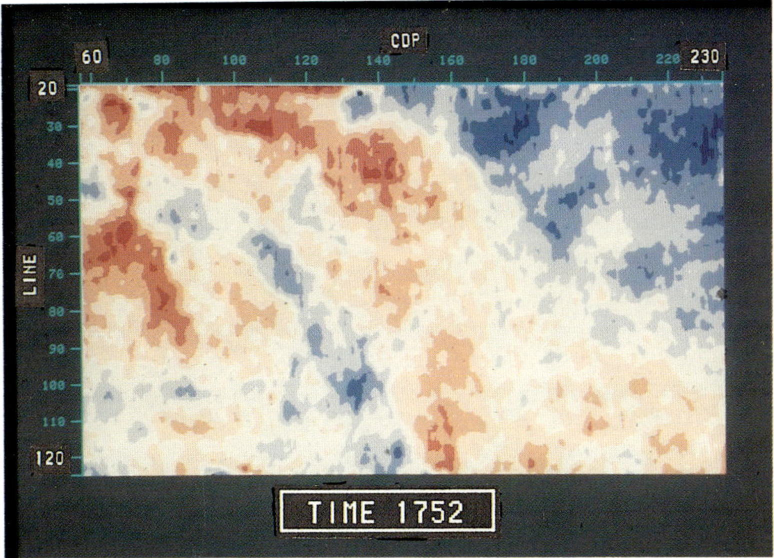

Fig. 29. Time slice amplitude section, clearly showing the geometry of the paleo-channel (see also Figure 19). The strong negative amplitudes are due to gas content in the reservoir.

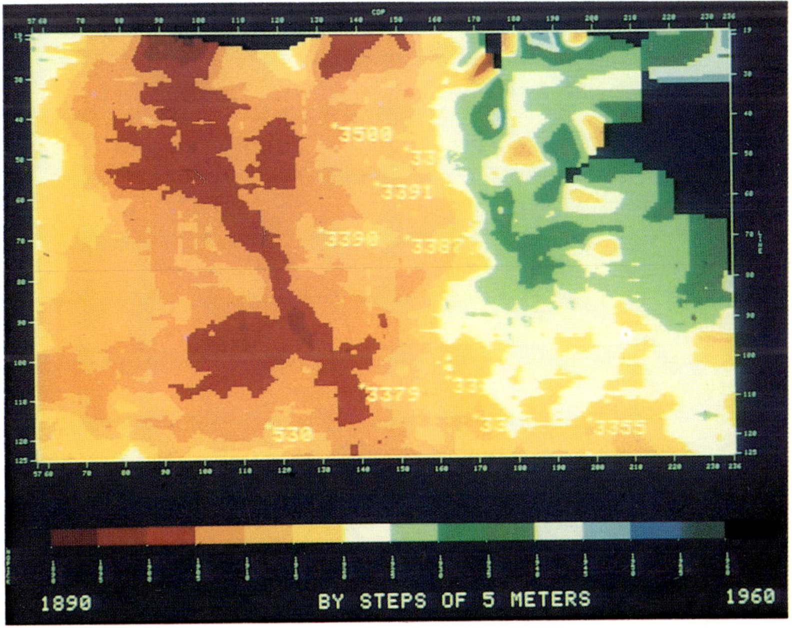

Fig. 30. AC-9 depth map. Closure of the reservoir is 20 m. The structure is elongated and follows the path of the paleo-river.

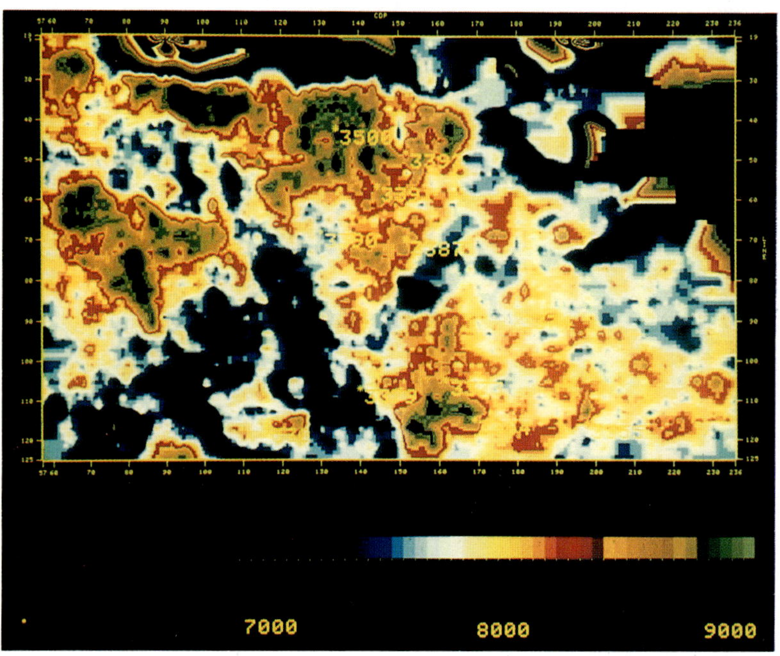

Fig. 31. AC-9 reservoir impedance map. Low-impedance values
correlate with hydrocarbon content (see Figure 33).

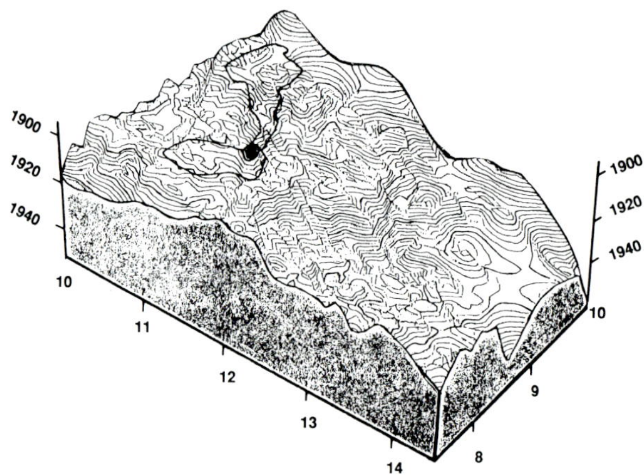

Fig. 32. 3-D diagram of reservoir AC-9. The pay zone
is located on the structural top. The total difference in
level is around 50 m.

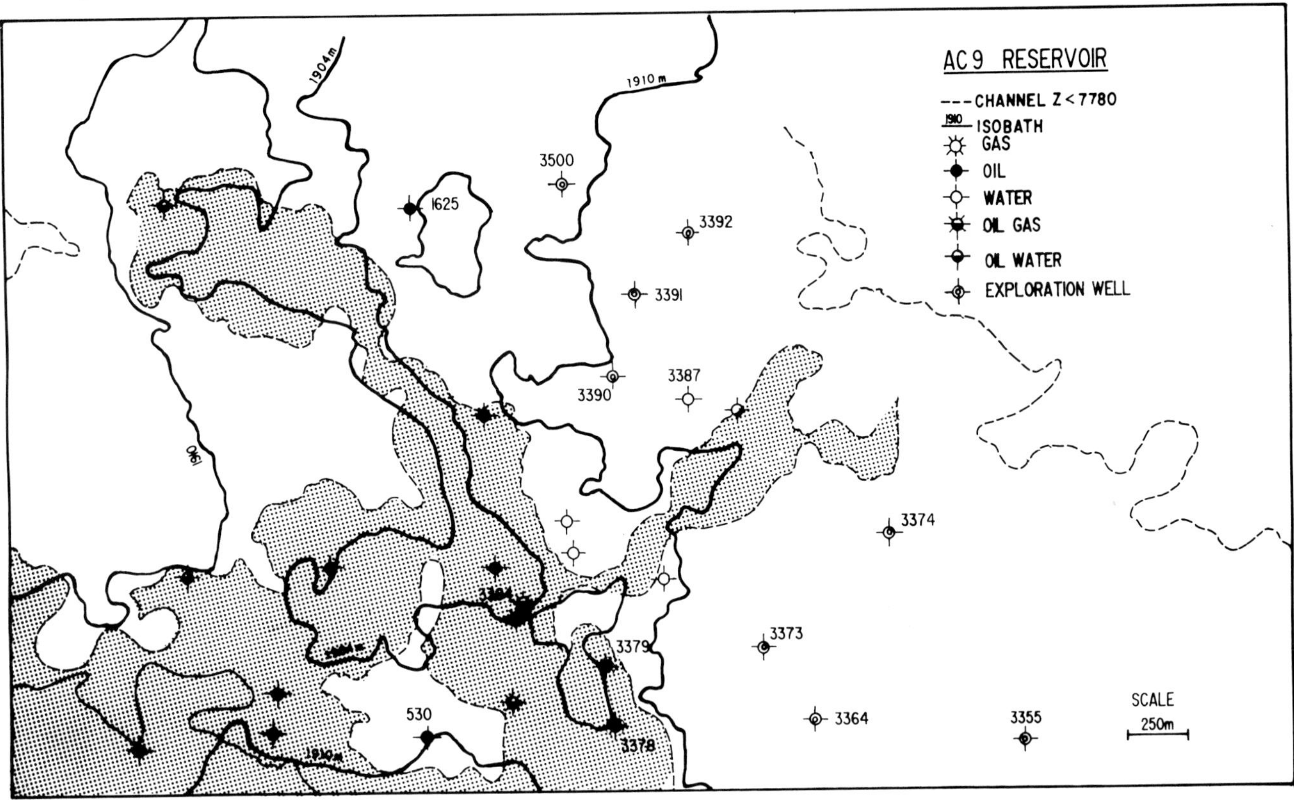

Fig. 33. Relationship between depth, impedance, and production. Impedances lower than 7780 have been shaded and the correlation with production wells is clear. The wells in the south are producing gas locally.

low-impedance areas, with a maximum of 26.5 percent at well 3384. The corresponding saturation is 57.3 percent. The reservoir map (Figure 33) shows oil wells aligned with the axis of the channel. Given the low closure values and the fairly high porosities, water injection likely would modify considerably the fluid distribution.

Study of Reservoir AC-7,8

The AC-7,8 reservoirs are continuous over the entire survey area and relatively homogeneous. Study of wells 3500, 530, and 3364 (Figure 34) gives the geologic succession shown in Table 3. Shales in the two reservoirs vary considerably and are difficult to detect (because of insufficient definition of the SP and resistivity curves).

The AC 7,8 levels are characterized by abnormally low seismic-impedance values. Because of the uniform thickness of the sealing Alymsk shales, this low impedance can be attributed to gas. Closure is around 12 m. Comparison of the depth and impedance maps (Figures 35 and 36) indicates that the 1888 m contour

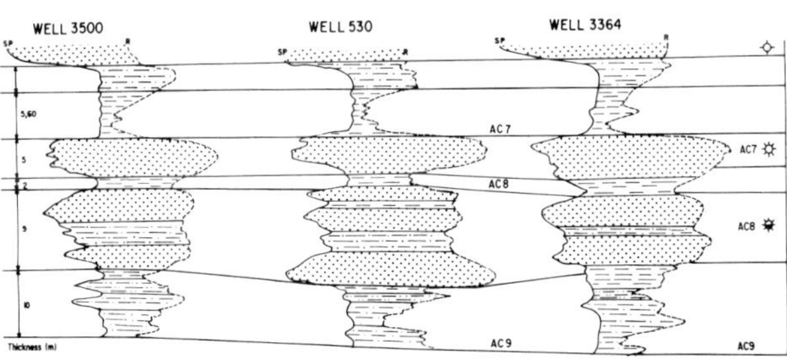

Fig. 34. AC-7,8 well logging characteristics. This is a multilayered gas-bearing sandstone reservoir. The combination of thin beds with high gas concentrations explains the difficulty for seismic in giving a detailed study. However, the homogeneous range of porosity (24 percent) of both productive layers allows an overall study of gas content.

Table 3. Thickness of AC members

Sandstones		
Alymsk shales (seal)	10 m	negative coefficient
Sandstones of reservoir AC-7	5 m gas and oil	negative coefficient
Fine layers of intercalated shales		
Shales	2 m	
Sandstone reservoir AC-8	9 m gas and oil	
Fine layers of intercalated shales		
Fine layers of shale and sandstone		

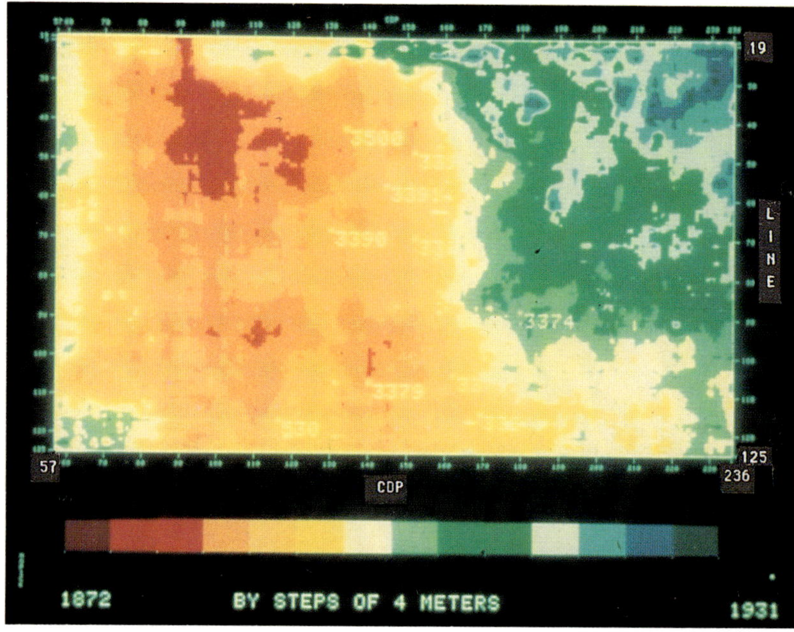

Fig. 35. AC-7–8 reservoir depth map. The north-to-south structural trend is always visible. The 1889 m contour is a determining factor in understanding the impedance map.

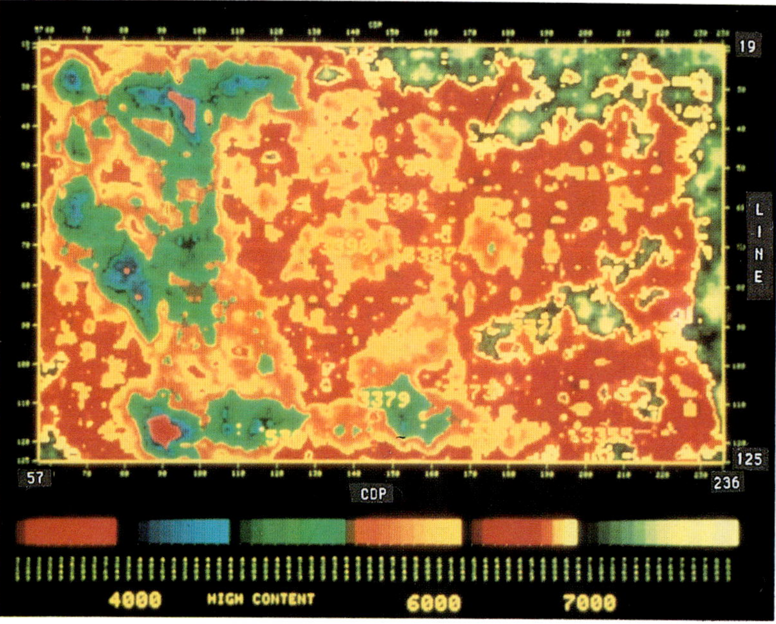

Fig. 36. AC-7–8 impedance map. Neither layer can be distinguished. Values lower than 6300 can be explained only by gas content (minimum value of Pimm shales). Correlation with the depth map is very instructive; low impedance values have been shifted westward. The structural top and central area show higher values despite having the same depth value.

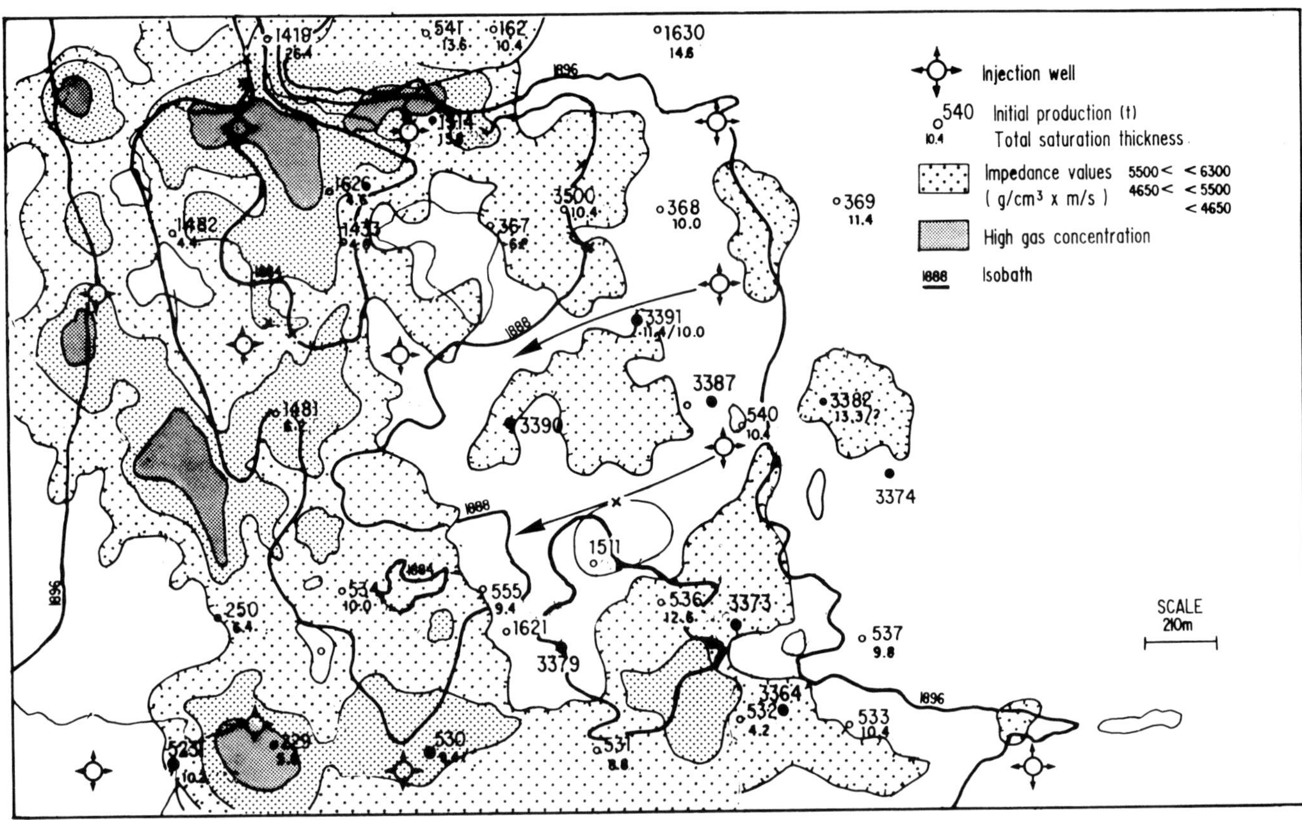

Fig. 37. AC-7,8 interpretative map. Significant depth and impedance values are shown. Injection wells are indicated and initial production showed a uniform distribution on the structural top (1896 m). Injection in the reservoir (540) may indicate a westward sliding of the gas slick accompanied by preferential water circulation along the northeast-to-southwest axis.

correlates with the envelope of impedance values less than 6000, and the structural tops and impedance minima do not coincide. Impedance values can be classified into three groups: (1) below 6300 g/cm^3 m/s, (2) between 6300 and 7000, and (3) between 7000 and 8000. These values can be compared with those of Pimm shales (between 6000–7600) with an average of 6800, and with sandstone impedance values between 7800-9000, with an average of 8500. The AC-9 sandstones show values over 8000. Values under 6300 can only be assigned to a factor other than shale content, this factor being the gas content.

Porosity values are 20–24 percent for AC-7 and 21–23 percent for AC-8. The maximum values for AC-7 are localized in three areas: (1) around well 3373 (25.6 percent) and 3379 (24 percent), (2) around well 530 (24.3 percent), and (3) around well 3391 (24 percent). The maximum values for AC-8 are localized similarly: (1) around well 3373 (23.7 percent) and 3379 (23.3 percent), (2) around 530 (23.7 percent), and (3) around 3387 and 3391 (23.6 percent). Oil and gas saturation values for level AC-7 show a comparable phenomenon: highly differentiated values of 40 per-

cent at 3374 to 66 percent at 3364. Given the small changes in level, we may suppose that the gas was originally located in high structural areas in the high-porosity zones and that injection into low areas (Figure 37) caused progressive movement of the gas toward the west (explaining noncompatibility with the depth map) with the gas remaining in higher concentrations in the high structural areas (1626, 523) and areas located farther from the injection wells (3379).

Conclusion

The study of impedance and well logging results on this 3-D block improved understanding of the sedimentation process and the petrophysical characteristics of the reservoirs. In this case, with two-component sedimentation (pure sandy shale series), a close relationship between lithology and density or velocity parameters can be assumed. Appropriate color assignment helps in understanding depositional phenomena. In the case of reservoirs BC-2,3; AC-9; and AC-7,8, a comparison of depth, impedance, and production maps is helpful for reservoir engineering, mapping of the rock types, and better siting of injection wells.

Renoux and Etienne

Acknowledgment

We thank the Ministry of Petroleum in the CIS (former USSR) for permission to publish this paper.

References for General Reading

Bazanov, E. A., Gurari, F. G.,, Kozyrcy, V. D., Naliukin, V. D., Pritula, Yu. A., Rovnin, L. I., Rudevich, M. Ya., Trofimuk, A. A., and Sarkisyan, S. G., 1975 and 1976, Gas and oil bearing province of Siberia: Proc. 7th World Petroleum Congress, Mexico, **2**, 110–120.

Gogonenkov, G. N., Elmanbovich, S. S., Kirsanov, V. V., Mikhailov, Yu. A., 1984, Integrated interpretation of seismic and well logging data in the detailed phase of oil and gas exploration and in the search for stratigraphic traps: Geophys. Transactions, **30**, No. 2, 177–200.

Golienko, G. B., 1982, Some problems of the tectonic in the central part of the west Siberian plate: Internat.Geological Review, **24**, No. 7, 780–784.

Markovskiy, N. I., 1981, Influence of an ancient river network on formation of accumulations of oil and gas in Western Siberia: Internat. Geological Review, **23**, No. 7, 781–784.

Trofimuk, A. A., and Karogodin, Yu. N., 1982, The Bazhenovo suite, a unique natural oil reservoir: Internat. Geology Review., **24**, No. 3, 249–252.

Ushatinskiy, I. N., 1982, Lithology and petroleum prospectus of Jurassic-Neocomian bituminous deposits of Western Siberia: Internat. Geology Review, **24**, No. 10, 1211–1221.

Petroleum Industry in the U. S. S. R., Petroleum geology of the sedimentary basins: Beicip and Robertson Research International Ltd., **1**, 89–114.

Chapter 6: Reservoir Surveillance
Geophysical Methods for Reservoir Surveillance

*J. H. Justice**

Introduction

For many years, geophysics has played a role of paramount importance in the quest for new oil and gas fields around the world. Spectacular successes may be attributed to its use, and much effort has been devoted to developing new and better geophysical technology to be used in this search. Today, attention is turning to the careful exploitation of existing reservoirs where enormous resources still remain to be recovered. The potential in heavy oils alone is on the order of tens of billions of barrels.

Most enhanced oil recovery (EOR) operations are not well understood in detail, and often many years of monitoring production are required to begin to piece together a reasonably accurate reservoir description. In many situations an accurate description may never be achieved, and many production strategies are simply hit-or-miss. If recovery from existing reservoirs could be improved by even a few percent, while at the same time improving our understanding of the reservoirs, the effort might be well worth the cost.

To be able to see what has not been seen before is the dream which has stimulated many scientific advancements and much geophysical research has focused on the problem of imaging the structure of the earth's crust in sufficient detail to be able to predict where hydrocarbons may be found. Today new potential is being realized in geophysical imaging which may allow a more detailed knowledge of the inner structure and even physical state of a reservoir, and this knowledge may permit monitoring of changes in time. In particular, changes which are induced by EOR programs may be monitored. Geophysics stands today at the threshold of a new era in which new applications and new directions for research will be found. The benefits to be derived from new uses of geophysical technology are significant.

New applications of geophysical technology to reservoir surveillance are making use of some existing technology and are stimulating the development of new technologies. Some of the existing technology which is beginning to find use includes high resolution surface seismic imaging, particularly three-dimensional (3-D), multicomponent recording, and vertical seismic profiling (VSP). Some new technologies under development utilize boreholes for passive and active monitoring and include crossborehole tomography. New developments in rock physics are beginning to provide information from seismic data which goes beyond structural imaging. In particular, direct interpretation of the effects of some EOR processes and imaging of the variations in elastic parameters within the reservoir are becoming possible. By monitoring these parameters in time, changes introduced by production operations may become accessible for analysis and these analyses may aid in planning field development and operation.

Seismic technology may be supplemented with other techniques derived from electromagnetic and potential field theory. Electrical surveys and controlled-source audio-magnetotelluric (CSAMT) surveys have been considered for reservoir monitoring. Electromagnetic data can give information on resistivity variations which may be diagnostic of lithology changes or changes in fluid content. The tiltmeter is being used to monitor surface deformations induced by shallow EOR projects. Borehole gravimeters may find use in reservoir surveillance through their sensitivity to density variations in the reservoir. Density variation detectable by a borehole gravimeter may relate to variations in porosity or lithology.

The question now raised is what can be imaged or monitored and what will the resulting information tell about the reservoir? Seismic technology can give information about the velocities of both P- and S-waves, which may change as reservoir fluids move. Multicomponent data acquisition can yield information on shear wavefields, including S-wave birefringence, and may give clues to anisotropy associated with stress or fracturing. Seismic amplitudes carry information on attenuation in the medium. Phase variations may encode frequency selective attenuation and tuning effects which may be characteristic of se-

*Mobil Research and Development, Box 819047, Dallas, TX 75381.

quences of thin beds. Seismic data can provide images of these attributes, which can reveal variations within a reservoir and give clues about the fluids which may be present.

The study of rock properties is assuming a role of fundamental importance in relating variations in geophysical measurements to information about the state and composition of a reservoir. Much work needs to be done to establish links between geophysical measurements and their interpretation in terms of reservoir description. When geophysical measurements are repeated over some time span, observed changes may be linked more easily to changes in the reservoir than if just one set of measurements were made. A good example of this is the drop in P-wave velocity in heavy oils when heated, which may allow monitoring of a steam-injection program. "Difference" images between successive sets of measurements allow us to remove static (nontemporal) variations from consideration. Static variations in measurements would include those due to lithologic variations or variation in rock properties, such as porosity. Of course, even properties which might be expected to be static may become dynamic in some cases.

Probably the weakest link in using geophysical methods for reservoir surveillance is the lack of adequate research in rock physics to provide a foundation for interpretation. Our ability to make measurements exceeds our ability to interpret them. Difference imaging helps to narrow the possibilities in an interpretation, but only more research in petrophysics can help to remove uncertainties.

While much research remains to be done, a few basic petrophysical results may be exploited today for seismic monitoring of reservoirs. A review of some of these follows.

The Physical Basis for Seismic Monitoring of Reservoirs

Velocity of Propagation of Seismic Waves

Seismic waves exhibit two modes of propagation, compressional (P) and shear (S). The velocity of propagation in an isotropic elastic medium is a function of the Lamé parameters and the rock density. The Lamé parameters may be expressed in terms of familiar parameters in rock mechanics, including the bulk modulus (incompressibility), shear modulus, and Poisson ratio. In view of these relationships, knowledge of P- and S-wave velocities can be used to infer mechanical properties.

In addition to deriving mechanical parameters of the medium directly from P-and S-wave velocities, other relationships between these quantities have significance for interpreting rock properties. The ratio of P-wave to S-wave velocities can provide a means of distinguishing between sandstones and carbonates, consolidated and unconsolidated sediments, and liquid-saturated and gas-saturated sands. This ratio shows a characteristic increase in gaseous zones. S- to P-wave traveltime ratios have been associated with porosity development in clean carbonates.

At a more empirical level, various reservoir properties may be indirectly related to changes in seismic velocity. A widely used example of such an empirical relationship is expressed in the Wyllie time-average equation,

$$\frac{1}{V} = \frac{\phi S_w}{V_w} + \frac{\phi(1 - S_w)}{V_h} + \frac{(1 - \phi)}{V_m},$$

in which ϕ is the porosity of the medium, V is the velocity of propagation in the bulk material; V_w is the velocity in water, V_h in hydrocarbon, and V_m in the rock matrix; and S_w is the water saturation. Although not strictly correct, relationships of this kind provide a useful guide to predicting velocity changes when parameters are changed.

As an example, this equation was used to predict changes in P-wave velocities due to a waterflood project (Dunlap et al., 1988). In this particular time-lapse application, the porosity was assumed to be unchanging and changes in velocity were attributed to the changes in the relative saturations of fluids resulting from the replacement of oil by water as the flood progressed.

In another application, given a spatially lithologically uniform reservoir in which fluid saturations do not change significantly, an image of P-wave velocities might be interpreted as an image of porosity variation.

In 1983, Conoco carried out surface seismic surveys over a steam flood project and noted an apparent change in the reflection coefficient of the steamed zone. Laboratory experiments by Amos Nur of Stanford University confirmed that P-wave velocities drop in sands saturated with heavy oils as the temperature increases (Nur, 1984). Time-lapse imaging can reveal the velocity reduction as the temperature increases due to steam or in-situ combustion EOR projects (Justice et al., 1989). By reconstructing both P- and S-wave velocities, changes in rock properties attributed to the EOR process may possibly be imaged. This represents a powerful tool for reservoir monitoring.

In one in-situ combustion EOR project (Greaves and Fulp, 1987) the evolution of the burn was monitored by surface 3-D seismic surveys. Several associated effects seemed to contribute to the changes associated with the burn, including the development of a gaseous zone above the burn (bright-spot effect) as well as the change in velocity and associated amplitude attenuation. A similar study in a heavy-oil reservoir involving repeated 3-D surveys over a steam flood, has been reported (Pullen et al., 1987) with what appears to be good delineation of the steam zone around the injection wells. Injection of a gas such as air or oxygen into a porous medium can also reduce seismic velocities (according to Biot theory) and may provide an additional effect which can be imaged in in-situ combustion programs.

Although not injected in gaseous form, reductions in seismic velocities in samples of Ottawa sand due to CO_2 injection have also been observed by Nur, an observation that might provide a basis for monitoring CO_2 injection EOR programs. Laboratory studies by Shell Oil predict a 4 percent change in seismic velocities due to CO_2 injection in one reservoir (Henley, 1988). The implication is that this effect would not be nearly as strong as the thermal effect in heavy oil reservoirs.

In addition to laboratory experiments to predict changes in rock properties as a result of production strategies, a second technology has been developed (Justice et al., 1989) using modified reservoir simulators. The change in physical parameters in a reservoir model are monitored using a reservoir simulator. By applying appropriate transforms relating physical parameters of the medium to seismic velocities, we may infer the seismic response of the reservoir. This technique provides a powerful tool for understanding the seismic response of the reservoir and for planning geophysical surveillance programs. Use of the technique has indicated the possibility of monitoring certain production programs, including some waterfloods, with geophysical methods.

Obviously, there is reason to be optimistic about seismic monitoring of thermal EOR, both steam and in-situ combustion in heavy oil reservoirs and nonthermal EOR including water and CO_2 injection.

Amplitude Attenuation of Acoustic Waves

In experiments on velocity variations as a function of temperature, Nur and his students found that amplitude attenuation increases significantly as the temperature is increased. However, applications have not been reported.

Amplitude attenuation effects associated with propagation through fracture zones and through gas-saturated zones are also known and provide a further basis for seismic monitoring.

Seismic Anisotropy

Seismic wave anisotropy is the variation in velocity as a function of direction of propagation. A secondary effect of anisotropy is the splitting of S-waves into polarized components which propagate at different velocities. Seismic anisotropy can be associated with propagation through fracture zones. Thus inferring stress field or fracture orientation and fracture density may be possible (Martin and Davis, 1987).

The detection of fractured zones can be important in development drilling, and fracture orientation may play an important role in EOR programs. Fractures may provide high permeability pathways for fluid movement, and may divert a flood from its planned course or provide channels for early breakthrough. Surface multicomponent surveys to detect fracturing have had mixed success and more work remains to be done.

Data Acquisition Considerations

A reservoir is a unique object for study: its economic value is high, and it therefore justifies more detailed geophysical analysis than is used in exploring for new prospects. Most 3-D surveys currently are conducted for the purposes of field development or reservoir delineation, but 3-D surveys for EOR monitoring are beginning to be reported. Multicomponent seismic methods will be most easily justified in reservoir surveillance applications, although they may also find use in exploration. Finally, high-resolution imaging of all kinds finds justification in the reservoir environment. Seismic transmission tomography is being developed for EOR monitoring applications and reservoir description. High-resolution imaging places new demands on seismic data acquisition systems originally designed for two-dimensional seismic exploration.

Higher resolution requires higher frequency content and broadband recording as well as denser spatial sampling. Multicomponent and 3-D data acquisition call for a large number of recording channels, often 1000 or more. New borehole seismic tools with high resolution capabilities are required with broad bandwidths and sample rates no greater than a fraction of a millisecond. Borehole tools must be designed for severe conditions because corrosive fluids, high pressures, and high temperatures are not uncommon. Dewar-flask technology, of course, is available and widely used, but other solutions would be welcome. Ability to withstand borehole temperatures in excess of 350°F (175°C) for at least a few hours is certainly desirable. Multielement tools may minimize time in the borehole as well as minimizing lost production. High data rates over many channels will be required from tools operating, ideally, from standard seven-conductor wirelines.

A number of new seismic sources are being developed and improvements will probably come rather rapidly in many areas. Tools capable of radiating high levels of broadband acoustic energy laterally into the formation from a cased borehole without damage will find many applications. They should be able to withstand high temperatures and pressures, have short firing-cycle times, and minimal maintenance requirements. Orientable or polarized sources for S- and P-waves will also find important applications.

New electrical and electromagnetic survey tools, such as those used in controlled-source audio-magnetotelluric (CSAMT) surveying, will probably find use in reservoir surveillance. Borehole applications may require nonconducting casing for completed wells for electrical and EM borehole monitoring.

Data-Processing Considerations

Much of our processing technology finds direct application in reservoir surveillance. New types of surveys, however, require development of new systems. In particular, seismic tomographic reconstruction and multicomponent acquisition require new software and system development.

Interactive workstations and systems which incorporate artificial intelligence or pattern recognition capabilities, perhaps using new developments in neural networks, can be envisioned. Ultra-high-speed computing algorithms employing large numbers of parallel processors will find use, along with new developments in operating systems, compilers, and data management.

Field computing systems will benefit from microminiaturization and parallel processing, and much more data reduction and analysis will become possible as the data are being acquired. Permanent instrumentation, with on-site computing capabilities may become commonplace for reservoir surveillance, making this activity a routine part of the field engineer's information base.

Finally, data processing will be extracting more information from data sets than ever before. Utilizing the full seismic wavefield, many physical measurements, including the elastic parameters of the medium as a function of space and time, may be deducible. This information may be supplemented with electrical and EM parameters of the medium and borehole data, as well as density from gravity measurements. Systems for interpretation of integrated data sets may then allow a reasonably full and accurate assessment of the state of the reservoir and its probable future states. Reservoir simulators, predicting the performance of production programs operating in a feed-back loop, will continually update the reservoir model, resulting in highly accurate reservoir predictions.

Looking Ahead

As the developments, which have been incompletely surveyed here, become reality, the reservoir will become a well understood entity. The net result will be optimal management of the recovery process. The economic benefits will be substantial, with many billions of barrels of oil and trillions of cubic feet of gas being added to recoverable reserves.

References

Dunlop, K.N.B., King, G.A., and Breitenbach, E.A., 1988, Monitoring of oil/water fronts by direct measurement: Soc. Petr. Eng. Paper 18271, Soc. Petr. Eng. Ann. Mtg.

Greaves, R.J., and Fulp, T.J., 1987, Three-dimensional seismic monitoring of an enhanced oil recovery process: Geophysics, 52, 1175–1187.

Henley, D.C., 1988, Differential tomography of a CO_2 flood: Soc. Expl. Geophys. Interwell Seismic Surveying Workshop.

Justice, J.H., Vassiliou, A.A., Singh, S., Ranganayaki, R.P., Cunningham, P.S., Solanki, J.J., and Allen, K.P., 1989, Recent developments in geophysics for reservoir characterization: 59th Ann. Internat. Mtg., Soc. Expl. Geophys., Expanded Abstracts, 549–550.

Justice, J.H., Vassiliou, A.A., Singh, S., Logel, J.D., Hansen, P.A., Hall, B. R. , Hutt, P.R., Solanki, J.J., 1989, Acoustic tomography for monitoring enhanced oil recovery: The Leading Edge, 8, No. 2, 12–19.

Martin, M.A., and Davis, T.L., 1987, Shear-wave birefringence: A new tool for evaluating fractured reservoirs: The Leading Edge, 6, No. 10, 22–28.

Nur, A.M., 1984, Seismic monitoring of thermal enhanced oil processes: 54th Ann. Internat. Mtg., Soc. Expl. Geophys., Expanded Abstracts, 377–340.

Pullen, N. E., Jackson, R. K., Matthews, L.W., Hirsche, W.K., den Boer, L.D., 1987, 3-D imaging of heat zones in an athabasca tar sands thermal pilot: 57th Ann. Internat. Mtg., Soc. Expl. Geophys., Expanded Abstracts, 391–394.

Robertson, J.D., 1987, Carbonate porosity from S/P travel-time ratios: Geophysics, 52, 1346–1354.

Aspects of Rockphysics in Seismic Reservoir Surveillance

Zhijing Wang and Amos Nur[‡]*

Introduction

Seismic reflection methods have been used in the past mostly to delineate structures which might bear hydrocarbons and relatively little use has been made for the determination of the rock properties of direct interest to hydrocarbon recovery (e.g., porosity, permeability) and reservoir surveillance.

Because of the growing realization that reservoirs are more heterogeneous than assumed in the past, a major shift in the use of seismic methods is taking place. One of the central aspects of this shift is the need to better understand the relation between the seismic properties of reservoir and reservoir related rocks, their production properties (porosity, permeability), and their state (mineralogy, saturation, pore pressure, etc.). Improved understanding is needed for the evaluation of stratigraphic traps, fracture detection, and determining the spatial distribution of porosity and permeability. Reservoir complexity typically is related to the spatial heterogeneity in porosity, permeability, clay content, fracture density, etc., that cannot be inferred at a satisfactory level of detail from well testing data, logs, or cores, but, hopefully, may be obtained from seismic measurements.

During the past several years, we have carried out a series of laboratory investigations on the effects of enhanced oil recovery (EOR) processes on acoustic velocities in both reservoir fluids and rocks. Surprisingly large changes were discovered with temperature and pressure changes and with CO_2 flooding and hydrocarbon solvent flooding. These results imply that reservoir parameters may be detected in-situ using seismic measurements to map the spatial heterogeneity of porosity and permeability, detect anomalous pore pressure and its temporal variations, detect subsurface fractures, track thermal fronts, monitor the movement of gas caps, and track the fluid fronts in CO_2 flooding, hydrocarbon solvent flooding, and possibly water flooding.

Reservoir surveillance using seismic methods has become increasingly attractive for many investigators.

Some in-situ investigations on seismic monitoring of thermal (in-situ combustion and steam flooding), hydrocarbon miscible, and water floodings have been carried out and reported (Britton et al., 1983; Greaves and Fulp, 1987; Pullin et al., 1987; den Boer and Matthews, 1988; King, 1988; Macrides et al., 1988; Hirsche and Sedgwick, 1989). In-situ investigations were, or should have been, based on the rock-physical results of laboratory measurements because whether seismic methods effectively can be used to monitor a specific EOR or production process depends on the sensitivity of seismic wave velocities and/or amplitudes to this process. Seismic velocities in reservoir rocks depend in complex ways on many reservoir parameters and factors, and a specific EOR or production process may cause more velocity changes in one reservoir than in another.

In this paper we summarize some of the results of our studies on seismic wave velocities in reservoir fluids and rocks containing hydrocarbons under conditions relevant to thermal EOR, CO_2 flooding, hydrocarbon solvent flooding, gas drive, and other EOR and production processes.

Velocity in Reservoir Fluids

Seismic properties of reservoir fluids are of great importance in reservoir descriptions and production assessments using seismic methods. Unfortunately, they often have been neglected, despite the fact that one purpose of exploration and reservoir management is to delineate reservoir fluids and assess fluid saturations. They are needed particularly in the interpretation of reservoir seismic surveys and acoustic logs, in production and EOR assessment, management, and monitoring, and in seismic modeling. Without knowing and understanding the seismic properties of the reservoir fluids, correct interpretation and understanding of field results will be difficult and sometimes impossible. In this section, we summarize some laboratory results on the measurement of seismic velocities in reservoir fluids.

Velocity in Reservoir Gases

Seismic velocities in reservoir gases are relatively unimportant to geophysicists. Under reservoir condi-

*Chevron Oil Field Research, 1300 Beach Blvd., P.O. Box 446, LaHabra, CA. Formerly Core Laboratories.
[‡]Department of Geophysics, Stanford University.

tions, the velocities in both hydrocarbon and nonhydrocarbon gases range from about 200 to 400 m/s. They increase as temperature increases, decrease as pressure increases in low-pressure ranges, and decrease as the molecular weight or density increases. In contrast to those in liquids, seismic velocities in gases can be calculated fairly accurately from gas theory and models such as equations of state (Batzle and Wang, 1992). Since gases are highly compressible, the velocity changes in dry rocks upon gas saturation usually are negligible.

Velocity in Water

Water has very complex properties. Seismic velocities in water behave very differently from those in hydrocarbons as temperature changes. Figure 1 shows the velocities in pure water as a function of both temperature and pressure. At atmospheric pressure, the velocity increases as temperature increases to about 73°C, and then as temperature increases further, the velocity decreases. At higher pressures, the velocity maximum shifts to higher temperatures and the velocity becomes less dependent on temperature. When water contains impurities (e.g., salt), the velocity is slightly higher.

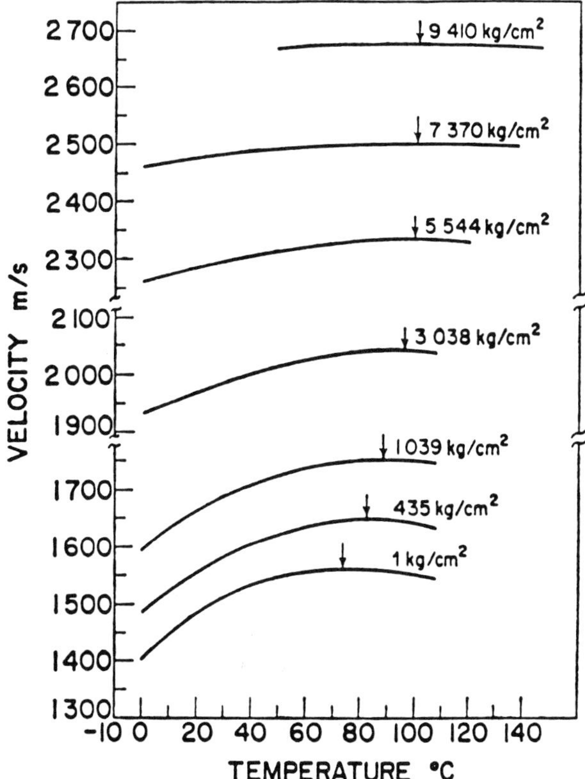

Fig. 1. Velocity in pure water as a function of temperature and pressure (Lawson and Hughes, 1963).

Velocity in Light Hydrocarbons

The compressional velocities in 26 pure hydrocarbon samples (13 n-alkanes, 10 1-alkenes, and 3 cycloparaffins) were reported in Wang and Nur (1987) as a function of temperature at atmospheric pressure, with accuracy of about 0.1 percent. In contrast to the velocities in hydrocarbon gases, the velocities in all these liquid hydrocarbons decrease approximately linearly with increasing temperature (Figure 2), with the data closely fitted by the simple function

$$V = V_o - b(T - T_o), \qquad (1)$$

where V and V_o are the velocities at temperatures T and T_o, and b, the temperature coefficient of velocity, is a constant for a given liquid hydrocarbon.

The approximately linear variation of velocities with temperature has been observed in many liquids. For example, velocities in nitrobenzene, methyl acetate, ethyl acetate, and diethylene glycol monoethyl ether all show linear decreases with increasing temperature in the range of 25° to 110°C (Rao and Rao, 1959). Similar velocity decreases were shown in Bradley (1963) for Dow Corning 200 (TM) fluid, in Chavez et al. (1981) for liquid trichlorofluoromethane, and in Hunter et al. (1963) for liquid mercury.

The measured velocity decreases versus temperature in the 26 hydrocarbons are very large, amounting to 35 to 45 percent as the temperature increases by 100°C. The exact magnitude depends on molecular weight. Velocities in heavier hydrocarbons are less sensitive to temperature changes than those in light hydrocarbons. The temperature coefficient b in equation (1) is fitted by the empirical relation

$$\frac{1}{b} = 0.306 - \frac{7.6}{M}, \qquad (2)$$

where M is molecular weight.

In light crude oils, the velocities have essentially the same relationship with temperature as the pure hydrocarbons shown above, although the absolute change with temperature varies among oils. Figure 3 shows the measured velocities in a 34° API oil. An approximately linear relationship also exists between the velocity and temperature.

Velocity in Heavy Hydrocarbons

Velocities in heavy hydrocarbons respond to temperature and pressure changes a little differently than those in light hydrocarbons. Figure 4 shows that the velocity in heavy crude A (10° API) decreases rapidly as temperature increases at different pressures but the velocity decrease is nonlinear. This velocity decrease apparently is related to the composition of the heavy oils, which usually contain some solid and semi-solid particles such as asphalteens and waxes, and when

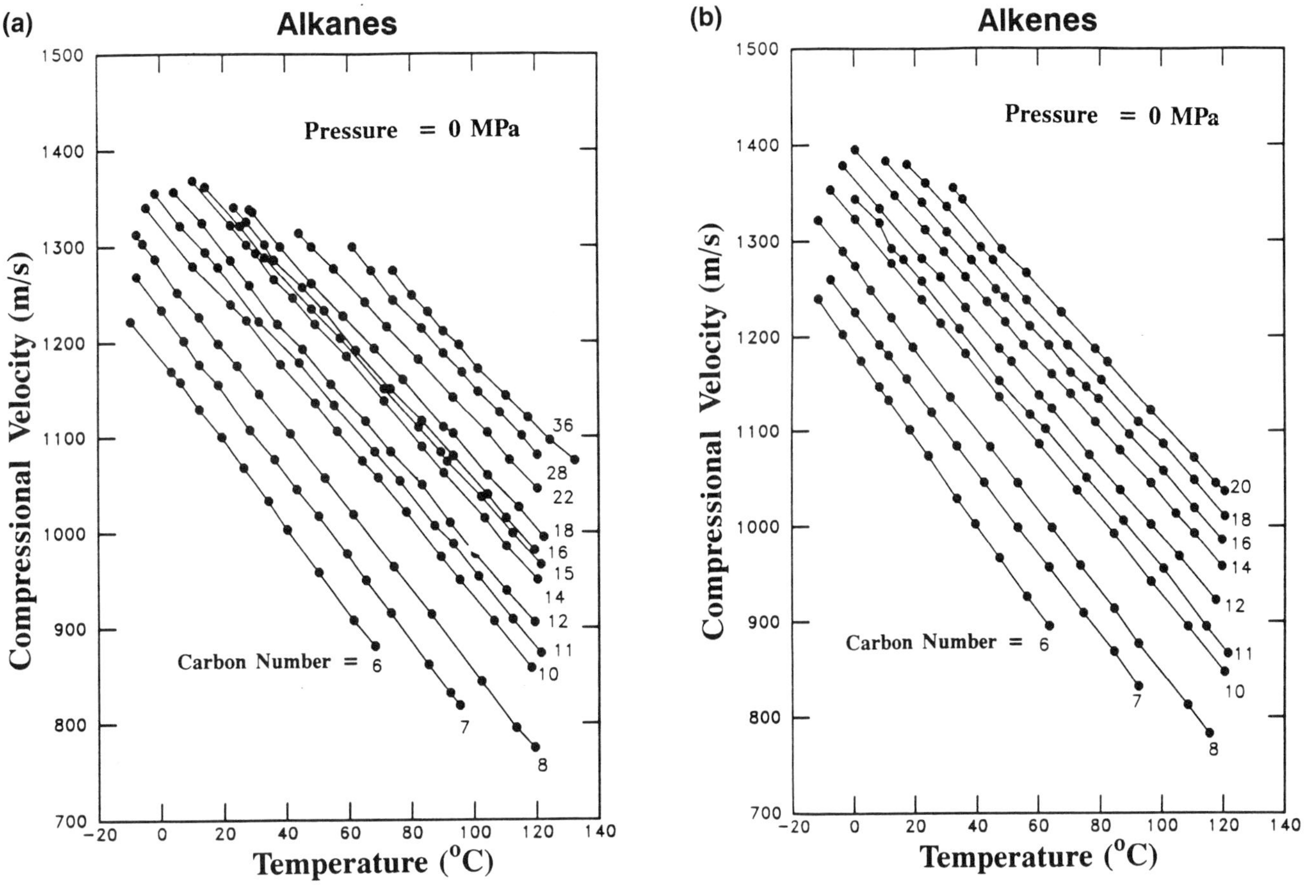

(a) Alkanes

(b) Alkenes

Fig. 2. Velocities in normal alkanes (a) and 1-alkenes (b) versus temperature.

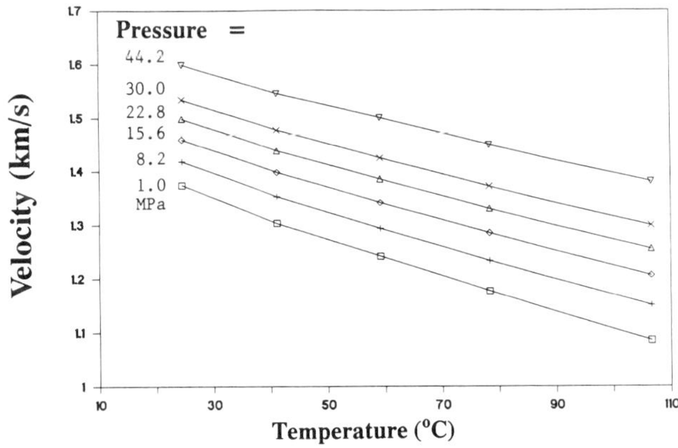

Fig. 3. Velocity in a 34° API crude oil versus temperature at different pressures (Wang et al., 1988a).

melting, these materials cause faster velocity decreases.

Figure 5 shows the velocities in a tar (11° API) and a crude oil (22° API) recovered from steam flooding pilot projects. In both materials, marked velocity decreases occur with increasing temperature. These velocity decreases are caused mainly by the increasing compressibility with increasing temperature.

Velocity in Live Oils

Velocity in a live oil (gas-saturated oil) was measured as a function of pressure at two temperatures as reported in Wang et al. (1988a). Figure 6 shows that the velocity decreases linearly with decreasing pressure. As the pressure reached the bubble point (3000 psig or 20.7 MPa at 72°C), the velocity did not drop sharply as expected from theory, suggesting that the bubbles were too small to affect the velocity. The pressure was then released to let the saturated gases escape, and later, the velocity was measured again at the same temperatures versus pressure in the remainder of the live oil. The velocity again had a linear relationship with pressure but was substantially higher than before. This means that dissolved gases in crude oils can lower the velocity significantly (above 15 percent).

Velocity-Temperature-Molecular Weight

A simple functional relation is obtained when the velocities in alkanes and alkenes are plotted as a function of the inverse of the molecular weight (Figure 7) revealing an approximately linear relationship. As seen in Figure 7, the velocity data fit straight lines to within 1.5 percent. Therefore, equation (1) can be extended to include both temperature T and molecular weight M,

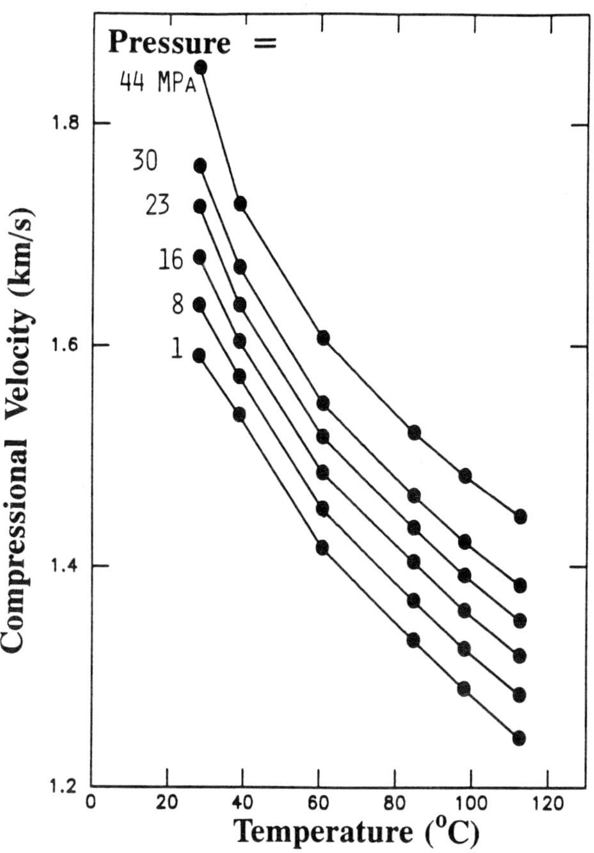

Fig. 4. Velocity in a 10° API crude oil versus temperature at different pressures (Wang et al., 1988a).

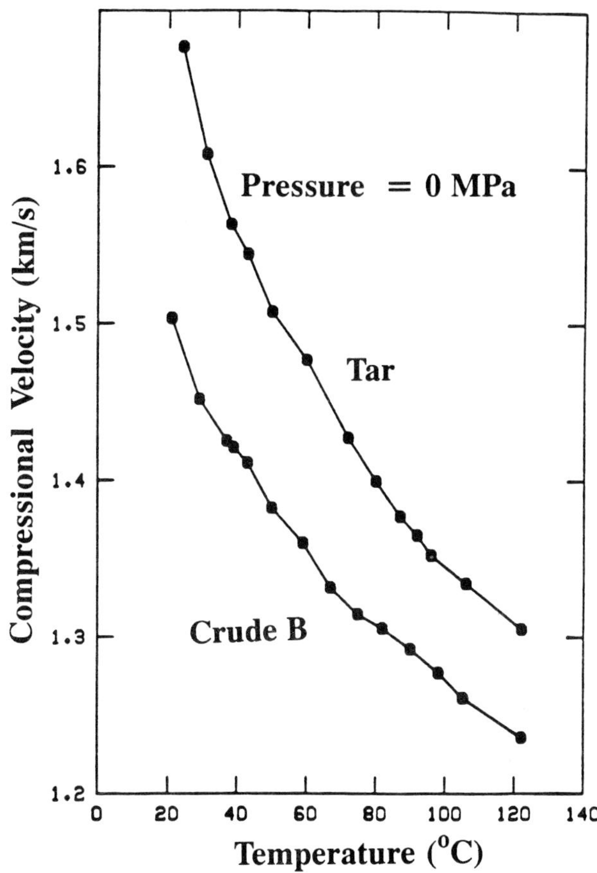

Fig. 5. Velocities in a tar and a crude oil recovered from steam flood pilot projects versus temperature (Wang and Nur, 1986).

$$V = V_o - b(T - T_o) - a\left(\frac{1}{M} - \frac{1}{M_o}\right), \qquad (3)$$

where M_o is a reference molecular weight. The coefficient b is generally a function of molecular weight, and a is a function of temperature.

We interpolated the coefficients a and b from the velocities in alkanes and alkenes and calculated seismic velocities in 5 crude oils whose average molecular weights were known. The results are shown in Table 1. For three light oils (oils E, F, and G), the calculated velocities (V_C) fit the measured (V_M) within 2 percent, but for heavy oils C and D, the deviations are 7 percent and 5 percent. The misfit may be caused by the inaccuracy of determining the molecular weights and the impurities and nonfluid materials.

Velocity-Temperature-Pressure-API Gravity

Experimental results from Wang et al. (1988a) show that velocities in crude oils decrease as the API gravities increase (as densities decrease) (Figure 8). The velocity data as a function of both temperature and pressure in 10 crude oils with API gravities ranging from 5 to 62 degrees were regressed to:

$$V = V_o + \beta_1 T + \beta_2 P + \beta_3 PT, \qquad (4)$$

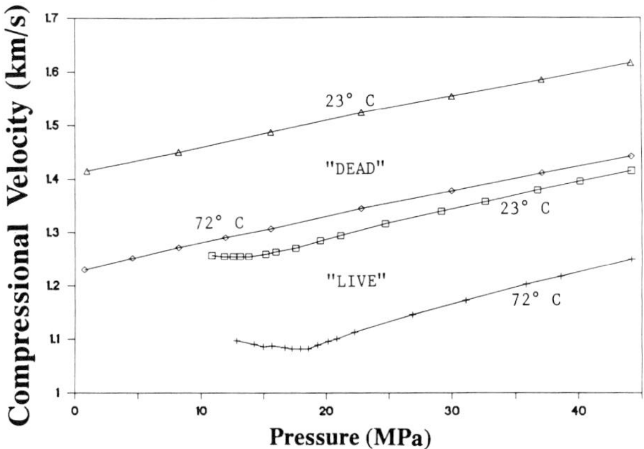

Fig. 6. Velocities in a live oil versus pressure at two temperatures. Here "live" means the oil is saturated with gas and "dead" after the saturated gas has escaped.

where β_1, β_2, β_3, and V_o are the regression coefficients. The correlation factors are greater than 0.99 for most of the oils. Figure 9 shows the calculated velocities, along with those measured by Han (1988, pers. comm.), in a 16° API oil. The calculated velocities deviate from the measured ones by about 3 percent. Examples from Wang (1988) show similar results in other oils.

Note that, unlike gases, velocities in liquids can not be calculated accurately from theory. The calculated velocity values using theory typically deviate from those measured by 20 to 100 percent (Wang, 1988).

Effect of Temperature on Velocities in Heavy Oil Sand and Tar Sand

In 1984 Tosaya et al. (1984) first noticed surprisingly large velocity decreases with increasing temperature in cores of heavy oil and tar sands (with 28 to 37 percent porosities) from Venezuela, California, and Canada. In some cores such velocity decreases reached 40 percent and more with temperature increasing from 25° to 150°C. Because such changes were not anticipated, a systematic study was undertaken to identify their causes and understand what controls their magnitude.

We conducted a series of laboratory investigations on rocks and sands saturated with heavy hydrocarbons. Compressional (P-) and shear (S-) wave velocities were measured in sandstones and clean sands saturated with paraffin wax, heavy crudes, water, air, and light hydrocarbons (Wang and Nur, 1986; 1987; 1988a, b; Nur and Wang, 1987; Wang, 1988; Wang and Nur, 1990). Compressional and shear velocities in wax-saturated sandstones and clean sands are significantly higher than in cores saturated with water in the

Table 1. Comparison of the measured (V_M) with the calculated (V_C) velocities by equation (3) in five crude oils with known molecular weights.

Oil Name	API Gravity	Molecular Weight	22°C V_M	22°C V_C	50°C V_M	50°C V_C	75°C V_M	75°C V_C
Oil G	43	193	1315	1321	1212	1205	1123	1115
Oil F	34	243			1268	1253	1185	1159
Oil E'	~23	~290			1308	1291	1214	1192
Oil C	10	504					1362	1265
Oil D	10.5	506					1340	1268

Pressure = 0 psig, Velocity Unit: m/s

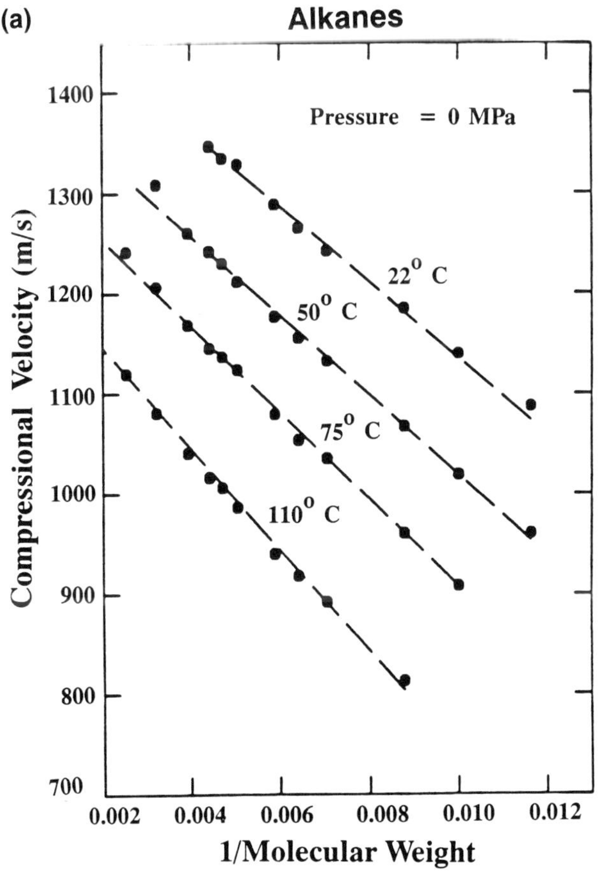

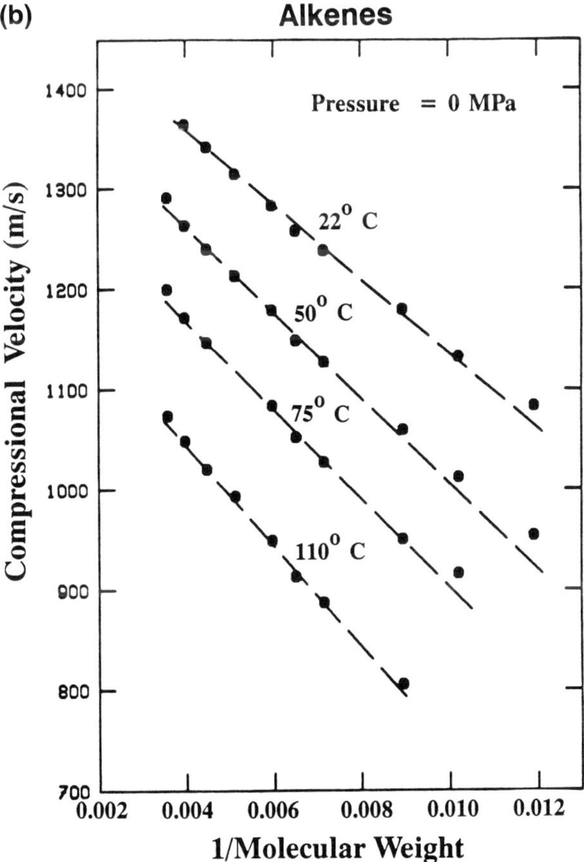

Fig. 7. Velocities in normal alkanes (a) and 1-alkenes (b) versus the inverse of the molecular weight at different temperatures (Wang and Nur, 1987).

temperature interval of 20° to 45°C, due to the higher effective elastic moduli of the wax-saturated rocks. The effect is particularly large for the *S*-wave velocity.

The *P*-wave velocities increased in sandstones and sands saturated with pure hydrocarbons as the molecular weight of the hydrocarbons increased. There was almost no change in *S*-wave velocities. However, both *P*- and *S*-wave velocities decreased by about 6 to 9 percent as the temperature was increased by 100°C.

Results also were obtained on the effects of temperature and oil content on velocities in tar sand samples, having about 30 percent porosities, from California and Venezuela. Figure 10 shows that in heavy oil-saturated sands velocities are extremely sensitive to temperature. A large decline in *P*-wave velocity (up to 40 percent) suggests that velocity may have application in mapping heated oil zones around injection wells in steam floods or fire floods. Samples of Kern River sand and Venezuelan sand having 50 percent of the pore space filled with heavy oil show similar velocity behavior with temperature changes but with smaller magnitude of the velocity decrease than samples having 100 percent pore space filled with heavy oil.

Figures 11a and 11b show that both *P*- and *S*-wave velocities in a Berea sandstone sample (with 19 percent porosity) saturated with a 10° API oil decrease rapidly as temperature increases. The velocity decreases range from 10 percent at high differential pressures to 20 percent at low differential pressures.

In Figure 12, the *P*-wave velocity decreases by about 5 percent in an unconsolidated Ottawa sand sample (with 37 percent porosity) saturated with air or water, as the temperature increases from 22° to 122°C. In contrast, in the same temperature interval, the velocity decreases by up to 35 percent when the same sand sample is mixed with a tar recovered from a steam flood pilot. Figure 12 also shows that the velocity is not sensitive to the tar percentage in this tar

sand. When the sand is mixed with 10.7 percent tar (by weight), the pore space of the sand sample is 70 percent filled. This means that the sample is grain supported, so that velocity at low temperature is high. The pore space of the sand sample is completely filled when the sand contains 20 percent tar (weight). This tar sand sample is partially tar-supported, and consequently, the velocity at low temperature is lower than that of the sand with 10.7 percent tar. At higher temperatures, both tar sand samples may become tar-supported due to the high thermal expansivity of the tar, so that the velocities are close.

As the 10.7 percent tar sand sample is saturated with water, the velocity becomes less sensitive to temperature changes. However, the velocity decrease is still much higher than that in a loose sand sample saturated with either gas or water.

The large decrease of the *P*-wave velocity in tar sands is caused not only by the compressibility increase, but also the decrease in the shear modulus, as the temperature increases. Tar at room temperature is semi-solid [with viscosity in the order of 1000 Pa.s (10^6 cp)], so that the tar samples under pressure have finite shear moduli which are much higher (by orders) than the shear moduli of either the tar or the unconsolidated sand. The *S*-wave velocities in tar sands at room temperature and 10 MPa differential pressure are around 900 to 1250 m/s, depending on the tar content. As the temperature increases, the tar melts so that the shear modulus and thus the *S*-wave velocity decrease; this contributes to the fast decrease in the *P*-wave velocities. In consolidated rocks, the magnitude of the *P*-wave velocity decrease is smaller.

Table 2 summarizes the effect of temperature on *P*-wave velocities in hydrocarbons and rocks saturated with hydrocarbons, air, and water. Generally speaking, in hydrocarbon-saturated rocks, as temperature increases by 100°C, the compressional velocities de-

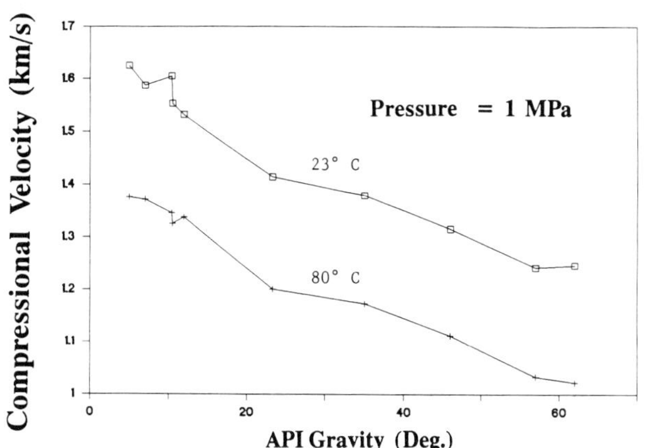

Fig. 8. Velocity versus API gravity of 10 crude oils at two temperatures.

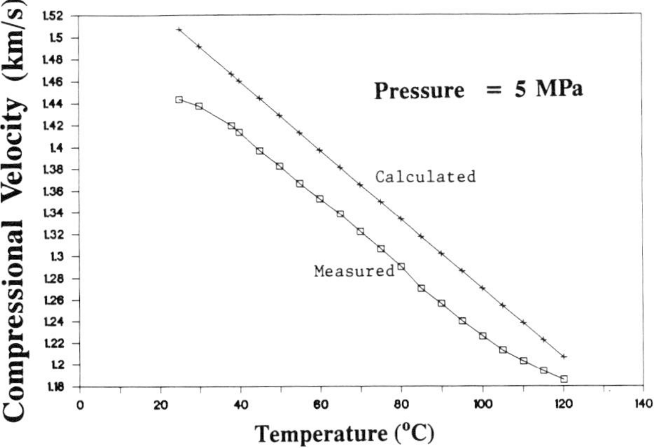

Fig. 9. Measured velocity versus the calculated using the regression coefficients in equation (4) in a 16° API oil.

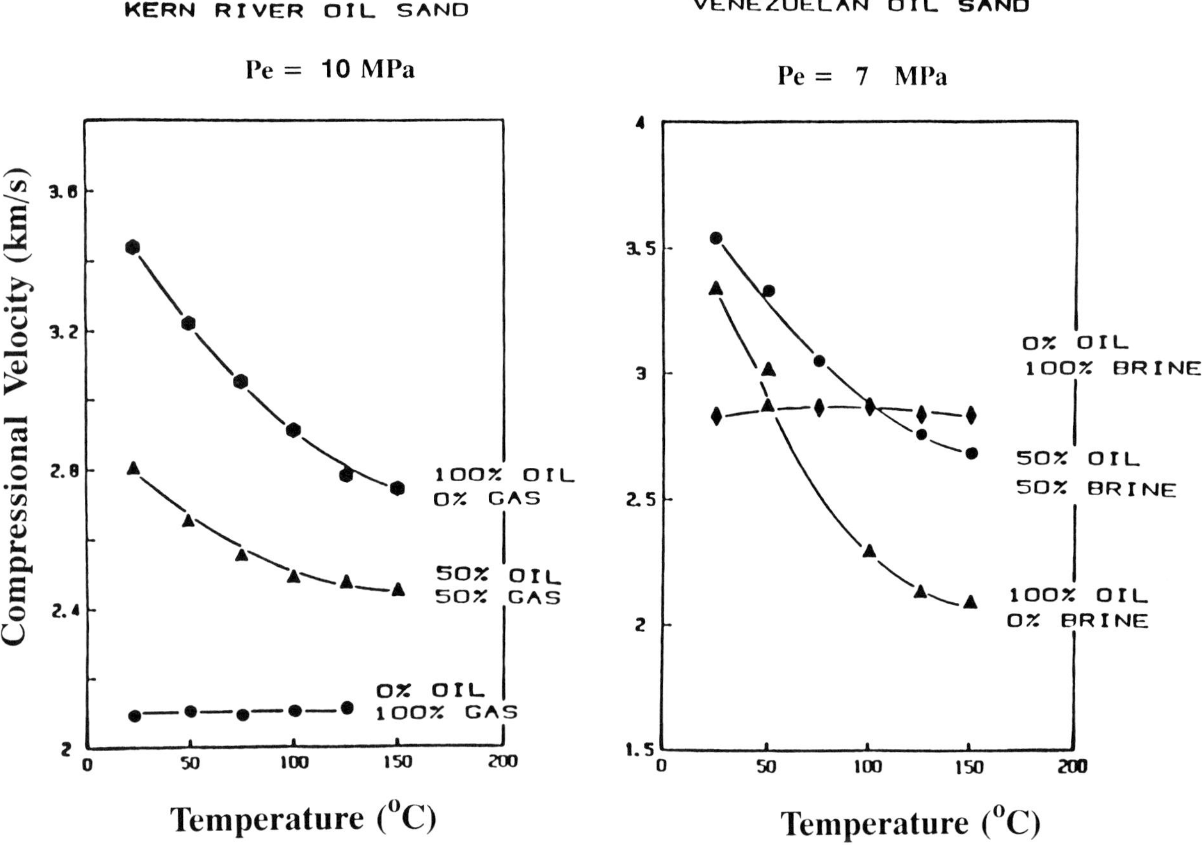

Fig. 10. Compressional velocities in Kern River and Venezuelan heavy oil sands versus temperature and oil content (Tosaya et al., 1984).

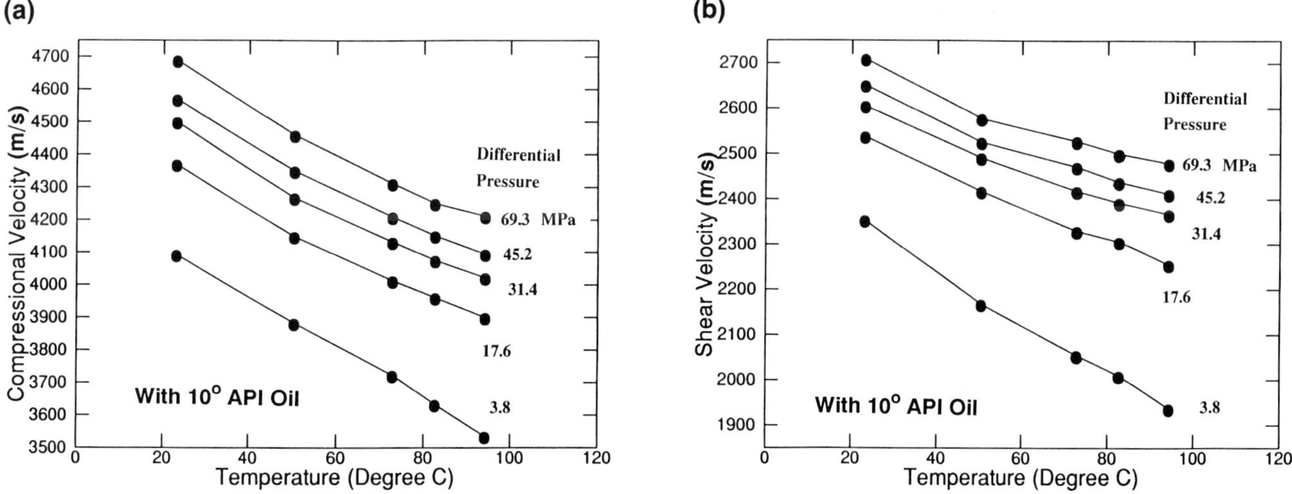

Fig. 11. Compressional (a) and shear (b) velocities in a Berea sandstone sample saturated with a 10° API oil versus temperature.

crease by 9 to 40 percent, depending on the composition and viscosity of the hydrocarbons. In consolidated rocks saturated with heavy hydrocarbons, decreases range from 9 to 22 percent, and from 15 to 40 percent in unconsolidated sands saturated with hydrocarbons. In contrast, velocity decreases in either water- or air-saturated samples are substantially less. These large temperature effects suggest that seismic methods should be successful in monitoring thermal EOR fronts, including hot water, fire, and steam floods.

Effect of CO₂ Flooding on Velocities in Rocks

To investigate the potential of using seismic methods to monitor carbon dioxide (CO_2) flooding, *P*- and *S*-wave velocities were first measured in sandstones and unconsolidated sands (with porosities ranging from 6 to 37 percent) saturated with air and with n-hexadecane ($C_{16}H_{34}$). Afterward, the velocities were measured again after the n-hexadecane-saturated cores were flooded with CO_2 both below and above the critical temperature of CO_2 (Wang and Nur, 1988c).

Table 2. Decreases of compressional velocities in hydrocarbons and rocks with hydrocarbons, air, and water, respectively, as temperature increases from 22° to 122°C.

| | Compressional Velocity Decreases in 22° - 122°C | |
in	Absolute (m/s)	(%)
Pure Hydrocarbons	350 - 450	20 - 30
Heavy hydrocarbons	270 - 800	15 - 40
Sand with Air	80	5
Sand with Water	110	7
Sand with Pure H.C.	280 - 320	15 - 17
Sand with Heavy H.C.	300 - 700	15 - 30
Sandstone with Air	130 - 160	3 - 6
Sandstone with Water	200 - 260	5 - 7
Sandstone with Pure H.C.	200 - 400	6 - 8
Sandstone with Heavy H.C.	300 - 900	9 - 22

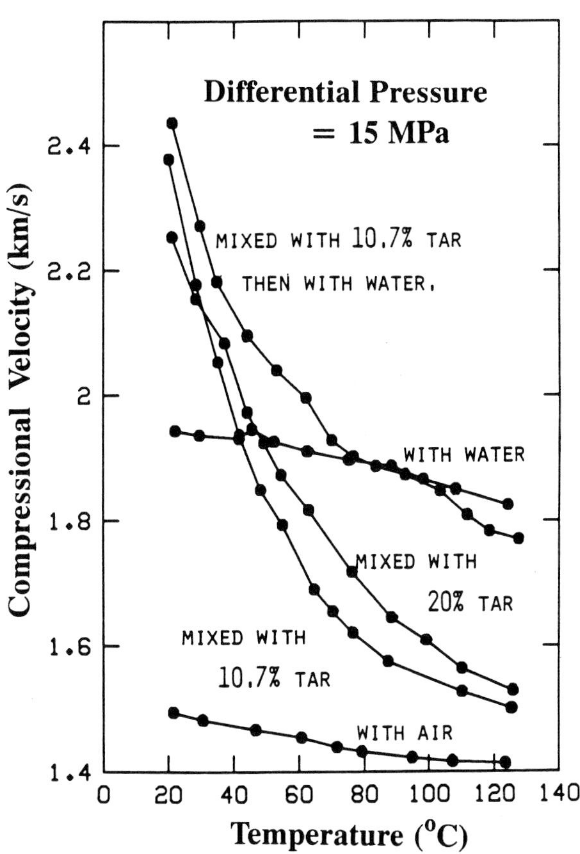

Fig. 12. Compressional velocity in Ottawa unconsolidated sand saturated with air, water, and mixed with tar, respectively.

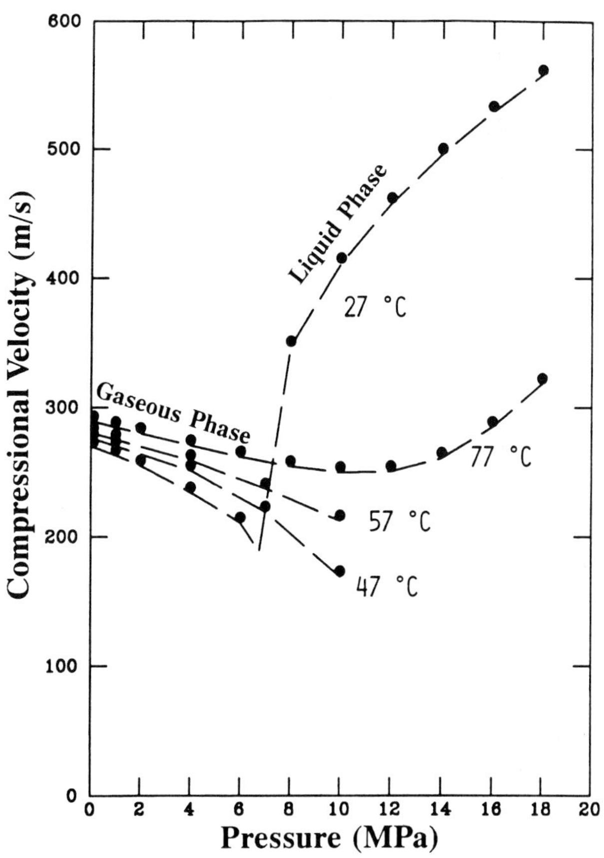

Fig. 13. Velocity in carbon dioxide (CO_2) versus pressure at different temperatures.

Carbon dioxide has a critical temperature of 31°C (88°F) at a pressure of 7.4 MPa (1070 psi). Above the critical temperature, CO_2 behaves as a gas with density increasing with increasing pressure. The density of CO_2 can be very high at high pressure below the critical temperature, even higher than that of many reservoir hydrocarbons (Holm and Josendal, 1982). For instance, at 21°C (70°F) and 6.9 MPa (1000 psi), CO_2 has a density of 0.8 g/cc. Such a high density will decrease both the P- and S-wave velocities when CO_2 replaces the hydrocarbons in a rock's pore spaces.

Velocity in CO_2 alone is very low compared with that in most liquid hydrocarbons. Figure 13 shows the P-wave velocity in CO_2 as a function of both temperature and pressure. Above the critical temperature, the velocity in CO_2 is very low and only nominally dependent on either pressure or temperature. Below the critical temperature, the velocity depends on the phase of CO_2. When CO_2 is in the liquid phase, the velocity is a strong function of pressure, increasing rapidly with increasing pressure. However, the velocity in liquid CO_2 is still much lower than that in water

or liquid hydrocarbons. When CO_2 is in the gaseous phase, the velocity decreases with increasing pressure in low-pressure ranges and increases with increasing pressure at higher pressures, resembling the behavior of most gases.

The low velocity and high density of CO_2 may make possible the use of seismic methods to monitor the CO_2 flooding process in-situ. When CO_2 replaces pore hydrocarbons, the compressional velocity decreases substantially while the shear velocity is less affected. The increase of the overall rock density also will decrease both P- and S-wave velocities. When the CO_2 is in the liquid phase, its velocity is higher but its density is also very high, so that the overall velocity in the rock still decreases dramatically.

Figures 14a and 14b show the P- and S-wave velocities in a Beaver sandstone sample (with 6 percent porosity) saturated with $C_{16}H_{34}$ and then flooded with CO_2 at a constant overburden pressure of 20 MPa. Both the P- and S-wave velocities in the hydrocarbon-saturated core decrease with increasing pore pressure. The P-wave velocity decreases markedly after CO_2 is

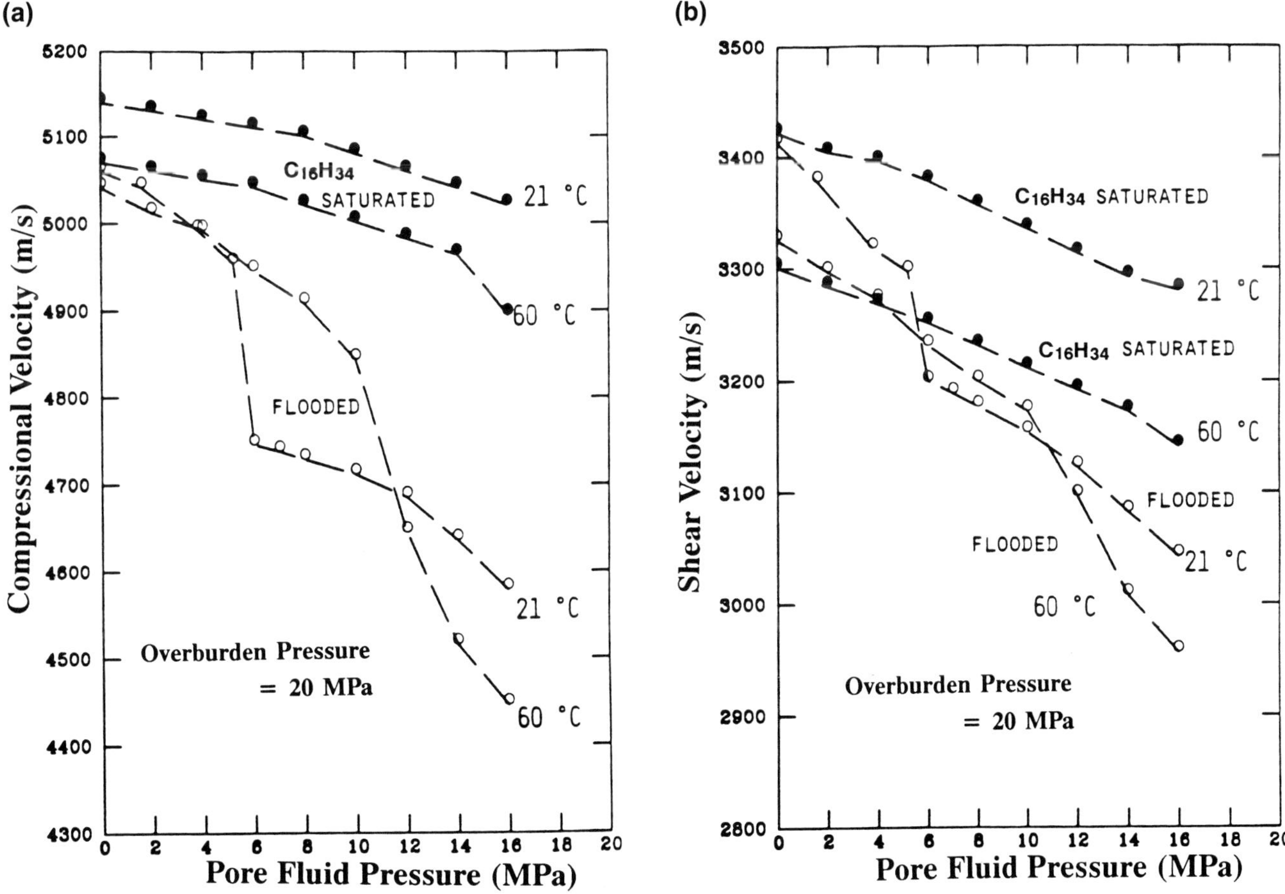

Fig. 14. Compressional (a) and shear (b) velocities in a Beaver sandstone saturated with $C_{16}H_{34}$ and then flooded with CO_2.

flooded through the core, the velocity decrease being approximately 8 percent. At 21°C as the pore fluid pressure decreases to below 6 MPa, the P-wave velocity increases sharply as the CO_2 changes its phase from liquid to gas, suggesting that both the low velocity and the high density of CO_2 contribute to the velocity decrease.

The CO_2 flooding of the Beaver sandstone core also decreases the S-wave velocity (Figure 14b). At 21°C, the S-wave velocity decreases by as much as 240 m/s or 7.3 percent at a pore fluid pressure of 16 MPa upon CO_2 flooding. As with P-waves, the S-wave velocity also increases as the CO_2 in the pore spaces transforms from liquid to gas. This increase is caused by the combined effect of density and viscosity decrease and effective pressure increase.

The P-wave velocity in an unconsolidated Ottawa sand sample with 37 percent porosity saturated with $C_{16}H_{34}$ is relatively insensitive to pore fluid pressure change (Figure 15). However, CO_2 flooding dramatically decreases the P-wave velocity, by about 470 m/s or 24 percent at a differential pressure of 20 MPa. The velocity lowering is even larger at lower differential pressures. The large effect of the CO_2 flooding on the

P-wave velocity is caused in part by the characteristics of the unconsolidated sand. Since the bulk modulus of unconsolidated air-saturated sand is very low, liquid saturation can increase the bulk modulus greatly. When the hydrocarbon-saturated sand is flooded with CO_2, the sand becomes partially saturated, and the P-wave velocity becomes even lower than that for the air-saturated sand (Murphy, 1982).

Figure 16 summarizes the P-wave velocity decreases caused by CO_2 flooding in 7 sandstones and 1 unconsolidated sand having various porosities for three different pore fluid pressures. In low porosity samples, the decrease can be as high as over 11 percent, while in high porosity samples, the decrease is only around 5 percent. In unconsolidated sand, the velocity decreases up to 30 percent. The relation we observe is predicted by the Gassmann relation (Gassmann, 1951) and Biot theory (Biot, 1956a,b): as the porosity increases, the velocity difference between gas- and liquid-saturated rocks decreases due to the increased overall density, provided that the bulk modulus of the dry rock does not decrease dramatically fast, which is generally true for most sandstones. For low porosity rocks, liquid saturation increases the bulk

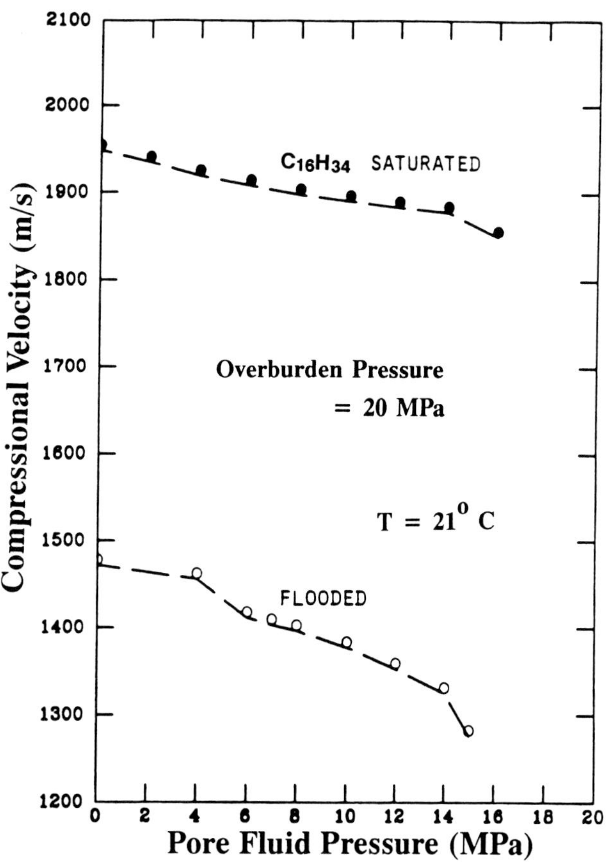

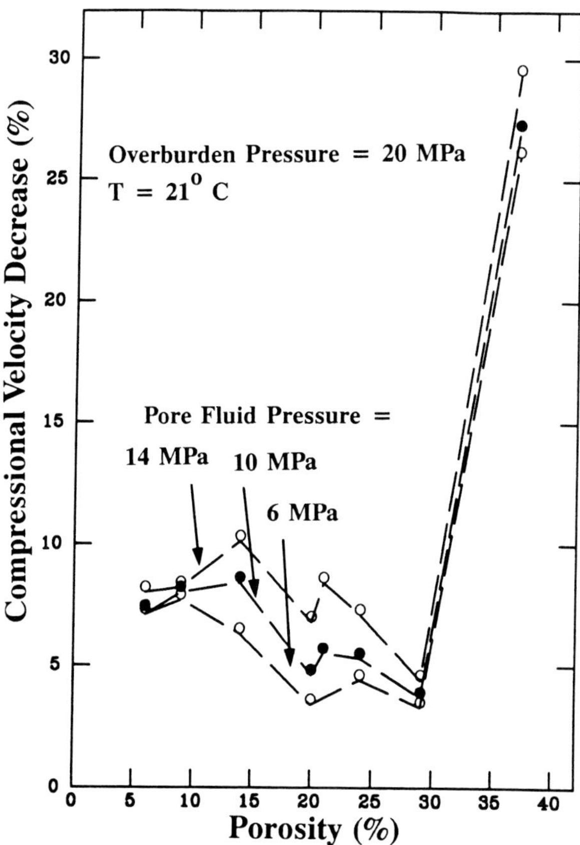

Fig. 15. Compressional velocity in Ottawa unconsolidated sand saturated with $C_{16}H_{34}$ and then flooded with CO_2.

Fig. 16. Decreases of compressional velocities caused by CO_2 floodings in 7 consolidated sandstones and 1 unconsolidated sand.

modulus greatly but has little effect on the density, and consequently P-wave velocity increases by a large percentage. For high porosity rocks, liquid saturation increases both the bulk modulus and the overall density, so that P-wave velocity does not increase as much. CO_2 flooding is the opposite process to liquid saturation.

The effect of CO_2 flooding on P-wave velocities calculated by the Gassmann relation gives the same magnitude of velocity decrease but the absolute velocity values do not match the measured values exactly (deviations of 2–3 percent).

Effect of Hydrocarbon Solvent Flooding on Velocities in Rocks

The effect of hydrocarbon solvent flooding should be, theoretically, similar to that of CO_2 flooding, with slight differences because the density of hydrocarbon solvent is lower than that of CO_2 at high pressures. When the injected solvent is in the gaseous phase, P-wave velocity will decrease due to the gas saturation. When the solvent is in the liquid phase, P-wave velocity also will decrease since the solvent has much higher compressibility than the indigenous oil.

The hydrocarbon solvents used in miscible flooding pilot projects are composed of light hydrocarbons, mostly methane and ethane with small portions of heavier hydrocarbons up to pentane. The composition determines that the velocity of the solvent, ranging from 300 to 600 m/s (depending on temperature and pressure), is much lower than the velocities in most reservoir oils. When a reservoir oil is replaced by low-velocity solvents, the P-wave velocity in the overall rock should drop considerably.

A series of laboratory experiments are being carried out to study hydrocarbon solvent flooding effects (Hirsche and Sedgwick, 1989). Cores from several carbonate reservoirs were selected for laboratory tests, with porosity values ranging from 3 to 26 percent and permeability values from 0.1 to 7000 mD. Figure 17 shows the P-wave velocities in a dolomite core sample from the Keg River formation which has a porosity of 18 percent and permeability of 6710 mD. The velocity was first measured in the dry core (with air in pores), then in the same core saturated with a mineral oil which has velocity and density close to the reservoir crude, and finally the core was flooded with hydrocarbon solvent. The solvent flooding decreases the P-wave velocity substantially (from 6 to 15 percent), especially at low differential pressures. At a differential pressure of 20.7 MPa (3000 psi), the velocity decrease is 12 percent, a magnitude of velocity decrease that should be measurable by seismic methods.

Figure 18 shows the P-wave velocities at 89°C in a limestone core also from the Keg River formation which has porosity of 16 percent and permeability of

90 mD. The velocity decrease upon solvent flooding ranges from 6 to 17 percent. The velocity in the solvent-flooded core is even lower than that in the dry rock (with air) for this particular core, which may be caused by the high density and high compressibility of the solvent.

So far about 80 carbonate cores have been tested. The decrease in P-wave velocities caused by hydrocarbon solvent flooding under reservoir conditions ranges from 2 to 20 percent. In contrast, S-wave velocities are much less affected. This wide range of velocity decreases is caused by differences in pore geometry and structure of the rocks. For rocks containing mostly round or high aspect-ratio pores, the compressibility of the pores is low, so that change in pore fluid compressibility causes only nominal change in the overall compressibility. In this case, both miscible and immiscible floodings have only small effect on the P-wave velocity. For rocks containing mostly cracks or pores of low aspect ratios, the compressibil-

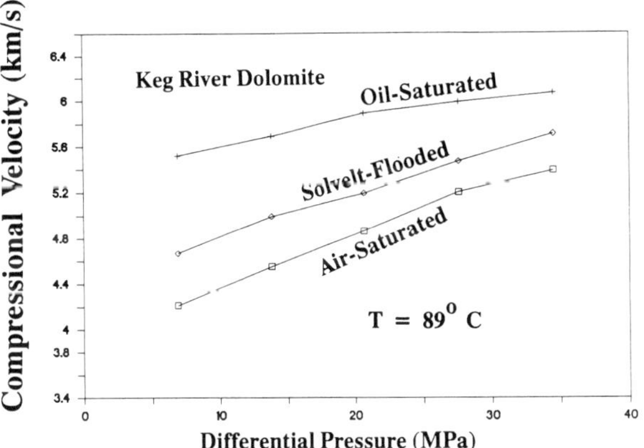

Fig. 17. Compressional velocity in a Keg River dolomite sample saturated with air, oil, and then flooded with a hydrocarbon solvent, respectively.

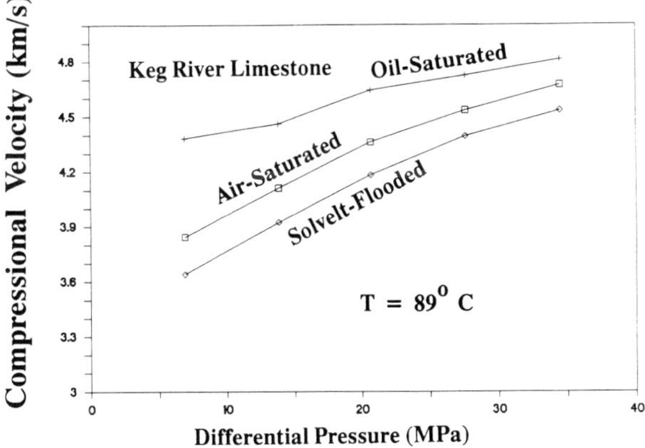

Fig. 18. Compressional velocity in a Keg River limestone sample saturated with air, oil, and then flooded with a hydrocarbon solvent, respectively.

ity of the cracks/pores is very high, so that a change of pore fluid compressibility contributes greatly to the overall compressibility. In this case, flooding has a large effect on the *P*-wave velocity. This emphasizes the importance of laboratory measurements before committing to a seismic monitoring project in-situ.

The laboratory results also reveal that, for every core tested, the effect of solvent flooding on *P*-wave velocities is larger at low differential pressures than that at high differential pressures. At low differential pressures, the thin pores and cracks are still open so that the rock is sensitive to pore fluid changes. At high differential pressures, most thin pores and cracks are closed and what are left open are the round and high aspect-ratio pores, so that the rock is less sensitive to pore fluid changes. The dependence of velocity decrease on differential pressure suggests that seismic monitoring of miscible and immiscible floods (including CO_2 floods, hydrocarbon solvent floods, and gas drive) likely will be successful in shallow reservoirs.

Effect of Pore Fluid Pressure and Saturation on Velocities in Rocks

Some main factors which control *P*- and *S*-wave velocities in porous rocks are confining pressure, pore fluid pressure, and hydrocarbon saturation. Most EOR processes involve fluid injection into the reservoir. The injection pressure always exceeds the pore fluid pressure.

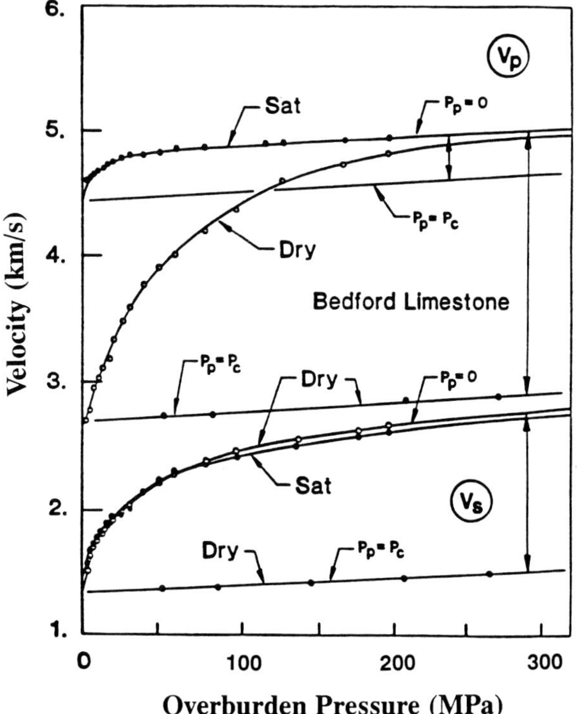

Fig. 19. Compressional and shear velocities in Bedford limestone versus overburden and pore pressures and saturation (Nur and Simmons, 1969).

Figure 19 shows pressure effects for a limestone; this pattern is typical for many sedimentary rocks. The *P*- and *S*-wave velocities at dry and liquid-saturated conditions depend very differently on the overburden and pore fluid pressures. The velocities in dry rock increase markedly with increasing overburden pressure. The magnitude of this increase generally is determined by the pore aspect ratios: velocities increase more in rocks with low aspect ratio pores (or cracks) than with high aspect ratio (or round) pores.

When a sample is saturated with water at room pressure, a large increase in *P*-wave velocity is observed, whereas *S*-wave velocity is changed only nominally. From laboratory observations, in rocks with velocities much dependent on differential pressure when saturated with gas, the increase of *P*-wave velocity upon liquid saturation also is high, especially in low differential pressure ranges. Many low porosity sedimentary rocks belong to this category. In these rocks, *S*-wave velocities usually increase, not decrease, as the rocks are saturated with liquid, especially in low differential pressure ranges. This *S*-wave velocity increase usually is very small for saturation with low-viscosity fluids and can be as high as 2–5 percent for saturation with high-viscosity fluids like heavy oils.

The effect of pore pressure is to counteract that of overburden pressure. Consequently, both *P*- and *S*-wave velocities in a rock saturated with fluids are much lower when the pore pressure is equal to the overburden pressure than when it is equal to atmospheric pressure. The strong dependence of velocities on pressure and saturation is confined to low differential pressures. At differential pressures above 100 to 200 MPa, velocities show only a small increase with further increasing differential pressures.

The reason for the effect of overburden pressure is that pressure deforms the most compliant part of the pore space (e.g., microcracks and loose grain contacts) and thus increases the stiffness of the rock, i.e., the effective bulk and shear moduli. High pore fluid pressure mechanically opposes the closing of cracks and grain contacts, thus leading to low effective moduli and velocities.

The influence of the pore fluid, as separate from its pressure, is related to its compressibility. When pore fluid is relatively incompressible (e.g., brine), the effective bulk modulus of the rock is high so that the *P*-wave velocity is high. In contrast, when the pore fluid is changed from gas to liquid, the shear modulus does not change so that the *S*-wave velocity only nominally changes, because the pore fluid viscosity is low.

Both *P*- and *S*-wave velocities show very little sensitivity to partial saturation in the liquid saturation interval of 0 to 80 percent. When liquid saturation is close to 100 percent, *P*-wave velocity increases markedly (e.g., Murphy, 1982). *S*-wave velocity is always insensitive to the degree of liquid saturation. These results indicate that, unfortunately, velocity measurements can not yield information on the degree of partial liquid saturation in reservoir rocks.

Figure 20a shows that the *P*-wave velocity is much higher when the Berea sandstone core with 19 percent porosity is saturated with a 10° API heavy oil than when saturated with either distilled water or normal decane ($C_{10}H_{22}$) (Wang et al., 1988b). This contradicts the Gassmann (1951) relation and Biot (1956a,b) theory because the velocity and density of the heavy oil are very close to those of water. A possible explanation is that since the heavy oil has very high viscosity (30 Pa.s at room conditions), the viscous oil in the pore space is physically and/or chemically bound to the grains, particularly when the thin pores or cracks are filled with the oil, so that both the bulk and shear moduli are high. As the temperature increases, both *P*- and *S*-wave velocities decrease rapidly due to the oil's compressibility increase and viscosity decrease.

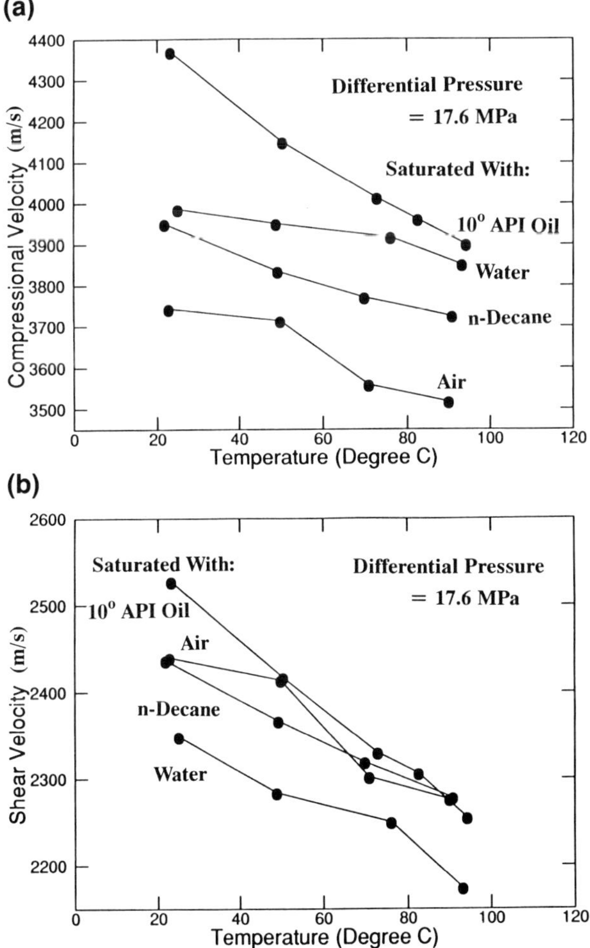

Fig. 20. Compressional (a) and shear (b) velocities in a Berea sandstone sample saturated with a 10° API oil, water, normal decane, and air, respectively.

The *S*-wave velocity is lower when the core is saturated with either water or decane than when saturated with air, whereas it increases upon heavy oil saturation, which may be caused by the viscosity and density differences of the pore fluids (Figure 20b).

Seismic Reservoir Surveillance: A Rockphysics View

The Rockphysics Basis

Because an average of nearly two-thirds of the original oil in place cannot be recovered by conventional recovery methods, enhanced oil recovery (EOR) methods hold great promise in most hydrocarbon reservoirs. However, one of the problems faced by EOR schemes is the need to determine, with accuracy not yet possible, the spatial and temporal distributions of the reservoir properties and the injected fluids. Of particular interest and promise are changes which occur during production and thermal and miscible/immiscible floods. In all EOR operations, it is especially important to determine the direction of propagation, shape, rate of movement, and spatial heterogeneity of the heat or fluid front. To accomplish such determinations, we would like to monitor reservoirs continuously throughout their volume.

Chemical and radioactive tracers have been used in monitoring of fluid break-throughs in EOR processes. However, tracers can not locate the position of the front of the injected fluids until they arrive at the monitor wells. Also, chemical tracers may cause polymerization of reservoir fluids and chemical precipitation which can reduce the effective permeability of the reservoir rocks, and radioactive tracers may cause environmental and safety problems. In contrast, seismic methods potentially can yield distribution maps of the injected fluids and reservoir temperature without any damages or dangers.

Whether seismic methods can be used in reservoir surveillance and in monitoring of EOR processes depends mainly on the magnitude of the velocity and/or attenuation changes caused by such processes. Little information has been obtained on the effect of EOR processes on seismic wave attenuation, but the velocity effects have been studied and look very promising. Velocity changes related to EOR processes in hydrocarbon reservoirs may be caused by:

(1) Temperature changes and the presence of gases in thermal EOR processes such as in-situ combustion, steam floods, and hot water injection. Both increases in temperature and the presence of gases decrease seismic velocities.

(2) Changes in the compressibility of the pore fluids and thus in the compressibility of the overall rock, such as in CO_2 floods, hydrocarbon solvent floods, and gas drive. In these EOR pro-

cesses, the injected fluids displace the liquid hydrocarbons in the reservoir and consequently increase the overall compressibility of the reservoir rock and lower the velocity.

Other possible factors affecting velocity are:

(1) Pore fluid pressure changes because of fluid injection. Zones of high pore fluid pressure should have lower seismic velocities.

(2) Production processes such as water intrusion, gas cap formation and movement, and formation fluid pressure depletion.

Applications of Rock Physics in Reservoir Surveillance

Seismic monitoring of thermal floods. One of the most important conclusions of our previous studies is that velocities in hydrocarbon-bearing rocks show large changes with changing temperature (Wang and Nur, 1986; 1987; 1988b; 1990). Changes in P-wave velocities range from 6 percent decrease in light hydrocarbon saturated rocks up to over 40 percent decrease in heavy hydrocarbon-saturated rocks as the temperature increases by 100°C (Table 2). Such velocity decreases are so large that they should cause noticeable seismic effects.

Tracking thermal fronts is important because the migration of the steam or hot water away from injection wells is controlled by permeability heterogeneities, and even modest heterogeneities or anisotropy can cause major economic losses. Thus it is important to determine the details of the shape and rate of movement of fire or steam fronts.

The very large magnitudes of the effects of temperature measured in laboratory samples strongly suggest that seismic monitoring of thermal EOR fronts, including hot water, fire, and steam floods, should be successful. Indeed, the seismic method may provide a thermometer to map the spatial distribution of heat in a heavy oil reservoir.

Monitoring CO_2 floods. In CO_2 floods, the injected CO_2 displaces hydrocarbons in reservoir rocks. Because CO_2 has very high compressibility and density, substitution of CO_2 for the liquid hydrocarbons in the pore spaces of reservoir rocks increases the compressibility and decreases the P-wave velocity. Thus seismic methods can offer promise in monitoring of CO_2 floods, tracking the CO_2 fronts, and mapping flooded zones.

Monitoring hydrocarbon solvent floods. In hydrocarbon solvent floods, mixtures of light hydrocarbons (mostly methane and ethane) are injected into the reservoir to displace in-situ oil. Hydrocarbon solvent flooding decreases compressional velocities in reservoir rocks and the magnitude of the velocity decrease may be high enough (5 to 16 percent under reservoir conditions) that seismic methods can be used in monitoring of hydrocarbon solvent floods (Hirsche and Sedgwick, 1989).

Monitoring gas drive and gas cap. Gas drives include using enriched hydrocarbon gas, high-pressure hydrocarbon gas, and nitrogen gas to displace oil in reservoirs. Because all these gases are highly compressible, the compressibilities of both the pore fluid and the overall rock increase substantially, and the P-wave velocity should decrease. This velocity decrease should be high enough to be detected seismically.

The effects of free gas have been applied extensively as a hydrocarbon indicator (as in "bright-spot" method). These effects can be used to monitor gas caps, particularly when a gas cap forms, grows, moves, or shrinks as recovery progresses. For instance, in high gas-to-oil ratio (GOR) reservoirs, a gas cap may form and grow as the fluid pressure is lowered by production.

Monitoring water floods. The most common secondary recovery process is water flooding in which water injected in some wells enhances oil flow through producing wells. As with all floods, the water may be channeled along more permeable pathways and thus may bypass hydrocarbons. The tracking and mapping of a water front would be of great economic value.

Several effects must be considered in order to monitor a water flood using seismic methods: most importantly, the P-wave velocity is lower in an oil with dissolved gases (live oil) than in water (Wang et al., 1988a). Injected water should cause P-wave velocity to increase and perhaps seismic methods may be able to detect such velocity increases and thus track water fronts. The second effect is associated with the change of temperature. Because the injected water is often much colder than the ambient reservoir temperature, some cooling of the residual oil in the flooded zone may take place, associated with an increase in velocities. Whether this effect is large enough to be detectable seismically depends on the magnitude of the temperature change due to water flooding, on the sensitivity of the velocity in the specific reservoir rock to temperature, and on the rate at which the injected water is warmed up after injection. Finally, the water is injected at higher pressure than the ambient formation pressure. This increase in pore pressure should lead to a decrease in seismic velocities. However, because most water floods are fairly deep, this effect is usually small. Systematic study is needed on the effects of water flooding on velocities in a variety of reservoir rocks.

Direct detection of liquid hydrocarbons. Our experimental results (Wang and Nur, 1987) show that there may be sufficient differences between P-wave velocities in rocks saturated with light hydrocarbons and with water, when the reservoir temperature is sufficiently high (50°C) to detect the hydrocarbons. Furthermore, in heavy hydrocarbon reservoirs at high temperatures (i.e., caused by thermal EOR process), the differences between the P-wave velocities in hydrocarbon- and water-saturated rocks are even larger. This implies, for example, that at high temperatures, seismic waves might be reflected not only from gas-liquid but also from water-oil interfaces in reservoirs.

Even more intriguing is the potential for using the hydrocarbon-indicator concept to directly detect liquid hydrocarbon reservoirs. This potential is based on our experimental results which show that P-wave velocities are measurably higher in heavy than in light hydrocarbon-saturated rocks. Therefore, reflection from interfaces between light and heavy oil-saturated layers obtained in high-resolution seismic measurements could be used to determine hydrocarbon types in reservoirs.

Monitoring foam injections. Foams are accumulations of gas bubbles, separated from each other by thin films of liquid. The injection of foam into reservoir rocks saturated with oil creates a large number of resillient interfaces, which exerts a piston-like force on the oil. The "quality" of a foam is defined as the ratio of the contained gas volume to the total foam volume. Such "quality" in reality can reach to 0.96, indicating that foams can contain a large amount of gases. The compressibility of the reservoir rocks should increase dramatically as the oil is displaced by foam and, consequently, the P-wave velocity should decrease. Although at present no laboratory results have been reported on the effect of foam on velocities in rocks, theoretically seismic methods should be able to detect the velocity changes caused by foam injection and therefore potentially can be used to monitor foam injections.

Production monitoring and 4-D seismology. As discussed, seismic methods, if sufficiently precise, could be used to monitor various EOR and production processes and to delineate oil saturation boundaries in reservoirs. Consequently, repeated 3-D seismic surveys in reservoirs undergoing EOR can be used to generate 4-D (3-D plus time) seismograms. These time-lapse seismograms will yield the temporal as well as spatial distributions of oil, temperature, injected fluids, and the progress of production in a reservoir. They should offer information about subsurface reservoir conditions which never before has been seen. They may give reservoir engineers the most valuable information they can get regarding the management of EOR projects (Nur and Wang, 1987; Wang and Nur, 1988b).

Importance of Laboratory Rock-Physical Measurements

Laboratory rock-physical measurements are very important and may be vital in seismic surveillance. All theories and models involve simplifications and assumptions, and rocks are so complicated that important parameters may be left out if there are no accompanying laboratory measurements. We suggest that cores from a reservoir be tested in the laboratory before seismic monitoring surveys are undertaken. Doing laboratory measurements first may save a large amount of money.

From our laboratory experiences, the magnitude of P-wave velocity changes upon flooding often varies between cores from different reservoirs and different wells. In general, monitoring an EOR process requires at least 4 to 5 percent velocity change (an impedance change of about 10 percent). In some reservoirs seismic methods may not be suitable. However, unconsolidated sand reservoirs are almost always excellent candidates for seismic monitoring of any EOR processes.

An important caveat about laboratory measurements of seismic velocities is that the frequencies used in the laboratory necessarily are much higher than those of seismic waves. The length of a rock sample must be greater than half of the wavelength, which means that for cores the wave frequencies must be in the ultrasonic range (usually around 0.1 to 1 MHz). However, both theory (e.g., Kjartansson, 1980) and experiments (e.g., Spencer, 1981; Jones, 1986) indicate that seismic wave velocities in dry rocks are nearly independent of frequency in the range from seismic to ultrasonic frequencies, and in most liquid-saturated rocks, the velocities at ultrasonic frequency are 1 to 5 percent higher than at seismic frequency. Our analysis has shown that the *temperature dependence* of the velocities in rocks with heavy hydrocarbons is not sensitive to wave frequencies (Wang and Nur, 1988d). In other EOR processes such as CO_2 floods, hydrocarbon solvent floods, and gas drive, the measured velocity changes are relative values, so that the observed laboratory velocity decreases should be about the same as those in the seismic frequency band.

Conclusion

Laboratory results show that seismic velocities are sensitive to various EOR processes such as in-situ combustion (fire flooding), steam flooding, hot water injection, CO_2 flooding, hydrocarbon solvent flooding, and gas drive. In thermal EOR processes, the velocities decrease dramatically in rocks impregnated with heavy hydrocarbons as the temperature increases. The magnitude of the velocity decrease depends on many factors such as the consolidation, pore structure and geometry, mineralogy, overburden and pore pressure of the rocks, and the composition, viscosity, and API gravity of the pore fluids. Seismic methods can be used in reservoir surveillance of thermal EOR processes.

Injected CO_2 and hydrocarbon solvents decrease P-wave velocities in rocks saturated with liquid hydrocarbons under reservoir conditions. The magnitude of the velocity decrease varies, ranging from 2 to 20 percent in consolidated rocks and over 25 percent in unconsolidated sands under reservoir conditions. Seismic methods may be suitable for monitoring some floods. As a rule of thumb, unconsolidated sand reservoirs almost always are good candidates for seismic reservoir surveillance.

Velocities in reservoir fluids also are important in reservoir surveillance and description. The velocities in hydrocarbons (including pure hydrocarbons and crude oils) are related systematically to the molecular weight, API gravity, pressure, temperature, and composition of the hydrocarbons. Empirical equations can be used to calculate velocities in hydrocarbons as a function of temperature and pressure if the API gravities are known.

Laboratory measurement plays an important role in seismic reservoir surveillance. Because seismic velocities depend in complex ways on many parameters and factors and because rock properties vary so much, it is necessary to carry out laboratory measurements before seismic monitoring.

The richness of the seismic effects reported in this paper, especially the sensitivity of velocity to temperature, CO_2 flooding, and hydrocarbon solvent flooding, clearly indicate a future direction of reservoir seismology: efforts to describe reservoirs in more detail and to monitor recovery processes.

Acknowledgment

We thank Dr. J. Justice for inviting us to contribute to this book. We are grateful to Dr. R. E. Sheriff for editing and reviewing the manuscript.

References

Batzle, M. L. and Wang, Z., 1992, Seismic properties of pore fluids: in press, Geophysics.

Biot, M. A., 1956a, b, Theory of propagation of elastic waves in a fluid saturated porous solid, I. Low frequency range; II. Higher frequency range: J. Acoust. Soc. Am., **28**, 168–191.

Bradley, D. L., 1963, Velocity of sound in a Dow Corning 200 fluid as a function of temperature and pressure: J. Acoust. Soc. Am., **35**, 1565–1567.

Britton, M. W., Martin, W. L., Leibrecht, R. J., and Harmon, R. A., 1983, Street Ranch pilot test of fracture-assisted steamflood technology: J. Petr. Tech., **35**, 511–522.

Chavez, M., Tsumura, R., and del Rio, F., 1981, Speed of sound in saturated liquid trichlorofluoromethane: J. Chem. Eng. Data, **26**, 1–2.

den Boer, L. D. and Matthews, L. W., 1988, Seismic characterization of thermal flood behavior: Proc. of the 4th Internat. Conf. on Heavy Crude and Tar Sands, **4**, 613–627.

Gassmann, F., 1951, Elastic waves through a packing of spheres: Geophysics, **16**, 673–685.

Greaves, R. J. and Fulp, T. J., 1987, Three-dimensional seismic monitoring of an enhanced oil recovery process: Geophysics, **52**, 1175–1187.

Hirsche, W. K. and Sedgwick, G., 1989, Miscible flood monitoring using seismic methods: Canadian Soc. Expl. Geophys. and Canadian Soc. Petr. Geol. Joint Conf., Abstracts, 65.

Holm, L. W. and Josendal, V. A., 1982, Effect of oil composition on miscible-type displacement by carbon dioxide: Soc. Petr. Eng. J., 97–98.

Hunter, J. L., Welch, T. J., and Montrose, C. J., 1963, Excess absorption in mercury: J. Acoust. Soc. Am., **35**, 1568–1570.

Jones, T. D., 1986, Pore fluids and frequency-dependent wave propagation: Geophysics, **51**, 1939–1953.

King, G. A., 1988, The application of seismic methods for reservoir description and monitoring: Presented at the Soc. Expl. Geophys. and Chinese Petr. Soc. Joint Prod. Geophys. Mtg., Daqing, China.

Kjartansson, E., 1980, Attenuation of seismic waves in rocks and applications in energy exploration: Ph.D. Thesis, Stanford University.

Lawson, A. W. and Hughes, A. J., 1963, High pressure properties of water, in Bradley, R. S., Ed., High pressure physics and chemistry, vol. 1: Academic Press.

Macrides, C. G., Kanasewich, E. R., and Bharatha, S., 1988, Multiborehole seismic imaging in steam injection heavy oil recovery projects: Geophysics, **53**, 65–75.

Murphy, III, W. F., 1982, Effect of microstructure and pore fluids on the acoustic properties of granular sedimentary materials: Ph.D. thesis, Stanford University.

Nur, A. and Simmons, G., 1969, The effect of saturation on velocity in low porosity rocks: Earth Plan. Sci. Let., **7**, 183–193.

Nur, A. and Wang, Z., 1987, In-situ seismic monitoring EOR: The petrophysical basis: Proc. of Ann. Internat. Mtg., Soc. Petr. Eng., Φ, 307–314.

Pullin, N., Matthews, L., and Hirsche, W. K., 1987, Techniques applied to obtain very high resolution 3-D seismic imaging at an Athabasca tar sands thermal pilot: Geophysics: The Leading Edge, **6**, 10–15.

Rao, K. S. and Rao, B. R., 1959, Study of temperature variation of ultrasonic velocities in some organic liquids by modified fixed-path interferometer method: J. Acoust. Soc. Am., **31**, 439–441.

Spencer, J. W., 1981, Stress relaxations at low frequencies in fluid saturated rock: Attenuation and modulus dispersion: J. Geophys. Res., **86**, 1803–1812.

Tosaya, C., Nur, A., Aronstam, P., and Da Prat, G., 1984, Monitoring of thermal EOR fronts by seismic methods: Presented at the 1984 Soc. Petr. Eng. Calif. Reg. Mtg.

Wang, Z., 1988, Wave velocities in hydrocarbons and hydrocarbon saturated rocks—with applications to EOR monitoring: Ph.D. Thesis, Stanford University.

Wang, Z. and Nur, A., 1986, Effect of temperature on the seismic wave velocities in sandstones and sands with hydrocarbons: 56th Ann. Internat. Mtg., Soc. Expl. Geophys., Expanded Abstracts, 3–5.

Wang, Z. and Nur, A., 1987, Wave velocities in hydrocarbons and hydrocarbon saturated rocks: 57th Ann. Internat. Mtg., Soc. Expl. Geophys., Expanded Abstracts, 1–3.

Wang, Z. and Nur, A., 1988a, The effects of temperature on velocities in rocks with heavy hydrocarbons: SPE Res. Eng., **3**, 158–164.

Wang, Z. and Nur, A., 1988b, Seismic velocities in tar sands: The basis for in-situ recovery monitoring: Proc. of the 4th Internat. Conf. on Heavy Crude and Tar Sands, **4**, 601–611.

Wang, Z. and Nur, A., 1988c, Effect of CO_2 flooding on wave velocities in rocks: Proc. of Sympos. of the 6th SPE/DOE Joint Mtg. on Enhan. Oil Recov., 275–288; also SPE Res. Engin., 1989, Nov., 429–436.

Wang, Z. and Nur, A., 1988d, Velocity dispersion and the ''local flow'' mechanism: 58th Ann. Internat. Mtg., Soc. Expl. Geophys., Expanded Abstracts, 548–550.

Wang, Z. and Nur, A., 1990, Wave velocities in hydrocarbon saturated rocks: Experimental results: Geophysics, **55**, 723–733.

Wang, Z., Nur, A., and Batzle, M., 1988a, Acoustic velocities in petroleum oils: Proc. of Internat. Mtg., Soc. Petr. Eng., Ω, 571–586; also J. Petr. Tech., 1990, Feb., 192–200.

Wang, Z., Nur, A., and Batzle, M., 1988b, Effect of different pore fluids on velocities in rocks: 58th Ann. Internat. Mtg., Soc. Expl. Geophys., Expanded Abstracts, 928–930.

3-D Seismic Monitoring of an In-Situ Thermal Process: Athabasca Canada

*L. Matthews**

Introduction

Vast resources of untapped tar sands and heavy oil reserves are waiting to be developed when the technology becomes available to mobilize this highly viscous hydrocarbon both efficiently and economically. A good deal of the ultimate success of heavy oil in-situ thermal recovery is necessarily related to preliminary knowledge of the reservoir's characteristics, together with reliable imaging of the subsequent subsurface heat movement throughout the life of the project.

The volume of heavy oil in Alberta is conservatively estimated at 267×10^9 cubic meters (1.68 trillion barrels) with only about 7 percent being accessible with surface mining (Energy Resources Conservation Board, 1987). The Athabasca area contains the largest of the Alberta tar sand deposits with estimated reserves of 209×10^9 m^3. Most in-situ projects are confined to reservoirs with oil that is immobile at ambient temperatures and pressures. With the proper addition of heat, however, many of these reservoirs can become producible at commercial rates. There are currently 6 commercial and 20 experimental in-situ projects operating in Alberta. Each reservoir has its own unique geology and basic reservoir parameters. It is therefore extremely important that the maximum knowledge, not only concerning the reservoirs' lithologic characteristics, but also the heat movement over time be known in order to maximize production.

Many investigators have indicated encouraging potential for imaging these heat zones (Britton et al., 1983; den Boer and Matthews, 1988; Greaves and Fulp, 1987; Pullin et al., 1987; Macrides et al. 1988). The techniques that are currently being evaluated are microseismic, vertical seismic profiling (VSP), tomography, and high-resolution 3-D seismic. With the exception of microseismic, all of these monitoring methods rely on significant changes in acoustic impedance as a direct result of heating the reservoir. High-resolution 3-D seismic produces the most complete areal coverage and dense subsurface sampling while VSP and tomography can add positive event correlation and vertical detail.

*Western Geophysical, Suite 600, 321 6th Ave. S.W., Calgary, T2P 3H3, Canada.

Recording identical surveys during the life of a pilot can provide a measure of the project's progress and allow ongoing modification of this process. Until recently, this type of information has been available only by using well data, which are extremely sparse when the very complex nature of most reservoirs is considered.

An overall methodology for monitoring heat movement over time is very important and should consist of the following:

Rock properties analysis,
Computer modeling,
Seismic field testing,
Stratigraphic interpretation from 3-D base survey, and
Heat zone imaging from 3-D monitor surveys.

Just such a monitoring scheme, spanning a three year period, was employed at an Athabasca thermal pilot operated by Amoco Canada with Amoco's partners, Alberta Oil Sands Technology and Research Authority (AOSTRA) and Petro-Canada.

The Gregoire Lake In-situ Steam Pilot (GLISP) site, located in northeastern Alberta, Canada, approximately 50 km south of Fort McMurray, was designed to test a steam stimulation process. The well configuration consists of a central injector H-6 surrounded by three equidistant producers: H-3, H-4, and H-5, and three observation wells: H0-7, H0-8, and H0-9. The pilot location and well pattern are illustrated in Figure 1. The geology of the GLISP site, Figure 2, is characterized by a 50 m thick section of discontinuous unconsolidated Cretaceous McMurray (MCMR) tar sands, containing an 8° API oil, resting on a carbonate Devonian (DVNN) unconformity approximately 240 m below the surface. Four high-resolution 3-D seismic surveys were conducted; the base 3-D survey was acquired in April 1985 prior to steam injection and the monitor surveys were recorded in January 1987, April 1987, and November 1987 subsequent to steaming in the production and injection wells.

Rock Properties

The basis of the seismic thermal monitoring technique is the temperature dependence of acoustic velocity and/or seismic amplitude which is observed while heating samples saturated with heavy oil from room temperature to 150°C. Wang and Nur (1986) attribute the bulk of the decrease in acoustic velocity to a decrease of the tar viscosity caused by melting. Changing the confining pressure does not produce as marked a difference in the observed velocity decrease.

Wang and Nur show that differences in porosity, pore geometry, rock mineralogy, and consolidation, combined with differences in chemical composition and saturation of the hydrocarbons occupying the pore spaces, make the acoustic velocity decrease profile for each rock unique. Thus, the effectiveness of seismic thermal monitoring depends on accurate measurement of these data for representative samples obtained at the actual field location.

Samples from the GLISP site showed an approximate 30 percent decrease in compressional velocity when being heated from 20 to 150°C. A representative velocity-temperature profile can be seen in Figure 3.

The significance of this change in velocity is that:

The top and bottom of the heated tar sand will act as acoustic reflectors.

Seismic reflections below the heated zone will be delayed in time, a phenomenon referred to as "pushdown".

Computer Modeling

In order to simulate the seismic response of heated tar sands, computer modeling was conducted, combining both sonic logs and rock properties information. Figure 4 is a model showing the seismic response of a wedge of 75°C hot tar sand increasing in thickness from 0 to 20 m. The model demonstrates that (1) the boundaries between the hot and cold tar sand produce significant reflections, and (2) pushdown is observed

immediately below the heated tar sand on the Devonian reflection.

Based on this model, one might expect to resolve a heated tar zone as thin as 5 m with dominant frequencies as high as 175 Hz (Kallweit and Wood, 1982). These estimates of potential resolution, however, are based on a very simplistic model and do not take noise into account.

Field Testing

To obtain high resolution seismic data, a comprehensive set of field tests was conducted to establish the field acquisition parameters that would produce data having exceptionally high signal-to-noise (S/N) ratio over a broad frequency band. Single-Sensor SM-11 30-Hz seismometers were used for tests. Both surface and permanently buried spreads were employed.

The near-surface weathered layer is generally a complex low-velocity absorptive medium of rapidly changing thickness. Nowhere are these characteristics more apparent than in unfrozen muskeg. An environment like this would be expected to act as a wave guide for high-amplitude, low-velocity noise; produce large static variations; and act as a severe high-cut filter. A demonstration of these effects is shown in Figure 5, which compares data from surface and subsurface seismometers for unprocessed field profiles recorded from the same 50 g charge. The dramatic improvement in S/N ratio, high-frequency content, and static shifts for the profile recorded with subsurface phones is very apparent. A synthetic trace generated from the H-3 sonic log, using a 20-150 Hz zero-phase wavelet, exhibits a remarkable correlation with the unprocessed field data obtained with the buried phones. These results indicate the importance of locating the geophones and shots below the muskeg layer in this area. Next to burying the seismometers,

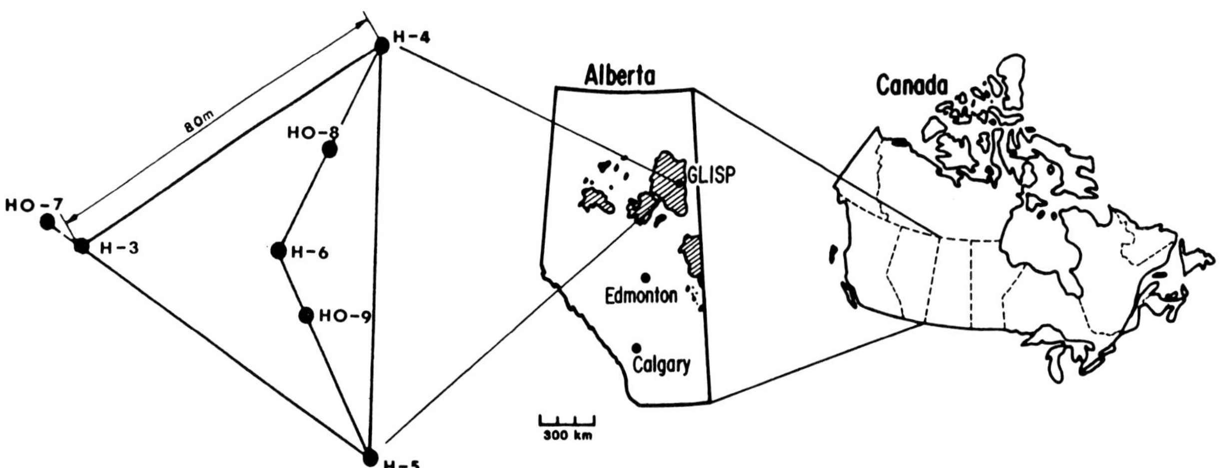

Fig. 1. Distribution of heavy oil resources in Alberta, GLISP pilot site location and well configuration.

charge size had the greatest impact on the high-frequency content of the field data. Charges ranging from 1 g (cap) to 1000 g were tested. Two criteria influenced the final charge size selection of 18 g. As the source size was increased above 50 g, progressively more detrimental pulse broadening was observed. While the 1 g charge appeared to emphasize the high-frequency content, the overall signal amplitude at the Devonian was suspect. A charge size of 18 g provided a good compromise between wavelet breadth and S/N ratio.

For all of the buried-phone profiles a very low velocity (260 m/s), high-amplitude noise wave was observed on the near traces. Processing proved unable to remove this distortion from the short-offset ranges.

Test results indicated that the most important field parameters were (1) Seismometer depth—should be buried below the muskeg, (2) Charge size—should be small (1 to 50 g), and (3) Recording offset—should be beyond the noise train. Based on the analysis of the field tests, the following recording parameters were judged optimal for the subsequent 3-D surveys:

Seismometer depth—single SM-11 (30 Hz) cemented at 13 m depth,
Charge size—18 g in cement at 18 m depth,
Recording offset—50 to 150 m, and
CMP multiplicity—6 to 12 fold.

3-D Field Acquisition: Base Survey

Field acquisition of the base 3-D survey was conducted in April 1985. Figure 6 illustrates the well configuration and associated recording geometry. At

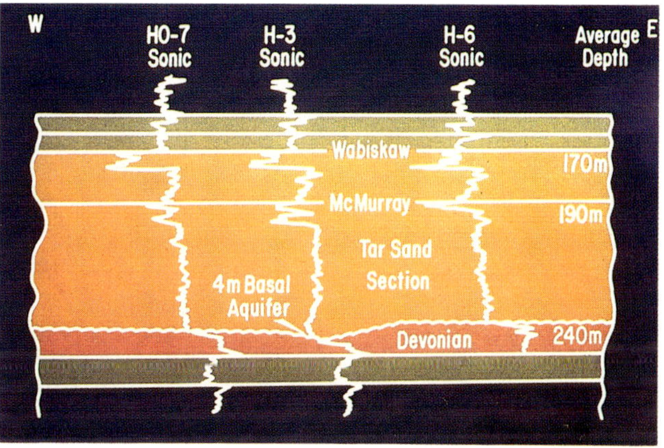

Fig. 2. Geology of the GLISP pilot site.

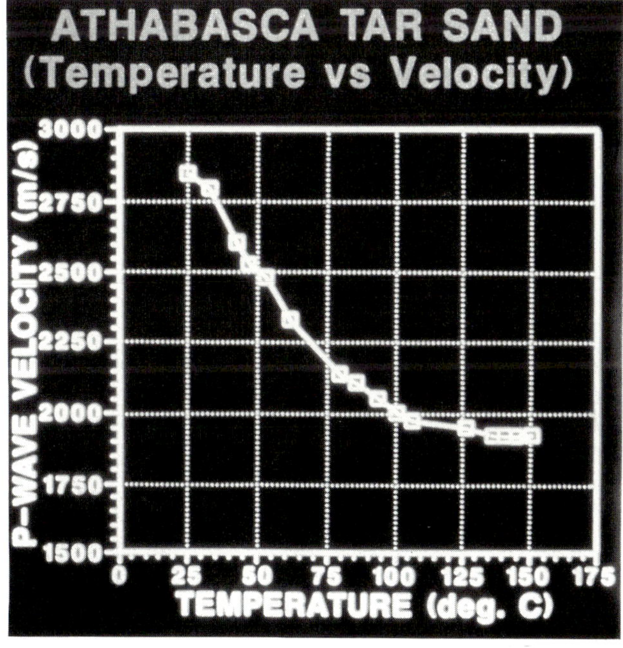

Fig. 3. Representative velocity versus temperature profile.

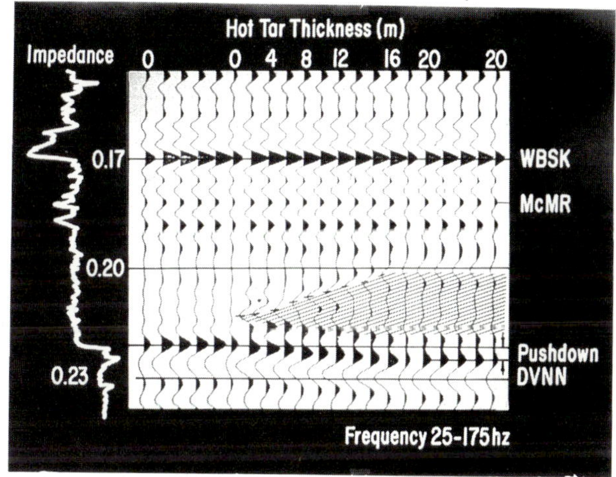

Fig. 4. Seismic response of a wedge of 75°C hot tar sand increasing in thickness from 0 to 20 m.

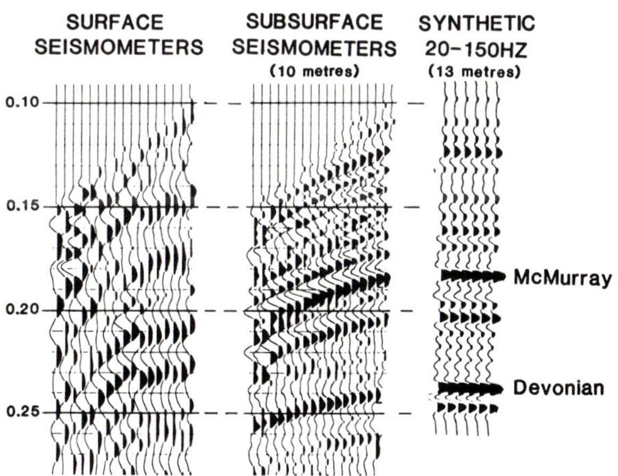

Fig. 5. Comparison of H-3 synthetic with surface-phone and buried-phone shot records generated by a 50 g source at 13 m depth.

the time the base survey was recorded only the H-3 and H0-7 wells had been drilled. The survey area was 168 m by 196 m with 4 m by 4 m common midpoint (CMP) bins. Acquisition resulted in a minimum of 12 fold coverage except near the perimeter of the pilot area. The downhole seismometer locations are designated by the symbol "⊕" and shot points by the symbol "X". To access the shot holes for future monitor surveys, they were cased to total depth with 4 inch PVC. To protect the geophone leads between surveys, a length of 4 inch PVC surface casing was used with a protective top cap. A 120 channel SERCEL 348 telemetry system recorded the 3-D survey. The sampling interval was 1 ms. Details of the base survey were given in Pullin et al. (1986).

The final processing flow was established after extensive parameter testing. The most significant steps were Q compensation for attenuation losses caused by dispersion, surface-consistent deconvolution, 3-D post-stack migration, and bandpass frequency filtering.

Time differences between the H0-7 VSP first arrivals and the integrated sonic log were used to derive a Q model for the area. Application of an inverse Q correction, based on this model, proved useful as a data preconditioner for subsequent deconvolution (Hale, 1982).

Improved waveform consistency on the output data was obtained by using surface-consistent deconvolution. In contrast to normal single-channel deconvolution techniques, this process exploits the statistical advantage of many traces to estimate the wavelet's autocorrelation and attributes systematic portions to contributions from the near-surface at each source and receiver location (Tanner and Coburn, 1981).

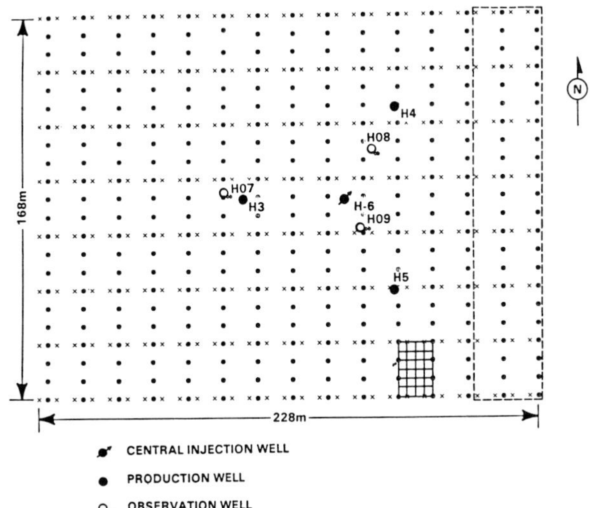

Fig. 6. Plat showing well configuration, source and receiver locations, and a portion of the stacking cell grid.

Post-stack 3-D migration proved most useful for improving spatial resolution because migration shrinks the Fresnel zone. The Devonian unconformity reflection on the migrated data demonstrated a narrow sinuous channel feature, which was not apparent on the unmigrated seismic. This channel was verified by the H0-7 and H-3 wells.

Post-stack, quality-control filter panels indicated that some signal frequencies as high as 240 Hz were recovered. The final optimal frequency filter passband chosen for the 3-D data was 25-220 Hz.

Monitor Surveys

The monitor survey area was 168 m × 228 m, which included an expansion to the east. This expansion was necessary because of an alteration to the originally planned well pattern. Each of the monitor surveys employed this acquisition configuration and were recorded in January, April, and November of 1987.

Monitor 1 (January 1987) was acquired after a four week steam and soak period at the three production wells H-3, H-4, and H-5. At the conclusion of Monitor 1 the central injection well, H-6, underwent a short period of hot water injection before being converted to steam. By the time the second 3-D monitor occurred in April, steam had been injected continuously into H-6 for 10 weeks. Monitor 3, recorded in early November 1987, was completed after another period of continuous steam injection in the central well.

The processing sequence employed was identical to that used for the base 3-D survey in order to minimize the creation of artificial differences not heat related. The exception to this sequence was the application of all-pass, phase-correcting filters to the monitors, when it was observed that the residual wavelet phase of all surveys was not identical. The probable reason for this variation was the seasonal difference in ground conditions affecting the ghosting behavior of the recorded wavelet.

Interpretation

As mentioned previously, two effects were important in the interpretation of the heated zones on the monitor surveys: (1) pushdown and (2) character changes occurring within the tar sand section. Both of these effects can be observed when comparing the base (pre-steam) and the monitor (post-steam) surveys.

Pushdown Mapping

If the seismic wavefront travels through a heated interval, then all the reflections from horizons below this interval are delayed, due to the velocity decrease caused by heating. To map this pushdown, reflectors above and below the reservoir were timed on both the base and monitor surveys. Traveltimes through the

McMurray interval were consistently delayed on the monitor surveys near the injection wells, as expected. The color coded pushdown maps, generated after Monitors 1 and 2, are shown in Figure 7. Monitor 1 shows a pattern that correlates to the injection well locations (H-3, 4, and 5). Monitor 2, on the other hand, shows expanding heat affected areas around H-3, 4, and 5 as well as the development of additional heating at the central H-6 injection site. These maps, while indicative of the areal extent of heating, do not provide the vertical distribution and are not directly correlatable to the temperature and/or thickness of the heated zone. Pushdown is dependent on two variables, thickness and velocity. Since velocity is a function of temperature, use of pushdown alone to predict thickness without first knowing temperature and vice versa is not possible.

Maps of the average tar sands velocity after the first and second monitor survey were obtained through a traveltime inversion procedure documented in Lines et al. (1989). The procedure established a depth model by image ray conversion of time to depths following the base survey. Having established reflector depths from the base survey, the traveltime changes for the monitor surveys were used to calculate velocity changes by employing simple distance-time-velocity relationships. The results obtained by use of this inversion technique corroborated those determined by the pushdown technique.

Character Change Analysis

Seismic amplitude is directly proportional to the change in acoustic velocity and density at an interface. Further, as the tar sands are heated, the velocity decreases by an amount directly related to the magnitude of the temperature change. Taken together, these phenomena mean that a heated zone, if thick enough, should show a change in seismic amplitude when compared to the base presteam survey. One example of this effect can be observed in Figure 8. The event at 0.2 s on the monitor survey, postulated to be the result of steam injected into H-5, is not as pronounced on the base data set.

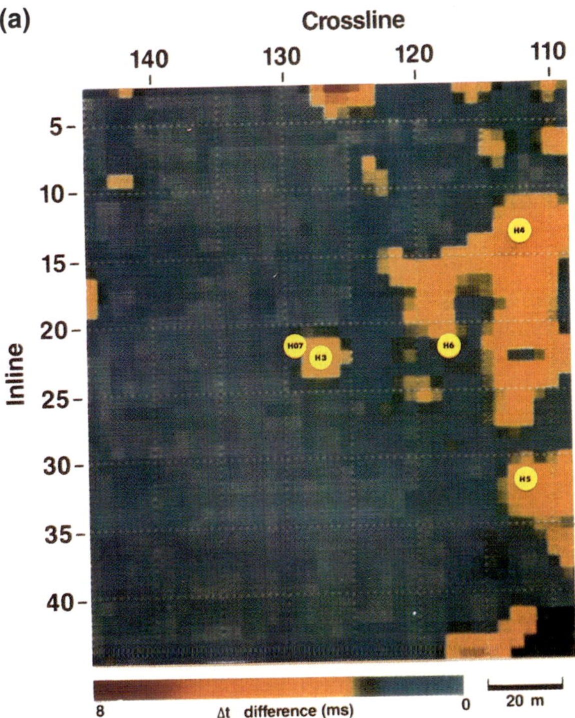

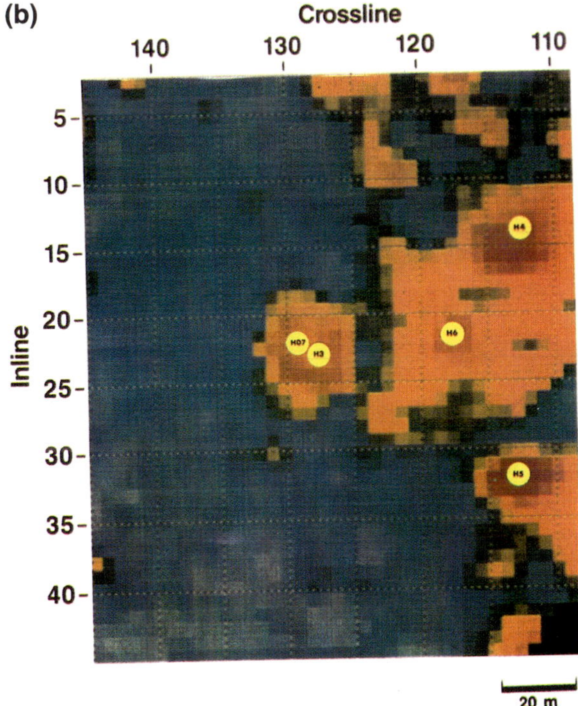

Fig. 7. Difference maps. (a) Baseline—monitor 1 (pushdown). (b) Baseline—monitor 2 (pushdown).

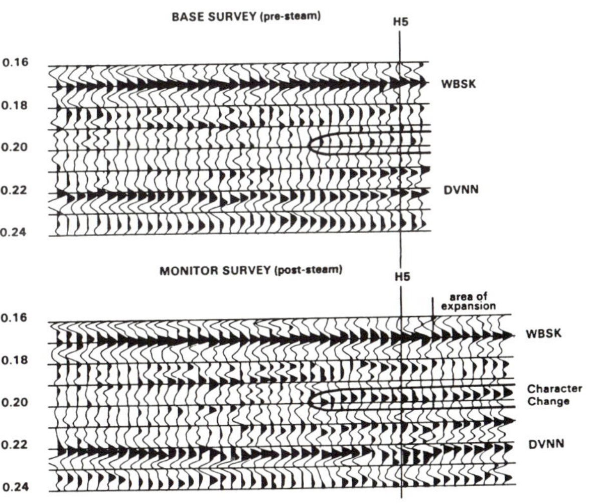

Fig. 8. Example of seismic character change associated with pre- and post-steam surveys.

Inversion

As seismic amplitude is directly linked to acoustic changes, all 3-D data sets were mathematically inverted and color-coded to provide a pseudo-acoustic impedance (velocity times density) log for each seismic trace as a function of depth. This method of converting seismic reflection data to pseudo-acoustic impedence logs has been described in Lindseth (1979).

Simplifying assumptions were made about the density in order to separate the velocity from acoustic impedance. The 220 Hz maximum signal frequency for these data permitted the resolution of layers between 2 and 3 m thickness. The inverted data volume for each monitor survey were independently subtracted from the base survey. The resulting 3-D volumes of velocity-difference data were then sliced by horizontal and vertical planes to reveal the distribution of heat in the subsurface. Figures 9 and 10 show the cross sections obtained by vertically slicing the velocity difference volumes for Monitor Survey 1 and 2 in an east-west direction, through wells H-3 and H-6. The tar interval extends from the Wabiskaw, at 170 m, to the Devonian, at 240 m. Differences are shaded so that larger magnitudes, which represent zones at higher temperature, are darker. The H-3 and H-6 injector positions are indicated by vertical lines on both sections.

The velocity difference section for Monitor 1 indicates that a relatively small portion of the tar zone has been heated. The velocity anomalies appear to be small, isolated, and disconnected. However, two im-portant observations can be made. First, the significant velocity anomalies are confined to the tar interval, suggesting that artifacts due to the differencing process have been minimized. Second, the vertical distribution of velocity anomalies suggests that two layers of preferred permeability exist at depths of 200 and 215 m, respectively. This is significant because steam was injected over an interval below the 215 m level, implying that heat has migrated upward, an observation substantiated by observation-well temperature profiles (Lowe, 1989).

The velocity difference section for Monitor 2 indicates dramatic heating effects. Again, the majority of the anomalies present are confined to the tar interval. Anomalies of lower magnitude, lying below the Devonian level, suggest that some residual pushdown artifacts may be present. Once again, there appear to be two main layers of heat concentration. However, in this case, the anomalies are large, laterally elongated, and well connected. Moreover, connections between anomalies in the first layer at 200 m and the second layer at 215 m suggest that vertical pathways of preferred permeability exist and are related to a fracturing program conducted at the well location. In addition, the magnitude of the velocity anomalies present suggest that the decrease in velocity exceeds 30 percent.

Figures 11 and 12 show the result obtained by slicing the velocity difference volumes for Monitor 1 and 2 horizontally at a depth of 200 m. This depth corresponds with an approximate 10 m interval of low oil saturation that is acting as a thermal thief zone. In both cases, the areal distributions of heat at this depth correlate well with those obtained independently from

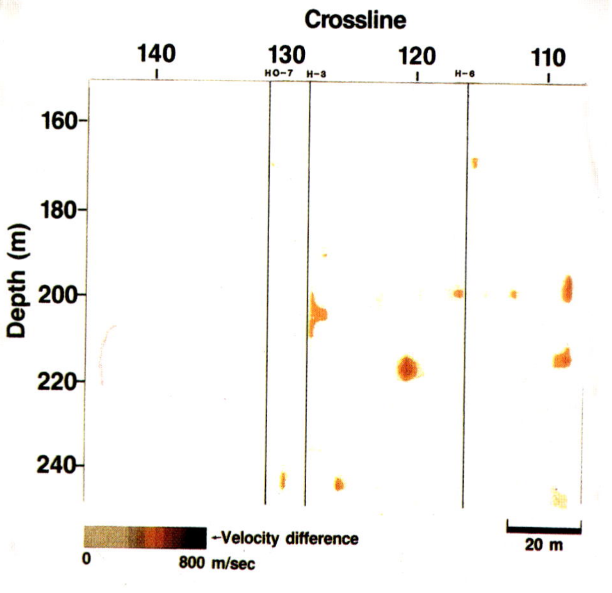

Fig. 9. Velocity difference profile in depth after monitor 1.

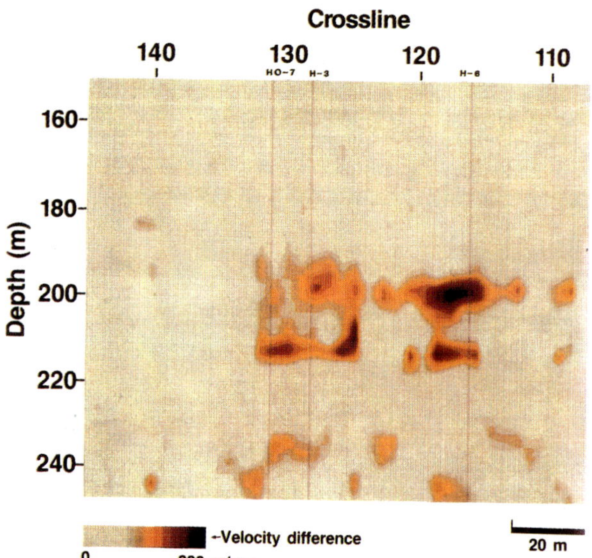

Fig. 10. Velocity difference profile in depth after monitor 2.

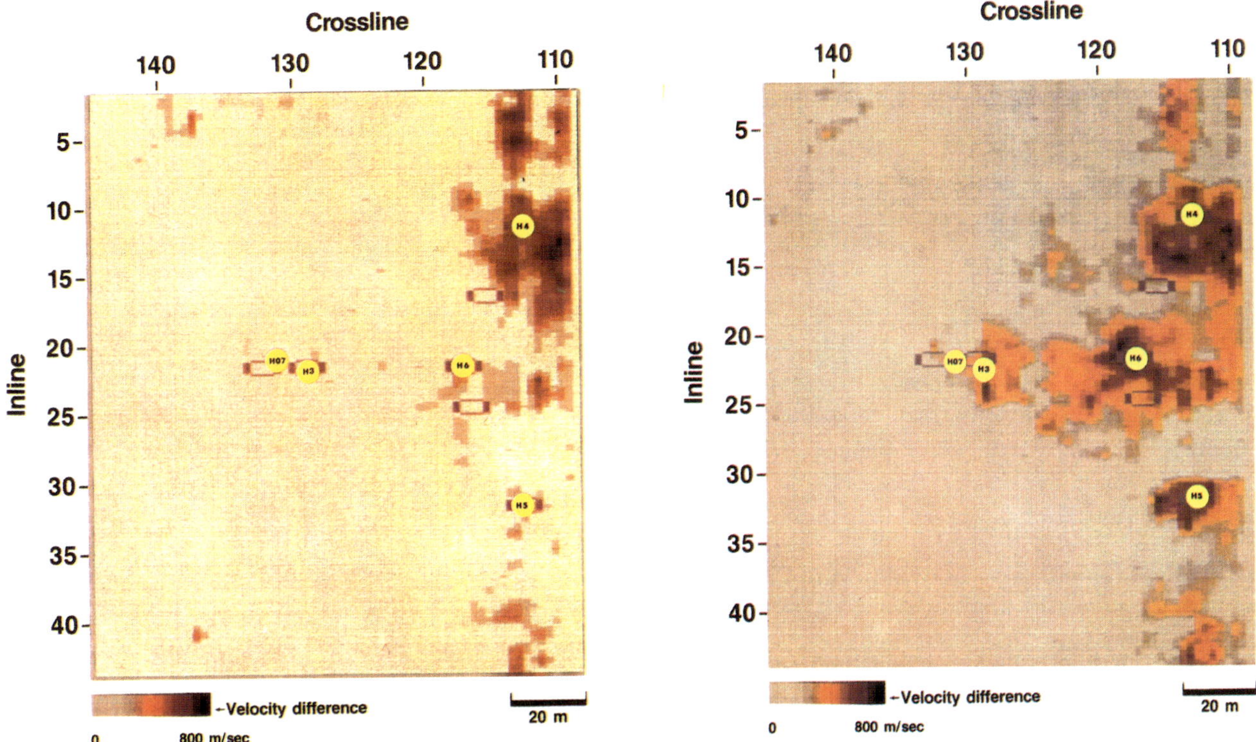

Fig. 11. Velocity difference depth slice after monitor 1. (200 m)

Fig. 12. Velocity difference depth slice after monitor 2. (200 m)

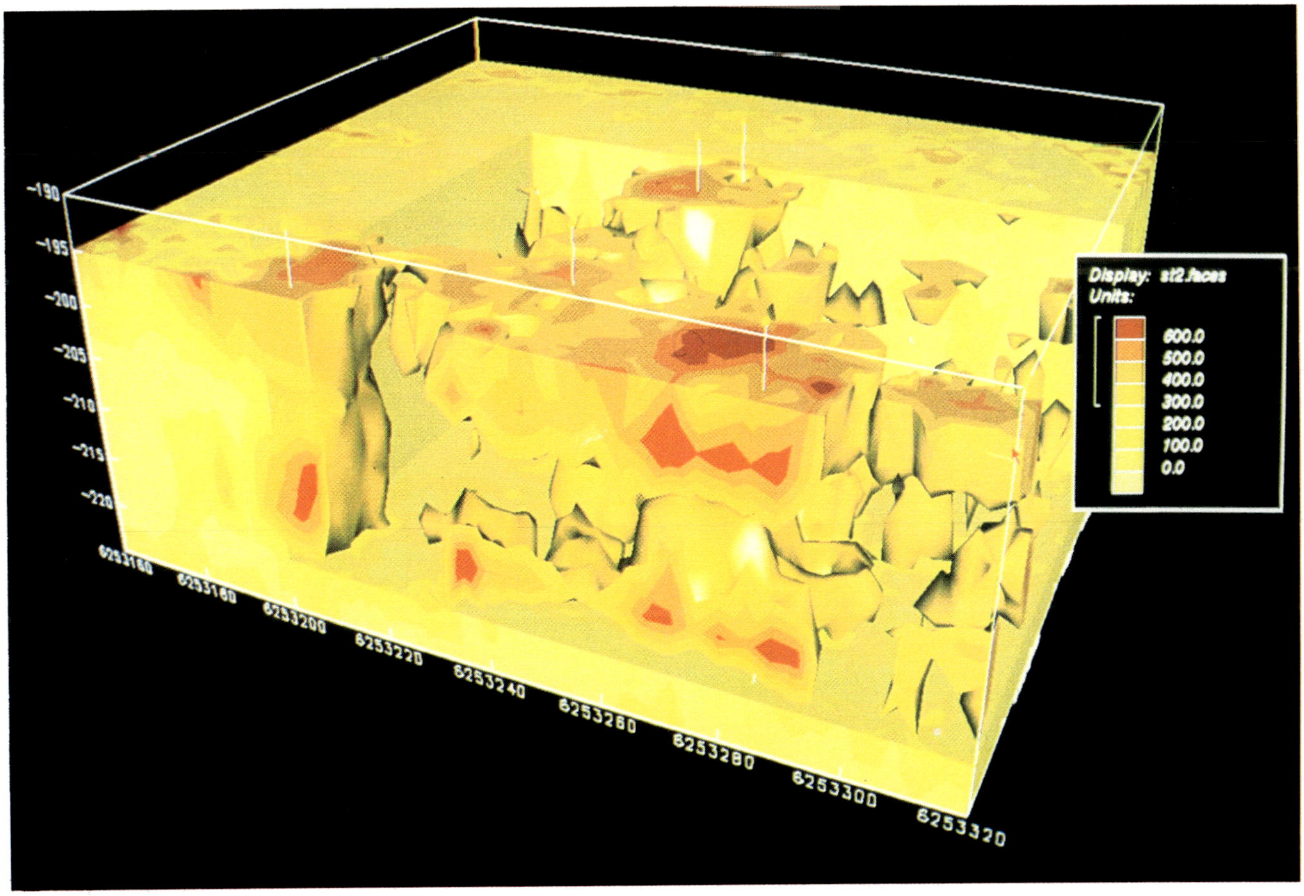

Fig. 13. 3-D volume display of velocity difference between base survey and monitor survey number 2. Units are in meters per second.

pushdown measurements for the entire tar interval. This correlation substantiates the velocity-depth distributions obtained from seismic inversion. Further, by combining a sequence of such slices, it is possible to reconstruct a true 3-D image of the heat zones as they exist in the subsurface. Figure 13 is such a volume display and demonstrates the interpretational advantage gained by employing advanced 3-D visualization display techniques.

Conclusions

Laboratory testing shows that a significant decrease in acoustic velocity occurs as the result of heating rock samples saturated with heavy hydrocarbons. The magnitude of this decrease is sufficient to allow seismic monitoring of thermal enhanced oil recovery processes. Seismic data have been shown to be useful in mapping the in-situ location of heated tar zones. With careful use of acquisition and processing parameters, repeat seismic surveys can provide valuable information about the effects of steaming operations at a reasonable cost. Inverted seismic data and sophisticated data display techniques yield significant new insights about the reservoir parameters, thus leading to a better understanding of the patterns of the heat-front movement. The integration of this data into the design and operation of steam injection projects should help in improving their future viability.

Acknowledgments

The author gratefully acknowledges numerous significant contributions made by all those employees within Amoco Production Company and Western Atlas International who participated in the many aspects of this project. In particular, a special thanks to Norm Pullin for the original concept, as well as Amoco Canada Petroleum Company Ltd., Alberta Oil Sands Technology and Research Authority (AOSTRA), and Petro Canada Resources for kindly releasing this material.

References and Additional Reading

Britton, M. W., Martin, W. L., Leibrecht, R. J., and Harmon, R. A., 1983, Street Ranch pilot test of fracture-assisted steamflood technology: J. Petr. Tech., 35, 511–522.

den Boer L. D., and Matthews, L. W., 1988, Seismic characterization of thermal flood behavior: Proc. of the 4th Internat. Conf. of Heavy Crude and Tar Sands.

Energy Resources Conservation Board, 1987, Alberta's reserve of crude oil, oil sands, gas, natural gas liquids and sulphur: ERCB Report ST88-18.

Greaves, R. J., and Fulp, T. J., 1987, Three-dimensional seismic monitoring of an enhanced oil recovery process: Geophysics, 52, 1175–1187.

Hale, D., 1982, Q-adaptive deconvolution: Stanford Exploration Project, Report Number 30, 133–158.

Kallweit, R. S., and Wood, L. C., 1982, The limits of zero-phase wavelets: Geophysics, 47, 1035–1046.

Lines, L. R., Jackson, R., and Covey, J. D., 1989, Seismic velocity models for heat zones in Athabasca tar sands: 59th Ann. Internat. Mtg., Soc. Expl. Geophys, Expanded Abstracts, 751–754.

Lindseth, R., 1979, Synthetic sonic logs—a process for stratigraphic interpretation: Geophysics, 44, 3–26.

Lowe, K., 1989, Thermal zone monitoring in an Athabasca oil sands pilot: Presented at 1989 Soc. Petr. Eng. Ann. Mtg.

Macrides, C. G., Kanasewich, E. R., and Bharatha, S., 1988, Multiborehole seismic imaging in steam injection heavy oil recovery projects: Geophysics, 53, 65–75.

Nur, A., 1982, Seismic imaging in enhanced recovery: SPE/DOE 10680.

Pullin, N., Matthews, L., Hirsche, K., 1986, Techniques applied to obtain very high resolution 3-D seismic imaging at an Athabasca tar sands thermal pilot: Presented at the 56th Ann. Internat. Mtg., Soc. Expl. Geophys.; published, 1987, The Leading Edge, 6, No. 12, 10–15.

Taner, M. T., and Coburn, K., 1981, Surface-consistent deconvolution: Presented at 51st Ann. Internat. Mtg., Soc. Expl. Geophys.

Wang, Z., and Nur, A., 1986, Effect of temperature on wave velocities in sands and sandstones with heavy hydrocarbons: Soc. Petr. Eng., Paper 15646.

Three-Dimensional Seismic Monitoring of an Enhanced Oil Recovery Process[1]

Robert J. Greaves and Terrance J. Fulp**

Introduction

Improving the efficiency of reservoir production can increase proven reserves. The final stages in the production of a field are enhanced oil recovery (EOR) processes. Effective management of EOR processes requires detailed reservoir description and observations of the volume of the reservoir being swept by the process. High-resolution 3-D reflection seismic surveying can be an effective tool in obtaining reservoir description, and, as demonstrated by this case study, can in some cases actually map the EOR process as it proceeds.

In this case study, 3-D seismic reflection data were used to monitor the propagation of a pilot in-situ combustion (fire-flood) process. Three identical 3-D seismic surveys were recorded over the pilot site at preburn, midburn, and postburn times. In this way, the combusion propagation was monitored over (calendar) time.

Acquisition and computer processing of the data were identical for each set of survey data, so that a direct comparison of the individual data sets could be made. To facilitate interpretation, the attributes of the seismic traces were calculated using Hilbert transform techniques as described by Taner and Sheriff (1977). Reflection strength, in this paper referred to as "envelope amplitude," was then used in the analysis of the reflection seismic data. In a unique application of reflection seismic data, the envelope amplitude traces from the preburn data volume were subtracted from their counterpart traces in the midburn and postburn data volumes, generating "difference volumes."

The combustion process substantially increased in-situ temperature and gas saturation in those sections of the reservoir affected by the burn. Both seismic velocity and density of the reservoir were changed. Zones with altered properties were detected by anomalous amplitude responses in the reflection from the top of the reservoir and from a limestone formation directly below the reservoir. The direction of the combustion propagation and estimates of its volume were based on the interpretation of these anomalies as observed in the difference volumes. The interpretation was supported by data available from monitor wells and postburn coring.

Background

Three 3-D seismic surveys were shot over a period of 15 months. The first (preburn) survey was recorded several months previous to ignition of the combustion process. The second (midburn) survey was recorded four months after ignition, and the final (postburn) survey was shot ten months after ignition.

The objectives of the seismic program were to

(1) detect a change in seismic reflection character attributable to the combustion process,
(2) determine the direction of burnfront propagation, and
(3) determine the volume of reservoir swept by the combustion process.

The basic premise was that an increase in gas saturation in the reservoir formation would produce measurable changes in reflection amplitude. Bright spots and dim spots, caused by anomalous gas concentrations, are well-known phenomena in exploration seismology. Increased gas saturation in the parts of the reservoir reached by the combustion process was expected to create bright spots and dim spots in the shadow zone (Sheriff, 1980). The 3-D data would be used to map that progression in time.

The EOR program consisted of a five-well pilot test covering a very small portion of the Holt Field in north-central Texas. The test consisted of four production wells separated by 90 m (300 ft) with a central injection well (Figure 1). The engineering objective was to propagate the combustion process from the injection well radially outward, creating and flushing an increased oil saturation zone, the oil bank, toward the production wells. Although the concept is simple, the implementation is quite difficult and is very sensitive to the details of the reservoir geology.

The reservoir is the Holt sand, a 12 m (40 ft) thick sandstone capped by a 2.5 m (8 ft) thick limestone

[1]Originally published in Geophysics, **52**, 1175–1187.
*ARCO Oil and Gas Company, Research and Technical Services, 2300 West Plano Parkway, Plano, TX 75075.

encased as a unit in thick shale (Figure 2). The sand is silty and laced with shale stringers and some calcite cementation zones. In this part of the field, the sand occurs at about 500 m (1650 ft) and dips to the north at 10 degrees. A thin limestone occurs about 45 m (150 ft) below the reservoir and is identified as the Palo Pinto limestone. From extensive core analysis, the horizontal permeability of the sand was found to be several times greater than the vertical permeability. Numerous fractures were observed, and an average orientation of N27E was measured. Although a detailed model of the reservoir and the burn process was not constructed, some effects of the process were anticipated. The combustion process would primarily propagate updip (to the south) due to the differing fluid densities. Propagation would primarily occur laterally within the reservoir from the initiation points. Vertical propagation would be limited to fractures or other natural permeability pathways. Finally, propagation might be further guided to the southwest along fracture-induced permeability pathways.

Acquisition and Processing

Several factors guided the choices of acquisition parameters:

(1) the target area was very small (90 m × 90 m);
(2) the target was relatively shallow (500 m);
(3) the data collection was to be repeated as identically as possible; and
(4) the amplitudes and spatial extents of the seismic anomalies would probably be quite small.

Simple seismic modeling based on well logs and the anticipated effect of increased gas saturation indicated that *detection* of the burnfront would be straightforward. However, very high-resolution seismic data would be required to map the lateral extent of the process and determine the net burn volume. A calculation of the resolution limit was made based on Widess (1973), and the center frequency required to resolve 7.5 m (25 ft) vertically was determined to be 100 Hz. The necessary resolution was felt to be achievable, given the shallow depth of the reservoir, if proper acquisition and processing techniques were applied.

The data collection array consisted of a modified 3-D patch geometry. Figure 1 shows the positions of shotpoints and receiver stations within the test area. This patch style of survey allowed the shot pattern to be arranged as necessary, such that the CMP data were collected with high fold over the area of primary interest (Figure 3) even though surface access was limited by buildings, wells, pipelines, etc. Furthermore, the geophones could be permanently installed at each receiver location to guarantee that the receiver array would be duplicated in each survey. A modification of the simple patch geometry was made to account for the migration of reflection points updip, by extending shot and receiver locations down-dip (to the north).

The receiver group spacing was 6 m (20 ft) with a single high-frequency (40 Hz) marsh geophone comprising each group. Each of the 182 receivers was buried 6 m (20 ft) below the surface. The 165 shotpoints were distributed along crossed lines with a 12 m (40 ft) spacing between individual shots. Each shot consisted of a 2.5 kg (3 lb) dynamite charge buried at 23 m (75 ft). The recording system used was a 192-channel GUS-BUS with sampling interval of 1 ms and band-pass recording filters set at 50 Hz low-cut and 320 Hz antialias. The burial of both shots and receivers improved the signal-to-noise (S/N) ratio of the recorded data by eliminating air-wave noise and substan-

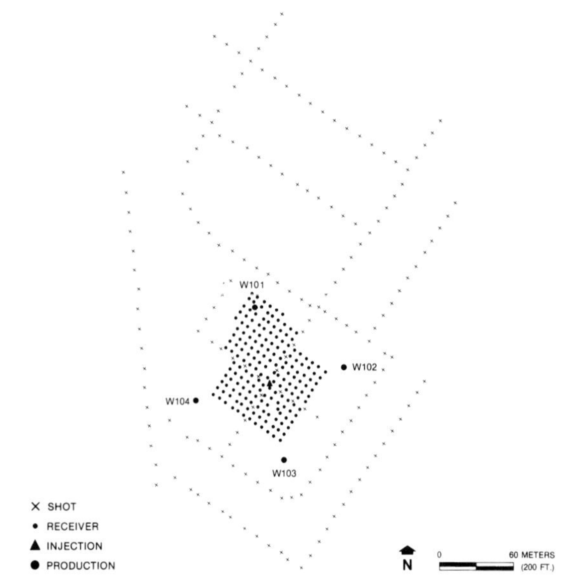

Fig. 1. 3-D seismic survey shot and receiver geometry with locations of production and injection wells.

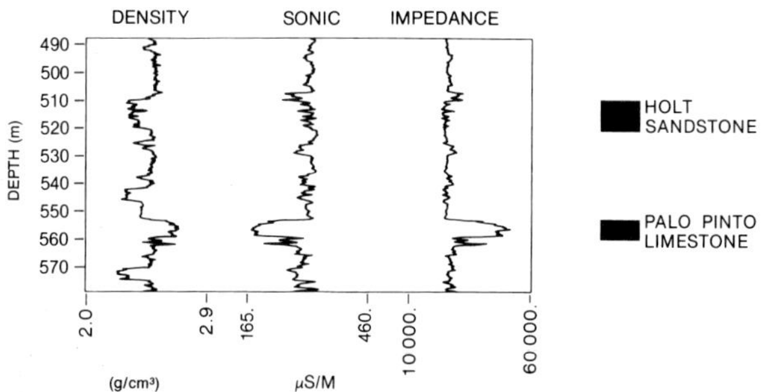

Fig. 2. Example of sonic and density logs with calculated impedance for the stratigraphic section including the Holt sandstone (reservoir) and the Palo Pinto limestone.

tially reducing the amplitude of the surface-wave noise. The 50 Hz low-cut filter was chosen at this high level to eliminate surface-wave noise further and to preserve the dynamic range of the recording system for digitization of the desired high-frequency signal. For recording shallow reflection seismic data, it is especially important to eliminate surface-wave noise that can seriously degrade the quality of the shallow data window. The small, deeply buried shots, the high-frequency phones and the low-cut filter all combined to eliminate this problem. The resulting frequency range of the recorded data was substantially higher than the range of standard exploration seismic, as shown in Figure 4a, and yet the range retained the minimum two-octave bandwidth considered necessary for high resolution.

The computer processing of the 3-D data sets used a standard sequence designed for 3-D CMP data. Throughout the processing, extra care was taken to retain true relative amplitude and the maximum usable frequency range. The traces were gathered into 3 × 3 m (10 × 10 ft) CMP bins. The statics and normal-movement (NMO) corrections were quite small due to the simple geologic structure and small area of the test. For more structurally complex geology, it would have been more difficult to make proper velocity adjustments because the patch geometry has the disadvantages of uneven fold and offset distribution. Three-dimensional surface-consistent statics were computed and were found to be on the order of 2 to 3 ms. Normal-movement corrections were applied using a datumed root-mean-square (rms) velocity function derived from the well control. Standard spiking deconvolution was applied before stack. A phaseless deconvolution technique was applied to balance further the usable spectrum of each stacked trace. In the final processing step, the data were migrated using an *f-k* migration algorithm and the velocity function derived

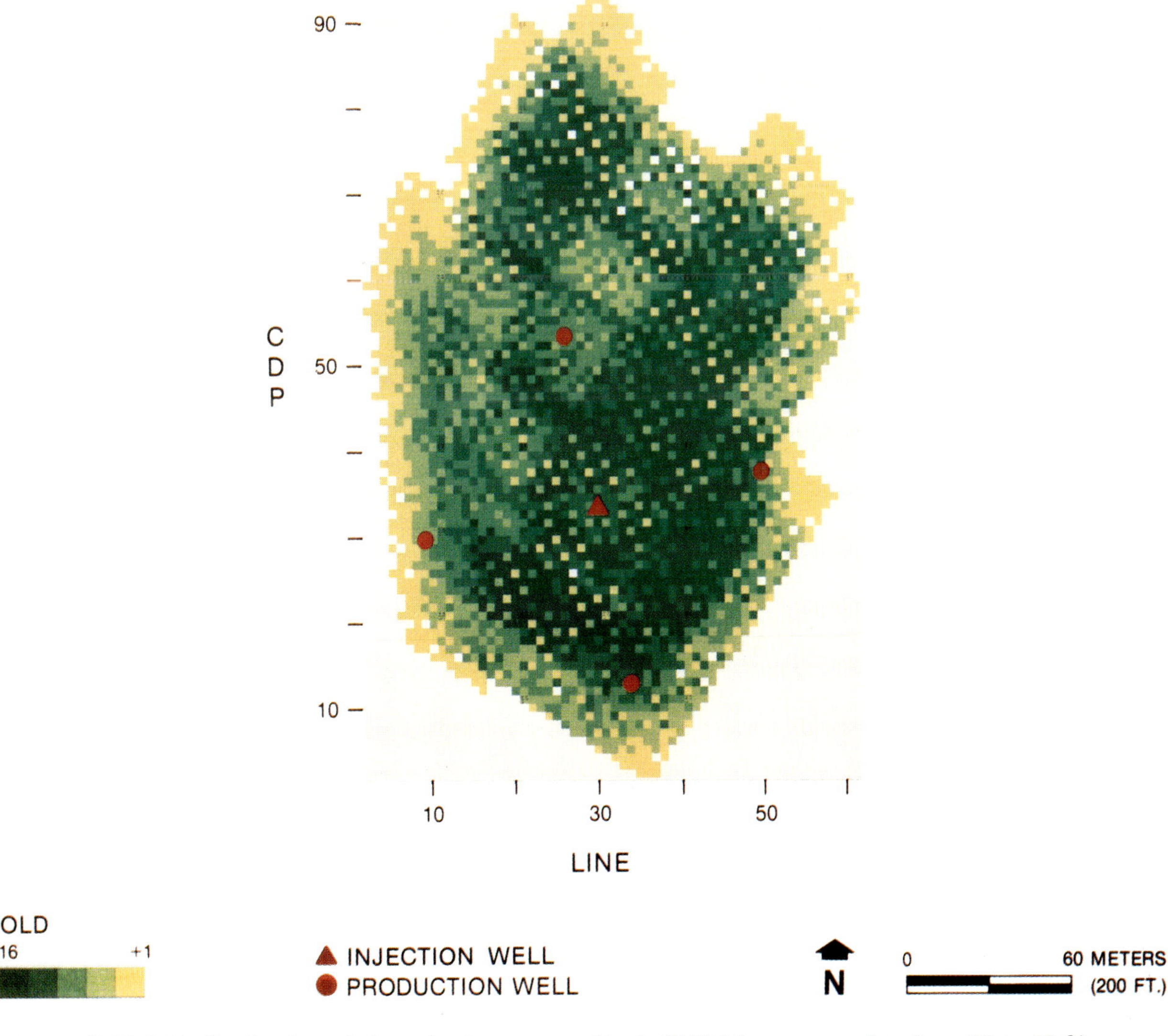

Fig. 3. CMP fold distribution of the seismic surveys. Each CMP bin covers a 3 × 3 m (10 × 10 ft) area.

from the sonic logs. This approach to migration was deemed adequate, given the localized area of interest and the simple velocity structure. The frequency spectrum of a fully processed trace, windowed in the reflection zone of interest, indicates that the 40–180 Hz bandwidth of the recorded data was enhanced during processing and the center frequency of 100 Hz was obtained (Figure 4b).

As postprocessing steps, the data were properly phase-corrected using well control, and the seismic attributes were calculated for each data set. To remove the geologic structure from the reflectors of interest, static adjustments were made, thereby allowing horizon views to be sliced from the 3-D data volume. Finally, in a unique step, the preburn horizon envelope amplitude at each level of interest was subtracted from the corresponding values in the midburn and postburn data volumes. The preburn data were used as the baseline seismic expression relative to which change was observed. Anomalies in the difference volumes were then interpreted directly.

Observed Anomalies

Bright Spots

Comparison of the envelope amplitudes of the reflection event at the top of the Holt sand reservoir revealed an increase in amplitude, a "bright spot," which developed after the combustion process was initiated. In Figure 5, a 2-D north-south section, line

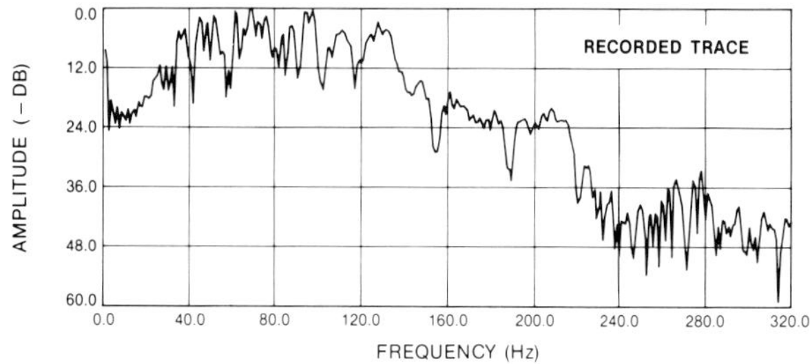

a. Pre-processed seismic trace from a shot record.

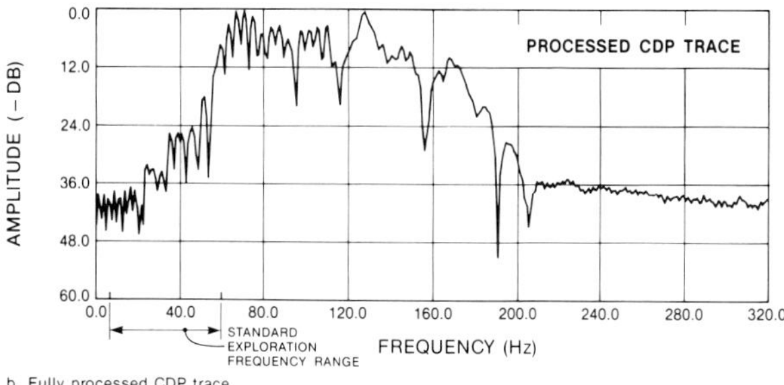

b. Fully processed CDP trace.

Fig. 4. An example of the power spectra of the trace data before and after processing. The spectra are for the window from 0.350–0.750 s which includes the reflection sequence from the Holt sandstone, to and including the Palo Pinto limestone.

14, is shown as it appeared at preburn, midburn, and postburn times. The reflection from the top of the Holt sand is identified as a trough occurring at about 385 ms. At this horizon, the envelope amplitudes at preburn time compared to midburn time show a zone of increased amplitude near well W104, with maximum change between CMP 16 and CMP 30. By postburn time, the bright spot had increased in lateral extent from CMP 16 to CMP 36, but it had not increased in maximum amplitude relative to midburn time.

Horizon slices at the top of the Holt sand, from the envelope amplitude difference volumes, are displayed in Figures 6a and 6b. The midburn difference shows a positive amplitude anomaly in the southwestern side of the data. This corresponds to the bright-spot development observed in line 14, Figure 5, at midburn time. Another, smaller bright spot is located to the southeast of the injection well at line 43, CMP 21. The difference amplitude at postburn time, Figure 6b, shows that the bright spot has grown to cover most of the area within the production wells, the midburn peak to the southwest has shifted downdip toward well W104, and the maximum amplitude of the difference anomaly has increased by about 10 percent.

Dim Spots

The strong reflection centered at 410 ms in line 14 (Figure 5) is identified as the Palo Pinto limestone. In line 14, a slight decrease in envelope amplitude occurs in the shadow of the bright spot centered around CMP 22. At midburn time, the decrease in amplitude is about 10 percent, but by postburn time, the decrease is nearly 25 percent, as marked by the change from deep orange and red shading to yellow.

A similar display of another north-south section, line 33 in Figure 7, shows a more substantial dim spot. This anomaly does not coincide with any bright spot at the Holt level at midburn time and only a modest Holt bright spot at postburn time. The dim-spot anomaly, pointed out by the arrows within the figure, is also stronger at midburn time than at postburn time. This lack of spatial coincidence (between bright and dim spots) is important in the interpretation of the results as described below.

The difference slice at the Palo Pinto reflection (Figure 8) clearly shows this anomaly. The dim spot at midburn time, Figure 8a, covers much of the pilot area with two negative-amplitude anomalies. One peak is located at the injection well, but the stronger peak lies about 30 m (100 ft) south-southwest of the injection well. The anomalies do not coincide with bright spots in the Holt reflection. A lower amplitude lobe of the dim spot extending to the southwest edge of the data does correlate with the maximum bright spot observed at midburn time. The dim spot observed at postburn time is lower in amplitude and extends over a signifi-cantly smaller portion of the pilot area. The two peaks of the midburn anomaly have merged into a ridge extending approximately southwest-northeast across the injection well with the larger area and peak of the anomaly to the southwest of the injection well.

Interpretation

Combustion Model

A simple model of the in-situ combustion process, based on combustion tube experiments, was described by Tadema (1959). The combustion process within the

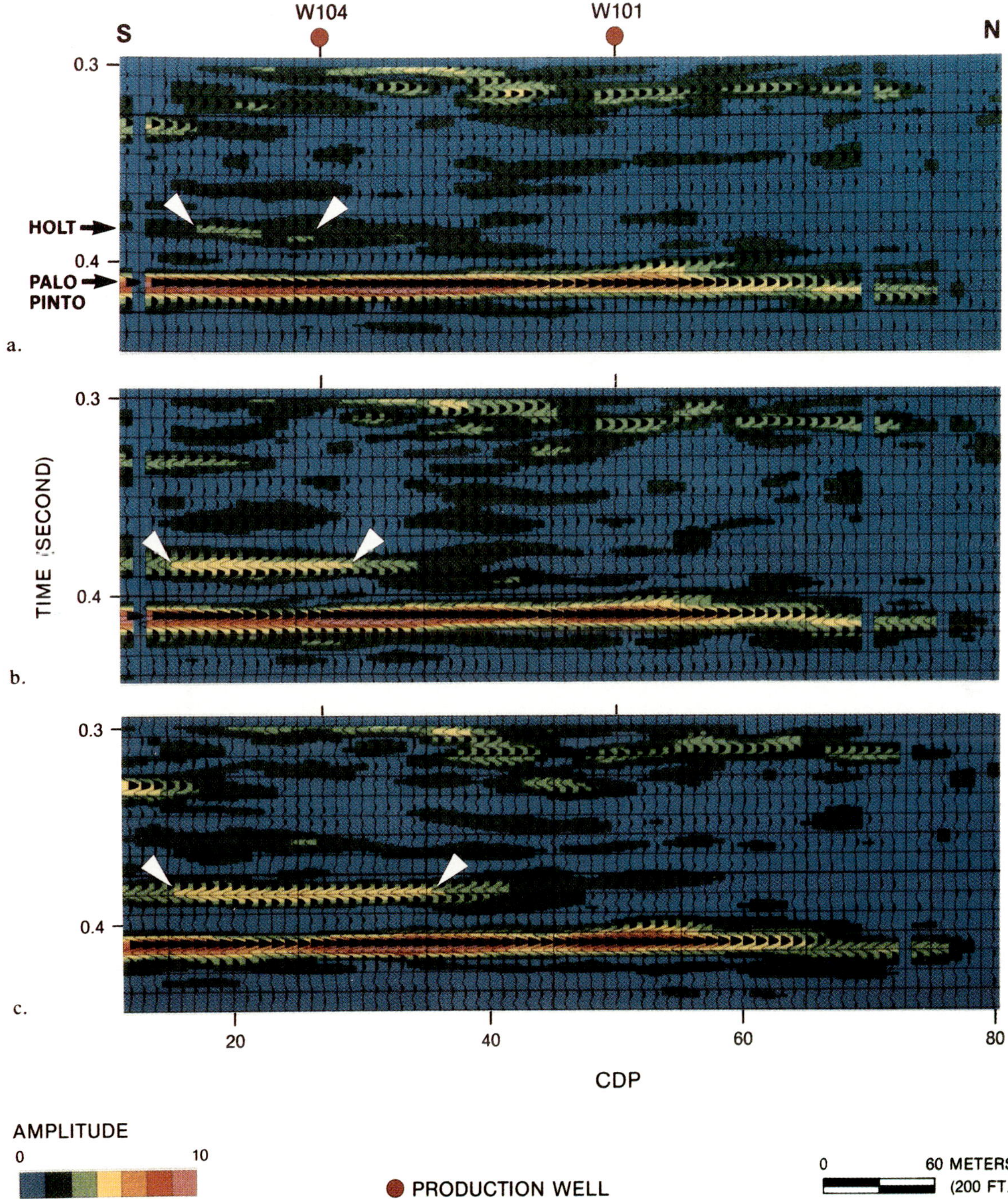

Fig. 5. Line 14, from (a) preburn, (b) midburn, and (c) postburn 3-D seismic data volumes. The reflection wiggle traces are overlain by a color scale of the calculated envelope amplitude. Dip was removed by static shifts before display. A bright spot was created (see arrows) at the top of the Holt sandstone by midburn time (b), and it increases in extent by postburn time (c). A dim spot in the reflection from the Palo Pinto limestone formed just below the peak of the bright spot.

reservoir can be divided into various zones, with each zone defined by its relative temperature and fluid saturations. The "combustion zone" propagates through the reservoir and is defined by maximum oxidation of the heaviest, or immobile, hydrocarbons. In its wake is left the "clean-burnt sand," a hot reservoir matrix with high gas saturation. Ahead of the combustion zone are several zones at lower temperatures and with distinctive percentages of oil, water, and gases until at some distance the original reservoir temperature and fluid mixture are encountered. Of particular interest are (1) that the clean-burnt sand zone has been subjected to very high temperatures, and (2) that combustion gas, as well as some injection gas, are forced ahead of the combustion zone. If this model were expanded into three dimensions, it would consist of a series of concentric rings which propagate radially from the injection well. The model is quite simple and does not at all account for geologic complexities, but it is useful as a starting point for the interpretation of the seismic anomalies in terms of the physical process of in-situ combustion.

Generalizing this model to the Holt sand reservoir, it is important to point out that the Holt sand reservoir, at the initiation of the burn, had little to no gas saturation.

Detection

As stated, the first objective of the seismic program was to detect a change in reflection character attributable to the combustion process. The bright spots and dim spots are considered true combustion-caused anomalies for the following reasons. First, the changes do occur in the reflection from the reservoir and the reflection just below it, as expected. Searches through the difference data volumes substantially above and below the reservoir reflection showed no extended, coherent anomalies. Second, the background noise level of the difference data volumes is substantially lower than the observed anomalies. Figures 6 and 8

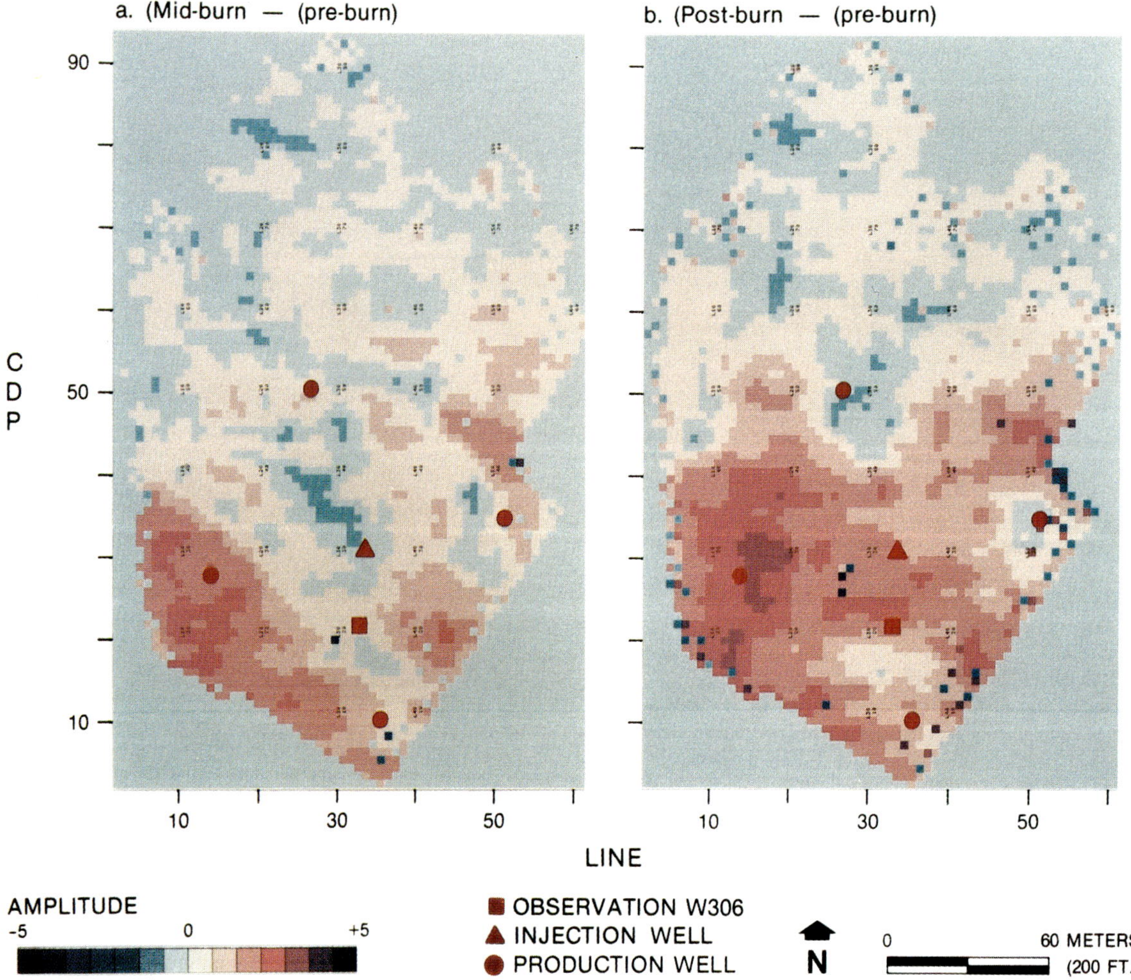

Fig. 6. The difference in envelope amplitude at the top of the Holt sandstone (0.385 s) displayed in horizontal time-slice form. Bright spots occur as positive anomalies. Well locations are marked for position in the subsurface at the top of the Holt sandstone.

show this by the amplitudes to the north of the injection well. Most importantly, the seismic anomalies were confirmed by well-log and core data. At the time postburn seismic data were collected, several cores and logs through the burned zone were collected.

In Figure 9, density and sonic traveltime logs show-ing the reservoir sand from a preburn observation well, W306, are compared to similar logs collected in two postburn boreholes. Within the zones of clean-burnt sand, the logs show substantial decreases in both density and velocity.

The combustion process was expected to increase the gas saturation with consequent changes in density

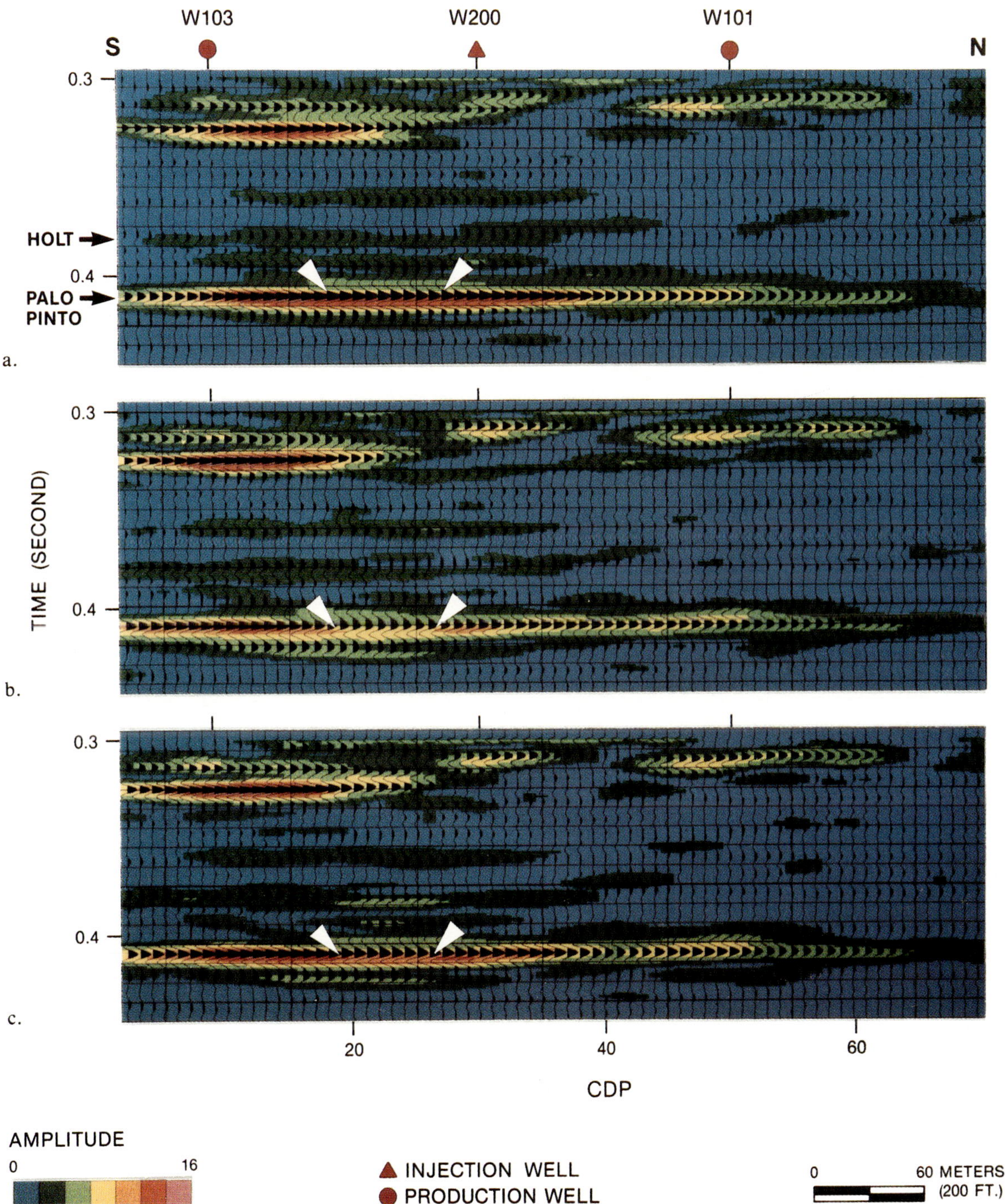

Fig. 7. Line 33, from (a) preburn, (b) midburn, and (c) postburn 3-D seismic data volumes. The reflection wiggle traces are overlain by a color scale of the calculated envelope amplitude. Dip was removed by static shifts before display. A dim spot was created (see arrows) in the Palo Pinto reflection (0.410 s) by midburn (b), but it decreases somewhat by postburn (c).

and velocity. Comparison of log density values showed decreased density in burned zones averaging about 5 percent. This density decrease can be fully accounted for by a change from 100 percent fluid-filled pores to partial gas saturation. The sonic-log velocities measured in burned zones decreased 15 percent to 35 percent, averaging 25 percent. This decrease in velocity is much greater than can be accounted for by increased gas saturation in the original pore space.

Ultrasonic measurements made on preburn core showed a 3 to 4 percent decrease in velocity, going from 100 percent water saturation to 100 percent gas saturation. Although this is a smaller decrease than that reported by Domenico (1976), a similar result was reported by Frisillo and Stewart (1980). In our case, the larger effect on velocity was due to permanent alteration of the rock matrix by the very high forma-

tion temperatures. Ultrasonic measurements showed a 25 percent decrease in velocity for cores heated to 700°F. The velocity decrease may be due to weakening of the rock by oxidation of organics and an alteration of clays. Therefore, the observed velocity decrease is the combined effect of changes in fluid saturation and of damage to the rock matrix.

The effect of increasing gas saturation on the seismic response is a nonlinear relation. As shown by Domenico (1974), it is the first few percentage increases in gas saturation (up to about 10 percent) that affect the seismic impedance the most. Further increases in gas saturation change the impedance very little. Therefore, if combustion gas is forced ahead of the burn zone in sufficient volume to increase gas saturation even a few percent, what is observed as a bright spot is both the clean-burnt zone and the zones ahead of the

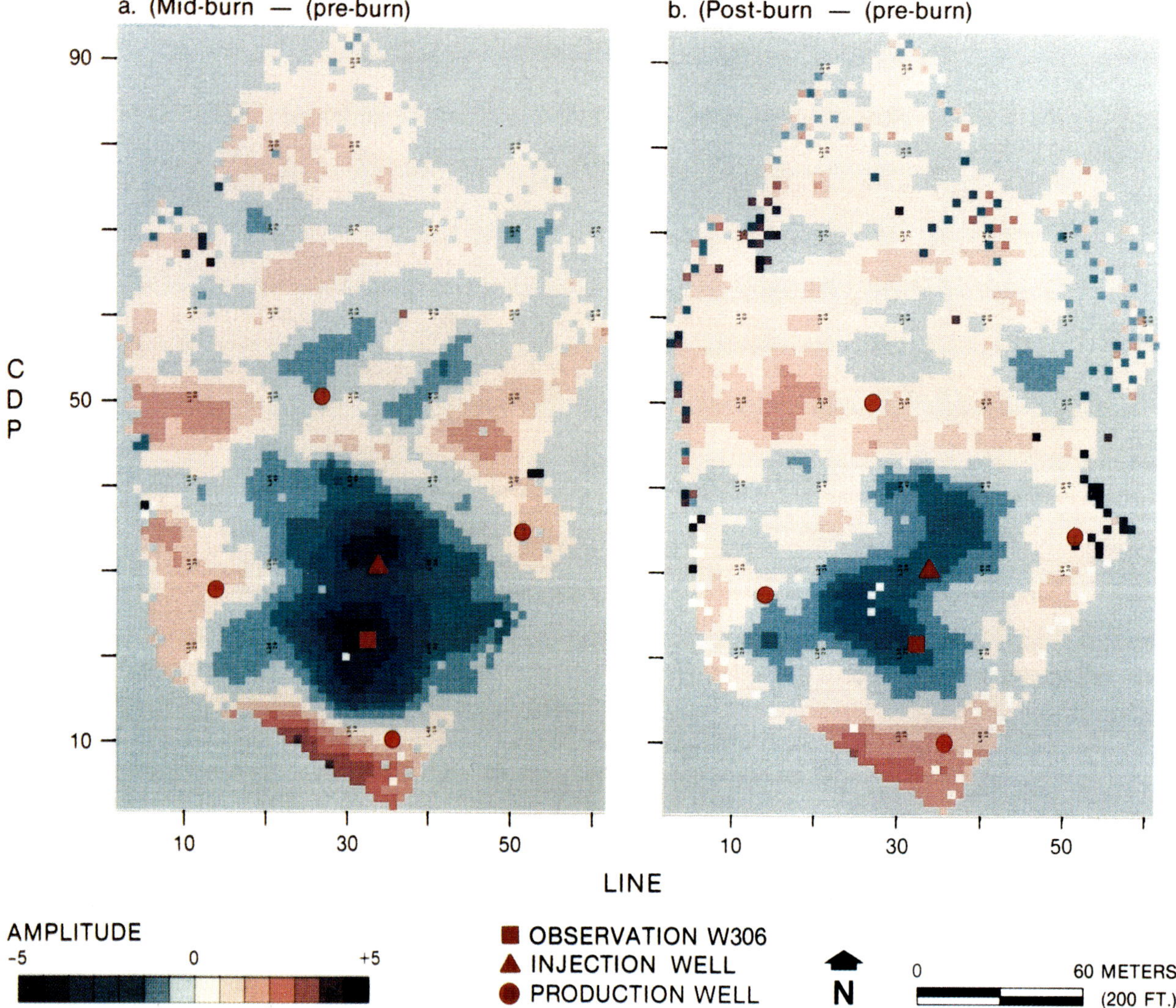

Fig. 8. The difference in envelope amplitude at the Palo Pinto reflection (0.410 s) displayed in horizontal time-slice form. Dim spots occur as negative anomalies. Well locations are marked for position in the subsurface at the top of the Holt sandstone.

combustion point reached by steam and combustion gas.

Certainly the phenomenon of attenuation is even more complex. Similar to seismic impedance, it is the initial change from 100 percent water saturation that increases the attenuation substantially. However, unlike impedance, attenuation decreases as the rock reaches 100 percent gas saturation (Frisillo and Stewart, 1980). Therefore, in the initial stages of the process, the dim spot will reflect both the clean-burnt zone and the zone reached by combustion gases. However, as 100 percent gas saturation is reached, the dim spot will more likely be an indication of just the clean-burnt zone.

Propagation

Interpretation of the positions of the bright-spot and dim-spot anomalies over calendar time provides a reasonable description of the combustion propagation.

First, it is quite clear from Figures 8a and 8b that the area around well W101 (to the north of the injection well) was not affected by the combustion process. This well was the only production well in which large quantities of gases were not observed. Therefore, this well was either too far downdip to be reached by the combustion process or was isolated by permeability barriers. One can also see from Figure 8 that the process did propagate to the southwest of the injection well, probably guided by the fracture system. The location of the strongest dim spot at midburn time shown in Figure 8a was verified by an observation well W306, which recorded the highest formation temperature at the time of the midburn survey.

The postburn dim spot (Figure 8b) decreased in amplitude and lateral extent compared to midburn time. It appears that the combustion process reached its maximum lateral propagation within the production area by midburn time or soon after. In a full-scale EOR

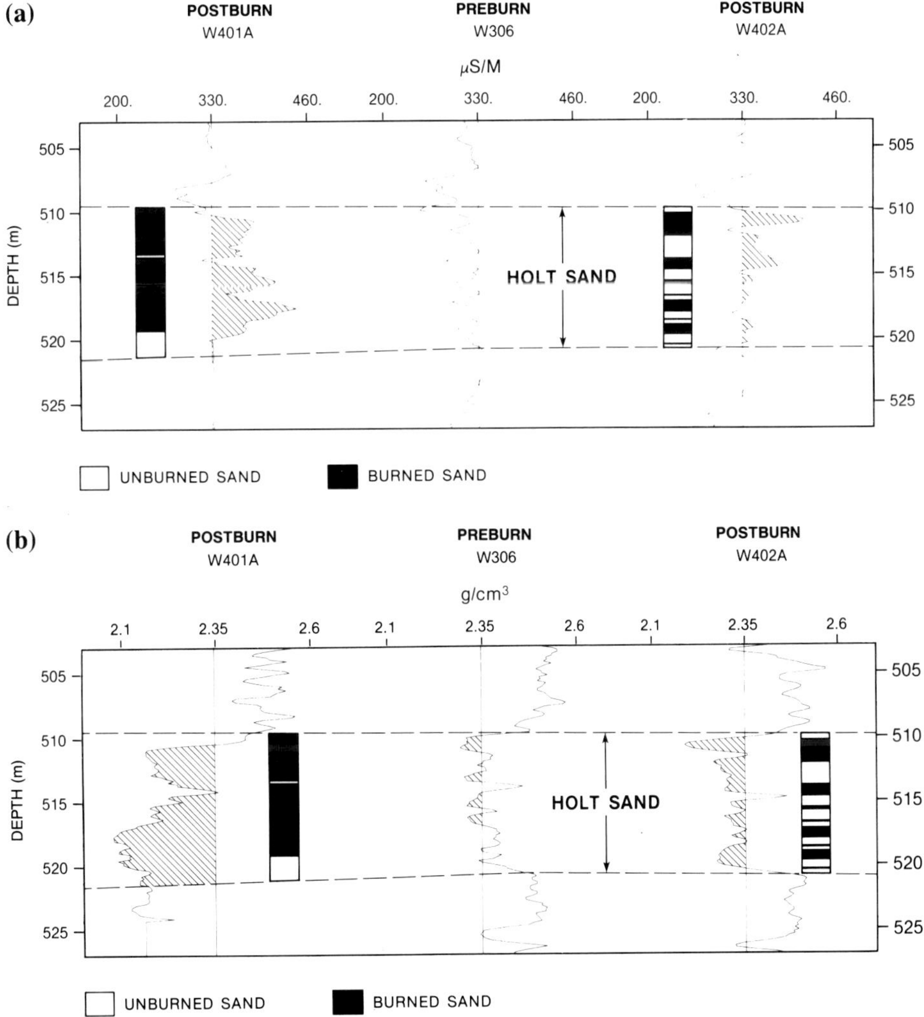

Fig. 9. Comparison of sonic preburn well W306 and postburn wells W401A and W402A. (a) Sonic logs; (b) density logs.

project, this knowledge would be crucial in adjusting the program to sweep the reservoir more efficiently.

Although the midburn bright spot (Figure 6a) also shows that the process moved to the southwest, it does not extend back to the injection well. Therefore, the combustion gas most likely propagated laterally within the reservoir until it encountered a vertical permeability pathway which allowed the gas to stream to the top of the reservoir. Once established, this pathway also allowed the burn to move to the top of the reservoir. Up to that point, the successful part of the burn was contained in the middle of the reservoir.

The postburn bright spot (Figure 6b) increased in area from midburn time. A major fault (providing the southern closure to the field) is located approximately 300 m (1000 ft) to the south of the test. This fault probably blocked further southward propagation of the combustion gases, forcing them back along the top of the reservoir toward the injection well. If sustained injection of gas from midburn to postburn time had continued to fuel the combustion process, that process would have moved out beyond the production area and the area of seismic coverage.

Burn volume

The final objective of this study was to estimate the volume of reservoir swept by the combustion process. Although the data do not have the spatial resolution to map the detailed distribution of the process, we have attempted to interpret the decreased amplitudes in the Palo Pinto reflection as estimates of burn thickness. The mechanisms of the attenuation are not separable, but several factors are certainly important: pore fluid state and interaction with the rock matrix, formation temperature, matrix velocity and density, and the increased reflectivity due to gas saturation in the reservoir.

A simple mathematical approach was chosen (after Waters, 1978) in which amplitude is expressed as

$$A = A_o e^{-\alpha z}, \qquad (1)$$

where A_o is the initial amplitude of the propagating wavelet, α is the attenuation parameter, and z is the propagation distance. If two seismic waves are considered identical except that one has passed through a zone Δz where the attenuation is different, then the reflection amplitude from a level past the zone of attenuation can be compared directly to find Δz. For this study, observed seismic waves are the before-burn and after-burn data traces, and Δz is the estimate of burn thickness. After taking the natural logarithms and accounting for two-way propagation, the equation becomes the linear relation

$$\Delta z = \frac{1}{2(\alpha_B - \alpha_A)} \ln\left(\frac{A_A}{A_B}\right). \qquad (2)$$

The reservoir was cored in twelve locations within the test pattern at postburn time. The cores confirmed that the burn had occurred in somewhat vertically isolated zones within the reservoir. The net burn thickness observed in each core was compared to the logarithm of the ratio for appropriate seismic amplitudes at the CMPs corresponding to the bottom-hole location of each core. Figure 10a shows the comparison of net burn thickness to the logarithm of midburn amplitude over preburn amplitude, and Figure 10b shows the same relationship for postburn data. Using least-squares estimation, a line was fit through the data points (lines are forced through the origin). The slope of either of these lines could be used in equation (2) to estimate burn thickness at all other CMPs. Since the midburn data (Figure 10a) appeared to fit the simple model better, the midburn data were used to estimate burn thickness. This reemphasizes the belief that burn

(a)

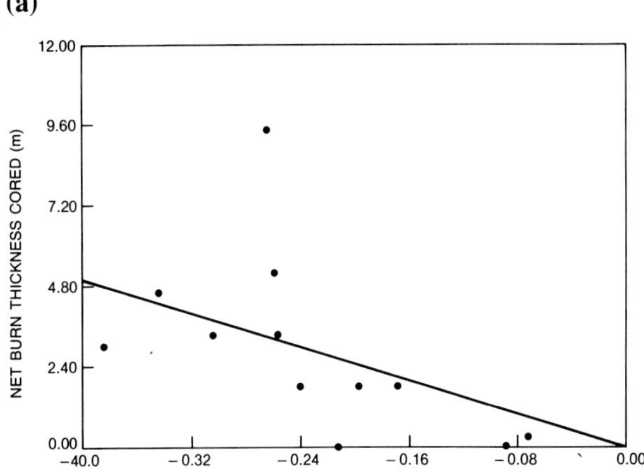

(b)

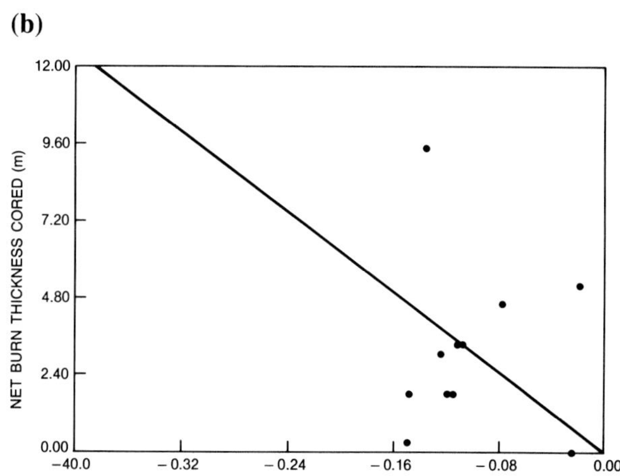

Fig. 10. Net burn thickness from postburn cores versus natural logarithm of the ratio of dim-spot (Palo Pinto) amplitudes. The lines are least-squares fits to the data points. (a) Midburn/preburn amplitudes; (b) postburn/preburn amplitudes.

propagation, at least within the production area, had ceased by midburn time. The postburn dim spot is more likely a map of formation damage only, suggesting that a more complicated model is needed to explain the attenuation due to alteration of the rock fabric.

Using the slope of the line in Figure 10a and equation (2), the midburn data were converted to an estimate of net burn thickness (Figure 11). Overlain on that estimate is a computer-generated contour map based on core data. A good correlation is observed, and the correlation could have been even better if there had been core data to the southwest. This also implies that even without the core data for calibration, a good estimate of relative burn thickness could have been made using only the seismic data. The observation of seismic attenuation is a useful approach in mapping certain recovery processes. Resolution could be improved utilizing borehole-to-borehole techniques.

Conclusions

Reflection seismic surveying can be used to monitor the progress of some EOR processes. In this case study, a fireflood process was detected, its propagation direction and extent were determined, and an estimate of net burn volume was made.

The 3-D seismic data detected the burn zone and showed that the gas propagated predominately updip to the southwest. A dim spot observed in a reflector just below the reservoir level was interpreted as a map of areal extent of the burned zone. A region of maximum net burn thickness was located about 30 m (100 ft) from the initiation point of the burn. Comparison of the midburn and postburn dim spots led to the

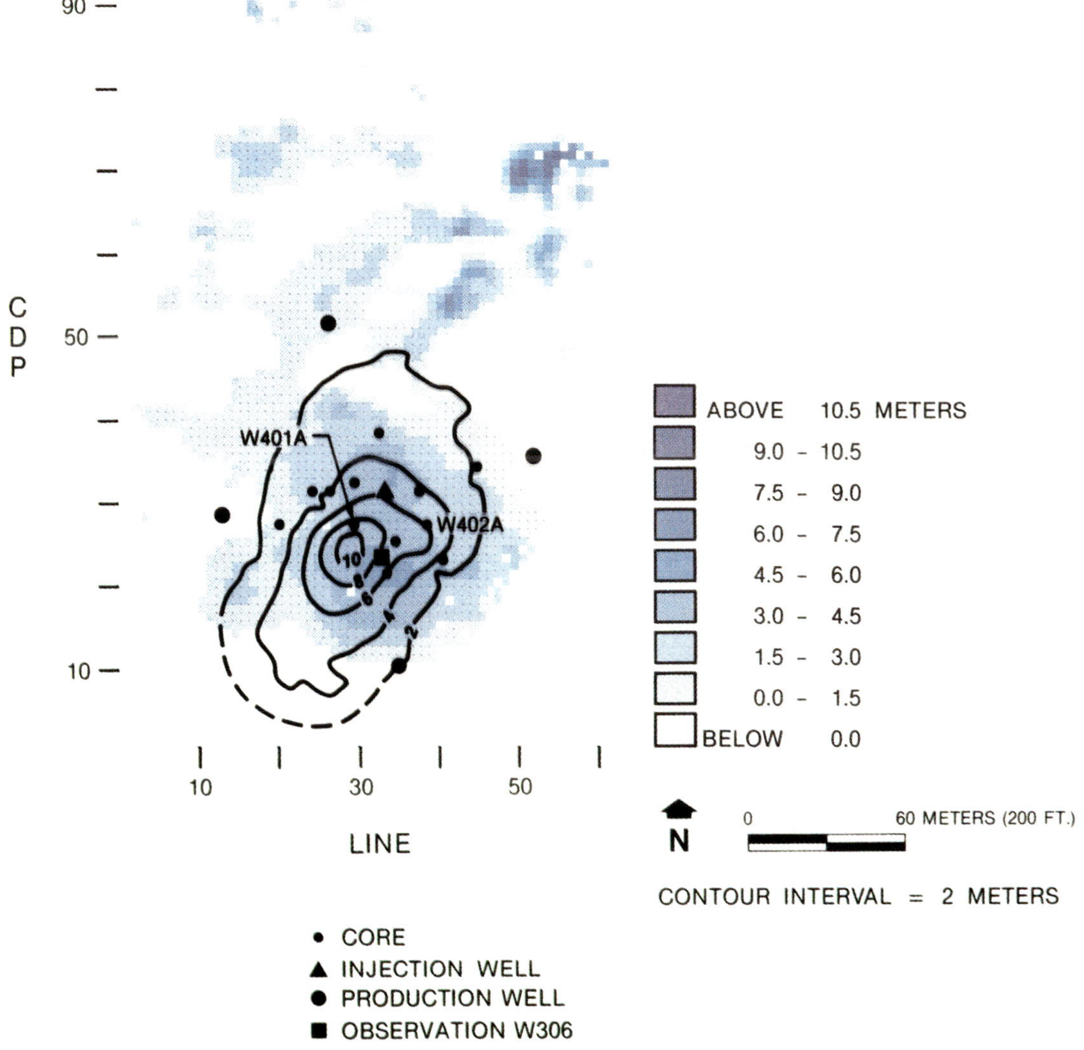

Fig. 11. Burn thickness calculated from midburn dim-spot amplitudes using equation (2) and the slope of the line in Figure 10a for the calibration constant. Overlain is a line contour map of net burn thickness observed in cores.

conclusion that the majority of the reservoir swept by the combustion process occurred in the first few months after ignition. The shape, orientation, and volume of the burn interpreted from the seismic data were confirmed by temperature monitor wells and postburn coring.

It was concluded that the attenuation increase, due to high-temperature alterations of the reservoir rock and pore fluid changes, was the best seismic indicator of the combustion process.

The subtraction of the baseline (preburn) data from the midburn and postburn data for interpretation of dynamic anomalies proved to be a very powerful technique. The subtraction technique has great potential for detecting anomalous seismic responses related to active reservoir processes.

Acknowledgments

The authors would like to thank ARCO Oil and Gas Company for allowing us to publish these results. Production Engineering Research provided the financial support for this project. In Geophysical Support, L. F. Konty and D. R. Paschal were instrumental in survey planning. L. J. Hix and P. W. Wise added their expertise in data acquisition. Data processing was designed by R. Chen and S. A. Svatek. M. L. Batzle made the petrophysical measurements on core samples and provided core descriptions. J. D. Robertson contributed support and ideas during the interpretation of the data sets.

References

Domenico, S. N., 1974, Effect of water saturation on seismic reflectivity of sand reservoirs encased in shale: Geophysics, **39,** 759–769.

———— 1976, Effect of brine-gas mixture on velocity in an unconsolidated sand reservoir: Geophysics, **41,** 882–894.

Frisillo, A. L., and Stewart, T. J., 1980, Effect of partial gas/brine saturation on ultrasonic absorption in sandstone: J. Geophys. Res., **85,** 5209–5211.

Sheriff, R. E., 1980, Seismic stratigraphy: Internat. Human Res. Dev. Corp., 185–198.

Tadema, H. J., 1959, Mechanism of oil production by underground combustion: Proc. 5th World Petr. Congress, sec. II, 279–287.

Taner, M. T., and Sheriff, R. E., 1977, Application of amplitude frequency, and other attributes to stratigraphic and hydrocarbon determination: Am. Assn. Petr. Geol. Memoir **26,** 301–302.

Waters, K. H., 1978, Reflection seismology: John Wiley and Sons, 203–207.

Widess, M. B., 1973, How thin is a thin bed?: Geophysics, **38,** 1176–1180.

Acoustic Tomography in Reservoir Surveillance

J. H. Justice, A. A. Vassiliou,* M. E. Mathisen,* S. Singh,[‡] P. S. Cunningham,[‡] P. R. Hutt[§]*

Introduction

Tomographic reconstruction techniques are being successfully applied to reservoir imaging. Several types of tomographic images can now be acquired and reconstructed which add significantly to our knowledge of reservoir rock and fluid properties, as well as reservoir structure and dynamics.

The first important application of this technology resulted from rock properties research on heavy oil sands. Nur (1984) reported that acoustic velocities and attenuation in sands saturated with heavy oils are strongly dependent on the degree of heating. Although the explanation for the observed effect was not clear, it was immediately apparent that this might provide the basis for monitoring flood fronts in thermal enhanced oil recovery (EOR) projects in heavy oil reservoirs. Since 1984, projects to test the theory and to verify the results predicted by the laboratory experiments have been quite promising and interest in geophysical monitoring of EOR processes is growing steadily.

EOR procedures are used when primary production (that in which the well flows under its own pressure, or in which a pump alone is used) has reached a very low rate, but where sufficient oil remains to justify more expensive procedures. In most EOR projects, a liquid or gas is injected under high pressure, the purpose being to push the oil bank toward producing wells and/or to reduce the viscosity of the oil so that it moves through the reservoir more easily. Production wells located around an injector well then produce the additional oil being swept by the "flood". Since hydrocarbon reservoirs are often heterogeneous and structurally complex, fluid flow paths cannot always be accurately predicted, and finding a way to monitor these processes to ensure their economic efficiency becomes important.

*Mobil Research and Development Corp., Box 819047, Dallas, TX 75381.
[‡]Mobil Exploration and Production Services, Inc., formerly Mobil Research and Development Corp.
[§]Mobil North Sea, Ltd., formerly Mobil Research and Development Corp.

Some of the first attempts to test the geophysical approach to EOR monitoring focused on the use of high-resolution surface seismic surveys. According to the laboratory results, thermal EOR processes might change acoustic impedance in the zone being stimulated, resulting in a change in the reflectivity. Surface seismic surveys, if successful, could be used to monitor the areal distribution and movement (and sweep efficiency) of the flood. However, the critical question of vertical conformance (vertical distribution of the flood) would go largely unanswered using only this approach, due to the low vertical resolution of surface seismic data.

As a result, considerable interest developed in the possible application of crosshole tomography to provide information on vertical conformance. Crosshole tomography constructs an image of the vertical plane connecting the wells using a multitude of measurements from various source locations in one well to a multitude of receiver locations in another well. Thus crosshole tomography, together with surface seismic data, might provide a relatively complete picture of the distribution of the injected fluid. The information obtained could be used to construct a more accurate description of the reservoir, resulting in a more complete understanding of the reservoir's characteristics and behavior. The resulting information could be used for planning further development, for taking corrective action, and for increasing the efficiency of the EOR process. The final result would be EOR programs which are better designed and more efficient, resulting in increased production at lower cost per barrel. In addition, recoverable reserves which might formerly have been missed could be identified and appropriate action could be taken.

With initial successes credited to the use of tomographic imaging, it was soon realized that the images, which reveal structural, stratigraphic, or lithologic details of the reservoir between wells, have great value in their own right. When reservoir cross-sections are reconstructed from well log data, they represent only best guesses based on limited information and lack the confirmation which tomographic imaging can provide.

Because costly development programs are based on this kind of analysis, tomographic confirmation of reservoir structure becomes extremely important. Great demands are placed on our ability to correctly interpret a tomographic image which reconstructs the seismic velocity or attenuation field. Because many factors affect the parameters which are imaged, additional information must be brought to bear on the interpretation process. Though much can be done, research in rock physics has been inadequate to answer all questions related to the relationship between parameters and seismic measurements, and much more work is needed. While the use of geophysics for reservoir surveillance is still in its infancy, the work is stimulating new research in areas such as rock properties, and answers to many new questions are beginning to emerge.

In this paper we consider the application of crosshole tomography to monitoring thermal EOR processes in heavy oil reservoirs and look at the use of reservoir simulators for understanding the seismic response of oil recovery programs.

Tomographic Image Reconstruction for Geophysical Applications

Two geophysical measurements, traveltime, which is used to reconstruct the unknown velocity field, and attenuation, used to construct a Q map, can be represented as path integrals. They therefore satisfy the requirements for tomographic image reconstruction in accordance with the work of Radon (Deans, 1983). We shall take the traveltime measurements as the measurements of interest, because they are more readily obtained, even though our comments will apply in either case.

The algebraic formulation of the straight-ray reconstruction problem is well known. We assume that the region to be imaged has been partitioned into cells and that the slowness in each cell is specified by a single parameter P_j. The resulting equation in matrix form is

$$\mathbf{T} = \underset{\sim}{\mathbf{S}}\,\mathbf{P}, \qquad (1)$$

where $\mathbf{T}$ is the vector containing the traveltime for each raypath, $\mathbf{P}$ is the vector containing the slowness in each cell, and $\underset{\sim}{\mathbf{S}}$ is the raypath matrix composed of elements S_{ij}, where

$$S_{ij} = \text{arc length of the segment of raypath } i \text{ in cell } j. \qquad (2)$$

The matrix $\underset{\sim}{\mathbf{S}}$ in equations (1) and (2) is generally not square, and usually the number of raypaths exceeds the number of cells so that the problem is overdetermined. A variety of algorithms is available for solving the nonsquare system. Although the algebraic reconstruction technique (ART) and simultaneous reconstruction technique (SIRT) algorithms have been mentioned often in the literature, they do not represent good choices for solving equation (1), even if computing power is limited. They suffer from a number of disadvantages and are usually modified in many ways in an attempt to improve their performance. Though we have worked with many versions of ART and SIRT, we cannot recommend either for convergence or accuracy. Better choices would be conjugate gradients and singular-value decomposition. The latter, which can be highly accurate, is more costly computationally than the former. Comparison of results using different algorithms is given later.

Acoustic waves do not travel along straight raypaths, as x-rays are assumed to do in medical applications. In thermal EOR projects, velocity variations within the affected zones can be of the order of several tens of percent and only reconstruction algorithms which account for a variable index of refraction should be considered.

Let us now turn our attention to tomography in which refraction/diffraction effects are accounted for. Various types of approximate reconstruction algorithms have been suggested to extend straight-ray algorithms, but none has met with much success.

We have obtained good results with an algebraic approach in which we linearize the problem, propagate the wavefield through the current estimate of the slowness field, and then compute a very accurate estimate of the Jacobian matrix to update the slowness field. It is extremely important that the Jacobian matrix be accurately computed, and probably one of the major weaknesses of other approaches to diffraction tomography is that inaccurate approximations were used. Generally we take a uniform model as the starting estimate, an approach which cannot be expected to work well if a sufficiently accurate wave-propagation algorithm is not used. We cannot recommend the use of ray tracing for this reason.

With first-arrival tomographic reconstruction algorithms, it is extremely important that the forward-modeling algorithm account for the actual observed arrivals. These arrivals may result from critical refraction, for example, and unless such events are computed in the forward-modeling algorithm, accurate reconstruction cannot be obtained. When small diffractors are present (a small steam zone in a young EOR project, for example), ray tracing cannot be used to reconstruct an accurate image, and the wave equation must be the basis for reconstruction. Recently our work has focused on an extremely fast algorithm for computing wavefronts of the full acoustic wavefield at closely spaced increments in time. Rays can then be traced along orthogonal trajectories backward (in time) through the resulting wavefronts. This procedure automatically finds the minimum-time ray(s) connecting a source to a receiver position. Since these rays, in general, do not satisfy the ray equations, they could not be found by more cumbersome classical ray tracing. We use this procedure for raypath analysis only. We do not recommend the use of tomographic reconstruction algorithms based on ray tracing for seismic imaging. We continue our discussion in the context of ray tracing, however, for conceptual reasons.

In some cases, ray bending and focusing is so severe that portions of the imaged region are not illuminated by minimum time raypaths. These portions will still be reconstructed with limited accuracy if the starting model is uniform, since minimum time rays will illuminate these regions until the velocity-field reconstruction is sufficiently accurate that the seismic energy no longer penetrates them, after which additional iterations will not result in further updates in these zones. In order to obtain a quantitatively accurate reconstruction of them, minimum traveltime data are no longer sufficient and traveltimes corresponding to raypaths for later arrivals must be included. In these cases, the full elastic wavefield becomes very important and the complexity of the reconstruction problem increases significantly.

We have investigated a large number of alternative error measures for use with our reconstruction algorithm. While the familiar least-square error criterion (L_2 norm), which is usually used in reconstruction, can be used, it is not recommended if the data set is noisy or contains outliers. Although the minimum mean absolute error (L_1 norm) is often suggested in such cases, our experience with both the Karmarker and Simplex algorithms resulted in excessive run times and cost. More important, these algorithms did not achieve good reconstructions unless they were constrained with essentially exact bounds on the unknown velocity field. For these reasons, neither Karmarker nor Simplex algorithms can be recommended.

A reasonable alternative, which combines the computational advantages of the L_2 norm with the robustness to outliers of the L_1 norm would be to choose an L_p norm, computed using reweighted least squares (Scales et al., 1988) and with p chosen close to one in value. Our experience with this approach has resulted in very accurate reconstructions which are sensitive to outliers. The computational advantages of the L_2 norm are essentially realized in this approach and this is our recommended procedure if the data are likely to be corrupted with noise, as they generally are.

We have also carried out some evaluations of the total least-squares (TLS) error criterion (Golub and Van Loan, 1980) for tomographic reconstruction. This work was motivated by the fact that any model-based reconstruction procedure is subject to inaccuracies in the governing physical equation. These result from the use of data generated by forward modeling using some form of the wave equation and accurate wave-propagation data. The TLS procedure allows for corrections to the assumed physical process as well as to the data which it generates. While the TLS error criterion did little to improve straight-ray tracing, it did slightly improve the accuracy of reconstructions in the curved-ray case. Further research on the TLS might be warranted.

In interpreting tomograms, it is important to understand that the propagating energy which arrived "in phase" was not confined to a single raypath, but rather is the summed contribution of an infinity of paths, both inside and outside the plane of reconstruction, which forms a volumetric element, which we refer to as a "leaf", a three-dimensional (3-D) Fresnel zone. An accurate reconstruction back-propagates observed data into the intersection of the leaf with the plane of reconstruction. The contribution to the leaf which lies out of the plane of reconstruction is imaged in the plane. As a result, the resolution of a tomogram must be understood in a 3-D context. An anomaly which is included in the leaf can contribute to the reconstructed image in the plane. Because we solve an overdetermined system using an L_p norm, unless an anomaly belongs to a large number of source-receiver Fresnel zones (leafs), the anomaly may or may not influence the reconstructed image. The limits of resolution of a tomogram should be interpreted in a 3-D context. A plot of the intersection of a leaf (associated with a single source-receiver pair) with the plane of reconstruction for a homogeneous medium is shown in Figure 1. Since the actual shape of the leaf is determined by the resolving power of the wavelet, the individual Figures 1(a), (b), and (c) show how spatial resolution is related to temporal resolution. A resolving time of 0.4 ms means that two wavelets in this class which arrive less than 0.4 ms apart cannot be resolved. The leaf represents the envelope of the in-the-plane minimum-time wavepaths which contribute "in phase" to the first arrival. The leaf provides a convincing argument for not using reconstruction algorithms based on single rays. The leaf also provides a nice vehicle for understanding the limits of tomographic image resolution. Other factors being constant, the size of the leaf is determined by the dominant frequency content of the data, with the leaf size decreasing as the dominant frequency increases. In general, resolution will be variable across an actual tomogram due to the complex structure of the leafs. It is possible to construct a measure for resolution and to produce overlays which show resolution for individual tomograms.

Well deviations, particularly deviations out of the image plane, must be accounted for if accurate reconstructions are to be obtained. Essentially all wells are deviated enough to cause error in tomographic reconstructions. Our reconstruction code does not actually compute an image in a single plane, but rather in a volume element which includes the entire section of each wellbore and the associated raypaths (or leafs) used to reconstruct the image.

Model Studies: Numerical Modeling

Before considering some actual examples of seismic tomography, consider a rather common type of reservoir model which is frequently encountered. The particular model is drawn from an actual reservoir and will be used to illustrate a few of the points we have made. The model, shown in Figure 2, is a sequence of layers with alternating high and low velocities. Such a model not only produces significant ray bending, but also exhibits first arrivals which result from critical refractions. Data was generated by full finite-difference acoustic wave-equation modeling so that all propagation paths are accounted for. Data for reconstruction was recorded at 20 borehole receiver locations from 20 borehole source locations. The velocity reconstruction using first-arrival traveltimes with the iterative algorithm discussed earlier is shown in Figure 3.

To illustrate the effect of applying straight-ray algorithms, we applied the standard algebraic algorithm for straight-ray reconstruction (Dynes and Lytle, 1979). Examples were run using ART as well as the more robust conjugate-gradient algorithm. Neither result is acceptable and both illustrate the kind of artifacts commonly associated with this type of reconstruction. Straight-ray results with the ART algorithm are shown in Figure 4. Not only is the tomogram qualitatively incorrect, it is also significantly in error quantitatively. We have noted this defect with ART algorithms even on tomograms where the raypaths are nearly straight. It is difficult to know how to iterate with ART, and quite variable results can be obtained depending on how it is used.

We mentioned earlier that the least-square error (L_2) criterion is not the only possible criterion and we referred to alternative criteria, including minimum mean absolute error (L_1) using Karmarker and Simplex algorithms, and L_p error criteria using reweighted least squares. To illustrate our remarks concerning error measures, we generated the steam-flood EOR model shown in Figure 5. Data were generated with both straight-ray and curved-ray forward modeling so

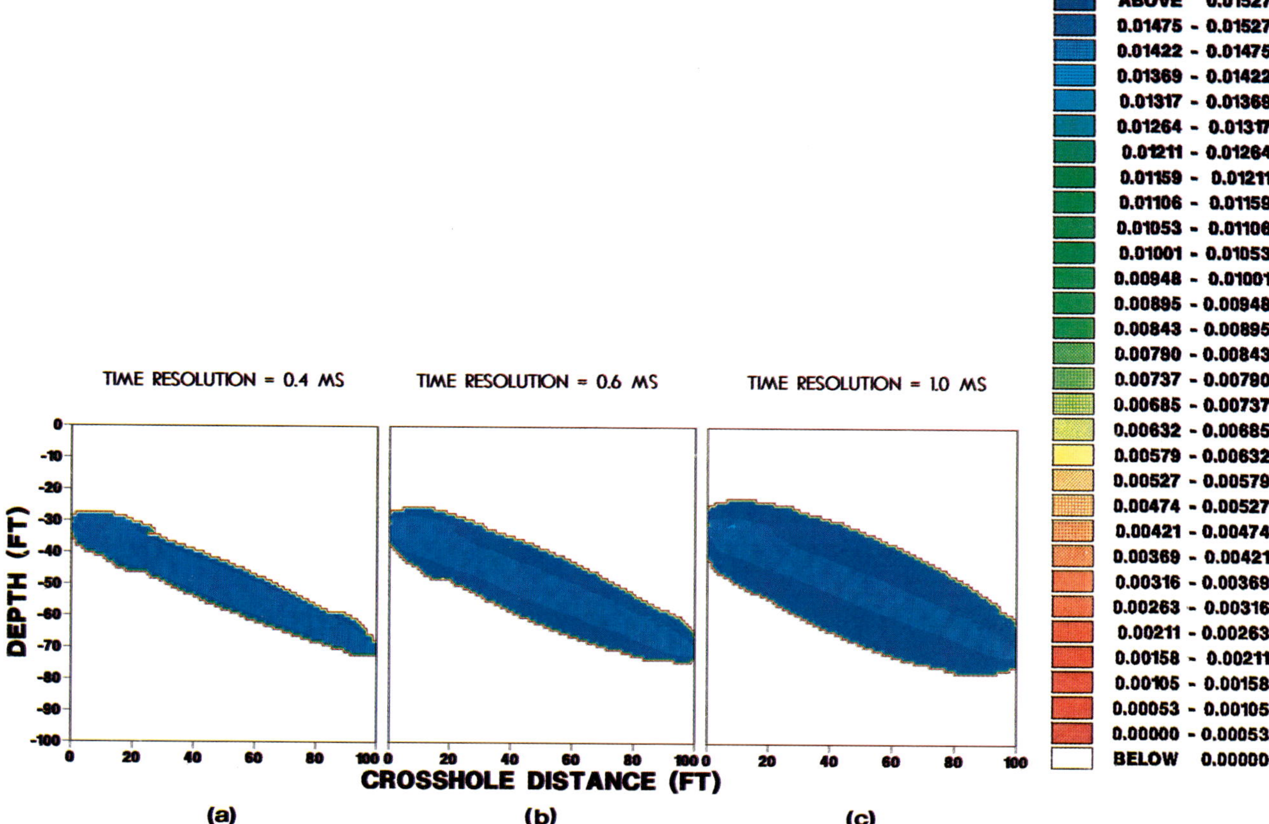

Fig. 1. Intersection of a leaf with plane of reconstruction for a homogeneous medium. (a) Time resolution 0.4 ms, (b) 0.6 ms, (c) 1.0 ms.

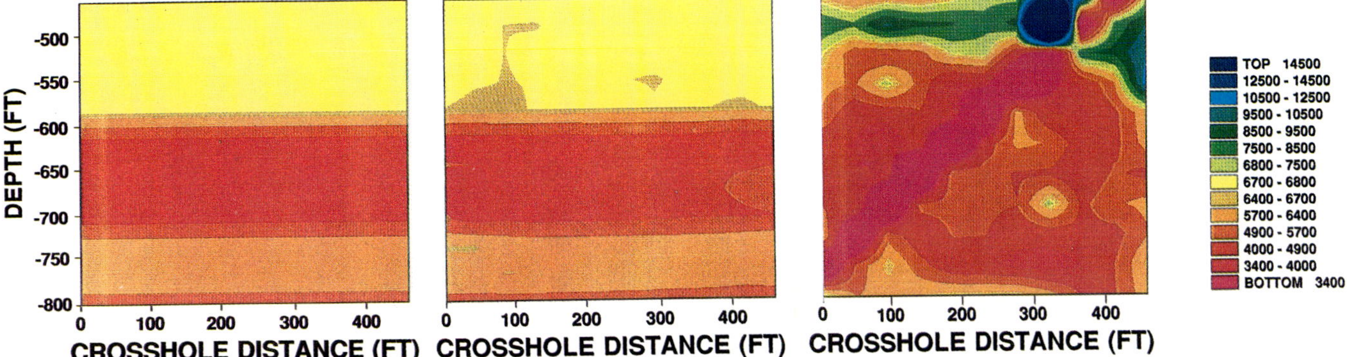

Fig. 2. Horizontally layered model with alternating high and low velocities.

Fig. 3. Reconstruction for 20 receiver locations from 20 borehole source locations.

Fig. 4. Reconstruction using straight-ray ART algorithm.

that straight-ray or curved-ray inversion could be made exact.

To evaluate the Karmarker and Simplex algorithms, we used the simplest case of straight-ray data and reformulated the reconstruction equations (1) as a linear-programming problem with constraints. Results produced extreme (nonphysical) values for the slowness. The reconstructions with very tight constraints on the velocities are shown in Figure 6 (Karmarker) and Figure 7 (Simplex). These results are best evaluated by comparing them with the L_2 reconstruction shown in Figure 8.

To test the L_p error criteria, we used a real data set with very high signal-to-noise ratio. Some of the data were corrupted by adding a high level of random noise (outliers). Without the outliers, the L_2 error criterion gives an excellent reconstruction. The reconstruction using the L_2 norm with the corrupted data is shown in Figure 9. The L_p norm algorithm, with p chosen to be 1.02, was then applied to the same corrupted data set and the result is shown in Figure 10. The resulting reconstruction is very close to the L_2 reconstruction obtained with the uncorrupted data set. Use of the reconstruction algorithm with p near 1.0 gains the robustness to outliers of the L_1 norm while avoiding the computational burden and severe constraints required for the Karmarker and Simplex algorithms.

The use of the TLS criterion is illustrated using the simplified reef model shown in Figure 11. This model produces significant ray focusing through the reef high-velocity zone. The L_2 curved-ray reconstruction is shown in Figure 12. To improve this reconstruction with the L_2 norm would require more data than was used. The TLS reconstruction using the same data set is shown in Figure 13; it is more faithful to the original model, suggesting that it might be preferred to the L_2 reconstruction.

Finally, we consider the case of imaging a small diffractor whose dimensions are smaller than the seismic wavelength. The diffractor is shown in Figure 14 as a small velocity anomaly near the center of an otherwise homogeneous material. Rays traced with the ray equations from sources (on the left) to receivers (on the right) would either miss the diffractor or penetrate it rather than diffract from it, as required by wave theory. The back-propagated traveltimes could not accurately reconstruct the diffractor and the tomogram would be inaccurate. Figure 15 shows the reconstruction with straight-ray imaging and Figure 16 shows the result with wave-equation imaging. The diffractor is nicely resolved in the latter example.

Physical Models

Numerical modeling, which can tell us much about the characteristics of image reconstruction algorithms, cannot substitute for real data. Perhaps the closest we can come to modeling field data is to build scaled physical models with known properties and then acquire and process data from these models. A simple layered model with four layers, shown in Figure 17, is constructed from silicone rubbers with the velocities and dimensions shown. Data were acquired using piezoelectric transducers. The survey included 55 receiver locations as shown in the figure. The scaled dimensions give an interwell distance of 141 m, an interval of 6 m between stations in the boreholes, and a data sampling rate of 0.5 ms. The model survey simulated a less than state-of-the-art field survey. The reconstructed image shown in Figure 18 illustrates the accuracy achievable with appropriate tomographic reconstruction algorithms.

(Text continued on page 329)

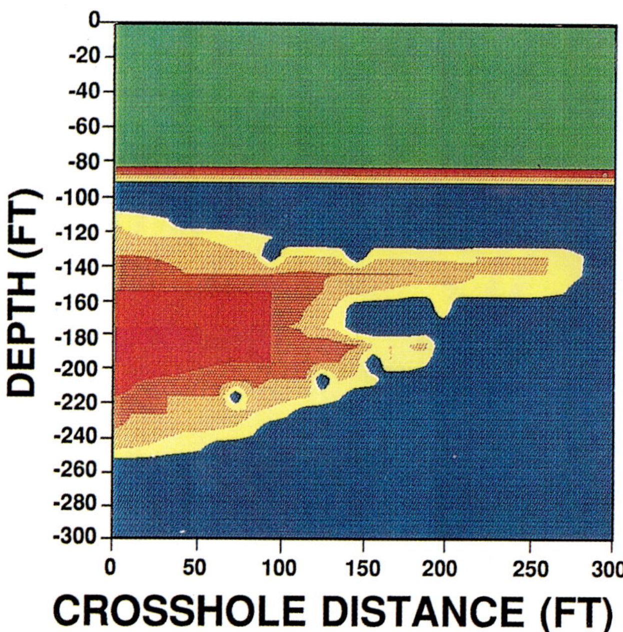

Fig. 5. Steam-flood EOR model.

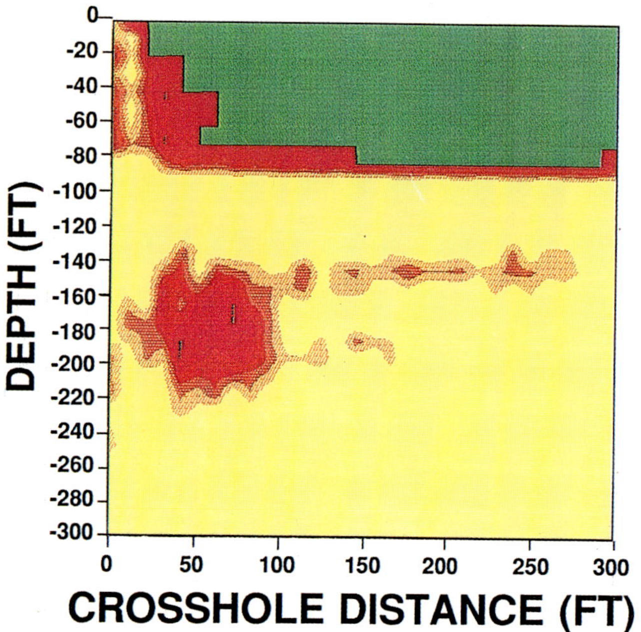

Fig. 6. Steam-flood EOR model reconstruction using Karmarker algorithm with tight constraints on velocity values.

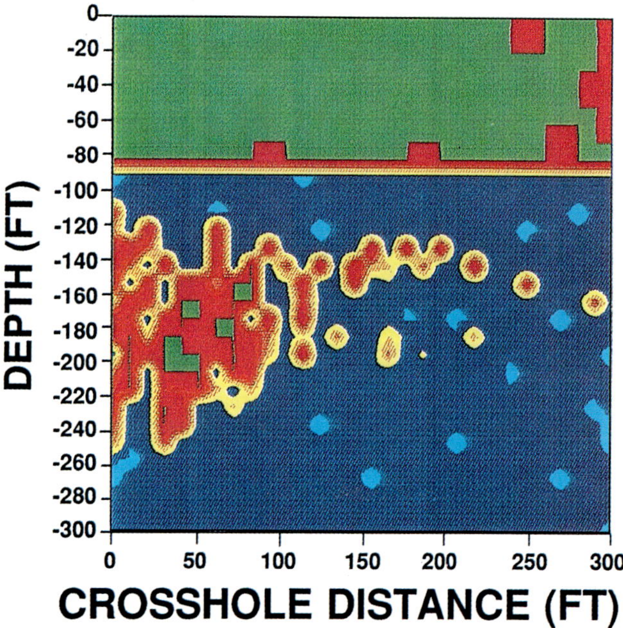

Fig. 7. Steam-flood EOR model reconstruction using Simplex algorithm with tight constraints on velocity values.

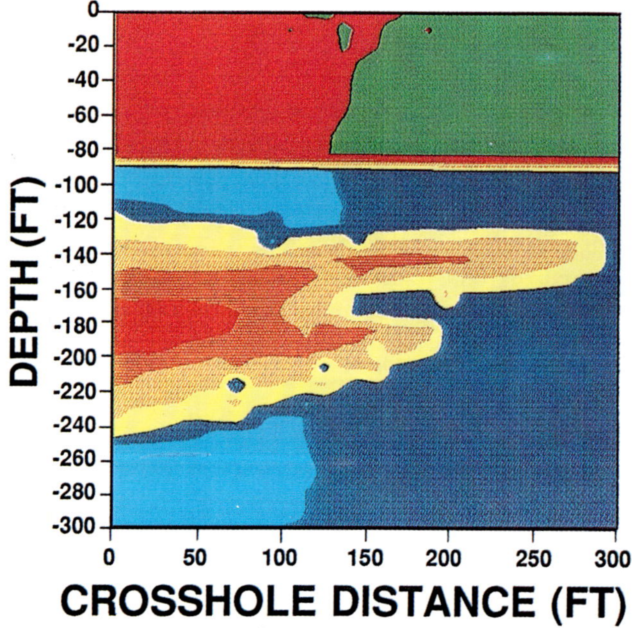

Fig. 8. Steam-flood EOR model reconstruction using L_2 criterion.

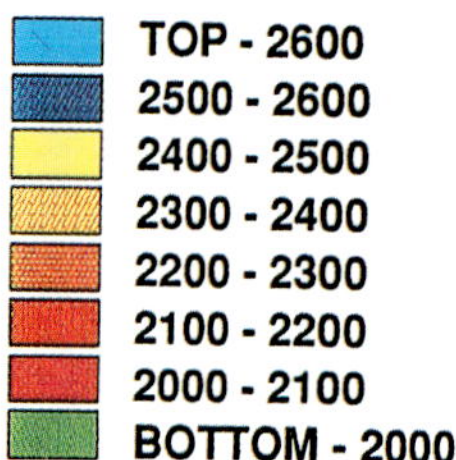

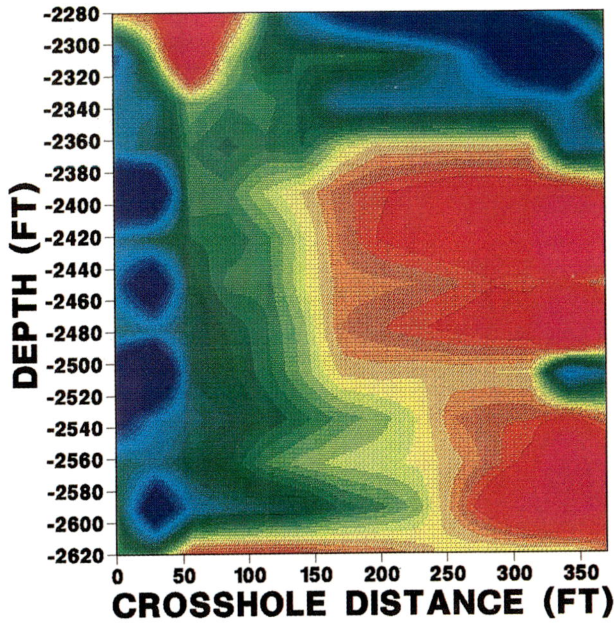

Fig. 9. real data corrupted with addition of random noise; reconstruction using L_2 criterion.

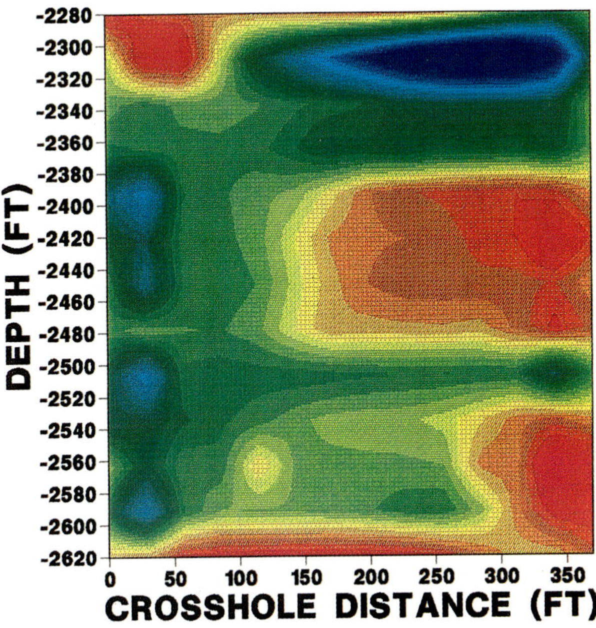

Fig. 10. Real data corrupted with addition of random noise; reconstruction using L_p criteria with p chosen to be 1.02.

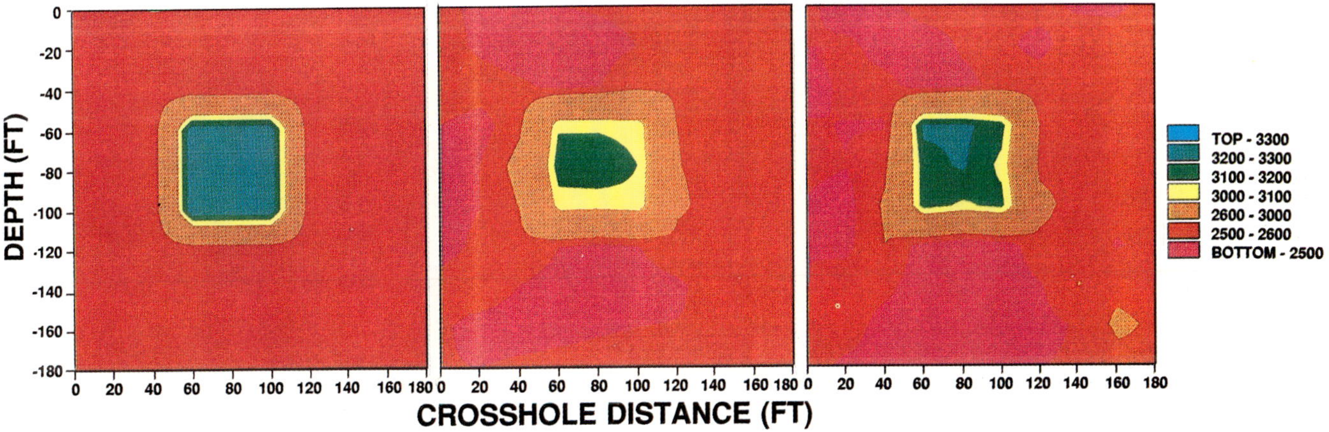

Fig. 11. Simplified reef model.

Fig. 12. Simplified reef model: L_2 reconstruction.

Fig. 13. Simplified reef model: Total least-squares (TLS) reconstruction.

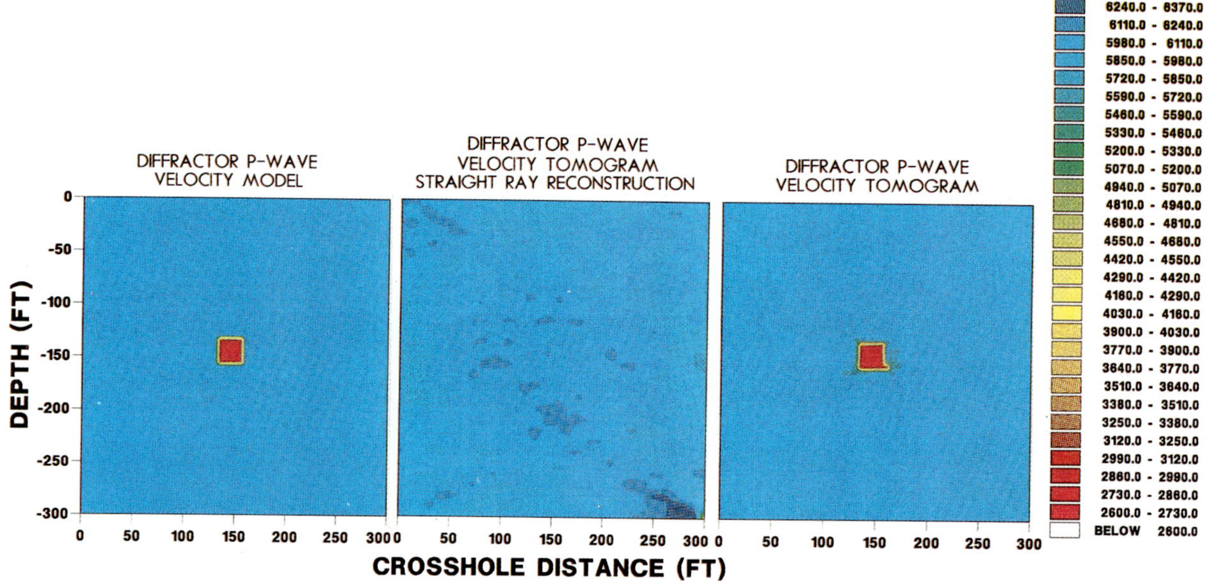

Fig. 14. Small diffractor model.

Fig. 15. Small diffractor model: Straight-ray reconstruction.

Fig. 16. Small diffractor model: Wave-equation imaging.

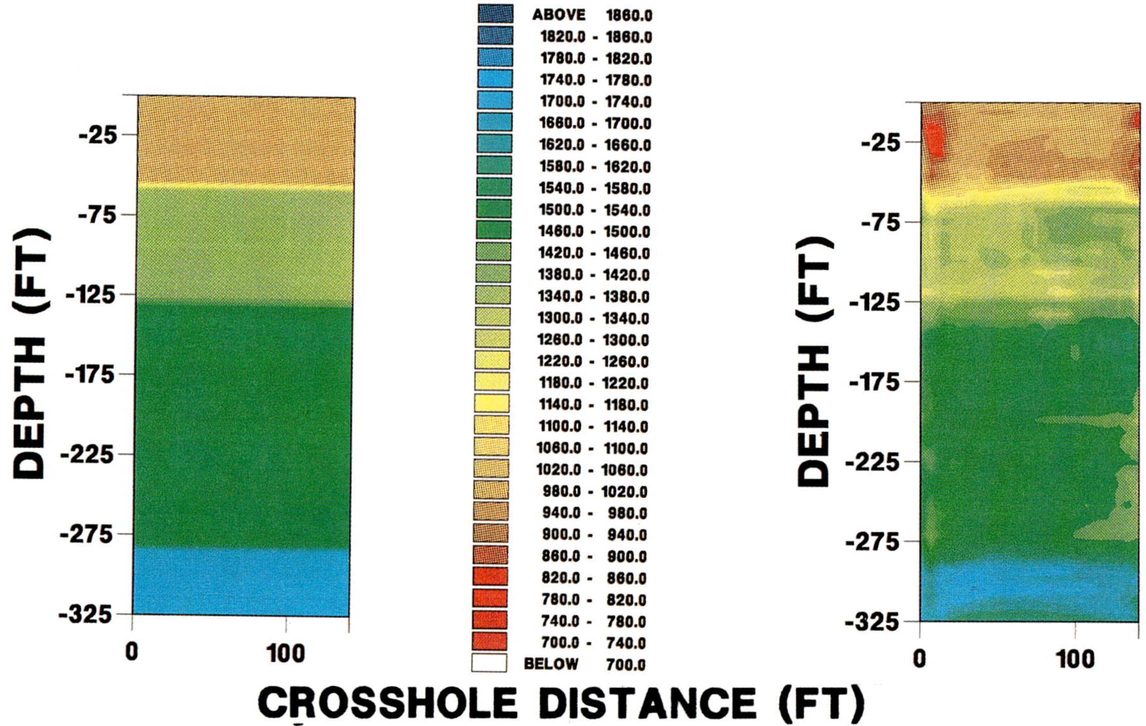

Fig. 17. Four-layer physical model made of silicone rubbers.

Fig. 18. Four-layer physical model: Reconstructed image.

Applications to Enhanced Oil Recovery Monitoring

Laboratory results reported in 1984 (Nur, 1984) showed a significant drop in velocity of sands saturated with heavy oils as the temperature was increased. The effect has now been observed in-situ and provides a means of monitoring thermal EOR processes in heavy oil reservoirs. Tomographic surveys are repeated at regular intervals and the changes in the reconstructed velocity field between successive tomograms is attributed to the effects of heating (cooling), assuming that no other changes occurred in the reservoir. If both *P*- and *S*-wave tomograms can be obtained, they can be combined to reconstruct tomograms of elastic parameters, including incompressibility (bulk modulus), shear modulus, and Poisson ratio. The elastic parameters are derivable only if each of the *P*- and *S*-wave tomograms is accurate. They should not be attempted with inaccurate reconstruction algorithms.

Our first field example involves repeated tomographic surveys across an EOR project. The position of the line relative to the injector and production wells is shown in Figure 19. The first survey in October 1987, a few months after injection had begun, is taken as the base survey, although lowered velocities associated with the injection fluid are already beginning to be imaged (Figure 20). The second survey in March 1988 is shown in Figure 21. A third survey, in October 1988, completes the example and is shown in Figure 22. The successively increased effects of the injection fluid and the consistency of the images in time are easily seen. Difference tomograms are useful in highlighting the relative changes and they indicate regions where the injection fluid may be concentrating or bypassing. These examples illustrate clearly the value of the tomographic survey as a monitoring tool for an EOR project.

The second example illustrates that single surveys across EOR projects, when calibrated with available well data, can also be used to document important variations in fluid distribution and reservoir properties. Single survey interpretations are important because they provide a timely response to operating units in the field and help to clarify problems and identify objectives for subsequent surveys.

As shown in Figure 23, well logs define variations in stratigraphy, facies, structure, temperature, and fluid properties at the survey wells. An unconformity, which is defined by both electrical resistivity and sonic logs, occurs at the top of the reservoir. Beneath the unconformity, sand stratigraphy and sedimentary facies are documented at each well by the resistivity logs. Log-signature variations, shaded in yellow, correspond to changes in sand sedimentary facies. Correlations between wells are illustrated by dashed lines.

The uppermost "amalgamated braided-channel sand facies" is, according to core-description and core-analysis data, the best reservoir unit. Sonic logs define a low-velocity "desaturated zone" at the top of the sand. This zone, which is shaded in red, occurs within the amalgamated sand braided-channel facies except at the top of the reservoir where it extends into the underlying meandering sand channels. Velocities in this zone are lowest at the structurally high end of the reservoir. Temperature logs document a temperature maximum near the top of the reservoir at 1200–1300 ft. The higher temperature documents that steam, which is injected into the base of the sand at a location between the two tomography wells, is rising to the top of the formation.

To facilitate crosshole interpretation, the tomographic velocity data are divided into five velocity fields which correlate with the log-defined contacts. Confidence in the accuracy of the velocity fields is increased by the strong correlation between the tomographic and sonic-log data. This is illustrated for the whole tomogram by the overlap of tomographic velocities measured at the borehole (plotted as red dots) with the sonic logs.

The desaturated zone at the top of the reservoir is imaged by the lower velocity fields displayed in orange and red. The red field, 4000–5000 ft/s, occurs in the amalgamated braided-channel facies at the structurally

EOR INJECTION PATTERN

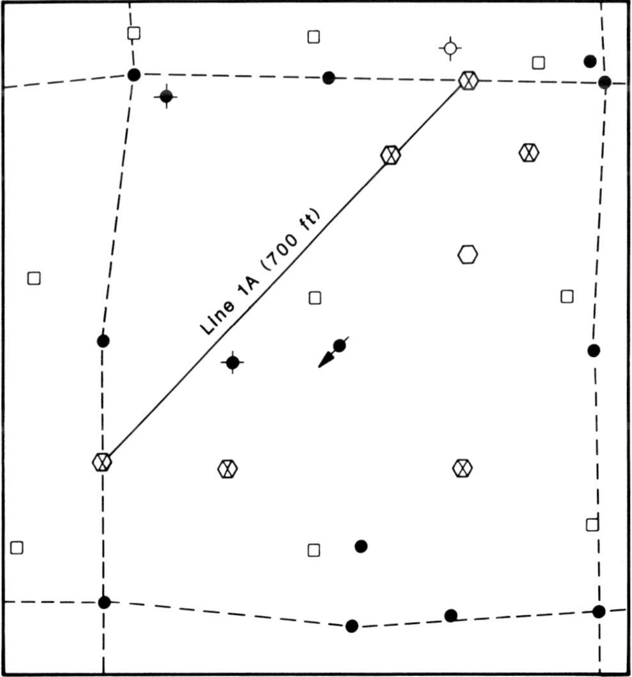

Fig. 19. Location of thermal EOR survey. Injection well indicated by arrow symbol, production wells by solid dots, observation wells by *x*.

high side of the tomogram. The orange field, 5000–6200 ft/s, also occurs in the amalgamated braided-channel facies but, at the high end of the reservoir (left), extends across facies boundaries to deeper units. The velocities are lowest adjacent the red field and increase down structure. These sands have been significantly influenced by the steam process and are referred to as air sands (Jones, 1990).

Most sands have velocities that are between 6200 and 8000 ft/s. The yellow velocity field, 6200–7000 ft/s, is formed by heavy oil sands that have been partially affected by the steam process. It is the broadest velocity field shown and cuts across several sand-facies contacts. Heavy oil sands, which have not been significantly affected by the steam process, are imaged by a high-velocity field, 7000–8000 ft/s, displayed in green at the base of the formation. These correspond in a qualitative way to hot and cold saturated zones, as discussed in Jones (1990). Since the temperature variations as defined on the logs are relatively small (approximately 20°F), the presence of gas is probably the dominant factor causing the lower velocities in the yellow field.

The highest velocity field, 8000–9000 ft/s, shown in blue near the base of the formation, is controlled by lithology rather than fluid content. Log and core data indicate that the higher velocities are caused by a sequence of thin carbonate-cemented sands.

Tomographic Imaging in Reservoir Characterization

While monitoring enhanced oil-recovery processes will continue to be an important application of seismic tomography, particularly due to its potential impact on the economics of the projects, the use of tomographic imaging for reservoir characterization can be equally important. Single tomograms can be used to image crosshole geology and reservoir properties in almost any reservoir where images can be obtained. Interpretation of these tomograms is complex, and requires that we correctly relate the imaged velocity or attenuation fields to the rock and fluid properties in the reservoir. Generally, all types of information must be integrated to reach an accurate interpretation.

As an example, we consider a clastic and carbonate midcontinent reservoir. As shown in Figure 24, changes in tomographic velocities correlate with formation boundaries and Paleozoic carbonate reservoir zones defined by neutron logs. The contacts are clearly defined at the wellbore and extend, for the most part, directly to the other well. The irregular surface at the top of the upper reservoir (light blue velocity field) may represent a paleo karst surface, known to be a waterflood thief zone. Reservoir heterogeneity is documented by variations in the neutron-log signature of the lower reservoir and the discontinuous nature of the velocity fields in the tomogram. The uniform velocity field corresponding to the upper

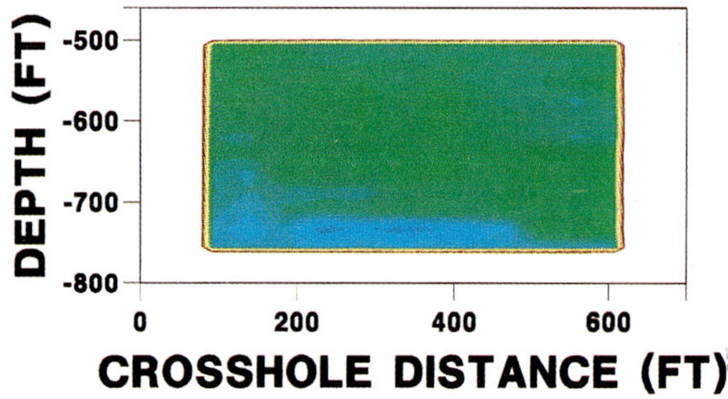

Fig. 20. First survey in October 1987 shortly after injection began.

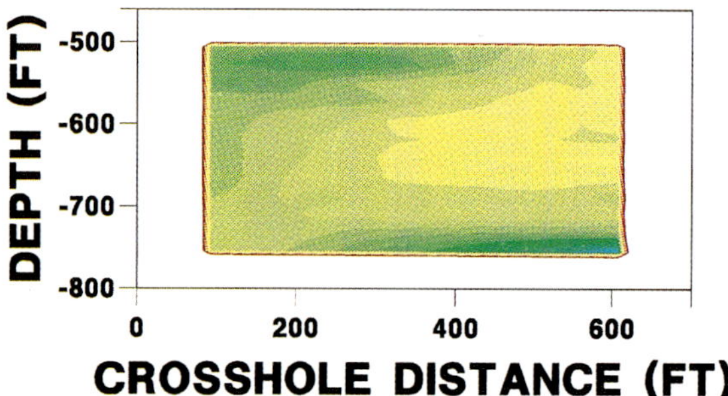

Fig. 21. Second survey in March 1988.

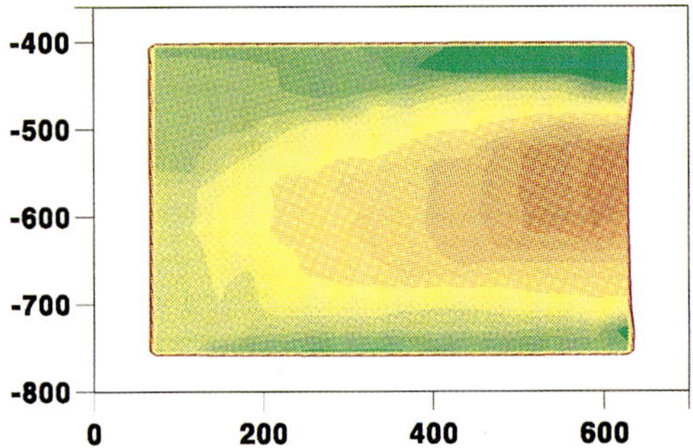

Fig. 22. Third survey in October 1988.

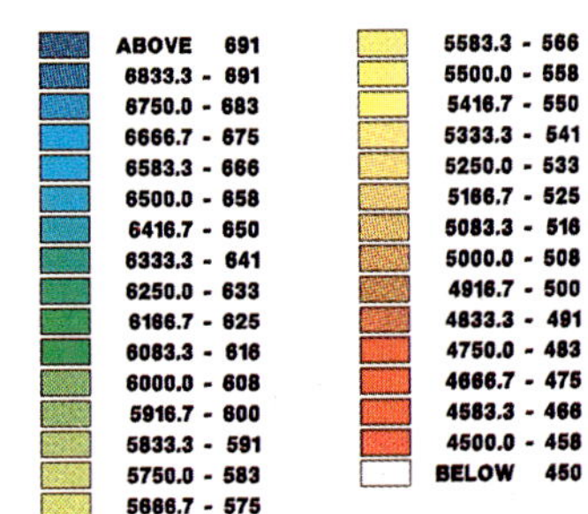

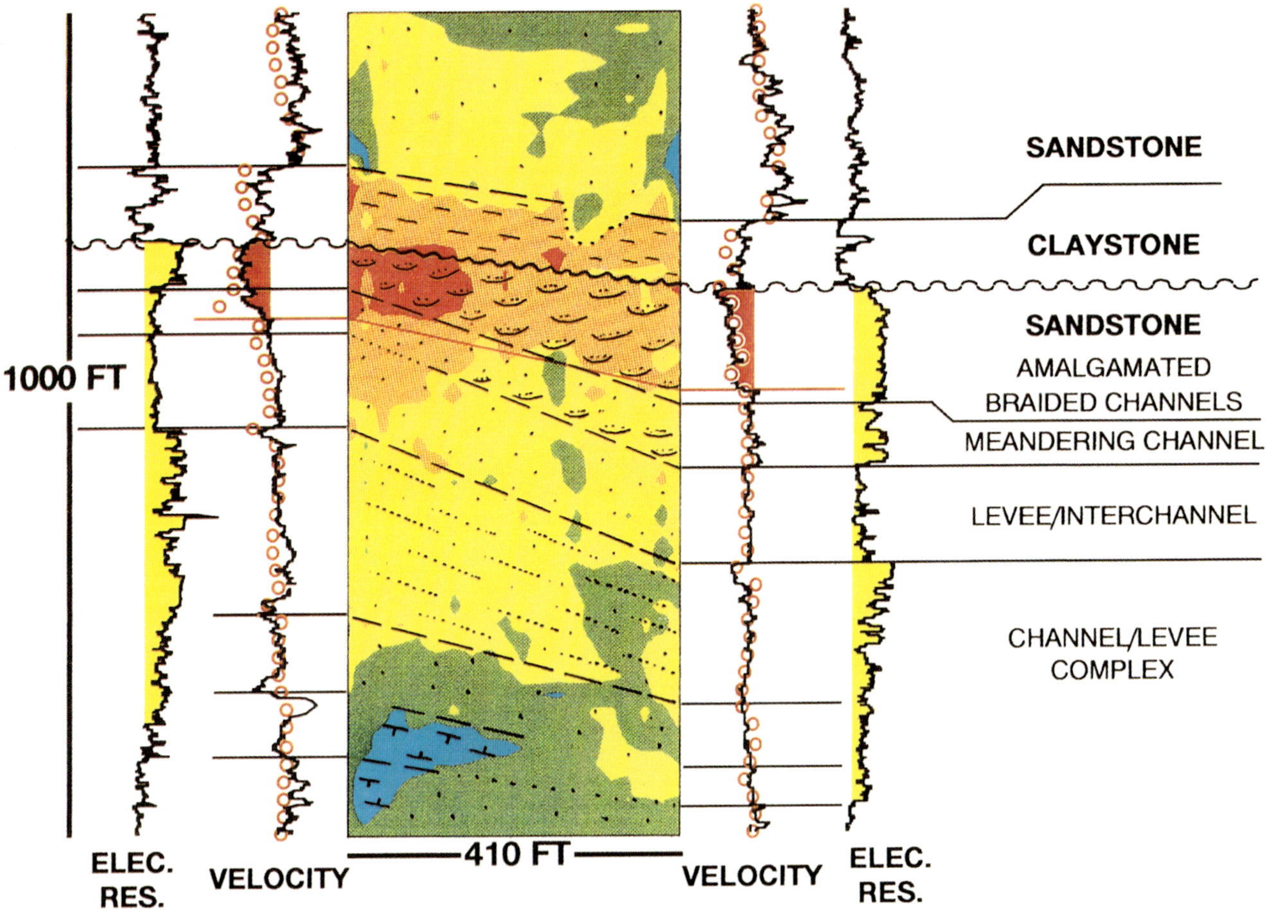

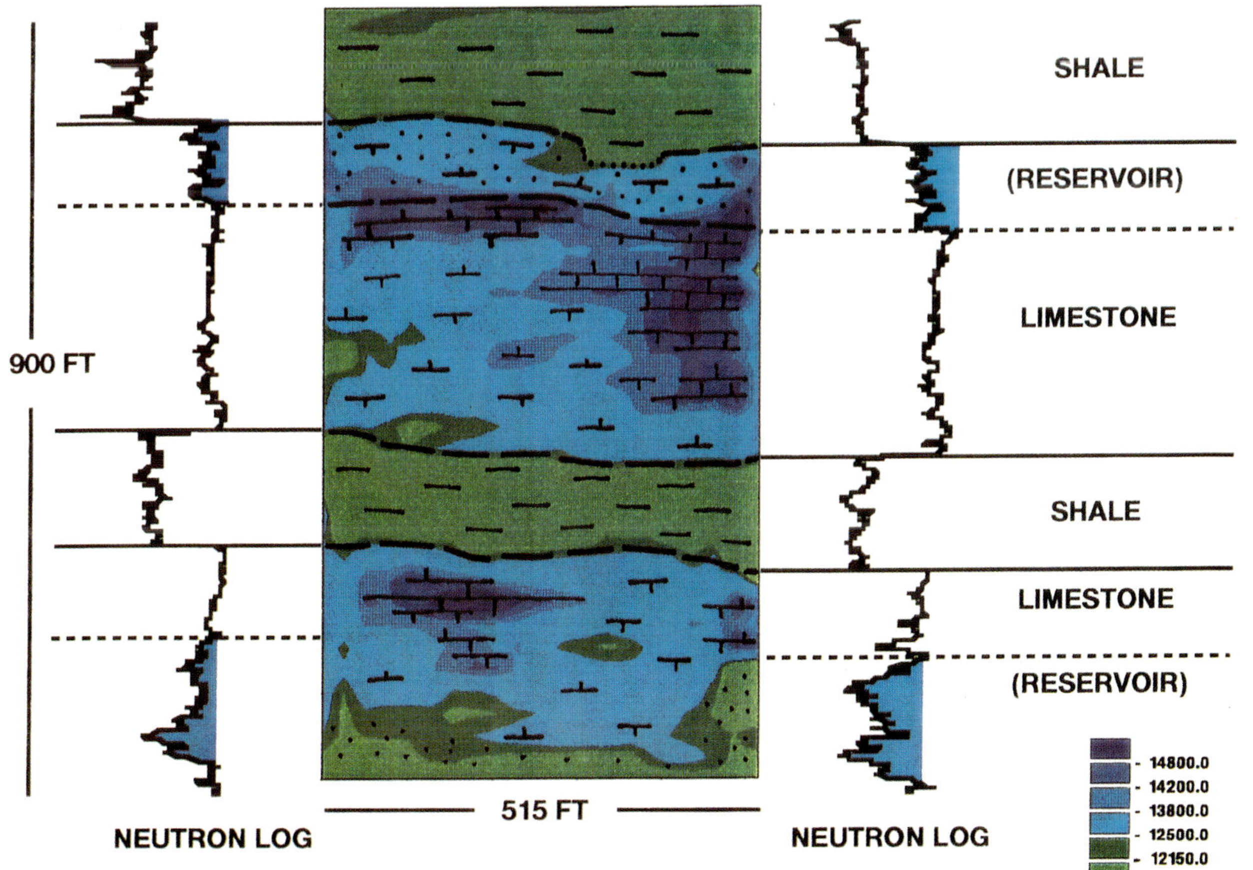

Fig. 23. Interpretation of *P*-wave tomogram involving heavy oil sands.

Fig. 24. *P*-wave tomogram involving shale-carbonate sequence.

reservoir, in contrast, connects wells with similar log signatures and documents a more homogeneous reservoir. A normal fault with 70 ft of displacement is documented in Figure 25, a 3-D tomogram across a Pennsylvanian sandstone reservoir. The tomograms not only define the precise location of the fault but also document variations in velocity which can be correlated with variations in reservoir quality defined by logs and cores.

Understanding Tomographic Imaging in EOR Monitoring

We presented examples documenting tomographic imaging for reservoir surveillance in thermal EOR programs in heavy oil reservoirs. This application was based on the observation (Nur, 1984) that there is a change in the velocity associated with heating sands saturated with heavy oils. If this is the only effect, then tomographic imaging for EOR monitoring might be limited. However, certain fundamental physical parameters affect the elastic S- and P-wave velocity fields in predictable ways when perturbed. If we can predict how a particular EOR program will perturb

these parameters, we can estimate effects and determine the basis for a monitoring program. This information can be used both in planning surveillance programs and in interpreting the resulting images.

The observations which we have just made suggest a new approach to understanding the seismic response of EOR programs (including waterflooding) using reservoir simulators (Justice et al., 1989). Over the years, powerful simulation programs have been (and are being) developed. By monitoring the changes in the fundamental parameter set associated with these simulators and applying appropriate transforms, we can obtain a rather complete understanding of the seismic response of a particular reservoir to a particular EOR program. The basis for monitoring one type of program in one reservoir may be quite different from monitoring the same program in a different reservoir. This observation is providing a strong foundation for planning and interpretation of surveillance programs.

As an example, let us consider a simple reservoir model of a homogeneous reservoir which has been uniformly depleted by primary production and which

Fig. 25. 3-D tomogram across a Pennsylvanian sandstone reservoir showing normal fault with 70 ft throw.

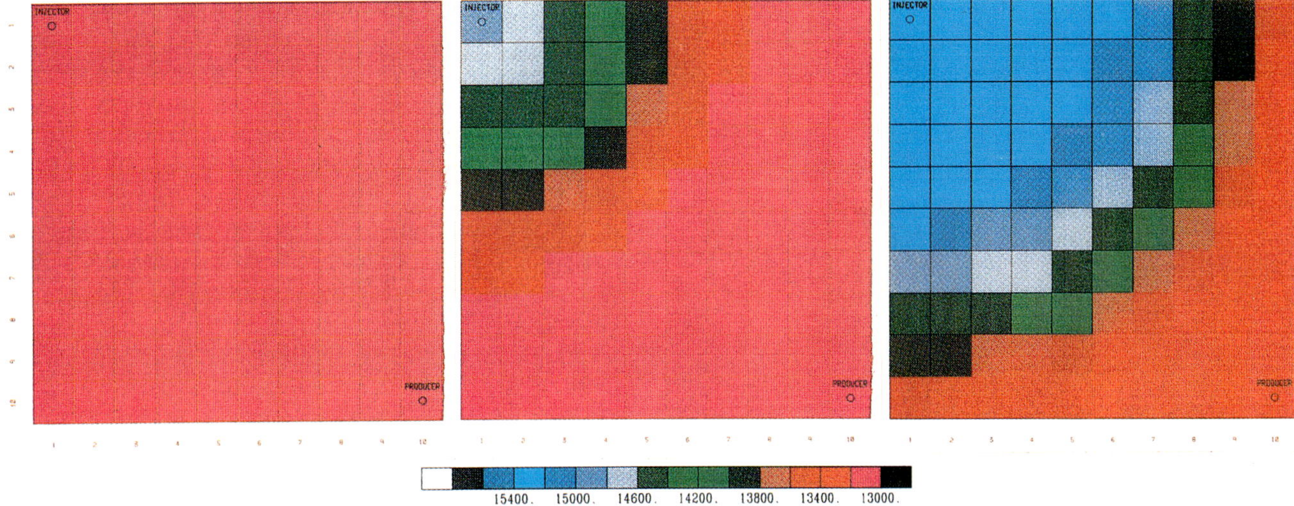

Fig. 26. Model of waterflood in homogeneous reservoir.

Fig. 28. Model of waterflood in homogeneous reservoir after 11 months.

Fig. 27. Model of waterflood in homogeneous reservoir after 9 months.

has a constant *P*-wave velocity field. The model is shown in Figure 26 with the *P*-wave velocity field color coded. We now begin the simulation of a waterflood with the injector in the upper left hand corner and the producer in the lower righthand corner. Two successive images are shown in Figures 27 and 28 as the flood progresses. The sharply defined change in the *P*-wave velocity field tracks the movement of the flood, indicating that it could be monitored with seismic surveillance. The physical phenomenon which underlies this particular set of images is the change from pressures below the bubble point of the reservoir gas to pressures above the bubble point as the flood progresses. Gas going into solution produces the observed changes in the velocity field. Another reservoir might behave quite differently.

Commentary

Tomographic imaging is beginning to play a very important role in reservoir characterization, surveillance, and analysis. Important applications include monitoring enhanced oil-recovery programs and imaging to better understand reservoir structure and dynamics as an aid in planning reservoir development programs. Tomographic image reconstruction based on ray tracing is generally inadequate to produce accurate or usable reconstructions. This inadequacy is even more true when reconstructions are based on the solution of underdetermined systems, as has been commonly reported in the literature. Nonuniqueness, which is not necessarily an inherent feature of tomographic reconstruction, can be introduced by incorrect imaging and is often the result of demanding more resolution than the data justify. Limitations on resolu-

tion, however, are inherent and are influenced by survey geometry and the bandwidth of the data. Important advances in our ability to produce high-resolution tomographic images will come as improvements in data quality become possible. New borehole seismic tools and data recording systems being developed will play a major role in increasing current limits on resolution.

Tomographic imaging is surely destined to play a major role in providing critical data required for optimum reservoir management. This role may involve anything from optimizing the efficiency of an enhanced oil-recovery program to planning development drilling and production strategy of a field under development. New work in rock physics, theoretical analysis, data-acquisition tools and procedures, computer modeling, and image-reconstruction techniques will allow more accurate interpretations made on a timely basis. These interpretations in turn will play an important role in reservoir management decisions. The net result should be increased profitability and an increase in our net recoverable reserves.

Appendix A: Tomographic Image Reconstruction for Diffracting Media; Traveltime Inversion

Consider a region to be imaged which is rectangular and assume that sources can be distributed along one side of the rectangle with receivers distributed along one or two of the other sides. The vertical sides correspond to boreholes and the upper side to the earth's surface. The rectangle is partitioned in some way into grid cells or pixels and the slowness in each cell depends on one or more parameters associated with that cell, or with the nodes of the partition. A traveltime measurement T_i is indexed by the source and receiver pair for which it was recorded. The unknown parameters are also indexed and are represented by the parameter vector $\mathbf{P}$. The measurement (traveltime) vector $\mathbf{T}_0$ corresponds to an initial estimate of the parameter set $\mathbf{P}_0$. We may now compute a first-order Taylor expansion of $\mathbf{T}$ about $\mathbf{P}_0$:

$$\mathbf{T} \approx \mathbf{T}_0 + \partial\mathbf{T}/\partial\mathbf{P}|_{\mathbf{P}_0}\,(\mathbf{P} - \mathbf{P}_0). \tag{A-1}$$

Define the Jacobian matrix $\mathbf{J}$ by the equation

$$\partial\mathbf{T}/\partial\mathbf{P}|_{\mathbf{P}_0} = \mathbf{J}(\mathbf{P}_0). \tag{A-2}$$

We may then write the first-order approximation to the change in the unknown parameter set as

$$\Delta\mathbf{T} = \mathbf{J}(\mathbf{P}_0)\Delta\mathbf{P}. \tag{A-3}$$

Proceeding with the Newton approximation method, we iterate on this equation to obtain the successive update equations,

$$\Delta T_k = \mathbf{J}(P_k)\Delta P_{k+1}. \tag{A-4}$$

We note that the matrix $\mathbf{J}$ is computed using the previous estimate of the parameter set.

We are now faced with the problem of solving equation (A-4) at each iteration. We may, for example, use singular-value decomposition, but for most problems, a conjugate-gradient method with an L_p norm would be preferred. Partial singular-value decomposition may also be used (Justice et al., 1988).

For the sake of discussion, we assume that a singular-value decomposition (SVD) is chosen, then a minimum norm L_2 solution can be obtained by computing the singular-value decomposition of the matrix $\mathbf{J}$,

$$\mathbf{J} = \mathbf{U}_q\Delta_q\mathbf{V}_q^T, \tag{A-5}$$

where $\mathbf{U}$ and $\mathbf{V}$ are orthogonal matrices of eigenvectors $\mathbf{J}\mathbf{J}^T$ and $\mathbf{J}^T\mathbf{J}$, and Δ is the diagonal matrix of singular values. It is assumed that null singular values have been removed from equation (A-5) and that the resulting decomposition uses only q nonzero singular values and their corresponding singular vectors (see Aki and Richards, 1980, for details). We may now define a generalized inverse $\mathbf{J}^+$ for the matrix $\mathbf{J}$ using the equation

$$\mathbf{J}^+ = \mathbf{V}_q\Delta_q\mathbf{U}_q^T. \tag{A-6}$$

The updated estimate of the parameter set, for this iteration, is then given by

$$\Delta P_{k+1} = \mathbf{J}^+(P_k)\Delta T_k. \tag{A-7}$$

The algorithm discussed here can be modified for full wavefield inversion as well as for tomographic reconstruction, but it is not the purpose of this paper to consider nontomographic inversion. The algorithm operates directly on the modeling equation at each iteration. Approximations to the wave equation can be economically used in the initial iterations but the approximate full wave equation should always be used in the final iterations. Ray tracing itself should not be used since the process breaks down when inhomogeneities exceed about 10 percent of the background velocity (which is common), with a correlation length on the order of a wavelength or less. Further, as was pointed out earlier, traveltime data must be backpropagated into the full Fresnel zone associated with the source and receiver. The data cannot be correctly backpropagated if confined to single raypaths joining source and receiver. For a discussion of tomographic image reconstruction with attenuation data, see Justice and Vassiliou (1990).

References and Further Reading

Aki, K., and Richards, P. G., 1980, Quantitative seismology, Vols. I, II: W. H. Freeman Co.

Bois, P., La Porte, M., Lavergne, M., and Thomas, G., 1972, Well-to-well seismic measurements: Geophysics, **37**, 42–83.

Deans, S. R., 1983, The Radon Transform and some of its applications: John Wiley & Sons.

Dines, K. A., and Lytle, R. J., 1979, Computerized geophysical tomography: Proc. Inst. Elec. Electron. Eng., **67**, 1065–1073.

Golub, G. H., and Van Loan, C. F., 1980, An analysis of the total least squares problem: SIAM J. Num. Anal., 17, 883–887.

Herman, G. T., 1980, Image reconstruction from projections: Academic Press.

Ivansson, S., 1986, Seismic borehole tomography- theory and computational methods: Proc. Inst. Elec. Electron. Eng., **74**, 328–338.

Jones, J., 1990, Why cyclic steam predictive models get no respect: Soc. Petr. Eng. Paper 20022, 95–104.

Justice, J. H., Vassiliou, A. A., 1990, Acoustic tomography in hydrocarbon reservoirs: Proc. Inst. Elec. Electron. Eng., **78**, 711–722.

Justice, J. H., Vassiliou, A. A., Nguyen, D. T., 1988, Geophysical diffraction tomography: Signal Processing, **15**, 227–235.

Justice, J. H., Vassiliou, A. A., Yale, D. P., Bansal, P. P., and Lin, J. C., 1989, 59th Ann. Internat. Mtg., Soc. Expl. Geophys., Expanded Abstracts, 760–762.

Nur, A. M., 1984, Seismic monitoring of thermal enhanced oil processes: 34th Ann. Internat. Mtg., Soc. Expl. Geophys., Expanded Abstracts, 337–340.

Scales, J., Gersztenkorn, A., and Treitel, S., 1988, Fast L_p solution of large, sparse, linear systems: J. Comp. Phys., **75**, 314–333.

Tarantola, 1987, Inverse problem theory- Methods for data fitting and model parameter estimation: Elsevier Science Publ.

Vassiliou, A. A., Justice, J. H., and Guinzy, N. J., 1989, A comparison of norms for non-diffraction and diffraction tomography: Acoustical Imaging, **17**: Plenum Press, 627–638.

Chapter 7: Alternative and Emerging Technologies
Alternative Technologies

*Thomas L. Dobecki**

Introduction

Chapter 7 deals with "alternative technologies" in the sense of applications which do not involve controlled-source seismic reflection. There are investigative methods besides seismic reflection which have reservoir geophysics applications. This paper concentrates upon geophysical methodology which has been effective, excluding controlled-source seismics. Some of these, as will be shown, are truly "seismic" but are so far removed from typical seismic applications that we must consider them as alternative technologies.

Alternative technologies can be used with significant effectiveness where controlled-source seismics (a) are not effective or are impractical for a specific target, or (b) are substantially more expensive but not substantially more useful. Fortunately, there are quite a few circumstances where such is true and, hence, there are many applications for alternative methodologies. What we can measure are any characteristics of a reservoir or reservoir process which involve modification of such diverse physical parameters as in situ strain, temperature, fluid salinity, bulk density, magnetic susceptibility, porosity, fluid pressure, and the distribution of fluids. Each of these, in turn, may have some significance owing to natural reservoir character or due to artificially induced modification, as during secondary or enhanced production operations. In many situations, controlled source seismic methods are not sensitive to these, yet they still must be mapped in order to provide insight into reservoir characteristics or process performance, and in many circumstances, they are the only means possible. They will be described in terms of their abilities to map such things as orientation and geometry of induced hydraulic fractures, position and trend of thermal and chemical enhanced oil recovery (EOR) processes, progress of reservoir depletion, natural fracturing, and direct hydrocarbon indication.

This paper describes applications by technique and not by application. This format is chosen because

*McBride-Ratcliff and Associates, 7220 Langtry, Houston, TX 77040-6698.

many of these methods find application in reservoir description, delineation, or process monitoring.

Finally, I have limited my discussion to only those methodologies which have a direct view of the reservoir or process being monitored. As such, I have not discussed geothermal or geochemical measurements which have an indirect view of the reservoir or process but do not have the resolving power required by the objectives of delineation, description, or surveillance. Also, I have not discussed nuclear or chemical tracers.

Elastic (Deformation) Methods

These elastic methods might qualify as true seismic methods in that they involve transmission and detection of elastic deformations. However, the time scales involved are beyond the realm of the seismic methods described in accompanying chapters in this book. Wave periods are measured in terms of minutes to weeks and sample rates are in terms of minutes instead of milliseconds. Included in such methods are tiltmeter surveys and tidal strain surveys. The only applications of these surveys have been to process monitoring applications.

Tiltmeters

Tiltmeters are biaxial (horizontal) sensors which respond to ground surface tilt much like a carpenter's level (and actually function on much the same principle). Thus, if such a device is mounted at the earth's surface or in a shallow borehole and if a reservoir activity (e.g.,an EOR process) causes the ground to move so that the device "tilts", the angle of tilt can be determined by the vector sum of the output from the biaxial sensor. Numerous processes cause uplift and subsequent tilt response. The most popular applications of tiltmeters have been the mapping of induced hydraulic fractures (Pollard and Holzhausen, 1979; Evans, 1983; Evans and Holzhausen, 1983; Davis, 1983) and thermal recovery processes (Holzhausen et al., 1985). In the former, the opening of a hydraulic fracture deforms the ground surface in a pattern dependent upon the geometry of the fracture. Figure 1 shows the qualitative patterns of ground surface deformation due to an induced, vertical planar fracture and a horizontal fracture. Thermal recovery methods cause a swelling of affected reservoir regions which also results in a surface deformation. Whereas the tilt response to the opening and closing of an induced hydraulic fracture may extend over a few minutes to several hours duration, thermal process signals often develop over several weeks. An interesting effect has been observed in monitoring Canadian steam injections (e.g., Holzhausen et al., 1985; Singhal and Card, 1988) in that overall reservoir swelling occurring over about six weeks has superimposed short time scale tilt events which are related to hydraulic fracturing generated by the actual steam injection. Thus, the tilt field not only describes the gross shape and trend tendencies of the steam injection but also the manner by which steam is being distributed into the reservoir.

The magnitude of uplift is dependent upon fracture and process size, depth, and the elastic characteristics

Fracture Height = 400 ft. (122 m.)

Fracture Width = 0.1 inches (2.54 mm.)

Fracture Length = 750 ft.

Poisson's Ratio = 0.25

Fracture Volume = 450 bbls. (71.5 m3)

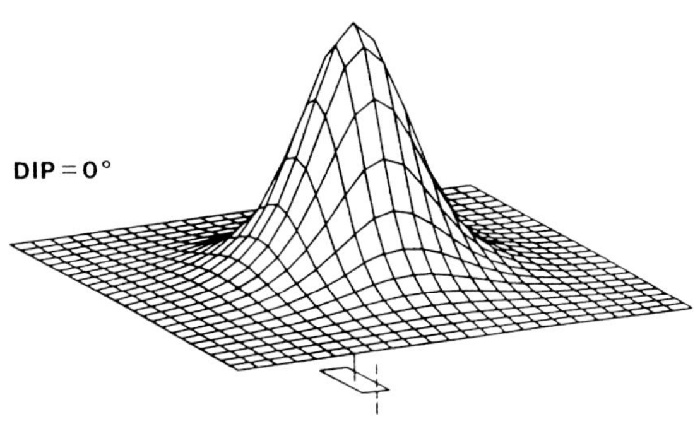

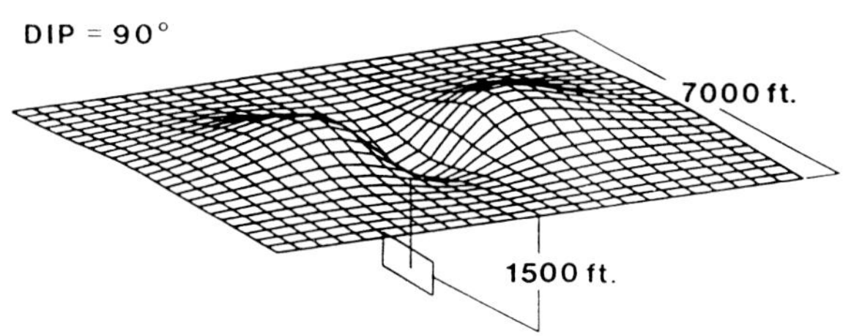

Fig. 1. Exaggerated illustration of ground surface deformation above a horizontal (upper) and a vertical hydraulic fracture of finite length.

of the rock mass and is measured in terms of millimeters to tens of millimeters. Precise leveling could measure such deformation but it is time consuming whereas an array of tiltmeters can immediately produce the deformation pattern. Tiltmeters have sensitivities on the order of 10^{-9}–10^{-10} radians, so very small tilts can be detected. Typical tilt signals for relatively shallow (< 2000 m) processes are on the order of 10^{-4}–10^{-6} radians, but this level decreases rapidly with depth. Extraneous noise sources which limit the depth capabilities of such surveys include tidal forces, thermal (sunrise/sunset) expansion of surface soils, rain loading, barometric effects, and surface cultural sources. These, particularly tidal effects, may have levels in excess of 10^{-4} radians which could swamp a small signal. Some of the more predictable noise sources such as tides can effectively be removed from the tilt time series. This is not the case with more random sources such as rainfall loading. The net effect is that tiltmeters are restricted to mapping processes which are < 2000 m in depth. Mapping deeper processes are best dealt with by utilizing near-field measurements such as the passive borehole seismic technique described later in this chapter.

Tidal Strain

If a wellbore intersects a fluid filled fracture (natural or induced), then a description of the fracture can be derived by an analysis of changes in the fluid level (or bottom hole pressure) within the wellbore as the fracture is alternately compressed and dilated by passage of tidal forces. The amplitude and phase of the diurnal and semi-diurnal components of the water level variations depend upon the shape and size of the fracture, how compressible the fracture is, and the orientation (strike and dip) of the fracture (Bower, 1983). This theory has been applied to the analysis of bedrock fracture patterns (Bower, 1983) and to the orientation of induced hydraulic fractures (Hanson, 1982). Associated problems include: (a) analysis is valid for single plane fractures only, (b) a test well must be shut in for several weeks to obtain adequate statistics, and (c) other effects, such as variable ocean loading on the continental shelf for near-coastal sites may be overwhelming contributors to the signal. Because this method involves measurement at or near the process depth (near-field), there does not appear to be a practical depth limitation to its application.

Microseismics and Acoustic Emissions (MS/AE)

If some activity within a reservoir creates microseismic events, then the volume of the reservoir in which that activity is occurring can be delineated by analysis of the resulting seismic events. Such activities might include fluid flow within a fracture system, periodic strain accumulation and release due to gas build up and flow, thermally induced failure in overburden materials above a fireflood, rock breakage during hydraulic fracturing, subsidence due to fluid withdrawal, or even massive gas flow in a blown-out well. Under these circumstances, we sit back and listen to nature. We control neither the character nor the timing of the source and merely monitor activity and, hopefully, plot locations of discrete events or calculate parameters about the source region. We, however, are dealing with seismic events of only Richter magnitude of -3 (negative three) or so, which are small events indeed.

Early development of downhole seismic arrays to monitor hydraulic fracture growth was in support of the Los Alamos Hot Dry Rock geothermal project (Albright and Pearson, 1982) although parallel development for porous, petroleum environment reservoirs was being carried out by Sandia National Laboratory (Seavey, 1982). Some commercial applications of hydraulic fracture mapping in oil and gas fields, to depths of over 3500 m, utilizing a single well site (as opposed to subsurface arrays) have been reported (Dobecki, 1983; Lacy, 1984).

A single-well determination is possible by using an oriented triaxial borehole seismometer ("TABS"– Lacy, 1984) and analyzing the particle motion polarization to determine the seismic source location and, therefore, fracture position. Figure 2 shows such a microseismic signal generated during a hydraulic fracture test. This triaxial signal is typical of fracture generated signals (sharp arrivals, 200–400 Hz peak frequency, relative phase shift in particle motion at S-wave arrival). Data handling includes analysis of relative P-wave arrival times or S-P times if several

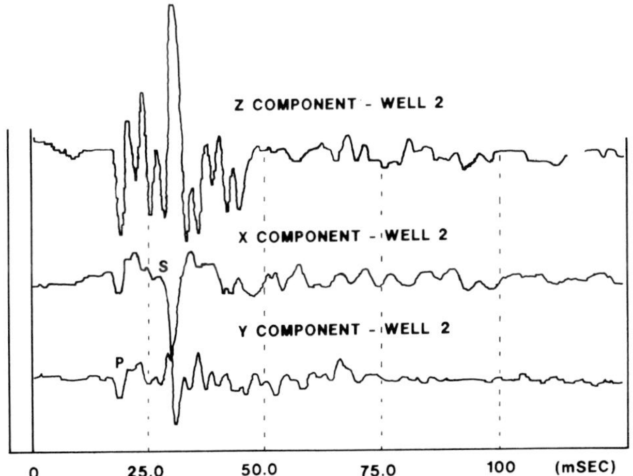

Fig. 2. Microseismic signals generated by a hydraulic fracture. Shown are signals detected by a three component VSP tool in a well near the injection well. Note the P to S phase change seen by the triaxial device.

sensors are in place, or merely an analysis of the P-wave polarization direction for a single sensor. While principally a tool for determining fracture-plane azimuth, continuing research is looking into gaining additional information regarding fracture geometry (height, length, and dip). When several observation wells are available, arrays of sensors enable more precise location of events and hence a better picture of the three-dimensional (3-D) fracture geometry.

Similar applications have been applied to map an in situ combustion process in a Canadian heavy oil field (Dusseault and Nyland, 1982), an in situ combustion and gasification experiment in a Wyoming coal seam (Beckham, 1978), and hydraulic fracturing in coalbed methane production (Dobecki, 1988). It is suspected that such microseismic investigations might be applied to reservoir surveillance. Lacy (1984) has suggested that such measurements may also be useful in describing open natural fracture sets within a producing reservoir.

Serious consideration has been given to the installation of surface seismometer arrays for process monitoring by Sandia and Gas Research Institute (GRI) experiments (Maurer and Mauk, 1987). Low signal levels and long-path attenuation coupled with very high surface-noise generated by the process hardware (pumps, blenders, etc.) caused these tests to be largely unsuccessful. Stanford University (1987) utilized surface seismometers to monitor seismic radiation caused when the fracture pump stroke travels down the well and couples into the fracture. This low-frequency (frequency of the pump) pressure pulse causes the fracture to radiate energy which can be detected at the surface.

A variation on this idea of coupling a pressure pulse into a fracture which has found commercial application is the so-called "hydraulic impedance" method (Holzhausen and Egan, 1987; Holzhausen et al., 1991). This concept employs the introduction of a pressure transient, such as generated by rapid closing or opening of a valve on a pressurized wellhead. The wellbore resonates as a pipe organ and the nature of the resonant pressure behavior is governed by the geometry of the wellbore plus any features that intersect it. A computerized pressure sensor records the pressure time history at the valve and the plots are fitted to numerical simulations containing the actual well geometry and trial fracture geometries. Besides the obvious potential value in being able to describe the geometry (but not azimuth) of induced fractures, the method should also find potential application to reservoir development by describing natural fracture systems (depths and dimensions) if these are intersected by the wellbore.

Gravitational Methods

It would seem that there are at least three applications of the gravity method which are suited to reservoir development purposes. These are: (1) repeat (iterant) surveys to monitor changes in reservoir mass, (2) borehole gravity, and (3) joint inversion. The last item, joint inversion, does not view gravity as a stand-alone tool, but as a component of an intelligent combination of geophysical data sets (Ramo and Bradley, 1982; Lines et al., 1988). The purpose of this combination is to produce a better, constrained (i.e., one which deals with all data) interpretation of the subsurface by reducing the number of possible alternatives. The usual combination includes gravity and seismic as both have explicit dependence on density. Rather than detail how such a "marriage" of data sets can produce better interpretations (a combination which is still underutilized), this section will concentrate on the other two areas where gravity is a stand-alone tool.

To imply that repeating gravity surveys would be useful requires that a redistribution of mass is occurring as a function of time. It would be easy to project that repeat gravity surveys over fields being subjected to secondary and EOR processes would be appropriate if either the injected fluids (CO_2 or steam, for example) or fluids created by the process (in situ combustion gasscs) arc of substantially different density than the natural reservoir fluids. However, if this were the case, then other techniques, such as seismic reflection or electromagnetics would also be effective with, perhaps, greater resolution.

Where gravity is a unique as well as a higher resolution technique is in borehole gravimeter (BHGM) applications. The BHGM gains resolution by being brought closer to the target under investigation. The BHGM, by measuring at discrete depths within the wellbore, determines the vertical gradient of the gravity field which can yield in situ density (LaFehr, 1983). Density logs only "see" a few centimeters into the borehole wall while the BHGM samples a much larger subsurface volume.

Circumstances which would cause substantial, and economically interesting, density changes in the subsurface would include gas accumulations, intensely fractured regions, and thermally hot zones, to name but a few. Not many case histories exploiting this are available. One (Shultz, 1989) deals with a theoretical development using the BHGM to monitor redistribution of fluids after the reservoir is pumped. The advantage of the BHGM in such an application would be the removal of the reservoir depth limitation of a surface gravity survey.

Perhaps part of the low utilization of the BHGM is due to the high costs of such a survey, both in terms of the contractor's fee and the down-time of the well, particularly if it is a producing well. Both could be reduced by new generations of BHGM's and borehole gravity gradiometers (BGGM) which are in various stages of research development. These would reduce the amount of time required to be spent in the well without sacrificing (and hopefully improving) data accuracy.

Magnetics

Quoting from Nettleton (1971):

" . . . the sources of magnetic anomalies are within the basement rocks. Therefore, magnetic interpretation usually begins at the base of the sedimentary section . . ."

Indeed, geophysicists, save those in mining or environmental pursuits, assume that the targets sought are transparent to magnetic investigation. Magnetics rarely are used for reservoir development purposes unless some process, natural or artificial, has concentrated magnetic minerals within the reservoir. An exception is trying to locate the position of a blowout from a relief well as it is being drilled. Here, the magnetic survey is being guided toward the iron in the casing (Jones et al., 1985).

The only reported example I have found deals with mapping the proppant distribution in a hydraulic fracture (Wood et al., 1983). "Proppant" is the term used for materials mixed and injected with the fracturing fluid during a hydraulic fracture operation. When the driving pressure is released, the proppant remains in the fracture and holds it open. The proppant is usually sand; however, in a research experiment sponsored by GRI, a special iron-based proppant was substituted. The propped fracture, then, would be a concentration of magnetic material and therefore detectable by a magnetic survey. A test of this concept was run for a very shallow fracture (15 m). An array of highly sensitive, superconducting (SQUID) magnetometers emplaced at the surface recorded magnetic field fluctuations during the fracture test. Anomalous residual (after correcting for tilting of the SQUIDs by the opening of the fracture and variation in the primary magnetic field) magnetic variations on the order of 1–2 nanotesla were observed. The test was only marginally successful considering the shallow test depth, and the method has been virtually abandoned as being impractical.

Electrical/EM

It might be considered that of all the alternative methods which could be applied to reservoir geophysics projects, electrical and electromagnetic (EM) methods offer the greatest potential utility. EM methods involve wave propagation phenomena and are, by association, more familiar to seismologists. The conductivity of a reservoir is almost completely a function of porosity and saturant chemistry, so reservoir description, including fluids and contacts between different fluids, can be realized (e.g., Whightman et al., 1983). Electrical conductivity in a fractured reservoir is known to be anisotropic (Taylor and Jansen, 1988) such that some measure of fracture characterization may be gained through mapping conductivity anisotropy. Conductivity can also be very dependent upon temperature, and thermal EOR process monitoring could be accomplished through conductivity mapping. Finally, numerous oil fields are geochemically active environments causing pyritic plumes or halos over such oil fields which themselves may be detected (Spies, 1983), offering insight into the delineation of such reservoirs.

The bulk of published reports on such applications have been reported in the literature of the Soviet Union, long a leading proponent of electrical applications. A summary of Soviet applications, plus a listing of appropriate references is provided in Spies (1983). Spies classifies the applications of electrical methods as (1) soundings for structural mapping (exploration), (2) sounding to measure the presence of oil directly (reservoir delineation or description), (3) sounding to measure the geoelectric properties of the oil bearing horizon (reservoir description), and (4) detection of a geochemical plume or halo above or surrounding the deposit (reservoir delineation?). Some question is attached to the last classification because plume formation reflects the permeability distribution of the surrounding sediments. Thus mapping a halo might be more indicative of fault or fracture concentrations than the shape of a reservoir.

Daniels (1977), Dobecki (1980), Lytle (1982), and Yang and Ward (1985) have all presented mathematical developments of dc electrical resistivity and/or induced-polarization models for measurements in and between boreholes or a borehole and the surface, in the search for anomalous conductors or resistors of various regular shape. As stated in Dobecki (1980):

"Modeling of in-situ energy extraction processes (e.g., coal gasification and oil shale retorting) has been examined by assuming the affected zones may be described by oblate or prolate spheroids of lowered electrical resistivity. The purpose of the model study is to determine if such subsurface bodies may be monitored and defined by electrical resistivity measurements made in boreholes away from the reaction zone."

The sense is that, if an EOR process is characterized by a growing conductive mass, then its presence and approximate size or shape can be determined electrically. The injection of brines into a petroleum reservoir provides an obvious resistivity contrast. The injection of CO_2 or polymers does not suggest an obvious contrast, but input is needed from the petrophysicist. Lee and Wayland (1982) as well as Daily (1984) show laboratory results which verify dramatic drops in resistivity as a reservoir is heated and pyrolized, as would occur during in situ combustion. Thus, some EOR processes should develop significant conductivity targets making the theoretical modeling studies applicable.

Similar thinking has led other investigators to utilize surface EM surveys to try to map EOR processes. Among these are Bartel (1982) and Bartel and Wayland (1981) for steamflood and fireflood mapping with controlled-source audiomagnetotellurics (CSAMT), and Bartel et al. (1985) for CSAMT mapping of underground coal gasification. Singhal and Card (1988), in addition to their tiltmeter study, also used EM and self-potential (SP) to monitor a steam flood in Alberta. The sense of using SP is that electrokinetic and electrochemical potentials may be generated by moving high temperature fluids in a shallow reservoir. A similar study for a thermal flood is provided in Dorfman et al. (1977).

Lytle et al. (1981) used tomographic inversion of cross-borehole EM to map the changing position (rate and shape) of a conductive fluid front injected into a reservoir. This was mainly a feasibility study which demonstrated that such methods can (a) track fronts and (b) determine reservoir properties (e.g., permeability) by tracking the advance rates of fronts through repeat surveys. Similar tomographic studies have been reported in Daily (1984) for mapping an in-situ oil-shale retort, in Somerstein et al. (1984) also for oil-shale retorting, and in Laine (1987) for mapping a steam flood. Asch and Morrison (1989) present a similar development for monitoring resistivity changes in a subsurface volume using a combination of downhole and surface electrodes instead of in a cross-borehole mode. With the current emphasis on crosshole seismic tomography surveys, crosshole electrical and EM surveys must also be considered as wells of opportunity become available. The principal shortcoming is, of course, that electrical and EM measurements cannot be made within or between metal-cased boreholes.

A novel approach to using electrical methods for the mapping of hydraulic fractures developed at Sandia National Laboratories (Bartel et al., 1976) applies the mise-à-la-masse method, established in mining geophysics, to fracture mapping. Referring to Figure 3, current is applied at the level of fracture initiation within the borehole before fracturing is induced. Surface equipotentials should be circular for such a simple point source model. As the fracture grows, the fracture fluid (usually KCl) is conductive and carries current. The point source, therefore, becomes a time-varying line source, which distorts the surface equipo-

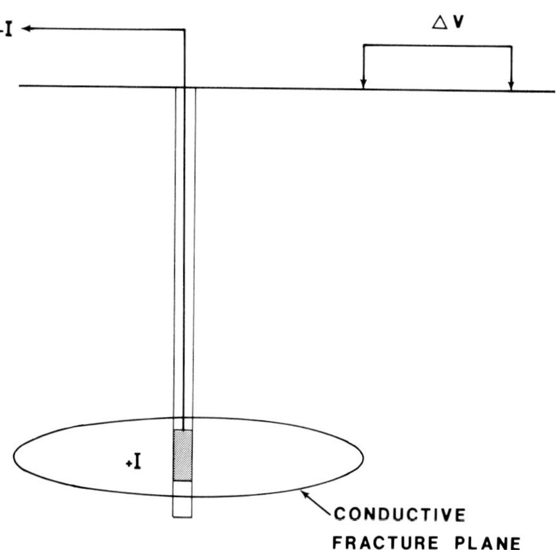

Fig. 3. Principle of the induced current method of hydraulic fracture mapping. A downhole current electrode couples into a conductive fracture as the fracture grows.

tentials (e.g., Figure 4) in a manner which allows some description of the induced fracture. This method has been used successfully for fracture tests at 2500 m process depth. As with the tiltmeter, surface measurement of induced electrical potentials gets more difficult with depth. Unless the length of the induced fracture is about the same or greater than its depth, the measurement can only provide fracture azimuth information (i.e., no reliable length or height information). The electrical signature from a fracture at depths over about 1000 m becomes insensitive to length or height unless the fracture is large. The distortion of potentials caused by surface metal hardware (pipelines, casing, grounding mats, etc.) impose practical, logistical limitations as well.

A final development to be noted is in the processing of EM data. Levy et al. (1988) present a means of processing magnetotelluric (MT) data such that the resulting sections (e.g., Figure 5) are pseudo-impulse-response sections very much in appearance like seismic spike sections. The amplitudes of the spikes are due to EM reflection and transmission coefficients at layer interfaces. Such processing may find application by geophysicists who are more accustomed to time sections than to sounding curves. A side thought is that MT is important in areas where seismic does not

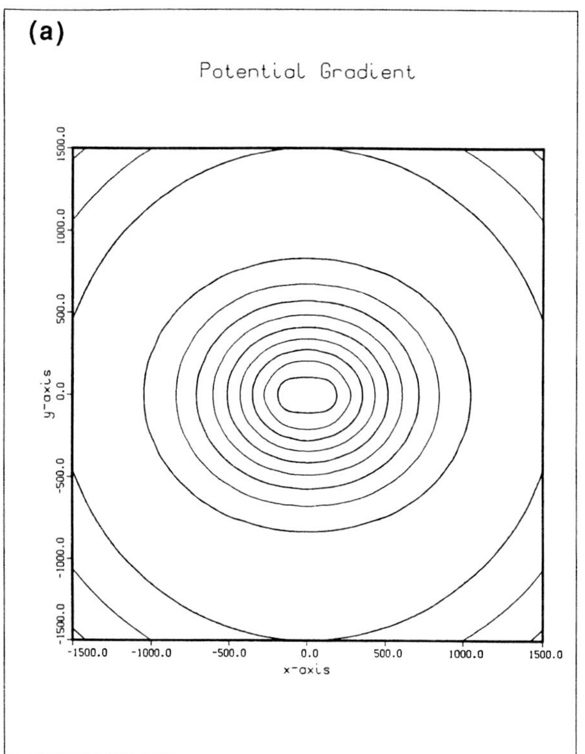

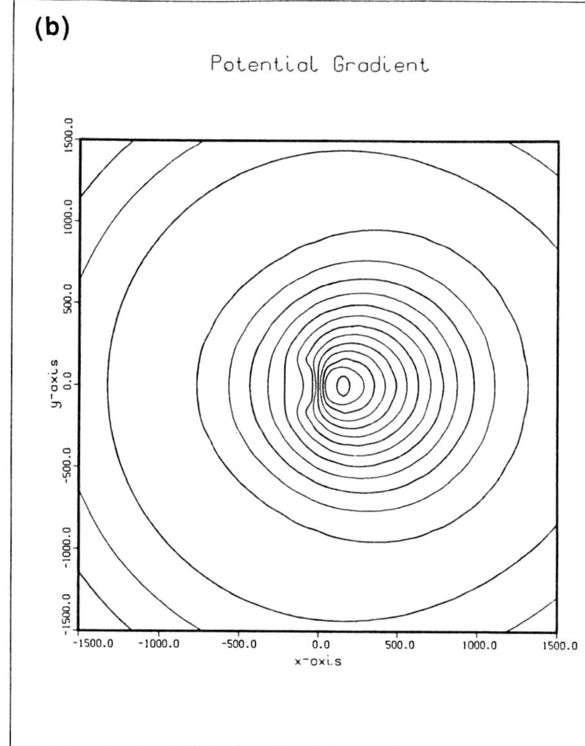

Fig. 4. Two examples of the resultant electrical potential gradient field for a current carrying hydraulic fracture. The fracture is 1500 m deep, 1500 m long, and extends along the x-axis. In Case (a), the fracture is symmetric (i.e., extends 750 m each way from the well). In Case (b) the fracture is asymmetric (1000 m toward $+x$ and 500 m toward $-x$). Model generated by L.C. Bartel, Sandia National Laboratory.

work well, i.e., high-resistivity surface layers such as volcanics and carbonates.

Several important caveats are coupled to the use of electrical and EM methods for purposes cited. Measurements cannot be made from or to steel cased wells and the presence of casings, surface pipelines, etc. (cultural noise) make surface measurements in existing oil fields very difficult if not impossible. Finally, the sequence of resistivities in beds above and below the pay zone, plus considerations of the width-to-depth ratio for an electrical target (geologic noise) have a strong influence on the anomaly generated (Garg and Keller, 1984) and, under the most favorable cultural-noise conditions, circumstances not connected to the target may render electrical surveys inapplicable. Certainly, the resolution (lateral as well as vertical) of EM methods is generally poor compared with seismic.

Remote Sensing

While all geophysical surveys are forms of remote sensing, this term usually refers to mapping from air- or space-borne platforms by way of electromagnetic interrogation. The range of spectral sensors employed extends from gamma-ray spectroscopy to far-ultraviolet scanners, near-ultraviolet and visible-spectrum

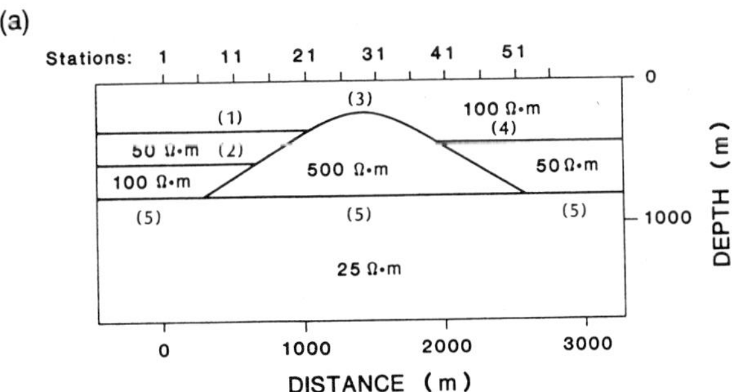

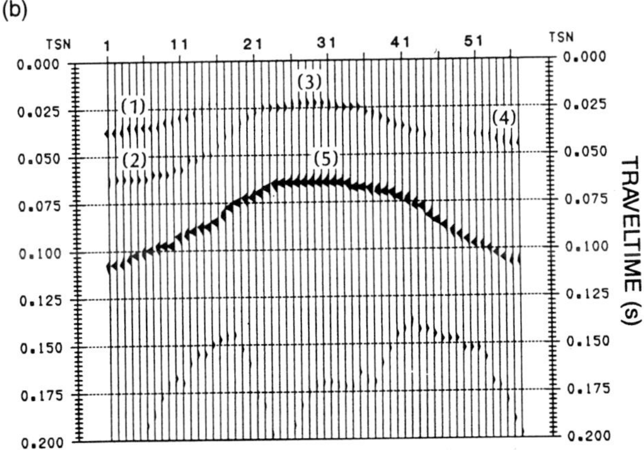

Fig. 5. A 2-D conductivity structure simulating a resistive salt dome buried in a layered medium is shown in (a). The EM reflectivity time section obtained by inverting the TE mode data is shown in (b). From Levy et al., 1988.

cameras, middle-infrared and far-infrared scanners, radar, and airborne EM systems (e.g., INPUT™). Some scanners have wide spectral response, from far ultraviolet to far infrared.

Normally, remote sensing is employed as a "broad brush" tool; i.e., a means of obtaining wide, regional coverage at fairly low cost (e.g., University of Texas, 1988). Aside from airborne EM systems, minimal penetration into the earth is realized, so these methods are, basically, surficial mapping tools. Remote sensing cannot describe or delineate petroleum reservoirs or processes directly. Geochemical anomalies which affect vegetation and therefore the spectral image are observed by remote sensing (Feder, 1985), but the same caveat noted in the section on electrical and EM regarding their association with permeable zones rather than with the whole reservoir applies. Remote sensing does see features which can be related to a reservoir, especially when integrated with other (e.g., seismic or surface EM) geophysical data (John Everett, Earth Satellite Corp., pers. comm.). For example, fracturing at the earth's surface may be fairly distinct on remote sensing data sets. If the trend and pattern of surficial fracturing can be extrapolated to reservoir depths, then an independent means of determining how to tie fault structures from one seismic line to another may be gained. In fracture-controlled oil fields, such as Devonian shales in the central and eastern U.S. or some Gulf coast fields, comparison of remotely sensed fractures with other geophysical data sets and production records can sometimes explain why some wells are winners and others are not, and where the next potential winner should be drilled (Weaver and Cress, 1984; Reynolds, 1987).

Regarding reservoir surveillance applications, Feder and Vixo (1987) discuss a nighttime, far-infrared survey over an East Texas steamflood (460 m depth) obtained through regular, repeat overflights. The heating pattern was quite distinct and changed with time. The authors allow that rock and soil are indeed terrible conductors of thermal energy and that visualizing a dynamic process through insulating cap rock and overburden is precluded over a useful time frame; therefore, reservoirs must be in fairly direct hydraulic communication and the observed thermal anomalies must be due to convective groundwater flow in overlying fractures. They further speculate that in situ combustion would be even more dramatic and that CO_2 injection might be observed as a moving cold anomaly.

Comments

In spite of a broad variety of geophysical methods available, few outside of seismic have become standard tools for reservoir applications. Some (e.g., tilt-meters and passive borehole seismic) have found commercial application, and the potential for growth in electromagnetics seems promising. Developments which have to occur before any significant increase in the use of alternative geophysical methods will happen include:

(1) An increase in the library of petrophysical data dealing with a wide variety of fluids (e.g., CO_2, polymers, surfactants, gels, etc.), temperature effects on these fluids, and their behavior in a porous reservoir rock. If we can't predict the geophysical effect, we can't pick the proper tool.

(2) An increase in the dialog among geophysicists, reservoir engineers, process engineers, and geologists. In much the same way as the dialog between geophysicists and geologists created an "explorationist", there needs to be a move toward creating a "productionist" class. Geophysicists can't pick the right tool if they don't know which targets are the important ones.

(3) Education. There are other geophysical phenomena in addition to elastic wave propagation. Although the heavy investment in seismic has been warranted, expertise in sister methods is suffering for seismic's prosperity. As more data and better communication become established, perhaps there will be new emphasis on revitalizing understanding alternative technologies.

References and General Reading

Albright, J.N. and Pearson, C.F., 1982, Acoustic emissions as a tool for hydraulic fracture location: Experience at the Fenton Hill Hot Dry Rock site: Soc. Petr. Eng. J., **22**, 523–530.

Asch, T. and Morrison, H.F., 1989, Mapping and monitoring electrical resistivity with surface and subsurface electrode arrays: Geophysics, **54**, 235–244.

Bartel, L.C., 1982, Modeling and analysis of the CSAMT geophysical technique results to map oil recovery processes: 57th Tech. Conf., Soc. Petr. Eng., Am. Inst. Min. Metall. Eng., New Orleans, Paper SPE 11192.

Bartel, L.C. and Wayland, J.R., 1981, Results from using the CSAMT geophysical technique to map oil recovery processes: 56th Tech. Conf., Soc. Petr. Eng., Am. Inst. Min. Metall. Eng., Paper SPE 10230.

Bartel, L.C., Davidson, G.S., Jacobson, R.D., and Uhl, J.E., 1985, Results from using the CSAMT technique to monitor the Tono UCG experiment: In Situ, **9**, 293–328.

Bartel, L.C., McCann, R.P., and Keck, L.J., 1976, Use of potential gradients in massive hydraulic fracture mapping and characterization: 51st Tech. Conf., Soc. Petr. Eng. of Am. Inst. Min. Metall. Eng., Paper SPE 6090.

Beckham, L.W., 1978, Estimating the extent of an in situ coal gasification process using a passive seismic technique: Report SAND78–0145, Sandia National Laboratory.

Bower, D.R., 1983, Bedrock fracture parameters from the interpretation of well tides: J. Geophys. Res., **88**, 5025–5035.

Daily, W., 1984, Underground oil-shale retort monitoring using geotomography: Geophysics, **49**, 1701–1707.

Daniels, J.J., 1977, Three-dimensional resistivity and induced polarization modeling using buried electrodes: Geophysics, **42**, 1006–1009.

Davis, P.M., 1983, Surface deformation associated with a dipping hydrofracture: J. Geophys. Res., **88**, 5826–5834.

Dobecki, T.L., 1980, Borehole resistivity curves near spheroidal masses: Geophysics, **45**, 1513–1522.

Dobecki, T.L., 1983, Hydraulic fracture orientation using passive borehole seismics: Ann. Tech. Conf., Soc. Petr. Engr., Am. Inst. Min., Metall., Petr. Eng., Paper SPE 12110.

Dobecki, T.L., 1988, Microseismic investigation of fracture height growth – Black Creek and Mary Lee treatments, Warrior Basin, Alabama: Final report submitted to Taurus Exploration Inc., Birmingham, Alabama.

Dorfman, M.H., Oskay, M.M., and Gaddis, M.P., 1977, Self-potential profiling — a new technique for determination of heat movement in a thermal oil recovery flood: 52nd Ann. Tech. Conf., Soc. Petr. Eng., Am. Inst. Min. Metall. Eng., Paper SPE 6790.

Dusseault, M.B. and Nyland, E., 1982, Fireflood microseismic monitoring – rock mechanics implications: *in* Issues in rock mechanics, 23rd Sympos. Soc. Min. Eng., Amer. Inst. Min. Metall. Petr. Eng.

Evans, K., 1983, On the development of shallow hydraulic fractures as viewed through the surface deformation field: Part I—Principles: J. Petr. Tech., **35**, 406–410.

Evans, K. and Holzhausen, G., 1983, On the development of shallow hydraulic fractures as viewed through the surface deformation field: Part II—Case histories: J. Petr. Tech., **35**, 411–420.

Feder, A.M., 1985, Contemporary remote sensing for hydrocarbon exploration, development—with case histories: Oil and Gas J., **83**, No. 38, 160–171.

Feder, A.M. and Vixo, D.L., 1987, Remote sensing in drilling and production applications: World Oil, **204** (March), 58–60.

Garg, N.R. and Keller, G.V., 1984, Synthetic electric sounding surveys over known oil fields: Geophysics, **49**, 1959–1967.

Hanson, J.M. and Owen, L.B., 1982, Fracture orientation by the solid earth tidal strain method: Ann. Tech. Conf. Soc. Petr. Eng., Am. Inst. Min. Metall. Eng., Paper SPE 11070.

Holzhausen, G.R. and Egan, H.N., 1987, Detection and control of hydraulic fractures in water injection wells: Calif. Regional Tech. Conf. Soc. Petr. Eng., Am. Inst. Min. Metall. Eng., Paper SPE 16362.

Holzhausen, G.R., Card, C.C., Raisbeck, J.M., and Dobecki, T.L., 1985, Hydraulic fracture growth during steam stimulation in a single-well test: Calif. Regional Tech. Conf. of the Soc. Petr. Eng., Am. Inst. Min. Metall. Eng., paper SPE 13619.

Holzhausen, G.R., Egan, H.N., and Baker, G.S., 1991, Hydraulic impedance – safe, low-cost well testing: Petr. Eng. Int., January.

Jones, D.L., Hoehn, G.L., and Kuckes, A.F., 1985, Improved magnetic model for determination of range and direction to a blowout well: Ann. Tech. Conf. Soc. Petr. Eng., Am. Inst. Min. Metall. Eng., Paper SPE 14388.

Lacy, L.L., 1984, Comparison of hydraulic fracture orientation techniques: Ann. Tech. Conf. Soc. Petr. Eng., Am. Inst. Min. Metall. Eng., Houston, Paper SPE 13225.

LaFehr, T.R., 1983, Rock density from borehole gravity surveys: Geophysics, **48**, 341–356.

Laine, E.F., 1987, Remote monitoring of the steam-flood enhanced oil recovery process: Geophysics, **52**, 1457–1465.

Lee, D.O. and Wayland, J.R., 1982, Measurement of formation resistivity changes induced by in situ combustion: 57th Tech. Conf., Soc. Petr. Eng., Am. Inst. Min. Metall. Eng., Paper SPE 11051.

Levy, S., Oldenburg, D., and Wang, J., 1988, Subsurface imaging using magnetotelluric data: Geophysics, **53**, 104–117.

Lines, L.R., Schultz, A.K., and Treitel, S., 1988, Cooperative inversion of geophysical data: Geophysics, **53**, 8–20.

Lytle, R.J., 1982, Resistivity and induced-polarization probing in the vicinity of a spherical anomaly: Inst. Elect. and

Electron. Eng., Trans. Geosci. Remote Sensing, **GE-20**, 493–499.

Lytle, R.J., Lager, D.L., Laine, E.F., Salisbury, J.D., and Okada, J.T., 1981, Fluid-flow monitoring using electromagnetic probing: Geophys. Prosp., **29**, 627–638.

Maurer, K.D. and Mauk, F.J., 1987, Seismic wave motion for a new model of hydraulic fracture with an induced low-velocity zone: J. Geophys. Res., **92**, 9293–9309.

Nettleton, L.L., 1971, Elementary gravity and magnetics for geologists and seismologists: Monograph 1, Soc. Expl. Geophys.

Pollard, D.D. and Holzhausen, G.R., 1979, On the mechanical interaction between a fluid-filled fracture and the earth's surface: Tectonophysics, **53**, 27–57.

Ramo, A.O. and Bradley, J.W., 1982, Bright spots, milligals, and gammas: Geophysics, **47**, 1693–1705.

Reynolds, S.M., 1987, Photogeology: unique system for detecting more and better drillable prospects: Oil and Gas J., **85**, No. 13, 120–124.

Schultz, A.K., 1989, Monitoring fluid movement with the borehole gravity meter: Geophysics, **54**, 1267–1273.

Seavey, R.W., 1982, Borehole seismic unit: Report SAND82–0373, Sandia National Laboratory.

Singhal, A.K. and Card, C.C., 1988, Monitoring of steam stimulation in the McMurray formation, Athabasca deposit, Alberta: J. Petr. Tech., **40**, 483–490.

Somerstein, S.F., Berg, M., Chang, D., Chung, H., Johnson, H., Richardson, B., Pizzicara, J., and Winfield, W.S., 1984, Radio-frequency geotomography for remotely probing the interiors of operating mini- and commercial-sized oil-shale retorts: Geophysics, **49**, 1288–1300.

Spies, B.R., 1983, Recent developments in the use of surface electrical methods for oil and gas exploration in the Soviet Union: Geophysics, **48**, 1102- 1112.

Stanford University, 1987, Adaptive signal processing for seismic monitoring of hydraulic fractures: Report GRI-87/0013, Gas Research Institute,.

Taylor, R. and Jansen, J., 1988, Azimuthal resistivity techniques to delineate zones of high secondary porosity in fracture controlled aquifers *in* Proc. 2nd Natl. Indoor Action Conf. Aquifer Restoration, Ground Wat. Monitoring, and Geophys. Methods: Assn. of Gr. Water Scient. Engr. and US EPA EMSL.

University of Texas, 1988, Relationship between radar lineaments, geologic structure, and in situ stress in East Texas: Report GRI-88/0093, Gas Research Institute,.

Weaver, K.A. and Cress, W.D., 1984, The use of lineament mapping and electromagnetic surveying for detecting natural fractures in the Devonian shale: Eastern Reg. Mtg., Soc. Petr. Eng., Am. Inst. Min. Metall. Eng., Paper SPE 13358.

Wightman, E., Kaufman, A.A., and Hoekstra, P., 1983, Mapping gas-water contacts in shallow producing formations with transient EM: 53rd Ann. Internat. Mtg., Soc. Expl. Geophys., Expanded abstracts, 59–60.

Wood, M.D., Parkin, C.W., Yotam, R., Hanson, M.E., Smith, M.B., Abbott, R.L., Cox, D., and O'Shea, P., 1983, Fracture proppant mapping by use of surface superconducting magnetometers: Symp. on low permeability gas reservoirs, Soc. Petr. Eng., Am. Inst. Min. Metall. Eng. and US Dept. of Energy, Paper SPE/DOE 11612.

Yang, F.W. and Ward, S.H., 1985, Single-borehole and cross-borehole resistivity anomalies of thin ellipsoids and spheroids: Geophysics, **50**, 637– 655.

Emerging Geophysical Technologies

*Wulf Massell**

Introduction

This book discusses geophysical methods and how their use improves the interdisciplinary reservoir model. A reservoir model should summarize our understanding of how and where oil and gas are distributed so that they can be produced and recovered efficiently. Previous chapters have introduced the three-dimensional (3-D) surface seismic method, vertical seismic profiling (VSP), and crosswell seismic techniques with numerous examples. A primary objective of these methods is to define the 3-D shape and distribution of reflectors that describe the subsurface lithology distribution. As resolution and repeatability improve we seek qualitative and quantitative information to characterize and identify the lithology between reflectors. We are challenged to explain the detail we see in our data, especially on crosswell seismic tomograms, and we find ourselves re-examining the principles and fundamentals of wave behavior and how they relate to the heterogeneity and dynamics of rock/fluid systems.

It is tempting to predict that future improvements in technology will be incremental and will have an evolutionary rather than a revolutionary impact. However, I believe that we are at a threshold of being able to integrate a critical mass of diverse technologies that will deliver a significant surge in oil recovery efficiency. A short historical review will set the stage for my predictions of future developments in geophysics.

The Historical Impact of New Technologies

Many oil discoveries can be linked to the introduction of new technologies. A plot of oil finding efficiency versus the cumulative number of exploratory wells drilled in the United States is shown in Figure 1. The plot is from Lindseth's 1990 SEG Distinguished Lecture Tour (Lindseth, 1990). It is not sensitive to market pressures of supply and demand that control the number of wells that are drilled. We might expect that the new oil discovery rate should decline geometrically, but this is not what we see. Lindseth points out that the surges in oil-finding efficiency correlate with

*E&P Imaging Corporation, 4002 El James Drive, Spring, TX 77388.

the introduction of new technologies. In his words: "Each new petroleum technology operates something like a sieve, sifting out the reservoirs it reveals and causing a temporary surge in discoveries which later subsides until the next wave of technology arrives."

Several surges are visible in Figure 1. In the 1920s, gravity and early seismic refraction methods ushered in a rash of oil field discoveries associated with Gulf Coast salt domes. The seismic reflection method introduced in the 1930s guided exploration drilling where subsurface geology could not be inferred from surface maps. A sequence of improvements led to the use of magnetic tape to record seismic field traces in the late 1950s. Magnetic tape was introduced to keep from having to reshoot seismic lines, but it quickly led to analog playback centers which performed static corrections, real-time normal moveout, basic signal processing, and also produced the first record sections.

Playback centers also made possible the common-midpoint (CMP) method. The idea had previously been patented by W. Harry Mayne, although the playback technology needed for implementation wasn't available. The CMP method produces better reflection quality than is possible on single-fold profiles. The combination of play-back centers and CMP technology boosted the sagging oil discovery rate in the early 1960s. At about that time digital recording was introduced and digital computers started to replace the analog playback centers. With increased dynamic range and the versatility of digital processing, seismic sections started to look like geologic sections. Digital technology sustained a declining oil discovery efficiency, culminating with the Prudhoe Bay discovery in 1969.

Seismic processing to preserve reflection character and amplitude in the mid 1970s led to the hunt for "bright spots" and "flat spots." Seismic interpreters began to recognize stratigraphic sequences and depositional systems. The power of an integrated approach to interpretation spurred the industry to form exploration teams; geology and geophysics became better integrated and a new breed of "explorationists" emerged.

In the mid 1970s 3-D seismic methods provided a new view of the subsurface; complex structures and depositional features could now be examined in an internally consistent 3-D seismic data volume. Petroleum engineers began to utilize 3-D interpretations to design development and recovery programs and thus a third member joined the geologist-geophysicist teams. Today, 3-D surveying accounts for more than half of the expenditure for seismic services in the Gulf of Mexico and the North Sea. Most early 3-D surveys were intended to support reservoir development programs in the 1980s. Speculative 3-D surveys are now being shot preceding government lease sales. Although the United States is the most mature oil exploration province in the world, continued improvements in technology will sustain the depletion curve well into the next century.

A Look to the Future

By the year 2000 we will be able to drill exploration and development wells with a confidence and accuracy not previously known. Just as magnetic resonance imaging (MRI) gives visual access to the exterior and interior of human joints, we will follow the progress of a well in real time within a 3-D seismic volume and examine the rock formations with a borehole televiewer. Drillers on the rig floor, as well as staff in the home office, will be watching drilling progress on a 3-D image on a color CRT. Borehole seismic sources, especially ones that can be deployed with the drill string still in place, will update the data base by adding higher resolution images of the geologic section just ahead of the bit. An areal array of a thousand or more surface geophones and receivers in nearby wells will record a large number of diverse raypaths to update and improve the subsurface image. The database will be updated in real time to truly look ahead of the bit.

Seismic interpretation today, whether of 2-D or 3-D data, involves picking horizons, faults, or sedimentary features on vertical or horizontal 2-D sections. These picks are control points for maps that lead to a comprehensive subsurface interpretation model. The interpreter's view of the data is confined to the opaque 2-D field of pixels, or picture elements, that comprise

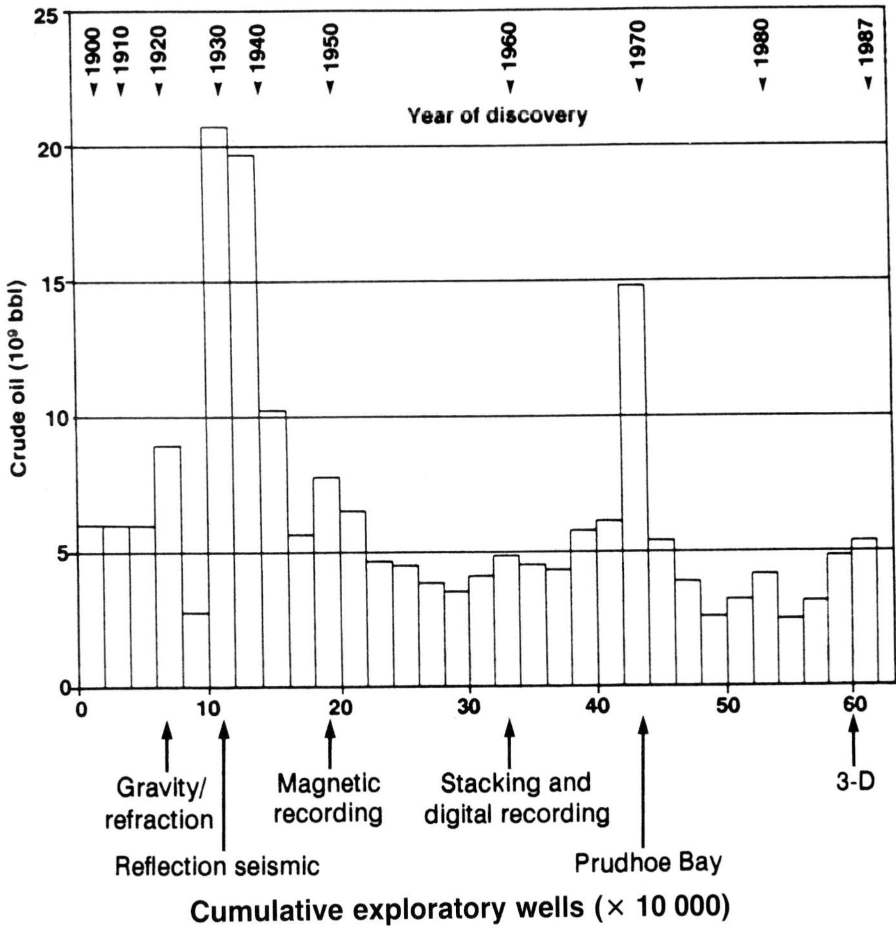

Fig. 1. Oil-finding efficiency indicated by graph of new oil found per well versus number of wells drilled. (After Lindseth, 1990.)

the seismic section. To represent 3-D data in three dimensions we can use a 3-D raster of voxels or volume elements. Voxel representation can include varying degrees of transparency that can be defined universally according to, for example, the seismic amplitude. This kind of graphic display of uninterpreted data is a new emerging field known as scientific visualization. Ideally, the interpreter would be able to interact with the data and selectively remove some voxels that impede a view of the interior of the data volume. A method to search neighboring voxels that share certain data attributes could direct a volume-fill algorithm similar to the area-fill features available in computer-graphics paint programs. Such a "predicate bug" algorithm can be used as an automatic picker to extract whole reflection units. A feature that might appear insignificant on a 2-D seismic section may be a hint of a cross-cutting artifact that is a clue to useful geologic information. Voxel viewing allows the interpreter to see such features without first having to "pick" them.

All seismic data have a 3-D character at some stage during processing. Easy access to computer graphic volume visualization will provide new insight to migration focussing and a better understanding of AVO (amplitude versus offset) variations and noise patterns on 2-D data. Field quality control can benefit if the observer can see source-generated noise in a 3-D display of source location, offset, and time. When interpreters are given visual access to the interior of "unpicked" 3-D data volumes, they will discover far more about the subsurface than by picking event surfaces. To benefit from seismic visualization we must develop tools for the user to interact with the data. These tools must be so simple and intuitive that they become natural extensions of an inquiring user.

Much of the recent growth in 3-D can be linked to the higher performances of digital processors, the availability of larger memories, and innovative processing algorithms. Nearly interactive versions of 2-D prestack depth migration algorithms are already available on supercomputers. Interpreters will have real-time parameter control within the next several years to optimize parameter choices through interactive seismic data processing. The effect will be similar to the introduction of magnetic tapes in field recording; instead of "instant replay", tomorrow's interpreters will have "instant reprocess." They will be able to alter a process step and change the velocity field or the statics parameters, or to swap out entire algorithms, and instantly observe the effect on the final image.

Such capabilities are already possible on supercomputers but they, like 747 jumbo jets, are neither priced nor designed to be flown interactively by a single passenger. Cost-effective, application-specific parallel computers already surpass supercomputer performance at a price that is comparable to today's seismic workstations, but the most important ingredient, the software, is just beginning to be developed. We continue to depend on the computer industry to solve issues, such as large capacity data storage media, higher I/O rates, and faster processors.

Emerging Technologies

Seismic reflection data provide a spatial framework between wells to which geologic interpretation must conform. An interpreter examines the seismic data for clues that reveal the depositional environment, probable lithology, and pore-fluid distribution. We will always want improved seismic resolution to close the correlation gap between well-log lithology and that inferred from seismic data. Developing technologies that improve downhole seismic imaging may provide the resolution needed to link the two data types.

The desired degree of resolution should be guided by the intended use of the reservoir model. There is always a tradeoff between having too much detail data and the amount of detail that can be usefully included in a reservoir model. A systematic method of extracting a model of any desired scale from too much data, or to interpolate when we lack enough information, is possible using geostatistical techniques (Wolf, 1990).

Geostatistical methods provide convenient ways to analyze and correlate large spatial data distributions. Commonly used tools to measure the spatial variation between data points include variograms, krieging, cokrieging, and conditional simulation. These measures combine to produce maps of spatial variation and accompanying measures of the map's reliability. In addition to simplifying complex data distributions, geostatistics can "fill-in" missing data where spatial sampling is too coarse. Methods of integrating well-log and seismic data to extend porosity and permeability maps using variograms and cokrieging have been reported by several authors (Fogg and Lucia, 1990; Doyen, 1988; Uzcategui and Del Pino, 1987; Thadani et al., 1987). The current challenge is to tie the relatively "soft" information derived from seismic data with "hard" information from well-logs. Geostatistics is likely to play an increasingly important role in building reservoir models as interpreters are given easier access to, and become more familiar with, its tools.

Borehole Seismic Receivers

Improved resolution of seismic images requires increased signal bandwidth as well as a larger number and greater diversity of raypaths. Two active areas of research are the development of multilevel clamped receivers and controllable downhole sources. Multilevel receivers reduce acquisition time and make VSP and crosswell measurements more cost effective.

Most downhole receivers can be used only in open hole environments. A number of research groups use multiple hydrophone receiver strings. Pressure- sensitive hydrophones, however, do not provide the directional sensitivity of clamped three-component geophones (although some directivity can be achieved during processing the outputs of individual elements). The alternative of clamped receiver sondes at multiple levels in the borehole represents a formidable increase in operational complexity. Each sonde must be clamped individually allowing sufficient cable slack between sondes to prevent taut-cable vibration. Such a tool will likely be limited to use in cased holes.

A multilevel receiver tool is currently being developed by the National Laboratories and the Oil Industry[1]. The prototype design features individually clamped three-component receiver "shuttles" spaced at 50 ft and suspended below a telemetry cartridge operating on a standard seven-conductor wireline. The tool will relay 16 channels of multiplexed data to the surface from data buffers. Receiving elements will be high-temperature piezoelectric accelerometers that operate to 200°C. A prototype tool is scheduled for field tests in late 1991. Planned improvements include a 96-channel capability and a sampling interval of 1/4 ms (Sandia and OYO, 1990).

Borehole Seismic Sources

A variety of borehole seismic sources are in use for interwell and VSP applications, each with its own strengths and weaknesses. It is unlikely that any one source will fill all of the needs for reservoir delineation and characterization. The firing rate and mobility of sources is especially important in crosswell tomography where resolving power depends on both the number and diversity of raypaths. The surviving tools

[1]This cooperative group is known as the Crosswell Seismic Forum, an operating unit of the Oil Recovery Technology Partnership. Contact: R.J. Hanold at Los Alamos or D. Northrup, Sandia National Laboratories.

and methods will be those that significantly reduce the cost per unique raypath.

Crosswell seismic surveys tend to require shorter raypaths than reverse VSP surveys. Most downhole sources generate signals in excess of 1000 Hz; however, because the earth attenuates high frequencies, not all sources are suitable for long travelpaths and only a few can be used in open-hole environments. A wide-bandwidth, controllable downhole source that can be used while drilling without interfering with the drilling process is needed. Several methods have been tested, though none delivers significant signal bandwidth. Table 1 summarizes existing downhole source types and their behavior. Many are prototype versions. The most consistent performers are air guns, the hydraulic vibrator, and explosive devices. The borehole airgun of Bolt Technology has been the most reliable downhole source. Both single guns and arrays are used. The guns require high pressure air hoses connected to compressors or bottled gas at the surface. Air pressure must exceed the ambient mud pressure in the borehole by at least 1500 psi, a requirement which limits operation to less than 8000 ft depths.

A single-axis downhole hydraulic vibrator operating on an umbilical feed has been developed and tested by Chevron Oil Company (Paulsson, 1988; Paulsson et al., 1990; Lo et al., 1990). This prototype source has performed satisfactorily in numerous VSP and crosswell profiling experiments although its operating depth is limited by the umbilical hydraulics. Plans are to build a new model for deep environments that will use a self-contained hydraulic system (Paulsson, 1990). The proposed tool will be a tri-axial generator, producing independent *P*-, *SH*-, or *SV*-waves. It will use a variety of sweep modes between 5 and 1440 Hz and be capable of a maximum output of 6000 pounds. An electric downhole motor-pump operating on a conventional wireline to depths to 20 000 ft and temperatures to 400°F will power the hydraulic drive.

Table 1. Downhole Seismic Sources

Frequency	Effective transmission distance		
	<100 ft	<500 ft	<1000 ft
High-frequency <1000 Hz	Sonic tools Mechanical hammers Sparkers Magnetostrictive Cylindrical bender Explosive Airguns Imploders	Sparkers Magnetostrictive Bender Explosive Airguns Imploders	Bender Explosive Airguns Imploders
Low-frequency: <500 Hz	Vibrators	Vibrators	Vibrators

Two other downhole vibrators are being developed. One is the Swept Frequency Borehole Source built by Western Atlas (Kennedy et al., 1988). This device creates an acoustic resonance between two widely separated bladders in a cased borehole. A pulsing "spitter" valve injects clear water midway between the bladders, causing the trapped drilling fluid to resonate at a frequency determined by the inter-bladder distance. A resonating frequency is achieved by a servo system that monitors the standing wave and controls the frequency of the water spitter. The frequency radiated into the surrounding earth is determined when the distance between the bladders is one-half wavelength. A programmed sweep of frequencies can be generated by gradually changing the distance between the bladders.

The second type is a pneumatic vibrator that operates on a high pressure air hose. This tool can operate in dry or fluid-filled boreholes. The device was developed at Sandia National Laboratories and is licensed to Santerra Corporation (Santerra Corp., 1990). The driving gas is vented to the surface and the tool does not have the depth limitation of air guns. Prototype tests indicate this is a workable source although it has less energy than the hydraulic vibrator. Both the pneumatic and hydraulic vibrators must be clamped to the borehole wall and require cased or smooth wellbores.

Electromechanical sources such as magnetostrictive and piezoelectric transducers produce limited seismic energy and thus have a limited effective range of several hundred feet. In hard rock terrain these devices have transmitted surprisingly wideband signals over large distances. Examples are the Cylindrical Bender (Balogh et al., 1988; Harris, 1988) and the Sparker and Gas Plasma sources (Owen et al., 1988). These tools are intended primarily for crosswell work.

A recent report compares the downhole performance of chemical explosives with a perforating gun, an air gun, and a water gun in both open and cased holes and in various lithologies (Chen et al., 1990). An array of 32 small explosive charges separated by 10 ft was suspended below an electronic control module. Each charge was attached to a cable as the array was lowered into the well so that no wired charges were stored above ground. The experiment demonstrated that controlled explosives in a well generate a more desirable signal with broader bandwith useful over larger distances than nonexplosive sources.

An attractive proposed source manufactures the explosive material in the borehole as needed. This source will be developed in a joint project proposed to the Crosswell Seismic Forum by Exxon Production Research and Los Alamos National Laboratories (Exxon and Los Alamos National Lab., 1990). This source is a repeating binary chemical device that will not require special cabling or clamping. The design uses the "slapper" detonating technology developed at Los Alamos. The borehole tool will manufacture and detonate a controlled amount of explosive at a rate of one explosion every 2–5 s. The amount of explosive will be adjustable to avoid damaging the wellbore casing or cement bond. An upper chambered subassembly will house sufficient chemical components to produce thousands of explosions (Figure 2). Plans are to inject the explosive mix into the borehole fluid between the upper chemistry assembly and a lower

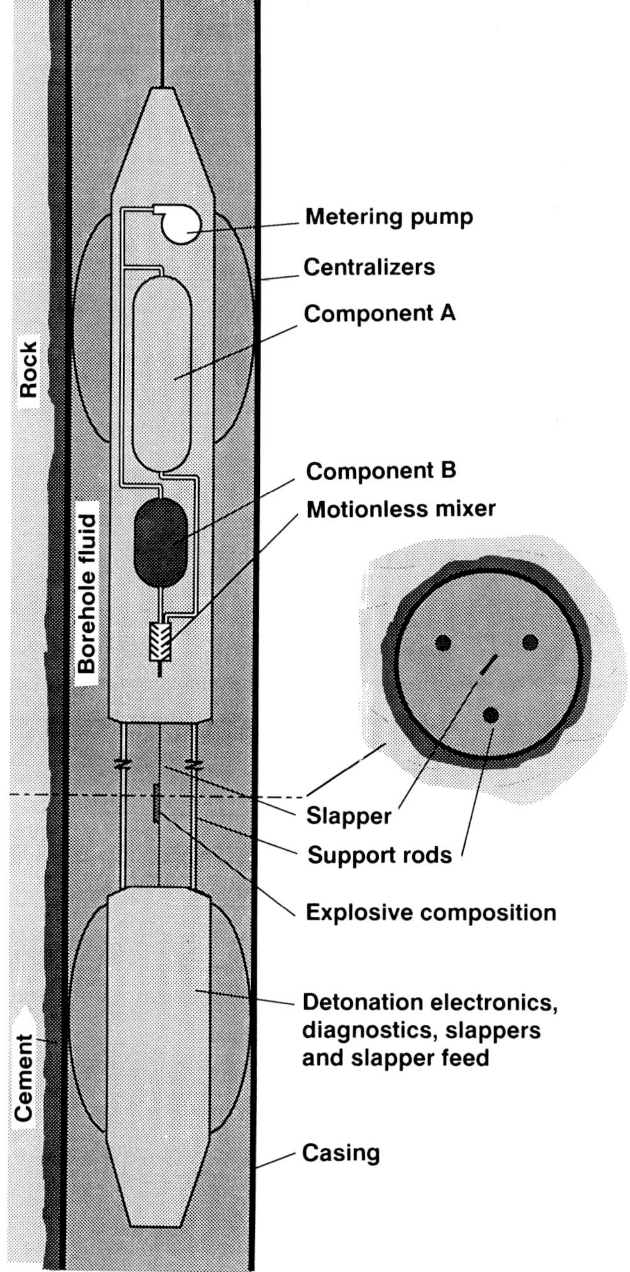

Fig. 2. Proposed borehole source using slapper technology.

detonating assembly, the two assemblies being rigidly connected by three supporting rods. The lower unit will house a roll of "slapper detonators" that will be fed into the open space somewhat like a roll of caps in a toy cap gun. The slapper uses electrical power concentrated at a rapidly bursting conductor bridge to propel a thin disc at high velocity into the nearby explosive charge. Because the explosive is manufactured at the rate at which it is used, no more than a couple of grams will be available for detonation at any one time. The explosive is soluble in water and oil so that undetonated explosive will be absorbed by the drilling fluid and be desensitized quickly if it is not fired. Primary explosives are not used and accidental detonation is unlikely since a high-power electrical pulse is required to detonate the slapper. The tool is designed to work on a standard seven-conductor wireline. Charge size will be accurately controlled and range from 1 to 10 grams, representing 1.5 to 15 kJ of energy. The tool will be in continuous motion and fire at 2–5 s intervals. More than 10 000 one-gram equivalent charges will be contained in the tool. Casing deformation can be controlled to less than the API specification for casing drift. The two chemicals are stable to 135°C at pressures up to 6500 psi. The proposed research will test the repeatability and firing rate of this tool in the wellbore environment.

A repeatable, controllable explosive device for downhole seismic studies represents a giant leap forward. The bandwidth of explosives exceeds that of other seismic sources. By keeping the energy output at a low nondestructive level, recordings of successive "pops" can be summed for signal enhancement and random-noise attenuation.

Recording While Drilling

Two additional borehole sources are designed to be used while the drill string is still in the hole. One of these (Rector et al., 1988), utilizes the drill bit noise as the seismic signal. When the surface trace is correlated with a "pilot" signal that travels up the drill string, the result is similar to a conventional VSP trace. This source is being marketed under the trade name Tomex™ by Western Atlas International, Inc. Published records show a favorable comparison between Tomex results and walkaway VSPs. Users of Tomex report privately that signal quality, although good, varies with the hardness of the rock being drilled.

Another downhole source, the Seismic Telemetry Logging tool (STL), is in experimental development by Klaveness Research Co. (Klaveness, 1989). The controlling mechanism is in a section of the drill collar located just above the drill bit. The source is impulsive and is armed and fired remotely without tripping the bit out of the hole. Klaveness calls this device a "pinger." Drilling must be stopped (such as when adding a pipe joint) during recording; the source is then activated and "fired" with a cycle time of about 20 s. Several "pings" can be recorded from the same depth for stacking. Both the STL and Tomex sources are entirely mechanical requiring no electronic components in the wellbore. No reflection data from the STL source have been published.

Reverse VSP

The future for reverse VSP is exciting. With a large areal array of recording stations on the surface we should achieve high-resolution images of the subsurface around the well, including reflectors below the bit, while drilling. A prestack migration algorithm running on a parallel processing workstation could update a seismic image as a new VSP record is made. The greatest challenge is to improve the energy level and bandwidth of downhole sources.

In addition to prestack reflection imaging, reverse VSP enables tomographic reconstruction of direct-wave traveltimes to provide a migration velocity field which will enhance computed images. An order of magnitude improvement in spatial resolution can be expected over conventional 3-D surface-to-surface surveys. Thus, reverse VSPs should produce real-time, high-resolution, 3-D velocity and acoustic-impedance logs prior to running other logs.

Full-Waveform Logging Tools

The number of sources and receivers in full-waveform logging tools is increasing. Downhole arrays of receivers and longer spacings allow deeper penetration into formations. The range of source frequencies is also expanding. The possibility of direct measurement of Stoneley waves for permeability determinations is especially intriguing. Seismic waves in well bores excite fluid motion in fractures and pore spaces that intersect the well and affect Stoneley wave propagation. Stoneley wave attenuation due to energy dissipation in fractures has been observed in laboratory studies (at ultrasonic frequencies up to 250 kHz). The studies relate observed attenuation to the well radius, fracture size, and the lithologic properties of the medium (Tang et al., 1989).

Surface Seismic Methods: Exploration 3-D

Reduced costs for 3-D seismic services will encourage their use for exploration (exploration 3-D). Some contractors offer 3-D surveys using 480 fixed receiver channels on land for less than $25 000/mile^2, including processing. The current average rate of $50 000/mile^2 using rolling CMP swath techniques still limits use.

S-Wave Data

S-wave studies using a horizontal vibrator source have been used to identify lithologies with some success. One obstacle to inversion is our inability to generate all three components of particle motion in one pass of a seismic source. Lithology inversion requires high quality data and presently the best *S*-wave data requires sources and receivers to be below the weathering zone. While no near-term solution is in sight, three-component receiver stations with "onboard" electronics that digitize and multiplex onto one transmission channel are being marketed (Gassaway, 1990). The motivation to use these devices depends on their cost and developments in how to deal with the data.

Three-component geophones permit higher resolution, both on the surface and downhole. Both types of measurements are needed for lithology mapping. Increasing the fold and the number of channels, coupled with the use of telemetry and denser seismic grids, can only improve the information derivable from seismic data.

S-Wave Birefringence Fracture Studies

S-wave measurements are being used to quantify fracture orientation and density. *S*-waves split into two events of different polarizations traveling at different velocities while traversing an anisotropic medium. The degree of *S*-wave splitting correlates to fracture intensity, while the orientation of polarization relates to fracture orientation (Crampin and Lynn, 1987; Davis and Lewis, 1990; Ebrom, 1990). The contribution by Ebrom and Sheriff in this chapter gives more details of the method.

Full-Waveform (Vector) Recording

Using single-station, three-component geophones instead of arrays of vertical geophones can simplify field procedures. Shorter station intervals will be required to resolve surface noise interference problems. This will generate more data, result in higher processing costs, and, as a reward, yield higher resolution. Compensating decreasing computer costs, higher data I/O rates, and reduced data storage costs will likely keep expenses constant while subsurface resolution is increased.

The conversion of *P*-wave to *S*-wave energy at reflection boundaries can be used to infer differences in Poisson's ratio between the lithologies on either side of reflecting surfaces; this is the basis for amplitude variation with offset (AVO) studies and is used to indicate the presence of gas.

The complexity of dealing with several reflected waves that travel at different velocities can be overwhelming. Incremental progress is being made in unraveling these phenomena. VSP and crosswell surveys use three-component receivers to record the full vector wavefield. Many companies are pursuing research and public reports are now emerging. A 3-D seismic survey utilizing *P*- and *S*-wave vibrators, multicomponent receivers, and a three-component VSP tool was conducted by the Reservoir Characterization Project at the Colorado School of Mines in the Silo Field in Wyoming. The survey detected fracture variations that correlated with production from the Cretaceous Niobrara Chalk at a depth of 2400 m. Structural features in the Niobarra are known to relate to increases in fracture density and correlate with *S*-wave splitting and polarization. A report on this project by Lewis, Davis, and Vuillermoz is included in this chapter.

Full waveform recording and processing will undoubtedly reveal more subsurface information and will thus impact porosity, permeability, and fluid content measurements. However, we have some way to go before the practical methods and theoretical underpinnings are in place to exploit full-wavefield seismology.

3-D Time-Lapse Data

Reservoir monitoring using time-lapse (or differential) seismology involves comparing 3-D surveys spaced in calender time (Greaves and Fulp, 1987). Differencing data from successive surveys suppresses the effects of geologic heterogeneities and focuses on changes such as those due to fluid flow. Successive snapshots and "difference" volumes, sections, and slices have been used to map EOR processes. Monitoring of temperature changes as a result of steam and fire flooding has been discussed in Chapter 6. This technique may be useful to monitor gas-cap volumes, particularly the blowdown of gas caps that typically occurs at the end of an oil reservoir's economic life. Changes in seismic velocities associated with temperature change have also been used in tracking cold injection fluids using tomographic cross-well seismic methods (Nur, 1984). We may be able to use seismic measurements to monitor the pressure and temperature variations associated with injection and production programs. Differential seismology may also permit measurements of changes in in-situ stress distributions over the life of a reservoir.

Computer Developments: Silicon Chips

The silicon chip is sometimes credited as the singular and most significant invention in our lifetime. A microchip combines millions of components that operate in a few billionths of a second. Silicon chip technology continues to deliver astounding improvements in computational efficiencies as well as larger and cheaper computer memories without increasing their cost. As chips become smaller they not only run cooler and faster, but as we approach the scale of electron interaction, the number of accidents drops, defects decline, and nothing wears out. Improved performance through architectural changes in chip design has been a key issue in reduced instruction set computers (RISC architectures). The industry's confidence in designing and building new chips is trending toward application- specific computers that discard the excess baggage of general-purpose logic in favor of silicon compilers designed for special applications. Thus it is becoming possible to harness the power of supercomputers in special-purpose workstations.

Parallel Processing

Parallel-computing architectures are well suited to the simultaneous and repetitive tasks in seismic data processing. A variety of architectures and experimental systems are being introduced, ranging from several to a very large number of processing elements. Two classes of machines, single-instruction stream/multiple-data stream (SIMD) and multiple-instruction stream/multiple-data stream (MIMD) are emerging. In SIMD machines each processor and its associated memory module receives the same instruction set from a central controller. All processing elements operate on different data from independent data streams. Various static or dynamic interconnections, or topologies, between processing elements are possible. In MIMD machines processors can run independent programs and are connected to either shared or distributed memory through high-speed routing networks. Processing interaction in a MIMD machine is often synchronized by a central controller.

The processing speeds possible with parallel computer processors can surpass that of supercomputers at a significant fraction of their cost. A major bottleneck to harnessing this power, however, is the currently low data-transfer rate, or bandwidth, between critical input/output components of a processing system. Processing power can leave computers "data-starved" because the input rates to the process flow can't keep up with the processors.

Processor speeds have steadily increased nearly a thousand-fold during the last twenty years while improvement in tape I/O rates have barely increased by one order of magnitude. It takes about 1 ms to read or write a one-second segment (250 samples) of a seismic trace to or from tape.[2] A moderate-sized 3-D data set (360 lines of 375 6-s traces, 1500 samples/trace, or about 810 MB) on 6 tapes requires about 14 minutes active tape read time to transfer. With an acquisition fold of 60, on 360 tapes, 15 hours is required to read this data set on one tape drive. Faster magnetic disks have evolved to become the primary medium for on-line seismic data storage as processors outperformed the I/O capabilities of multiple tape drives. As

disk system performance began to lag increasing processor speeds, DRAM (dynamic random access memory) chips, the standard for internal computer memories, assembled in large-capacity mass-memory units (several gigabytes of external DRAM), emerged to feed supercomputers. Each of these storage technologies lead to increased data densities. Following this trend another order of magnitude improvement in memory densities is predicted for this decade (National Research Council, 1988).

Although variations of semiconductor, optical, and molecular-engineered technologies are being evaluated as alternatives, the dominant medium for off-line data storage is still the magnetic tape. A logical extension to rapid on-line data storage is the vertical recording technique. The retrieval speed of magnetic disks is limited by the plater speed. A number of conventional disks, eight for example, might record each bit of an eight-bit byte simultaneously. (The read/write rate of such a unit is eight times that of a single disk). Additional gains can be had by breaking up a block of data into multiple blocks and then writing each of these to a separate device simultaneously. This is known as "disk striping." These methods have a multiplying effect on improving transfer rates and offer a short-term, though expensive, solution.

Once the data transfer problem is solved, the power of parallel computing will begin to impact methods of seismic data processing. The first truly interpretive processing will likely combine striped-disk systems feeding parallel processors that are ported to a VIDEO-RAM cache to monitor the progress of seismic imaging in response to real-time user-controlled parameter decisions. Bits and pieces of the needed software are emerging and new algorithms will undoubtedly replace the old batch-oriented serial processing routines.

Workstations and Interpretive Processing

Seismic data processing today is a sequence of batch processes that is frequently interrupted to insert parameter decisions based on previous results. As computers become faster, memory cheaper, and data storage larger, seismic processing will become a geologic interpretation exercise with real-time visual feedback to guide parameter decisions. The decreasing costs of parallel computer architectures will enable true human-in-the-loop data processing. The current complexities of seismic data processing will melt away to become intuitive procedures for the interpreter. When interpreters become the "consumers" of tomorrow's computer technologies, they will make the fullest use of their workstations. The subsurface interpreter in the year 2000 will not have to be a computer scientist or a data processing expert.

Our current use of workstations is most often a mere adaptation of interpretation technology that was ap-

---

[2]A standard 2400 ft tape reel at 6250 bpi density holds approximately 160 Mbytes of formatted seismic data. At 200 ips a tape drive will operate at nearly 1 MB/s.

plied to 2-D paper-sections in the past. Our goal then was to "make a map." Conventional interpretation emphasized picking reflecting horizons, faults, or sedimentary features on 2-D displays. With the advent of scientific visualization tools, interpreters of tomorrow will have visual access to data "out of the plane of the section" while examining data "in the section." In just a few years the interpreter will visually navigate through an unpicked 3-D seismic data volume using methods that are just as comfortable as examining a 3-D object held in one's hand. The "object" might be an enlarged portion or an entire 3-D data set. Currently available data gloves can monitor the orientation and movements of one's fingers while a computer draws the same hand and allows it to interact with other data on a CRT screen (Lanier, 1989). This method is superior to using a 2-D joy stick or a mouse for interacting with a computer. A new Nintendo game allows a player to put on a "power glove" and reach into the storyline on the screen to "punch it out" with an opponent boxer (Birmingham, 1990). Data selection and manipulation tools borrowed from today's researchers in "artificial reality" will give interpreters access to not only examine and manipulate 3-D data but also to leave a quantitative trail of interpretations that synthesizes the complex data set into a manageable subsurface model.

Tom Furness at the University of Washington is exploring the notion of an air traffic controller using his hand to reach into a visual 3-D space where he can "touch" the 3-D icon of an airplane to establish radio communication with the pilot. Furness, who used to build head-mounted displays for fighter pilots, believes that within this decade "full-motion laser imagery as sharp as a photograph will be projected directly onto a person's retina, doing away with display screens altogether" (Stewart, 1991). Just a little of this emerging visualization technology can make a significant difference in the way we interact with 3-D seismic data.

Shared Data Bases

Perhaps the single most significant emerging technology, the one that will have the most lasting impact, is the formation of interdisciplinary teams who have both the motivation and the means to apply their different skills to map, describe, and monitor a reservoir. Integrated teamwork requires a shared data management and visualization system that can be accessed by each member of the team. Geophysicists, geologists, and engineers already use desk-top computers and workstations specialized for their profession's applications; their interpretations must be shared with the team, who must have easy access and the ability to exchange interpretations. Systems are becoming hardware transparent.

As computer hardware and operating systems become more sophisticated the user interface to these must become simpler and "friendlier." The consumer's first introduction to a user-friendly interface was the Macintosh from Apple Computer, Inc. The point-and-click mechanical "mouse" and the Macintosh "look and feel" set a standard that many other computer companies have since emulated. The UNIX operating system is emerging as a standard that promises to free the user from hardware pecularities. The importance of a well designed graphics user interface (GUI) to integrate the work between professionals cannot be over emphasized. Opportunities for innovation in describing and integrating reservoir interpretations will expand significantly once acceptable GUIs emerge.

Interpretation Improvements: Porosity/Permeability Mapping

Attempts to relate porosity to permeability have had mixed success. In select instances porosity variations have been correlated to seismic response. Seismic data allow mapping relationships away from a well. Additional information about geologic structure, stratigraphy, and formation diagenisis will improve determinations of seismically mapped porosity.

To correlate porosity with permeability, we must measure fluid movement. Time-lapse seismic studies performed at intervals of six months to several years holds the best hope in tracking fluid movements. Temperature changes caused by injected fluids influence seismic velocity and can be used to map changes over periods of time.

Amplitude Versus Offset Seismic Response

The rock properties that can be extracted from zero-offset seismic data are (1) subsurface geometry and (2) the gradients of P-wave impedance (Backus, 1987). The observation made in the 1970s that gas accumulations can produce bright spots called attention to the forgotten fact that P-wave reflectivity involves density as well as velocity contrast. A paper at the 1982 SEG convention (Ostrander, 1984) focused attention on how the amplitude of reflected P-waves varies with incidence angle and is related to the gradient of S-wave impedance. For incidence angles up to about 30 degrees it is possible to detect anomalously low Poisson's ratios related to gas-bearing rocks. Poisson's ratio, density, and seismic velocities are all related to lithology, pore-fluid content, depth, and age of a rock. Bright spots are now routinely analyzed using AVO (amplitude versus offset) analysis. AVO analysis when tied to well information has enabled interpreters to extend lithology information away from a well with more confidence.

Conclusion

Seismic data processing attempts to image the output of a few receivers on the surface of the earth to reconstruct the position and characteristics of subsurface reflectors. The imaging algorithms are designed to reconstruct specular reflections, i.e., they require that we can image the source as it would appear on the shiny surface of a geologic reflector. If a subsurface element is not properly "illuminated" by the source then it cannot be "seen" by the receiver. The amplitude of a diffuse reflector is not likely to survive the noise levels encountered in 2-D data processing. However, the 3-D seismic method may be more forgiving if we know the velocity structure of the subsurface before we begin imaging. By combining the transmission properties of the earth derived from tomographic imaging techniques we may be able to reinforce the validity of an interpretation model.

Since digital recording and data processing were introduced to the geophysical industry our technology has evolved rapidly. We've been forced to continuously re-examine the principles and methods that we use only to discover that we can do better. Geophysics applied to reservoir development provides a unique opportunity to see predictions in action and re-examine results in light of new knowledge. We're suddenly presented with more subsurface control than we've ever had, and the opportunity of integrating a diverse variety of physical data into our technology. There will be many more questions than answers at first. It is safe to predict that the interpretation skills of an integrated exploitation team will feed back into each of the contributing professions far more wisdom than can be generated if we were to work in isolation.

References

Backus, Milo M. ,1987, Amplitude versus offset: A review: 57th Ann. Internat. Mtg., Soc. Expl. Geophys., Expanded Abstracts, 359–364.

Balogh, W.T., Owen, T.E., and Harris, J.M., 1988, New piezoelectric transducer for hole-to-hole seismic applications: 58th Ann. Internat. Mtg., Soc. Expl. Geophys., Expanded Abstracts, 155–157.

Birmingham, J., 1990, Boys and their toys: Continental Profiles, June 1990, 30 + 8 (Continental Airlines complimentary passenger magazine).

Chen, S.T., Ericksen, E.A., and Miller, M.A., 1990, Experimental studies of downhole seismic sources: Geophysics, 55, 1645–1651.

Crampin, S and Lynn, H., 1987, Shear wave VSP's: A powerful new tool for fracture and reservoir description: Soc. Petr. Eng. Mtg., SPE Paper 16866.

Davis, T., and Lewis, C. , 1990, Reservoir characterization by 3-D, 3-C seismic imaging, Silo Field, Wyoming: The Leading Edge, 9, no. 11, 22–25.

Doyen, P.M., 1988, Porosity from seismic data: A geostatistical approach: Geophysics, 10, 1263.

Ebrom, D, 1990, Acquisition and interpretation of multicomponent seismic data in isotropic and azimuthally anisotropic media: PhD Thesis, Univ. Houston.

Exxon and Los Alamos National Lab., 1990, Binary explosive seismic source development: A cooperative proposal to Crosswell Seismic Forum.

Fogg, G. G., and Lucia, F. J., 1990, Reservoir modelling of restricted platform carbonates: Geologic/geostatistical characterization of interwell-scale reservoir and heterogeneity, Dune Field, Crane County, Texas: Report of Investigation No. 190, Bureau of Economic Geology, Univ. Texas at Austin.

Gassaway, G.S., 1990, Three-component particle motion filter: 60th Ann. Internat. Mtg., Soc. Expl. Geophys., Expanded Abstracts, 944–947.

Gilder, G., 1989, Microcosm: the quantum revolution in economics and technology., Simon & Shuster.

Greaves, R.J., and Fulp, T.J., 1987, Three-dimensional seismic monitoring of an enhanced oil recovery process: Geophysics, 52, 1175–1187.

Harris, J.M., 1988, Crosswell seismic measurements in sedimentary rocks: 58th Ann. Internat. Mtg., Soc. Expl. Geophys., Expanded Abstracts, 147–150.

Kennedy, W., Wiggins, W., and Aronstam, P., 1988, Swept frequency borehole source for inverse VSP and cross-borehole surveying: 58th Ann. Internat. Mtg., Soc. Expl. Geophys., Expanded Abstracts, 158–160.

Klaveness, A., 1989, Emerging technology in borehole geophysics with multiple applications in drilling, production, exploration, and enhanced recovery: SEG 5th Ann. Gulf Coast Expl. Develop. Mtg. Expo.

Lanier, J., 1989, Virtual environments and Interactivity: windows to the future: Panel Session at SIGGRAPH '89 (American Association of Computing Machinery Special Interest Group on Computer Graphics) in Computer Graphics, 23, no. 5, 7–18.

Lindseth, R., 1990, The new wave in exploration geophysics: Geophysics: The Leading Edge, 9, no. 12, 9–15.

Lo, T.W., Inderwiesen, P.L., Howlett, D.L., Melton, D.R., Livingston, N.D., Paulsson, B.N.P., Fairborn, J.W., 1990, The McKittrick Cross-Well Seismology Project, Part I. Data Acquisition and Tomographic Imaging: 60th Ann. Internat. Mtg., Soc. Expl. Geophys., Expanded Abstracts, 30–33.

National Research Council, 1988, Global trends in computer technology and their impact on export control.

Nur, A. M., 1984, Seismic monitoring of a thermal enhanced oil process: 54th Ann. Internat. Mtg., Soc. Expl. Geophys., Expanded Abstracts, 337–340.

Ostrander, W.J., 1984. Plane-wave reflection coefficients for gas sands at nonnormal angles-of-incidence: Geophysics, 49, 1637–1648.

Owen, T.E., Balogh, W.T., and Peters, W.R., 1988, Arc discharge pulse source for borehole seismic applications: 58th Ann. Internat. Mtg., Soc. Expl. Geophys., Expanded Abstracts, 151–154.

Paulsson, B.N.P., 1988, Three-component downhole seismic vibrator: 58th Ann. Internat. Mtg., Soc. Expl. Geophys., Expanded Abstracts, 139–142.

Paulsson, B.N.P., 1990, Proposal for the development of a three-component vibratory seismic source, by Chevron Oil Field Research Company, Sandia National Laboratories, and Los Alamos National Laboratories to the Crosswell Seismic Forum.

Paulsson, B.N.P., Fairborn, J.W., Cogley, A.L., Howlett, D.L., Melton, D.R., Livingston, N., 1990, The McKittrick Cross-Well Seismology Project, Part I. Data Acquisition and Tomographic Imaging: 60th Ann. Internat. Mtg., Soc. Expl. Geophys., Expanded Abstracts, 26–29.

Rector, J.W., Marion, B.P., and Widrow, B., 1988, Use of drill-bit energy as a downhole source: 58th Ann. Internat. Mtg., Soc. Expl. Geophys., Expanded Abstracts, 161–164.

Sandia and OYO Geospace, 1990, Advanced borehole receiver project. Crosswell seismic forum of the oil recovery technology partnership.

Santerra, Bolt, and Sandia, 1990, Proposal for seismic source development by Santerra Corporation, Bolt Technology Corporation, and Sandia Laboratories: Proposal to the Crosswell Seismic Forum.

Stewart, D, 1991, Through the looking glass into an artificial world—via computer: Smithsonian, **21**, No. 10, 36–43.

Tang, X.M., C.H. Cheng,, and M. N. Toksöz, 1989, Stoneley wave propagation in a fluid-filled borehole with a vertical fracture: 59th Ann. Internat. Mtg., Soc. Expl. Geophys., Expanded Abstracts, 30–32.

Thadani, S.D., Alabert, F., and Journel, A.G., 1987, An integrated geostatistical/pattern recognition technique for characterization of reservoir spatial variability: 57th Ann. Internat. Mtg., Soc. Expl. Geophys., Expanded Abstracts, 372–375.

Uzcategui, O., and Del Pino, E., 1987, Seismic inversion for mapping porosity and thickness by using SVD and geostatistical methods: 57th Ann. Internat. Mtg., Soc. Expl. Geophys., Expanded Abstracts, 369–371.

Wolf, D.J., 1990, Mathematics in geostatistics: A tutorial for geoscientists without the Ph.D. in mathematics (Geostatistics for poets): 60th Ann. Internat. Mtg., Soc. Expl. Geophys., Expanded Abstracts, 336–338.

Anisotropy and Reservoir Development

Dan Ebrom and R. E. Sheriff[‡]*

Fracturing is an important factor in fluid flow in many hydrocarbon reservoirs. Knowledge of the orientation of fractures is necessary to locate production and injection wells so that hydrocarbons are drained efficiently, and to avoid premature breakthrough of flushing fluids and bypassing of hydrocarbons. Fracturing affects seismic traveltimes and causes velocity anisotropy, i.e., differences in seismic velocity when measured in different directions. Measurements of anisotropic velocity effects may offer a means of determining fracture intensity and orientation, and thus permeability anisotropy.

The effect of anisotropy on P-wave traveltimes is usually small and may not be distinguishable from heterogeneity effects, but for S-waves (shear waves) it is apt to be distinctive. S-waves exhibit splitting (birefringence), that is, dependence of velocity on the direction of polarization, a phenomenon which is not caused by heterogeneities. The degree of splitting gives a rough measure of the fracture intensity, and the polarization of the faster S_1-wave is generally parallel to the strike of the fractures. If all three components of particle motion are recorded, then the directions of the principal axes can be determined by coordinate rotation and used to infer fracture orientation. Reflection strength differences between the events for fast S-waves (S_1-) and slow S-waves (S_2-) may give direct indications of zones of fracture intensity. In concept, fracture orientation can also be inferred from SH-wave velocity analysis if lines are acquired at different azimuths.

Introduction to Anisotropy

Anisotropy is a general term referring to changes in physical property measurements resulting from changes in either direction or polarization. Seismic velocity anisotropy is a common characteristic of many earth materials, but it is generally neglected in seismic surveying. Most reflection work is done near vertical angles of incidence, where the effects of P-wave stratigraphic (layer-induced) anisotropy are

difficult to differentiate from other complicating factors. However, fracture-induced S-wave splitting is apt to be most easily detected for near-vertical raypaths, and so surface seismic holds promise as an appropriate tool for detecting this anisotropy.

A number of different types of anisotropy can be derived from first principles. Depending upon how they are grouped, eight different types of anisotropy, generally named after equivalent crystal types, are possible (Figure 1). At seismic wavelengths, however, the only anisotropy types reported thus far are transverse isotropy (hexagonal anisotropy), orthorhombic anisotropy, and monoclinic anisotropy. The two important types of transverse isotropy are (1) that with a nearly vertical symmetry axis (''polar anisotropy'') and (2) that with a nearly horizontal axis (''azimuthal anisotropy''). Orthorhombic anisotropy is equivalent to a superposition of polar anisotropy and azimuthal anisotropy. Monoclinic anisotropy can be produced by superimposing tilted fractures on a layered medium (Schoenberg and Muir, 1989). The stress-strain relationships for transverse isotropy require five independent elastic moduli (sometimes called ''stiffness'' or ''compliance'' components), nine for orthorhombic anisotropy, and thirteen for monoclinic anisotropy, compared with only two for the isotropic case.

Polar Anisotropy

Polar anisotropy (or ''layering anisotropy'') is the strongest anisotropy commonly encountered. It arises from layering, which can occur on many different scale lengths. If there are more than roughly 10 layers in one wavelength, a layered medium acts as a homogeneous anisotropic unit (Melia and Carlson, 1984; Helbig, 1984; Ebrom et al., 1990b, Carcione et al., 1991).

At the individual crystal scale, stresses cause preferred orientations of crystal growth and realignment with velocity anisotropy consequences (Silver, 1988). At the clastic grain scale, platey clay fragments (preferentially aligned by gravity during deposition) produce anisotropic shale rocks (Jones and Wang, 1981). At the sedimentary scale, interleaved sand, silt, and shale units produce anisotropic formations. Horizontal velocity usually exceeds vertical velocity.

*Allied Geophysical Laboratories, University of Houston, Houston, TX 77204-4231.
[‡]Geosciences Dept., Univ. Houston, Houston, TX 77204-5503.

356 **Ebrom and Sheriff**

The percent difference between the largest and smallest velocities is a common measure of anisotropic intensity. Laboratory *P*-wave anisotropy has been reported as high as 20 percent (Jones and Wang, 1981). *S*-wave anisotropy can be as large as 40 percent in laboratory specimens (ibid). Winterstein (1986) interpreted shear anisotropies of 25 percent from *S*-wave stacking-velocity/depth-conversion anomalies at seismic frequencies. *P*-wave stacking-velocity/depth-conversion anomalies are much smaller than for *S*-waves (ibid.). Polar anisotropy can distort structural mapping beneath massive shale units (Banik, 1984). More accurate measures of anisotropy, such as the "Thomsen parameters" ε, δ, and γ (Thomsen, 1986b), are better quantitative predictions of velocities at intermediate angles.

Azimuthal Anisotropy

Azimuthal anisotropy is the anisotropy type of special interest for reservoir engineering. In its simplest form, azimuthal anisotropy is mathematically identical to polar anisotropy, except for a symmetry axis rotation (Figure 2). Azimuthal anisotropy is most commonly caused by the preferred orientation of mechanical flaws (microcracks or fractures) and, in tight petroleum reservoirs, these flaws may control reservoir permeability. In an azimuthal sense, the horizontal symmetry axis is perpendicular to the more-or-less vertical fracture orientation.

Lynn and Thomsen (1990) interpreted a 17 percent difference between vertical S_1 and S_2 velocities in a seismic experiment over a fractured unit in Pennsylvania.

Brodov et al. (1990) observed a positive correlation between the magnitude of *S*-wave azimuthal anisotropy and production in a fractured unit. Vertical birefringence of 4 percent was associated with poor production (3 m^3, 20 barrels, of oil per day) whereas birefringence of 7 percent was associated with better production (14 m^3, 100 barrels, of oil per day). Brodov et al. derived their anisotropy estimates by combining multicomponent surface seismic with multicomponent walkaway VSPs. (VSP data are more likely than

SYSTEM	UNIT CELL	ANGLES	EDGES	ELASTIC CONSTANTS
TRICLINIC		$\alpha \neq \beta \neq \gamma \neq 90°$	$a \neq b \neq c$	21
MONOCLINIC		$\alpha = \gamma = 90°$ $\beta > 90°$	$a \neq b \neq c$	13
ORTHORHOMBIC		$\alpha = \beta = \gamma = 90°$	$a \neq b \neq c$	9
TRIGONAL		$\alpha = \beta = \gamma \neq 90°$	$a = b = c$	7
TETRAGONAL		$\alpha = \beta = \gamma = 90°$	$a = b \neq c$	7
HEXAGONAL		$\alpha = \beta = 90°$ $\gamma = 120°$	$a = b \neq c$	5
CUBIC		$\alpha = \beta = \gamma = 90°$	$a = b = c$	3

Fig. 1. Unit cells of different crystal classes and the number of elastic constants needed to characterize an anisotropic velocity field (far right-hand column). Isotropy (not shown) requires only two elastic constants, the Lamé constants λ and μ.

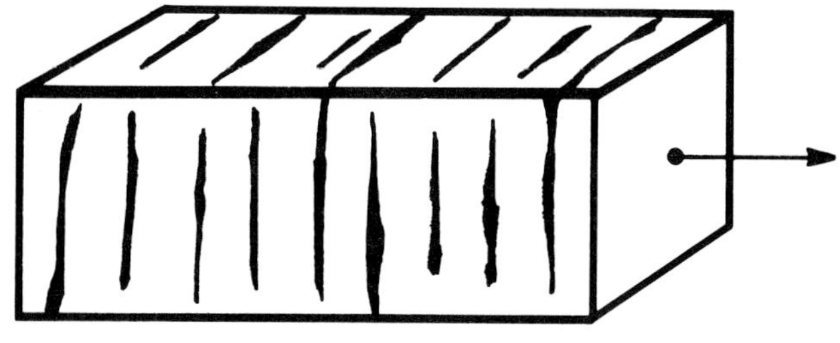

Azimuthal Anisotropy

Horizontal axis of symmetry

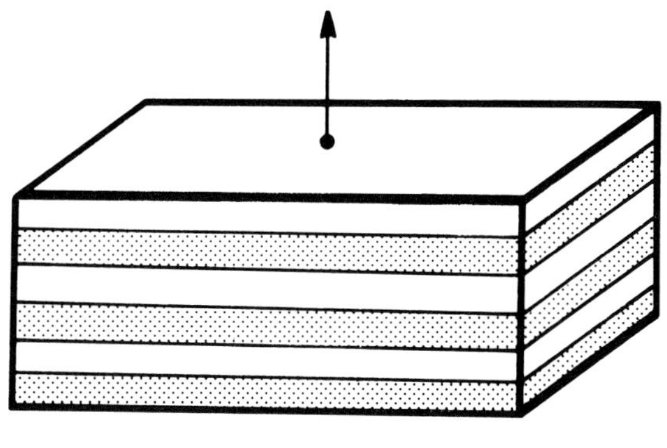

Polar Anisotropy

Vertical axis of symmetry

Fig. 2. Geologic systems producing azimuthal anisotropy (above) and polar anisotropy (below). Both are transversely isotropic (requiring five elastic constants), but their symmetry axes differ by 90°. Azimuthal anisotropy relates to fracture permeability, whereas polar anisotropy can be responsible for distortion of mapped horizons.

surface seismic data to encounter a geometrical phenomenon known as the "shear-wave window", which complicates interpretation; see Booth and Crampin, 1985). This positive correlation between oil production and the magnitude of S-wave birefringence is consistent with experiments performed by Tatham et al. (1987), who found that S-wave birefringence was roughly proportional to the number of fractures per unit length.

Currently, seismic measurements cannot distinguish between many small-aperture fractures and a few large-aperture fractures. Schoenberg and Douma (1988) have shown that, at long wavelengths, the compliance added by fractures can be produced by very different combinations of fracture dimensions.

Orthorhombic and Monoclinic Anisotropy

Orthorhombic anisotropy arises because a vertical fracture system has been superimposed upon a horizontally layered system. This is the expected situation in a fractured reservoir, for example. VSP data from the Paris Basin have been successfully interpreted using an orthorhombic model (Bush and Crampin, 1987; MacBeth, 1990). Layering anisotropy is usually much stronger than fracture anisotropy so that the overall effect is difficult to distinguish from polar anisotropy. Examples of monoclinic anisotropy have been observed in the field (Crampin et al., 1980 and Winterstein and Meadows, 1990a).

S-Wave Splitting

In anisotropic media, except for special symmetry cases, the particle motion in the passage of a seismic wave is generally neither parallel nor perpendicular to the propagating ray. Three wavefronts arise: one, with particle motion nearly collinear with the ray, travels fastest; this wavefront is the quasi-P or qP-wave. Two other wavefronts, with particle motions roughly orthogonal to the ray and to each other, travel at slower speeds; these are quasi-S waves. The faster of the two is termed S_1 and the slower S_2. Phase and group velocities differ. It is phase velocity that is given by elasticity theory but group velocity that is involved in traveltime measurements (see Postma, 1955).

Unlike the isotropic case where an S-wave can have any orientation perpendicular to the direction of travel, for a given ray direction in anisotropic media, only two distinct polarizations can be supported. (There are some exceptions in particular symmetry directions.) These two polarizations have different velocities and hence different arrival times if they originated together. This phenomenon is called "S-wave splitting" or "birefringence". The name originates from the observation that an S-wave of arbitrary polarization entering an anisotropic region spontaneously splits into a faster and a slower wave, each in

one of the natural orientations of the media (called the "principal axes").

For transversely isotropic media, the S-wave polarization with particle motion parallel to the symmetry planes has a velocity field that is elliptical in form, that is, the wavefronts have ellipsoidal shape. However, the qS-wave with particle motion perpendicular to fracturing (layering) has a velocity field that is never elliptical (except in the degenerate case of isotropy).

The second S-wave and the P-wave are coupled in a quadratic equation when determining phase velocities (Postma, 1955).

Micro-Crack Models

Elastic theories of azimuthal anisotropy based on microcracks have been developed in Hudson (1982) from earlier work in Garbin and Knopoff (1975). They predict velocity curves that are strikingly different for P-waves when cracks are gas-filled, rather than liquid-filled (Figure 3), but S-wave curves are not qualitatively different. This is not surprising, since the fluid in a microcrack will not affect the rigidity of a rock whereas it will affect the compressibility. Hudson's theory neglects the possibility of pore volumes that interconnect with the micro-cracks. If there is interconnection, then liquid-filled cracks may behave as if they were filled with gas (Thomsen, 1986a).

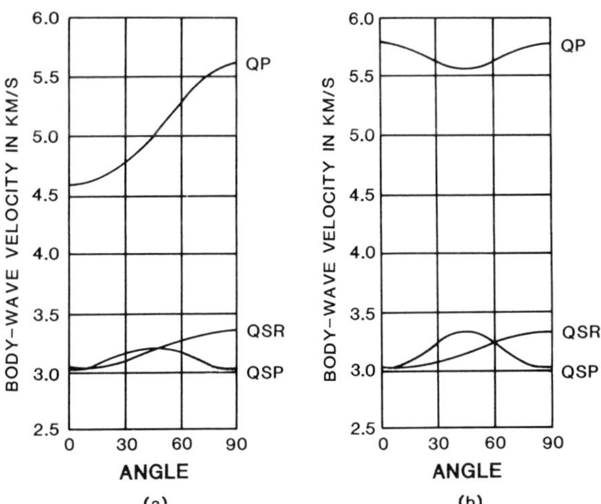

Fig. 3. Velocities of qP- (quasi-P) and the two qS-waves as functions of angle, when micro-cracks are gas-filled (on left) and liquid-filled (on right). qSP is the quasi-shear wave with particle motion roughly perpendicular to the micro-cracks, and qSR parallel to the micro-cracks. (Zero degrees is perpendicular to the fractures and 90° is parallel to the fractures.) Thomsen (1986a) suggests that both gas-filled and liquid-filled micro-cracked mediums will have velocity variations of the form seen on the left if the micro-cracks are connected to pore spaces (after Crampin, 1984a).

How to Measure Anisotropy

S-wave splitting offers a possible means for determining fracture intensity and fracture orientation. The time difference between S_1 and S_2 arrivals depends on the product of the thickness of the anisotropic unit and the intensity of fracturing. The particle motion of the faster *S*-wave yields the direction parallel to fracturing, and the particle motion of the slower *S*-wave yields the direction perpendicular to fracturing.

In order to determine particle motions, it is necessary to record at least two and preferably all three components of particle wave motion. The analysis of 3-D particle motion from three-component recording is called "hodogram analysis". It is essentially a coordinate rotation so that the axes correspond with the principal axes of the medium. The principal anisotropy axes in individual units have been interpreted from multicomponent data (Squires et al., 1989, and Davis and Lewis, 1990).

An alternative to hodogram analysis is to record *SH*-waves on several lines at different azimuths. Both S_1 and S_2 wavefronts will be observed (with different amplitudes) for all azimuths except for those perpendicular and parallel to the fracture orientation, when only one of the *S*-wave types will arrive (assuming a flat reflector, background isotropy, and neglecting out-of-the-plane arrivals). Thus, for two particular line orientations, reflectors beneath an anisotropic member generate only one event (Lynn and Thomsen, 1990).

Another alternative is to record on a seismic line both *SH*- and *SV*-waves from both *SH*- and *SV*-sources. There will be little *SH*-energy from *SV*-sources if the anisotropic principal axes are oriented with the seismic line. Likewise little *SV*-energy will be seen from *SH*-sources. Rotation of sources and receivers (done in processing) so that cross-energy is minimized (Alford rotation) determines the orientations of the principal axes (Alford, 1986; Thomsen, 1988).

In Figures 4 and 5 we see a practical implementation of these Alford rotations. After multicomponent surface *S*-wave data were acquired in Texas (over the Austin chalk), Alford rotation was used to convert an *SV*/*SH* (in-line/crossline) data set into an S_1/S_2 (fast/slow) data set. The data show that only the S_2-section is sensitive to the presence of fractures. Reflector "dim-outs" on the S_2-section correlate with the presence of fractures. The data suggest that surface seismic can be used to directly determine the location of subsurface fracture zones.

Figure 6 shows physical model data examples (Ebrom et al., 1990a). An *SH*-line parallel to fractures will show a wave that traveled at S_2 velocity (because particle motion is perpendicular to the fractures, even though the ray parallels the fractures). An *SH*-line perpendicular to the fractures has S_1 velocity for the near offsets (but at far offsets it tends to the S_2 velocity). The traveltime-offset curve for the line parallel to the fractures obeys the normal-moveout equation whereas other lines exhibit some abnormal moveout due to the angular dependence of the velocity field. Calculation of the velocity over the interval between events above and below the anisotropic unit could be used to determine the fracture orientation independently of hodograms.

An often-mentioned property of *S*-waves is that they "have no memory", i.e., the recorded polarization corresponds to the anisotropy axes of the last region encountered. In spite of this limitation, recent work by

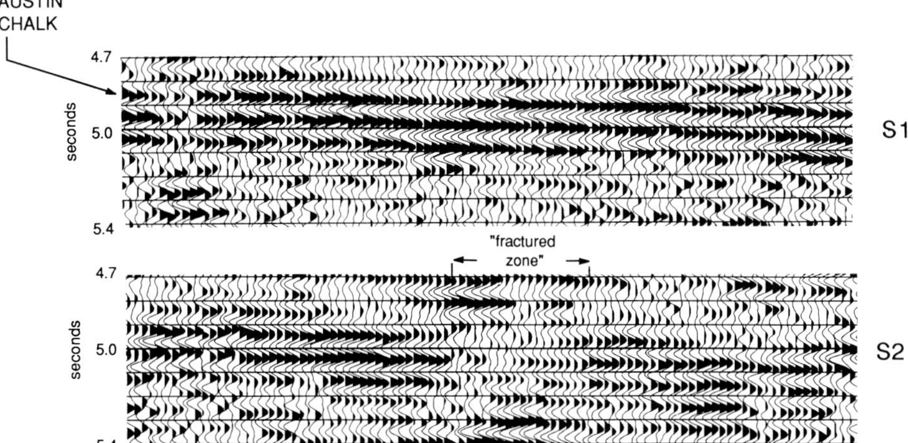

Fig. 4. Fast (S_1) and slow (S_2) *S*-wave sections for a mile of the Austin Chalk in Burleson County, Texas. Note that the top of the chalk is at 4.85 s on the fast section but at 4.9 s on the slow section. The S_1 reflection from the Austin Chalk is continuous across the entire line. The S_2 reflection from the Austin Chalk "dims out" in localized parts of the line, indicating zones of intense fracturing (from Mueller, 1991).

Winterstein and Meadows (1990b) suggests that layer-stripping techniques applied to multicomponent data can recover the *S*-wave polarizations from increasing times, and thus indicate where the stress field orientation changed with depth. Winterstein and Meadows found a major anisotropy axis shift associated with an unconformity separating Pliocene and Miocene units, an example of seismic measurements relating to geologic features.

Research Directions

While it is believed that *S*-wave anisotropy axes relate to permeability anisotropy, rather than to present- or paleo-stress fields, this has not been firmly established. Microcrack-induced anisotropy and frac-ture-induced anisotropy can yield equivalent compliance, and microcracks may not relate to formation permeability. Recent work (Ebrom et al., 1990b) suggests that dispersive effects at frequencies circa 500 Hz may distinguish between these two families of flaws and these frequencies may be achievable in cross-borehole tomographic studies. However, if interest is only in fracture orientation rather than fracture intensity, problems may not arise since microcracks and major fractures are apt to be aligned together.

A horizontal well should be oriented perpendicular to fractures to test noncommunicating fracture zones. Azimuthal anisotropy offers a possible solution to determining the fracture orientation.

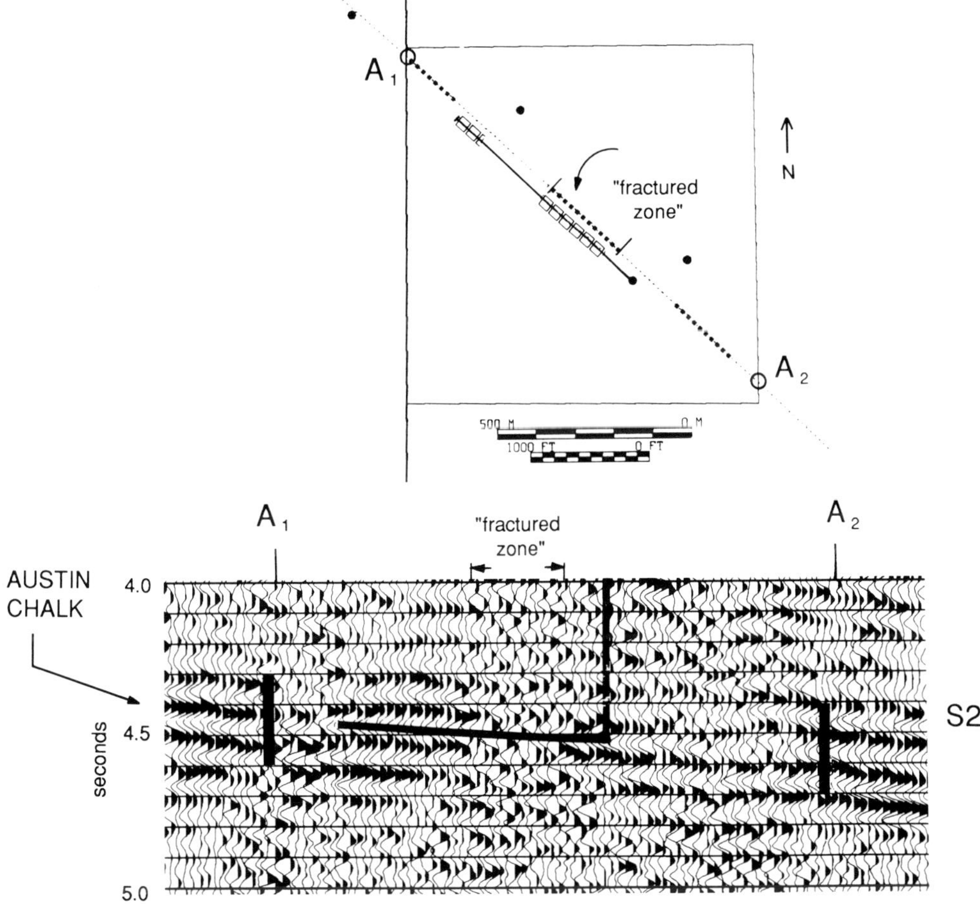

Fig. 5. Slow (S_2) *S*-wave sections across an oil lease in Fayette County, Texas. Points A1 and A2 indicate the limits of the lease on the seismic line. Small black boxes correspond to interpreted zones of Austin Chalk reflector "dim-out" on the S_2 section. A vertical well was drilled and then kicked horizontally for 2000 ft (as indicated at 4.5 s). Analysis of cuttings indicated calcite-filled fractures over the horizontal zones depicted by open squares on the wellbore trace. Note the correlation between surface seismic predictions of fracturing (small black squares) and borehole indications. The horizontal well was drilled less than 500 ft from a vertical borehole that has produced less than 20 000 barrels of oil (50 BOPD initial production); the new well has produced 90 000 barrels to date (500 BOPD initial production), with an estimated production of 150 000 barrels. (From Mueller, 1991.)

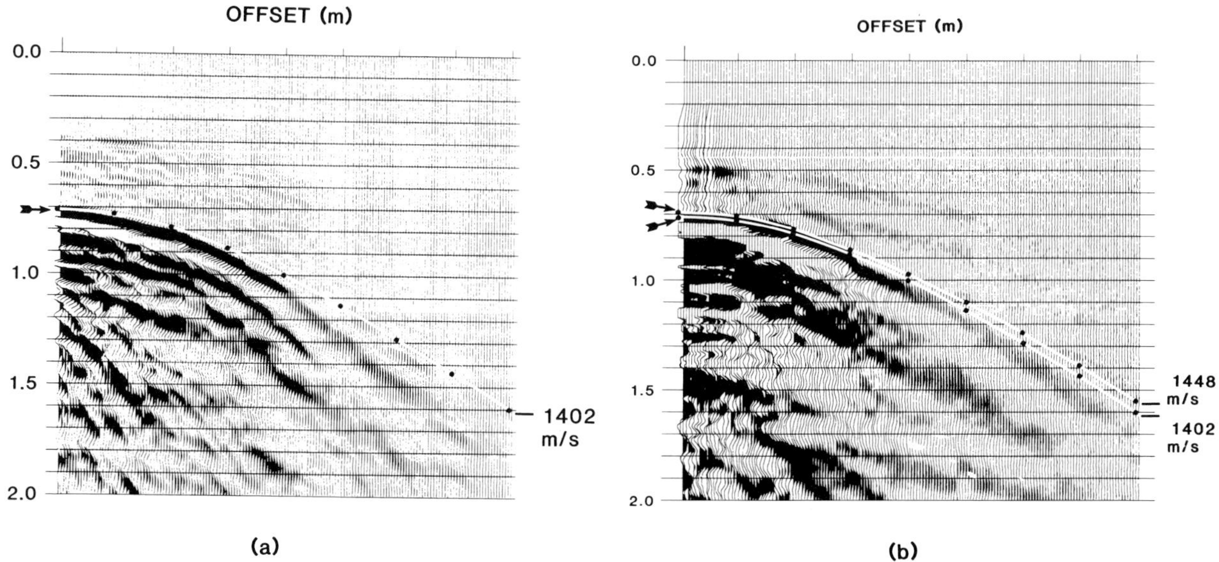

Fig. 6. Physical-model *SH* data acquisition along seismic lines running parallel and perpendicular to a vertical fracture set. (a) Parallel to the fractures; the single NMO velocity is S_2 (1402 m/s); (b) perpendicular to the fractures showing abnormal moveout; the near-offset data give the S_1 velocity (1448 m/s) whereas the far-offset data give the S_2 velocity (1402 m/s).

Acknowledgments

Appreciation is expressed to Stuart Crampin, whose suggestions have been incorporated. Crampin (1989) and Winterstein (1990) have (independently) published glossaries of anisotropic terminology, to which interested readers are directed. Crampin (1981) has also written a brief description of the mathematics involved, which makes a nice companion piece to Postma's (1955) classic tutorial. We also thank Leon Thomsen for his suggestions; Thomsen (1988) offers an intuitive tutorial exposition. Thanks are also due to Mike Mueller for his review and for Figures 4 and 5.

References and General Reading

Alford, R. M., 1986, Shear data in the presence of azimuthal anisotropy: Dilley, Texas: 56th Ann. Internat. Mtg., Soc. Expl. Geophys., Expanded Abstracts, 476–479.

Banik, N. C., 1984, Velocity anisotropy of shales and depth estimation in the North Sea basin: Geophysics, 49, 1411–1419.

Booth, D. C., and Crampin, S., 1985, Shear-wave polarizations on a curved wavefront at an isotropic free surface: Geophys. J. Roy. Astr. Soc., 53, 467–496.

Brodov, L., Kuznetzov, V., Tikhonov, A., Cliet, C., Marin, D., and Michon, D., 1990, Measurements of azimuthal anisotropy parameters for reservoir study: 4th Internat. Workshop on Seismic Anisotropy, Edinburgh, Collected Abstracts, 14.

Bush, I., and Crampin, S., 1987, Observations of EDA and PTL anisotropy in shear-wave VSP: 57th Ann. Internat. Mtg., Soc. Expl. Geophys., Expanded Abstracts, 646–659.

Carcione, J. M., Kosloff, D., and Behle, A., 1991, Long-wave anisotropy in stratified media: a numerical test: Geophysics, 56, 245–254.

Crampin, S., McGonigle, R. and Bamford, D., 1980, Estimating crack parameters from observations of *P*-wave velocity anisotropy: Geophysics, 45, 345–360.

Crampin, S., 1981, A review of wave propagation in anisotropic and cracked elastic media: Wave Motion, 3, 343–391.

Crampin, S., 1984a, An introduction to propagation in anisotropic media: Geophys. J. R. Astr. Soc., 76, 17–28.

Crampin, S., 1984b, Effective elastic constants for wave propagation through cracked solids: Geophys. J. R. Astr. Soc., 76, 135–145.

Crampin, S., 1989, Suggestions for a consistent terminology for seismic anisotropy: Geophys. Prosp., 37, 753–770.

Davis, T. L., and Lewis, C., 1990, Reservoir characterization by 3-D, 3-C seismic imaging, Silo field, Wyoming: The Leading Edge, 9, 22–25.

Ebrom, D. A., Tatham, R. H., Sekharan, K. K., McDonald, J. A., and Gardner, G. H. F., 1990a, Hyperbolic traveltime analysis of first arrivals in an azimuthally anisotropic medium: A physical modeling study: Geophysics, 55, 185–191.

Ebrom, D. A., Tatham, R. H., Sekharan, K. K., McDonald, J. A., and Gardner, G. H. F., 1990b, Dispersion and anisotropy in laminated versus fractured media: an experimental comparison: 60th Ann. Internat. Mtg., Soc. Expl. Geophys., Expanded Abstracts, 1416–1419.

Garbin, H. D., and Knopoff, L., 1975, Elastic moduli of a medium with liquid-filled cracks: Q. Appl. Math., 33, 301–303.

Helbig, K., 1984, Anisotropy and dispersion in periodically layered media: Geophysics, 49, 364–373.

Hudson, J. A., 1980, Overall properties of a cracked solid: Math. Proc. Camb. Phil. Soc., 88, 371–384.

Jones, L. E. A. and Wang, H. F., 1981, Ultrasonic velocities in Cretaceous shales from the Williston basin: Geophysics, 46, 288–297.

Lynn, H. B. and Thomsen, L. A., 1990, Reflection shear-wave data collected near the principal axes of azimuthal anisotropy: Geophysics, 55, 147–156.

Macbeth, C., 1990, Inversion of shear-wave polarizations for anisotropy using three-component offset VSPs: 60th Ann. Internat. Mtg., Soc. Expl. Geophys., Expanded Abstracts, 1404–1406.

Martin, M. A. and Davis, T. L., 1987, Shear-wave birefringence: a new tool for evaluating fractured reservoirs: The Leading Edge, 6, 10, 22–28.

Melia, P. J., and Carlson, R. L., 1984, An experimental test of *P*-wave anisotropy in stratified media: Geophysics, **49**, 374–378.

Mueller, M., 1991, *S*-wave studies of fracturing on the Austin Chalk: Geophys. J. Internat., in press.

Postma, G. W., 1955, Wave propagation in a stratified medium: Geophysics, **20**, 780–806.

Schoenberg, M. and Douma, J., 1988, Elastic wave propagation in media with parallel fractures and aligned cracks: Geophys. Prosp., **36**, 571–590.

Schoenberg, M. and Muir, F., 1989, A calculus for finely layered media: Geophysics, **54**, 581–589.

Silver, P. and Chan, W., 1988, Constraints on continental anisotropy from shear-wave splitting in SKS: 3rd Internat. Workshop on Seismic Anisotropy, Collected Abstracts.

Squires, S. G., Kim, C. D. Y. and Kim, D. Y., 1989, Interpretation of total wavefield data over Lost Hills Field, Kern County, California: Geophysics, **54**, 1420–1429.

Tatham, R. H., Matthews, M. D., Sekharan, K. K., Wade, C. J., and Liro, L. M., 1987, A physical model study of shear-wave splitting and fracture intensity: 57th Ann. Internat. Mtg., Soc. Expl. Geophys., Expanded Abstracts, 642–645.

Thomsen, L. A., 1986a, Elastic anisotropy due to aligned cracks: EOS, **67**, 1207.

Thomsen, L. A., 1986b, Weak elastic anisotropy: Geophysics, **51**, 1954–1966.

Thomsen, L. A., 1988, Reflection seismology over azimuthally anisotropic media: Geophysics, **53**, 304–313.

Winterstein, D. F., 1986, Anisotropy effects in *P*-wave and *SH*-wave stacking velocities contain information on lithology: Geophysics, **51**, 661–672.

Winterstein, D. F., 1989, Velocity anisotropy terminology for geophysicists: Geophysics, **55**, 1070–1088.

Winterstein, D. F., and Meadows, M. A., 1990a, Shear-wave polarizations and subsurface stress directions at Lost Hills Field: 60th Ann. Internat. Mtg., Soc. Expl. Geophys., Expanded Abstracts, 1431–1434.

Winterstein, D. F., and Meadows, M. A., 1990b, Changes in shear-wave polarization azimuth with depth in Cymric and Railroad Gap Oil Fields: 60th Ann. Internat. Mtg., Soc. Expl. Geophys., Expanded Abstracts, 1435–1438.

Three-Dimensional Multicomponent Imaging of Reservoir Heterogeneity, Silo Field, Wyoming[1]

Catherine Lewis, Thomas L. Davis[‡], and Claude Vuillermoz[§]*

Introduction

The internal heterogeneity of a petroleum reservoir has a critical influence on its production performance. Where facies changes and structural deformation divide the reservoir into a myriad of compartments, open fractures provide conduits for fluid flow. Unfortunately, the spatial distribution of fractures in a reservoir is impossible to predict from the limited sampling of well information alone.

Three-dimensional (3-D) seismic surveys have been used to map reservoirs for production, but compressional-wave seismic data are not very useful for fracture description because *P*-waves are relatively insensitive to fractures. Instead, recent studies have indicated that shear waves may be sensitive to fractures because the fractures can introduce anisotropy to the reservoir (Crampin, 1987). A shear wave entering the fractured anisotropic layer splits into two waves traveling with different velocities. The two waves have approximately perpendicular polarizations, and from the polarizations we can derive the orientation of the fractures. From the difference in traveltimes of the two waves, we can get a gross estimate of the density of fracturing.

To evaluate the ability of shear waves to image fractures at depth, the Reservoir Characterization Project at Colorado School of Mines conducted an investigation of a fractured reservoir using 3-D multicomponent seismic data at Silo Field (Figure 1). There, oil production from chalks of the Cretaceous Niobrara Formation is extremely variable, depending on fracture development. Acquired in 1987, this survey may be the first 3-D multicomponent experiment ever conducted.

Processing multicomponent seismic data creates unique challenges that are not obstacles in conventional *P*-wave processing. A critical step of shear-wave processing is the rotation of the components into a fracture coordinate system, defined by axes parallel and perpendicular to the fracture strike (Alford, 1986; Thomsen, 1988). Rotation analysis becomes problematic when the survey is in three dimensions, because wave polarizations as well as velocities may vary with raypath azimuth. We sorted the shear-wave data by shot-to-geophone azimuth to look for changing polarizations characteristic of orthorhombic anisotropy, a type of anisotropy that may be expected from the combination of vertical fractures and horizontal sedimentary layering. Instead, we identified a simpler type of anisotropy that results from vertical fractures alone (Lewis, 1989).

We made an integrated interpretation of reservoir heterogeneity using fracture-identification logs and the multicomponent survey. The 3-D compressional-wave survey resolved small-scale structural features that influenced fracturing of the Niobrara reservoir. We associated the relative magnitudes of anisotropy determined from delays between the two shear-wave components with the variable production performance and fracture density in the field.

Previous Studies

Over the last decade, the recognition of shear-wave splitting and the ability to record the full waveform have stimulated a growing interest in seismic anisotropy. Shear-wave splitting has been observed in vertical seismic profiles (Crampin and Bush, 1986; Johnston, 1986). In two single-component, shear-wave seismic lines running perpendicular to each other, Lynn and Thomsen (1986, 1990) demonstrated that shear-wave splitting due to anisotropy could be observed on surface seismic data. Recognizing that the interference of split shear waves could destroy the coherency of shear-wave data, Alford (1986) introduced a coordinate rotation to separate the split shear waves. Garotta and Granger (1988) introduced energy ratios in rotation analysis to focus converted-wave

[1]Presented at the 59th Annual International Meeting, Society of Exploration Geophysicists. Published in Geophysics, **56**, 2048–2056.

*Exxon Production Research Co., P.O. Box 2189, Houston, TX 77252.
[‡]Geophysics Department, Colorado School of Mines, 1500 Illinois Ave., Golden, CO 80401.
[§]CGG American Services, Inc., 2500 Wilcrest, Houston, TX 77042.

data. Thomsen (1988) describes rotation algorithms in more detail. With two components of shear-wave data on a 2-D shear-wave line at Silo Field, Martin and Davis (1987) related azimuthal anisotropy determined from split shear waves to fracturing in the chalk reservoir. In a recent application of shear-wave technology, Squires et al. (1989) reported variations in anisotropy in association with a changing fracture orientation across an anticline.

Geologic Setting of Silo Field

Silo field produces oil from naturally fractured chalks in the Upper Cretaceous Niobrara Formation. Across the field, the sediments dip about one-half degree to the southwest. Over forty wells have tested oil at Silo Field, but economic production depends on fracture density sufficient to provide fluid conductivity in the reservoir. Cores taken in the Niobrara indicate that fracturing is variable. In the Golden Buckeye 9-1 in section 9 of T15N-R64W, less than 2 km southeast of the survey area, fracturing is minimal. In contrast, the Combs 1 in section 35 of T16N-R65W, about 1 km west of the survey, had poor core recovery due to intense fracturing.

The stratigraphic column at Silo Field is estimated to be 4000 m thick, including 3000 m of Cretaceous sediments. Over 2000 m of marine Pierre Shale overlie the Niobrara. The Shannon, Sussex, and Rocky Ridge are thin sandstone units within the Pierre Shale, which generate seismic reflections within the shale.

Acquisition of 3-D Multicomponent Survey

The 3-D multicomponent survey covers a little over six square kilometers in the heart of Silo Field (Figure 1). It is a true 3-D survey, because all possible azimuths of raypath direction were recorded. The vibrators moved through a stationary grid of geophones. Both the compressional- and shear-wave surveys had eight lines of geophones, in linear arrays, laid out in an east-west direction, with north-south lines for the vibrators (Figure 2; Lewis, 1989). During the acquisition, the geophones were frozen in the ground, covered by up to three feet of snow.

Four components of shear-wave motion were recorded. The shear-wave vibrators had two perpendicular orientations, east-west and north-south, with two orientations of horizontal geophones. We refer to east-west as the x-direction and to north-south as the y-direction, and to the four components as xx, xy, yy, and yx, indicating their source and receiver orientations, respectively. The sampling interval was 4 ms for the entire survey.

Processing of 3-D Multicomponent Survey

The P-wave survey was processed using a modern processing sequence, with a common midpoint (CMP) bin size of 25 m by 25 m. The survey was migrated with one-pass 3-D 45-degree finite-difference migration. Processing of the shear-wave survey required much more innovation.

A critical step of shear-wave processing is the rotation of the components from the acquisition coordinate system into a fracture coordinate system, defined by axes parallel and perpendicular to the fracture strike at the reservoir level. Rotation separates the reflections and maximizes the time delay between the faster and slower shear waves. Rotation analysis was complicated by the possibility that the shear-wave polarization directions could depend on the direction of travel in the earth.

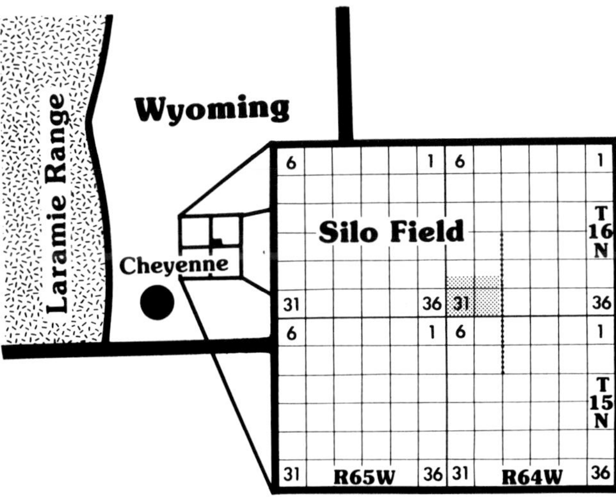

Fig. 1. Location of 3-D multicomponent seismic survey over Silo Field, along with 2-D multicomponent line interpreted by Martin and Davis (1987).

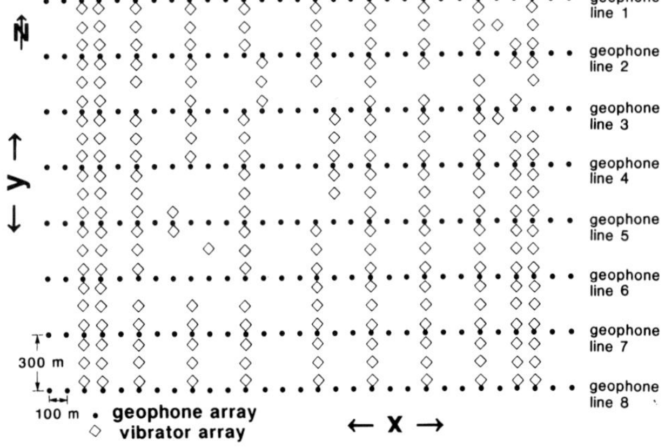

Fig. 2. Acquisition layout of shear-wave survey. Vibrators moved along north-south lines through a stationary grid of eight geophone lines. All directions of shot-to-geophone azimuth were covered. The x-direction is east-west, and the y-direction is north-south.

In processing the shear-wave survey before rotation into the fracture coordinates, any process that might distort the wavelet shape or relative amplitude among the four components was carefully avoided (Lewis, 1989). The amplitude adjustment was one factor for all four components. Residual statics were calculated on deconvolved data, but deconvolution and *f-k* filtering were avoided in the rotation analysis and rotation because these processes might change the relative amplitudes among the four components.

Rotation Analysis of Shear-Wave Survey

To look for a dependence of the polarization direction on raypath azimuth, a 3-D algorithm was implemented in the rotation analysis. The familiar azimuthal anisotropy is extensive dilatancy anisotropy (EDA), which results from a vertical system of parallel microcracks (Crampin, 1987). On a larger scale, these cracks are tectonic fractures that are of interest in reservoir analysis. In contrast, horizontal layering in a sedimentary basin is characterized by transverse isotropy, produced by periodic thin layering (PTL). Crampin (1988) showed on theoretical grounds that horizontal layering and vertical fractures can combine to produce a complicated pattern known as orthorhombic anisotropy.

The equal area projections shown in Figure 3 illustrate the polarization patterns of shear waves for the cases of EDA and orthorhombic anisotropy (Crampin, 1988; Booth et al., 1986). The inner circle is the outer limit of the shear-wave window, at about 35 degrees, inside which the shear-wave arrivals are free of mode conversions at the free surface (Evans, 1984). The hachures represent the polarization direction of the first-arriving shear wave. In the case of EDA anisotropy, the first-arriving shear wave is always polarized parallel to the fracture orientation within the shear-wave window. With orthorhombic anisotropy, however, polarization of the first-arriving shear wave may change by 90 degrees even near zero offset. A seismic line running in a direction perpendicular to the major axis of anisotropy, or oblique to it, would encounter one of two difficulties: either the shear-wave line would lose coherency because rapidly changing polarizations near the shear-wave singularity (indicated by the dot on the lower hemisphere projection) were being stacked, or the rotation analysis would indicate a polarization direction that would be an incorrect interpretation of fracture orientation.

At Silo Field, the shear-wave survey includes all azimuths, since each shot was recorded into all eight lines of geophones. For the 3-D rotation analysis, the offset was limited to the shear-wave window, where the arriving shear waves are free from interference from mode conversions at the free surface. By dividing the survey into ten "pie slices" of shot-to-geophone

azimuth, parallel and subparallel polarizations were isolated, except for the azimuth containing the shear-wave singularity (Figure 4). The four shear-wave components were stacked by common geophone and by azimuth since polarizations were not being mixed within the azimuths. Then rotation analysis was performed on each of the geophone-azimuth stacks.

Rotations of the four components of the geophone-azimuth stacks were displayed by five 18-degree incre-

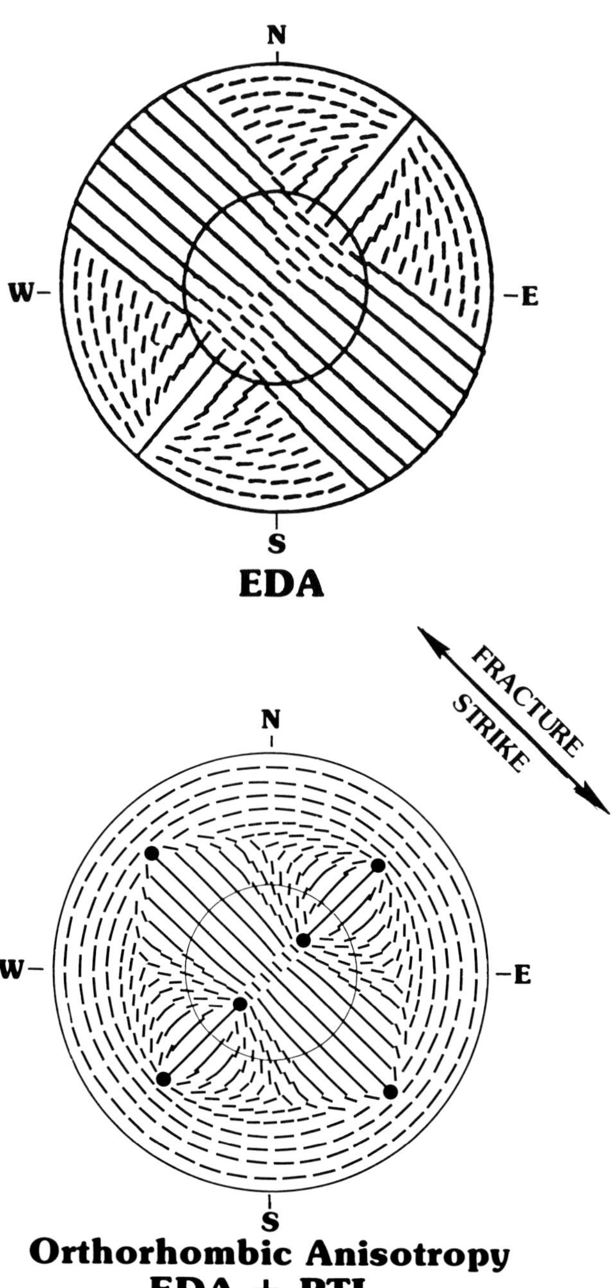

Orthorhombic Anisotropy
EDA + PTL

Fig. 3. Equal area projections of EDA and orthorhombic anisotropy, with fractures oriented northwest. Hachures and the polarization directions of the first-arriving shear wave (Crampin, 1988).

ments (Figure 5). The wiggle traces recorded from an east-west source on an east-west receiver, designated *xx*, are shown rotated 18, 36, 54, 72, and 90 degrees; likewise, the rotated traces recorded from an east-west source on a north-south receiver are designated *xy*, etc. The reflection between 3.2 and 3.4 s is from the Shannon Member of the Pierre Shale, and the Niobrara reflection occurs between 3.7 and 3.9 s. High-amplitude, low-frequency noise that obliterates the reflections in certain azimuths was identified as ground-roll.

Interpretation of the rotation angle for the Silo Field data was made with careful examination of the ground-roll content in each azimuth. Because of the severity of the problem, analysis was performed on the strongest shear-wave reflection, which is the Shannon, and the rotation of the Niobrara reflection was evaluated and compared to the Shannon. An energy-ratio analysis, comparing the energy of the in-line components to the crossline components, measured the rotation angle of each "pie slice" of azimuth to a precision of 1 degree (Gurch, 1989).

The S1 polarization directions from the rotation analysis, plotted within each azimuth, are shown for geophone line 4 (Figure 6). The azimuths that are contaminated with groundroll are marked with "GR." The overall S1 direction is northwest, and the S2 direction is northeast. The first observation from the rotation analysis is that orthorhombic anisotropy is not present in the Silo Field area. Since azimuth 9 is the sector containing the principal axis of anisotropy,

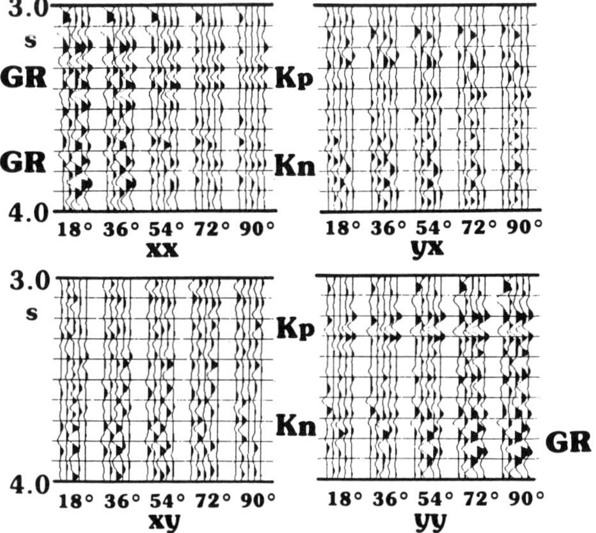

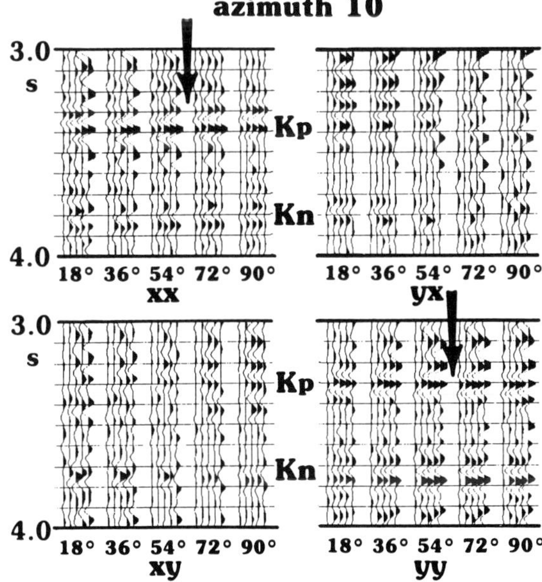

Fig. 5. Window containing Niobrara and Shannon reflection in the geophone-azimuth stacks. The four components are displayed, rotated by 18-degree increments, for azimuths 6 and 10 of geophone line 4. Arrows point to the rotation containing the highest coherent amplitudes on the *xx* and *yy* components of azimuth 10, which indicates the approximate polarization of the shear waves. Groundroll (GR) can be recognized as large incoherent amplitudes in azimuth 6.

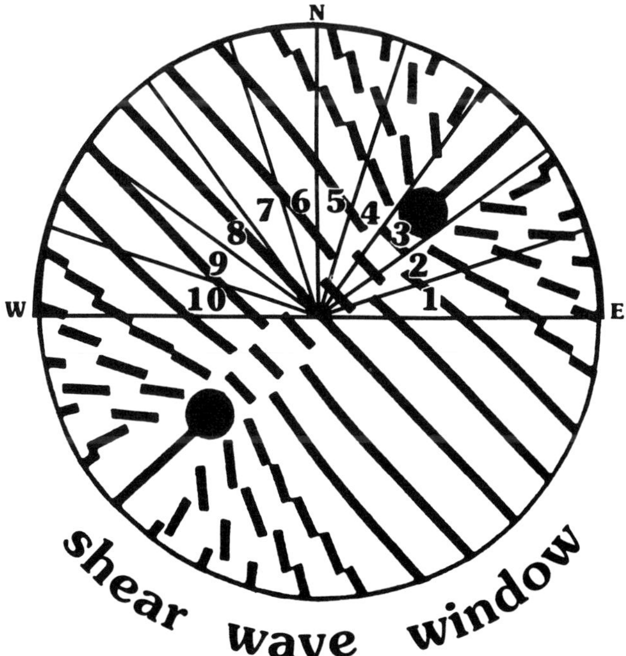

Fig. 4. "Pie slices" of raypath azimuth stacked to look for orthorhombic anisotropy. Each pie slice isolated parallel or subparallel polarizations, except for the one containing the singularity, indicated by the dot.

toward the northwest, azimuth 4 would be the sector that would contain the minor axis (Figure 4). Although the calculated rotation angle for azimuth 4 is distorted by groundroll on one trace, the S1 event is still evident on the rotated *yy*-component, indicating an S1 polarization still to the northwest. This result, parallel polarization indicating the EDA anisotropy, was substantiated by all the other geophone lines.

Azimuths 5 and 6 are examples of how the rotation angle can be corrupted by the presence of groundroll (Figure 5). The calculated angles are 73 and 36 degrees, respectively, and examination of the rotated traces leads to the conclusion that the energy-ratio calculation has keyed in on the groundroll amplitude.

Because azimuths, 1, 2, 3, 8, and 9 are the azimuths that are free of groundroll in the Shannon window from 3.1 to 3.5 s, they were averaged to determine a rotation angle on geophone line 4. Clean azimuths are more difficult to find in the Niobrara interval, 3.6 to 4.0 s. Azimuths 8 and 10 are free of groundroll, and they indicate a rotation angle that parallels the Shannon rotation angle.

Rotation angles for all the other geophone lines were determined in the same manner. The rotation angles calculated from energy ratios among groundroll-free azimuths in the same geophone line vary by about 10 to 15 degrees. Because the anisotropy at Silo Field is

not the orthorhombic but the EDA pattern of anisotropy, the true polarization direction must be the same in all azimuths. This suggests that the noise level inherent in the data is ±8 degrees. The rotation angles vary between geophone lines from 54 to 62 degrees. Since the variation of the rotation angle between geophone lines is at the noise level of the analysis, a single rotation angle of 58 degrees counterclockwise was used for the entire survey. The component oriented N58°W is referred to as the S1 component, with earlier arrival times for the shear-wave events. The component polarized N32°E is the S2 component.

Post-Rotation Processing

After rotation of the shear-wave survey into its natural coordinates, a conventional processing flow was implemented to complete the processing. The cross-components, *xy* and *yx*, were dropped from the processing at that stage since they are noise. The two components of shear-wave data were CMP-stacked in 50-m square bins. Fold increases toward the center of the survey: average fold is 45, with a range from 1 to 60. Predictive deconvolution with a 70 ms gap was used to remove the groundroll from the rotated shear components.

The post-rotation velocity analysis of the shear-wave survey required careful attention. It is common knowledge that a change in moveout velocity can move the onset time of an event up or down on the stacked section (e.g., Iverson, 1990). Since delay times between the split shear components are significant to the interpretation of anisotropy and fracture intensity, care was taken to make sure that the velocities correctly flattened the shear-wave events in the midpoint gathers.

Reservoir Heterogeneity at Silo Field

The structural interpretation of the Niobrara reservoir was important as the framework for making the subsequent interpretation of fracture distribution. Nine wells in Silo Field penetrate the Niobrara within the survey area. Although the area is basically flat, the Niobrara reservoir is dissected by listric normal faults with small displacements (Figure 7). These faults developed over the edge of a deeper salt-withdrawal feature.

Figure 8 is a line through the 3-D survey showing the salt-solution collapse and fill structure in the Permian and Mesozoic section. The location of the line is shown on an isopach of the deep Permian interval from which the salt was dissolved (Figure 9). Other isopachs indicate that salt dissolution began during the early Jurassic and continued into the Cretaceous after Niobrara deposition (Lewis, 1989). Notice how the overlying reflections, including the Niobrara, drape into the depression. As the chalks were lithified and

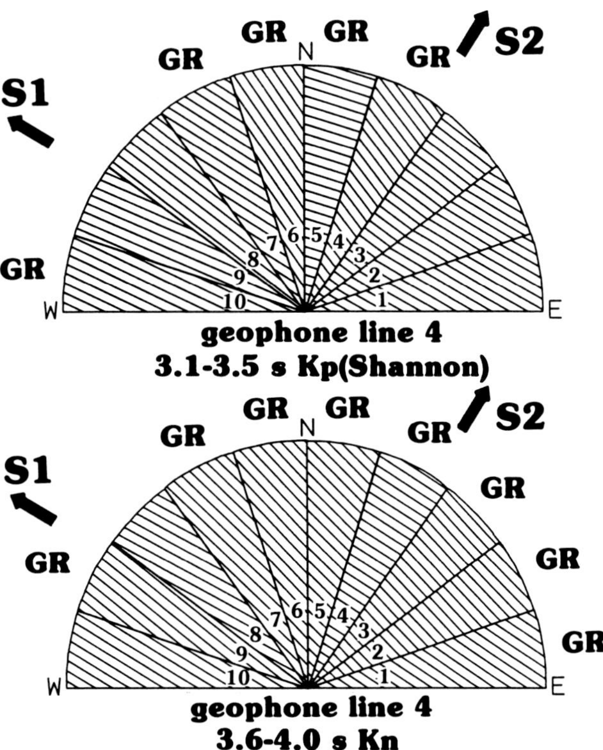

Fig. 6. Polarizations of S1 wave in "pie slices" of raypath azimuth. Azimuths contaminated with groundroll (GR) were eliminated from the interpretation. The overall direction of the S1 polarization is N58°W.

became brittle, this subtle structural deformation fractured the reservoir.

The reflection from the top of the Permian Wolfcamp, an interval known to contain evaporites, is truncated where salt has been removed (Figure 8). In the Berry #1 well, only about 11 km to the west in section 13 of T16N-R66W, 10 m of anhydrite were cored in the Wolfcamp (Doyle, 1987). The two lobate areas of salt removal appear as thin regions in the southwest corner of the survey (Figure 9). In all, a thickness of about 100 m of Permian salt was removed.

Niobrara Structure

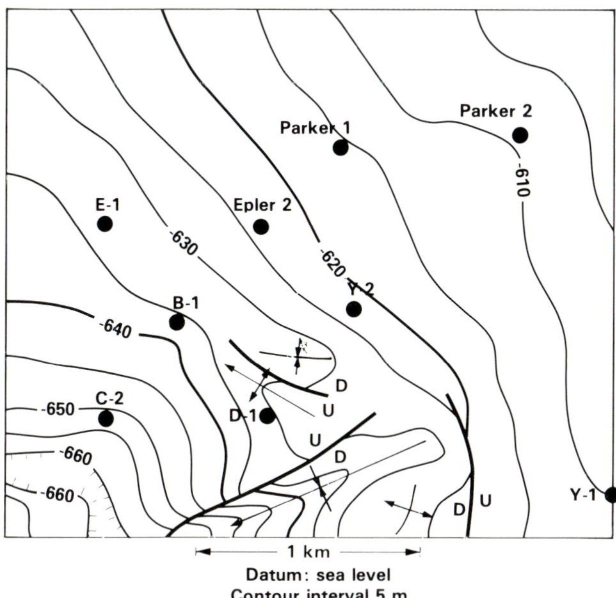

Fig. 7. Niobrara structure map. The Niobrara dips gently to the southwest, and it is draped and faulted over the edge of a deeper salt-withdrawal structure.

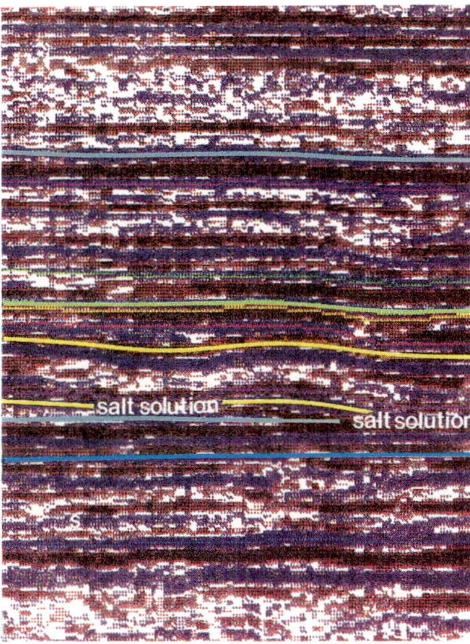

Fig. 8. P-wave line, extracted from the 3-D survey, showing the structural collapse and sedimentary filling where salt was dissolved from the Permian section. Note the truncation of the Wolfcamp reflector, which is known from core to contain anhydrite in the area near Silo Field.

Fractures in Silo Field Wells

The integrated interpretation of the reservoir heterogeneity culminated in the interpretation of the 3-D shear-wave survey in light of fracture-identification well logs from the survey area and the production history of the wells. A three-component VSP was critical to making the seismic ties between the shear- and compressional-wave surveys (O'Rourke, 1986).

Figure 10 is a line extracted from the shear-wave 3-D survey that passes through two wells in the field, with the S1 component polarized N58°W, and the S2 component polarized N32°E. On the S1 component of the shear-wave survey, the reflection from the top of the Niobrara arrives at almost 3.7 s, and it is about 80 ms delayed on the S2 component. The Dakota Group and Mississippian are reflectors below the Niobrara.

Deep Permian Isopach

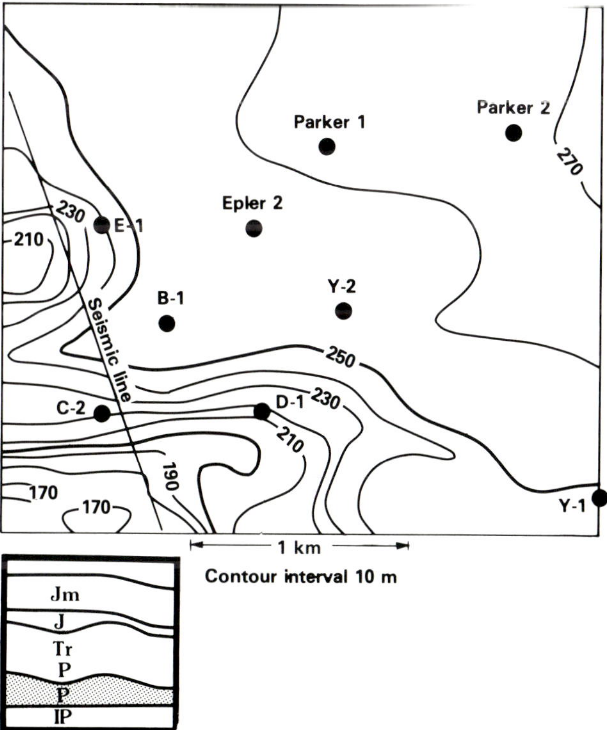

Fig. 9. Isopach of deep Permian interval from which salt was dissolved. Two lobate areas of thinning in the southwest corner indicate the areas of salt removal. The location of the P-wave seismic line in Figure 8 is shown, and the isopached interval is shaded on the thumbnail sketch of the seismic interpretation.

368
Lewis et al.

Dipmeter and full waveform sonic logs provide a partial understanding of the fracture distribution of the reservoir at the well locations, but another indication of the fracture development is the production history of the wells. However, the productivity of an individual well depended on the completion technique that was used as well as natural fracture development. Prior to the introduction of horizontal drilling, the best producers at Silo Field were the Goertz B-1 and Goertz C-2 wells.

Full waveform sonic logs provide a measure of comparison for the amount of fracturing in the wells. Core in the Niobrara from the Goertz B-1 is very fractured. The Goertz C-2 has the same full-waveform signature as the Goertz B-1, indicating similar fracturing (Figure 11). Dipmeter and full waveform sonic logs from the Goertz D-1 well show well-developed fracturing also (Figure 12). In contrast, the Parker 2 has full waveform sonic and dipmeter logs that indicate minimal fracturing in the Niobrara.

Dipmeter logs point to an average fracture orientation of N50°W to N55°W in the survey area. The Goertz D-1 dipmeter reads between N30°W and N65°W, and the Goertz C-2 well has fractures between N40°W and N70°W. The Parker 1 dipmeter reads

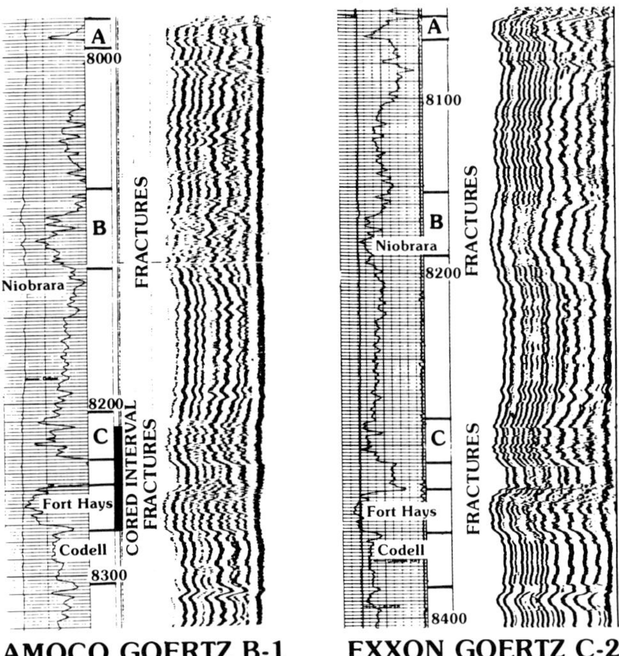

Fig. 11. Full waveform sonic logs for the Goertz B-1 and Goertz C-2 wells, showing evidence of well-developed fractures. The fractured interval in core from the Goertz B-1 corresponds to the amplitude loss seen on the waveforms.

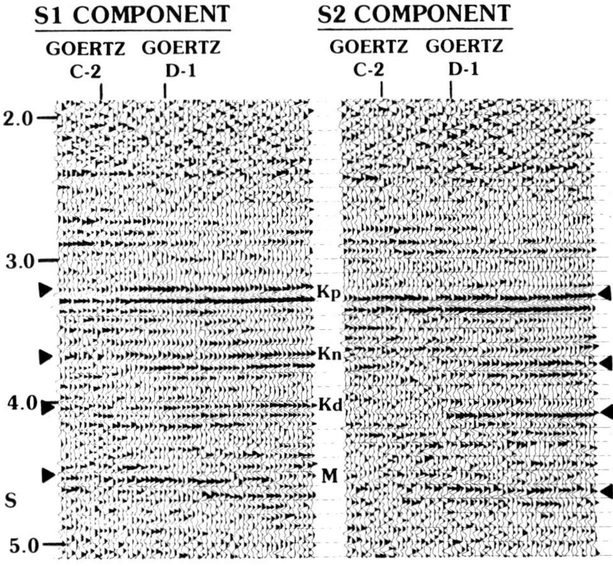

Fig. 10. Two components of shear-wave data for an east-west line passing through the Goertz C-2 and Goertz D-1 wells. The S1 component is polarized N58°W, in a direction parallel to the predominant strike of the fractures. The S2 component is polarized in the perpendicular direction, N32°E. Note the later arrival times of the events on the S2 section. The strongest reflector (*Kp*) is the Shannon Member of the Pierre. The Niobrara (*Kn*) has about 80 ms of delay between the sections. The anisotropy was calculated for the reservoir from the interval times between the Niobrara and the Dakota (*Kd*).

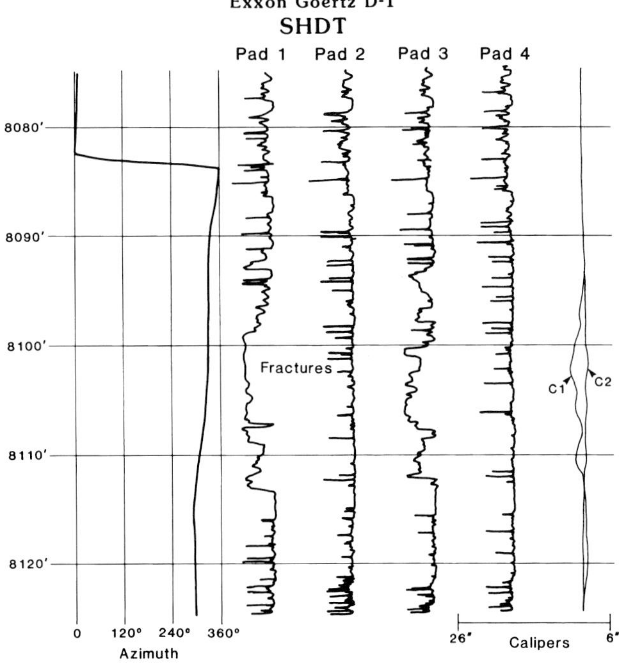

Fig. 12. Dipmeter from the Goertz D-1 well. Low resistivity on tracks 1 and 3 is from fractures with an orientation between N30°W and N65°W.

N35°W to N70°W, with a clear reading of N55°W over 3 m of one of the chalks.

Fracture Identification From Azimuthal Anisotropy

The few points of well control on fracture behavior in the reservoir were enhanced laterally by a picture drawn from the stacked 3-D shear-wave survey. The split shear waves provided a qualitative measure of the development of open fractures because the fracture density of an interval is proportional to its anisotropy (Lynn and Thomsen, 1990). Anisotropy was measured as the delay between the split shear-wave components, normalized by the shorter interval time:

$$\frac{\Delta ts2 - \Delta ts1}{\Delta ts1},$$

where $\Delta ts2$ is the interval time of the S2 component and $\Delta ts1$ is the interval time of the S1 component. In this equation, S1 refers to the component with the overall earlier arrival times on the seismic section. If $\Delta ts2$ is less that $\Delta ts1$ in a specific interval, the equation gives negative anisotropy. This means that the principal axis for that interval is perpendicular to the regional axis. In other words, for Silo Field the principal axis is toward the northwest for positive anisotropy and toward the northeast for negative anisotropy.

The anisotropy was measured for the reservoir interval between the reflections from the top of the Niobrara and the Dakota, and it is proportional to the fracture density. The anisotropy of the Niobrara interval ranges from −7 to +8 percent, calculated from delays up to 28 ms (Figure 13). The 5 percent anisotropy at the Goertz B-1 and the Goertz C-2 wells can be directly related to their productivity from natural fractures oriented to the northwest. According to Petroleum Information, these wells have cumulative productions of 260 Mbbl and 178 Mbbl, current to

March 1990, respectively. Northeast of these wells is an area of slightly weaker anisotropy that corresponds to less rewarding production. The Parker 1 well is steadily producing a small amount of oil, for a cumulative production of 29 Mbbl. In contrast, the Goertz E-1 well, with almost zero anisotropy at its location, produced only 2 Mbbl of oil before being abandoned. The Parker 2 well was a marginal producer, now abandoned, and it occurs in an area of negative anisotropy.

Although the levels of anisotropy correlate to the production from fractures in an overall sense, the noise in the shear-wave survey still affects some measurements. The Epler 2 well, located between the Goertz B-1 and Parker 1, has produced 54 Mbbl of oil. It falls at the edge of a small pocket of weak anisotropy located within a larger area of strong positive anisotropy, and its 1 percent anisotropy is misleading since it is a moderately good producer for the field. The Goertz D-1 well occurs in an area of 4 percent anisotropy, and its dipmeter log shows clear fracturing in the Niobrara. It was unsuccessful in production, probably because it was drilled three years after the Goertz B-1 and three other wells in the section had already depleted the area. The State of Wyoming Y-2 hole was lost in an experimental completion, but its 5 percent anisotropy indicates that it might have been a good producer.

Discussion of Fracture Interpretation

At the Niobrara level, the dominant fracture direction is N50°W to N55°W. Dipmeter logs indicate this, and the rotation of the shear-wave data agrees with the well information. The dominantly positive anisotropy indicates that overall the fracture strike is to the northwest, because the faster split shear wave is polarized in that direction (Figure 13). The most intense fracturing, associated with better production in the southwest corner of the survey, seems to be related to the salt-withdrawal structure.

The survey area is on trend with the northwest-oriented fractures identified to the southeast on a 2-D multicomponent line (Martin and Davis, 1987, Figure 1). Anisotropy on the 2-D line reaches a maximum of 3 percent in the Niobrara interval, compared with 8 percent on the 3-D survey. Wells near the 2-D line are marginal producers, in contrast with the profitable Goertz B-1 and Goertz C-2 wells associated with the large values of positive anisotropy. The negative anisotropy in the northeast corner of the 3-D survey ties with the area of negative anisotropy in the center of the 2-D line.

The lateral variability, or spatial wavelength, of the reservoir heterogeneity resulting from variations in fracture intensity is just a few hundred meters. Every one of the nine wells drilled in the survey area missed

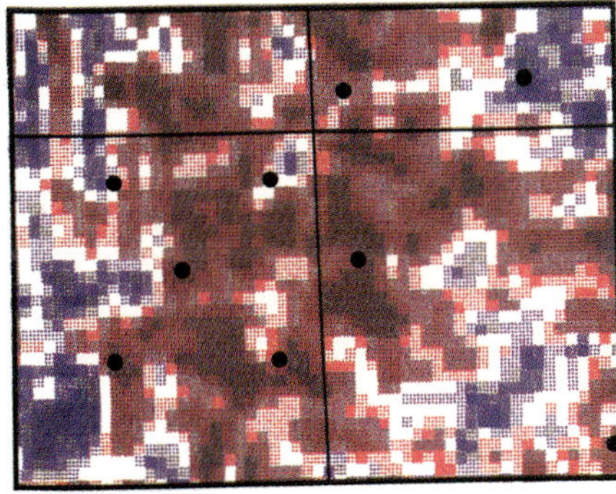

Fig. 13. Map of anisotropy in the Niobrara interval. The wells are identified in Figures 7 and 9, and section lines are shown. Note that the spatial wavelength of the anisotropy is less than the typical well spacing. None of the wells penetrate the areas of largest anisotropy (8 percent).

the most intense fracturing, indicated by the strongest areas of positive anisotropy in the Niobrara interval. If this survey had been available at the time of field development, the Goertz B-1, Goertz C-2, Goertz D-1, and State of Wyoming Y-2 wells could have been better located or deviated to horizontal in order to tap the high-permeability areas of greater fracture density.

Conclusions

Integrated interpretation of the multicomponent seismic survey resulted in a picture, derived from the split shear waves, of the internal heterogeneity of a fractured reservoir. Large positive values of anisotropy, up to 5 percent, associated with an overall fracture orientation of N58°W, were tied to the major producers at Silo Field, and negative or low magnitudes of anisotropy were found in areas of unsuccessful production. The 3-D survey resolved a spatial wavelength of fracture distribution that is less than the well spacing in the field. In similar fields under development, this technique potentially could be used to describe the reservoir in advance of drilling.

The 3-D rotation analysis of the shear-wave data provided redundancy in determining the fracture orientation and the pattern of anisotropy. Analysis of the shear-wave polarizations is very sensitive to the presence of groundroll, but azimuths containing groundroll were carefully eliminated in the rotation analysis. The orthorhombic pattern of anisotropy, expected from the combined effects of vertical fractures and horizontal layering, was not found at Silo Field. A simpler pattern of anisotropy, resulting from the vertical fractures alone, was identified in the azimuthal rotation analysis.

Acknowledgments

We gratefully acknowledge the cooperation in research of CGG American Services, Inc., in the acquisition and processing of the multicomponent survey. We also thank the consortium who supported this research: Amoco, ARAMCO, ARCO Oil and Gas Company, British Petroleum, Canadian Hunter Exploration Ltd., CGG American Services, Inc., Chevron Oil Field Research Company, Conoco Inc., Digital Images, Elf Acquitaine, Exxon Production Research Company, Golden Geophysical, H & H Star Energy, Institut Francais du Petrole, Intevep, S.A., Japan National Oil Corporation, Marathon Oil Company, Meridian Oil Company, Mobil Research and Development Corporation, Oryx Energy Company, PETROBRAS, Pan-Canadian Petroleum Company, Santa Fe Minerals, Inc., Teledyne Exploration, Tenneco Oil Exploration and Production Company, Texaco Inc., UNOCAL, Union Pacific Resources Company, and Western Geophysical Research and Development. Bill Iverson at the University of Wyoming supplied production and well information, and Mike Gurch at Colorado School of Mines contributed to the 3-D rotation analysis. The VSP was processed by Tom O'Rourke at Colorado School of Mines in 1986.

We are grateful to Exxon Production Research Company for its support of the publication of this research.

References

Alford, R. M., 1986, Shear data in the presence of azimuthal anisotropy: Dilley, Texas: 56th Ann. Internat. Mtg., Soc. Expl. Geophys., Expanded Abstracts, 476–479.

Booth, D. C., Crampin, S., Chesnokov, E. M., and Krasnova, M. A., 1986, The effects of near-vertical parallel cracks on shear-wave polarizations: Geophys. J. Roy. Astr. Soc., **87**, 583–594.

Crampin, S., 1987, Geological and industrial implications of extensive dilatancy anisotropy: Nature, **328**, no. 6130, 491–496.

——1988, Nonparallel shear wave polarizations in sedimentary basins: 58th Ann. Internat. Mtg., Soc. Expl. Geophys., Expanded Abstracts, 1130–1132.

Crampin, S., and Bush, I., 1986, Shear waves revealed: Extensive-dilatancy anisotropy confirmed: 56th Ann. Internat. Mtg., Soc. Expl. Geophys., Expanded Abstracts, 481–484.

Doyle, J. D., 1987, Depositional environments of the Pennsylvanian and Lower Permian of the Hartville Uplift and adjacent area in eastern Wyoming, western Nebraska, and southwest South Dakota: Ph.D. Thesis, Colorado School of Mines, T-3140.

Evans, R., 1984, Effects of the free surface on shear wavetrains: Geophys. J. Roy. Astr. Soc., **76**, 165–172.

Garotta, R., and Granger, P. Y., 1988, Acquisition and processing of 3C X 3-D data using converted waves: 58th Ann. Internat. Mtg., Soc. Expl. Geophys., Expanded Abstracts, 995–997.

Gurch, M. L., 1989, Four-component rotation of 3-D shear wave data, Silo field, Wyoming: M.S. Thesis, Colorado School of Mines, T-3716.

Iverson, W. P., 1990, Apparent azimuthal shear anisotropy from normal-moveout differentials: Geophysics, **55**, 1637–1638.

Johnston, D. H., 1986, Azimuthal anisotropy: shear wave birefringence: 56th Ann. Internat. Mtg., Soc. Expl. Geophys., Expanded Abstracts, 464–466.

Lewis, C., 1989, Three-dimensional multicomponent imaging of reservoir heterogeneity, Silo Field, Wyoming: Ph.D. Thesis, Colorado School of Mines, T-3700.

Lynn, H. B., and Thomsen, L. A., 1986, Reflection shear-wave data along the principal axes of azimuthal anisotropy: 56th Ann. Internat. Mtg., Soc. Expl. Geophys., Expanded Abstracts, 473–476.

——1990, Reflection shear-wave data collected near the principal axes of azimuthal anisotropy: Geophysics, **55**, 147–156.

Martin, M. A., and Davis, T. L., 1987, Shear-wave birefringence: a new tool for evaluating fractured reservoirs:cm;0: The Leading Edge, **6**, no. 10, 22–28.

O'Rourke, T. J., 1986, Shear and compressional analysis of VSP in Silo Field, Denver-Julesberg basin: M.S. Thesis, Colorado School of Mines, T-3198.

Squires, S. G., Kim, C. D., and Kim, D. Y., 1989, Interpretation of total wave-field data over Lost Hills field, Kern County, California: Geophysics, **54**, 1420–1429.

Thomsen, L., 1988, Reflection seismology over azimuthally anisotropic media: Geophysics, **53**, 304–313.

Chapter 8: Economics
Economics and Reservoir Geophysics

J. E. Warren[*]

Recent years have seen major advances in geophysical technology: better data acquisition by digital recording, improved processing by supercomputers, and enhanced interpretation using interactive workstations with imaginative graphics, to name a few. The attention of applied geophysics is shifting from being almost exclusively interested in the discovery of hydrocarbons toward a focus on the understanding of reservoirs. The use of geophysics by reservoir engineers is increasing rapidly. Much current research and development is devoted to providing a tool for the reservoir engineer.

From the perspective of a reservoir engineer, the most significant geophysical advance has been the emergence of three-dimensional (3-D) seismic as a cost-effective tool for describing a reservoir's external and internal structure. A 3-D data set integrated with conventional well data (cuttings, cores, logs, and flow tests), vertical seismic profiles (VSP), amplitude variation with offset (AVO) information, and geologic concepts, can produce a much more accurate description of a reservoir's structural style, internal stratigraphy, continuity, and fluid content. The resulting consistent geologic model then, either on its own or as the basis for a numerical simulation model, can be used to specify appraisal-well sites to minimize dry holes, to estimate reserves and productivity more accurately, to determine the optimal production system (number and location of production/injection wells, platform sites, and fluid-handling capacity), and to monitor production performance (primary, secondary, and/or EOR).

To date, more than 1000 3-D surveys have been conducted worldwide and most of them have been demonstrably successful—both technologically and economically. For example, total costs for detailed 3-D acquisition, processing, and interpretation are typically \$40–50 000/km^2 offshore and \$60–80 000/km^2 onshore; thus offshore survey costs (\$160–200 per acre) are far less than lease acquisition costs in the Gulf of Mexico and much less than the cost of an appraisal well almost anywhere. One \$20 million ap-praisal well is equivalent to a 90 000 acre 3-D survey of an entire block in the U.K. North Sea.

The real benefits of 3-D reservoir seismic are diverse. For example,

(1) Onshore in the east Painter Reservoir field in the Rocky Mountain overthrust area, thirteen consecutive wells were drilled and completed successfully (\$4–5 million per well) following a \$1.6 million 3-D survey. As many as six wells were drilled simultaneously, greatly accelerating the development of the field.

(2) In the North Sea, most discoveries are now routinely followed by a 3-D survey to minimize the required number of delineation wells. By shortening the appraisal period, the lead time needed to achieve "first oil" (the beginning of positive cash flow) can be reduced substantially. If production begins one year earlier owing to this reduction, the benefit, at a discount rate of 20 percent per year. is equivalent to an increase of about \$2.00 per barrel in the sales price of the crude oil.

(3) Onshore in the Rio Urucu and the Leste du Urucu fields of the Solimoes basin, Brazil, the interpretation of a 3-D seismic survey suggested delineation-well locations that increased the estimated hydrocarbon reserves by more than 40 percent (from 84 to 120 million barrels of oil equivalent) over the original estimate based on 2-D data. Similarly, a study of ten major fields in the North Sea indicated an average increase in reserves of 36 percent (range: 7 percent to 81 percent) after 3-D surveys compared with prior 2-D estimates.

(4) Onshore in the Arauca field, Llanos basin, Colombia, a \$6 million 3-D survey demonstrated that the field was noncommercial at prevailing oil prices. If the survey had been conducted earlier, at least two of three unsuccessful delineation wells, based on a 2-D interpretation, at a cost of \$20 million per well, would probably not have been drilled.

(5) Offshore in the Gullfaks field, Norwegian North Sea, a \$2 million 3-D survey dramatically changed the structural model of the field and four successful oil wells were drilled in portions of the field that previ-

*Frontier Resources International, Houston, TX.

ously could not be mapped using the 2-D seismic data. This field was declared commercial with reserves of about 1500 million barrels. Another 3-D survey in the Cormorant field, UK North Sea, resulted in a major revision in the fault pattern. This alteration forced a drastic change in the development plan regarding the location of production and injection wells.

(6) Onshore in the Northeast Viboras field, Brooks County, Texas, an experimental 3-D survey was shot to evaluate various data-acquisition techniques before a proposed seven-well infill-drilling program was carried out; this field was chosen because of good subsurface control based on 40 wells, good exisiting 2-D seismic data, and a consistent structural and stratigraphic model of the reservoir. Interpretation of the 3-D data forced a re-evaluation of the reservoir model and a revision of the planned drilling program. The revised three-well infill program was profitable, including the 3-D cost; the discounted cash flow rate of return for the project was 160 percent per year. The original seven-well program probably would have been an economic disaster resulting in four dry holes, two noneconomic holes, and one marginal producer.

These examples show, both qualitatively and quantitatively, the value of 3-D seismic data. Other geophysical techniques (VSP, AVO, seismic attributes, geostatistics, crosswell tomography, time-lapse seismology, etc.) are less well established but very promising. Some that are discussed elsewhere in this book have paid substantial dividends.

The timing of information acquisition is clearly important. Data acquired too early are apt to not be of as good quality as data acquired later, because methods continually improve, but delayed information acquisition means that intermediate decisions will be made and costs incurred without the benefit of the information. Delays in the availability of information will delay income from the exploitation of the reservoir, decreasing the reservoir's present worth $\times$ value, and quite possibly resulting in facilities which have inappropriate capacities.

Since a model of a reservoir is needed to provide input for economic decision making throughout the life of the reservoir—from development through production to abandonment—3-D seismic data, other geophysical information, and engineering data (production rates, pressures, gas-oil ratios, water-oil ratios, etc.) are needed to ensure the continuous evolution of the reservoir model. However, the value of the information obtained from any data-gathering process must be economically weighed against the real cost of the data, including the deferment of cash flow. The following paper analyzes the decision-making process per se; it emphasizes the effect of uncertainty and attempts to quantify the value of information.

Economics and Decisions

*J. E. Warren**

Introduction

Although an exploration program may be deemed successful after a petroliferous reservoir has been discovered, the discovery satisfies only the necessary condition for success; the sufficient condition is that the discovered resource be economic to develop (Megill, 1971 and Newendorp, 1975). Consequently, the development decision is of paramount importance to the future health of the exploration program. More importantly, since the decision involves an expenditure that is two or more orders of magnitude greater than that for exploration, it may be critical to the survival of the corporation itself. Later in the life of the reservoir, similar decisions may be required regarding in-fill drilling or enhanced recovery, but, because all of these decisions are fundamentally economic in nature, they can be treated in the same manner.

However, before any decision-making process can be invoked, one basic question must be answered: "Should the decision be made on the basis of existing information or should additional information be acquired?" This question cannot be answered adequately unless a method is available for estimating the value of the additional information acquired in terms of the economic measure used to make the decision. Consider the following schedule:

Appraisal Development Production

$$\Vert\blacktriangleleft\!\!-\!\!-\!\!-\!\!\blacktriangleright\Vert\blacktriangleleft\!\!-\!\!-\!\!-\!\!\blacktriangleright\Vert\blacktriangleleft\!\!-\!\!-\!\!-\!\!\blacktriangleright\Vert\text{time}\longrightarrow$$

| 0 | τ | $\tau+\theta$ | T |

If the decision is deferred, the additional cost is not only that for information acquisition or appraisal (reservoir delineation/description) but also that associated with the delay in production startup (onset of cash flow). Figure 1 summarizes the decision-making process in its simplest form (Lohrenz, 1988). The decision points have been identified (■) and probabilities have been assigned to each possible outcome (uncertainty) at each chance node (●). This diagram (decision tree) provides a framework for analyzing the information-buying decision. However, outcome evaluation, prob-

*Frontier Resources International, Houston, TX.

ability assignment, and procedure need further discussion (Holloway, 1979).

Outcome Evaluation

An expenditure, when contemplated, must be evaluated in a consistent and objective manner that recognizes corporate goals. It is not possible to fund every proposed project; therefore, the allocation of available funds implies competition among projects—development of newly discovered reserves, production purchase, corporate acquisition, or enhanced recovery from existing reservoirs. For evaluation, an economic model that faithfully interrelates the project parameters is required; it must be easy to use and to understand, and it must be sufficiently flexible to allow for judgement when data are lacking.

The first step in the evaluation of an investment opportunity is the calculation of the net cash flow stream; it represents the cash inflows and outflows over time. The net cash flow is equivalent to profit plus depreciation plus any other noncash charges for the period (usually taken to be one year), and it can be positive or negative. For the jth time increment,

$$(\text{net cash flow})_j = (\text{gross revenue})_j - (\text{royalty})_j -$$
$$(\text{taxes})_j - (\text{operating cost})_j - (\text{investment})_j.$$

Note that the net cash flow rate (stream) is defined on an after-tax basis since the investment represents money that has already passed through the tax system. If the net cash flows over a sequence of time intervals are summed, the result is the cumulative net cash flow (Figure 2). At some time $t = n\Delta t$:

$$(\text{cumulative net cash flow})_t = \sum_{j=1}^{n} (\text{net cash flow})_j.$$

Cash flow is generally negative during the development period. Income from production offsets costs at a time of maximum exposure (d); the pay-out time is m. The project life is complete when the net cash flow rate reaches zero at time T.

Present Worth

Any investment can be visualized as the purchase of a series of net cash flows in the future. Since the

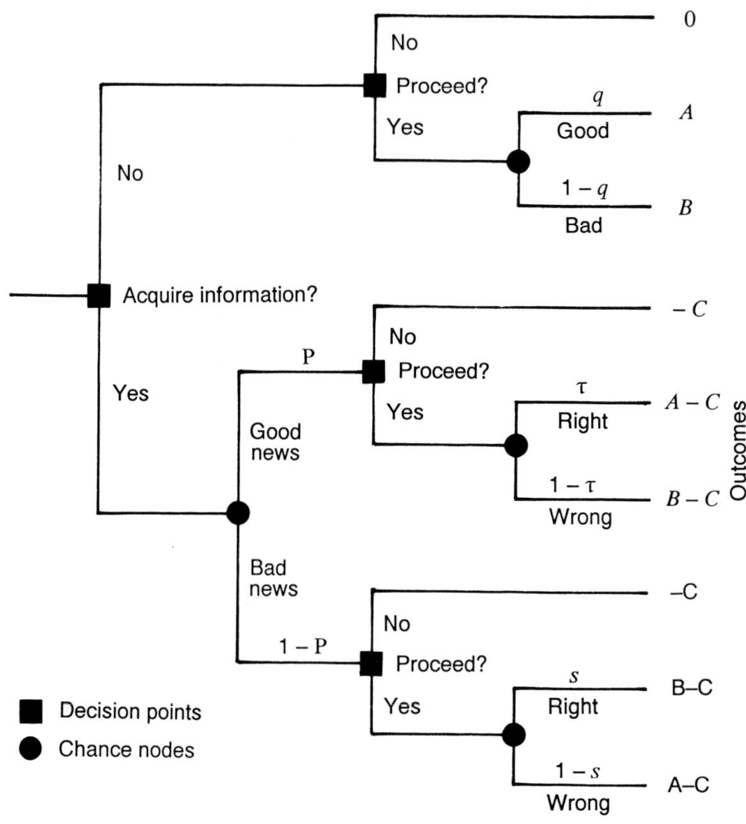

Fig. 1. Decision-making process. Assigned probabilities are p, q, τ, and s.

money invested could have been left in a bank to earn interest or employed in some other corporate project, it is necessary to consider the time value of money or the concept of the present-worth of future net cash flows. If an investment I_0 is made today at a fixed annual interest rate k, and the interest is compounded at fixed time intervals Δt, the future value of the investment at time $t = n\Delta t$, is given by the following[1]:

$$(\text{future value})_t = I_0(1 + k\Delta t)^n;$$

$$I_0 = (\text{future value})_t/(1 + k\Delta t)^n.$$

Defining k to be the minimum acceptable corporate rate of return ("hurdle rate"), the opportunity rate of

---

[1]Note that

$$\frac{\text{LIM}}{\Delta t \to 0}\left[\frac{1}{(1 + k\Delta t)}\right] = e^{-kt};$$

consequently, on a continuous basis,

$$I_0 = (\text{future value})_t e^{-kt}$$

and

$$\text{cummulative } I_0 = \int_0^t (\text{future value})_t e^{-kt} dt.$$

return (average return for all corporate projects) or the corporate discount rate, the present-worth of the cumulative future net cash flow from a project can be defined as follows:

$$(\text{present worth})_t = \sum_{j=1}^{n}\left[\frac{(\text{net cash flow})_j}{(1 + k\Delta t)^j}\right]; \quad t = n\Delta t$$

The project's history can be depicted by the present-worth cumulative net cash flow profile (Figure 3).

The project life is unaffected by discounting since it represents the economic limit for the project (where the actual cash flow rate becomes zero). The present worth of the project's cumulative net cash flow is an economic measure that permits this project to be compared to other corporate projects. The requirement that the present-worth of cumulative net cash flow is positive (≥ 0) ensures that the project satisfies the corporate constraint on investment. Its use implies that all positive net cash flows can be reinvested at the opportunity or "hurdle" rate of return. The project is evaluated on a stand-alone basis with no exchange of tax credits within the corporation; exploration expenditures up to the completion and testing of the discovery well are treated as "sunk costs" rather than investments—their only impact on the evaluation is through the tax element of the net cash flow stream by way of depreciation of tangible costs, offset against revenue for expensed costs and neglect of capitalized costs. (It is tacitly assumed that the present-worth benefits derived from capitalized costs and abandonment costs are canceled by the project's salvage value.) It should also be noted that the required rate of return k is actually an "apparent rate of return" since it includes inflation; the "real rate of return" is equal to $(k - i)$, where i is the inflation rate. For simplicity, let us assume that the inflation rate is negligible and that any cost escalation is also negligible.

Based on the schedule shown in the introduction, let us define the following elements to complete the specification of the economic model:

(present worth$_{\text{cumulative net cash flow}}$)$_k$

$$= \sum_{1}^{T} \left[\frac{(\text{net cash flow})_j}{(1 + k)^j} \right]$$

(present worth$_{\text{cost of development}}$)$_k$

$$= \sum_{1}^{\tau + \theta} \left[\frac{-(\text{net cash flow})_j}{(1 + k)^j} \right]$$

(present worth$_{\text{cost of information}}$)$_k$

$$= \sum_{1}^{\tau} \left[\frac{-(\text{net cash flow})_j}{(1 + k)^j} \right]$$

where τ = appraisal period (years), θ = development period (years), T = economic life (years); $\Delta t = 1.0$ (year).

The project is characterized by a present worth subject to the constraint present-worth of cumulative net cash flow ≥ 0. For ranking purposes, i.e., allocation of available funds among viable projects, let us define the following:

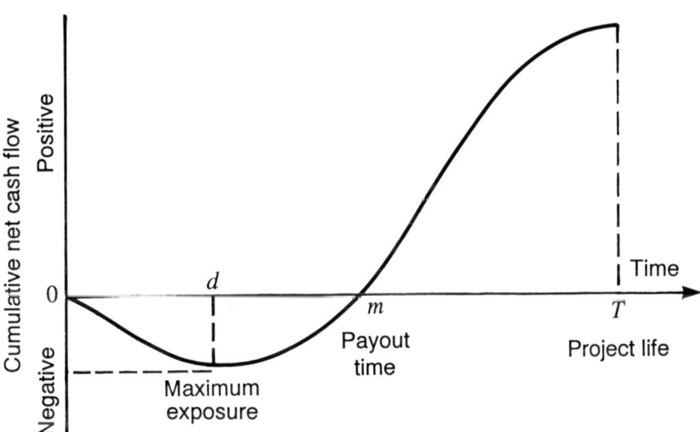

Fig. 2. Cumulative net cash flow.

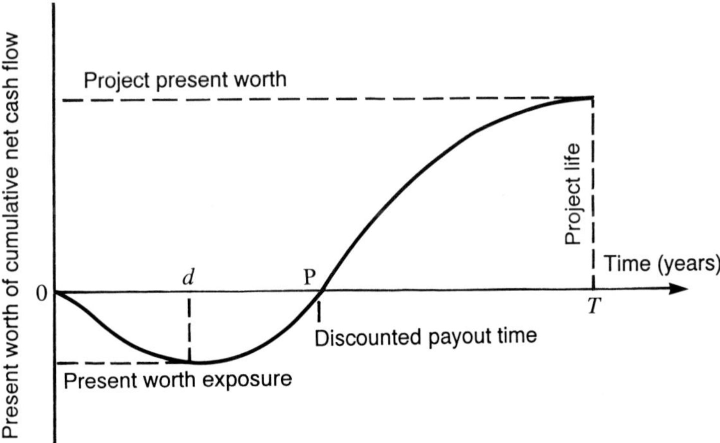

Fig. 3. Present worth of cumulative net cash flow profile: d equals time of maximum exposure, p equals time of payout, and T equals project life.

(profitability index)$_k$

$$= \frac{(\text{present worth}_{\text{cumulative net cash flow}})_k}{(\text{present worth}_{\text{cost of development}})_k + (\text{present worth}_{\text{cost of information}})_k}$$

Physical Model

The next step in the evaluation requires a physical model for the behavior of the reservoir. It must be simple enough to be imbedded in the economic model, but it must also be realistic. Let us assume that the production history shown in Figure 4 is an adequate representation of the performance of the reservoir (Warren, 1981). The production rate increases rapidly after start-up until reaching the capacity of the production-transport facilities q_0, and then remains at the maximum rate until it can no longer be sustained by the available well productivity; after this, the production rate declines until it reaches the economic limit q_{EC}. Let us assume that well productivity $q'(t)$ is linearly related to the hydrocarbons not yet produced; i.e., it can be expressed as follows:

$$q'(t) = q_0'\left[1 - \frac{Q(t)}{R}\right]$$

where $q'(t)$ = total productivity (STB/yr),

$\qquad q_0' = wq_{0i}'$ = total initial productivity (STB/yr),

$\qquad w$ = number of producing wells,

$\qquad q_{0i}'$ = initial productivity of an individual well (STB/yr),

$\qquad Q(t)$ = cumulative production (STB), and

$\qquad R$ = reserves (STB).

This equation defines a constant-percentage decline in productivity for a bounded, homogeneous reservoir. This assumption is not overly restrictive since the behavior equation exactly describes a liquid-expansion drive, adequately describes a solution-gas drive, and approximately describes a bottom-water or gas-cap drive subject to water-oil ratio or gas-oil ratio control. This equation is a good approximation for any type of productivity decline during the early life of the reservoir. Rarely is enough information available early enough to make a better choice. The actual production rate $q(t)$ is, thus, the following:

$$q(t) = \begin{cases} 0; & 0 \leq t \leq \tau + \theta \\ q_0 t/\varepsilon; & \tau + \theta < t < \tau + \theta + \varepsilon \\ q_0; & \tau + \theta + \varepsilon \leq t \leq \tau + \theta + t^* \\ q_0 e^{-\lambda(t-\tau-\theta-t^*)}; & \tau + \theta + t^* < t \leq T \end{cases}$$

where λ is the decline rate, and the project can be characterized as follows:

$\qquad q_0$ = initial productivity = $\alpha\lambda R$,

$\qquad t^*$ = time when production begins to decline =

$$[(1 - \alpha)/\alpha\lambda] + \varepsilon/2,$$

$\qquad T = t^* + (1/\lambda)\ \ln(q_0/q_{EC})$,

$\qquad Q_{EC}$ = cumulative economic production =

$$R[1 - \alpha\varepsilon^{-\lambda(T-t^*)}],$$

where α is the fraction of the initial productivity represented by q_0, the maximun production rate.

For simplicity, let us assume that $\varepsilon \rightarrow 0$ (this actually might be the case if the wells were predrilled, if pipeline capacity were not immediately available, or if certification were not completed promptly). Then, to be consistent with the discrete form of the definition of present worth, let us express the annual increments of production as follows:

$(\Delta Q)_j$

$$= \begin{cases} 0; & 0 \leq j \leq \tau + \theta \\ q_0; & \tau + \theta < j \leq \tau + \theta + t^* \\ q_0\left[\dfrac{1 - e^{-\lambda}}{\lambda}\right]e^{-\lambda(j-\tau-\theta-t^*-1)}; & \tau + \theta + t^* < j \leq T. \end{cases}$$

Given q_{0i}' and price/cost estimates q_{EC}, the production history is completely described by three parameters,

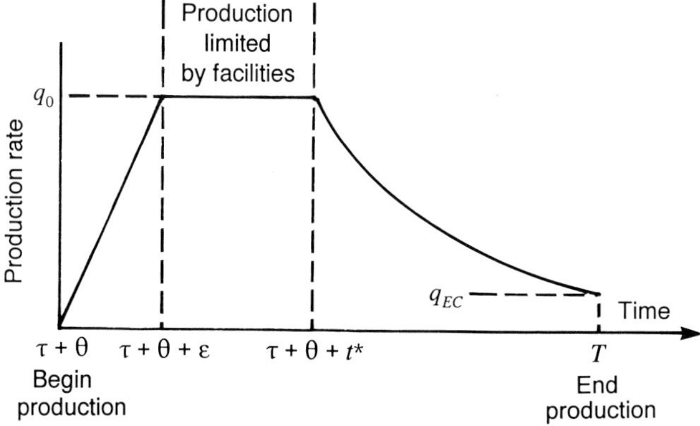

Fig. 4. Production history.

α, λ, and R. Of these, the first two are arbitrary design parameters; e.g.,

$$\alpha = q_0/q_0' = q_0/wq_{0i}'$$

$$\lambda = q_0'/R = wq_{0i}'/R.$$

Since the total initial well productivity q_0' is a function of the number of production wells w and the "plateau" production rate q_0 reflects the capacity of the production-transport system, α and λ can be varied independently until the present worth is maximized for a particular level of reserves R. Consequently, let us assume that the optimal combination of α and λ (or alternatively, w and q_0) has been determined (see Example 1).

Uncertainty

To this point, the evaluation has been treated as though it were completely deterministic. This is far from the truth; at the initial decision, $t = 0$, the data set probably consists of a structural/stratigraphic concept,

---

Example 1: Determining fraction of initial productivity α and number of wells w.

Assume the following:

R = reserves = 404×10^6 STB,

q_{oi}' = initial productivity/well = 10 000 STB/day/well,

q_{EC} = 14 300 STB/day,

ϵ = 0, and

k = discount rate = 0.15/year;

then we get:

(Cumulative present-worth of net cash flow)$_k$ ($ $\times$ 10^6)

α	$w = 25$	$w = 26$	$w = 27$
0.82	205.812	206.456	206.200
0.83	206.165	206.502	206.254
0.84	206.143	206.471	206.199

Optimal: $w = 26$ wells; $\alpha = 0.83$ (note the broad maximum).

q_0' = initial productivity = (26) (10 000) = 260 000 STB/day

q_0 = "plateau" rate = (0.83) (260 000) = 216 000 STB/day

λ = decline factor = (465) (260 000)/(404 $\times$ 106) = 0.235/year

t^* = time when production begins to decline = "plateau" period

= (1 − 0.83)/(0.83) (0.235) = 0.872 years

T = economic life = (0.872) + $\left[\ln(216\ 000)/(14\ 300)\right]$/(0.235) = 12.43 years

Q_{EC} = cumulative economic production = (404) $\times$ 10 $\left\{1 - (0.83)\ \exp\left[-(0.235)\ (12.43 - 0.872)\right]\right\}$

= 382 $\times$ 10^6 STB

---

the two-dimensional (2-D) seismic survey used to locate the discovery well, and a selection of logs, cores, fluid samples, and tests from the discovery well itself. This inventory suggests that significant uncertainty is present in our perception of the reservoir. In fact, it is this uncertainty that justifies a reservoir delineation and description program. If we assume that development costs, operating costs, and crude-oil prices are deterministic and that there will be no changes in the tax regime (certainly a sweeping and gratuitous assumption), the principal sources of uncertainty are the estimates for reserves R and for initial individual well productivities q_{0i}'. Of these, the latter is less worrisome since, in a continuous (or, at least, piece-wise continuous) reservoir, the model uses the total initial well productivity which is obtained by summing individual well productivities. Fortunately, the central limit theorem tends to reduce the variation in the sum (Warren, 1981). Therefore, let us assume that the uncertainty in the physical model is restricted to the estimate of the reserves.

The availability and the quality of information dictate the level of sophistication to be used to estimate reserves. The simplest approach, requiring the least information, is to make a volumetric estimate (Warren, 1981):

$$R = A \cdot H \cdot \rho$$

where R = reserves, A = area, H = net hydrocarbon thickness, and ρ = recovery factor.

Let us assume that three estimates, representing the 5, 50, and 95 percentiles, are available for each of the parameters in the equation for reserves (based on 2-D seismic and well data plus geology). Then, using a remarkably accurate nonparametric (distribution free) approximation (Keefer, 1980) for the expectation E:

$$E(x) = 0.630\ \bar{x} + 0.185(x_m + x_M)$$

$$E(x^2) = 0.630\ \bar{x}^2 + 0.185(x_m^2 + x_M^2)$$

where x_m = minimum, $P(<x_m) = 0.05$, $\bar{x}$ = median, $P(<x) = 0.50$, and x_M = maximum, $P(<x_M) = 0.95$. Since the three parameters are uncertain, their product will tend to be lognormally distributed and, the following approximation is acceptable:

$$\delta(R) = \left\{\ln\left[\frac{E(R^2)}{E^2(R)}\right]\right\}^{1/2}$$

where $E(R) = E(A) \cdot E(H) \cdot E(\rho)$ and $E(R^2) = E(A^2) \cdot E(H^2) \cdot E(\rho^2)$.

Finally, the continuous lognormal distribution, defined by $E(R)$ and $\delta(R)$, can be represented by three discrete values of R:

$$R_m = \text{minimum} = \bar{R}e^{-1.645\ \delta(R)}; \quad P(<R_m) = 0.05,$$

$$\tilde{R} = \text{median} = E(R)e^{-\delta^2(R)/2}; \quad P(<\tilde{R}) = 0.50,$$

$$R_M = \text{maximum} = \tilde{R}e^{1.645\,\delta(R)}; \quad P(<R_M) = 0.95.$$

Each chance node involving reserves requires evaluation for these three levels of reserves to preserve the uncertainty (see Example 2).

Probability Assignment

In the preceding sections, the evaluation of outcomes was based on an economic model with an imbedded physical model. A measure (present worth) for the outcome that satisfies corporate constraints was prescribed, and an attempt was made to preserve the uncertainty associated with the estimated reserves by introducing a three-point, nonparametric representation for its distribution. Now, the choices of alternatives must be related to the outcomes by the assignment of probabilities.

Types of Probability

Probability is a numerical statement regarding the liklihood (chance) that an event (outcome), or sequence of events, will occur when all of the possibilities are considered. Two types of probabilities are pertinent to this discussion[2]:

(1) Objective probability relies on empirical data from the past (long-run) or on the analysis of a discrete set of possible outcomes (intelligent gambling).
(2) Subjective probability is intuitive in nature; it may be based on experience, training, biases, policy and/or emotion, with not all of the possible outcomes necessarily known (i.e., there is no model).

Both types of probability will be used; the mix depends on the availability of data, the data quality, and the objectives of the analysis. While some people are very cynical about the use of probabilities (particularly subjective ones) in the decision-making process, it is better than doing nothing; the use of probability theory at least forces us to look at data and to quantify experience (Holloway, 1979).

Classes of Probability

Owing to the presence of uncertainty, an outcome may depend on one, or a combination, of the following:

At a chance node the outcome may be

[2]The classical method, based solely on logic, determines the degree to which evidence conforms to a model or hypothesis—after the fact. Theory becomes law.

"Mutually exclusive" meaning the occurrence of one event precludes the occurrence of any other; the probability of A or B occurring is $P(A + B)$:

$$P(A + B) = P(A) + P(B);$$

no memory is involved. Or

"Mutually exhaustive" meaning simply that all of the possibilities have been considered;

$$P(A+B) = P(A) + P(B) = 1.$$

With two chance nodes separated by a decision node, the probability may be

"Conditional" where the occurrence of one event depends entirely on the occurrence of the other, so memory is involved; the probability of A and B occurring is $P(AB)$:

$$P(AB) = P(B|A)P(A) = P(B)\,P(A|B).$$

With two chance nodes separated by a decision node with new information acquired at the first chance node, the probability is

"Bayesian" where the occurrence of one event alters our prior estimate of another set of events; if A occurs, the original set of probabilities, $P(B_j)$, is updated to give a new and improved set of probabilities, $P(B_j|A)$;

Example 2: Estimation of reserves involving uncertainty

Assume the following parameters for minimum, median, and maximum estimates:

	x_m	$\tilde{x}$	x_M	$E(x)$	$E(x^2)$
A (acres)	3800	7300	13 000	7710	67.5×10^6
H (feet)	120	200	400	222	57.5×10^3
p (STB/acre ft)	74	220	450	236	68.9×10^3

$$E(R) = \text{Expected reserves} = (7710)(222)(236) = 404 \times 10^6 \text{ STB}$$

$$E(R^2) = (67.5 \times 10^6)(57.5 \times 10^3)(68.9 \times 10^3) = 267 \times 10^9 \text{ STB}^2$$

Then,

$$\delta(R) = \ln\left\{E(R^2)/\left[E(R)^2\right]\right\}^{1/2} = \left\{\ln\left[(267\,000)/(404)^2\right]\right\}^{1/2} = 0.702$$

and

$$R_m = (0.316)\exp\left[-(1.645)(0.702)\right] = 100 \times 10^6 \text{ STB}; \quad P(<R_m) = 0.05$$

$$\tilde{R} = (404)\exp\left[-(0.702)^2 2\right] = 316 \times 10^6 \text{ STB}; \quad P(<\tilde{R}) = 0.50$$

$$R_M = (316)\exp\left[(1.645)(0.702)\right] = 1000 \times 10^6 \text{ STB}; \quad P(<R_M) = 0.95.$$

$$P(B_j | A) = \frac{P(A | B_j)\, P(B_j)}{\displaystyle\sum_{j=1}^{n} P(A | B_j)\, P(B_j)}$$

See Example 3.

"Variance reduction" is an alternative to modifying the probabilities by using the Bayes theorem, when the impact of additional information can be approximated by reducing the variance associated with the lognormal distribution of reserves while preserving its mean or expected value. The method of conjugate distributions can be used for revision, or a simpler approach, using Kolmogoroff's principle, can be used: i.e.,

$$\sigma_N^2(R) = \ln\left[\frac{e^{\sigma^2(R)} - 1}{N + 1} + 1\right]$$

where N = number of appraisal wells and $\sigma^2(R)$ = initial estimate (before appraisal). If experience allows the estimation of a relationship between geophysical work, such as a three-dimensional (3-D) seismic program, and the equivalent number of appraisal wells, this method can be used to approximate the benefits that would be obtained (see Example 4).

Procedure

Now that we have an economic model to evaluate the outcomes and have a basis for assigning probabilities, we can begin to reduce the decision tree. Working back from the outcomes, there are two simple rules to follow:

(1) Replace each chance node with its expected value, E(V).

$$E(V) = \sum_{j=1}^{n} (P_j V_j); \quad \sum_{j=1}^{n} P_j = 1.$$

(2) Replace each decision node by the largest expected value in an algebraic sense (it can be either positive or negative).

This procedure is repeated until, in the final step, we have the following:

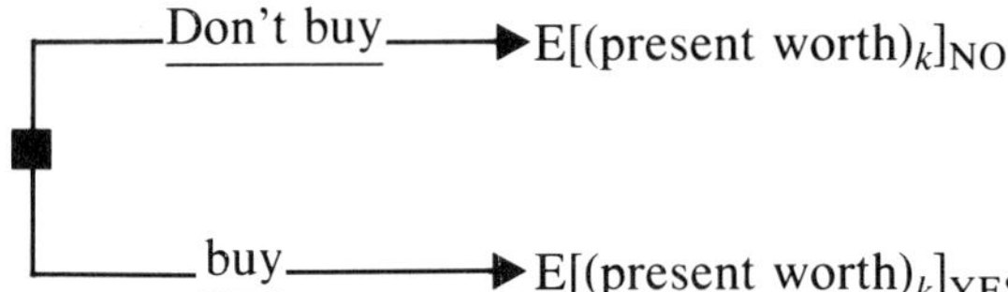

Obviously, the decision with the larger value is chosen. This sounds too simple—and, it is. There are other concepts that must be considered (see Example 5).

Example 3: Probabilities

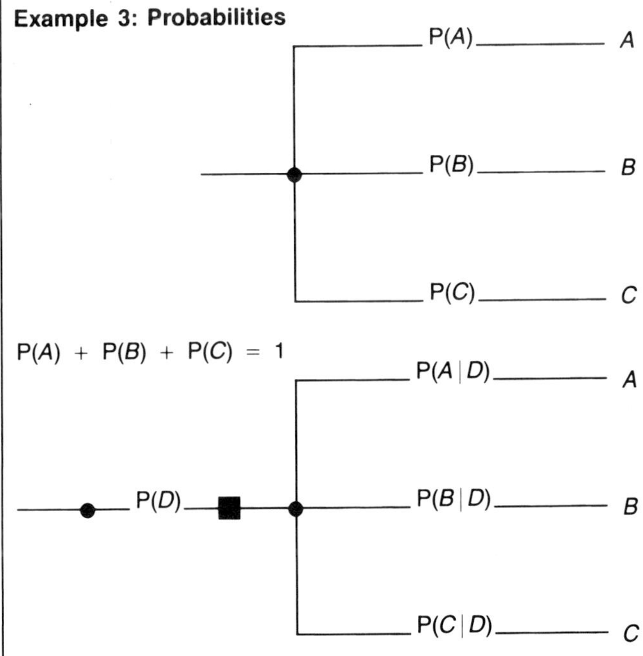

$P(A|D) + P(B|D) + P(C|D) = 1$ since $P(A|D) = P(D|A)\,P(A) / P(D)$; etc., and $P(D) = P(D|A)\,P(A) + P(D|B)\,P(B) + P(D|C)\,P(C)$.

Note that $P(D|A)$, $P(D|B)$, and $P(D|C)$ are estimates that are completely subjective or are based on past performance; they represent the perception of an expert, an expert program, or historical data.

Example 4: The effect of variance reduction

Consider the following (based on optimization from Example 1 and reserves from Example 2) as N additional delineation wells are drilled:

N	$\delta(R)$	$R_m \times 10^6$ STB	$\widetilde{R} \times 10^6$ STB	$R_M \times 10^6$ STB	$\bar{R} \times 10^6$ STB
0	0.702	100	316	1000	404
1	0.526	148	352	836	404
2	0.439	178	367	755	404
3	0.384	199	375	706	404

Assume 120 days/well and 10^7/well; this yields:

(Cumulative net present worth of cash flow) $_K$, ($ $\times$ 10^6)

N	R_m	$\widetilde{R}$	R_M	E(V)	(P_s)$_k$
0	−271	86.0	771	147	0.621
1	−176	132	625	166	0.730
2	−131	137	531	160	0.776
3	−105	132	464	149	0.801

Although there is a continuous increase in the probability (P_s)$_k$ with each additional delineation well, the expected value of the present worth decreases after the first delineation well owing to increased expenditure and deferral of cash flow from production.

Optimization

In the case of the original unmodified estimate for the reserves distribution, each of the three reserve levels used to characterize the distribution must be evaluated. However, it was assumed that an optimal development plan (α,λ) would be used in the evaluation. This seems inconsistent since only the distribution is known at the time of the decision. Therefore, let us assume that, for evaluation in the "zero information" case, the optimal production system (q_0 and w, number of wells) based on the expected reserves, $E(R)$, will be used for all three levels of reserves to furnish a lower limit for the project's $E[(\text{present worth})_k]$.

Similarly and (hopefully) consistently for the case of "perfect information", the optimal production system for each level of reserves will be determined and the evaluation will be performed with the system for the appropriate reserve level. It is assumed that this will provide an upper limit for the project's $E[(\text{present worth})_k]$ (see Example 6).

In the case of "imperfect information," let us assume that the level of reserves inferred by the information will be used to determine the optimal production system. All three levels of reserves will be evaluated on the basis of the optimal system; but, the expected value of present worth for that inference will be based on the adjusted probabilities (see Example 7).

Value of Information

The maximum value of the acquired information can be determined by evaluating the perfect-information and zero-information cases. If the difference is not sufficient to justify the cost of acquisition and the cost of delaying the development decision, nothing more is needed. Then the decision must be made between immediate development and abandonment (see Example 6). If the difference is large enough, the more realistic case of imperfect information must be investigated, albeit subjectively (Warren, 1983) (see Example 7).

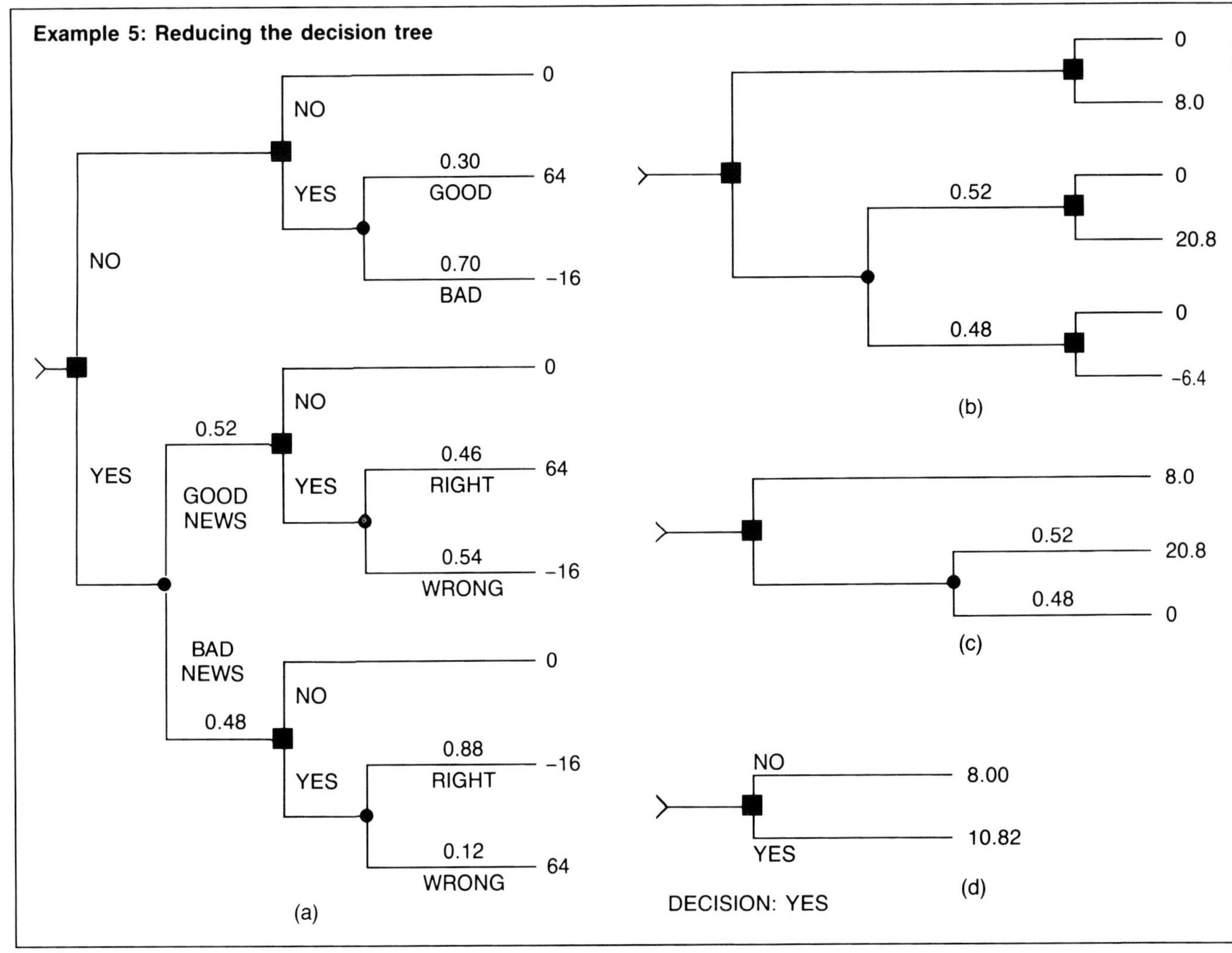

The difference in $E[(\text{present worth})_k]$ between the perfect- and zero-information cases is not the only measure that is significant; the uncertainty associated with the expected value is also important. Fortunately, the use of three-point approximation (5, 50, and 95 percentiles) allows us to take advantage of the fact that the percentiles of a transformed distribution can be obtained by transforming the percentiles of the original distribution. Then, the values of present worth for R_m and R_M represent the 90 percent confidence limit for the transformed distribution and can be used as a measure of uncertainty. Again, using the transformed-

---

Example 6: The value of information

For zero information:
Using the reserves distribution from Example 2, the optimal production system from Example 1, and the evaluations from Example 4,

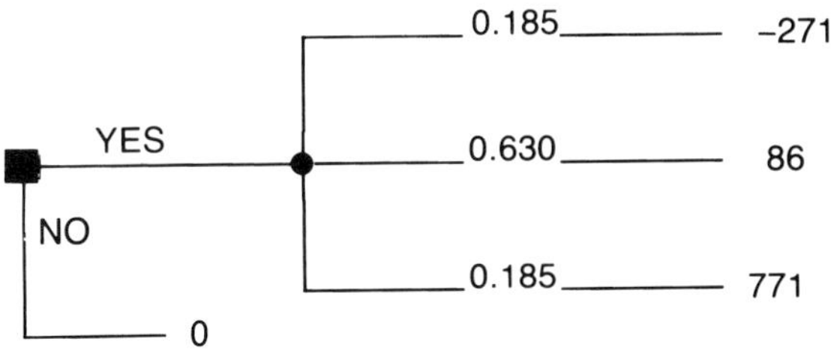

$E(V) = (0.185)\,(-271) + (0.630)\,(86) + (0.185)\,(771) = \$147 \times 10^6$.

For perfect information:
Optimizing the production system for each of the reserve levels, R_m, $\widetilde{R}$, R_M.

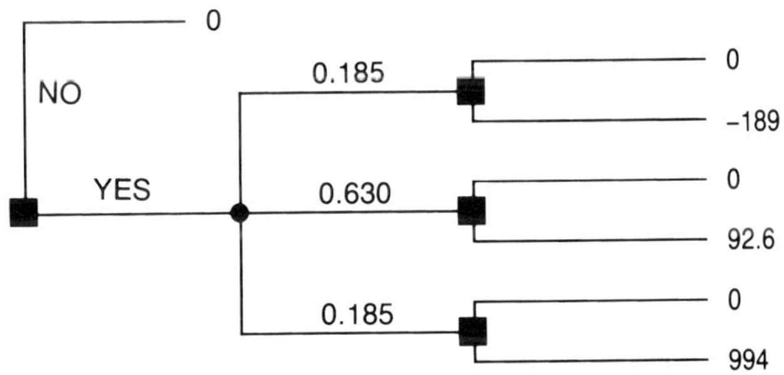

$E(V) = (0.185)(0) + (0.630)(92.6) + (0.185)(994) = \$242 \times 10^6$.

$\Delta p = $ value of perfect information $= 242 - 147 = \$95 \times 10^6$.

Note that the value assigned to Δp, $\$95 \times 10^6$, is in present-worth after-tax dollars; the actual expenditure will be larger and will depend on the tax regime. For example, assuming 70 percent tax rate and unit-of-production depreciation; the following results obtain for an information-acquisition program over one year:

Actual expenditure ($\times 10^6$)	Δp, ($\times 10^6$)
0	95
50	44
100	16
150	0

Consequently, up to $\$150 \times 10^6$ can be spent to acquire "perfect" information over one year before the zero option is more attractive.

Example 7: Imperfect information

Assume the following:
$$P(R'_m | R_m) = P(\widetilde{R}' | \widetilde{R}) = P(R'_M | R_M) = 0.70,$$
$$P(R'_m | \widetilde{R}) = P(R'_M | \widetilde{R}) = 0.20,$$
$$P(\widetilde{R}' | R_m) = P(\widetilde{R}' | R_M) = 0.15,$$
$$P(R'_m | R_M) = P(R'_M | R_m) = 0.10.$$
Since $P(R_m) = P(R_M) = 0.185$ and $P(\widetilde{R}) = 0.630$,

$P(R | R')$

R	R'_m	$\widetilde{R}'$	R'_M
R_m	0.3347	0.0718	0.0763
$\widetilde{R}$	0.3897	0.8564	0.3897
R_M	0.0763	0.0718	0.5340

and $P(R'_m) = P(R'_M) = 0.2245$ and $P(\widetilde{R}') = 0.5150$. Optimizing at each reserve level,**

Expected present worth, $E\left[\text{Present worth}(R | R')\right]_k$

	R_m	$\widetilde{R}$	R_m
R_m	−169	12.3	236.
$\widetilde{R}$	−236	92.6	684.
R_M	−585	−124	994.

Combining the probability $P(R | R')$ and the

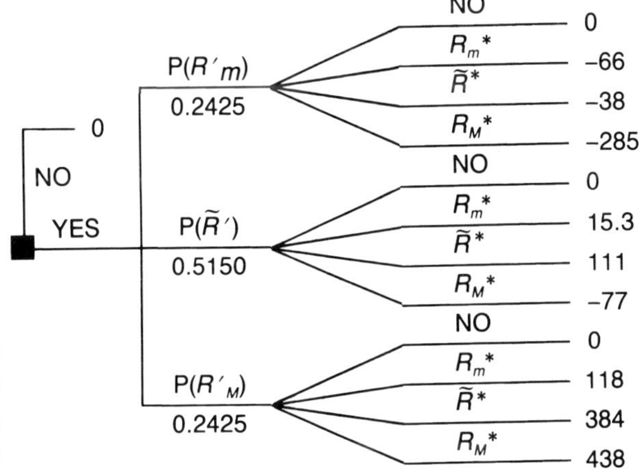

$E(V) = (0.2425)(0) + (0.5150((111) + (0.2425)(438) = \$163 \times 10^6$.

$\Delta I = $ value of imperfect information $= 164 - 147 = \$16 \times 10^6$.

Again, the actual expenditure could be as much as $\$17 \times 10^6$ over six months to equate with the "zero" option.

A prime indicates an imperfect estimate; an asterisk the basis for optimization.

distribution concept (Warren, 1981), it is possible to determine the probability that the required rate of return k will be achieved, $(P_s)_k$, since it is simply the probability that present worth is greater than zero; $E[(P_s)_k]$ must be used with imperfect information.

dure for reducing the decision tree to a single rational course of action results in an expected value for the decision per se, a measure of the uncertainty associated with it and a value for the information required.

Summary

This chapter describes a consistent, logical approach to the decision-making process. The need to consider the acquisition of additional information as an integral part of the process is emphasized. The idea of the decision tree, which provides the framework for analyzing decision making, requires specific consideration of outcome evaluation, probability assignment, and "tree reduction". The evaluation of outcomes includes uncertainty (reserves distribution) and the time value of money (present-worth concept) explicitly. The assignment of probabilities demands quantification of opinions—particularly, subjective estimates regarding the quality of information; and the procedure for reducing the decision tree to a single rational course of action results in an expected value for the decision per se, a measure of the uncertainty associated with it and a value for the information required.

References

Holloway, C. A. 1979, Decision making under uncertainty: Models and choice: Prentice-Hall Inc.

Keefer, D. L. 1980, Three-point approximations for judgmental probability distributions: Proc., ORSA-TIMS Nat. Mtg.

Lohrenz, J. 1988, Net value of information: J. Petr. Tech. (April)

Megill, R. E. 1971, An introduction to exploration economics: Petroleum Publ. Co.

Newendorp, P. D. 1975, Decision analysis for petroleum exploration: Petroleum Publ. Co.

Warren, J. E. 1981, The development decision: Frontier areas, Proc., Soc. Petr. Eng. Hydrocarbon economics and evaluation symposium.

Warren, J. E. 1983, The development decision: Value of information: Proc., Soc. Petr. Eng. Hydrocarbon economics and evaluation symposium.

Appendix: Glossary

*R. E. Sheriff**

abnormal pressure: Formation fluid pressure which differs appreciably from the normal **hydrostatic pressure**, which is the pressure produced by a column of the fluid extending to the surface. Pressure greater than normal (sometimes called **geopressure**) is associated with low seismic velocity and low formation density. Pressure may be as great as that of the weight of the overburden (**lithostatic pressure**).

absorption: See *attenuation* and *Q*.

acoustic impedance: The product of density and seismic velocity. Reflectivity depends on changes in acoustic impedance.

amplitude: The excursion of a seismic trace from its null value.

amplitude shadow zone: A region of lowered amplitude sometimes used as a *hydrocarbon indicator* (q.v.).

amplitude variation with offset (AVO): The variation in the amplitude of a seismic reflection with source-geophone distance. Depends on the velocity, density, and Poisson ratio contrasts at the reflectivity boundary. Used as a hydrocarbon indicator for gas

anisotropic: Having properties whose values depend on the direction of measurement.

appraisal well: A well intended to ascertain reservoir properties or limits rather than to produce hydrocarbons. Also called a **delineation well**.

arbitrary line: A vertical seismic section from a 3-D data volume that is not necessarily straight or in either line or crossline directions.

array sonic log: A sonic log that measures wave arrivals at several offset distances, permitting the detection and analysis of several waves. Schlumberger trade name. Halliburton's Full Wave Sonic System and Atlas Wireline DAC are similar.

arrival (seismic): See *event*.

aspect ratio: The ratio of the width of something (such as a fracture or a pore) to its length.

attenuation: A loss of amplitude. Seismic attenuation is caused by divergence (geometrical spreading), absorption (conversion of seismic energy into heat), and redistribution of energy (by reflection, peg-leg multiples, scattering, etc.).

attribute, seismic: A measurement made on seismic data, such as interval velocity, envelope amplitude, instantaneous phase, instantaneous frequency, dip magnitude, dip azimuth.

AVO: *Amplitude variation with offset* (q.v.).

azimuthal anisotropy: A change in properties with compass direction such as might be produced by vertical fracturing.

azimuthal VSP: See *vertical seismic profiling*.

azimuth/dip display: An encoded display of dip direction/magnitude, usually displayed by color encoding on an automatically tracked map.

band-pass: Frequencies within specified limits. The limits of frequencies passed are usually specified as those that are down by 3 dB (or 70 percent) from the peak value. **Band-reject** filters reject frequencies within a band.

basic wavelet: See *embedded wavelet*.

bin: One of a set of discrete areas into which a survey area is divided. Data are sorted among bins according to midpoint locations for unmigrated data or according to reflecting points for migrated data. After sorting and correcting for normal moveout, the data elements within each bin are summed (stacked) and divided by some function of the number of elements to obtain the output trace for the particular bin.

birefringence: Splitting of an *S*-wave into two waves that travel at different velocities through an anisotropic medium. Occurs when an *S*-wave enters a transversely anisotropic region (perhaps a fracture zone) not in the direction of the symmetry axis. Also called **shear-wave splitting** and **double refraction**.

borehole gravimeter: A remote reading gravimeter which can be lowered through a borehole. The difference between the gravity readings at two different depths gives the apparent density between the depths. The effective penetration is much larger than other borehole measurements.

*Dept. Geosciences, University of Houston, Houston, TX 77204–5503.

borehole gravity gradiometer: A measurement of the vertical gradient of gravitational acceleration. This can be measured indirectly by taking the difference between borehole gravity measurements at two depths.

borehole televiewer log: A well log wherein a pulsed, narrow acoustic (sonar) beam scans the borehole wall in a tight helix as the tool moves up the borehole. The traveltime and amplitude provide information about the borehole wall, such as revealing fractures, vugs, etc.

borehole-to-borehole measurements: See *crosshole method*.

boxcar filter: A filter which passes without attenuation all frequencies between certain limits while completely rejecting frequencies outside those limits.

boxcar frequency spectrum: The frequency-domain situation where the amplitude of each frequency component is the same within a passband and zero for all frequencies outside the passband.

breakthrough: The arrival at a producing well of a fluid injected into a formation to push hydrocarbons toward the producing well.

bright spot: An increase of amplitude assumed to be caused by hydrocarbon accumulation. See *hydrocarbon indicator*. Sometimes any local increase of amplitude on a seismic section for any reason.

bubble point: The temperature and pressure at which part of a liquid begins to convert to gas.

capillary pressure: The pressure required to displace one fluid with another with which it is not miscible.

chemical enhanced recovery: Increased recovery of hydrocarbons because of injecting a chemical (solvent or surfactant) into the reservoir.

cokriging: A geostatistical technique for estimating one or more spatially distributed random variables, utilizing both the spatial autocorrelations and crosscorrelations of those variables. It is a method of bivariate least-squares prediction filtering that is essentially a bivariate version of Wiener filtering. See Journel and Huijbregts (1978).

common-midpoint (CMP) method: A recording-processing method where each source is recorded at a number of geophone locations and each geophone records a number of source locations. These data are corrected for normal moveout and combined (stacked) to provide a CMP section that simulates the traces that would be recorded by a coincident source and geophone. The method produces considerable attenuation of several types of noise.

communication: The situation where permeability is continuous between two wells; usually established by observing if pressure changes in one well affect pressure in the other well.

complex trace analysis: A mathematical calculation of envelope amplitude, instantaneous frequency, instantaneous phase. See Taner et al. (1979).

conditional simulation: A Monte Carlo technique for generating multiple images of a reservoir that highlight possible physical property variations (of porosity, permeability, etc.).

coning: The funneling down of gas or the funneling up of water into a hydrocarbon zone because of excessive pressure differential near a well.

convolutional model: A concept that a seismic trace can be represented by the convolution of an embedded wavelet with the reflectivity, with noise added. A consequence of the concept that each reflected wave causes its own effect at each geophone (or hydrophone) independent of what other waves are affecting the geophone at that time, and that the response is simply the sum (linear superposition) of the effects of all the waves.

correlation coefficient: See *crosscorrelation function*.

critical point: The pressure-temperature combination above which the distinction between gas and liquid no longer exists. The **critical pressure** is the pressure needed to condense a vapor at its critical temperature. The **critical temperature** is the highest temperature at which a fluid can exist as a liquid and above which its vapor cannot be liquified regardless of the amount of pressure applied.

crosscorrelation function: A measure of the similarity of two waveforms, of the degree of linear relationship between them, or of the extent to which one is a linear function of the other. Normalized crosscorrelation is also called **correlation coefficient**.

crosshole method: 1. Technique for investigating the region between two or more boreholes by measuring the transit times and/or amplitudes of seismic waves from a source located in one borehole to geophones in other boreholes. Usually implies **crosshole tomography**, the reconstruction of an object from wave projections. Three-component geophones may be used. 2. A technique for resistivity or electromagnetic measurements between boreholes, used for fracture and cavity detection and in reservoir studies.

CSAMT: Controlled-source audio-magnetotelluric technique.

Darcy's Law: A statement that the fluid flow rate is proportional to the permeability, cross-sectional area, pressure gradient, and inversely proportional to the viscosity.

delineation well: A well intended to ascertain reservoir properties or limits rather than to produce hydrocarbons. Also called an **appraisal well**.

detectable limit: The minimum thickness for a bed to give a seismic reflection; approximately 1/30 the dominant wavelength for excellent data. Also called limit of visibility. Compare *resolvable limit*.

detuned amplitude: Seismic amplitude measurements after correcting for the *tuning effect* (q.v.).

diagenesis: Any chemical, physical, or biological change undergone by sediment after its deposition (except for weathering and metamorphism). It embraces processes such as compaction, cementation, leaching, hydration, recrystalization, replacement, dolomitization, reworking, authigenesis, bacterial action, and concretion formation. Diagenesis may destroy or create porosity (secondary porosity) or affect permeability.

difference map, difference section: The result of subtracting one map or section from another. Where the two maps were acquired at different times, the difference (a **time-lapse map**) shows the changes that have occurred. Where the two maps are of different horizons, the difference (**isochron** or **isopach** map) represents the thickness of the intervening interval.

differential pressure: The difference between the lithostatic pressure (the pressure of the overlying column of rock) and the hydrostatic pressure (the pressure of the interstitial fluid).

diffraction tomography: See *tomography*.

dim spot: A local decrease of the amplitude of a seismic event. Where a significant positive acoustic impedance contrast occurs in the absence of hydrocarbons, the presence of hydrocarbons may lessen the contrast and hence the amplitude of a reflection.

dip/azimuth display: An encoded display of dip magnitude/direction, usually displayed by color encoding on an automatically tracked map.

divergence: See *attenuation*.

diurnal and semidurnal components: Components with a periodicity of about 24 or 12 hours. Tidal effects have such periods because of the Earth's rotation.

DMO (dip-moveout processing): A seismic processing operation to correct for the fact that, for dipping reflections, the component traces of a CMP gather do not involve a common reflecting point. DMO effectively corrects for the reflection-point smear that results when reflectors dip. Events with various dips stack with the same velocity after DMO. DMO can be performed in a number of ways.

double refraction: See *birefringence*.

downhole source: A source of seismic energy that can be used within a borehole. See *vertical seismic profiling*.

drive: The energy source that causes interstitial fluids to flow. Natural drives include expansion of a gas cap, gas coming out of solution as the pressure is lowered (solution drive), water drive, and gravity drive. Natural drives are supplemented in secondary and enhanced recovery efforts; see *secondary recovery*, *enhanced oil recovery*, and *reservoir drive*.

dynamic range: The ratio of the maximum reading to the minimum reading (often noise level) which can be recorded by and read from an instrument without change of scale.

effective permeability: The effective permeability of fluids depends on their relative saturations, that is, the presence of one fluid effectively changes the permeability to another fluid.

electrical resistivity: Resistivity is resistance per unit distance. Apparent resistivity is the resistivity of a uniform semi-infinite medium that gives the observed resistivity. Resistivity measurements have to be corrected for the geometry of the measuring system.

electromagnetic method (EM method): A method in which the magnetic and/or electric fields associated with natural or artificially generated subsurface currents are measured. A multitude of methods are used for measurements in both time and frequency domains using both natural and artificial sources.

EM: Electromagnetic. See *electromagnetic method*.

embedded wavelet: The wavelet shape which would result from reflection of an actual wavetrain by a single sharp interface with positive reflection coefficient. See *convolutional model*. Often called **equivalent wavelet** or **basic wavelet**.

enhanced oil recovery (EOR): Stimulating production by injecting materials other than gas or water, such as miscible fluids, surfactants or other chemicals, and steam, or setting fire to the hydrocarbons in a reservoir. Includes steam injection, steam flooding, miscible fluid injection, surfactant flooding, alkaline chemical injection, polymer flooding, fire flooding.

EOR: *Enhanced oil recovery* (q.v.).

equivalent wavelet: See *embedded wavelet*.

event: A lineup on a number of traces which indicates the arrival of new seismic energy, denoted by a systematic phase or amplitude change on a seismic record; an **arrival**. May indicate a reflection, refraction, diffraction, or other type of wavefront.

excitation-at-the-mass method: See *mise-à-la-masse method*.

exploration 3-D: A widely spaced form of 3-D acquisition with lines spaced 4-5 times as far apart as required by the sampling theorem, relying on interpolation to give the data sampling required for migration. While the resulting survey lacks resolution, it reveals broad structural aspects and is generally preferable to 2-D methods.

fault slice: A slice through a three-dimensional data volume along a curved surface parallel to a fault plane. Used to ascertain closure against the fault and secondary faulting.

flat spot: A horizontal seismic reflection attributed to an interface between two fluids such as gas and water or gas and oil. See *hydrocarbon indicator*.

flip-flop acquisition: The use of two seismic ships to carry out a marine survey, where each ship alternately acts as the source for both its own and the other ship's streamer.

flood: Injection of a fluid into a reservoir to flush oil out of the reservoir. Injection of water (**waterflood**) is the most common, but gas, steam, CO_2, surfactants, etc. are also done. Combustion may also be initiated within a reservoir supported by oxygen or air injection (**fireflood**). See *enhanced oil recovery*.

folding frequency: See *Nyquist frequency*.

formation microscanner: A logging tool that produces a result similar to a *borehole televiewer log* (q.v.).

formation testing: The gathering of data on a formation to determine its potential productivity. A drill-stem test involves lowering a tool to the bottom on a string of drill pipe, setting a packer to isolate a formation from the formations above and below. A port on the tool is opened to allow the trapped pressure below the packer to bleed into the drill pipe, gradually exposing the formation to atmospheric pressure and allowing the well to produce.

forward modeling: Calculating the results to be expected from a model, often called simply **modeling**. The opposite process is called inverse modeling or *inversion* (q.v.).

4-D survey: *Time-lapse seismology* (q.v.) using 3-D data. The fourth dimension is calender time.

Fourier theorem: A waveshape can be synthesized by the superposition of cosine waves of different frequencies, amplitudes, and phase shifts.

frequency-domain method: An *induced polarization* or *electromagnetic method* (q.v.) where frequency is the variable.

Fresnel zone: The portion of a reflector from which reflected energy can reach a detector within one-half period of the first reflected energy. Reflected energy from such a zone interferes constructively. The Fresnel zone limits horizontal resolution with unmigrated seismic data.

fringe area: The extra area around the periphery of the area to be surveyed that must be covered because of the inward movement of dipping reflectors and focusing of seismic amplitude dispersed over Fresnel zones during 3-D migration.

gamma: A nanotesla, a unit of measure of magnetic fields.

gas drive: See *reservoir drive*.

geometrical spreading: See *attenuation*.

geophone: A seismic detector, usually of the moving-coil type; used in land work. Hydrophones are used at sea.

geophone group: A number of geophones (or hydrophones), often 6 to 24, that are hard-wired together to effectively form a large geophone. The use of groups discriminates against random energy and waves that are traveling with large horizontal components. Groups of sources can be used in the same way.

geopressure: See *abnormal pressure*.

geostatistics: Application of statistical estimation techniques to spatially correlated random variables for geological/geophysical applications. Involves techniques for interpolation and extrapolation of physical measurements using correlation and probability concepts.

gravity survey: Measurement of the acceleration of gravity over an area or with depth in a borehole.

ground roll: A horizontally traveling surface wave that is apt to be especially strong when surface sources (such as the vibroseis) are used. The most important noise contributor on land.

halo: An anomalous ring around a feature. The mechanism for creation of reported halos around hydrocarbon fields is in dispute. Apparent halos are also created by residualizing methods.

hodogram: Trace of the motion of a vector; used in studying seismic waves from three-component recording.

horizon slice: Display of the data from a three-dimensional set of data interpreted to lie on the same reflecting horizon, thus showing areal variations in amplitude or other attributes. Compare *time slice* (q.v.).

horizon tracking: Picking a seismic section or a 3-D data volume at a chosen point on the seismic waveform for a particular horizon, often made automatically by a picking algorithm.

horizontal section: A *time slice* (q.v.); compare *horizon slice*.

huff and puff: A production technique whereby a fluid is alternately injected into a borehole for a period of time and then the well is produced.

hydrocarbon indicator: Seismic effect that indicates a local hydrocarbon accumulation. Indicators include an amplitude increase (**bright spot**) or decrease (**dim spot**), polarity reversal, or a wave-shape change (**phasing**), a change in the frequency content (especially a local lowering of frequency), a horizontal event reflected from a gas-water, gas-oil, or oil-water contact (**flat spot**), a decrease in amplitude below the accumulation (**amplitude shadow zone**), lower velocity than laterally equivalent sediments producing an apparent sag in lower reflections because of increased time in transmitting the accumulation (**velocity sag** or **push-down**), and a **low-frequency shadow** immediately underneath a reservoir.

hydrofracturing: Fracturing subsurface reservoirs by injecting fluid under very high pressure. The fluid may contain proppants to hold the fractures open after the pressure is released.

hydrophones: Pressure-sensitive detectors used underwater.

hydrostatic pressure: The pressure of the interstitial fluid in a rock, normally the pressure of a column of fluid extending to the surface.

induced polarization: An electrical exploration method involving measurement of the slow decay of voltage in the ground following the cessation of an excitation current pulse (time-domain method) or low-frequency variations of earth impedance (frequency-domain method).

INPUT: A time-domain electromagnetic survey system in which measurements are made during the off period between source pulses.

in situ: Material in its original location. Used in connection with measurements of properties of material which do not involve moving the material (and risk altering it thereby).

instantaneous frequency: See *complex-trace analysis*.

instantaneous phase: See *complex-trace analysis*.

intelligent interpolation: Automatic determination of values between control points, usually based on character considerations similar in concept to that expected by interpreters. Used specifically for interpolating between widely spaced or irregularly spaced seismic lines onto a 3-D grid to provide the sampling required for 3-D migration and interpretation.

interactive: A process in which a human is involved on-line. At an **interactive work station** a human inputs instructions and the computer responds with a display of the results of executing the instructions. "Interactive" implies that the computer is on-line and that the operator waits at the terminal to get the response, so that he can modify the instructions if the response is not satisfactory.

interference test: A test to determine the connectivity of a reservoir by observing the changes in one well when the pressure in another well changes.

interpretation: Deriving a simple, plausible geologic model that is compatible with observed data. The model is never unique and involves a sequence of somewhat arbitrary choices.

inversion: Determining the cause from observation of effects. Inversion has to satisfy constraints that are imposed. The manufacture of a *seismic log* (q.v.) from seismic data is an example of one-dimensional inversion.

inverted VSP: *Reverse VSP* (q.v.).

isotropy: Having properties whose values are the same regardless of the direction of measurement.

joint inversion: *Inversion* (q.v.) to satisfy measurements of two or more different kinds.

kriging: A spatial-domain statistical method for determining the best linear unbiased estimate of a value at an unknown point based on spatially distributed measurements and some known control points. Analagous to time-domain Wiener filtering.

L_1, L_2 norm: Criteria for error minimization. L_1 relates to minimizing the mean absolute error and L_2 to the mean square error.

limit of separability, visibility: See *resolvable limit*, *detectable limit*.

lithostatic pressure: The pressure of the overlying column of rock.

live oil: Oil that is saturated with gas.

longitudinal wave: *P-wave* (q.v.).

magnetic susceptibility: A measure of the degree to which a substance may be magnetized.

magnetotelluric method (MT): A method in which orthogonal components of the electric and magnetic fields induced by natural primary sources are measured simultaneously as a function of frequency. Inversion to a resistivity distribution is a difficult and nonunique operation.

maximum likelihood: The most probable (value). The concept that different probability models generate different samples and that any given sample is more likely to have come from some models than from others. The method requires specification of a probability model, determination of a formula for the likelihood function, and maximization of the likelihood function.

microseism: Feeble earth tremors due to natural causes such as wind, water waves, etc.

midpoint: The surface location midway between the source and detector locations.

migration: A process to relocate and focus seismic data at the locations of the reflectors or diffracting points that cause them. Conventional (two-dimensional) migration is based on observed in-line dip components; dip components perpendicular to the lines are not measured and thus are not allowed for. Three-dimensional migration, possible where three-dimensional data are available, allows for true dip. Important in displaying seismic structure and in making faults and subtle stratigraphic features sharper and more evident, and in attenuating spatial noise. *DMO* (q.v.) is sometimes regarded as a migration process.

minimum-phase wavelet: A wavelet wherein the energy builds up as rapidly as possible after the wavelet begins. Most field data are nearly minimum phase

miscible: The ability of two fluids to mix to form one phase. Oil and water are immiscible, oil and gas are miscible at high pressure and/or temperature.

miscible drive: A method of enhanced recovery in which solvents or gases (as propane, LPG, natural gas, carbon dioxide, or a mixture) are injected into the reservoir to reduce interfacial forces between oil and water in the pore channels and thus displace oil from the reservoir rock.

mise-à-la-masse method: An electrical-exploration method in which one current electrode is positioned in a conducting mineral either in outcrop or in a borehole. The other current electrode is a great distance away and the potential electrodes are moved about with the objective of mapping the mineral deposit. Also called **excitation-at-the-mass method**.

Monte-Carlo method: A mathematical method whereby a calculation is repeated many times using random values. The result gives a statistical estimate for a solution.

MWD: Measurement while drilling

ninety-degree wavelet: A wavelet where the phase of each component is 90 degrees. The resulting wavelet is antisymmetric.

noise: Any unwanted energy. **Seismic noise** consists of energy other than primary reflections; includes microseisms, source-generated noise, multiples, instrument noise, harmonic distortion, etc.

normal moveout (NMO): The variation of reflection arrival time because of sourcepoint-to-geophone distance (**offset**). The additional time required for energy to travel from a source to a flat, reflecting bed and back to a geophone at some distance from the source point compared with the time to return to a geophone at the source point.

Nyquist frequency: A frequency associated with sampling, equal to half the sampling frequency. Also called **folding frequency**. Frequencies greater than the Nyquist frequency alias as lower frequencies from which they are indistinguishable. The **Nyquist theorem**, also called the **sampling theorem**, states that aliasing will not occur if all frequencies are less than the Nyquist frequency.

octave: The interval between two frequencies having a ratio of 2 (or ½).

offset: (1) The distance from the source-point to a geophone, or more commonly to the center of a geophone group. (2) In vertical seismic profiling, the horizontal distance between source and receiver. The distance from source to well head may be different if the well is not vertical.

offset VSP: *Vertical seismic profiling* (q.v.) using a source located some distance away (offset) from the point that is vertically over the recording position in the borehole.

orthorhombic symmetry: A type of symmetry such as might result from vertical fractures of a horizontally layered media.

overpressuring: The situation where the interstitial pressure is greater than that of a column of interstitial fluid extending to the surface. Produces a decrease in seismic velocity that sometimes can be used to map the overpressured zone and predict the degree of overpressuring.

passive seismic: Seismic techniques that do not use an artificial source of seismic energy. Such techniques are used for mapping hydrofractures and other features.

patch: A large geophone group up to several hundred feet across in both directions, containing several hundred geophones.

patch shooting: See *shooting through the patch*.

permeability: A measure of the ease with which a fluid can pass through the pore spaces of a formation. The permeability constant k is expressed by *Darcy's law* (q.v.).

petrophysics: Study of the relationships among the physical properties of rocks. Specifically, studies of how porosity, permeability, etc. relate to seismic velocity, electrical resistivity, temperature, etc.

phase: **1.** The argument of a wave. If the representation of a wave is, for example, A sin $(\kappa x{-}\omega t)$, the argument $(\kappa x{-}\omega t)$ is the phase. **2.** The angle of lag or lead of a cosine wave with respect to a reference. **3.** Any portion of a nonhomogeneous system that is bounded by a surface and may be mechanically separated from the other phases. The three phases of H_2O, for example, are ice, liquid water, and steam. **4. Instantaneous phase** is one of the results of *complex trace analysis* (q.v.).

phasing: A change in waveshape, which is sometimes a *hydrocarbon indicator* (q.v.). Results from the change in reflectivity because of a change in pore fluid.

Poisson's ratio: The ratio of the fractional decrease in width to the fractional increase in length for a body under one-dimensional tension. See *Vs/Vp*.

polar assymmetry: Symmetry such as what might result from horizontal layering, where horizontal properties differ from vertical properties but there is no variation with azimuth.

polarity: The condition of being positive or negative.

polarity reversal: The change in the polarity of a reflection from being positive where the pore fluid is brine to being negative where the pore fluid is a hydrocarbon accumulation, or vice-versa; one of several posssible hydrocarbon indicators. Also called **phase change** and related to *phasing* (q.v.).

polarity standard: The SEG standard for causal seismic data specifies that the onset of a compression from an explosive source is represented by a negative number, that is, by a downward deflection when displayed graphically; this standard is historically based, so that refraction first arrivals break downward. For a zero-phase wavelet, a positive reflection coefficient is represented by a central peak, normally plotted black on a variable area or variable density display; this convention is called **positive standard polarity** and the reverse convention is **negative standard polarity** or **reverse polarity**. Polarity standards are not specified for wavelets other than minimum-phase or zero-phase ones.

polarized S-waves: *S-waves* (q.v.) where particle motion is restricted to a particular orientation (e.g., horizontal (SH) or vertical (SV) planes), or to a particular relationship between motions in orthogonal directions (eliptically polarized).

porosity: The volume fraction of space not occupied by the rock matrix.

precise leveling: Determining relative elevation levels to very high precision, usually to detect changes in levels with time as a result of subsidence, fault movement, etc.

pressure test: Measurements of the changes in formation pressure at a borehole as production conditions are changed, for example, the pressure changes following shut-in of the borehole being tested.

pressure transient testing: Observing changes in the pressure of surrounding boreholes with time following change in the pressure of one borehole, to determine the connectivity of formations.

primary porosity: The porosity remaining from that present at deposition.

primary recovery: Use of the natural energy present in a reservoir to drive fluids to producing boreholes.

proppant: The structurally strong materials injected into fractures opened during hydraulic fracturing to prevent collapse of the fracture upon release of the fracturing pressure.

proximity survey: See *offset VSP* and *salt proximity survey*.

pseudo-log: See *seismic log*.

P-waves: Compressional waves (acoustic or sonic) that involve particle motion in the direction of wave travel.

Q: The ratio of 2π times the peak energy to the energy dissipated in a cycle. The ratio of 2π times the power stored to the power dissipated. The seismic Q of rocks is of the order of 50 to 300. Q is related to other measures of absorption:

$$1/Q = \alpha V/\pi\nu = \alpha\lambda/\pi = h\,T/\pi = \delta/\pi,$$

where V, ν, λ, and T are, respectively, velocity, frequency, wavelength, and period. α is the **absorption coefficient**, h the **damping factor**, δ the **logarithmic decrement**.

raypaths: Lines perpendicular to wavefronts (in isotropic media) along which (it is assumed) seismic energy travels.

reflecting point: For a particular source-geophone combination, the point on a reflecting surface where raypaths from the source and to the geophone make equal angles with the reflector. For horizontal reflectors with horizontally uniform velocities, the reflecting point lies below the midpoint.

reflection coefficient: See *reflectivity*.

reflection point smear: The consequences of common-midpoint traces not involving the same reflection point for dipping reflectors. This effect can be remedied by *DMO processing* (q.v.) or migration before stacking.

reflection tomography: See *tomography*.

reflectivity: **Reflection coefficient**; the ratio of the amplitude of the displacement of a reflected wave to that of the incident wave. For normal incidence on an interface which separates media of densities ρ_1 and ρ_2 and velocities V_1 and V_2, the reflection coefficient for a plane wave incident from medium 1 is

$$R = (\rho_2 V_2 - \rho_1 V_1)/(\rho_2 V_2 + \rho_1 V_1).$$

A negative reflection coefficient implies phase inversion, that a compression is reflected as a rarefaction. In the more general case of a plane-wave incident at an angle, both reflected P- and S-waves and transmitted P- and S-waves will be generated. The amplitude of each of these waves may be found from the *Zoeppritz's equations* (q.v.).

reservoir delineation: Defining the limits of a hydrocarbon accumulation and barriers to fluid flow.

reservoir depletion: Loss of *reservoir drive* (q.v.).

reservoir description: Determining the thickness, porosity, permeability, and other properties of reservoirs and their spatial and directional variations.

reservoir drive: The energy that causes reservoir fluids to flow out of the reservoir rock and into a well bore. **Gas drive** depends on the expansion of the gas in a reservoir to provide the driving energy; the pressure is reduced as a reservoir is produced. **Solution gas drive** depends on gas coming out of solution in the oil as the pressure is reduced. **Water drive** depends on water pressure to force hydrocarbons into the wellbore, and depends on the connectivity of the reservoir with a surrounding aquifer. Water drive is much more efficient at driving oil than is gas drive. Natural reservoir drives are supplemented by injecting gas, water, or other substances in secondary recovery and *enhanced oil recovery* (q.v.).

reservoir geophysics: The use of geophysical methods to assist in delineating, describing, or monitoring a hydrocarbon reservoir.

reservoir simulation: Use of a computer or physical model of a reservoir to test how the reservoir will perform as production or stimulation proceeds.

reservoir surveilance: Monitoring reservoir changes because of movement of the reservoir fluids.

resolution: The ability to separate effects, usually expressed as the ability to separate two features which are very close together.

resolvable limit: The thinnest layer for which reflections from top and base of the layer can be distinguished and are correctly located. A thickness of about one quarter-wavelength. Also called **limit of separability**. Compare *detectable limit*.

retrograde condensation: The formation of liquid droplets in a gas as a well is produced and the pressure drops. Some hydrocarbons exist naturally above their critical temperatures in the reservoir; as a result, when pressure is decreased, instead of expanding to form a gas, they may condense to form a liquid.

reverse VSP: A *vertical seismic profile* (q.v.) made with the source in the borehole and geophones on the surface.

rms velocity: Root-mean-square velocity. For a series of parallel layers of velocity V_i, where the traveltime for seismic energy perpendicularly through each is t_i, the rms velocity is

$$V_{\mathrm{rms}} = [(\Sigma\ V_i^2 t_i)/(\Sigma\ t_i)]^{1/2}.$$

rms velocity is often used erroneously for **stacking velocity**, the velocity determined from velocity analysis based upon normal-moveout measurements. Stacking velocity approaches V_{rms} as the offset approaches zero if velocity layering and reflectors are parallel and the layers are isotropic.

S_1, S_2: Designations of seismic S-waves traveling at the highest and lowest (respectively) velocities because of anisotropy.

salt proximity survey: A survey to determine the location of the salt-sediment interface at a salt-dome flank. Normally involves recording offset and azimuthal VSP surveys in a borehole into the salt or near the salt flank from various surface locations. May also involve borehole gravimeter, long-spaced electrical log measurements, or other types of measurements.

sampling theorem: See *Nyquist frequency*.

seal: A seal or cap rock through which fluid flow is so small that even during long geologic time only a small amount of fluids will have passed through. Cap rocks have permeabilities of the order of 10^{-6}–10^{-8} darcies and very high capillary pressures.

secondary porosity: Porosity developed subsequent to deposition because of solution, chemical replacement, or other processes.

secondary recovery: A method of recovering additional hydrocarbons by injecting liquids or gases into the reservoir to sustain reservoir energy.

Seiscrop section: A *time slice* (q.v.).

Seislog: See *seismic log*.

seismic lithologic modeling (SLIM): An iterative technique whereby a seismic velocity and density model is changed until synthetic seismic data calculated from it matches observed data within some specified tolerance.

seismic log: A calculation of acoustic impedance or velocity as functions of depth or time, based on reflection amplitude measurements. Also called **synthetic sonic log, G-log, Seislog, synthetic acoustic impedance, Saile section, seismic lithologic modeling (SLIM)**. A one-dimensional inversion.

self potential (SP): See *SP*.

sender: *Transmitter* (q.v.).

SH-wave: An S-wave where particle motion is in a horizontal plane.

shear wave: *S-wave* (q.v.).

shear-wave splitting: See *birefringence*.

shooting through the patch: A land 3-D acquisition technique wherein sources are moved around and through a distribution of geophone groups laid out on the ground.

Shuey approximation: An approximation of the Zoeppritz equations used with amplitude-offset (AVO) studies. See Shuey (1985).

side lobe: A buildup of wavelet energy that does not indicate the correct arrival. Seismic wavelets generally involve several cycles, especially when the bandwidth is limited.

signal: That which we wish to measure.

signature: The aspect of a waveshape which makes it distinctive; character

simulation: Representing a real situation by a model; specifically, representing a hydrocarbon reservoir by a computer or physical model to determine how to produce hydrocarbons from the reservoir efficiently.

Snell's Law: A statement that wavefronts and raypaths bend when the velocity changes. If the angles of incidence and reflection are i and r at an interface separating media of velocities V_i and V_r, $\sin i/V_i = \sin r/V_r$.

solution gas drive: See *reservoir drive*.

sonic log: A well log of the traveltime (transit time) for acoustic waves per unit distance, which is the reciprocal of the *P*-wave velocity. Also called **acoustic-velocity log** and **continuous-velocity log**. Usually measured in microseconds per foot. Used for porosity determination by the *time-average equation* (q.v.). The interval transit time is integrated down to the borehole to give total traveltime. Some logs display the complete waveform rather than just the arrival time of the first energy, from which the velocity of other waves can also be determined. Amplitudes measured from some sonic logs are used to determine the quality of cement bonds or for fracture detection.

sonic waveform log: A *sonic log* (q.v.) in which the complete waveform is recorded rather than just the first arrival of energy.

source group: A number of sources used simultaneously or a sequence of records that are stacked together. Used to discriminate against random energy and waves that are traveling with large horizontal components, the same reasons as for the use of geophone groups.

sources, seismic: Explosives and vibrators are mainly used for land operations, and airguns for most marine work.

SP (self potential): The natural ground voltage observed between nearby nonpolarizing electrodes, either on the surface of the ground or in a borehole. Self potentials may be generated by minerals, shale contacts, streaming fluids, temperature changes, etc. Also called **spontaneous potential**.

stacking: Combining traces from different records to build up signal strength and attenuate noise. The **stacking velocity** is the velocity calculated from normal-moveout measurements assuming a model. Sometimes erroneously called "*rms velocity*" (q.v.). See *common-midpoint method*.

standard polarity: See *polarity standard*.

stochastic: Random; a value determined from a specified distribution by chance. Opposite of deterministic.

strain: A deformation in size or shape, such as results from application of a stress (force/unit area).

streamer: A long cable, typically about 2 miles (3+ km) long, that is towed behind a seismic ship. The streamer contains the hydrophone detectors.

superconducting (SQUID) magnetometers: A sensitive magnetometer which detects magnetic field changes by means of a superconducting loop containing Josephson junctions.

SV-wave: An *S*-wave where particle motion is in a vertical plane.

swath method: A type of *three-dimensional survey* (q.v.) in which a series of sources located among planted geophones results in acquiring CMP data over a strip of terrain. The entire setup is then moved to an adjacent area to record over another strip and so the three-dimensional volume of data are built up.

S-wave: A seismic wave that involves particle motion perpendicular to the direction of wave travel. There are two such directions, in consequence of which there can be two independent modes of *S*-waves. *S*-waves that involve just one of these are polarized; see *polarized S-waves*. Also called a shear wave, transverse wave.

S-wave splitting: See *birefringence*.

synthetic seismogram: The seismic data that would be obtained for a particular geologic model with an assumed sonic wavelet. The models are usually based on well log information.

synthetic sonic log: See *seismic log*.

TABS: triaxial borehole seismometer or three-component geophone: See *three-component geophone*.

televiewer: See *borehole televiewer log*.

thermal enhanced recovery: Recovery of hydrocarbons as a consequence of increasing the temperature of a reservoir by steam injection or igniting combustion within the reservoir.

three-component geophone (TABS): A sensor that records the three mutually orthogonal components of particle velocity. Provides the information for determining the type of wave responsible for an event and its direction of approach.

3-D: Three-dimensional.

three-dimensional migration: See *migration*.

three-dimensional (3-D) survey: A survey involving collection of seismic data over an area with the objective of determining spatial relations in three dimensions, as opposed to their apparent components along lines of survey. Various arrangements are used in such surveys. The data from such a survey constitute a volume which can be displayed in different ways, such as vertical sections along *arbitrary lines, time slices, horizon slices, fault slices* (see individual entries).

three-D log: See *sonic waveform log*.

throats: The connections between pores in a rock matrix.

tidal strain: The semidiurnal and diurnal changes that result from changes at any location in the direction of attraction of the moon and sun.

tiltmeter: An instrument to measure tilting such as might result from production-associated subsidence, fault movement, etc.

time-average equation: The empirical equation used to calculate porosity from velocity measurements:

$$(1/V) = \phi/V_f + (1 - \phi)/V_m$$

where ϕ is the porosity, V_f is the velocity of the interstitial fluid, and V_m is the velocity of the rock matrix material, or, in terms of specific transit times for the rock, fluid, and matrix materials, Δt, Δt_f, Δt_m:

$$\Delta t = \phi \Delta t_f + (1 - \phi)\Delta t_m$$

time-domain method: An *induced polarization* or *electromagnetic method* (q.v.) where time is the variable.

time-lapse seismology: Repeating a seismic survey to determine the changes that have occurred in the interval, such as may be caused by enhanced oil recovery efforts. Results are sometimes displayed as difference sections or maps. When using multiple 3-D surveys run at different times, this is sometimes called a **4-D survey**, the fourth dimension being time.

time slice: A display of the seismic measurements (usually amplitude) corresponding to a single arrival time (or single depth) for a grid of data points; a horizontal slice or section through a volume of 3-D data. Also called a **horizontal section** or **Seiscrop section**. Compare *horizon slice*.

time tie: 1. Two or more sets of data that should have the same arrival times, such as traces at line intersections or seismic data and a synthetic seismogram. Migrated 2-D lines may not tie at line intersections. 2. To relate reflection events to contacts seen in wells. Used as both verb and noun.

tomography: A method for finding the velocity and attenuation distribution from a multitude of observations using combinations of source and receiver locations. Space is divided into cells and the data are expressed as line integrals along raypaths through the cells. **Transmission tomography** involves borehole-to-borehole or surface-to-borehole observations (or surface-to-surface observations, as in refraction seismic work). **Reflection tomography** involves surface-to-surface observations (as in conventional reflection seismic work). A velocity (and sometimes attenuation factor) is assigned to each cell and traveltimes (and amplitudes) are calculated by tracing rays through the model. The results are compared with observed times (and amplitudes), and the model is then perturbed and the process repeated iteratively to minimize errors. **Diffraction tomography** involves least-time travel through the grid of cells without attempting to obey Snell's law. Tomographic solutions are often determined iteratively.

tracking: Picking an event on a seismic section at a workstation. The tracking may be manual or automatic. Autotracking may be 2-D (along a seismic section) or 3-D (carried throughout a volume). The interpreter needs to verify the results of autotracking and occassionally intervene because autotracking algorithms encounter difficulties at discontinuities (as at faults) and where waveshape changes (perhaps because of interference).

transit time: The traveltime of a wave over a unit of distance. See *sonic log*.

transmission tomography: See *tomography*.

transmitter (Tx): In resistivity and IP surveying, a current waveform generator. Also called a **sender**.

transverse isotropy: A situation where directional properties are the same in any direction about an axis but different at different angles to the axis. The most common case of anisotropy, often resulting from layering or parallel fracturing.

triaxial (borehole) seismometer (TABS): An assembly of three mutually orthogonal (borehole) geophones.

tuning effect: Constructive or destructive interference between the wavelets from two or more closely spaced reflectors. The composite wavelet exhibits amplitude and phase effects that depend on the time delays between the successive reflection events and the magnitude and polarity of their associated reflection coefficients, and also on the shape of the embedded wavelet.

tuning thickness: The thickness of a layer so that interference of the reflections from the top and base produces a maximum amplitude for reflections of opposite polarity. Occurs when the layer thickness is 1/4 of the dominant wavelength.

undershooting: Seismic surveying using a source on one side of a property into a spread on the opposite side in order to obtain subsurface coverage under the property itself. Used when the surface of the property is inaccessible (such as on a line crossing a river) or to get data beneath some feature whose presence might introduce intolerable uncertainties if measured through it (such as to map beneath a salt dome).

velocity sag: A time delay because of the low velocity of overlying sediments. Sometimes a *hydrocarbon indicator* (q.v.).

vertical seismic profiling (VSP): Measurements of the response of a geophone at various depths in a borehole to sources on the surface. In a conventional **zero-offset VSP** the source is near the wellhead so that travelpaths are nearly vertical. In an **offset VSP** the location of the surface source is an appreciable horizontal distance from the receiver; this results in obtaining reflection points from a region to the side of the borehole as the geophone depth is changed. A **walkaway VSP** achieves a similar result by moving the source while leaving the geophone at the same depth. An **azimuthal VSP** involves sources located at different azimuths from the borehole geophone. A **directional hole VSP** is one also run in a nonvertical borehole. A **reverse VSP** involves interchanging source and geophone, using a *downhole source* (q.v.).

vibroseis method: A seismic method whereby a vibrator injects energy into the ground at low intensity but over a long period of time (7 to 35 s), and subsequent correlation processing effectively shortens it into a compact wavelet.

VSP-to-CMP transform: A rearrangement of offset vertical seismic profile data to locate it at reflecting points, assuming horizontal reflectors. The data still require either migration or a positioning by iterative model-based ray tracing if dips are appreciable.

Vs/Vp: The ratio of velocities of S-wave to P-wave. This ratio is sensitive to lithology or interstitial fluid. The ratio is about 1/2 for most rocks, is larger for gas saturation, and zero for fluids. Measurements of Poisson's ratio are equivalent

$$\frac{v_S}{v_P} = \left[\frac{0.5 - \sigma}{1 - \sigma}\right]^{1/2}, \quad \sigma = \frac{0.5 - \left(\frac{v_S}{v_P}\right)^2}{1 - \left(\frac{v_S}{v_P}\right)^2}.$$

walkaway VSP: A *vertical seismic profile* (q.v.) performed by moving source points to progressively larger horizontal distances from the receiver.

water drive: See *reservoir drive*.

wavefronts: Surfaces in the earth that are all undergoing the same sense of motion at any given time in response to a seismic wave.

wavelength: The distance between two successive portions of a wave that undergo motion in the same sense. Usually indicated by the symbol λ.

wavelet: The shape of a reflection from a single, sharp, isolated interface.

wavelet processing: A type of deconvolution that attempts to ascertain the shape of the embedded wavelet, find an operator to change it to a more interpretable wavelet, and apply this operator to the data. Also used to make sure the embedded wavelets in all the elements of a gather are the same, so that high frequency elements are not lost in stacking, and to improve the tie between data recorded in different ways, or to tie synthetic seismograms to field data.

wave period: The time between corresponding successive phases; the reciprocal of frequency.

wettable: Having sufficiently low surface tension that a liquid spreads over the surface of a solid.

workstation: A computer terminal at which interpretation (or processing) can be performed on an interactive basis. See *interactive*.

zero-phase wavelet: An embedded wavelet where the phase of each component is zero; such a wavelet is symmetrical with the reflecting interface at the center of symmetry, where the amplitude is maximum.

Zoeppritz's equations: Equations which express the partition of energy when a plane wave impinges on an interface at an arbitrary angle. In the general case for an interface between two solids, four waves are generated: reflected P-wave and S-wave and transmitted P-wave and S-wave. The partition of energy among these is found from boundary conditions which require continuity of normal and tangential displacement and stress at the boundary. The *Shuey approximation* (q.v.) to the Zoeppritz equations is often used in studies of *amplitude variation with offset* (q.v.) studies.

zonation of a reservoir: Vertical subdivision of a producing zone into subzones that may produce somewhat independently because of permeability differences or separation by units with limited permeability.

References

Journel, A. G., and Huijbregts, C. J., 1978, Mining Geostatistics: Academic Press Inc.

Shuey, R. T, 1985, A simplification of the Zoepritz-equations: Geophysics, **50**, 609–614.

Taner, M. T., Koehler, F., and Sheriff, R. E., 1979, Complex trace analysis: Geophysics, **44**, 1041–1063.

Index